ESSENTIAL

*i*GENETICS

PETER J. RUSSELL

Reed College

Benjamin
Cummings

San Francisco Boston New York
Capetown Hong Kong London Madrid Mexico City
Montreal Munich Paris Singapore Sydney Tokyo Toronto

Senior Project Manager:	Peggy Williams
Production Editors:	Larry Lazopoulos, Jamie Sue Brooks
Composition:	TSI Graphics, Inc.
Illustrators:	Steve McEntee, Electronic Publishing Services, Inc.
Instructional Media Designer:	Margy Kuntz
Designer, Text and Cover:	Jennifer Dunn
Copyeditor:	Janet Greenblatt
Proofreaders:	Martha Ghent, Carla Breidenbach
Indexer:	Barbara Littlewood
Manufacturing Supervisor:	Pam Augspurger
Marketing Manager:	Josh Frost
Cover Printer:	The Lehigh Press, Inc.
Text Printer and Binder:	Courier Book Companies, Inc., Kendallville

ISBN 0-8053-4697—X

www.aw.com/bc

Contents

CHAPTER *24*

Preface

Overview of the Text

The structure of DNA was first described in 1953, and since that time genetics has become one of the most exciting and ground-breaking sciences. A continual flood of discoveries not only expands our understanding about heredity but also affects our daily lives in areas ranging from disease therapy to courtroom evidence. Experimentally, the development of gene cloning techniques in the 1970s revolutionized the way we look at genes and their expression, and the development of PCR in the 1980s enabled a second revolution, enhancing our abilities to examine genes at the molecular level. In the past 10 or so years, the rapid advances in molecular techniques have made it possible to consider sequencing entire genomes to identify all genes and to study the organization of genes in the chromosomes. At this writing in 2002, the genomes of many viruses, many bacteria, and some eukaryotes have been completely sequenced. Most importantly, a working draft of the human genome sequence has been accomplished, and efforts are under way to refine the sequence and interpret its contents. Complete genome sequences enable us to focus on genomes rather than individual genes and therefore to ask more complex questions about gene expression. As we move into the postgenomic era, then, our knowledge about genes and gene functions will increase enormously. Thus, it is a very important time to learn about the basic concepts of genetics, and this textbook has been written to teach both the classic and modern molecular aspects of this subject.

Essential iGenetics, First Edition, is a shorter and less complex version of my *iGenetics* textbook. While *iGenetics* has a "molecular first" organization, this new textbook has the traditional "Mendel first" organization. As such, it is a replacement for my *Fundamentals of Genetics,* Second Edition, textbook. *Essential iGenetics* was written using the more comprehensive text as a starting point for the overall treatment of the subject, and paying attention to reducing the complexity of the discussions. In addition, great care has been taken to keep the text accessible to students by making it easy to read, with a consistent level of coverage and a logical progression of ideas. Thus, *Essential iGenetics,* First Edition, is an ideally suitable text for students who have a limited background in biology and chemistry. The length of this new text is much shorter than more comprehensive texts, making it ideal not only for one-semester courses in genetics, but also for one-quarter courses.

Essential iGenetics reflects the dynamic nature of the field of genetics. *Essential iGenetics* emphasizes an experimental, inquiry-based approach, with solid treatment of many research experiments that have contributed to our knowledge of genetics. In this way, students are exposed to the processes of science, learning about the formulation and study of scientific questions in a way that will be of value in their study of genetics and in all areas of science.

Essential iGenetics contains pedagogical features such as "Principal Points," "Keynotes," "Summaries," and "Analytical Approaches for Solving Genetics Problems," designed to be useful learning tools for students of genetics. Problem solving is a major feature of the book, and the end-of-chapter "Questions and Problems" have been consistently praised by class testers. Essential iGenetics maintains a format that allows instructors to use the chapters out of sequence to accommodate various teaching approaches.

Essential iGenetics includes the following features:

- The text has a traditional Mendel first organization. The text coverage has the following flow: Mendelian genetics and its extensions, gene mapping, and non-Mendelian inheritance; DNA structure, replication, and gene expression; DNA cloning and manipulation, applications of recombinant DNA technology, and the analysis of genomes; control of gene transcription and the genetics of cancer; DNA mutation and repair, transposable elements, and chromosomal mutations; and population genetics, quantitative genetics, and molecular evolution. As mentioned earlier, the chapters can readily be used in any sequence to fit the needs of individual instructors.
- Twenty-four interactive activities called *iActivities* have been designed to promote interactive problem solving. Found on the *Essential iGenetics* website, these activities are based on case studies presented at the beginnings of the chapters. An example from Chapter 14 is the analysis of DNA microarray results for a fictional patient with breast cancer to determine gene expression differences and then determine which drugs

would be useful for treating her cancer. I have checked all *iActivities* at every stage of their development to help ensure accuracy and quality. A brief description of the *iActivity* appears in the text at the appropriate time in chapter, urging students to try it out.

- Fifty narrated animations on the *iGenetics* CD-ROM help students visualize challenging concepts or complex processes, such as meiosis, DNA replication, translation, restriction mapping, and gene mapping. As with the *iActivities,* I have been involved with the entire development of the animations, outlining them, editing the storyboards, helping describe the steps for the artists, and working closely with the animators until the animations were complete. We have made a special effort to base the animations on the text figures so that students do not have to think about the processes in a different graphic format. These animations are of very high quality, showing a level of detail not typical of animations that are supplements to texts. A media flag with the title of the animation appears next to the discussion of that topic in the chapter.

- The material on recombinant DNA technology and the manipulation of DNA is covered over three chapters. Recombinant DNA Cloning Technology (Chapter 13) discusses how to clone DNA, how to make and screen recombinant DNA libraries, how to analyze genes and gene transcripts, how to sequence DNA, and how to amplify DNA using PCR. "Applications of Recombinant DNA Technology" (Chapter 14) describes how the molecular tools presented in Chapter 13 can be applied to study biological processes, test for genetic disease mutations, isolate human genes, fingerprint DNA, develop gene therapy approaches, develop commercial products, and engineer plants genetically. "Genome Analysis" (Chapter 15) assembles in one chapter an overview of the analysis of genomes, including how genomes are sequenced completely, a summary of the properties of key genomes that have been sequenced, how genome sequences are analyzed, and how transcriptional and translational profiles can be analyzed for many genes simultaneously.

- All the molecular aspects of genetics are included so that the book reflects our current understanding of genes at the molecular level.

- Human examples are used extensively throughout the text, and discussions include our current molecular understandings of various human genetic diseases. Human genes mentioned in the text are keyed to the OMIM (Online Mendelian Inheritance in Man) online database of human genes and genetic disorders at http://www3.ncbi.nlm.nih.gov/Omim/, where the most up-to-date information is available about the genes.

- The Suggested Reading section contains references to papers that were key to the development of concepts and references to current research in the areas being discussed.

Organization and Coverage

The four major areas of genetics—transmission genetics, molecular genetics, population genetics, and quantitative genetics—are covered in 24 chapters. Chapter 1 is an introductory chapter designed to summarize the main branches of genetics, explain the basic concepts of genetics (the molecular nature of genes, the transmission of genetic information, the expression of genes, and the sources of genetic variation), describe what geneticists do and what their areas of research encompass, introduce genetic databases and maps, and discuss the transmission of chromosomes from cell division to cell division and from generation to generation by the processes of mitosis and meiosis, respectively. This knowledge will help you understand your subsequent study of molecular genetics and transmission genetics.

The next six chapters deal with the transmission genetics. Chapters 2 and 3 present the basic principles of genetics in relation to Mendel's laws. Chapter 2 is focused on Mendel's contributions to our understanding of the principles of heredity, and Chapter 3 covers mitosis and meiosis in the context of animal and plant life cycles, the experimental evidence for the relationship between genes and chromosomes, and methods of sex determination. Mendelian genetics in humans is introduced in Chapter 2 with a focus on pedigree analysis and autosomal traits. The topic is continued in Chapter 3 with respect to sex-linked genes. The exceptions to and extensions of Mendelian analysis (such as the existence of multiple alleles, the modification of dominance relationships, gene interactions and modified Mendelian ratios, essential genes and lethal alleles, and the relationship between genotype and phenotype) are described in Chapter 4. In Chapter 5, gene mapping in eukaryotes is presented. In this chapter, we describe how the order of and distance between the genes on eukaryotic chromosomes are determined in genetic experiments designed to quantify the crossovers that occur during meiosis. We also discuss the more specialized analysis of genes by tetrad analysis, primarily in fungal systems, and the phenomenon of recombination in mitosis. In Chapter 6, we discuss the ways of mapping genes in bacteria and in bacteriophages, which take advantage of the processes of conjugation, transformation, and transduction. Fine structure analysis of bacteriophage genes concludes this chapter. In Chapter 7, we address the genetics of extranuclear genomes of mitochondria and chloroplasts. We cover the classic genetic experiments that are used to study non-Mendelian inheritance. We also discuss two contrasts to non-Mendelian inheritance: maternal effect and genomic imprinting.

The "molecular core" of *Essential iGenetics* is found in the next eight chapters which detail the current level of our knowledge about the molecular aspects of genetics and about DNA cloning and genome analysis. The first five of these chapters focus on the nature of the genetic material and the expression of genes. In Chapter 8, we cover the structure of DNA, presenting the classic experiments that revealed DNA and RNA to be genetic material and that established the double helix model as the structure of DNA, and the details of DNA structure and organization in prokaryotic and eukaryotic chromosomes. We cover DNA replication in prokaryotes and eukaryotes and recombination between DNA molecules in Chapter 9. In Chapter 10, we examine some aspects of gene function, such as the genetic control of the structure and function of proteins and enzymes and the role of genes in directing and controlling biochemical pathways. A number of examples of human genetic diseases that result from enzyme deficiencies are described to reinforce the concepts. The discussion of gene function in Chapter 10 enables students to understand the important concept that genes specify proteins and enzymes, setting them up for the next two chapters, in which gene expression is discussed. In Chapter 11, we discuss the first step in the expression of a gene: transcription. We describe the general process of transcription and then present the currently understood details of the transcription of messenger RNA, transfer RNA, and ribosomal RNA genes and the processing of the initial transcripts to the mature RNAs for both prokaryotes and eukaryotes. In Chapter 12, we describe the structure of proteins, the evidence for the nature of the genetic code, and a detailed expression of our current knowledge of translation in both prokaryotes and eukaryotes.

In the three remaining chapters of the molecular core of the book, we cover gene manipulation and genome analysis. In Chapter 13, we discuss recombinant DNA technology and other molecular techniques that are essential tools of most areas of modern genetics. There are descriptions of the use of recombinant DNA technology to clone and characterize genes and to manipulate DNA. Then, in Chapter 14 we discuss the applications of recombinant DNA technology in the analysis of biological processes, the diagnosis of human diseases, the isolation of human genes, forensics (DNA typing), gene therapy, the development of commercial products, and the genetic engineering of plants. In Chapter 15 we discuss genome analysis, focusing on the Human Genome Project for mapping and sequencing the complete genomes of humans and other selected organisms. The chapter describes how complete genomes are sequenced, discusses the features of a number of genomes that have been sequenced, and goes into the types of research scientists are engaging in to detail the global expression of genes in cells at the RNA and protein levels.

The next three chapters discuss the regulation of gene expression. Chapter 16 focuses on the regulation of gene expression in prokaryotes. In this chapter, we discuss the operon as a unit of gene regulation, the current molecular details in the regulation of gene expression in bacterial operons, and regulation of genes in bacteriophages. Chapter 17 focuses on the regulation of gene expression and development in eukaryotes, explaining how eukaryotic gene expression is regulated, stressing molecular changes that accompany gene regulation, short-term gene regulation in simple and complex eukaryotes, gene regulation in development and differentiation, and immunogenetics. Next, in Chapter 18 we discuss the relationship of the cell cycle to cancer and the various types of genes that, when mutated, play a role in the development of cancer. We also discuss the fact that cancer usually requires a number of independent mutational events to develop. Finally, we consider the induction of cancer by chemicals and radiation (carcinogens).

Some of the ways in which genetic material can change or be changed are described in Chapters 19-21. Chapter 19 covers the processes of gene mutation, the procedures that screen for potential mutagens and carcinogens (the Ames test), some of the mechanisms that repair damage to DNA, and some of the procedures used to screen for particular types of mutants. Chapter 20 presents the structures and movements of transposable genetic elements in prokaryotes and eukaryotes. Chromosomal mutations—changes in normal chromosome structure or chromosome number—are discussed in Chapter 21. Chromosomal mutations in eukaryotes and human disease syndromes that result from chromosomal mutations, including triplet repeat mutations, are emphasized.

In Chapters 22 and 23, we describe the genetics of populations and quantitative genetics, respectively. In Chapter 22, "Population Genetics," we present the basic principles in population genetics, extending our studies of heredity from the individual organism to a population of organisms. This chapter includes an integrated discussion of the developing area of conservation genetics. In Chapter 23, "Quantitative Genetics," we consider the heredity of traits in groups of individuals that are determined by many genes simultaneously. In this chapter we also discuss heritability: the relative extent to which a characteristic is determined by genes or by the environment. Both Chapters 22 and 23 include discussions of the application of molecular tools to these areas of genetics. Chapter 24, "Molecular Evolution," discusses evolution at the molecular level of DNA and protein sequences. The study of molecular evolution uses the theoretical foundation of population genetics to address two essentially different sets of questions: how DNA and protein molecules evolve and how genes and organisms are evolutionarily related.

Pedagogical Features

Because the field of genetics is complex, making the study of it potentially difficult, we have incorporated a number of special pedagogical features to assist students and to enhance their understanding and appreciation of genetic principles:

- Each chapter opens with an outline of its contents and a section called "Principal Points." "Principal Points" are short summaries that alert students to the key concepts they will encounter in the material to come.
- Throughout each chapter, strategically placed "Keynote" summaries emphasize important ideas and critical points allow students to check their progress.

- Important terms and concepts—highlighted in bold—are defined where they are introduced in the text. For easy reference, they are also compiled in a glossary at the back of the book.
- Some chapters include boxes covering special topics related to chapter coverage. Some of these boxed topics are *Genetic Terminology* (Chapter 2), *Elementary Principles of Probability* (Chapter 2); *Equilibrium Density Gradient Centrifugation* (Chapter 9); *Cloning of a Sheep* (Chapter 17); and *Hardy, Weinberg, and the History of Their Contribution to Population Genetics* (Chapter 22).
- Each chapter has a "Summary" following the presentation of the concepts, further reinforcing the major points that have been discussed.
- With the exception of the introductory Chapter 1, all chapters precede the set of questions and problems with a section titled "Analytical Approaches for Solving Genetics Problems." Genetics principles have always been best taught with a problem-solving approach. However, beginning students often do not acquire the necessary experience with basic concepts that would enable them to attack assigned problems methodically. In the "Analytical Approaches" sections, typical genetic problems are talked through in step-by-step detail to help students understand how to tackle a genetics problem by applying fundamental principles.
- The problem sets that close the chapters include approximately 475 questions and problems designed to give students further practice in solving genetics problems. The problems for each chapter represent a range of topics and difficulty. The answers to questions marked by an asterisk (*) can be found at the back of the book, and answers to all questions are available in a separate supplement, the *Study Guide and Solutions Manual*.
- Comprehensive and up-to-date suggested readings for each chapter are listed at the back of the book.
- Special care has been taken to provide an extensive, accurate, and well cross-referenced index.

Supplements

For Students

Study Guide and Solutions Manual (0-8053-4707-0)
Prepared by Bruce Chase of the University of Nebraska, the *Study Guide and Solutions Manual* has detailed solutions for all the problems in the text and contains the following features for each chapter: chapter outline of text material, key terms, suggestions for analytical approaches, problem-solving strategies; and 1,000 additional questions for practice and review. It also includes a review of important terms and concepts; a "Thinking Analytically" section, which provides guidance and tips on solving problems and avoiding common pitfalls; additional questions for practice and review; and questions that relate to chapter-specific animations and *iActivities*.

The Genetics Place (www.geneticsplace.com)
This online learning environment offers interactive learning activities, practice quizzes, links to related Web sites, a syllabus manager, and a glossary. A subscription to the Genetics Place is included free with every new copy of the text.

For Instructors

Instructor's Guide (0-8053-4702-X)
This guide presents sample lecture outlines, teaching tips for the text, and media tips for using and assigning the media component in class.

PowerPoint CD-ROM (0-8053-4699-6)
This cross-platform CD-ROM features all illustrations from the text, which can be edited and customized for lecture presentation. You can edit labels, import illustrations and photos from other sources, and export figures into other programs, including PowerPoint, or for use on the Web.

Transparency Acetates (0-8053-4705-4)
Approximately 175 full-color figures from the text are included.

Printed Test Bank (0-8053-4704-6)
Prepared by Holly Ahern of Adirondack Community College, the test bank includes approximately 1,100 multiple-choice and true-or-false questions. The entire set of questions is also available in Macintosh and Windows test-generating software (0-8053-4546-9).

Acknowledgments

Publishing a textbook and all its supplements is a team effort. I have been very fortunate to have some very talented individuals working with me on this project.

For their help in honing the text, I thank all of the reviewers and class testers involved in this edition.

I would like to thank the following individuals for their talents and efforts in crafting some of the chapters in the text: Drs. Ann Sodja and Tim Hinnman (Wayne State University) for their work on Chapter 17, "Regulation of Gene Expression in Eukaryotes;" Dr. Andrew

Clark (Pennsylvania State University) for consulting on the generation of Chapter 22, "Population Genetics;" Dr. Edmund Brodie III (Indiana University) for his work on Chapter 23, "Quantitative Genetics;" and Dr. Dan E. Krane (Wright State University) for writing Chapter 24, "Molecular Evolution."

Thank you also to Bruce Chase (University of Nebraska, Omaha) for the end-of-chapter questions he has contributed. And, thank you to both Janet Greenblatt and Bruce Chase for their excellent work on putting together the *Study Guide and Solutions Manual*. I also thank Mary Healey (Springfield College) for providing some of the end-of-chapter questions and problems. I also thank Barbara Littlewood for generating an excellent, comprehensive index for this book.

A number of talented individuals worked with me to develop the *iGenetics* website and have contributed to its success: Margy Kuntz (*iActivities*), who did an excellent job researching this subject matter and then authoring highly creative and rich *iActivities* that are designed to enhance critical thinking in genetics; Dr. Todd Kelson (Ricks College; animation storyboards); Dr. Hai Kinal (Springfield College, animation storyboards); Dr. Robert Rothman (Rochester Institute of Technology; animation storyboards); Steve McEntee (*iActivity* art development, art style for the animations and text art); Kristin Mount (animations); Richard Sheppard (animations); Eric Stickney (animations); Holly Ahern (Adirondack Community College; website quiz questions); and the staff of Six Red Marbles and Binary Labs.

I am grateful to the literary executor of the late Sir Ronald A. Fisher, F.R.S.; to Dr. Frank Yates, F.R.S.; and to Longman Group Ltd. London, for permission to reprint Table IV from their book *Statistical Table for Biological, Agricultural and Medical Research* (Sixth Edition, 1974).

I would like to thank Larry Lazopoulos, Jamie Sue Brooks, and TSI Graphics, Inc. for their handling of the production phase of the book.

Finally, I wish to thank those at Benjamin Cummings who helped to make *Essential iGenetics* a reality. Most especially, I thank Peggy Williams, Senior Project Editor, for her dedication and devotion to making the textbook and its supplements of the highest quality possible. I am particularly appreciative of Peggy's hard work, organizational efforts, and sense of humor throughout this project.

Peter J. Russell

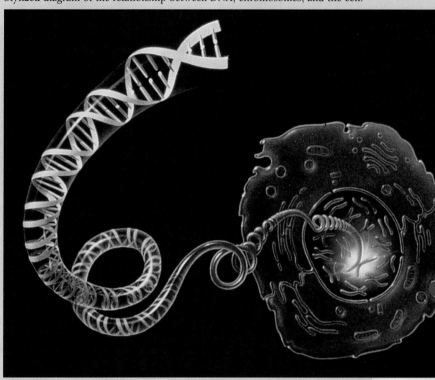

Stylized diagram of the relationship between DNA, chromosomes, and the cell.

1

Genetics: An Introduction

PRINCIPAL POINTS

Genetics often is divided into four major subdisciplines: transmission genetics, which deals with the transmission of genes from generation to generation; molecular genetics, which deals with the structure and function of genes at the molecular level; population genetics, which deals with heredity in groups of individuals for traits determined by one or only a few genes; and quantitative genetics, which deals with heredity of traits in groups of individuals where the traits are determined by many genes simultaneously. Depending on whether the goal is to obtain a fundamental understanding of genetic phenomena or to overcome problems in society or exploit discoveries, genetic research is considered basic or applied, respectively.

Eukaryotes are organisms whose genetic material is located in a membrane-bounded nucleus within the cells. The genetic material is distributed among several linear chromosomes. Prokaryotes, by contrast, lack a membrane-bounded nucleus.

Diploid eukaryotic cells have two haploid sets of chromosomes, one set coming from each parent. The members of a pair of chromosomes, one from each parent, are called homologous chromosomes. The complete set of chromosomes in a eukaryotic cell is called its karyotype.

1

Mitosis is the process of nuclear division in eukaryotic cells represented by M in the cell cycle (G_1, S, G_2, and M). Mitosis results in the production of daughter nuclei that contain identical chromosome numbers and are genetically identical to one another and to the parent nucleus from which they arose.

Meiosis occurs in all sexually reproducing eukaryotes. A specialized diploid cell (or cell nucleus) with two sets of chromosomes is transformed through one round of DNA replication and two rounds of nuclear division into four haploid cells (or four nuclei), each with one set of chromosomes.

Meiosis generates genetic variability through the processes by which maternal and paternal chromosomes are reassorted in progeny nuclei and through crossing-over between members of a homologous pair of chromosomes.

i DO YOU HAVE YOUR MOTHER'S EYES AND MUSICAL ability? Your father's nose and sense of humor? How can you determine which of these traits you have inherited from your parents—or what traits you might pass on to your children? The answers lie in the tiny segments of DNA known as genes and the way those genes are transmitted from parent to child. In this chapter, you will have a chance to review the basic concepts of genetics and to learn more about what geneticists do and how they work. Then, in the iActivity, you will discover some of the different methods geneticists might use to determine if a specific trait is inherited.

Welcome to the study of **genetics,** the science of heredity. Genetics is concerned primarily with understanding biological properties that are transmitted from parent to offspring. The subject matter of genetics includes heredity; the molecular nature of the genetic material; the ways in which genes, which determine the characteristics of organisms, control life functions; and the distribution and behavior of genes in populations.

Genetics is central to biology because gene activity underlies all life processes, from cell structure and function to reproduction. Learning what genes are, how genes are transmitted from generation to generation, how genes are expressed, and how gene expression is regulated is the focus of this book. Genetics is expanding so rapidly that it is not possible to describe everything we know about it between these covers. The important principles and concepts are presented carefully and thoroughly; readers who want to go further can consult the suggested readings at the end of the text or search for research papers using the Internet—for example, by searching the PubMed database supported by the National Library of Medicine, National Institutes of Health, at http://www.ncbi.nlm.nih.gov/PubMed/.

We assume that you have a general understanding of genetics. This chapter presents a brief introduction to genetics and also discusses the transmission of chromosomes from cell division to cell division and from generation to generation by the processes of mitosis and meiosis, respectively. This knowledge will help you understand your subsequent study of genes and the results of genetic crosses.

Classical and Modern Genetics

Humans recognized long ago that offspring tend to resemble their parents. Humans have also performed breeding experiments with animals and plants for centuries: classical genetic engineering, if you will. However, the principles of heredity were not understood until the mid-nineteenth century, when Gregor Mendel analyzed quantitatively the results of crossing pea plants that varied in easily observable characteristics. The importance of his findings was not recognized in Mendel's lifetime, but when the principles of heredity were rediscovered at the turn of the twentieth century, it was realized that Mendel's work constituted the foundation of modern genetics.

Since the turn of the twentieth century, genetics has been an increasingly powerful tool for studying biological processes. An important approach used by many geneticists is to isolate mutants affecting a particular biological process and then, by comparing the mutants with normal strains, to obtain an understanding of the process. Such research has gone in many directions, such as analyzing heredity in populations, analyzing evolutionary processes, identifying the genes that control the steps in a biological process, mapping the genes involved, determining the products of the genes, and analyzing the molecular features of the genes, including the regulation of the genes' expression.

Research in genetics was revolutionized in 1972, when Paul Berg constructed the first recombinant DNA molecule in vitro, and in 1973, when Herbert Boyer and Stanley Cohen cloned a recombinant DNA molecule for the first time. Kary Mullis's development in 1986 of the polymerase chain reaction (PCR) to amplify specific segments of DNA spawned a second revolution. Recombinant DNA technology, PCR, and other molecular technologies are leading to an ever-increasing number of exciting discoveries that are furthering our knowledge of basic biological functions and will lead to improvements in the quality of human life. Already the complete genomic DNA sequences have been determined for a number of organisms, including humans, heralding the exciting field of *genomics*. As scientists analyze the data, we can expect major contributions to our biological knowledge. For example, we will know about every gene in the human genome: where it is in the genome, its sequence, and its regulatory sequence. Such knowledge undoubtedly will lead to a better understanding of human genetic diseases and contribute to their cures. The science fiction scenario of each of us having our DNA genome sequence on a chip that we carry with us may well become a reality. However, knowledge about our genome will raise social and ethical concerns that must be resolved carefully. For example, if a person has a genome sequence that includes a gene with the potential to cause a life-shortening disease, to what extent should that information be private? These questions notwithstanding, this is a very exciting time to be a student of genetics.

Figure 1.1

DNA. **(a)** Three-dimensional molecular model of DNA. **(b)** Stylized diagram of the DNA double helix.

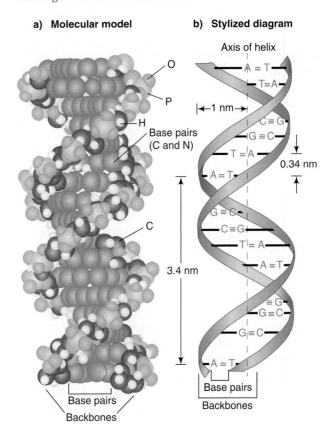

a) **Molecular model** b) **Stylized diagram**

Basic Concepts of Genetics

In this section, we summarize some of the important concepts and processes of genetics so that they are clear to you as we develop the field of genetics in more detail.

DNA, Genes, and Chromosomes

The genetic material of all living organisms, both eukaryotes and prokaryotes, is **DNA (deoxyribonucleic acid;** see Chapter 8). DNA is also the genetic material of many viruses that infect prokaryotes and eukaryotes. A number of other viruses have **RNA (ribonucleic acid)** as genetic material. DNA is made up of two chains (also called strands); each chain consists of building blocks called **nucleotides,** each of which consists of the sugar deoxyribose, a phosphate group, and a **base.** The arrangement of the nucleotides in the chains forms a double helix (Figure 1.1).

There are four bases in DNA: adenine (A), guanine (G), cytosine (C), and thymine (T). In RNA, uracil (U) occurs in place of thymine. The sequence of these bases within a strand determines the genetic information stored in that strand. **Genes,** which Mendel called factors, are specific sequences of nucleotides. These sequences describe the traits that are passed on from parent to offspring.

In the cell, the genetic material is organized into structures called **chromosomes** (see Chapter 8). *Chromosome* means "colored body" and is so named because these threadlike structures are visible under the light microscope only after they are stained with dyes. Many but not all prokaryotes have a single, usually circular chromosome. In eukaryotes, the nucleus contains a number of linear chromosomes, each consisting of a single DNA molecule complexed with protein. Different organisms have different numbers of chromosomes. The total amount of genetic material in a given organism's nuclear chromosomes is called the **genome.**

Outside the nucleus, eukaryotes have DNA in the mitochondria (in animals and plants) and in chloroplasts (in plants; see Chapter 7). This extranuclear DNA differs in structure between species, but in many cases it is circular.

Transmission of Genetic Information

Genes determine such things as what we look like, how we grow, and what genetic diseases we may have. But how are such traits passed on from parents to offspring? As mentioned earlier, Gregor Mendel was the first to determine the laws of heredity (see Chapter 2). Mendel performed a series of careful breeding experiments with the garden pea. In brief, he picked strains of peas that differed in particular characteristics (also called *traits*). For example, the pea seeds were either smooth or wrinkled, and the flowers were either purple or white (Figure 1.2). Then he made genetic crosses, counted the number of times the traits appeared in the progeny, and interpreted the results. (This basic experimental design is still used in gene transmission studies today.) From these kinds of data, Mendel was able to conclude that inherited characteristics are determined by factors, or genes, and that each organism contains two copies of each gene: one inherited from its mother and one from its father. Mendel also hypothesized that alternative versions of genes, what we now call **alleles,** account for variations in inherited characters. For example, the gene for pea seed color exists in two versions, one for yellow (allele Y) and the other for green (allele y).

An organism having a pair of identical alleles for a trait is said to be **homozygous** for the trait (e.g., YY or yy for the seed color trait), whereas an organism having two different alleles for a gene is said to be **heterozygous** (e.g., Yy for the seed color trait). The complete genetic makeup of an organism is its **genotype.** All the observable properties an organism has are its **phenotype.** The

genotype alone is not responsible for the phenotype; rather, the genotype interacts with the environment—the external environment and the internal environment—to produce the phenotype. Thus, two individuals with identical genotypes (identical twins, for example) are not necessarily identical in phenotype.

Mendel considered the factors that controlled the phenotypes he studied in abstract terms. He correctly deduced that the factors segregated randomly into the gametes (Mendel's first law, the principle of segregation) and that the two factors controlling one trait assorted independently from the two factors controlling another trait (Mendel's second law, the principle of independent assortment) (see Chapter 2). Recall the thinking behind the principle of segregation as we look briefly at the following cross of two homozygotes:

P (parental) generation	YY	$\times$	yy
Parental phenotype	yellow seeds		green seeds
Haploid gametes	Y		y

Fusion of the gametes produces the following:

F₁ generation	Yy
F₁ phenotype	Yellow seeds because Y allele is **dominant** to y allele; can also say that y is **recessive** to Y

When F₁ plants are crossed:

F₁ × F₁ Yy × Yy

F₁ gametes Y y Y y

The F₂ generation is produced by random fusion of the gametes as follows:

F₂ generation

	Y	y
Y	YY	Yy
y	Yy	yy

The result is a genotypic ratio of 1 YY : 2 Yy : 1 yy; because Y is dominant to y, the phenotypic ratio is 3 yellow : 1 green. We will learn more about Mendel's experiments and his principles of heredity in Chapter 2.

It was almost two decades after Mendel's death (in 1884) that the material basis of gene segregation from generation to generation was shown. In 1902, Walter Sutton and Theodor Boveri proposed the chromosome theory of heredity, which posited that genes are on

Figure 1.2

Example of easily distinguishable alternative traits: purple-flowered *(left)* vs. white-flowered *(right)* pea plants.

Figure 1.3

Transcription. The DNA separates locally into single strands, and RNA polymerase makes an RNA copy of one of the DNA strands.

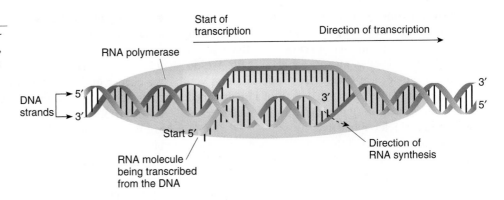

chromosomes and that the segregation patterns of genes can be explained entirely by the segregation of chromosomes from generation to generation (see Chapter 3). This is generally accepted today, with gene segregation paralleling the segregation of chromosomes in meiosis.

Expression of Genetic Information

Genes control all aspects of the life of an organism, encoding the products responsible for development, reproduction, and so forth. The process by which a gene produces its product and the product carries out its function is known as **gene expression.** But what products are expressed? And how are these products related to specific traits?

In 1941, George Beadle and Edward Tatum, working with mutations of the fungus *Neurospora crassa* ("new-ross-pore-a crass-a") (orange bread mold) that imposed nutritional requirements on the organism, showed that there was a firm relationship between genes and enzymes (a special subset of proteins; see Chapter 10). Cells function through myriad biochemical pathways, with each step catalyzed by one or more enzymes. Their studies of mutants elegantly showed that each step in the biochemical pathways they studied was under genetic control. Because the steps are catalyzed by enzymes, they proposed the **one gene–one enzyme hypothesis.** This hypothesis has been modified to the *one gene–one polypeptide hypothesis* because not all proteins are enzymes and not all proteins consist of only one polypeptide.

Beadle and Tatum showed that genes provide the instructions for making specific proteins. This process consists of two main steps: transcription (see Chapter 11) and translation (see Chapter 12). In **transcription,** the DNA separates locally into single strands, and an enzyme makes an RNA copy of one of the strands of the DNA

molecule (Figure 1.3). The enzyme that catalyzes the synthesis of a new RNA molecule using the DNA nucleotide pair sequence as a template is **RNA polymerase.** In prokaryotes, three classes of RNA are made: **messenger RNA (mRNA), transfer RNA (tRNA),** and **ribosomal RNA (rRNA).** Eukaryotes encode these three classes of RNA and **small nuclear RNA (snRNA).** The RNA molecules are essential for cell function in both prokaryotes and eukaryotes. The mRNAs specify the amino acid sequences of proteins, which are important structural and functional components of cells.

The process by which the base sequence information in mRNA is converted into an amino acid sequence in proteins is called **translation.** Translation occurs on **ribosomes** (Figure 1.4), large complexes of rRNA molecules and proteins. The base pair information that potentially specifies the amino acid sequence of a protein is called the **genetic code.** Each amino acid is specified by a three-nucleotide sequence of the mRNA; this sequence is called a **codon.** Other DNA sequences specify where the RNA copy is to stop.

Figure 1.4

A ribosome, the organelle on which translation of mRNA (protein synthesis) takes place. Two views of the three-dimensional model of the *E. coli* ribosome. The large subunit is shown in red, the small subunit in yellow.

At any one time, only some of the genes in a particular genome are active, and in complex multicellular organisms, only a specific set of genes are active in each tissue and organ. How is all this accomplished? We do not have anywhere near a complete understanding yet, but at the general level, this is the result of a finely tuned array of gene regulation signals determining which genes are active and which are inactive.

Our comprehension of the regulation of gene expression began in bacteria, when François Jacob and Jacques Monod in 1961 proposed an **operon** model to explain the regulation of the expression of genes that encode enzymes needed to metabolize lactose (a sugar; see Chapter 16). In their model, a genetic switch is involved. When the switch is set one way, transcription of the genes encoding the enzymes is blocked; when the switch is set the other way, transcription of the genes can take place. This model has been shown to be a general description of many gene systems in bacteria and their viruses.

Genetic switches also regulate gene expression in eukaryotes, but those switches are different from and typically more complex than those identified in bacteria (see Chapter 17). Moreover, many mechanisms are eukaryote-specific. Unquestionably, although much has been learned about gene regulation in eukaryotes, much remains to be learned.

Sources of Genetic Variation

Genetics has shown us that many of the differences between organisms are the result of differences in the genes they carry. These differences have resulted from the evolutionary process of **mutation** (a change in the genetic material; see Chapter 19), **recombination** (exchange of genetic material between chromosomes; see Chapters 5 and 6), and **selection** (the favoring of particular combinations of genes in a given environment; see Chapter 22).

A mutation is any heritable alteration in the genetic material. Mutations can occur spontaneously or be induced experimentally by the use of mutagenic agents such as radiation or chemicals. As we will learn in Chapter 19, living cells have a variety of systems that repair damage to the genetic material caused by mutagens. Thus, mutations are changes in the genetic material that become fixed because they remain uncorrected by the repair systems.

Recombination, the exchange of genetic material between chromosomes, is brought about by enzymes that cut and rejoin DNA molecules. In eukaryotes, recombination is a common event in meiosis when the physical exchange of homologous chromosomes occurs by crossing-over (see later in this chapter). Uncommonly, crossing-over also may occur in mitosis. In prokaryotes, which do not undergo meiosis or mitosis, recombination still occurs by crossing-over whenever two DNA molecules with identical or nearly identical sequences become aligned.

Selection was discovered by Charles Darwin in the mid-nineteenth century. The main consequence of selection is a change in the frequencies of genes affecting the trait or traits on which selection acts. As a result of selection, different genotypes contribute alleles to the next generation not strictly in proportion to their number in the population but in proportion to the selective advantage they have.

These three mechanisms of mutation, recombination, and selection individually and collectively produce new genetic variations that are essential for evolution.

Geneticists and Genetic Research

The material presented in this book is the result of an incredible amount of research done by geneticists working in many areas of biology. Geneticists use the methods of science in their studies. As researchers, geneticists typically use the **hypothetico-deductive method of investigation.** This consists of making *observations*, forming *hypotheses* to explain the observations, making experimental *predictions* based on the hypotheses, and finally *testing* the predictions. The last step provides new observations, producing a cycle that leads to a refinement of the hypotheses and perhaps, eventually, to the establishment of a theory that attempts to explain a set of observations.

As in all other areas of scientific research, the exact path a research project will follow cannot be predicted precisely. In part, the unpredictability of research makes it exciting and motivates the scientists engaged in it. The discoveries that have revolutionized genetics were not planned; they developed out of research in which basic genetic principles were being examined. Barbara McClintock's work on the inheritance of patches of color on corn kernels is an excellent example (see Chapter 20). After accumulating a large amount of data from genetic crosses, she hypothesized that the appearance of colored patches was the result of the movement (transposition) of a DNA segment from one place to another in the genome. Only many years later were these DNA segments—called *transposons* or *transposable elements*—isolated and characterized in detail. (A more complete discussion of this discovery and of Barbara McClintock's life is presented in Chapter 20.) We know now that transposons are ubiquitous, playing a role not only in the evolution of species but also in some human diseases.

The Subdisciplines of Genetics

Geneticists often divide genetics into four major subdisciplines. **Transmission genetics** (sometimes called classical genetics) is the subdiscipline dealing with how genes and genetic traits are transmitted from generation to generation and how genes recombine. Analyzing the pattern of trait transmission in a human pedigree or in crosses of experimental organisms is an example of a transmission genetics study. **Molecular genetics** is the subdiscipline dealing with the molecular structure and function of genes. Analyzing the molecular events involved in gene expression is an example of a molecular genetics study. **Population genetics** is the subdiscipline that studies heredity in groups of individuals for traits that are determined by one or only a few genes. Analyzing the frequency of a disease-causing gene in the human population is an example of a population genetics study. **Quantitative genetics** also considers the heredity of traits in groups of individuals, but the traits of concern are determined by many genes simultaneously. Analyzing the fruit weight and crop yield in agricultural plants are examples of quantitative genetics studies. Although these subdisciplines help us think about genes from different perspectives, there are no sharp boundaries between them. Increasingly, for example, population and quantitative geneticists analyze molecular data to determine allelic frequencies in large groups. Historically, transmission genetics developed first, followed by population genetics and quantitative genetics and then molecular genetics.

Through the products they encode, genes influence all aspects of an organism's life. Understanding transmission genetics, population genetics, and quantitative genetics will help you understand such things as population biology, ecology, evolution, and animal behavior. Similarly, understanding molecular genetics is useful when you study such topics as neurobiology, cell biology, developmental biology, animal physiology, plant physiology, immunology, and, of course, the structure and function of genomes.

Basic and Applied Research

Genetics research, and research in general, may be either basic or applied. In **basic research,** experiments are done to gain an understanding of fundamental phenomena, whether or not the knowledge gained leads to any immediate applications. In **applied research,** experiments are done with an eye toward overcoming specific problems in society or exploiting discoveries. Basic research was responsible for most of the facts we discuss in this book. For example, we know how the expression of many prokaryotic and eukaryotic genes is regulated as a result of basic research on model organisms such as the bacterium *Escherichia coli*, the yeast *Saccharomyces cerevisiae*, and the fruit fly *Drosophila melanogaster*. The knowledge obtained from basic research is used largely to fuel more basic research.

Applied research is done with different goals in mind. In agriculture, applied genetics has contributed significantly to improvements in animals bred for food (such as reducing the amount of fat in beef and pork) and

in crop plants (such as increasing the amount of protein in soybeans). A number of diseases are caused by genetic defects, and great strides are being made in understanding the molecular bases of some of those diseases. Drawing on knowledge gained from basic research, applied genetic research may involve developing rapid diagnostic tests for genetic diseases and producing new pharmaceuticals for treating diseases.

There is no sharp dividing line between basic and applied research. Indeed, in both areas researchers use similar techniques and depend on the accumulated body of information when building hypotheses. For example, **recombinant DNA technology**—procedures that allow molecular biologists to splice a DNA fragment from one organism into DNA from another organism and to clone (make many identical copies of) the new recombinant DNA molecule—has had a profound effect on both basic and applied research (see Chapters 13–15). Many biotechnology companies owe their existence to recombinant DNA technology as they seek to clone and manipulate genes in developing their products. In the area of plant breeding, recombinant DNA technology has made it easier to introduce traits such as disease resistance from noncultivated species into cultivated species. Such crop improvement traditionally was achieved using conventional breeding experiments. Figure 1.5a shows the Flavr Savr tomato, which was genetically engineered to spoil more slowly. These tomatoes can be picked later than normal tomatoes, when they have developed more flavor. In animal breeding, recombinant DNA technology is being used in the beef, dairy, and poultry industries, for example, to increase the amount of lean meat, the amount of milk, and the number of eggs. In medicine, the results are equally impressive. Recombinant DNA technology is being

Figure 1.5

Examples of products developed as a result of recombinant DNA technology. (a) The Flavr Savr tomato. This tomato has been genetically engineered so that its rate of spoilage is much slower than that of normal tomatoes; as a result, it can be picked at a later stage of ripeness and has more flavor than a normal tomato. **(b)** Humulin, human insulin for insulin-dependent diabetics.

a)

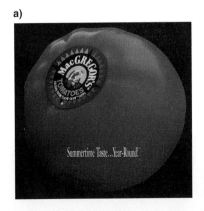

b)

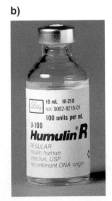

used to produce many antibiotics, hormones, and other medically important agents, such as clotting factor and human insulin (marketed under the name Humulin; Figure 1.5b), and to diagnose and treat a number of human genetic diseases. In forensics, *DNA typing* (also called *DNA fingerprinting* or *DNA profiling*) is being used in paternity cases, criminal cases, and anthropological studies. In short, the science of genetics is currently in an exciting and dramatic growth phase, and there is still much to discover.

iActivity You are a researcher investigating different ways to determine if the ability to taste a chemical called phenylthiocarbamide (PTC) is inherited in the iActivity *A Question of Taste* on the website.

KEYNOTE

Genetics can be divided into four major subdisciplines: transmission genetics, molecular genetics, population genetics, and quantitative genetics. Depending on whether the goal is to obtain a fundamental understanding of genetic phenomena or to overcome problems in society or exploit discoveries, genetic research is considered basic or applied, respectively.

Genetic Databases and Maps

In this section, we talk about two important resources for genetic research: genetic databases and genetic maps. Genetic databases have become much more sophisticated and expansive as tools for computer analysis have been developed and Internet access to databases has become routine, and constructing genetic maps has been part of genetic analysis for about 100 years.

Genetic Databases. The amount of information about genetics has increased dramatically. No longer can we learn everything about genetics by going to a college or university library; the computer now plays a major role. For example, a useful way to look for genetic information using the Internet is to enter key words into search engines such as Google (http://www.google.com). Typically, a vast number of hits are listed, some useful and some not. Therefore, it is difficult to summarize all genetic databases that are useful in this section. You must search for yourself and be critical about what you find. However, we can consider a set of important and extremely useful genetic databases at the National Center for Biotechnology Information (NCBI) World Wide Web site (http://www.ncbi.nlm.nih.gov). NCBI was created in 1988 as a national resource for molecular biology information, and its role is to "create public databases, conduct research in computational biology, develop software

tools for analyzing genome data, and disseminate biomedical information—all for the better understanding of molecular processes affecting human health and disease."

Here are some of the search tools available at the NCBI site:

- BLAST is a tool used to compare a nucleotide sequence or protein sequence with all sequences in the database to find possible matches. This is useful, for example, if you have sequenced a new gene and want to find out whether anything similar has been sequenced previously. Genes with related functions may be listed in the databases, allowing you to focus your research on the function of the gene you are studying.
- GenBank is the National Institutes of Health (NIH) genetic sequence database. This database is an annotated collection of all publicly available DNA sequences (more than 11 billion bases as of February 2001). You search GenBank by entering terms in the search window. For example, if you are interested in the human disease cystic fibrosis, you enter "cystic fibrosis" to find all sequences that have been entered into GenBank that include those two words in the annotations.
- PubMed is a search tool for accessing literature citations and abstracts and for linking to sites with electronic versions of journal articles (sometimes free, sometimes for a fee or by subscription). You search PubMed by entering terms, either subject words or author names.
- OMIM (Online Mendelian Inheritance in Man) is a database of human genes and genetic disorders authored and edited by Dr. Victor A. McKusick and his colleagues. You search OMIM by entering terms in a search window; the result is a list of linked pages, each with a specific OMIM entry number. The pages have details about the gene or genetic disorder, including genetic, biochemical, and molecular data, along with an up-to-date list of references. We refer to OMIM entries (and give the OMIM entry number) throughout the book each time we discuss a human gene or genetic disease.

A powerful feature of the NCBI databases is that they are linked, enabling the user to move smoothly between them and integrate the knowledge obtained in each of them. For example, a literature citation found in PubMed will have links to sequences in nucleotide and protein databases.

Genetic Maps. Since 1902, much effort has been made to construct **genetic maps** (Figure 1.6) for the commonly used experimental organisms in genetics. Genetic maps are like road maps that show the relative locations of towns along a road; that is, they show the arrangements of genes along the chromosomes and the genetic distances between the genes. The position of a gene on the

Figure 1.6

Example of a genetic map, here some of the genes on chromosome 2 of the fruit fly, *Drosophila melanogaster*.

(From *Principles of Genetics* by Robert Tamarin. Copyright © 1996. Reproduced with permission of The McGraw-Hill Companies.)

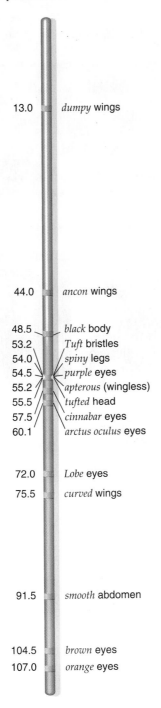

in the two original parents switches (i.e., recombines; see Chapter 5).

The goal of constructing genetic maps has been to obtain an understanding of the organization of genes along the chromosomes—for example, to inform us whether genes with related functions are on the same chromosome and, if they are, whether they are close to each other. Genetic maps have also proved very useful in efforts to clone and sequence particular genes of interest and, more recently, as part of genome projects, the efforts to obtain the complete sequences of genomes.

Model Organisms

Geneticists use various organisms in their research. The principles of heredity were first established in the nineteenth century by Gregor Mendel's experiments with the garden pea. Since Mendel's time, many organisms have been used in genetic experiments. Among the qualities that make an organism well suited for genetic experimentation are the following:

- The genetic history of the organism involved must be well known.
- The organism must have a short life cycle so that a large number of generations occur within a short time. In this way, researchers can obtain data readily over many generations. Fruit flies, for example, produce offspring in 10 to 14 days.
- A mating must produce a large number of offspring.
- The organism should be easy to handle. For example, hundreds of fruit flies can easily be kept in small bottles.
- Most importantly, genetic variation must exist between the individuals in the population so that the inheritance of traits can be studied. The more marked the variations, the easier the genetic analysis.

Cellular Reproduction

Recall from your previous studies that reproduction is the basis of heredity. Because genes are located on chromosomes, the segregation of genes from generation to generation parallels the segregation of chromosomes from generation to generation. To prepare us for our discussion of gene inheritance, we now discuss the basic organizational features of eukaryotic cells (including how they differ from prokaryotic cells), the general structure of eukaryotic chromosomes, and the transmission of chromosomes from cell division to cell division and from generation to generation by the processes of mitosis and meiosis, respectively.

Eukaryotic Cells

Eukaryotes (meaning "true nucleus") are organisms with cells within which the genetic material (DNA) is

map (and on the chromosome to which the map relates) is called a **locus** or **gene locus.** The genetic distances between genes on the same chromosome are calculated from the results of genetic crosses by counting the frequency of recombination, that is, the percentage of the time among the progeny that the arrangement of alleles

Figure 1.7

Eukaryotic organisms that have contributed significantly to our knowledge of genetics. (a) *Saccharomyces cerevisiae* (a budding yeast). (b) *Drosophila melanogaster* (fruit fly). (c) *Caenorhabditis elegans* (a nematode). (d) *Arabidopsis thaliana* (thalecress, a member of the mustard family). (e) *Mus musculus* (mouse). (f) *Homo sapiens* (human). (g) *Neurospora crassa* (orange bread mold). (h) *Tetrahymena* (a protozoan). (i) *Paramecium* (a protozoan). (j) *Chlamydomonas reinhardtii* (a green alga). (k) *Pisum sativum* (a garden pea). (l) *Zea mays* (corn). (m) *Gallus* (chicken).

Figure 1.8

Eukaryotic cells. Cutaway diagrams of **(a)** a generalized higher plant cell and **(b)** a generalized animal cell, showing the main organizational features and the principal organelles in each.

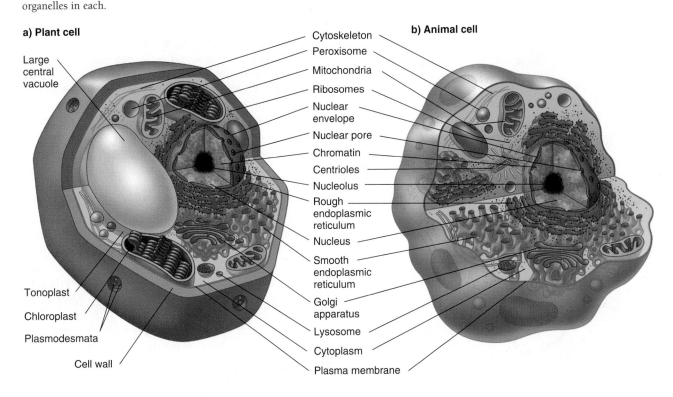

a) Plant cell

Large central vacuole

Tonoplast

Chloroplast

Plasmodesmata

Cell wall

Cytoskeleton

Peroxisome

Mitochondria

Ribosomes

Nuclear envelope

Nuclear pore

Chromatin

Centrioles

Nucleolus

Rough endoplasmic reticulum

Nucleus

Smooth endoplasmic reticulum

Golgi apparatus

Lysosome

Cytoplasm

Plasma membrane

b) Animal cell

located in the **nucleus,** a discrete structure bounded by a nuclear envelope. Eukaryotes can be unicellular or multicellular. Many eukaryotic organisms are used in genetic research. In genetics today, a great deal of research is done with six eukaryotes (Figure 1.7a–f): *Saccharomyces cerevisiae* ("sack-a-row-my-seas serry-vee-see-eye," a yeast), *Drosophila melanogaster* ("dra-soff-ee-la muh-lano-gaster," fruit fly), *Caenorhabditis elegans* ("see-no-rab-dye-tiss elly-gans," a nematode worm), *Arabidopsis thaliana* ("a-rab-ee-dop-sis thal-ee-ah-na," a small weed of the mustard family), *Mus musculus* ("muss muss-cue-lus," mouse), and *Homo sapiens* ("homo say-pee-ens," human). Humans are included not because they meet the criteria for an organism well suited for genetic experimentation but because ultimately we want to understand as much as we can about human genes and their expression. With this understanding, we will be able to combat genetic diseases and gain fundamental knowledge about our species' development and evolution.

Over the years, research with the following seven eukaryotes has also contributed significantly to our understanding of genetics (Figure 1.7g–m): *Neurospora crassa* (orange bread mold), *Tetrahymena* ("tetra-hi-me-na," a protozoan), *Paramecium* ("para-me-see-um," a protozoan), *Chlamydomonas reinhardtii* ("clammy-da-moan-as rhine-heart-ee-eye," a green alga), *Pisum sativum* ("pea-zum sa-tie-vum," garden pea), *Zea mays* (corn),

and *Gallus* (chicken). Of these, *Tetrahymena, Paramecium, Chlamydomonas,* and *Saccharomyces* are unicellular organisms, and the rest are multicellular.

You probably learned about many features of eukaryotic cells in your introductory biology course. Figure 1.8 shows a generalized higher plant cell and a generalized animal cell. Surrounding the cytoplasm of both plant cells and animal cells is the *plasma membrane.* Plant cells, but not animal cells, have a rigid cell wall outside the plasma membrane. The nucleus of eukaryotic cells contains DNA complexed with proteins and organized into a number of linear structures called *chromosomes.* The nucleus is separated from the rest of the cell—the cytoplasm and associated organelles—by the double membrane of the nuclear envelope. The membrane is selectively permeable and has pores about 20 to 80 nm (nm = nanometer = 10^{-9} meter) in diameter that allow materials to move between the nucleus and the cytoplasm. For example, *ribosomes,* which function in the cytoplasm to synthesize proteins, are assembled within the nucleus in the **nucleolus** and reach the cytoplasm through the pores.

The cytoplasm of eukaryotic cells contains many different materials and organelles. Of special interest to geneticists are the *centrioles, endoplasmic reticulum (ER), ribosomes, mitochondria,* and *chloroplasts.* Centrioles (also called basal bodies) are found in the cytoplasm of

Figure 1.9

Cutaway diagram of a generalized prokaryotic cell.

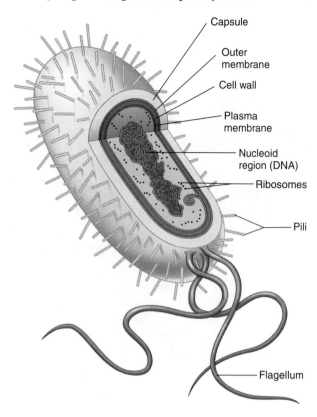

Many plant cells contain chloroplasts, large, triple-membraned, chlorophyll-containing organelles involved in photosynthesis (see Figure 1.8a). Chloroplasts also contain DNA that encodes some of the proteins that function in the chloroplast and some components of the chloroplast protein synthesis machinery.

In contrast to eukaryotes, prokaryotes (meaning "prenuclear") do not have a nuclear envelope surrounding their DNA (Figure 1.9); this is the major distinguishing feature of prokaryotes. Included in the prokaryotes are all the bacteria, which are spherical, rod-shaped, or spiral-shaped organisms; most are single-celled, although a few are multicellular and filamentous. The shape of a bacterium is maintained by a rigid cell wall located outside the cell membrane. Prokaryotes are divided into two evolutionarily distinct groups: the Bacteria and the Archaea. The Bacteria are the common varieties found in living organisms (naturally or by infection), in soil, and in water. Archaea often are found in much more inhospitable conditions, such as hot springs, salt marshes, methane-rich marshes, or the ocean depths. Bacteria generally vary in size from about 100 nm in diameter to 10 μm in diameter (μm = micrometer = 10^{-6} meter) and 60 μm in length; one species, the surgeonfish symbiont *Epulopiscium fishelsoni*, is 60 × 800 μm, fully a million times larger than *E. coli*.

In most cases, the prokaryotes studied in genetics are members of the Bacteria group. The most intensely studied is *E. coli* (Figure 1.10), a rod-shaped bacterium common in human intestines. Studies of *E. coli* have significantly advanced our understanding of the regulation of gene expression and the development of molecular biology. *E. coli* is also used extensively in recombinant DNA experiments.

nearly all animal cells (see Figure 1.8b), but not in most plant cells. In animal cells, a pair of centrioles is located at the center of the centrosome, a region of undifferentiated cytoplasm that organizes the spindle fibers that function in mitosis and meiosis, processes discussed later in this chapter.

The ER is a double-membrane system, continuous with the nuclear envelope, that runs throughout the cell. Rough ER has ribosomes attached to it, giving it a rough appearance, whereas smooth ER does not. Ribosomes bound to rough ER synthesize proteins to be secreted by the cell or to be localized in the cell membrane or particular organelles within the cell. The synthesis of proteins other than those distributed via the ER is performed by ribosomes that are free in the cytoplasm.

Mitochondria (singular, mitochondrion; see Figure 1.8) are large organelles surrounded by a double membrane; the inner membrane is highly convoluted. Mitochondria play a crucial role in processing energy for the cell. They also contain DNA that encodes some of the proteins that function in the mitochondrion and some components of the mitochondrial protein synthesis machinery.

Figure 1.10

Colorized scanning electron micrograph of *Escherichia coli*, a rod-shaped eubacterium common in human intestines.

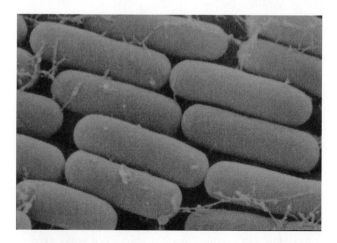

Eukaryotic Chromosomes

The genetic material of eukaryotes is distributed among multiple, linear chromosomes (see Chapter 8); the number of chromosomes, with rare exceptions, is characteristic of the species. Many eukaryotes have two copies of each type of chromosome in their nuclei, so their chromosome complement is said to be **diploid,** or 2N. Diploid eukaryotes are produced by the fusion of two gametes (mature reproductive cells specialized for sexual fusion), one from the female parent and one from the male parent. The fusion produces a diploid **zygote,** which then undergoes embryological development. Each gamete has only one set of chromosomes and is said to be **haploid** (N). The complete complement of genetic information in a haploid chromosome set is called the *genome*. Two examples of diploid organisms are humans, with 46 chromosomes (23 pairs), and *Drosophila melanogaster,* with 8 chromosomes (4 pairs). By contrast, laboratory strains of the yeast *S. cerevisiae,* with 16 chromosomes, are haploid.

In diploid organisms, the members of a chromosome pair that contain the same genes and that pair at meiosis are called **homologous chromosomes;** each member of a pair is called a **homologue,** and one homologue is inherited from each parent. Chromosomes that contain different genes and that do not pair during meiosis are called **nonhomologous chromosomes.** Figure 1.11 illustrates the chromosomal organization of haploid and diploid organisms.

In animals and in some plants, male and female cells are distinct with respect to their complement of **sex chromosomes,** the chromosomes in many eukaryotic organisms that are represented differently in the two sexes. One sex has a matched pair of sex chromosomes, and the other sex has an unmatched pair of sex chromosomes or a single sex chromosome. For example, human females have two X chromosomes (XX), whereas human males have one X and one Y (XY). Chromosomes other than sex chromosomes are called **autosomes.**

Under the microscope, we can see that chromosomes differ in size and morphology (appearance) within and between species. Each chromosome often has a constriction along its length called a **centromere,** which is important for the behavior of the chromosomes during cellular division. The location of the centromere in one of four general positions in the chromosome is useful in classifying eukaryotic chromosomes (Figure 1.12). A **metacentric chromosome** has the centromere in approximately the center, so the chromosome appears to have two approximately equal arms. **Submetacentric chromosomes** have one arm longer than the other; **acrocentric chromosomes** have one arm with a stalk and often with a "bulb" (called a *satellite*) on it; and **telocentric chromosomes** have only one arm because the centromere is at the end. Chromosomes also vary in relative size. Chromosomes of mice, for example, are all similar in length, whereas those of humans have a wide range of relative lengths. Chromosome length and centromere position are constant for each chromosome and help in identifying individual chromosomes.

Figure 1.11

Chromosomal organization of haploid and diploid organisms.

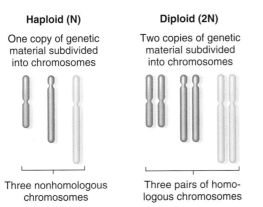

Figure 1.12

General classification of eukaryotic chromosomes as metacentric, submetacentric, acrocentric, and telocentric, based on the position of the centromere.

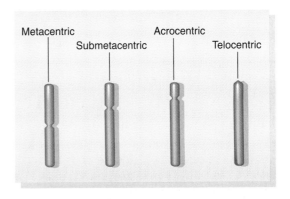

A complete set of all the metaphase chromosomes in a cell is called its **karyotype** ("carry-o-type"; literally, "nucleus type"). Metaphase chromosomes (see p. 16) are used because that is when chromosomes are in their most condensed (most compact) form during mitotic cell division and therefore are easy to see under the microscope after staining. The karyotype is species-specific, so a wide range of number, size, and shape of chromosomes is seen among eukaryotic organisms. Even closely related organisms may have quite different karyotypes.

Figure 1.13 shows the karyotype for the cell of a normal human male. It is customary, particularly with human chromosomes, to arrange chromosomes in order according to size and position of the centromere. This karyotype shows 46 chromosomes: two pairs of each of the 22 *autosomes* ("other chromosomes"; i.e., chromosomes other than the *sex chromosomes*) and one of each of the X and Y sex chromosomes (which differ greatly in size). In a human karyotype, the chromosomes are numbered for easy identi-

fication. Conventionally, the largest pair of homologous chromosomes is designated 1, the next largest 2, and so on. Although chromosome 21 is smaller than chromosome 22, it is called 21 for historical reasons. As shown in Figure 1.13, chromosomes with similar morphologies are arranged under the letter designations A through G.

Based on size and morphology alone, it is hard to distinguish the different chromosomes unambiguously when they are stained evenly. Fortunately, a number of procedures stain certain regions, or *bands*, of the chromosomes more intensely than other regions. Banding patterns are specific for each chromosome, enabling us to distinguish each chromosome in the karyotype clearly. One of these staining techniques is called G banding. In this procedure, chromosomes are treated with mild heat or proteolytic enzymes (enzymes that digest proteins) to digest the chromosomal proteins partially and then stained with Giemsa stain to produce dark bands called G bands (see Figure 1.13). In humans, approximately 300 G bands can be distinguished in metaphase chromosomes, and approximately 2,000 G bands can be distinguished in chromosomes from the prophase stage of mitosis. Conventionally, drawings (*ideograms*) of human chromosomes show the G banding pattern. Furthermore, a standard nomenclature has been established for the chromosomes based on the banding patterns so that scientists can talk about gene and marker locations with reference to specific regions and subregions. Each chromosome has two arms separated by the centromere. The smaller arm is designated p and the larger arm is designated q. Numbered regions and numbered subregions are then assigned from the centromere outward; that is, region 1 is closest to the centromere. For example, the breast cancer susceptibility gene *BRCA1* is at location 17q21, meaning that it is on the long arm of chromosome 17 in region 21. Subregions are indicated by decimal numerals after the region number; for example, the cystic fibrosis gene spans subregions 7q31.2–q31.3. That is, it spans both subregions 2 and 3 of region 31 of the long arm of chromosome 7.

Figure 1.13

G banding in a karyotype of human male metaphase chromosomes.

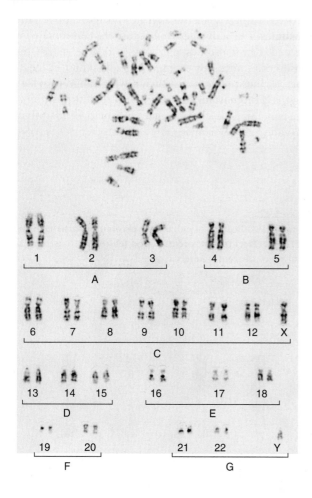

KEYNOTE

Diploid eukaryotic cells have two haploid sets of chromosomes, one set coming from each parent. The members of a pair of chromosomes, one from each parent, are called homologous chromosomes. Haploid eukaryotic cells have only one set of chromosomes. The complete set of chromosomes in a cell is called its karyotype. The karyotype is species-specific. Staining with particular dyes results in characteristic banding patterns, which has led to a numbering system for defining chromosome regions and subregions.

Mitosis

In both unicellular and multicellular eukaryotes, cellular reproduction is a cyclical process of growth, **mitosis** (*nuclear division,* or *karyokinesis*), and (usually) **cell division** (*cytokinesis*). The cycle of growth, mitosis, and cell division is called the **cell cycle.** In proliferating somatic cells, the cell cycle consists of two phases: the mitotic (or division) phase (M) and an interphase between divisions (Figure 1.14). Interphase consists of three stages: G_1 (gap 1), S, and G_2 (gap 2). During G_1 (the presynthesis stage), the cell prepares for DNA and chromosome replication, which take place in the S (synthesis) stage. In G_2 (the postsynthesis stage), the cell prepares for cell division, or the M phase. Put another way, chromosome replication takes place in interphase, and then mitosis occurs, resulting in the distribution of a complete chromosome set to each of two progeny nuclei.

The relative time spent in each of the four stages of mitosis varies greatly among cell types. In a given organism, variation in the length of the cell cycle depends primarily on the duration of G_1; the duration of S plus G_2 plus M is approximately the same in all cell types. For example, some cancer cells and early fetal cells of humans spend minutes in G_1, whereas some differentiated adult cells (such as nerve cells) spend years in G_1. Finally, some cells exit the cell cycle from G_1 and enter a quiescent, nondividing state called G_0.

During interphase, the individual chromosomes are elongated and are difficult to see under the light microscope. The DNA of each chromosome is replicated in the S phase, giving two exact copies, called **sister chromatids,** which are held together by the replicated but unseparated centromeres. (Because the centromeres have not separated, only one centromere structure is visible under the microscope.) More precisely, a **chromatid** is one of the two visibly distinct longitudinal subunits of all replicated chromosomes that become visible between early prophase and metaphase of mitosis. Later, when the centromeres separate, the sister chromatids become known as **daughter chromosomes.**

Mitosis occurs in both haploid and diploid cells. It is a continuous process, but for purposes of discussion, it is usually divided into four cytologically distinguishable stages called *prophase, metaphase, anaphase,* and *telophase.* Figure 1.15 shows the four stages in simplified diagrams. The photographs in Figure 1.16 show the typical chromosome morphology in interphase and in the four stages of mitosis in animal (whitefish early embryo) cells.

Prophase. At the beginning of **prophase** (see Figures 1.15 and 1.16b), the chromatids are very elongated. In preparation for mitosis they begin to coil tightly so that they appear shorter and fatter under the microscope. By late prophase, each chromosome, which was duplicated during the preceding S phase of interphase, can be seen to consist of two sister chromatids.

Many mitotic events depend on the *mitotic spindle* (spindle apparatus), a structure consisting of fibers composed of microtubules made of special proteins called *tubulins.* The mitotic spindle assembles outside the nucleus during prophase. In most animal cells, the centrioles (see Figure 1.8b) are the focal points for spindle assembly; higher plant cells usually lack centrioles, but they do have a mitotic spindle. Centrioles are arranged in pairs, and before the S phase, the cell's centriole pair replicates. Then, during mitosis, each new centriole pair becomes the focus of a radial array of microtubules called the aster. Early in prophase, the two asters are next to one another and close to the nuclear envelope; by late prophase, the two asters have moved far apart along the outside of the nucleus and are spanned by the microtubular spindle fibers.

Near the end of prophase, a substage called *prometaphase* occurs in which the nucleolus or nucleoli cease to be discrete areas within the nucleus and the nuclear envelope breaks down, allowing the spindle to enter the nuclear area. Specialized structures called **kinetochores** form on each face of the centromere of each chromosome and become attached to special microtubules called *kinetochore microtubules* (Figure 1.17).

Metaphase. Metaphase (see Figures 1.15 and 1.16c) begins when the nuclear envelope has completely disappeared. During metaphase, the kinetochore microtubules

Figure 1.14

Eukaryotic cell cycle. This cycle assumes a period of 24 hours, although great variation exists between cell types and organisms.

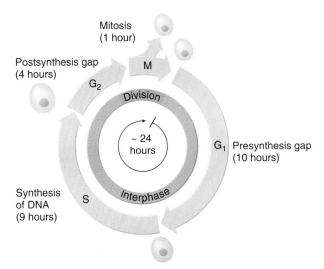

Figure 1.15

Interphase and mitosis in an animal cell.

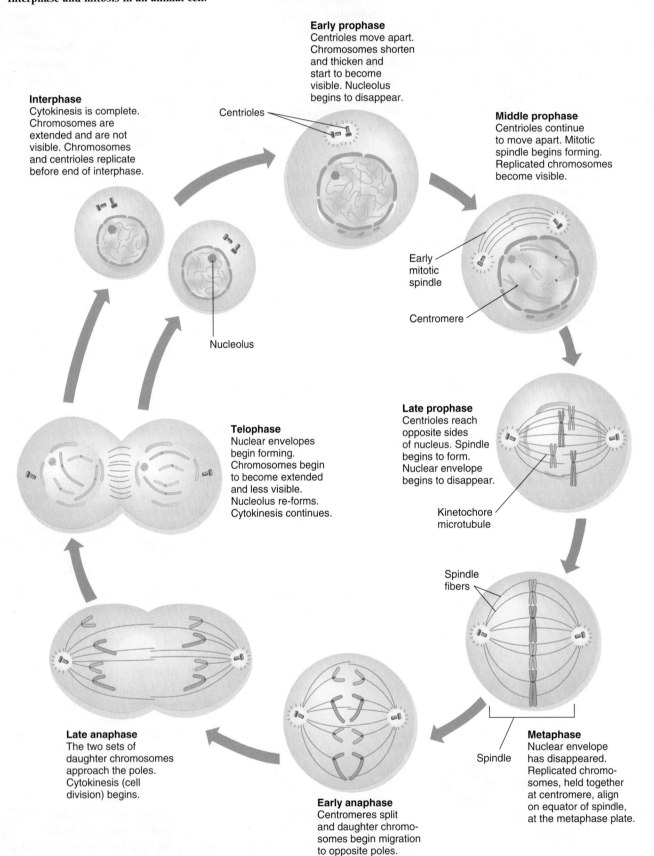

Interphase
Cytokinesis is complete. Chromosomes are extended and are not visible. Chromosomes and centrioles replicate before end of interphase.

Nucleolus

Early prophase
Centrioles move apart. Chromosomes shorten and thicken and start to become visible. Nucleolus begins to disappear.

Centrioles

Middle prophase
Centrioles continue to move apart. Mitotic spindle begins forming. Replicated chromosomes become visible.

Early mitotic spindle

Centromere

Late prophase
Centrioles reach opposite sides of nucleus. Spindle begins to form. Nuclear envelope begins to disappear.

Kinetochore microtubule

Spindle fibers

Metaphase
Nuclear envelope has disappeared. Replicated chromosomes, held together at centromere, align on equator of spindle, at the metaphase plate.

Spindle

Early anaphase
Centromeres split and daughter chromosomes begin migration to opposite poles.

Late anaphase
The two sets of daughter chromosomes approach the poles. Cytokinesis (cell division) begins.

Telophase
Nuclear envelopes begin forming. Chromosomes begin to become extended and less visible. Nucleolus re-forms. Cytokinesis continues.

Figure 1.16

Interphase and the stages of mitosis in whitefish early embryo cells. (**a**) Interphase. (**b**) Late prophase. (**c**) Metaphase. (**d**) Early anaphase. (**e**) Telophase.

a)

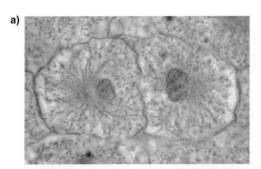

b)

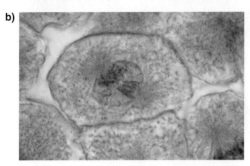

c)

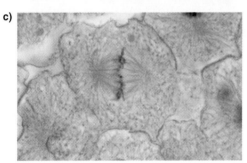

d)

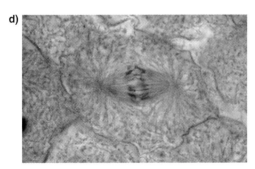

e)

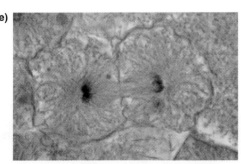

Figure 1.17

Kinetochores and kinetochore microtubules.

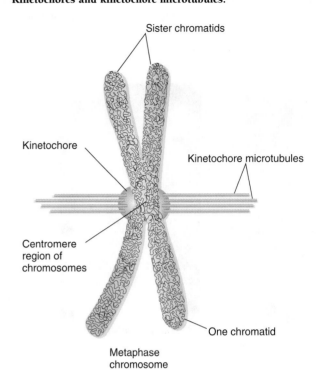

Sister chromatids

Kinetochore

Kinetochore microtubules

Centromere region of chromosomes

One chromatid

Metaphase chromosome

orient the chromosomes so that their centromeres become aligned in one plane halfway between the two spindle poles, with the long axes of the chromosomes at 90 degrees to the spindle axis. The plane where the chromosomes become aligned is called the metaphase plate.

Figure 1.18 shows electron micrographs of human chromosomes in metaphase. Note the highly condensed state of the sister chromatids. The micrograph in Figure 1.18c shows a human chromosome from which much of the protein has been removed. In the center of the chromosome is a dense framework of protein called a *scaffold,* which still has the form of the chromosome. The scaffold is surrounded by a halo of DNA that has uncoiled and spread outward.

Anaphase. Anaphase (see Figures 1.15 and 1.16d) begins when the joined centromeres of sister chromatids separate, giving rise to two daughter chromosomes. Once the paired kinetochores on each chromosome separate, the sister chromatid pairs undergo disjunction (separation), and the daughter chromosomes (the former sister chromatids) move toward the poles. In anaphase, the daughter chromosomes are pulled toward the opposite poles of the cell by the shortening kinetochore microtubules. As they are pulled, the chromosomes have characteristic shapes related to the location of the centromere along the chromosome's length. For example, a metacentric chromosome is V-shaped as the two roughly equal-length

Figure 1.18

Human metaphase chromosome. (a) Transmission electron micrograph of an intact chromosome. (b) Colorized scanning electron micrograph of an intact chromosome. (c) Transmission electron micrograph of an intact chromosome from which much of the protein has been removed; the DNA filaments have uncoiled.

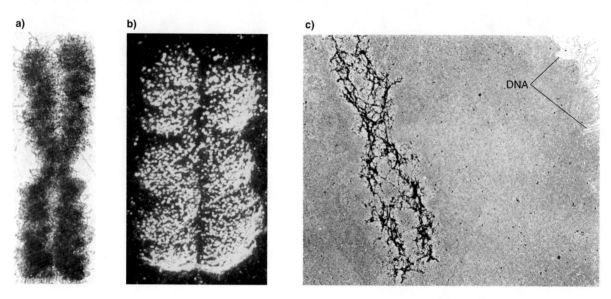

a)   b)   c)

DNA

chromosome arms trail the centromere in its migration toward the pole. Similarly, a submetacentric chromosome has a J shape with a long and short arm. Cytokinesis (cell division) usually begins in the latter stages of anaphase.

Telophase. During **telophase** (see Figures 1.15 and 1.16e), the migration of daughter chromosomes to the two poles is completed, and the two sets of daughter chromosomes are assembled into two groups at opposite ends of the cell. The chromosomes begin to uncoil and assume the elongated state characteristic of interphase. A nuclear envelope forms around each chromosome group, the spindle microtubules disappear, and the nucleolus or nucleoli re-form. At this point, nuclear division is complete: The cell has two nuclei.

Cytokinesis. **Cytokinesis** is division of the cytoplasm; usually it follows the nuclear division stage of mitosis and is completed by the end of telophase. Cytokinesis compartmentalizes the two new nuclei into separate daughter cells, completing the mitosis and cell division process (Figure 1.19). In animal cells, cytokinesis proceeds with the formation of a constriction in the middle of the cell; the constriction continues until two daughter cells are produced (see Figure 1.19a). By contrast, most plant cells do not divide by the formation of a constriction. Instead, a new cell membrane and cell wall are assembled between the two new nuclei to form a *cell plate* (see Figure 1.19b). Cell wall material coats each side of the cell plate, and the result is two progeny cells.

Gene Segregation in Mitosis. Mitosis maintains a constant amount of genetic material and a constant set of genes from cell generation to cell generation. Mitosis is a highly ordered process in which one copy of each duplicated chromosome segregates into both daughter cells. Thus, for a haploid (N) cell, chromosome duplication produces a cell in which each chromosome has doubled its content. Mitosis then results in two progeny haploid cells, each with one complete set of chromosomes (a genome). For a diploid (2N) cell, which has two sets of chromosomes (two genomes), chromosome duplication produces a cell in which each chromosome set has doubled its content. Mitosis then results in two genetically identical progeny diploid cells, each with two sets of chromosomes (two genomes). As a result, an equal distribution of genetic material occurs, and no genetic material is lost.

KEYNOTE

Mitosis is the process of nuclear division in eukaryotes. It is one part of the cell cycle (the M in G_1, S, G_2, and M), and it results in the production of daughter nuclei that contain identical chromosome numbers and are genetically identical to one another and to the parent nucleus from which they arose. Before mitosis, the chromosomes duplicate. Mitosis is usually followed by cytokinesis. Both haploid and diploid cells proliferate by mitosis.

Figure 1.19

Cytokinesis (cell division). **(a)** Diagram of cytokinesis in an animal cell. **(b)** Diagram of cytokinesis in a plant cell.

a) Animal cell   **b) Plant cell**

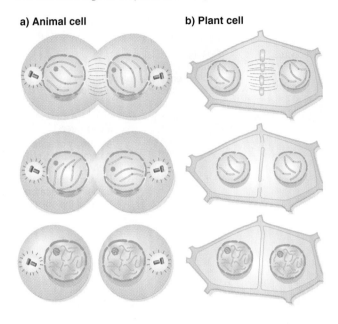

Meiosis

Meiosis consists of two successive divisions of a *diploid* nucleus after only one DNA replication (chromosome duplication) cycle. That original diploid nucleus contains one haploid set of chromosomes from the mother and one set from the father (with the exception of self-fertilizing organisms, such as many plants, in which case both sets of chromosomes come from the same parent). Meiosis occurs only at a special point in an organism's life cycle. In animals it results in the formation of haploid gametes (in **gametogenesis**), and in plants it results in the formation of haploid meiospores (in *sporogenesis*). (A meiospore undergoes mitosis to produce a gamete-bearing, multicellular stage called the gametophyte.) Before meiosis, homologous chromosomes replicate; during meiosis, they pair and then undergo two divisions—meiosis I and meiosis II—each consisting of a series of stages (Figure 1.20). Meiosis I results in reduction in the number of chromosomes from diploid to haploid (reductional division), and meiosis II results in separation of the sister chromatids. As a consequence, each of the four nuclei resulting from the two meiotic divisions receives one chromosome of each chromosome set (i.e., one complete haploid genome). In most cases, the divisions are accompanied by cytokinesis, so the result of meiosis of a single diploid cell is four haploid cells.

animation

Meiosis

Meiosis I: The First Meiotic Division.   Meiosis I, in which the chromosome number is reduced from diploid to haploid, consists of four stages: *prophase I, metaphase I, anaphase I,* and *telophase I.*

Prophase I.   As **prophase I** begins, the chromosomes have already duplicated (see Figure 1.20). Prophase I is divided into a number of substages. Except for the behavior of homologous pairs of chromosomes and crossing-over, prophase I of meiosis is very similar to prophase of mitosis.

In **leptonema** (early prophase I, the leptotene stage), the chromosomes have begun to coil. Once a cell enters leptonema, it is committed to the meiotic process. A key event in leptonema is *pairing* of homologous chromosomes, the loose alignment of homologous regions of the chromosomes. Then a most significant event begins: **crossing-over,** the reciprocal physical exchange of chromosome segments at corresponding positions along pairs of homologous chromosomes. If there are genetic differences between the homologues, crossing-over can produce new gene combinations in a chromatid. There is usually no loss or addition of genetic material to either chromosome because crossing-over involves reciprocal exchanges. A chromosome that emerges from meiosis with a combination of genes that differs from the combination with which it started is called a **recombinant chromosome.** Therefore, crossing-over is a mechanism that can give rise to *genetic recombination.*

In **zygonema** (early to mid-prophase I, the zygotene stage) a key event is **synapsis,** the formation of a very tight association of homologous chromosomes brought about by the formation along the length of the chromatids of a zipperlike structure called the **synaptonemal complex.** Because of the replication that occurred earlier, each synapsed set of homologous chromosomes consists of four chromatids and is called a **bivalent** or *tetrad.* The chromosomes are maximally condensed before the synaptonemal complex forms.

Following zygonema is **pachynema** (mid-prophase I, the pachytene stage). At the end of pachynema, the synaptonemal complex is disassembled, and the chromosomes have started to elongate.

In **diplonema** (mid- to late prophase I, the diplotene stage) the chromosomes begin to move apart. The result of crossing-over becomes visible during diplonema as a cross-shaped structure called a **chiasma** (plural, *chiasmata;* Figure 1.21). Because all four chromatids may be involved in crossing-over events along the length of the homologues, the chiasma pattern at this stage may be very complex.

Diplonema is followed rapidly in most organisms by the remaining stages of meiosis. However, in many animals, the oocytes (egg cells) can remain in diplonema for very long periods. In human females, for example, oocytes go through meiosis I up to diplonema by the seventh month of fetal development and then remain arrested in this stage for many years. At the onset of puberty and until menopause, one oocyte per menstrual cycle completes meiosis I and is ovulated. If the oocyte is

Figure 1.20

The stages of meiosis in an animal cell.

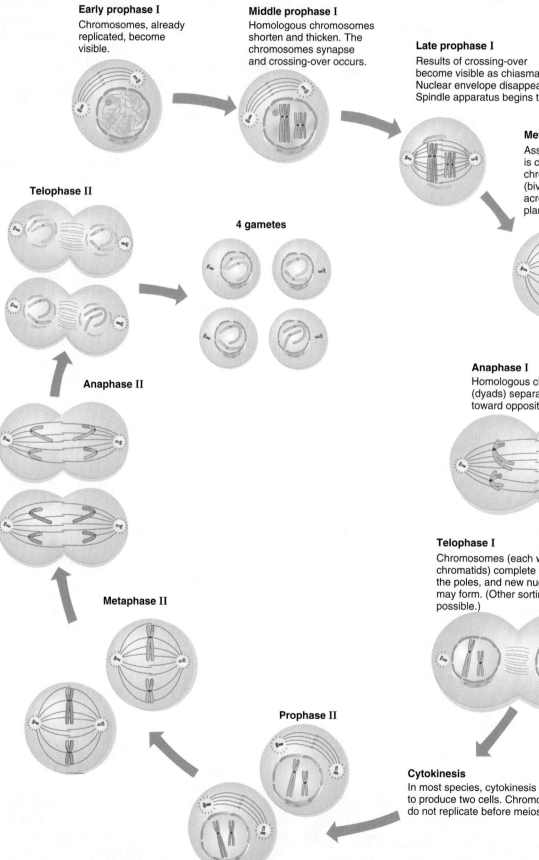

Early prophase I
Chromosomes, already replicated, become visible.

Middle prophase I
Homologous chromosomes shorten and thicken. The chromosomes synapse and crossing-over occurs.

Late prophase I
Results of crossing-over become visible as chiasmata. Nuclear envelope disappears. Spindle apparatus begins to form.

Metaphase I
Assembly of spindle is completed. Each chromosome pair (bivalent) aligns across the equatorial plane of the spindle.

Telophase II

4 gametes

Anaphase II

Anaphase I
Homologous chromosome pairs (dyads) separate and migrate toward opposite poles.

Telophase I
Chromosomes (each with two sister chromatids) complete migration to the poles, and new nuclear envelopes may form. (Other sorting patterns are possible.)

Metaphase II

Prophase II

Cytokinesis
In most species, cytokinesis occurs to produce two cells. Chromosomes do not replicate before meiosis II.

Figure 1.21

Appearance of chiasmata, the visible evidence of crossing-over, in diplonema.

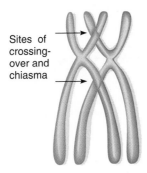

Sites of crossing-over and chiasma

fertilized by a sperm as it passes down the fallopian tube, it quickly completes meiosis II, and by fusion with a haploid sperm, a functional zygote is produced.

In **diakinesis** (late prophase), the nucleolus and nuclear envelope break down. Simultaneously, the spindle is assembled. The chromosomes can be counted most easily at this stage of meiosis.

The pairing, crossing-over, and synapsis phenomena described for prophase I apply to homologous chromosomes—namely, the autosomes. Even though the sex chromosomes are not homologous chromosomes, the mammalian Y chromosome has a small terminal portion called the *pseudoautosomal region* that is homologous to a portion of the X chromosome. In meiosis in males, crossing-over between the pseudoautosomal regions of the X and Y chromosomes is necessary for correct segregation of X and Y chromosomes as meiosis proceeds.

Metaphase I. By the beginning of **metaphase I** (see Figure 1.20), the nuclear envelope has completely broken down and the bivalents have become aligned on the equatorial plane of the cell. The spindle is completely formed now, and the microtubules are attached to the kinetochores of the homologues. Note particularly that the *pairs of homologues* (the bivalents) are found at the metaphase plate. In contrast, in mitosis of most organisms, replicated homologous chromosomes (sister chromatid pairs) align *independently* at the metaphase plate.

Anaphase I. In **anaphase I** (see Figure 1.20), the chromosomes in each bivalent separate, so the chromosomes of each homologous pair disjoin and migrate toward opposite poles, the areas in which new nuclei will form. (At this stage, each of the separated chromosomes is called a *dyad*.) This migration assumes that maternally derived and paternally derived centromeres segregate randomly to each pole (except for the parts of chromosomes exchanged during the crossing-over process) and that at each pole there is a haploid complement of replicated centromeres with associated chromosomes. *At this time, the segregated sister chromatid pairs remain attached at their respective cen-* *tromeres. In other words, a key difference between mitosis and meiosis I is that sister chromatids remain joined after metaphase in meiosis I, whereas in mitosis they separate.*

Telophase I. In **telophase I** (see Figure 1.20), the dyads complete their migration to opposite poles of the cell, and (in most cases) new nuclear envelopes form around each haploid grouping. In most species, cytokinesis follows, producing two haploid cells. Thus, meiosis I, which begins with a diploid cell that contains one maternally derived and one paternally derived set of chromosomes, ends with two nuclei, each of which is haploid and contains one mixed-parental set of dyads. After cytokinesis, each of the two progeny cells has a nucleus with a haploid set of dyads.

Meiosis II: The Second Meiotic Division. The second meiotic division is very similar to a mitotic division (see Figure 1.20).

In **prophase II,** the chromosomes condense.

In **metaphase II,** each of the two daughter cells organizes a spindle apparatus that attaches to the centromeres that still connect the sister chromatids. The centromeres line up on the equator of the second-division spindles.

During **anaphase II,** the centromeres split and the chromatids are pulled to the opposite poles of the spindle. One sister chromatid of each pair goes to one pole, and the other goes to the opposite pole. The separated chromatids are considered chromosomes in their own right.

In the last stage, **telophase II,** a nuclear envelope forms around each set of chromosomes, and cytokinesis takes place. After telophase II, the chromosomes become more elongated and are no longer visible under the light microscope.

The end products of the two meiotic divisions are four haploid cells (gametes in animals) from one original diploid cell (see Figure 1.20). Each of the four progeny cells has one chromosome from each homologous pair of chromosomes. Moreover, these chromosomes are not exact copies of the original chromosomes because of crossing-over. Figure 1.22 compares mitosis and meiosis.

Gene Segregation in Meiosis. Meiosis has three significant results:

1. Meiosis generates haploid cells with half the number of chromosomes found in the diploid cell that entered the process because two division cycles follow only one cycle of DNA replication (S period). Fusion of haploid nuclei restores the diploid number. Therefore, through a cycle of meiosis and fusion, the chromosome number is maintained in sexually reproducing organisms.

2. In metaphase I, each maternally derived and paternally derived chromosome has an equal chance of aligning on one or the other side of the equatorial metaphase plate. As a result, each nucleus generated by meiosis has some combination of maternal and paternal chromosomes.

Figure 1.22

Comparison of mitosis and meiosis in a diploid cell.

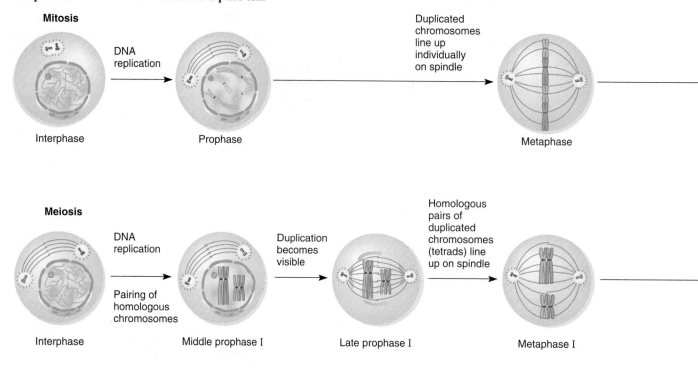

The number of possible chromosome combinations in the haploid nuclei resulting from meiosis is large, especially when the number of chromosomes in an organism is large. Consider a hypothetical organism with two pairs of chromosomes in a diploid cell entering meiosis. Figure 1.23 shows the two possible combinations of maternal and paternal chromosomes that can occur at the metaphase plate. Understanding this concept is useful when we consider gene segregation in Chapter 2.

The general formula for the number of possible chromosome arrangements at the metaphase plate in meiosis is 2^{n-1}, where n is the number of chromosome pairs. Similarly, the general formula for the number of possible chromosome combinations in the nuclei resulting from meiosis is 2^n. In *Drosophila*, which has four pairs of chromosomes, the number of possible combinations in nuclei resulting from meiosis is 2^3, or 8; in humans, which have 23 chromosome pairs, more than 4 million combinations are possible. Therefore, because there are many gene differences between the maternally derived and paternally derived chromosomes, the nuclei produced by meiosis are genetically quite different from the parental cell and from one another.

3. The crossing-over between maternal and paternal chromatid pairs during meiosis I generates still more variation in the final combinations. Crossing-over occurs during every meiosis, and because the sites of crossing-

Figure 1.23

The two possible arrangements of two pairs of homologous chromosomes on the metaphase plate of the first meiotic division. Paternal chromosomes are shown in yellow and green, maternal chromosomes in purple and tan.

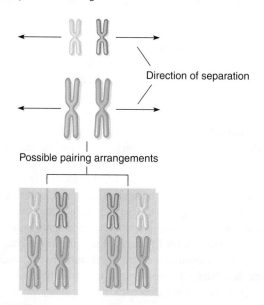

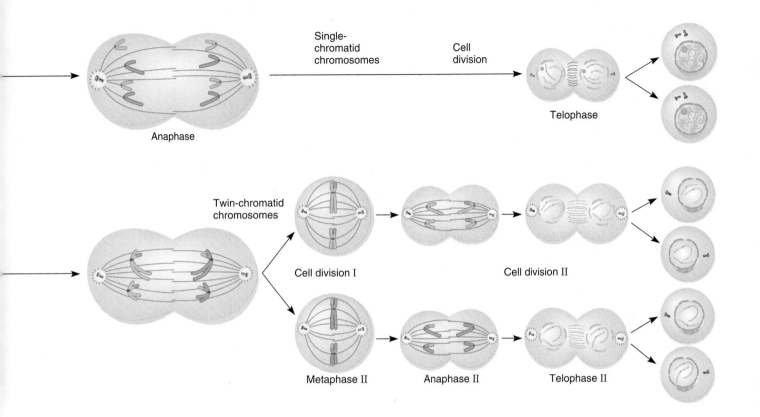

Anaphase

Single-chromatid chromosomes

Cell division

Telophase

Twin-chromatid chromosomes

Cell division I

Metaphase II

Anaphase II

Cell division II

Telophase II

over vary from one meiosis to another, the number of different kinds of progeny nuclei produced by this process is extremely large. Given the genetic features of meiosis, this process is of critical importance for understanding the behavior of genes.

The events that occur in meiosis are the bases for the segregation and independent assortment of genes according to Mendel's laws, discussed in Chapter 2.

KEYNOTE

Meiosis occurs in all sexually reproducing eukaryotes. It is a process by which a specialized diploid (2N) cell or cell nucleus with two sets of chromosomes is transformed, through one round of chromosome replication and two rounds of nuclear division, into four haploid (N) cells or nuclei, each with one set of chromosomes. In the first of two divisions, pairing, crossing-over, and synapsis of homologous chromosomes occur. The meiotic process, in combination with fertilization, results in the conservation of the number of chromosomes from generation to generation. It also generates genetic variability through the various ways in which maternal and paternal chromosomes are combined in the progeny nuclei and by crossing-over (the physical exchange of chromosome segments at corresponding positions along pairs of homologous chromosomes).

Meiosis in Animals and Plants. Lastly, we discuss briefly the role of meiosis in animals and plants.

Meiosis in Animals. Most multicellular animals are diploid through most of their life cycles. In such animals, meiosis produces haploid gametes, fusion of two haploid gametes produces a diploid zygote when their nuclei fuse in fertilization, and the zygote then divides by mitosis to produce the new diploid organism; this series of events, involving an alternation of diploid and haploid phases, is **sexual reproduction.** Thus, the gametes are the only haploid stages of the life cycle. Gametes are formed only in specialized cells. In males, the gamete is the sperm, produced through a process called **spermatogenesis;** in females, the gamete is the egg, produced by **oogenesis.** Spermatogenesis and oogenesis are illustrated in Figure 1.24.

In male animals, the sperm cells (spermatozoa) are produced within the testes. The testes contain the primordial germ cells (*primary spermatogonia*), which, via mitosis, produce *secondary spermatogonia*. Spermatogonia transform into *primary spermatocytes (meiocytes)*, each of which undergoes meiosis I and gives rise to two *secondary spermatocytes*. Each secondary spermatocyte undergoes meiosis II. The results of these two divisions are four haploid *spermatids* that eventually differentiate into the male gametes, the spermatozoa.

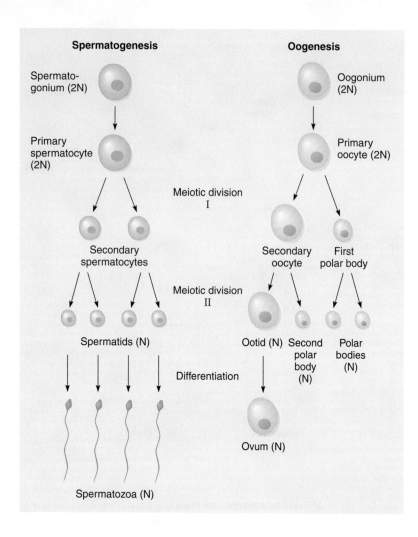

Spermatogenesis

Spermato-
gonium (2N)

Primary
spermatocyte
(2N)

Meiotic division
I

Secondary
spermatocytes

Meiotic division
II

Spermatids (N)

Differentiation

Spermatozoa (N)

Oogenesis

Oogonium
(2N)

Primary
oocyte (2N)

Secondary   First
oocyte   polar body

Ootid (N) Second   Polar
           polar   bodies
           body   (N)
           (N)

Ovum (N)

Figure 1.24

Spermatogenesis and oogenesis in an animal cell.

In female animals, the ovary contains the primordial germ cells (*primary oogonia*), which, by mitosis, give rise to *secondary oogonia*. These cells transform into **primary oocytes,** which grow until the end of oogenesis. The diploid primary oocyte goes through meiosis I and unequal cytokinesis to give two cells: a large one called the **secondary oocyte** and a very small one called the *first polar body*. In meiosis II, the secondary oocyte produces two haploid cells. One is a very small cell called a *second polar body;* the other is a large cell that rapidly matures into the mature egg cell, or **ovum.** The first polar body may or may not divide during meiosis II. The polar bodies have no function in most species and degenerate; only the ovum is a viable gamete. (In many animals, the cell that is actually fertilized is the secondary oocyte; however, nuclear fusion must await completion of meiosis by that oocyte.) Thus, in the female animal, only one mature gamete (the ovum) is produced by meiosis of a diploid cell. In humans, all oocytes are formed in the fetus, and one oocyte completes meiosis I each month in the adult female but does not progress further unless stimulated to do so by fertilization by a sperm.

Meiosis in Plants. The life cycle of sexually reproducing plants typically has two phases: the **gametophyte,** or haploid, stage, in which gametes are produced, and the **sporophyte,** or diploid, stage, in which haploid spores are produced by meiosis.

In angiosperms (the flowering plants), the flower is the structure in which sexual reproduction occurs. Figure 1.25 shows a generalized flower containing both male and female reproductive organs, the **stamens** and **pistils,** respectively. Each stamen consists of a single stalk, the filament, on the top of which is an anther. From the anther are released the pollen grains, which are immature male gametophytes (gamete-producing phase). The pistil, which contains the female gametophytes, typically consists of the stigma, a sticky surface specialized to receive the pollen; the style, a thin stalk down which a pollen tube grows from the adhered pollen grain; and, at the base of the structure, the ovary, within which are the ovules. Each ovule encloses a female gametophyte (the embryo sac) containing a single egg cell. When the egg cell is fertilized, the ovule develops into a seed.

Figure 1.25
Generalized structure of a flower.

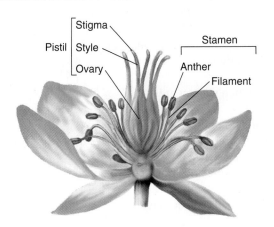

Among living organisms, only plants produce gametes from special bodies called gametophytes. Thus, plant life cycles have two distinct reproductive phases, called the **alternation of generations** (Figure 1.26). Meiosis and fertilization are the transitions between these stages. The haploid *gametophyte generation* begins with spores that are produced by meiosis. In flowering plants, the spores are the cells that ultimately become pollen and embryo sac. Fertilization initiates the diploid *sporophyte generation*, which produces the specialized haploid cells called spores, completing the cycle.

Summary

This chapter introduced the subject of genetics. Genetics often is divided into four main branches: transmission genetics, molecular genetics, population genetics, and quantitative genetics. Geneticists investigate all aspects of genes, and, depending on whether the goal is to obtain fundamental knowledge or to solve problems in society or exploit discoveries, genetic research is considered basic or applied, respectively.

This chapter also introduced some of the organisms that have been fruitful subjects of genetic studies. As the simplest cellular organisms, prokaryotes lack a membrane-bounded nucleus and have a single chromosome located in a nucleoid region. Eukaryotic organisms may be single-celled or multicellular and have several chromosomes located within a membrane-bounded nucleus.

Eukaryotic cells contain either one or two sets of chromosomes; the former are haploid cells and the latter are diploid cells. Both haploid and diploid cells proliferate by mitosis. Mitosis involves one round of DNA replication followed by one round of nuclear division, often accompanied by cell division. Thus, mitosis results in the production of daughter nuclei that contain identical chromosome numbers and that are genetically identical to one another and to the parent nucleus from which they arose.

Figure 1.26
Alternation of gametophyte and sporophyte generations in flowering plants.

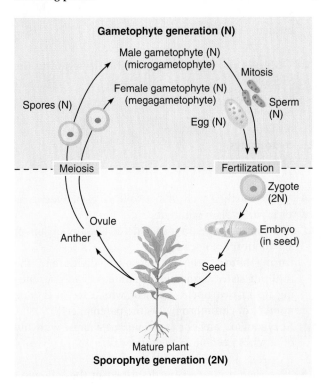

In all sexually reproducing organisms, meiosis occurs at a particular stage in the life cycle. Meiosis is the process by which a diploid cell (never a haploid cell) or cell nucleus undergoes one round of DNA replication and two rounds of nuclear division to produce four specialized haploid cells or nuclei. The products of meiosis are gametes or meiospores. Unlike mitosis, meiosis generates genetic variability in two main ways: through the various ways in which maternal and paternal chromosomes are combined in progeny nuclei and through crossing-over between maternally derived and paternally derived homologues to produce recombinant chromosomes with some maternal and some paternal genes.

Questions and Problems

Solutions to Questions and Problems marked with an asterisk are found at the end of this text, starting on page 564. Solutions to all problems are found in the Student Study Guide and Solutions Manual *that accompanies this text.*

*1.1 Interphase is a period corresponding to the cell cycle phases of
a. mitosis
b. S
c. $G_1 + S + G_2$
d. $G_1 + S + G_2 + M$

1.2 Chromatids joined together by a centromere are called
a. sister chromatids.
b. homologues.
c. alleles.
d. bivalents (tetrads).

*1.3 Mitosis and meiosis always differ in regard to the presence of
a. chromatids.
b. homologues.
c. bivalents.
d. centromeres.
e. spindles.

1.4 State whether each of the following statements is true or false. Explain your choice.
a. The chromosomes in a somatic cell of any organism are all morphologically alike.
b. During mitosis, the chromosomes divide and the resulting sister chromatids separate at anaphase, ending up in two nuclei, each of which has the same number of chromosomes as the parental cell.
c. At zygonema, any chromosome can synapse with any other chromosome in the same cell.

1.5 For each mitotic event described in the following table, write the name of the event in the blank provided in front of the description. Then put the events in the correct order (sequence); start by placing a 1 next to the description of interphase and continue through 6, which should correspond to the last event in the sequence.

Name of event		Order of event
	The cytoplasm divides and the cell contents are separated into two separate cells.	
	Chromosomes become aligned along the equatorial plane of the cell.	
	Chromosome replication occurs.	
	The migration of the daughter chromosomes to the two poles is complete.	
	Replicated chromosomes begin to condense and become visible under the microscope.	
	Sister chromatids begin to separate and migrate toward opposite poles of the cell.	

*1.6 Determine whether the answer to each of these questions is yes or no. Explain the reasons for your answer.
a. Can meiosis occur in haploid species?
b. Can meiosis occur in a haploid individual?

1.7 Select the correct answer. The general life cycle of a eukaryotic organism has the sequence
a. N→meiosis→2N→fertilization→N
b. 2N→meiosis→N→fertilization→2N
c. N→mitosis→2N→fertilization→N
d. 2N→mitosis→N→fertilization→2N

*1.8 Which statement is true?
a. Gametes are 2N; zygotes are N.
b. Gametes and zygotes are 2N.
c. The number of chromosomes can be the same in gamete cells and in somatic cells.
d. The zygotic and somatic chromosome numbers cannot be the same.
e. Haploid organisms have haploid zygotes.

1.9 All of the following happen in prophase I of meiosis except
a. chromosome condensation.
b. pairing of homologues.
c. chiasma formation.
d. terminalization.
e. segregation.

*1.10 Give the name of the stage of mitosis or meiosis at which each of the following events occurs.
a. Chromosomes are located in a plane at the center of the spindle.
b. The chromosomes move away from the spindle equator to the poles.

1.11 Given the diploid, meiotic mother cell shown here, diagram the chromosomes as they would appear

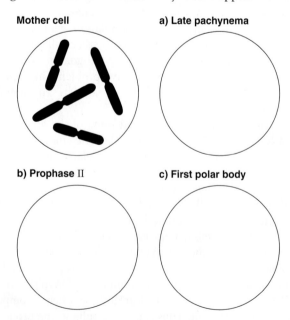

Mother cell   a) Late pachynema

b) Prophase II   c) First polar body

a. in late pachynema.
b. in a nucleus at prophase of the second meiotic division.
c. in the first polar body resulting from oogenesis in an animal.

1.12 The cells in the following figure were all taken from the same individual (a mammal). Identify the cell division events occurring in each cell and explain your reasoning. What is the sex of the individual? What is the diploid chromosome number?

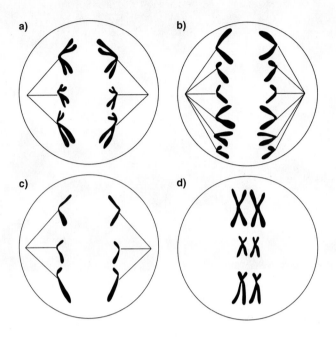

1.13 Does mitosis or meiosis have greater significance in the study of heredity? Explain your choice.

*1.14 Consider a diploid organism that has three pairs of chromosomes. Assume that the organism receives chromosomes A, B, and C from the female parent and A′, B′, and C′ from the male parent. To answer the following questions, assume that no crossing-over occurs.
a. What proportion of the gametes of this organism would be expected to contain all the chromosomes of maternal origin?
b. What proportion of the gametes would be expected to contain some chromosomes from both maternal and paternal origin?

*1.15 Normal diploid cells of a theoretical mammal are examined cytologically at the mitotic metaphase stage for their chromosome complement. One short chromosome, two medium-length chromosomes, and three long chromosomes are present. Explain how the cells might have such a set of chromosomes.

1.16 Is the following statement true or false? Explain your answer. "Meiotic chromosomes can be seen after appropriate staining in nuclei from rapidly dividing skin cells."

*1.17 Is the following statement true or false? Explain your answer. "All the sperm from one human male are genetically identical."

1.18 The horse has a diploid set of 64 chromosomes, and a donkey has a diploid set of 62 chromosomes. Mules (viable but usually sterile progeny) are produced when a male donkey is mated to a female horse. How many chromosomes will a mule cell contain?

*1.19 The red fox has 17 pairs of large, long chromosomes. The arctic fox has 26 pairs of shorter, smaller chromosomes.
a. What do you expect to be the chromosome number in somatic tissues of a hybrid between these two foxes?
b. The first meiotic division in the hybrid fox shows a mixture of paired and single chromosomes. Why do you suppose this occurs? Can you suggest a possible relationship between this fact and the observed sterility of the hybrid?

*1.20 At the time of synapsis preceding the reduction division in meiosis, the homologous chromosomes align in pairs, and one member of each pair passes to each of the daughter nuclei. In an animal with five pairs of chromosomes, assume that chromosomes 1, 2, 3, 4, and 5 have come from the father, and 1′, 2′, 3′, 4′, and 5′ have come from the mother. Assuming no crossing-over, in what proportion of the gametes of this animal will all the paternal chromosomes be present together?

2

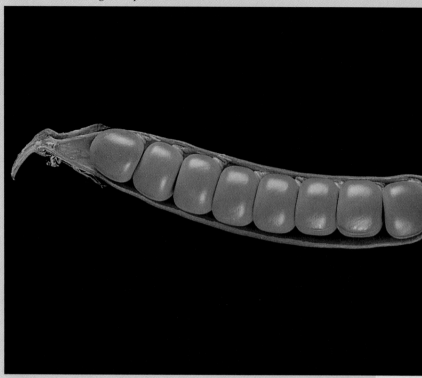

Mendelian
Genetics

PRINCIPAL POINTS

The genotype is the genetic makeup of an organism, whereas the phenotype is the observable properties (structural and functional) of an organism that are produced by the interaction between its genotype and the environment.

Genes provide the potential for the development of characteristics; this potential is affected by interactions with other genes and with the environment.

Mendel's first law, the principle of segregation, states that the two members of a gene pair segregate from each other in the formation of gametes.

To determine an unknown genotype (usually in an individual expressing the dominant phenotype), a cross is made between that individual and a homozygous recessive individual. This cross is called a testcross.

Mendel's second law, the principle of independent assortment, states that members of different gene pairs are transmitted independently of one another during the production of gametes.

Mendelian principles apply to all eukaryotes. The study of the inheritance of genetic traits in humans is complicated by the fact that controlled crosses cannot be done. Instead, human geneticists examine genetic traits by pedigree analysis—that is, by following the occurrence of a trait in family trees in which the trait is segregating.

ⓘ PEOPLE HAVE BRED ANIMALS AND PLANTS FOR specific traits for many centuries. But after Gregor Mendel developed his theory to explain the transmission of hereditary characteristics from generation to generation, breeding became an art form. Now people can use their knowledge of how a characteristic is passed from parent to offspring to produce food crops that are resistant to certain diseases, cows that produce more milk, dogs that are intelligent and gentle enough to make good guide dogs, and even furless cats. What were Mendel's experiments? What is the relationship between genes and traits? How can knowing the way in which characteristics are inherited allow people to breed for specific traits?

Later on you can try the iActivity for this chapter, which allows you to apply the knowledge you've gained in an effort to breed a very special pet.

Figure 2.1

Influences on the physical manifestation (phenotype) of the genetic blueprint (genotype): interactions with other genes and their products (such as hormones) and with the environment (such as nutrition).

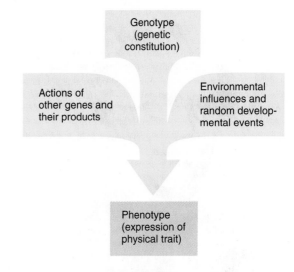

The understanding of how genes are transmitted from parent to offspring began with the work of Gregor Johann Mendel (1822–1884), an Augustinian monk. The goal of this chapter is for you to learn the basic principles of the transmission of genes by examining Mendel's work. Be aware that even though Mendel analyzed the segregation of hereditary traits, he did not know that genes control the traits or that genes are located on chromosomes or even that chromosomes existed.

Genotype and Phenotype

Before we begin our study of Mendel's work, we must distinguish between the nature of the genetic material and the physical characteristics that result from the expression of genes.

The characteristics of an individual that are transmitted from one generation to another are sometimes called **hereditary traits** (Mendel called them **characters**). These traits are under the control of **genes** (Mendel called them *factors*). The genetic constitution of an organism is called its **genotype,** and the **phenotype** is an observable characteristic or set of characteristics (structural and functional) of an organism produced by the interaction between its genotype and the environment.

Genes provide only the potential for developing a particular phenotypic characteristic. The extent to which that potential is realized depends both on interactions with other genes and their products and, in many cases, on environmental influences and random developmental events (Figure 2.1). A person's height, for example, is controlled by many genes, the expression of which can be significantly affected by internal and external environmental influences, such as the effects of hormones during puberty (an internal environmental influence) and the effects of nutrition (an external environmental influence). In other words, genes are a starting point for determining

the structure and function of an organism, and the route to the mature phenotypic state is highly complex and involves many interacting biochemical pathways.

It is important to understand that although the phenotype is the product of interaction between genes and environment, the contribution of the environment varies. In some cases, the environmental influence is great; in other cases, the environmental contribution is nonexistent. We will develop the relationship between genotype and phenotype in more detail as the text proceeds, but for our current task of examining Mendel's experiments, the important aspects just mentioned are sufficient.

KEYNOTE

The genotype is the genetic constitution of an organism. The phenotype is the observable manifestation of the genetic traits. The genes give the potential for the development of characteristics; this potential is often affected by interactions with other genes and with the environment. Thus, individuals with the same genotype can have different phenotypes, and individuals with the same phenotypes may have different genotypes.

Mendel's Experimental Design

The work of Gregor Johann Mendel (Figure 2.2) is considered the foundation of modern genetics. In 1843, he was admitted to the Augustinian monastery in Brno (now Brünn, Czech Republic). In 1854, he began a series of breeding experiments with the garden pea *Pisum sativum*

Figure 2.2
Gregor Johann Mendel, founder of the science of genetics.

Figure 2.3
Procedure for crossing pea plants.

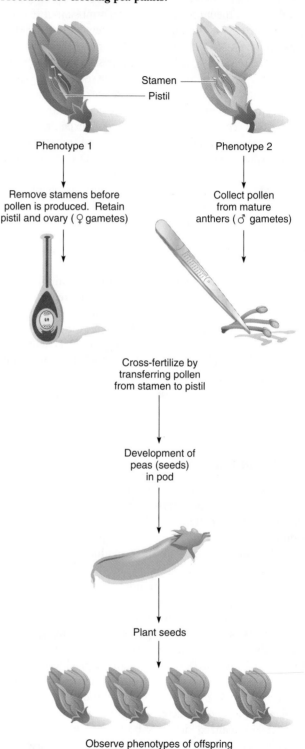

Stamen
Pistil

Phenotype 1   Phenotype 2

Remove stamens before pollen is produced. Retain pistil and ovary (♀ gametes)

Collect pollen from mature anthers (♂ gametes)

Cross-fertilize by transferring pollen from stamen to pistil

Development of peas (seeds) in pod

Plant seeds

Observe phenotypes of offspring

to learn something about the mechanisms of heredity. Probably as a result of his creativity, Mendel discovered some fundamental principles.

From the results of crossbreeding pea plants for different characteristics, such as seed shape, seed color, and flower color, Mendel developed a simple theory to explain the transmission of hereditary characteristics or traits from generation to generation. (Mendel had no knowledge of mitosis and meiosis, so he did not know that genes segregate according to chromosome behavior.) Although Mendel reported his conclusions in 1865, their significance was not fully realized until the late 1800s and early 1900s.

Mendel's experimental approach was effective because he made simple interpretations of the ratios of the types of progeny he obtained from his crosses and because he then carried out direct and convincing experiments to test his hypotheses. In his initial breeding experiments, he took the simplest approach of studying the inheritance of one trait at a time. (This is how you should work genetics problems.) He made carefully controlled matings (crosses) between pea strains that had obvious differences in heritable traits, and, most importantly, he kept very careful records of the outcomes of the crosses. The numerical data he obtained enabled him to do a rigorous analysis of the hereditary transmission of characteristics.

Generally, genetic crosses are done as follows. Two diploid individuals differing in phenotype are allowed to

produce haploid gametes by meiosis. Fusion of male and female gametes produces zygotes from which the diploid progeny individuals are generated. The phenotypes of the offspring are analyzed to provide clues to the heredity of those phenotypes.

Mendel did all his significant genetic experiments with the garden pea. The garden pea was a good choice because it fits many of the criteria that make an organism suitable for use in genetic experiments (see Chapter 1): It is easy to grow, bears flowers and fruit in the same year a seed is planted, and produces a large number of seeds. When the seeds are planted, they germinate to produce new pea plants.

Figure 2.3, which shows the procedure for crossing pea plants, begins with a cross section of a flower of the garden pea, showing the stamens (male reproductive organs) and pistils (female reproductive organs). The pea normally reproduces by **self-fertilization;** that is, the anthers at the ends of the stamen produce pollen (microspores of a flowering plant that germinate to form male [♂] gametophytes), which lands on the pistil (containing the female [♀] gametes) within the same flower and fertilizes the plant. This process is also called **selfing.** Fortunately, we can prevent self-fertilization of the pea by removing the stamens from a developing flower bud before their anthers produce any mature pollen. Then pollen taken from the stamens of another flower can be dusted onto the pistil of the emasculated flower to pollinate it.

Cross-fertilization, also called a **cross,** is the fusion of male gametes (in this case, pollen) from one individual and female gametes (eggs) from another. Once cross-fertilization has occurred, zygotes develop in the seeds (peas), which are then planted. The phenotypes of the plants that grow from the seeds are then analyzed.

Mendel obtained 34 strains of pea plants that differed in a number of traits. He allowed each strain to self-fertilize for many generations to make sure that the traits he wanted to study were inherited. This preliminary work ensured that he worked only with pea strains in which the trait under investigation remained unchanged from parent to offspring for many generations. Such strains are called **true-breeding** or **pure-breeding strains.**

Next, Mendel selected seven traits to study in breeding experiments. Each trait (or character) had two easily distinguishable, alternative appearances (phenotypes). Mendel studied the following character pairs (Figure 2.4):

Figure 2.4

Seven character pairs in the garden pea that Mendel studied in his breeding experiments.

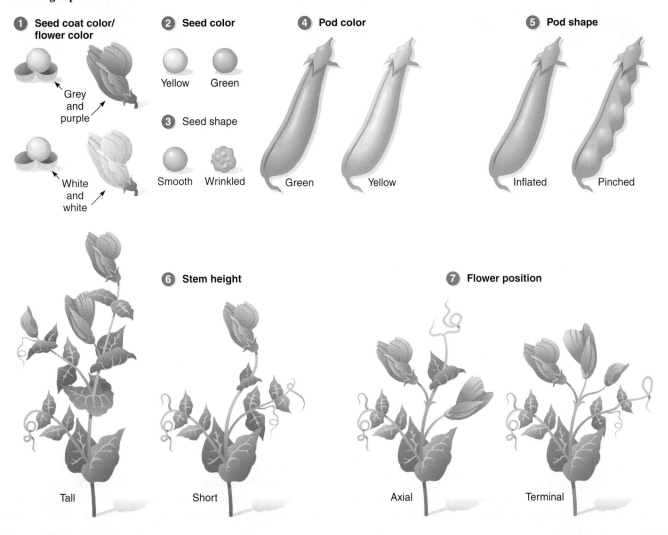

1. Flower and seed coat color (grey versus white seed coats, and purple versus white flowers; note that a single gene controls both these particular color properties of seed coats and flowers)
2. Seed color (yellow versus green)
3. Seed shape (smooth versus wrinkled)
4. Pod color (green versus yellow)
5. Pod shape (inflated versus pinched)
6. Stem height (tall versus short)
7. Flower position (axial versus terminal)

Monohybrid Crosses and Mendel's Principle of Segregation

Before we discuss Mendel's experiments, we must be clear on the terminology used in breeding experiments. The parental generation is called the **P generation.** The progeny of the P mating is called the **first filial generation,** or **F₁.** The subsequent generation produced by breeding together the F₁ offspring is the **F₂ generation.** Inter-breeding the offspring of each generation results in generations F₃, F₄, F₅, and so on.

Mendel first performed crosses between true-breeding strains of peas that differed in a single trait. Such crosses are called **monohybrid crosses.** For example, when he pollinated pea plants that gave rise only to smooth seeds[1] with pollen from a true-breeding variety that produced only wrinkled seeds, the outcome was all smooth seeds (Figure 2.5). The same result was obtained when the parental types were reversed; that is, when the pollen from a smooth-seeded plant was used to pollinate a pea plant that gave wrinkled seeds, the outcome was all smooth seeds. Matings that are done both ways—smooth female [♀] × wrinkled male [♂] and wrinkled female [♀] × smooth male [♂]—are called **reciprocal crosses.** Conventionally, the female is given first in crosses of plants. If the results of reciprocal crosses are the same, it means that the trait does not depend on the sex of the organism.

The significant point of this cross is that all the F₁ progeny seeds of the smooth × wrinkled reciprocal crosses were smooth; that is, they exactly resembled only one of the parents in this character rather than being a blend of both parental phenotypes. The finding that all offspring of true-breeding parents are alike is sometimes referred to as the *principle of uniformity in F₁*.

Next, Mendel planted the seeds and allowed the F₁ plants to self-fertilize to produce the F₂ seeds. Both smooth and wrinkled seeds appeared in the F₂ generation,

---

[1]Seeds are the diploid progeny of sexual reproduction. If a phenotype concerns the seed itself, the results of the cross can be seen directly by looking at the seeds. If a phenotype concerns a part of the mature plant, such as flower color, then the seeds must be germinated and grown to maturity before that phenotype can be seen.

Figure 2.5

Results of one of Mendel's breeding crosses. In the parental generation, he crossed a true-breeding pea strain that produced smooth seeds with one that produced wrinkled seeds. All the F₁ progeny seeds were smooth.

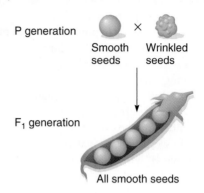

and both types could be found within the same pod. Typical of his analytical approach to the experiments, Mendel counted the number of seeds of each type. He found that 5,474 were smooth and 1,850 were wrinkled (Figure 2.6). The calculated ratio of smooth seeds to wrinkled seeds was 2.96:1, which is very close to a 3:1 ratio.

Mendel observed that although the F₁ progeny resembled only one of the parents in their phenotype, they did not breed true, a fact that distinguishes the F₁ plants from the parent they resembled. Moreover, the F₁ plants could produce some F₂ progeny with the parental phenotype that had disappeared in the F₁ generation. But how can a trait present in the P generation disappear in the F₁ generation and then reappear in the F₂ generation? Mendel concluded that the alternative traits in the cross—smooth or wrinkled seeds—were determined by **particulate factors.** He reasoned that these factors, which were transmitted from parents to progeny through the gametes, carried hereditary information. We now know these factors by another name: *genes.*

Since Mendel was examining a pair of traits (wrinkled and smooth), he surmised that each factor existed in alternative forms (which we now call **alleles**), each of which specified one of the traits. For the gene controlling pea seed shape, there is one form (allele) that results in a smooth seed and another allele that results in a wrinkled seed.

Mendel reasoned further that a true-breeding strain of peas must contain a pair of identical factors. Since the F₂ progeny exhibited both traits and the F₁ exhibited only one of those traits, each F₁ individual must have contained both factors, one for each of the alternative traits. In other words, crossing two different true-breeding strains brings together in the F₁ generation one factor from each strain. The eggs contain one factor from one strain and the pollen grains contain one factor from the other strain. Furthermore, because only one of the traits

Figure 2.6

The F₂ progeny of the cross shown in Figure 2.5. When the plants grown from the F₁ seeds were self-pollinated, both smooth and wrinkled F₂ progeny seeds were produced. Commonly, both seed types were found in the same pod. In his experiments, Mendel counted 5,474 smooth and 1,850 wrinkled F₂ progeny seeds for a ratio of 2.96:1.

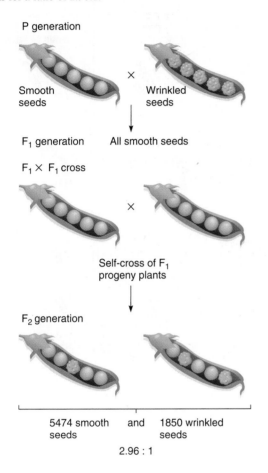

P generation

Smooth seeds   ×   Wrinkled seeds

F₁ generation   All smooth seeds

F₁ × F₁ cross   ×

Self-cross of F₁ progeny plants

F₂ generation

5474 smooth seeds   and   1850 wrinkled seeds

2.96 : 1

was seen in the F₁ generation, the expression of the missing trait must somehow have been masked by the visible trait; this masking is called *dominance*. For the smooth × wrinkled cross, the F₁ seeds were all smooth. Thus, the allele for smoothness is masking or **dominant** to the allele for wrinkledness. Conversely, wrinkled is said to be **recessive** to smooth because the factor for wrinkled is masked.

A simple way to visualize the crosses is to use symbols for the alleles, as Mendel did. For the smooth × wrinkled cross, we can give the symbol *S* to the allele for smoothness and the symbol *s* to the allele for wrinkledness. The letter used is based on the dominant phenotype, and the convention in this case is that the dominant allele is given the uppercase letter and the recessive allele the lowercase letter. (This convention was followed for many years, particularly in plant genetics. Now it is more conventional to base the letter assignment on the recessive phenotype. We will use this convention later.)

Using these symbols, the genotype of the parental plant grown from the smooth seeds is *SS* and that of the wrinkled parent is *ss*. True-breeding individuals that contain two copies of the same specific allele of a particular gene are said to be **homozygous** for that gene (Figure 2.7). When plants produce gametes by meiosis (see Chapter 1), each gamete contains only one copy of the gene (one allele); the plants from smooth seeds produce *S*-bearing gametes, and the plants from wrinkled seeds produce *s*-bearing gametes. When the gametes fuse during fertilization, the resulting zygote has one *S* allele and one *s* allele, a genotype of *Ss*. Plants that have two different alleles of a particular gene are said to be **heterozygous**. Because of the dominance of the smooth *S* allele, *Ss* plants produce smooth seeds (see Figure 2.7).

Figure 2.8 diagrams the smooth × wrinkled cross using genetic symbols; the production of the F₁ generation is shown in Figure 2.8a and that of the F₂ generation in Figure 2.8b. (In Figures 2.7 and 2.8, the genes are shown on chromosomes. Keep in mind that the segregation of genes from generation to generation follows the behavior of chromosomes.) The true-breeding, smooth-seeded parent has the genotype *SS*, and the true-breeding, wrinkle-seeded parent has the genotype *ss*. Since each parent is true-breeding and diploid (i.e., has two sets of chromosomes), each must contain two copies of the same allele. All the F₁ plants produce smooth seeds, and all are *Ss* heterozygotes.

The plants grown from the F₁ seeds differ from the smooth parent in that they produce equal numbers of two types of gametes: *S*-bearing gametes and *s*-bearing gametes. All possible fusions of F₁ gametes are shown in the matrix in Figure 2.8b, called a **Punnett square** after its originator, R. Punnett. These fusions give rise to the zygotes that produce the F₂ generation.

In the F₂ generation, three types of genotypes are produced: *SS, Ss,* and *ss*. As a result of the random fusing of gametes, the relative proportion of these zygotes is

Figure 2.7

Dominant and recessive alleles of a gene for seed shape in peas.

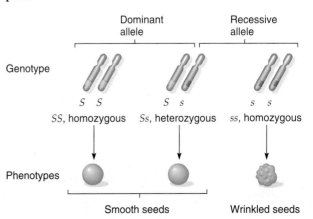

Dominant allele   Recessive allele

Genotype

S S   *S s*   *s s*

SS, homozygous   *Ss*, heterozygous   *ss*, homozygous

Phenotypes

Smooth seeds   Wrinkled seeds

Figure 2.8

The same cross as in Figures 2.5 and 2.6, using genetic symbols to illustrate the principle of segregation of Mendelian factors. (a) Production of the F_1 generation. **(b)** Production of the F_2 generation.

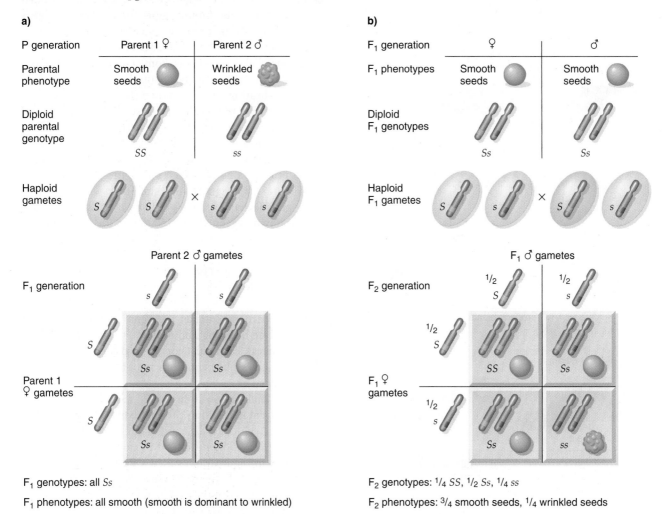

a)

b)

F_1 genotypes: all *Ss*

F_1 phenotypes: all smooth (smooth is dominant to wrinkled)

F_2 genotypes: ¼ *SS*, ½ *Ss*, ¼ *ss*

F_2 phenotypes: ¾ smooth seeds, ¼ wrinkled seeds

1:2:1, respectively. However, because the *S* factor is dominant to the *s* factor, both the *SS* and *Ss* seeds are smooth, and the F_2 generation seeds show a phenotypic ratio of 3 smooth : 1 wrinkled.

Mendel also analyzed the behavior of the six other pairs of traits. Qualitatively and quantitatively, the same results were obtained (Table 2.1). From the seven sets of crosses, he made the following general conclusions about his data:

1. The results of reciprocal crosses were always the same.
2. All F_1 progeny resembled one of the parental strains, indicating the dominance of one allele over the other.
3. In the F_2 generation, the parental trait that had disappeared in the F_1 generation reappeared. Furthermore, the trait seen in the F_1 generation was always found in the F_2 generation at about three times the frequency of the other trait.

The Principle of Segregation

From the sort of data we have discussed, Mendel proposed what has become known as **Mendel's first law,** the **principle of segregation:** *Recessive characters, which are masked in the F_1 generation from a cross between two true-breeding strains, reappear in a specific proportion in the F_2 generation.* In modern terms, this means that *the two members of a gene pair (alleles) segregate (separate) from each other during the formation of gametes.* As a result, half the gametes carry one allele, and the other half carry the other allele. In other words, each gamete carries only a single allele of each gene. The progeny are produced by the random combination of gametes from the two parents.

In proposing the principle of segregation, Mendel differentiated between the factors (genes) that determine

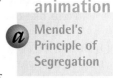

animation

Mendel's Principle of Segregation

Table 2.1   Mendel's Results in Crosses Between Plants Differing in One of Seven Characters

Character[a]	F₁	F₂ (Number) Dominant	Recessive	Total	F₂ (Ratio) Dominant : Recessive
Seeds: smooth versus wrinkled	All smooth	5,474	1,850	7,324	2.96:1
Seeds: yellow versus green	All yellow	6,022	2,001	8,023	3.01:1
Seed coats: grey versus white [b]	All grey	705	224	929	3.15:1
Flowers: purple versus white	All purple				
Flowers: axial versus terminal	All axial	651	207	858	3.14:1
Pods: inflated versus pinched	All inflated	882	299	1,181	2.95:1
Pods: green versus yellow	All green	428	152	580	2.82:1
Stem: tall versus short	All tall	787	277	1,064	2.84:1
Total or average		14,949	5,010	19,959	2.98:1

[a] The dominant trait is always written first.
[b] A single gene controls both the seed coat trait and the flower color trait.

the traits (the genotype) and the traits themselves (the phenotype). From a modern perspective, we know that genes are on chromosomes. The specific location of a gene on a chromosome is called its **locus** (or **gene locus;** plural, *loci*). Furthermore, Mendel's first law means that at the gene level, the members of a pair of alleles segregate during meiosis, and each offspring receives only one allele from each parent. Thus, **gene segregation** parallels the separation of homologous pairs of chromosomes at anaphase I in meiosis (see Chapter 1).

Box 2.1 (p. 36) presents a summary of the genetics concepts and terms we have discussed so far in this chapter. A thorough familiarity with these terms is essential to your study of genetics.

KEYNOTE

Mendel's first law, the principle of segregation, states that the two members of a gene pair (alleles) segregate (separate) from each other in the formation of gametes; half the gametes carry one allele, and the other half carry the other allele.

Representing Crosses with a Branch Diagram

The use of a Punnett square to represent the pairing of all possible gamete types from the two parents in a monohybrid cross (shown in Figure 2.8) is a simple way to predict the relative frequencies of genotypes and phenotypes in the next generation. In this section, however, we examine an alternative method, one that you are encouraged to master: the branch or fork diagram. (Box 2.2 discusses some elementary principles of probability that will help you understand this approach.) To use the branch diagram approach, it is necessary to know the dominance/recessiveness relationship of the allele pair so that the progeny phenotypic classes can be determined. Figure 2.9 illustrates the branch diagram analysis of the F₁ selfing of the smooth × wrinkled cross diagrammed in Figure 2.8.

The F₁ seeds from the cross in Figure 2.8 have the genotype *Ss*. In meiosis, we expect half the gametes to be *S* and half to be *s* (see Figure 2.9). Thus, ½ is the

Figure 2.9

Using the branch diagram approach to calculate the ratios of phenotypes in the F₂ generation of the cross in Figure 2.8.

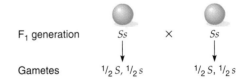

Box 2.1 Genetics Terminology

Alleles: alternative forms of a gene. For example, the alleles *S* and *s* represent the smoothness and wrinkledness of the pea seed. (Like gene symbols, allele symbols are underlined or *italicized*.)

Cross: a mating between two individuals, leading to the fusion of gametes.

Diploid: a eukaryotic cell or organism with two homologous sets of chromosomes.

Gamete: a mature reproductive cell that is specialized for sexual fusion. Each gamete is haploid and fuses with a cell of similar origin but of opposite sex to produce a diploid zygote.

Gene (Mendelian factor): the determinant of a characteristic of an organism. Gene symbols are underlined or *italicized*.

Genotype: the genetic constitution of an organism. A diploid organism in which both alleles are the same at a given gene locus is said to be **homozygous** for that allele. Homozygotes produce only one gametic type with respect to that locus. For example, true-breeding smooth-seeded peas have the genotype *SS*, and true-breeding wrinkle-seeded peas have the genotype *ss*; both are homozygous. The smooth parent is

homozygous dominant; the wrinkled parent is **homozygous recessive.**

Diploid organisms that have two different alleles at a specific gene locus are said to be **heterozygous.** Thus, F$_1$ hybrid plants from the cross of *SS* and *ss* parents have one *S* allele and one *s* allele. Individuals heterozygous for two allelic forms of a gene produce two kinds of gametes (*S* and *s*).

Haploid: a cell or an individual with one copy of each chromosome.

Locus (gene locus; plural = loci): the specific place on a chromosome where a gene is located.

Phenotype: the physical manifestation of a genetic trait that results from a specific genotype and its interaction with the environment. In our example, the *S* allele was dominant to the *s* allele, so in the heterozygous condition the seed is smooth. Therefore, both the homozygous dominant *SS* and the heterozygous *Ss* seeds have the same phenotype (smooth), even though they differ in genotype.

Zygote: the cell produced by the fusion of male and female gametes.

---

predicted frequency of each of these two types. But just as tossing a coin many times does not always give exactly half heads and half tails, the two gametes may not be produced in an exactly 1:1 ratio. However, the more chances (tosses), the closer the observed frequency will come to the predicted frequency.

From the rules of probability, we can predict the expected frequencies of the three possible genotypes in the F$_2$ generation. To produce an *SS* plant, an *S* egg must pair with an *S* pollen grain. The frequency of *S* eggs in the population of eggs is $\frac{1}{2}$, and the frequency of *S* pollen grains in the pollen population is $\frac{1}{2}$. Therefore, the expected proportion of *SS* smooth plants in the F$_2$ generation is $\frac{1}{2} \times \frac{1}{2} = \frac{1}{4}$. Similarly, the expected proportion of *ss* wrinkled progeny in the F$_2$ is $\frac{1}{2} \times \frac{1}{2} = \frac{1}{4}$.

What about the *Ss* progeny? Again, the frequency of *S* in one gametic type is $\frac{1}{2}$, and the frequency of *s* in the other gametic type is also $\frac{1}{2}$. However, there are two ways in which *Ss* progeny can be obtained. The first involves the fusion of an *S* egg with *s* pollen, and the second is a fusion of an *s* egg with *S* pollen. Using the product rule (see Box 2.2), the probability of each of these events occurring is $\frac{1}{2} \times \frac{1}{2} = \frac{1}{4}$. By using the sum rule (see Box 2.2), the probability of one or the other occurring is the sum of the individual probabilities, or $\frac{1}{4} + \frac{1}{4} = \frac{1}{2}$.

The prediction, then, is that one-fourth of the F$_2$ progeny will be *SS*, half will be *Ss*, and one-fourth will be *ss*, exactly as we found with the Punnett square in Figure 2.8. Either method—the Punnett square or the branch diagram—may be used with any cross, but as crosses become more complicated, the Punnett square becomes cumbersome.

Confirming the Principle of Segregation: The Use of Testcrosses

When formulating his principle of segregation, Mendel did a number of tests to ensure the correctness of his results. He continued the self-fertilizations at each generation up to the F$_6$ generation and found that in every generation, both the dominant and recessive characters were found. He concluded that the principle of segregation was valid no matter how many generations were involved.

Another important test concerned the F$_2$ plants. As shown in Figure 2.8, a ratio of 1:2:1 occurs for the genotypes *SS*, *Ss*, and *ss* for the smooth × wrinkled example. Phenotypically, the ratio of smooth to wrinkled is 3:1. At the time of Mendel's experiments, the presence of segregating factors that were responsible for the smooth and wrinkled phenotypes was only a hypothesis. To test his factor hypothesis, Mendel allowed the F$_2$ plants to self-pollinate. As he expected, the plants produced from wrinkled seeds bred true, supporting his conclusion that they were pure for the *s* factor (gene).

Selfing the plants derived from the F$_2$ smooth seeds produced two different types of progeny. One-third of the smooth F$_2$ seeds produced all smooth-seeded progeny, whereas the other two-thirds produced both smooth and wrinkled seeds in each pod in a ratio of 3 smooth : 1 wrinkled (Figure 2.10). For the plants that produced both seed types in the progeny, the actual ratio of smooth : wrinkled seeds was 3:1, the same ratio as seen for the F$_2$ progeny. These results support the principle of gene segregation. The random combination of gametes that form the zygotes of the original F$_2$ plants produces two genotypes that give rise to the smooth phenotype (see Figures

Box 2.2	**Elementary Principles of Probability**

A **probability** is the ratio of the number of times a particular event is expected to occur to the number of trials during which the event could happen. For example, the probability of picking a heart from a deck of 52 cards, 13 of which are hearts, is $P(heart) = {}^{13}\!/_{52} = {}^1\!/_4$. That is, we would expect, on the average, to pick a heart from a deck of cards once in every four trials.

Probabilities and the *laws of chance* are involved in the transmission of genes. As a simple example, let us consider the chance that a baby will be a boy (or a girl). Assume that an exactly equal number of boys and girls are born (which is not precisely true, but let's assume so for the sake of discussion). The probability that the child will be a boy is $^1\!/_2$, or 0.5. Similarly, the probability that the child will be a girl is $^1\!/_2$.

Now, a rule of probability can be introduced: the **product rule.** The product rule states that *the probability of two independent events occurring simultaneously is the product of each of their individual probabilities*. Thus, the probability that both children in families with two children will be girls is $^1\!/_4$. That is, the probability of the first child being a girl is $^1\!/_2$, the probability of the second being a girl is also $^1\!/_2$, and by the product rule, the probability of both children being girls is $^1\!/_2 \times {}^1\!/_2 = {}^1\!/_4$. Similarly, the probability of having three boys in a row is $^1\!/_2 \times {}^1\!/_2 \times {}^1\!/_2 = {}^1\!/_8$.

Another rule of probability, the **sum rule,** states that *the probability of occurrence of several mutually exclusive events is the sum of the probabilities of the individual events*. For example, if one die is thrown, what is the probability of getting a one and a six? The individual probabilities are calculated as follows: The probability of rolling a one, $P(one)$, is $^1\!/_6$, because there are six faces to a die. The probability of getting a six, $P(six)$, is also $^1\!/_6$, for the same reason. To roll a one or a six with a single throw of a die involves two mutually exclusive events, so the sum rule is used. The sum of the individual probabilities is $^1\!/_6 + {}^1\!/_6 = {}^2\!/_6 = {}^1\!/_3$. To return to our family example, the probability of having two boys or two girls is $^1\!/_4 + {}^1\!/_4 = {}^1\!/_2$.

2.8 and 2.9); the relative proportion of the two genotypes *SS* and *Ss* is 1:2. The *SS* seeds give rise to true-breeding plants, whereas the *Ss* seeds give rise to plants that behave exactly like the F$_1$ plants when they are self-pollinated in that they produce a 3:1 ratio of smooth : wrinkled progeny. *Mendel explained these results by proposing that each plant had two factors, whereas each gamete had only one. He also proposed that the random combination of the gametes generated the progeny in the proportions he found. Mendel obtained the same results in all seven sets of crosses.*

The self-fertilization test of the F$_2$ progeny proved a useful way to confirm the genotype of a plant with a given phenotype. A more common test to ascertain the genotype of an organism is to perform a **testcross,** a cross of an individual of unknown genotype (usually expressing the dominant phenotype) with a homozygous recessive individual to determine the unknown genotype.

Consider again the cross shown in Figure 2.8. We can predict the outcome of a testcross of the F$_2$ progeny showing the dominant, smooth-seed phenotype. If the F$_2$ individuals are homozygous *SS*, then the result of a testcross with an *ss* plant will be all smooth seeds. As Figure 2.11a shows, the Parent 1 smooth *SS* plants produce only *S* gametes. Parent 2 is homozygous recessive wrinkled, *ss*, so it produces only s gametes. Therefore, all zygotes are *Ss,* and all the resulting seeds have the smooth phenotype. In actual practice, then, if a plant with a dominant trait is testcrossed and only the dominant phenotype is seen among the progeny, then the plant must have been homozygous for the dominant allele. In contrast, heterozygous *Ss* F$_2$ plants testcrossed with a

homozygous *ss* plant give a 1:1 ratio of dominant to recessive phenotypes. As Figure 2.11b shows, the Parent 1 smooth *Ss* produces both *S* and *s* gametes in equal proportion, and the homozygous *ss* Parent 2 produces only *s* gametes. As a result, half the progeny of the testcross are *Ss* heterozygotes and have a smooth phenotype because of the dominance of the *S* allele, and the other half are *ss* homozygotes and have a wrinkled phenotype. In actual practice, then, if a plant with a dominant trait is testcrossed and the progeny exhibit a 1:1 ratio of dominant to recessive phenotype, then the plant must have been heterozygous.

In summary, testcrosses of the F$_2$ progeny from Mendel's crosses that showed the dominant phenotype resulted in a 1:2 ratio of homozygous dominant : heterozygous genotypes in the F$_2$ progeny. That is, when

Figure 2.10

Determining the genotypes of the F$_2$ smooth progeny of Figure 2.8 by selfing the plants grown from the smooth seeds.

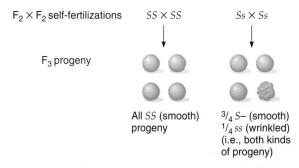

F$_2$ × F$_2$ self-fertilizations   *SS* × *SS*   *Ss* × *Ss*

F$_3$ progeny

All *SS* (smooth) progeny

$^3\!/_4$ *S*– (smooth)
$^1\!/_4$ *ss* (wrinkled)
(i.e., both kinds of progeny)

Figure 2.11

Determining the genotypes of the F₂ generation smooth seeds (Parent 1) of Figure 2.8 by testcrossing plants grown from the seed with a homozygous recessive wrinkled (ss) strain (Parent 2). (a) If Parent 1 is *SS*, then all progeny seeds are smooth. **(b)** If Parent 1 is *Ss*, then ½ of the progeny seeds are smooth and ½ are wrinkled.

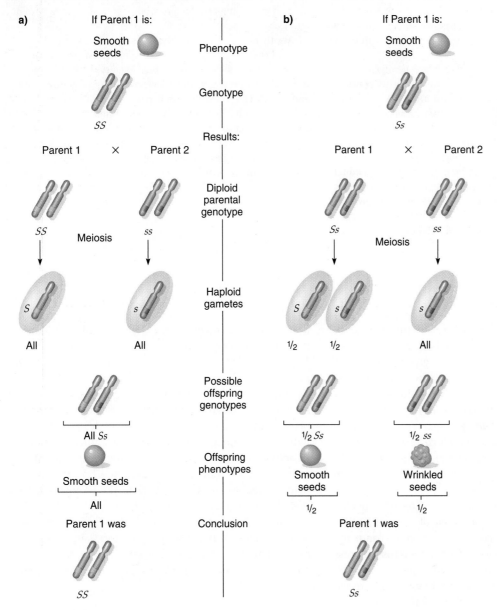

crossed with the homozygous recessive, one-third of the F₂ progeny with the dominant phenotype gave rise only to progeny with the dominant phenotype and were therefore homozygous for the dominant allele. The other two-thirds of the F₂ progeny with the dominant phenotype produced progeny with a 1:1 ratio of dominant phenotype to recessive phenotype and therefore were heterozygous.

K E Y N O T E

A testcross is a cross of an individual of unknown genotype, usually expressing the dominant phenotype, with a known homozygous recessive individual to determine the genotype of the unknown individual. The phenotypes of the progeny of the testcross indicate the genotype of the individual being tested.

Dihybrid Crosses and Mendel's Principle of Independent Assortment

The Principle of Independent Assortment

Mendel also analyzed a number of crosses in which two pairs of traits were simultaneously involved. In each case he obtained the same results. From these experiments he proposed what has become known as **Mendel's second law,** the **principle of independent assortment,** which states that *the factors for different traits assort independently of one another.* In modern terms, this means that *genes on different chromosomes behave independently in gamete production.*

animation

a Mendel's Principle of Independent Assortment

Consider an example involving smooth (*S*), wrinkled (*s*), yellow (*Y*), and green (*y*) seed traits (yellow is dominant to green). When Mendel made crosses between true-breeding smooth yellow plants (*SS YY*) and wrinkled green plants (*ss yy*), he got the results shown in Figure 2.12. All the F_1 seeds from this cross were smooth and yellow, as the results of the monohybrid crosses predicted. As Figure 2.12a shows, the smooth yellow parent produces only *S Y* gametes, which give rise to *Ss Yy* zygotes upon fusion with the *s y* gametes from the wrinkled green parent. Because of the dominance of the smooth and yellow traits, all F_1 seeds are smooth and yellow.

The F_1 plants are heterozygous for two pairs of alleles at two different loci. Such individuals are called dihybrids, and a cross between two of these dihybrids of the same type is called a **dihybrid cross.**

When Mendel self-pollinated the dihybrid F_1 plants to give rise to the F_2 generation (Figure 2.12b), he considered two possible outcomes. One was that the genes for the traits from the original parents would be transmitted together to the progeny. In this case, a phenotypic ratio of 3 smooth yellow : 1 wrinkled green would be predicted.

The other possibility was that the traits would be inherited independently of one another. In this case, the dihybrid F_1 plants would produce four types of gametes: *S Y, S y, s Y,* and *s y.* Given the independence of the two pairs of genes, each gametic type is predicted to occur with equal frequency. In $F_1 \times F_1$ crosses, the four types of gametes would be expected to fuse randomly in all possible combinations to give rise to the zygotes and hence the progeny seeds. All the possible gametic fusions are represented in the Punnett square in Figure 2.12b. In a dihybrid cross, there are 16 possible gametic fusions. The result is nine different genotypes, but because of dominance, only four phenotypes are predicted:

1 *SS YY,* 2 *Ss YY,* 2 *SS Yy,* 4 *Ss Yy*	= 9 smooth, yellow
1 *SS yy,* 2 *Ss yy*	= 3 smooth, green
1 *ss YY,* 2 *ss Yy*	= 3 wrinkled, yellow
1 *ss yy*	= 1 wrinkled, green

According to the rules of probability, if pairs of characters are inherited independently in a dihybrid cross, then the F_2 generation from an $F_1 \times F_1$ cross will give a 9:3:3:1 ratio of the four possible phenotypic classes. This ratio is the result of the independent assortment of the two gene pairs into the gametes and of the random fusion of those gametes.

This prediction was met in all the dihybrid crosses Mendel performed. In every case, the F_2 ratio was close to 9:3:3:1. For our example, he counted 315 smooth yellow, 108 smooth green, 101 wrinkled yellow, and 32 wrinkled green seeds, very close to the predicted ratio. To Mendel this result meant that the factors (genes) determining the specific, different character pairs he was analyzing were transmitted independently. This means that he rejected the possibility that the two traits were inherited together.

Figure 2.12

The principle of independent assortment in a dihybrid cross. This cross, actually done by Mendel, involves the smooth/wrinkled and yellow/green character pairs of the garden pea. **(a)** Production of the F_1 generation.

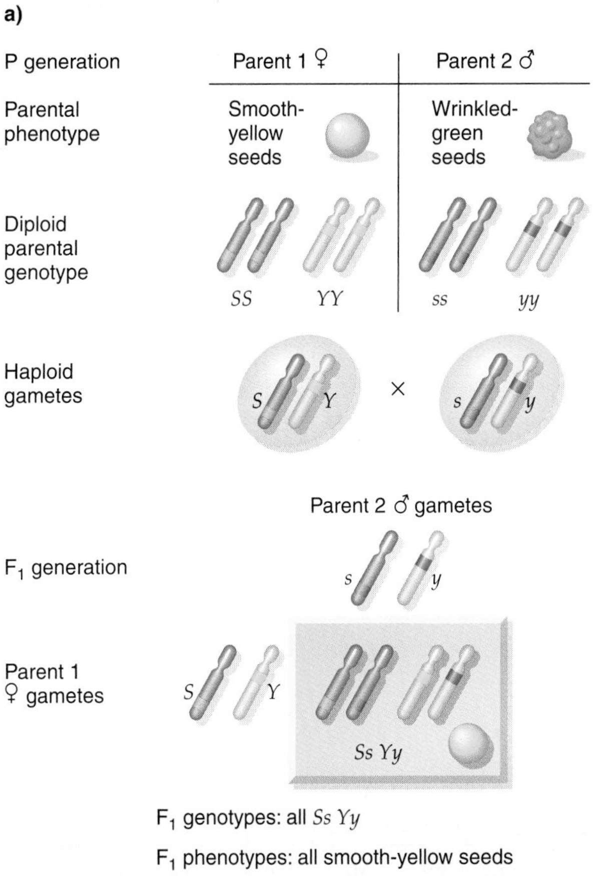

a)

P generation — Parent 1 ♀ | Parent 2 ♂

Parental phenotype — Smooth-yellow seeds | Wrinkled-green seeds

Diploid parental genotype — *SS*   *YY* | *ss*   *yy*

Haploid gametes — *S   Y* × *s   y*

Parent 2 ♂ gametes

F_1 generation — *s   y*

Parent 1 ♀ gametes — *S   Y*

Ss Yy

F_1 genotypes: all *Ss Yy*

F_1 phenotypes: all smooth-yellow seeds

Figure 2.12, continued

(b) The F_2 genotypes and the 9:3:3:1 phenotypic ratio of smooth yellow : smooth green : wrinkled yellow : wrinkled green, derived using the Punnett square. (Note that compared with previous figures of this kind, only one box is shown in the F_1 generation, instead of four. This is because only one class of gametes exists for Parent 2 and only one class for Parent 1. Previously we showed two gametes from each parent, even though those gametes were identical.)

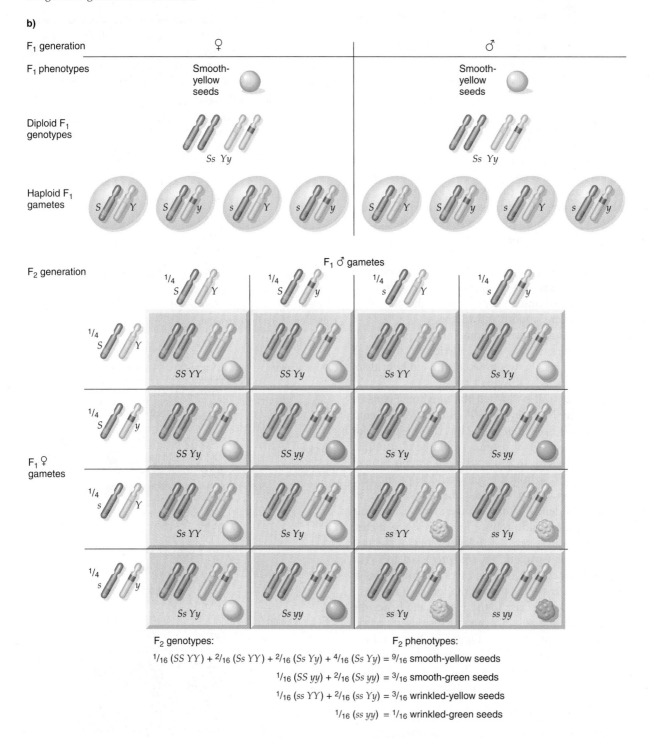

b)

F_2 genotypes:
$^1/_{16}$ (SS YY) + $^2/_{16}$ (Ss YY) + $^2/_{16}$ (Ss Yy) + $^4/_{16}$ (Ss Yy) = $^9/_{16}$ smooth-yellow seeds
$^1/_{16}$ (SS yy) + $^2/_{16}$ (Ss yy) = $^3/_{16}$ smooth-green seeds
$^1/_{16}$ (ss YY) + $^2/_{16}$ (ss Yy) = $^3/_{16}$ wrinkled-yellow seeds
$^1/_{16}$ (ss yy) = $^1/_{16}$ wrinkled-green seeds

KEYNOTE

Mendel's second law, the principle of independent assortment, states that genes for different traits assort independently of one another in gamete production.

Branch Diagram of Dihybrid Crosses

Rather than using a Punnett square, it is easier to get into the habit of calculating the expected ratios of phenotypic or genotypic classes by using a branch diagram to apply the laws of probability to the traits one at a time. *With practice, you should be able to calculate the probabilities of outcomes of various crosses just by using the laws of probability without drawing out the branch diagram. Diligently working problems really helps to hone this skill.*

Using the same example, in which the two gene pairs assort independently into gametes, we consider each gene pair in turn. Earlier we saw that an F_1 self of an *Ss* heterozygote gave rise to progeny of which three-fourths were smooth and one-fourth were wrinkled. Genotypically, the former class had at least one dominant *S* allele; that is, they were *SS* or *Ss*. A convenient way to signify this situation is to use a dash to indicate an allele that has no effect on the phenotype. Thus, *S–* means that phenotypically the seeds are smooth and genotypically they are either *SS* or *Ss*.

Now consider the F_2 progeny produced from a selfing of *Yy* heterozygotes; again, a 3:1 ratio is seen, with three-fourths of the seeds being yellow and one-fourth being green. Since this segregation occurs independently of the segregation of the smooth/wrinkled pair, we can consider all possible combinations of the phenotypic classes in the dihybrid cross. For example, the expected proportion of F_2 seeds that are smooth and yellow is the product of the probability that an F_2 seed will be smooth and the probability that it will be yellow, or $\frac{3}{4} \times \frac{3}{4} = \frac{9}{16}$. Similarly, the expected proportion of F_2 progeny that are wrinkled and yellow is $\frac{3}{4} \times \frac{1}{4} = \frac{3}{16}$. Extending this calculation to all possible phenotypes, as shown in Figure 2.13, we obtain the ratio of 9 *S– Y–* (smooth, yellow) : 3 *S– yy* (smooth, green) : 3 *ss Y–* (wrinkled, yellow) : 1 *ss yy* (wrinkled, green).

The testcross can be used to check the genotypes of F_1 progeny and F_2 progeny from a dihybrid cross. In our example, the F_1 plant is a double heterozygote, *Ss Yy*. This F_1 plant produces four types of gametes in equal proportions: *S Y*, *S y*, *s Y*, and *s y* (see Figure 2.12b). In a testcross with a doubly homozygous recessive plant—in this case, *ss yy*—the phenotypic ratio of the progeny is a direct reflection of the ratio of gametic types produced by the F_1 parent. In a testcross like this one, then, there will be a 1:1:1:1 ratio in the offspring of *Ss Yy* : *Ss yy* : *ss Yy* : *ss yy* genotypes, which means a 1:1:1:1 ratio of smooth yellow : smooth green : wrinkled yellow : wrinkled green phenotypes. This 1:1:1:1 phenotypic ratio is diagnostic of testcrosses in which the "unknown" parent is a double heterozygote.

In the F_2 generation of a dihybrid cross, there are nine different genotypic classes but only four phenotypic classes. The genotypes can be ascertained by testcrossing, as we have shown. Table 2.2 lists the expected ratios of progeny phenotypes from such testcrosses. No two patterns are the same, so here the testcross is truly a diagnostic approach to confirm genotypes.

 Go to the iActivity *Tribble Traits* on the website and discover how, as a Tribble breeder, you can choose the right combination of traits to produce the cuddliest creature.

Figure 2.13

Using the branch diagram approach to calculate the F_2 phenotypic ratio of the cross in Figure 2.12.

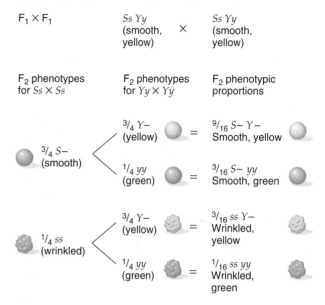

Trihybrid Crosses

Mendel also confirmed his laws for three characters segregating in other crosses. Such crosses are called **trihybrid crosses.** Here the proportions of F_2 genotypes and phenotypes are predicted with precisely the same logic used before—by considering each character pair independently. Figure 2.14 shows a branch diagram derivation of the F_2 phenotypic classes for a trihybrid cross. The independently assorting character pairs in the cross are smooth versus wrinkled seeds, yellow versus green seeds, and purple versus white flowers. There are 64 combinations of eight maternal and eight paternal gametes. Combination of these gametes gives rise to 27 different genotypes and 8 different phenotypes in the F_2 generation. The phenotypic ratio in the F_2 generation is 27:9:9:3:9:3:3:1.

Table 2.2	Proportions of Phenotypic Classes Expected from Testcrosses of Strains with Various Genotypes for Two Gene Pairs			
	Proportion of Phenotypic Classes			
Testcrosses	**A– B–**	**A– bb**	**aa B–**	**aa bb**
AA BB × aa bb	1	0	0	0
Aa BB × aa bb	1/2	0	1/2	0
AA Bb × aa bb	1/2	1/2	0	0
Aa Bb × aa bb	1/4	1/4	1/4	1/4
AA bb × aa bb	0	1	0	0
Aa bb × aa bb	0	1/2	0	1/2
aa BB × aa bb	0	0	1	0
aa Bb × aa bb	0	0	1/2	1/2
aa bb × aa bb	0	0	0	1

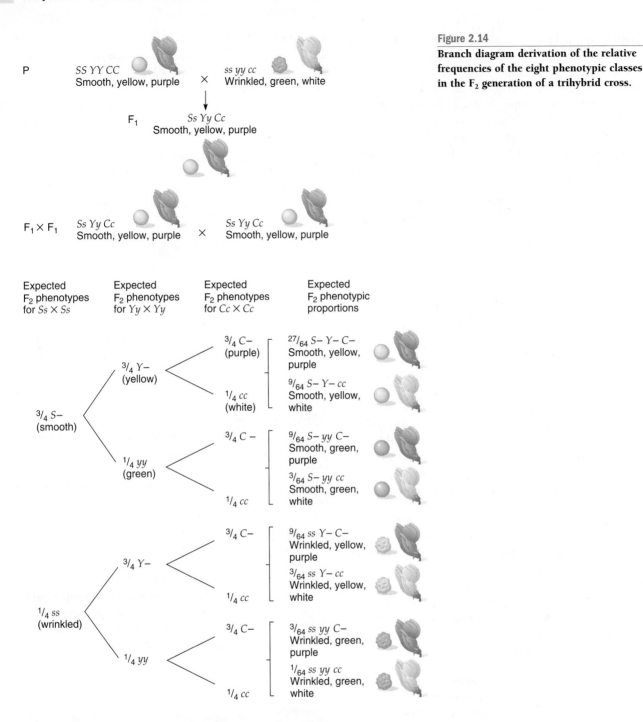

Figure 2.14

Branch diagram derivation of the relative frequencies of the eight phenotypic classes in the F₂ generation of a trihybrid cross.

Now that we have considered enough examples, we can make some generalizations about phenotypic and genotypic classes. In each example discussed, the F₁ generation is heterozygous for each gene involved in the cross, and the F₂ generation is generated by selfing (when possible) or by allowing the F₁ progeny to interbreed. In monohybrid crosses, there are two phenotypic classes in the F₂ generation; in dihybrid crosses, there are four; and in trihybrid crosses, there are eight. The general rule is that there are 2^n phenotypic classes in the F₂, where n is the number of independently assorting, heterozygous gene pairs (Table 2.3). (This rule holds *only* when a true dominant-recessive relationship holds for each of the

gene pairs.) Furthermore, we saw that there are 3 genotypic classes in the F₂ generation of monohybrid crosses, 9 in dihybrid crosses, and 27 in trihybrid crosses. A simple rule is that the number of genotypic classes is 3^n, where n is the number of independently assorting heterozygous gene pairs (see Table 2.3).

Incidentally, the phenotypic rule (2^n) can also be used to predict the number of classes that will come from a multiple heterozygous F₁ generation used in a testcross. Here the number of genotypes in the next generation will be the same as the number of phenotypes. For example, from *Aa Bb* × *aa bb*, there are four progeny genotypes (2^n, where n is 2)—*Aa Bb*, *Aa bb*, *aa Bb*, and *aa bb*—and

Table 2.3	Number of Phenotypic and Genotypic Classes Expected from Self-Crosses of Heterozygotes in Which All Genes Show Complete Dominance	
Number of Segregating Gene Pairs	**Number of Phenotypic Classes**	**Number of Genotypic Classes**
1^a	2	3
2	4	9
3	8	27
4	16	81
n	2^n	3^n

[a] For example, from $Aa \times Aa$, two phenotypic classes are expected, with genotypic classes of AA, Aa, and aa.

four phenotypes: both dominant phenotypes, the A dominant phenotype and b recessive phenotype, the a recessive phenotype and B dominant phenotype, and both recessive phenotypes.

The "Rediscovery" of Mendel's Principles

Mendel published his treatise on heredity in 1866 in *Verhandlungen des Naturforschenden Vereines* in Brünn, but it received little attention from the scientific community at the time. At the turn of the twentieth century, three researchers working independently on breeding experiments came to the same conclusions as Mendel. The three men were Carl Correns, Hugo de Vries, and Erich von Tschermak. Correns concentrated mostly on maize (corn) and peas, de Vries worked with a number of different plant species, and von Tschermak studied peas.

The first demonstration that Mendelism applied to animals came in 1902 from the work of William Bateson, who experimented with fowl. Bateson also coined the terms *genetics, zygote, F_1, F_2,* and **allelomorph** (literally, "alternative form," meaning one of an array of different forms of a gene), which other researchers shortened to *allele.* The term *gene* as a replacement for *Mendelian factor* was introduced by W. L. Johannsen in 1909.

Statistical Analysis of Genetic Data: The Chi-Square Test

Data from genetic crosses are quantitative. A geneticist typically uses statistical analysis to interpret a set of data from crossing experiments to understand the significance of any deviation of observed results from the results predicted by the hypothesis being tested. The observed phenotypic ratios among progeny rarely exactly match expected ratios owing to chance factors inherent in biological phenomena. A hypothesis is developed based on the observations and is presented as a **null hypothesis,** which states that there is no real difference between the

observed data and the predicted data. Statistical analysis is used to determine whether the difference is due to chance. If it is not, then the null hypothesis is rejected, and a new hypothesis must be developed to explain the data.

A simple statistical analysis used to test null hypotheses is called the **chi-square (χ^2) test,** which is a type of *goodness-of-fit test.* In the genetic crosses we have examined so far, the progeny seemed to fit particular ratios (such as 1:1, 3:1, and 9:3:3:1), and this is where a null hypothesis can be posed and where the chi-square test can tell us whether the data are consistent with that hypothesis.

To illustrate the use of the chi-square test, we will analyze theoretical progeny data from a testcross of a smooth yellow double heterozygote ($Ss\ Yy$) with a wrinkled green homozygote ($ss\ yy$) (see p. 41 and Table 2.2). (Additional applications of the chi-square test are given in Chapter 5.) The progeny data are as follows:

	154 smooth, yellow
	124 smooth, green
	144 wrinkled, yellow
	146 wrinkled, green
Total	568

We hypothesize that a testcross should give a 1:1:1:1 ratio of the four phenotypic classes if the two genes assort independently, and we use the chi-square test to test the hypothesis, as shown in Table 2.4.

First, in column 1, we list the four classes expected in the progeny of the cross. Then we list the observed (o) numbers for each phenotype using actual numbers, not percentages or proportions (column 2). Next, we calculate the expected (e) number for each phenotypic class, given the total number of progeny (568) and the hypothesis under evaluation (in this case a ratio of 1:1:1:1). Thus, in column 3 we list $1/4 \times 568 = 142$ for each phenotype. Now we subtract the expected number (e) from the observed number (o) for each class to find the difference, called the deviation value (d).

In column 5, the deviation squared (d^2) is computed by multiplying each deviation value in column 4 by

Table 2.4	Chi-Square Test Example				
(1)	**(2)**	**(3)**	**(4)**	**(5)**	**(6)**
Phenotypes	**Observed Number (o)**	**Expected Number (e)**	**d ($o - e$)**	**d^2**	**d^2/e**
smooth yellow	154	142	+12	144	1.01
smooth green	124	142	−18	324	2.28
wrinkled yellow	144	142	+2	4	0.03
wrinkled green	146	142	+4	16	0.11
Total	568	568	0		3.43

(7) $\chi^2 = 3.43$ (8) Degrees of freedom (df) = 3

| Table 2.5 | Chi-Square Probabilities | | | | | | | | | |

Probabilities

df	0.95	0.90	0.70	0.50	0.30	0.20	0.10	0.05	0.01	0.001
1	0.004	0.016	0.15	0.46	1.07	1.64	2.71	3.84	6.64	10.83
2	0.10	0.21	0.71	1.39	2.41	3.22	4.61	5.99	9.21	13.82
3	0.35	0.58	1.42	2.37	3.67	4.64	6.25	7.82	11.35	16.27
4	0.71	1.06	2.20	3.36	4.88	5.99	7.78	9.49	13.28	18.47
5	1.15	1.61	3.00	4.35	6.06	7.29	9.24	11.07	15.09	20.52
6	1.64	2.20	3.83	5.35	7.23	8.56	10.65	12.59	16.81	22.46
7	2.17	2.83	4.67	6.35	8.38	9.80	12.02	14.07	18.48	24.32
8	2.73	3.49	5.53	7.34	9.52	11.03	13.36	15.51	20.09	26.13
9	3.33	4.17	6.39	8.34	10.66	12.24	14.68	16.92	21.67	27.88
10	3.94	4.87	7.27	9.34	11.78	13.44	15.99	18.31	23.21	29.59
11	4.58	5.58	8.15	10.34	12.90	14.63	17.28	19.68	24.73	31.26
12	5.23	6.30	9.03	11.34	14.01	15.81	18.55	21.03	26.22	32.91
13	5.89	7.04	9.93	12.34	15.12	16.99	19.81	22.36	27.69	34.53
14	6.57	7.79	10.82	13.34	16.22	18.15	21.06	23.69	29.14	36.12
15	7.26	8.55	11.72	14.34	17.32	19.31	22.31	25.00	30.58	37.70
20	10.85	12.44	16.27	19.34	22.78	25.04	28.41	31.41	37.57	45.32
25	14.61	16.47	20.87	24.34	28.17	30.68	34.38	37.65	44.31	52.62
30	18.49	20.60	25.51	29.34	33.53	36.25	40.26	43.77	50.89	59.70
50	34.76	37.69	44.31	49.34	54.72	58.16	63.17	67.51	76.15	86.66

←——————————— | ———————————→
Fail to reject | Reject
at 0.05 level

Source: From Table IV in *Statistical Tables for Biological, Agricultural, and Medical Research* by Fisher and Yates, 6th ed., 1974. Reprinted by permission of Addison Wesley Longman Ltd.

itself. In column 6, the deviation squared is then divided by the expected number (e). The chi-square value, χ^2 (item 7 in the table), is the total of all the values in column 6. The more the observed data deviate from the data expected on the basis of the hypothesis being tested, the higher χ^2 will be. In our example, χ^2 is 3.43. The general formula is

$$\chi^2 = \Sigma \frac{d^2}{e}, \text{ where } \Sigma \text{ means "sum" and } d^2 = (o - e)^2$$

The last value in the table, item 8, is the degrees of freedom (df) for the set of data. The degrees of freedom in a test involving n classes is usually equal to $n - 1$. There are four phenotypic classes here, so in this case, df = 3.

The chi-square value and the degrees of freedom are next used to determine the probability (P) that the deviation of the observed values from the expected values is due to chance. The P value for a set of data is obtained from tables of chi-square values for various degrees of freedom. Table 2.5 is part of a table of chi-square probabilities. For our example—$\chi^2 = 3.43$ with 3 degrees of freedom—the P value is between 0.30 and 0.50. This is interpreted to mean that with the hypothesis being tested, in 30 to 50 out of 100 trials (that is, 30 to 50 percent of

the time), we could expect chi-square values of this magnitude or greater due to chance. We can reasonably regard this deviation as simply being a sampling, or chance, error. We must be cautious how we use this result, however, because a result like this does not tell us that the hypothesis is *correct*; it only indicates that the experimental data provide no statistically compelling argument against the hypothesis.

As a general rule, if the probability of obtaining the observed χ^2 values is greater than 5 in 100 (5 percent of the time, $P > 0.05$), then the deviation of expected from observed is not considered statistically significant, and the hypothesis being tested is not thrown out.

Suppose that in another chi-square analysis of a different set of data we obtained $\chi^2 = 15.85$ with 3 degrees of freedom. By looking up the value in Table 2.5, we see that the P value is less than 0.01 and greater than 0.001 ($0.001 < P < 0.01$). This means that from 0.1 to 1 time out of 100 (0.1 to 1 percent of the time), we could expect chi-square values of this magnitude or greater due to chance with the hypothesis being true. That this P value is less than 0.05 indicates that the results are not statistically consistent with the 1:1:1:1 hypothesis being tested because of the poor fit.

Mendelian Genetics in Humans

After the rediscovery of Mendel's laws, geneticists found that the inheritance of genes follows the same principles in all sexually reproducing eukaryotes, including humans. W. Farabee, in 1905, was the first to document a genetic trait in humans, *brachydactyly* (OMIM 112500 at http://www.ncbi.nlm.nih.gov/OMIM/), which results in abnormally broad and short fingers (Figure 2.15). By analyzing the trait in human families, Farabee learned that brachydactyly is inherited. The pattern of transmission of the abnormality over several generations led to the conclusion that the trait is a simple dominant trait. In this section, we explore some of the methods used to determine the mechanism of hereditary transmission in humans, and we learn about some inherited human traits.

Pedigree Analysis

The study of human genetics is complicated because controlled matings of humans are not possible for ethical reasons. The inheritance patterns of human traits usually are identified by examining the way the trait occurs in the family trees of individuals who clearly exhibit the trait. Such a study of a family tree, called **pedigree analysis,** involves carefully assembling phenotypic records of the family over several generations. The affected individual through whom the pedigree is discovered is called the **proband** (**propositus** if a male, **proposita** if a female).

Pedigree analysis has its own set of symbols; Figure 2.16 summarizes the basic symbols used here and elsewhere in the text. (The term *autosomal* was introduced in Chapter 1, and *sex-linked* is introduced in Chapter 3; they are included here for completeness.) Figure 2.17 presents a hypothetical pedigree to show how the symbols are assigned to the family tree.

The trait presented in Figure 2.17 is determined by a recessive mutant allele *a*. (Note that recessive mutant alleles may be rare or common in a population.) Generations are numbered with Roman numerals, and individuals are numbered with Arabic numerals; this makes it easy to refer to particular people in the pedigree. The trait in this pedigree results from homozygosity for the allele, in this case resulting from cousins mating. Since cousins

Figure 2.15

Photographs of (a) normal hands and (b) hands exhibiting brachydactyly.

a) **b)**

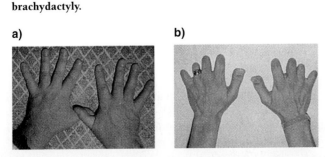

Figure 2.16

Symbols used in human pedigree analysis.

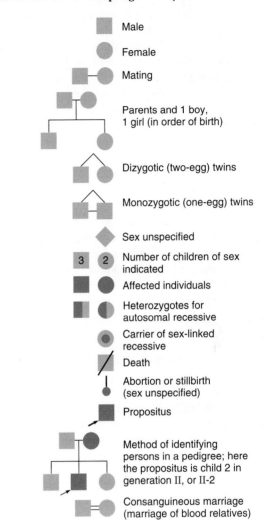

Figure 2.17

A human pedigree, illustrating the use of pedigree symbols.

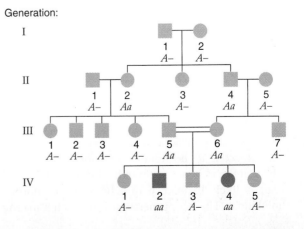

share a fair proportion of their genes, a number of alleles are homozygous in their offspring; in this case, one mutant recessive allele became homozygous and resulted in an identifiable genetic trait.

Gene symbols are included in this pedigree to show the deductive reasoning possible with such analysis. The trait appears first in generation IV. Since neither parent (the two cousins) had the trait, but they produced two children with the trait (IV-2 and IV-4), the simplest hypothesis is that the trait is caused by a recessive allele. Thus, IV-2 and IV-4 would both have the genotype *aa*, and their parents (III-5 and III-6) must have the genotype *Aa*. All other individuals who did not have the trait must have at least one *A* allele; that is, they must be *A–* (either *AA* or *Aa*). Since III-5 and III-6 are both heterozygotes, at least one of each of their parents must have carried an *a* allele. Furthermore, because the trait appeared only after cousins had children, the simplest assumption is that the *a* allele was inherited from individuals with bloodlines shared by III-5 and III-6. This means that II-2 and II-4 probably are both *Aa* and that one of I-1 and I-2 is *Aa* (perhaps both, unless the allele is rare).

Examples of Human Genetic Traits

Recessive Traits.   A large number of human traits are known to be caused by homozygosity for mutant alleles that are recessive to the normal allele. Such recessive mutant alleles produce mutant phenotypes because of a *loss of function* or modified function of the gene product resulting from the mutation involved.

Many serious abnormalities or diseases result from homozygosity for recessive mutant alleles. Two individuals expressing the recessive trait of albinism (deficient pigmentation: OMIM 203100) are shown in Figure 2.18a, and a pedigree for this trait is shown in Figure 2.18b. Individuals with albinism do not produce the pigment melanin, which protects the skin from harmful ultraviolet radiation. As a consequence, their skin and eyes are very sensitive to sunlight. Frequencies of harmful recessive mutant alleles usually are higher than frequencies of harmful dominant mutant alleles because heterozygotes for the recessive mutant allele are not at a significant selective disadvantage. Nonetheless, individuals homozygous for recessive mutant alleles usually are rare. In the United States approximately 1 in 17,000 of the white population and 1 in 28,000 of the African American population have albinism. Among the Irish, about 1 in 10,000 have albinism.

Here are some general characteristics of recessive inheritance for a rare trait:

1. Most affected individuals have two normal parents, both of whom are heterozygous. The trait appears in the F_1 generation because a quarter of the progeny are expected to be homozygous for the recessive allele. If the trait is rare, an individual expressing the trait is likely to mate with a homozygous normal individual. The next generation from such a mating

Figure 2.18

Albinism. (a) Two individuals with albinism: the blues musicians Johnny Winter (left) and Edgar Winter (right). **(b)** A pedigree showing the transmission of the autosomal recessive trait of albinism.

a)

b)   Pedigree

Generation:

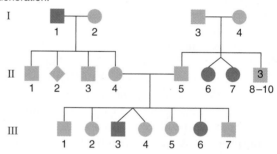

would be heterozygotes who do not express the trait. In other words, recessive traits often skip generations. In the pedigree in Figure 2.18b, for example, II-6 and II-7 must both be *aa*, which means that both parents (I-3 and I-4) must be *Aa* heterozygotes. I-1 is also *aa*, and so II-4 must be *Aa*. Since II-4 and II-5 produce some *aa* children, II-5 also must be *Aa*.

2. Matings between two normal heterozygotes should produce an approximately 3:1 ratio of normal progeny to progeny exhibiting the recessive trait. However, in the analysis of human populations (families), it is difficult to obtain a large enough sample to make the data statistically significant.

3. When both parents are affected, they are homozygous for the recessive trait. All their progeny will usually exhibit the trait.

Other examples of human recessive traits are cystic fibrosis (OMIM 602421: a lethal disease affecting pancreatic, lung, and digestive functions) and sickle-cell anemia (OMIM 141900: a disease resulting from defective hemoglobin).

Dominant Traits.   There are many known dominant human traits. Dominant mutant alleles may produce mutant phenotypes because of a *gain of function* of the gene product resulting from the mutation involved. In other words, the dominant mutant phenotype is a new property of the mutant gene rather than a decrease in its normal activity. Figure 2.19a illustrates one such trait, called woolly hair, in which an individual's hair is very tightly kinked, is very brittle, and breaks before it can grow very long. The best examples of pedigrees for this trait come from Norwegian families; one of these pedigrees is presented in Figure 2.19b. Since it is a fairly rare trait and since not all children of an affected parent show the trait, most woolly-haired individuals probably are heterozygous for the dominant allele involved rather than homozygous.

Dominant mutant alleles are expressed in a heterozygote when they are in combination with what is usually called the **wild-type allele,** the allele that predominates (is present in the highest frequency) in the population found in the "wild." Since many dominant mutant alleles that give rise to recognizable traits are rare, it is extremely unusual to find individuals homozygous for the dominant allele. An affected person in a pedigree is likely to be a heterozygote, and most pairings that involve the mutant allele are between a heterozygote and a homozygous recessive (wild type). Most dominant mutant genes that are clinically significant (i.e., cause medical problems) fall into this category.

Here are some general characteristics of dominant inheritance for a rare trait (refer to Figure 2.18b):

1. Every affected person in the pedigree must have at least one affected parent.
2. The trait usually does not skip generations.
3. On average, an affected heterozygous individual will transmit the mutant gene to half of his or her progeny. If the dominant mutant allele is designated A and its wild-type allele is a, then most crosses will be $Aa \times aa$. From basic Mendelian principles, half the progeny will be aa (wild-type) and the other half will be Aa and show the trait.

Other examples of human dominant traits are achondroplasia (OMIM 100800: dwarfism resulting from

Figure 2.19

Woolly hair. (a) Members of a Norwegian family, some of whom exhibit the trait of woolly hair. **(b)** Part of a pedigree showing the transmission of the autosomal dominant trait of woolly hair.

a)

b) Generation:

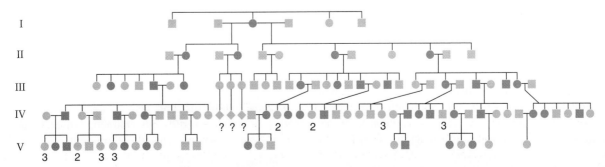

defects in long-bone growth), brachydactyly (malformed hands with short fingers), and Marfan syndrome (OMIM 154700: connective tissue defects potentially causing death by aortic rupture).

KEYNOTE

> Mendelian principles apply to humans and all other eukaryotes. The study of the human inheritance of genetic traits is complicated by the fact that no controlled crosses can be done. Instead, human geneticists analyze genetic traits by pedigree analysis.

Summary

In this chapter, we discussed fundamental principles of gene segregation and gene assortment. Genes are DNA segments that control the biological characteristics transmitted from one generation to another, that is, the hereditary traits. An organism's genetic constitution is its genotype, and the physical manifestation of a genetic trait is its phenotype. An organism's genes provide only the potential for the development of that organism's characteristics. That potential is influenced during development by interactions with other genes and with the environment. Individuals with the same genotype can have different phenotypes, and individuals with the same phenotype may have different genotypes.

The first person to obtain some understanding of the principles of heredity—that is, the inheritance of certain traits—was Gregor Mendel. From his breeding experiments with garden peas, Mendel proposed two basic principles of genetics. In modern terms, the principle of segregation states that the two members of a single gene pair (the alleles) segregate from each other in the formation of gametes. For each gene with two alleles, half of the gametes carry one allele, and the other half carry the other allele. The principle of independent assortment, proposed on the basis of experiments involving more than one gene, states that genes for different traits behave independently in the production of gametes. Both principles are recognized by characteristic phenotypic ratios—called Mendelian ratios—in particular crosses. For the principle of segregation, in a monohybrid cross between two true-breeding parents, one exhibiting a dominant phenotype and the other a recessive phenotype, the F_2 phenotypic ratio is 3:1 for the dominant : recessive phenotypes. For the principle of independent assortment, in a dihybrid cross, the F_2 phenotypic ratio is 9:3:3:1 for the four phenotypic classes.

The gene segregation patterns can be studied more definitively by determining the genotypes for each phenotypic class. This is done using a testcross, in which an individual of unknown genotype is crossed with a homozygous recessive individual to determine the unknown genotype. For example, in a monohybrid cross

resulting in an F_2 3:1 phenotypic ratio, the dominant class can be shown to consist of 1 homozygous dominant : 2 heterozygotes by using a testcross.

Geneticists have found that Mendelian principles of gene segregation apply to all eukaryotes, including humans. The study of the inheritance of genetic traits in humans is complicated because no controlled crosses can be done. Instead, human geneticists analyze genetic traits by pedigree analysis, that is, by examining the occurrences of the trait in family trees of individuals who clearly exhibit the trait. Many recessively inherited and dominantly inherited genetic traits have been identified by pedigree analysis.

Analytical Approaches for Solving Genetics Problems

The most practical way to reinforce Mendelian principles is to solve genetics problems. In this and all following chapters, we will discuss how to approach genetics problems by presenting examples of such problems and discussing their answers. These problems use familiar and unfamiliar examples and pose questions designed to get you to think analytically.

Q2.1 A purple-flowered pea plant is crossed with a white-flowered pea plant. All the F_1 plants produce purple flowers. When the F_1 plants are allowed to self-pollinate, 401 of the F_2 plants have purple flowers and 131 have white flowers. What are the genotypes of the parental and F_1 generation plants?

A2.1 The ratio of plant phenotypes in the F_2 generation is very close to the 3:1 ratio expected of a monohybrid cross. More specifically, this ratio is expected to result from an $F_1 \times F_1$ cross in which both are heterozygous for a specific gene pair. In addition, since the two parents differed in phenotype and only one phenotypic class appeared in the F_1 generation, it is likely that both parental plants were true-breeding. Furthermore, because the F_1 phenotype exactly resembled one of the parental phenotypes, we can say that purple is dominant to white flowers. Assigning the symbol P to the form of the gene that determines purpleness of flowers and the symbol p to the alternative form of the gene that determines whiteness of flowers, we can write the genotypes:

P generation: PP, for the purple-flowered plant
               pp, for the white-flowered plant
F_1 generation: Pp, which, because of dominance, is purple-flowered

We could further deduce that the F_2 plants have an approximately 1:2:1 ratio of $PP : Pp : pp$ by performing testcrosses.

Q2.2 Consider the three gene pairs Aa, Bb, and Cc, each of which affects a different character. In each case, the uppercase letter signifies the dominant allele and the lowercase letter the recessive allele. These three gene pairs assort independently of each other. Calculate the probability of obtaining:

a. an *Aa BB Cc* zygote from the cross *Aa Bb Cc × Aa Bb Cc*;

b. an *Aa BB cc* zygote from the cross *aa BB cc × AA bb CC*;

c. an *A B C* phenotype (i.e., having the dominant phenotype for each of the three genes) from the cross *Aa Bb CC × Aa Bb cc*;

d. an *a b c* phenotype (i.e., having the recessive phenotype for each of the three genes) from the cross *Aa Bb Cc × aa Bb cc*.

A2.2 We must break down the question into simple parts to apply basic Mendelian principles. The key is that the genes assort independently, so we must multiply the probabilities of the individual occurrences to obtain the answers.

a. First, we must consider the *Aa* gene pair. The cross is *Aa × Aa*, so the probability of the zygote being *Aa* is ²⁄₄, since the expected distribution of genotypes is 1 *AA* : 2 *Aa* : 1 *aa*. Then the probability of *BB* from *Bb × Bb* is ¼, and that of *Cc* from *Cc × Cc* is ²⁄₄, following the same sort of logic. Using the product rule (see Box 2.2), the probability of an *Aa BB Cc* zygote is ½ × ¼ × ½ = ¹⁄₁₆.

b. Similar logic is needed here, although we must be sure of the genotypes of the parental types, since they differ from one gene pair to another. For the *Aa* pair, the probability of getting *Aa* from *AA × aa* has to be 1. Next, the probability of getting *BB* from *BB × bb* is 0, so on these grounds alone we cannot get the zygote asked for from the cross given.

c. This question and the next ask for the probability of getting a particular phenotype, so we must start thinking about dominance. Again, we consider each character pair in turn. The probability of an *A* phenotype from *Aa × Aa* is ³⁄₄, from basic Mendelian principles. Similarly, the probability of a *B* phenotype from *Bb × Bb* is ³⁄₄. Lastly, the probability of a *C* phenotype from *CC × cc* is 1. Overall, the probability of an *A B C* phenotype is ³⁄₄ × ³⁄₄ × 1 = ⁹⁄₁₆.

d. The probability of an *a b c* phenotype from *Aa Bb Cc × aa Bb cc* is ½ × ¼ × ½ = ¹⁄₁₆.

Q2.3 In chickens, the white plumage of the leghorn breed is dominant over colored plumage, feathered shanks are dominant over clean shanks, and pea comb is dominant over single comb. Each of the gene pairs segregates independently. If a homozygous white, feathered, pea-combed chicken is crossed with a homozygous colored, clean, single-comb chicken, and the F₁ chickens are allowed to interbreed, what proportion of the birds in the F₂ generation will produce only white, feathered, pea-combed progeny if mated to colored, clean-shanked, single-combed birds?

A2.3 This example is typical of a question that presents the unfamiliar in an attempt to get at the familiar. The best approach to such questions is to reduce them to their simplest parts and, whenever possible, to assign gene symbols for each character. We are told which character is dominant for each of the three gene pairs, so we can use *W* for white and *w* for colored, *F* for feathered and *f* for clean shanks, and *P* for pea comb and *p* for single comb. The cross involves true-breeding strains and can be written as follows:

P generation: *WW FF PP × ww ff pp*
F₁ generation: *Ww Ff Pp*

Now the question asks the proportion of the birds in the F₂ generation that will produce only white, feathered, pea-combed progeny if mated to colored, clean-shanked, single-combed birds. The latter are homozygous recessive for all three genes, that is, *ww ff pp*, as in the parental generation. For the requested result, the F₂ birds must be white, feathered, and pea-combed and must be homozygous for the dominant alleles of the respective genes to produce only progeny with the dominant phenotype. What we are seeking, then, is the proportion of the F₂ chickens that are *WW FF PP* in genotype. We know that each gene pair segregates independently, so the answer can be calculated by using simple probability rules. We consider each gene pair in turn. For the white/colored case, the F₁ × F₁ is *Ww × Ww*, and we know from Mendelian principles that the relative proportion of F₂ genotypes will be 1 *WW* : 2 *Ww* : 1 *ww*. Therefore, the proportion of the F₂ chickens that will be *WW* is ¼. The same relationship holds for the other two pairs of genes. Since the segregation of the three gene pairs is independent, we must multiply the probabilities of each occurrence to calculate the probability for *WW FF PP* individuals. The answer is ¼ × ¼ × ¼ = ¹⁄₆₄.

Questions and Problems

*2.1 In tomatoes, red fruit color is dominant to yellow. Suppose a tomato plant homozygous for red is crossed with one homozygous for yellow. Determine the appearance of (a) the F₁ tomatoes, (b) the F₂ tomatoes, (c) the offspring of a cross of the F₁ back to the red parent, and (d) the offspring of a cross of the F₁ back to the yellow parent.

2.2 In maize, a dominant allele *A* is necessary for seed color, as opposed to colorless (*a*). Another gene has a recessive allele *wx* that results in waxy starch, as opposed to normal starch (*Wx*). The two genes segregate independently. Give phenotypes and relative frequencies for offspring resulting when a plant of genetic constitution *Aa WxWx* is testcrossed.

*2.3 F₂ plants segregate ³⁄₄ colored : ¼ colorless. If a colored plant is picked at random and selfed, what is the probability that both colored and colorless plants will be seen among a large number of its progeny?

*2.4 In guinea pigs, rough coat (R) is dominant over smooth coat (r). A rough-coated guinea pig is bred to a smooth one, giving eight rough and seven smooth progeny in the F_1 generation.

a. What are the genotypes of the parents and their offspring?

b. If one of the rough F_1 animals is mated to its rough parent, what progeny would you expect?

2.5 In cattle, the polled (hornless) condition (P) is dominant over the horned (p) phenotype. A particular polled bull is bred to three cows. Cow A, which is horned, produces a horned calf; polled cow B produces a horned calf; and horned cow C produces a polled calf. What are the genotypes of the bull and the three cows, and what phenotypic ratios do you expect in the offspring of these three matings?

*2.6 In jimsonweed, purple flowers are dominant to white. Self-fertilization of a particular purple-flowered jimsonweed produces 28 purple-flowered and 10 white-flowered progeny. What proportion of the purple-flowered progeny will breed true?

*2.7 Two black female mice are crossed with the same brown male. In a number of litters, female X produced 9 blacks and 7 browns, and female Y produced 14 blacks. What is the mechanism of inheritance of black and brown coat color in mice? What are the genotypes of the parents?

2.8 Bean plants may have different symptoms when infected with a virus. Some show local lesions that do not seriously harm the plant; others show general systemic infection. The following genetic analysis was made:

P   local lesions × systemic infection
F_1 all local lesions
F_2 785 local lesions : 269 systemic infection

What is probably the genetic basis of this difference in beans? Assign gene symbols to all the genotypes occurring in the genetic analysis. Design a testcross to verify your assumptions.

2.9 A normal *Drosophila* (fruit fly) has both brown and scarlet pigment granules in the eyes, which appear red as a result. Brown (bw) is a recessive allele on chromosome 2 that, in the homozygous condition, results in the absence of scarlet granules (so that the eyes appear brown). Scarlet (st) is a recessive allele on chromosome 3 that, when homozygous, results in scarlet eyes because of the absence of brown pigment. Any fly homozygous for recessive brown and recessive scarlet alleles produces no eye pigment and has white eyes. The following results were obtained from crosses:

P   brown-eyed fly × scarlet-eyed fly
F_1 red eyes (both brown and scarlet pigment present)
F_2 $9/16$ red : $3/16$ scarlet : $3/16$ brown : $1/16$ white

a. Assign genotypes to the P and F_1 generations.

b. Design a testcross to verify the F_1 genotype, and predict the results.

*2.10 Grey seed color (G) in garden peas is dominant to white seed color (g). In the following crosses, the indicated parents with known phenotypes but unknown genotypes produced the listed progeny. Give the possible genotypes of each female parent based on the segregation data.

Parents	Progeny		Female Parent
Female × Male	Grey	White	Genotype
grey × white	81	82	?
grey × grey	118	39	?
grey × white	74	0	?
grey × grey	90	0	?

*2.11 Fur color in the babbit, a furry little animal and popular pet, is determined by a pair of alleles, B and b. BB and Bb babbits are black, while bb babbits are white. A farmer wants to breed babbits for sale. True-breeding white (bb) female babbits breed poorly. The farmer purchases a pair of black babbits, and these mate and produce six black and two white offspring. The farmer immediately sells his white babbits and then consults with you for a breeding strategy to produce more white babbits.

a. If he performed random crosses between pairs of F_1 black babbits, what proportion of the F_2 progeny would be white?

b. If he crossed an F_1 male to the parental female, what is the probability that this cross will produce white progeny?

c. What would be the farmer's best strategy to maximize the production of white babbits?

2.12 In the jimsonweed, purple flower (P) is dominant to white (p), and spiny pods (S) are dominant to smooth (s). In a cross between a jimsonweed homozygous for white flowers and spiny pods and one homozygous for purple flowers and smooth pods, determine the phenotype of (a) the F_1 generation; (b) the F_2 generation; (c) the progeny of a cross of the F_1 back to the white spiny parent; and (d) the progeny of a cross of the F_1 back to the purple smooth parent.

*2.13 Cleopatra is normally a very refined cat. When she finds even a small amount of catnip, however, she purrs madly, rolls around in the catnip, becomes exceedingly playful, and appears intoxicated. Cleopatra and Antony, who walks past catnip with an air of indifference, have produced five kittens who respond to catnip just like Cleopatra. When the kittens mature, two of them mate and produce four kittens that respond to catnip and one that does not. When another of Cleopatra's daughters mates with Augustus (a nonrelative), who behaves just like Antony, three catnip-sensitive and two catnip-insensitive kittens are produced. Propose a hypothesis for the inheritance of catnip sensitivity that explains these data.

2.14 Using the information in problem 2.12, what progeny would you expect from the following jimsonweed crosses? You are encouraged to use the branch diagram approach.

a. *PP ss* × *pp SS*
b. *Pp SS* × *pp ss*
c. *Pp Ss* × *Pp SS*
d. *Pp Ss* × *Pp Ss*
e. *Pp Ss* × *Pp ss*
f. *Pp Ss* × *pp ss*

*2.15 In summer squash, white fruit (*W*) is dominant over yellow (*w*), and disk-shaped fruit (*D*) is dominant over sphere-shaped fruit (*d*). The following problems give the appearances of the parents and their progeny. Determine the genotypes of the parents in each case.

a. White, disk × yellow, sphere gives ½ white, disk and ½ white, sphere.
b. White, sphere × white, sphere gives ¾ white, sphere and ¼ yellow, sphere.
c. Yellow, disk × white, sphere gives all white, disk progeny.
d. White, disk × yellow, sphere gives ¼ white, disk; ¼ white, sphere; ¼ yellow, disk; and ¼ yellow, sphere.
e. White, disk × white, sphere gives ⅜ white, disk; ⅜ white, sphere; ⅛ yellow, disk; and ⅛ yellow, sphere.

*2.16 Genes *a*, *b*, and *c* assort independently and are recessive to their respective alleles *A*, *B*, and *C*. Two triply heterozygous (*Aa Bb Cc*) individuals are crossed.

a. What is the probability that a given offspring will be phenotypically *A B C*, that is, will exhibit all three dominant traits?
b. What is the probability that a given offspring will be genotypically homozygous for all three dominant alleles?

2.17 In garden peas, tall stem (*T*) is dominant over short stem (*t*), green pods (*G*) are dominant over yellow pods (*g*), and smooth seeds (*S*) are dominant over wrinkled seeds (*s*). Suppose a homozygous short, green, wrinkled pea plant is crossed with a homozygous tall, yellow, smooth one.

a. What will be the appearance of the F_1 generation?
b. If the F_1 plants are interbred, what will be the appearance of the F_2 generation?
c. What will be the appearance of the offspring of a cross of the F_1 back to its short, green, wrinkled parent?
d. What will be the appearance of the offspring of a cross of the F_1 back to its tall, yellow, smooth parent?

2.18 *C* and *c*, *O* and *o*, and *I* and *i* are three independently segregating pairs of alleles in chickens. *C* and *O* are dominant alleles, both of which are necessary for pigmentation. *I* is a dominant inhibitor of pigmentation. Individuals of genotype *cc* or *oo* or *Ii* or *II* are white, regardless of what other genes they possess. Assume that White Leghorns are *CC OO II*, White Wyandottes are *cc OO ii*, and White Silkies are *CC oo ii*. What types of offspring (white or pigmented) are possible, and what is the probability of each, from the following crosses?

a. White Silkie × White Wyandotte
b. White Leghorn × White Wyandotte
c. (Wyandotte-Silkie F_1) × White Silkie

2.19 Two homozygous strains of corn are hybridized. They are distinguished by six different pairs of genes, all of which assort independently and produce an independent phenotypic effect. The F_1 hybrid is selfed to give an F_2.

a. What is the number of possible genotypes in the F_2?
b. How many of these genotypes will be homozygous at all six gene loci?
c. If all gene pairs act in a dominant-recessive fashion, what proportion of the F_2 will be homozygous for all dominants?
d. What proportion of the F_2 will show all dominant phenotypes?

*2.20 The coat color of mice is controlled by several genes. The agouti pattern, characterized by a yellow band of pigment near the tip of the hairs, is produced by the dominant allele *A*; homozygous *aa* mice do not have the band and are nonagouti. The dominant allele *B* determines black hairs, and the recessive allele *b* determines brown. Homozygous $c^h c^h$ individuals allow pigments to be deposited only at the extremities (e.g., feet, nose, and ears) in a pattern called Himalayan. The genotype *C*– allows pigment to be distributed over the entire body.

a. If a true-breeding black mouse is crossed with a true-breeding brown, agouti, Himalayan mouse, what will be the phenotypes of the F_1 and F_2 generation?
b. What proportion of the black agouti F_2 mice will be of genotype *Aa BB Cc^h*?
c. What proportion of the Himalayan mice in the F_2 generation are expected to show brown pigment?
d. What proportion of all agoutis in the F_2 generation are expected to show black pigment?

2.21 In cocker spaniels, solid coat color is dominant over spotted coat. Suppose a true-breeding, solid-colored dog is crossed with a spotted dog, and the F_1 dogs are interbred.

a. What is the probability that the first puppy born will have a spotted coat?
b. What is the probability that if four puppies are born, all of them will have solid coats?

2.22 In the F_2 generation of his cross of red-flowered × white-flowered *Pisum*, Mendel obtained 705 plants with red flowers and 224 with white.

a. Is this result consistent with his hypothesis of factor segregation, from which a 3:1 ratio would be predicted?
b. In how many similar experiments would a deviation as great as or greater than this one be expected? (Calculate χ^2 and obtain the approximate value of *P* from Table 2.5.)

2.23 In tomatoes, cut leaf and potato leaf are alternative characters, with cut (*C*) dominant to potato (*c*). Purple stem and green stem are another pair of alternative characters, with purple (*P*) dominant to green (*p*). A true-breeding cut green tomato plant is crossed with a true-breeding potato purple plant, and the F_1 plants are allowed to interbreed. The 320 F_2 plants were phenotypically 189 cut purple, 67 cut green; 50 potato purple, and 14 potato green. Propose a hypothesis to explain the data, and use the χ^2 test to test the hypothesis.

*2.24 The simple case of just two mating types (male and female) is by no means the only sexual system known. The ciliated protozoan *Paramecium bursaria* has a system of four mating types, controlled by two genes (*A* and *B*). Each gene has a dominant and a recessive allele. The four mating types are expressed according to the following scheme:

Genotype	Mating Type
AA BB	A
Aa BB	A
AA Bb	A
Aa Bb	A
AA bb	D
Aa bb	D
aa BB	B
aa Bb	B
aa bb	C

It is clear, therefore, that some of the mating types result from more than one possible genotype. We have four strains of known mating type—"A," "B," "C," and "D"—but unknown genotype. The following crosses were made with the indicated results:

	Mating Type of Progeny			
Cross	A	B	C	D
"A" × "B"	24	21	14	18
"A" × "C"	56	76	55	41
"A" × "D"	44	11	19	33
"B" × "C"	0	40	38	0
"B" × "D"	6	8	14	10
"C" × "D"	0	0	45	45

Assign genotypes to "A," "B," "C," and "D."

*2.25 In bees, males (drones) develop from unfertilized eggs and are haploid. Females (workers and queens) are diploid and come from fertilized eggs. *W* (black eyes) is dominant over *w* (white eyes). Workers of genotype *RR* or *Rr* use wax to seal crevices in the hive; *rr* workers use resin instead. A *Ww Rr* queen founds a colony after being fertilized by a black-eyed drone bearing the *r* allele.
a. What will be the appearance and behavior of workers in the new hive, and what are their relative frequencies?
b. Give the genotypes of male offspring, with relative frequencies.

c. Fertilization normally takes place in the air during a nuptial flight, and any bee unable to fly would effectively be rendered sterile. Suppose that a recessive mutation, *c*, occurs spontaneously in a sperm that fertilizes a normal egg and that the effect of the mutant gene is to cripple the wings of any adult not bearing the normal allele *C*. The fertilized egg develops into a normal queen named Madonna. What is the probability that wingless males will be *found in a hive* founded two generations later by one of Madonna's granddaughters?
d. By one of Madonna's great-great-granddaughters?

*2.26 Consider the following pedigree, in which the allele responsible for the trait (*a*) is recessive to the normal allele (*A*).

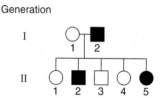

Generation

a. What is the genotype of the mother?
b. What is the genotype of the father?
c. What are the genotypes of the children?
d. Given the mechanism of inheritance involved, does the ratio of children with the trait to children without the trait match what would be expected?

2.27 For the following pedigrees A and B, indicate whether the trait involved in each case could be (a) recessive or (b) dominant. Explain your answer.

Pedigree A

Generation

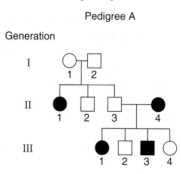

Pedigree B

Generation

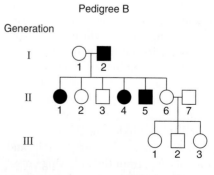

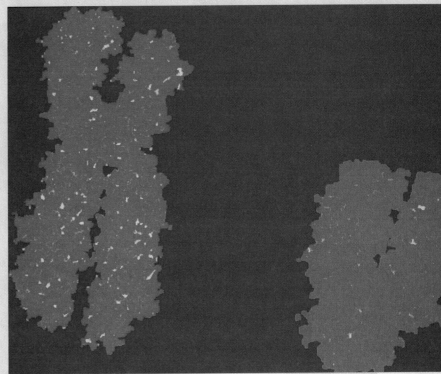

3

Chromosomal Basis of Inheritance, Sex Linkage, and Sex Determination

PRINCIPAL POINTS

The chromosome theory of inheritance states that genes are located on chromosomes.

Sex chromosomes are chromosomes in eukaryotes that are represented differently in the two sexes. In organisms with sex chromosomes, one sex is homogametic and the other is heterogametic.

Sex linkage is the physical association of genes with the sex chromosomes of eukaryotes. Such genes are referred to as sex-linked genes.

The correlation between gene segregation patterns and the patterns of chromosome behavior in meiosis supports the chromosome theory of inheritance.

In many eukaryotic organisms, sex determination is related to the sex chromosomes. In humans and other mammals, for example, the presence of a Y chromosome specifies maleness, and its absence results in femaleness. Several other sex determination mechanisms are known in eukaryotes.

In humans, the allele responsible for a trait can be inherited in one of five main ways: autosomal recessive, autosomal dominant, X-linked recessive, X-linked dominant, or Y-linked.

Since Mendel, many exceptions to his rules have been discovered.

i WHEN A CHILD IS BORN, THE FIRST QUESTION MOST people ask is "Is it a boy or a girl?" The answer, at the chromosomal level, depends on the sex chromosomes: two X chromosomes produce a girl, while an X and a Y chromosome produce a boy. But the genes contained on these chromosomes determine more than just the sex of an individual; they also are responsible for the inheritance of a number of other traits.

In this chapter, you will learn how chromosomes behave during nuclear division in eukaryotes, how sex is determined in humans and other organisms, and how certain traits are sex-linked in humans. After you have read this chapter, you can apply what you've learned by trying the iActivity, in which you will investigate the inheritance of deafness within a family.

On Mendel's foundation, early geneticists began to build genetic hypotheses that could be tested by appropriate crosses, and they began to investigate the nature of Mendelian factors. We now know that Mendelian factors are genes and that genes are located on chromosomes, and we start by considering the evidence for this association. In so doing, we will learn about the segregation of genes located on the sex chromosomes. Next, we will learn about various mechanisms of sex determination, and finally we will discuss sex-linked traits in humans. The goal of this chapter is to help you to learn how to think about gene segregation in terms of chromosome inheritance patterns.

Table 3.1	Chromosome Number in Various Organisms[a]
Organism	**Total Chromosome Number**
Human	46
Chimpanzee	48
Dog	78
Cat	72
Mouse	40
Horse	64
Chicken	78
Toad	36
Goldfish	94
Starfish	36
Fruit fly (*Drosophila melanogaster*)	8
Mosquito	6
Australian ant (*Myrecia pilosula*)	♂ 1, ♀ 2
Nematode	♂ 11, ♀ 12
Neurospora (haploid)	7
Sphagnum moss (haploid)	23
Field horsetail	216
Giant sequoia	22
Tobacco	48
Cotton	52
Potato	48
Tomato	24
Bread wheat	42
Yeast (*Saccharomyces cerevisiae*) (haploid)	16

[a]Except as noted, all chromosome numbers are for diploid cells.

Chromosome Theory of Inheritance

Around the turn of the twentieth century, cytologists had established that within a given species, the total number of chromosomes is constant in all cells, whereas the chromosome number varies considerably among species (Table 3.1). In 1902, Walter Sutton and Theodor Boveri independently recognized that the transmission of chromosomes from one generation to the next closely paralleled the pattern of transmission of genes from one generation to the next. To explain this correlation, they proposed the **chromosome theory of inheritance.** This theory states that the Mendelian factors—which we now know as genes—are located on chromosomes. In this section, we consider some of the evidence cytologists and geneticists obtained to support this theory.

Sex Chromosomes

The support for the chromosome theory of inheritance came from experiments that related the hereditary behavior of particular genes to the transmission of the **sex chromosomes,** the chromosomes in eukaryotes that are

represented differently in the two sexes. For many animals, the sex chromosome composition of an individual is directly related to the sex of the individual. The other chromosomes in eukaryotes, which are not represented differently in the two sexes, are the **autosomes.**

The sex chromosomes typically are designated the **X chromosome** and the **Y chromosome.** In humans and the fruit fly *Drosophila melanogaster,* for example, the female has two X chromosomes (she is XX with respect to the sex chromosomes), while the male has one X chromosome and one Y chromosome (he is XY). Figure 3.1a shows male and female *Drosophila,* and Figure 3.1b shows the chromosome sets of the two sexes. Because the male produces two kinds of gametes with respect to sex chromosomes (X or Y), and because the female produces only one type of gamete (X), the male is called the **heterogametic sex** and the female is called the **homogametic sex.** In *Drosophila,* the X and Y chromosomes are similar in size, but their shapes are different. (Note that in some organisms, the male is homogametic and the female is heterogametic.)

Figure 3.1

***Drosophila melanogaster* (fruit fly), an organism used extensively in genetics experiments. (a)** Female (left) and male (right). Top, adult flies; bottom, drawings of ventral abdominal surface to show differences in genitalia. **(b)** Chromosomes of *Drosophila melanogaster* diagrammed to show their morphological differences. A female (left) has four pairs of chromosomes in her somatic cells, including a pair of X chromosomes. The only difference in the male is an XY pair of sex chromosomes instead of two Xs.

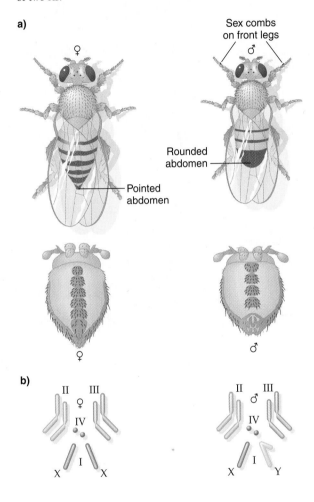

KEYNOTE

Sex chromosomes are chromosomes in eukaryotes that are represented differently in the two sexes. In many of the organisms encountered in genetic studies, one sex possesses a pair of identical chromosomes (the X chromosomes). The opposite sex possesses a pair of visibly different chromosomes. One is an X chromosome, and the other, structurally and functionally different, is called the Y chromosome. Commonly, the XX sex is female, and the XY sex is male. The XX and XY sexes are called the homogametic and heterogametic sexes, respectively.

Sex Linkage

Evidence to support the chromosome theory of heredity came in 1910 when Thomas Hunt Morgan reported the results of genetics experiments with *Drosophila*. Morgan received the 1933 Nobel Prize in Physiology or Medicine for his discoveries concerning the role played by the chromosome in heredity.

Morgan found in one of his true-breeding stocks a male fly that had white eyes instead of the brick-red eyes characteristic of the **wild type.** The term *wild type* refers to a strain, organism, or gene that is most prevalent in the "wild" population of the organism with respect to genotype and phenotype. For example, a *Drosophila* strain with all wild-type genes has brick-red eyes. Variants of a wild-type strain arise from mutational changes of the wild-type alleles that produce **mutant alleles;** the result is strains with mutant characteristics. Mutant alleles may be recessive or dominant to the wild-type allele; for example, the mutant allele that causes white eyes in *Drosophila* is recessive to the wild-type (red-eye) allele.

Morgan crossed the white-eyed male with a red-eyed female from the same stock and found that all the F_1 flies were red-eyed. He concluded that the white-eyed trait was recessive. Next, he allowed the F_1 progeny to interbreed and counted the phenotypic classes in the F_2 generation; there were 3,470 red-eyed and 782 white-eyed flies. The number of individuals with the recessive phenotype was too small to fit the Mendelian 3:1 ratio. (Later, he determined that the lower-than-expected number of flies with the recessive phenotype was the result of lower viability of white-eyed flies.) In addition, *Morgan noticed that all the white-eyed flies were male.*

Figure 3.3 diagrams the crosses. The *Drosophila* gene symbolism used is different from the symbolism we used for Mendel's crosses and is described in Box 3.1. You should understand the *Drosophila* gene symbolism before proceeding with this discussion. Note that the mother-son inheritance pattern presented in Figure 3.3

The pattern of transmission of X and Y chromosomes from generation to generation is straightforward. In Figure 3.2, the X is represented by a straight structure much like a slash mark, and the Y by a similar structure topped by a hook to the right. The female produces only X-bearing gametes, while the male produces both X-bearing and Y-bearing gametes. Random fusion of these gametes produces an F_1 generation with ½ XX (female) and ½ XY (male) flies.

Figure 3.2

Inheritance pattern of X and Y chromosomes in organisms where the female is XX and the male is XY. (a) Production of the F_1 generation. (b) Production of the F_2 generation.

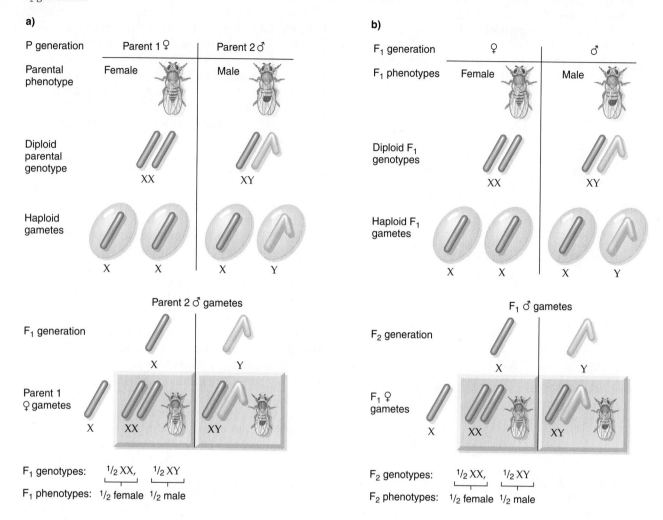

is the result of the segregation of genes located on a sex chromosome.

Morgan proposed that the gene for the eye color variant is located on the X chromosome. The condition of X-linked genes in males is said to be **hemizygous,** since the gene is present only once in the organism because there is no homologous gene on the Y. For example, the white-eyed *Drosophila* males have an X chromosome with a white-eye allele and no other allele of that gene in their genomes; these males are hemizygous for the white-eye allele. Since the white-eye allele of the gene is recessive, the original white-eyed male must have had the recessive allele for white eyes (designated w; see Box 3.1) on his X chromosome. The red-eyed female came from a true-breeding strain, so both of her X chromosomes must have carried the dominant allele for red eyes, w^+ ("w plus").

The F_1 flies are produced in the following way (see Figure 3.3a): The males receive their only X chromosome from their mother and hence have the w^+ allele and are red-eyed. The F_1 females receive a dominant w^+ allele from their mother and a recessive w allele from their father, so they are also red-eyed.

In the F_2 generation produced by interbreeding the F_1 flies, the males that received an X chromosome with the w allele from their mother are white-eyed; those that received an X chromosome with the w^+ allele are red-eyed (see Figure 3.3b). The gene transmission shown in this cross—from a male parent to a female offspring ("child") to a male grandchild—is called **crisscross inheritance.**

Morgan also crossed a true-breeding white-eyed female (homozygous for the w allele) with a red-eyed male (hemizygous for the w^+ allele) (Figure 3.4). This cross is the *reciprocal cross* of Morgan's first cross—white-eyed male × red-eyed female—shown in Figure 3.3. All the F_1 females receive a w^+-bearing X from their father and a w-bearing X from their mother (see Figure 3.4a).

Figure 3.3

The X-linked inheritance of red eyes and white eyes in *Drosophila melanogaster*.
The symbols *w* and *w*⁺ indicate the white- and red-eyed alleles, respectively. (**a**) A red-eyed female is crossed with a white-eyed male. (**b**) The F₁ flies are interbred to produce the F₂ flies.

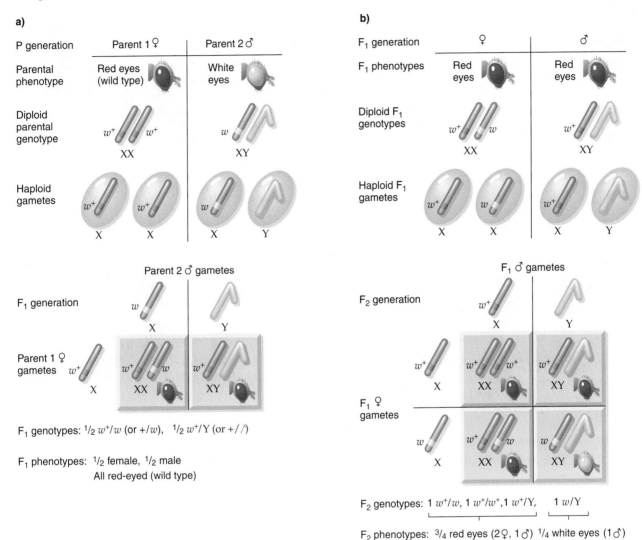

Consequently, they are heterozygous w^+/w and have red eyes. All the F₁ males receive a *w*-bearing X from their mother and a Y from their father, and so they have white eyes (see Figure 3.4a). This result is different from that of the cross in Figure 3.3.

Interbreeding of the F₁ flies (see Figure 3.4b) involves a *w*/Y male and a *w*⁺/*w* female, giving approximately equal numbers of male and female red- and white-eyed flies in the F₂. This ratio differs from the results obtained in the first cross, where an approximately 3:1 ratio of red-eyed : white-eyed flies was obtained and where none of the females and approximately half the

males exhibited the white-eyed phenotype. The difference in phenotypic ratios in the two sets of crosses reflects the transmission patterns of sex chromosomes and the genes they contain.

Morgan's crosses of *Drosophila* involved eye color characteristics that we now know are coded for by a gene found on the X chromosome. These characteristics and the genes that give rise to them are referred to as **sex-linked**—or, more correctly, as **X-linked**—because the gene locus is part of the X chromosome. *X-linked inheritance* is the term used for the pattern of hereditary transmission of X-linked genes. When

Figure 3.4

Reciprocal cross of that shown in Figure 3.3. (a) A homozygous white-eyed female is crossed with a red-eyed (wild-type) male. (b) The F_1 flies are interbred to produce the F_2 flies. The results of this cross differ from those in Figure 3.3 because of the way sex chromosomes segregate in crosses.

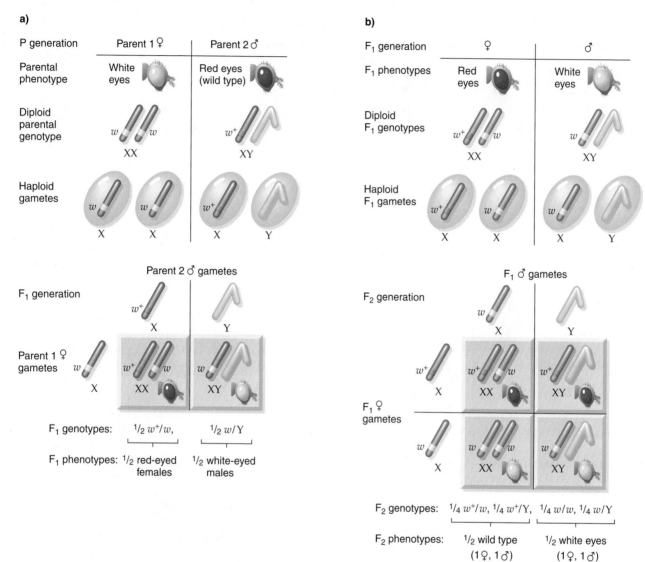

the results of reciprocal crosses are not the same, and different ratios are seen for the two sexes of the offspring, sex-linked characteristics may well be involved. By comparison, the results of reciprocal crosses are *always* the same when they involve genes located on the autosomes. Most significantly, Morgan's results strongly supported the hypothesis that genes were located on chromosomes. Morgan found many other examples of genes on the X chromosome in *Drosophila* and in other organisms, thereby showing that his observations were not confined to a single species. Later in this chapter, we discuss the analysis of X-linked traits in humans.

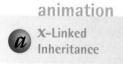

animation

X-Linked Inheritance

KEYNOTE

Sex linkage is the linkage of genes with the sex chromosomes of eukaryotes. Such genes, as well as the phenotypic characteristics these genes control, are called sex-linked genes. Genes only on the X chromosome are called X-linked genes. Morgan's pioneering work with the inheritance of sex-linked genes of *Drosophila* strongly supported but did not prove the chromosome theory of inheritance.

Box 3.1   **Genetic Symbols Revisited**

Unfortunately, no single system of gene symbols is used by geneticists; the gene symbols used for *Drosophila* are different from those used for peas in Chapter 2. The *Drosophila* symbolism is commonly, but not exclusively, used in genetics today. In this system, the symbol + indicates a wild-type allele of a gene. A lowercase letter designates a mutant allele of a gene that is *recessive* to the wild-type allele, and an uppercase letter is used for an allele that is *dominant* to the wild-type allele. *The letters are chosen on the basis of the phenotype of the organism expressing the mutant allele.* For example, a variant strain of *Drosophila* has bright orange eyes instead of the usual brick red. The mutant allele involved is recessive to the wild-type brick-red allele, and because the bright orange eye color is close to vermilion in tint, the allele is designated *v* and is called the vermilion allele. The wild-type allele of *v* is *v*⁺, but when there is no chance of confusing it with other genes in the cross, it is often shortened to +. In the Mendelian terminology used up to now, the recessive mutant allele would be *v*, and its wild-type allele would be *V*.

A conventional way to represent the chromosomes (instead of the way we have been using in the figures) is to use a slash (/). Thus, *v*⁺/*v* or +/*v* indicates that there are two homologous chromosomes, one with the wild-type allele (*v*⁺ or +) and the other with the recessive allele (*v*). The Y chromosome is usually symbolized as a Y or as a bent slash (/). Thus, Morgan's cross of a true-breeding red-eyed female fly with a white-eyed male could be written *w*⁺/*w*⁺ × *w*/Y or +/+ × *w*/\.

The same rules apply when the alleles involved are dominant to the wild-type allele. For instance, some *Drosophila* mutants, called *Curly*, have wings that curl up at the end rather than the normal straight wings. The symbol for this mutant allele is *Cy*, and the wild-type allele is *Cy*⁺, or + in the shorthand version. Thus, a heterozygote would be *Cy*⁺/*Cy* or +/*Cy*.

In the rest of the book, both the *A/a* (Mendelian) and *a*⁺/*a* (*Drosophila*) symbols will be used, so it is important that you be able to work with both. Since it is easier to verbalize the Mendelian symbols ("big *A*, small *a*"), many of our examples will follow that symbolism, even though the *Drosophila* symbolism in many ways is more informative. That is, with the *Drosophila* system, the wild-type and mutant alleles are readily apparent, since the wild-type allele is indicated by a +. The Mendelian system is commonly used in animal and plant breeding. A good reason for this is that after many years (sometimes centuries) of breeding, it is no longer apparent what the "normal" (wild-type) gene is.

Nondisjunction of X Chromosomes

Proof for the chromosome theory of inheritance came from the work of Morgan's student Calvin Bridges. Morgan's work showed that from a cross of a white-eyed female (*w/w*) with a red-eyed male (*w*⁺/Y), all the F₁ males should be white-eyed and all the females should be red-eyed. Bridges found rare exceptions to this result: about 1 in 2,000 of the F₁ flies from such a cross are either white-eyed *females* or red-eyed *males*.

To explain these exceptional flies, Bridges hypothesized that a problem had occurred with chromosome segregation in meiosis (see Chapter 1, pp. 18–20). Normally, homologous chromosomes (in meiosis I) or sister chromatids (in meiosis II or mitosis) move to opposite poles at anaphase (see p. 20); when this fails to take place, chromosome **nondisjunction** results. Nondisjunction can involve either autosomes or the sex chromosomes. For the crosses Bridges analyzed, occasionally the two X chromosomes failed to separate, so eggs were produced either with two X chromosomes or with no X chromosomes instead of the usual one. This particular type of nondisjunction is called **X chromosome nondisjunction.** Normal disjunction of the X chromosomes is illustrated in Figure 3.5a, and nondisjunction of the X chromosomes in meiosis I and meiosis II is shown in Figures 3.5b and 3.5c, respectively.

How can nondisjunction of the X chromosomes explain the exceptional flies in Bridges's cross? When nondisjunction occurs in the *w/w* female (Figure 3.6), two classes of exceptional eggs result with equal (and low) frequency: those with two X chromosomes and those with no X chromosomes. The XY male is *w*⁺/Y and produces equal numbers of *w*⁺- and Y-bearing sperm. When these eggs are fertilized by the two types of sperm, the result is four types of zygotes. Two types usually do not survive: the YO class, which has a Y chromosome but no X chromosome, and the triplo-X (XXX) class, which has three X chromosomes and the genotype *w/w/w*⁺ (see Figure 3.6).

The surviving classes are the red-eyed XO males (in *Drosophila*, the XO pattern produces a sterile male), with no Y chromosome and a *w*⁺ allele on the X, and the white-eyed XXY females (in *Drosophila*, XXY produces a fertile female), with a *w* allele on each X. The males are red-eyed because they received their X chromosome from their father, and the females have white eyes because their two X chromosomes came from their mother. This result is unusual, since sons normally get their X from their mother, and daughters get one X from each parent.

In sum, gene segregation patterns parallel the patterns of chromosome behavior in meiosis. In Figure 3.7, this parallel is illustrated for a diploid cell with two homologous pairs of chromosomes. The cell is genotypically *Aa Bb*, with the *A/a* gene pair on one chromosome and the *B/b* gene pair on the other chromosome. As the figure shows, the two homologous pairs of chromosomes

Figure 3.5

Nondisjunction in meiosis involving the X chromosome. (Nondisjunction of autosomal chromosomes and of all chromosomes in mitosis occurs in the same way.) **(a)** Normal X chromosome segregation in meiosis. **(b)** Nondisjunction of X chromosomes in meiosis I. (c) Nondisjunction of X chromosomes in meiosis II.

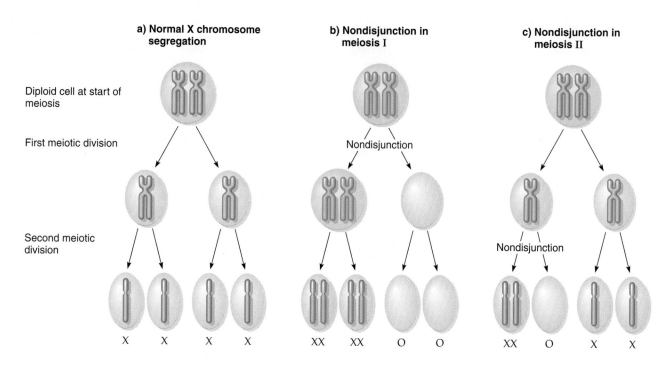

align on the metaphase plate independently, giving rise to two different segregation patterns for the two gene pairs. Since each of the two alignments, and hence segregation patterns, is equally likely, meiosis results in cells with equal frequencies of the genotypes *A B*, *a b*, *A b*, and *a B*. Genotypes *A B* and *a b* result from one chromosome alignment, and genotypes *A b* and *a B* result from the other alignment. In terms of Mendel's laws, we can see how the principle of segregation applies to the segregation pattern of one homologous pair of chromosomes and the associated gene pair, while the principle of independent assortment applies to the segregation pattern of both homologous pairs of chromosomes.

K E Y N O T E

An unexpected inheritance pattern of an X-linked mutant gene in *Drosophila* correlated directly with a rare event during meiosis, called nondisjunction, in which members of a homologous pair of chromosomes do not segregate to the opposite poles. The correlation between gene segregation patterns and the patterns of chromosome behavior in meiosis proved the chromosome theory of inheritance.

Sex Determination

In this section, we discuss some of the mechanisms of sex determination. In *genotypic sex determination systems*, sex is governed by the genotype of the zygote or spore; in *environmental sex determination systems*, sex is governed by internal and external environmental conditions.

Genotypic Sex Determination Systems

Genotypic sex determination, in which the sex chromosomes play a decisive role in the inheritance and determination of sex, may occur in one of two ways. In the **Y chromosome mechanism of sex determination** (seen in humans), the Y chromosome determines the sex of an individual. Individuals with a Y chromosome are genetically male, while individuals without a Y chromosome are genetically female. In the **X chromosome–autosome balance system** (seen in *Drosophila* and the nematode *Caenorhabditis elegans*), sex is determined by the *ratio between the number of X chromosomes and the number of sets of autosomes.* In this system, the Y chromosome has no effect on sex determination, but it is required for male fertility.

Figure 3.6

Rare primary nondisjunction during meiosis in a white-eyed female *Drosophila melanogaster* and results of a cross with a normal red-eyed male. Note the decreased viability of XXX and YO progeny.

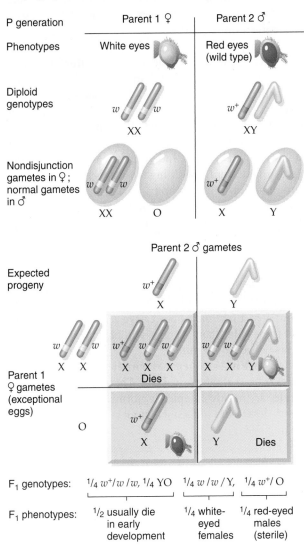

	Parent 1 ♀	Parent 2 ♂
P generation		
Phenotypes	White eyes	Red eyes (wild type)
Diploid genotypes	w w XX	w^+ XY
Nondisjunction gametes in ♀; normal gametes in ♂	w w XX O	w^+ X Y

Expected progeny

	Parent 2 ♂ gametes	
	w^+ X	Y
Parent 1 ♀ gametes (exceptional eggs) w w X X	w^+ w w X X X **Dies**	w w X X Y
O	w^+ X	Y **Dies**

F_1 genotypes:	¼ $w^+/w/w$, ¼ YO	¼ $w/w/$Y,	¼ $w^+/$O
F_1 phenotypes:	½ usually die in early development	¼ white-eyed females	¼ red-eyed males (sterile)

Figure 3.7 ▶

The parallel behavior between Mendelian genes and chromosomes in meiosis. This hypothetical *Aa Bb* diploid cell contains a homologous pair of metacentric chromosomes, which carry the *A/a* gene pair, and a homologous pair of telocentric chromosomes, which carry the *B/b* gene pair. The independent alignment of the two homologous pairs of chromosomes at metaphase I results in equal frequencies of the four meiotic products, *A B*, *a b*, *A b*, and *a B*, illustrating Mendel's principle of independent assortment.

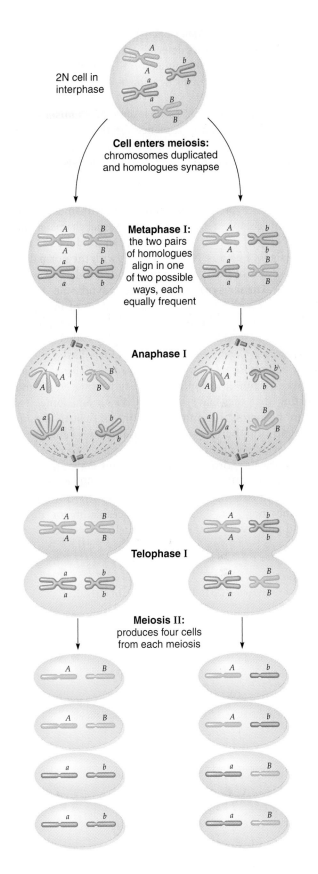

2N cell in interphase

Cell enters meiosis: chromosomes duplicated and homologues synapse

Metaphase I: the two pairs of homologues align in one of two possible ways, each equally frequent

Anaphase I

Telophase I

Meiosis II: produces four cells from each meiosis

Sex Determination in Mammals. In humans and other placental mammals, sex is determined by the Y chromosome mechanism of sex determination. In the absence of a Y chromosome, the gonads develop as ovaries.

Evidence for the Y Chromosome Mechanism of Sex Determination. Early evidence for the Y chromosome mechanism of sex determination in mammals came from studies in which nondisjunction in meiosis produced an abnormal sex chromosome complement. While sex determination in these cases follows directly from the sex chromosomes present, those individuals with unusual chromosome complements have many unusual characteristics.

Nondisjunction, for example, can produce XO individuals. In humans, XO individuals with the normal two sets of autosomes are female and sterile, and they exhibit **Turner syndrome.** A Turner syndrome individual is shown in Figure 3.8a, and her set of metaphase chromosomes is shown in Figure 3.8b. (A complete set of metaphase chromosomes in a cell is called its **karyotype;** see Chapter 1.) Such individuals have only one sex chromosome—an X chromosome. These females have a genomic complement designated as 45,X, indicating that they have a total of 45 chromosomes (sex chromosomes plus autosomes), in contrast to the normal 46, and that the sex chromosome complement consists of one X chromosome.

Turner syndrome individuals occur with a frequency of 1 in every 10,000 females born. Up to 99 percent of all 45,X embryos die before birth. Surviving Turner syndrome individuals have few noticeable major defects until puberty, when they fail to develop secondary sexual characteristics. They tend to be shorter than average, and they have weblike necks, poorly developed breasts, and immature internal sexual organs. They have a reduced ability to interpret spatial relationships, and they are usually infertile. All of these defects in XO individuals indicate that two X chromosomes are needed for normal development in females.

Nondisjunction can also result in the generation of XXY humans, who are male and have **Klinefelter syndrome** (Figure 3.9). About 1 in 1,000 males born have Klinefelter syndrome. These 47,XXY males have underdeveloped testes and are often taller than the average male. Some degree of breast development is seen in about 50 percent of affected individuals, and some show subnormal intelligence. Individuals with similar phenotypes are also found with higher numbers of X and/or Y chromosomes—for example, 48,XXXY and 48,XXYY. The defects in Klinefelter individuals indicate that one X and one Y chromosome are needed for normal development in males.

Some individuals have one X and two Y chromosomes; they have *XYY syndrome.* These 47,XYY individuals are male because of the Y. The XYY karyotype results from nondisjunction of the Y chromosome in meiosis. About 1 in 1,000 males born have XYY syndrome. They tend to be taller than average, and occasionally there are adverse effects on fertility.

About 1 in 1,000 females born have three X chromosomes instead of the normal two. These 47,XXX (triplo-X) females are mostly completely normal, although they are slightly less fertile and a small number have less than average intelligence.

Figure 3.8

Turner syndrome (XO). (a) Individual. **(b)** Karyotype.

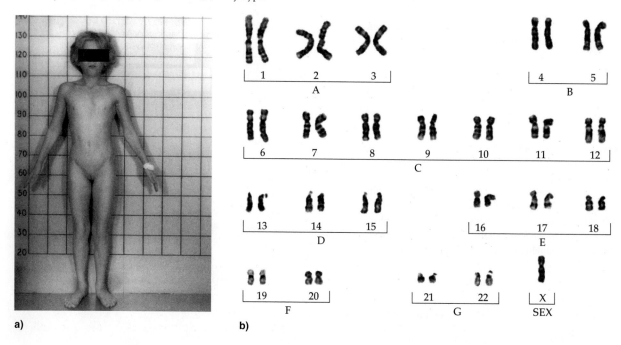

a)

b)

Figure 3.9
Klinefelter syndrome (XXY). (a) Individual. **(b)** Karyotype.

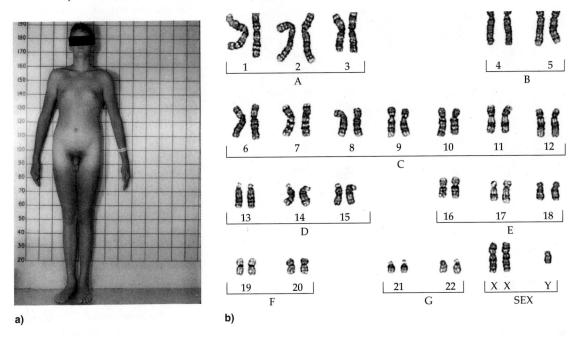

a)

b)

Table 3.2 summarizes the consequences of an unusual number of X and Y chromosomes in humans. In every case, the normal two sets of autosomes are associated with the sex chromosomes. The Barr bodies mentioned in the table are discussed next.

Dosage Compensation Mechanism for Extra X Chromosomes. Mammals can tolerate abnormalities in the number of sex chromosomes quite well, whereas, with rare exceptions, mammals with an unusual number of autosomes usually die. A **dosage compensation** mechanism in mammals compensates for X chromosomes in excess of one. The somatic cell nuclei of normal XX females contain a highly condensed mass of chromatin—named the **Barr body** after its discoverer, Murray Barr—not found in the nuclei of normal XY male cells. That is, somatic cells of XX individuals have one Barr body, and somatic cells of XY individuals have no Barr bodies (Figure 3.10 and Table 3.2). In 1961, Mary Lyon and Lillian Russell expanded this concept into what is now called the **Lyon hypothesis,** which proposed the following:

1. The Barr body is a highly condensed and (mostly) genetically inactive X chromosome (it has become "lyonized" in a process called **lyonization**).
2. The X chromosome inactivated is randomly chosen from the maternal and paternal X chromosomes in a process that is independent from cell to cell. (Once a maternal or paternal X chromosome is inactivated in a cell, all descendants of that cell inherit the inactivation pattern.)

Table 3.2	Consequences of Various X and Y Chromosome Abnormalities in Humans, Showing Role of the Y in Sex Determination	
Chromosome Constitution[a]	**Designation of Individual**	**Expected Number of Barr Bodies**
46,XX	Normal ♀	1
46,XY	Normal ♂	0
45,X	Turner syndrome ♀	0
47,XXX	Triplo-X ♀	2
47,XXY	Klinefelter syndrome ♂	1
48,XXXY	Klinefelter syndrome ♂	2
48,XXYY	Klinefelter syndrome ♂	1
47,XYY	XYY syndrome ♂	0

[a]The first number indicates the total number of chromosomes in the nucleus, and the Xs and Ys indicate the sex chromosome complement.

X inactivation occurs at about the sixteenth day following fertilization. Because of X inactivation, mammalian females heterozygous for X-linked traits are effectively genetic mosaics; that is, some cells show the phenotypes of one X chromosome, while the other cells show the phenotypes of the other X chromosome. This mosaicism is readily visible in, for example, the orange and black patches on calico cats (Figure 3.11). A calico cat is a female cat with the genotype *Oo B–*. That is, a calico is homozygous or heterozygous for the dominant *B*

a)   b)

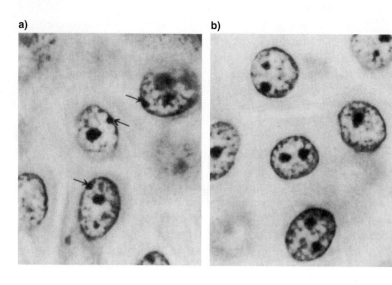

Figure 3.10

Barr bodies. (a) Nuclei of normal human female cells (XX), showing Barr bodies (indicated by arrows). **(b)** Nuclei of normal human male cells (XY), showing no Barr bodies.

allele of an autosomal gene for black hair. A calico is also heterozygous for an X-linked gene for orange hair. If the dominant *O* allele of that gene is expressed, orange hair results no matter what other genes for coat color the cat has. So orange and black patches are produced because of random X inactivation as the female develops. The orange patches are where the chromosome with the *O* allele was *not* inactivated, so that the active *O* allele masks the *B* allele(s), while the black patches are where the chromosome with the *O* allele *was* inactivated, allowing the *B* allele(s) to be expressed. (The white areas on calico cats are the result of the activity of yet another coat color gene that, when expressed, masks the expression of any other color gene, leaving white hairs. Very rarely, a calico cat will be male; it is an XXY cat with the appropriate coat color gene genotype.)

A similar, though less visible, phenotype is seen in human females heterozygous for an X-linked mutation that causes the absence of sweat glands (anhidrotic ectodermal displasia; OMIM 305100). That is, there is a mosaic of skin patches, some with sweat glands, others without.

The X inactivation process explains how mammals tolerate abnormalities in the number of sex chromosomes quite well, whereas, with rare exceptions, mammals with an unusual number of autosomes usually die. When lyonization operates in cells with extra X chromosomes, all but one of the X chromosomes typically become inactivated to produce Barr bodies; no such mechanism exists for extra autosomes. A general formula for the number of Barr bodies is the number of X chromosomes minus one. Table 3.2 lists the number of Barr bodies associated with the abnormal human X chromosome numbers we have discussed.

Three steps are involved in X inactivation: chromosome counting (determining the number of X chromosomes in the cell), selection of an X for inactivation, and X inactivation itself. In female somatic cells, the choice of which X chromosome is inactivated and which X chro-

mosome remains active is made at the *X-controlling element* (*Xce*), which is in the *X inactivation center* (*XIC* in humans; *Xic* in mice) of the X chromosome. A gene called *XIST* (humans)/*Xist* (mice), for *X* inactive specific transcripts, is also located within the *XIC/Xic* region. *XIST* is expressed from the *inactive* X rather than from the active X, which is the opposite of the expression pattern of other X-linked genes. The *Xist* gene is essential for X inactivation. *XIST/Xist* is transcribed to produce an RNA that is not translated. This RNA is stabilized during X inactivation and coats the X chromosome to be inactivated (from which it was transcribed), silencing

Figure 3.11
A calico kitten.

most of the genes. The X inactivation step itself is not well understood. It is clear, though, that inactivation is initiated from *XIC/Xic* and proceeds in both directions along the chromosome.

The Gene on the Y Chromosome That Determines Maleness.

The Y chromosome determines maleness in placental mammals. It uniquely carries an important gene (or perhaps genes) that sets the switch toward male sexual differentiation. The gene product is called **testis-determining factor,** and the corresponding hypothesized gene is the testis-determining factor gene. Testis-determining factor causes the tissue that will become gonads to differentiate into testes instead of ovaries. This is the central event in sex determination of many mammals; all other differences between the sexes are secondary effects resulting from hormone action or from the action of factors produced by the gonads. Therefore, sex determination is equivalent to testis determination.

The testis-determining factor gene was found by studying so-called *sex reversal* individuals, that is, males who are XX (instead of XY) and females who are XY (instead of XX). In the XX males, a small fragment from near the tip of the small arm of the Y chromosome had broken off during the production of gametes and become attached to one of the X chromosomes. A number of the XY females had deletions of the same region of the Y chromosome. These findings suggested that the testis-determining factor gene is in that small segment of the Y, and as a result, the human *SRY* gene and its mouse counterpart, *Sry,* were identified and cloned. One piece of evidence for why *SRY* and *Sry* are considered the testis-determining factor genes of humans and mice, respectively, comes from an experiment in which a DNA clone containing the *Sry* gene was introduced into XX mouse embryos by microinjection. Organisms or cells that have extra genes introduced into their genomes by genetic manipulation are called **transgenic organisms** or **transgenic cells,** and the introduced gene is called a **transgene.** The mice produced from these transgenic embryos were males with normal testis differentiation and subsequent normal male secondary sexual development. Thus, *Sry* alone is capable of causing a full phenotypic sex reversal in an XX transgenic mouse.

Sex Determination in *Drosophila*.

Drosophila melanogaster has four pairs of chromosomes: one pair of sex chromosomes and three pairs of autosomes. In this organism, the homogametic sex is the female (XX), and the heterogametic sex is the male (XY). However, an XXY fly is female and an XO fly is male, indicating that the sex of the fly is not the consequence of the presence or absence of a Y chromosome. In fact, the sex of the fly is determined by the ratio of the number of X chromosomes (X) to the number of sets of autosomes (A). Since *Drosophila* is diploid, a wild-type fly has two sets of autosomes, although abnormal numbers of sets can be produced as a result of nondisjunction. *Drosophila* exemplifies the *X chromosome–autosome balance system of sex determination.*

Table 3.3 presents some chromosome complements and the sex of the resulting flies. A normal female has two Xs and two sets of autosomes; the X:A ratio is 1.00. A normal male has a ratio of 0.50. If the X:A ratio is greater than or equal to 1.00, the fly is female; if the X:A ratio is less than or equal to 0.50, the fly is male. If the ratio is between 0.50 and 1.00, the fly is neither male nor female; it is an intersex. Intersex flies are variable in appearance, generally having complex mixtures of male and female attributes for the internal sex organs and external genitalia. Such flies are sterile.

Dosage compensation of X-linked genes also occurs in *Drosophila*. In this case, the dosage compensation mechanism is an upregulation of transcription of X-linked genes in males to equal the expression levels of the two X chromosomes in females.

Sex Determination in *Caenorhabditis*.

Sex determination in the nematode *Caenorhabditis elegans* also occurs by the X chromosome–autosome balance system. *C. elegans* has become a popular tool for developmental geneticists because an adult has only about a thousand cells, and the lineages of every one of those cells has been carefully defined from egg to adult. *C. elegans* has two sexual types: hermaphrodites and males. Most individuals are **hermaphroditic;** that is, they have both sex organs, an ovary and two testes. They make sperm when they are larvae and store those sperm as development continues. In adults, the ovary produces eggs that are fertilized by the stored sperm as the eggs migrate to the uterus. Self-fertilization in this way almost always produces more hermaphrodites. However, 0.2 percent of the time, males are

Table 3.3	Sex Balance Theory of Sex Determination in *Drosophila melanogaster*		
Sex Chromosome Complement	**Autosome Complement (A)**	**X:A Ratio**[a]	**Sex of Flies**
XX	AA	1.00	♀
XY	AA	0.50	♂
XXX	AA	1.50	Metafemale (sterile)
XXY	AA	1.00	♀
XXX	AAAA	0.75	Intersex (sterile)
XX	AAA	0.67	Intersex (sterile)
X	AA	0.50	♂ (sterile)

[a]If the X chromosome–autosome ratio is greater than or equal to 1.00 (X:A ≥ 1.00), the fly will be a female. If the X chromosome–autosome ratio is less than or equal to 0.50 (X:A ≤ 0.50), the fly will be male. Between these two ratios, the fly will be an intersex.

produced from self-fertilization. These males can fertilize hermaphrodites if the two mate, and such matings result in about equal numbers of hermaphrodite and male progeny because the sperm from males has a competitive advantage over the sperm stored in the hermaphrodite. Genetically, hermaphrodites are XX and males are XO; that is, an X chromosome–autosome ratio of 1.00 results in hermaphrodites, and a ratio of 0.50 results in males.

Dosage compensation of X-linked genes in *C. elegans* occurs by yet another mechanism. In this case, both X chromosomes in an XX hermaphrodite are transcribed, but each at half the rate of that of the single X chromosome in the XO male.

Sex Chromosomes in Other Organisms.   In birds, butterflies, moths, and some fish, the sex chromosome composition is the opposite of that in mammals. The male is the homogametic sex, and the female is the heterogametic sex. To prevent confusion with the X and Y chromosome convention, we designate the sex chromosomes in these organisms as Z and W: The males are ZZ and the females are ZW. Genes on the Z chromosome behave just like X-linked genes, except that hemizygosity is found only in females. All the daughters of a male homozygous for a Z-linked recessive gene express the recessive trait, and so on.

Plants exhibit a variety of arrangements of sex organs. Some species (the ginkgo, for example) have plants of separate sexes, with male plants producing flowers that contain only stamens and female plants producing flowers that contain only pistils. These species are called **dioecious** ("two houses"). Other species have both male and female sex organs on the same plant; such plants are said to be **monoecious** ("one house"). If both sex organs are in the same flower, as in the rose and the buttercup, the flower is said to be a *perfect flower*. If the male and female sex organs are in different flowers on the same plant, as in corn, the flower is said to be an *imperfect flower*.

Some dioecious plants have sex chromosomes that differ between the sexes, and a large proportion of these plants have an X-Y system. Such plants typically have an X chromosome–autosome balance system of sex determination like that in *Drosophila*. However, we see many other sex determination systems in dioecious plants.

Many species, particularly eukaryotic microorganisms, do not have sex chromosomes but instead rely on a *genic system* for the determination of sex. In this system, the sexes are specified by simple allelic differences at a small number of gene loci. For example, the yeast *Saccharomyces cerevisiae* is a haploid eukaryote that has two "sexes"—**a** and α—referred to as **mating types.** The mating types have the same morphologies, but crosses can occur only between individuals of opposite type. These mating types are controlled by the *MAT***a** and *MAT*α alleles, respectively, of a single gene.

Environmental Sex Determination Systems

In **environmental sex determination,** environmental factors (such as temperature) play a major role in determining the sex of progeny. In certain turtles, for example, eggs incubated above 32°C produce females, eggs incubated below 28°C produce males, and eggs incubated between these temperatures produce a mixture of males and females. No universal system for the temperature control of sex exists, however. The eggs of snapping turtles produce females either at 20°C or below or at 30°C or above and produce predominantly males at temperatures in between.

It is important to realize that while environmental factors trigger the particular sexual development pathway in these systems, the pathways themselves are under genetic control. Environmental sex determination mechanisms are much rarer than the genotypic mechanisms we have discussed up to now.

K E Y N O T E

Many eukaryotic organisms have sex chromosomes that are represented differentially in the two sexes; in humans and most other mammals, the male is XY and the female is XX. In other eukaryotes with sex chromosomes, the male is ZZ and the female is ZW. Sex determination commonly is related to the sex chromosomes. For humans and many other mammals, for instance, the presence of the Y chromosome confers maleness, and its absence results in femaleness. *Drosophila* and *Caenorhabditis* have an X chromosome–autosome balance system of sex determination: The sex of the individual is related to the ratio of the number of X chromosomes to the number of sets of autosomes. Several other sex-determining systems are known in eukaryotes, including genic systems, found particularly in simple eukaryotes, and environmental systems.

Analysis of Sex-Linked Traits in Humans

In Chapter 2, we introduced the analysis of recessive and dominant traits in humans; those traits were not sex-linked but were the result of alleles carried on autosomes. In this section, we discuss the analysis of X-linked and Y-linked traits in humans.

For the analysis of all pedigrees, whether the trait is autosomal or X-linked, collecting reliable human pedigree data is a difficult task. For example, often one has to rely on a family's recollections. Also, there may not be enough affected people to enable a clear determination of the mechanism of inheritance involved, especially when the trait is rare and the family is small. Further, the expression of a trait may vary, resulting in some individuals erroneously being classified as normal. Lastly, because the same mutant phenotype could result from mutations in more than one gene, it is possible that different pedigrees will indicate, correctly, that different mechanisms of inheritance are involved for the "same" trait.

> You are a genetic counselor helping a couple determine whether deafness could be passed on to their children in the iActivity *It Runs in the Family* on the website.

X-Linked Recessive Inheritance

A trait due to a recessive mutant allele carried on the X chromosome is called an **X-linked recessive trait.** At least 100 human traits are known for which the gene has been traced to the X chromosome. Most of the traits involve X-linked recessive alleles. The best known X-linked recessive pedigree is that of hemophilia A (OMIM 306700) in Queen Victoria's family (Figure 3.12). Hemophilia is a serious ailment in which the blood lacks a clotting factor, so a cut or even a bruise can be fatal to a hemophiliac. In Queen Victoria's pedigree, the first instance of hemophilia was in one of her sons. Since she passed the mutant allele on to some of her other children (carrier daughters), she must have been a carrier (heterozygous) herself. It is thought that the mutation occurred on an X chromosome in the germ cell of one of her parents.

In X-linked recessive traits, females usually must be homozygous for the recessive allele in order to express the mutant trait. The trait is expressed in males who possess only one copy of the mutant allele on the X chromosome. Therefore, affected males normally transmit the mutant gene to all their daughters but to none of their sons. The instance of father-to-son inheritance of a rare trait in a pedigree tends to rule out X-linked recessive inheritance.

Figure 3.12

X-linked recessive inheritance. (a) Painting of Queen Victoria as a young woman. **(b)** Pedigree of Queen Victoria (III-2) and her descendants, showing the inheritance of hemophilia. (Refer to Figure 2.16, p. 45, for an explanation of symbols used in pedigrees. In the pedigree shown here, marriage partners that were normal with respect to the trait may have been omitted to save space.) Since Queen Victoria was heterozygous for the sex-linked recessive hemophilia allele but no cases occurred in her ancestors, the trait may have arisen as a mutation in one of her parents' germ cells (the cells that give rise to the gametes).

a)

b)

Generation:

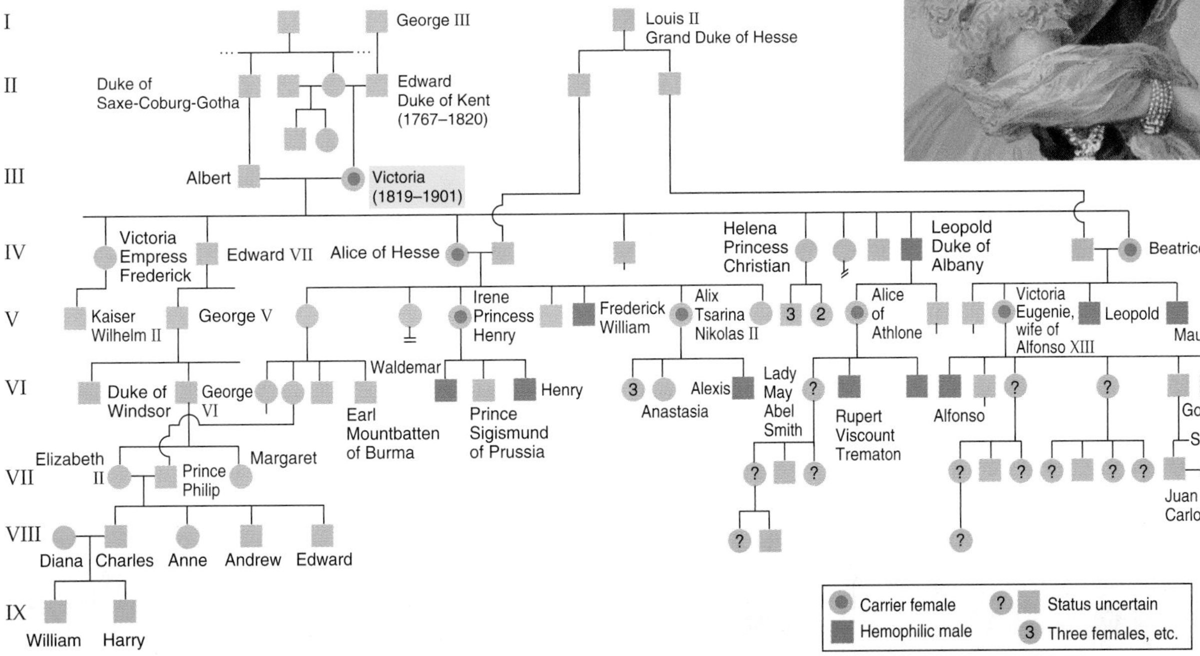

[Pedigree chart showing Queen Victoria's family and the inheritance of hemophilia across generations I through IX, with individuals including George III, Louis II Grand Duke of Hesse, Duke of Saxe-Coburg-Gotha, Edward Duke of Kent (1767–1820), Albert, Victoria (1819–1901), Victoria Empress Frederick, Edward VII, Alice of Hesse, Helena Princess Christian, Leopold Duke of Albany, Beatrice, Kaiser Wilhelm II, George V, Irene Princess Henry, Frederick William, Alix Tsarina Nikolas II, Alice of Athlone, Victoria Eugenie wife of Alfonso XIII, Leopold, Maurice, Duke of Windsor, George VI, Waldemar, Earl Mountbatten of Burma, Prince Sigismund of Prussia, Henry, Anastasia, Alexis, Lady May Abel Smith, Rupert Viscount Trematon, Alfonso, Gonzalo, Sophie, Elizabeth II, Prince Philip, Margaret, Juan Carlos, Diana, Charles, Anne, Andrew, Edward, William, Harry.]

Legend:
- ● Carrier female
- ■ Hemophilic male
- ? Status uncertain
- 3 Three females, etc.

Other characteristics of X-linked recessive inheritance are the following (refer to Figure 3.12):

1. For X-linked recessive mutant alleles, many more males than females should exhibit the trait because of the different number of X chromosomes in the two sexes.
2. All sons of an affected (homozygous mutant) mother should show the trait, since males receive their only X chromosome from their mothers.
3. The sons of heterozygous (carrier) mothers should show an approximately 1:1 ratio of normal individuals to individuals expressing the trait; that is, $a^+/a \times a^+/Y$ gives half a^+/Y and half a/Y sons.
4. From a mating of a carrier female with a normal male, all daughters will be normal, but half will be carriers; that is, $a^+/a \times a^+/Y$ gives half a^+/a^+ and half a^+/a females. In turn, half the sons of these carrier females will exhibit the trait.
5. A male expressing the trait, when mated with a homozygous normal female, will produce all normal children, but all the female progeny will be carriers; that is, $a^+/a^+ \times a/Y$ gives a^+/a females and a^+/Y (normal) males.

Other examples of human X-linked recessive traits are Duchenne muscular dystrophy (progressive muscle degeneration that shortens life) and two forms of color blindness.

X-Linked Dominant Inheritance

A trait due to a dominant mutant allele carried on the X chromosome is called an **X-linked dominant trait.** Only a few X-linked dominant traits have been identified.

One example of an X-linked dominant trait—faulty tooth enamel and dental discoloration (hereditary enamel hypoplasia; OMIM 130900)—is shown in Figure 3.13a; a pedigree for this trait is shown in Figure 3.13b. Note that all the daughters and none of the sons of an affected father (III-1) are affected and that heterozygous mothers (IV-3) transmit the trait to half their sons and half their

Figure 3.13

X-linked dominant inheritance. (a) The teeth of a person with the X-linked dominant trait of faulty enamel. **(b)** A pedigree showing the transmission of the trait for faulty enamel. This pedigree illustrates a shorthand convention that omits parents who do not exhibit the trait. Thus, it is a given that the female in generation I paired with a male who did not exhibit the trait.

a)

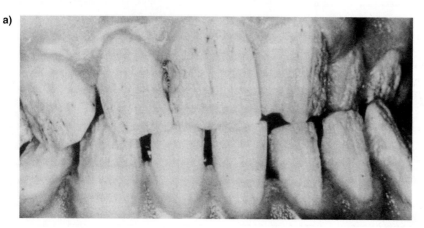

b) **Pedigree**

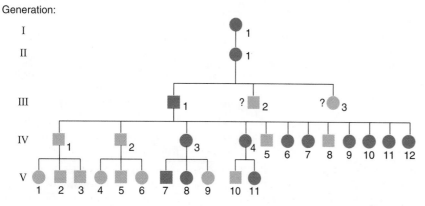

daughters. Other X-linked dominant mutant traits are webbing to the tips of the toes in a family in South Dakota (studied in the 1930s) and a severe bleeding anomaly called constitutional thrombopathy. In the latter (also studied in the 1930s), bleeding is not due to the absence of a clotting factor (as in hemophilia) but instead to interference with the formation of blood platelets, which are needed for blood clotting.

X-linked dominant traits follow the same sort of inheritance rules as the X-linked recessives, except that heterozygous females express the trait. In general, X-linked dominant traits tend to be milder in females than in males. Also, since females have twice the number of X chromosomes as males, X-linked dominant traits are more frequent in females than in males. If the trait is rare, females with the trait are likely to be heterozygous. These females will pass the trait on to half their male progeny and half their female progeny. Males with an X-linked dominant trait will pass the trait on to all of their daughters and to none of their sons.

Y-Linked Inheritance

A trait due to a mutant gene that is carried on the Y chromosome but has no counterpart on the X is called a **Y-linked**, or **holandric** ("wholly male"), **trait.** Such traits should be easily recognizable, since every son of an affected male should have the trait, and no females should ever express it. Several traits with Y-linked inheritance have been suggested. In most cases, the genetic evidence for such inheritance is poor or nonexistent. A number of genes on the Y chromosome have been identified, however, including the *SRY* gene for testis-determining factor mentioned earlier.

K E Y N O T E

> Controlled matings cannot be made in humans, so the study of inheritance in humans must rely on pedigree analysis. The data obtained from pedigree analysis enable geneticists to make judgments, with varying degrees of confidence, about whether a mutant gene is inherited as an autosomal recessive, an autosomal dominant, an X-linked recessive, an X-linked dominant, or a Y-linked allele.

Summary

Chromosome Theory of Inheritance

In Chapter 2, genes—the modern terms for Mendelian factors—were considered abstract entities that control hereditary characteristics. Cytologists working in the nineteenth century accumulated information about cell structure and cell division. The fields of genetics and cytology came together in 1902, when Sutton and Boveri independently hypothesized that genes are on chromosomes, "bodies" within the cell nucleus. The chromosome theory of heredity states that the patterns of chromosome transmission from one generation to the next closely parallel the patterns of transmission of Mendelian factors (i.e., genes) from one generation to the next.

Support for the chromosome theory of inheritance came from experiments that related the hereditary behavior of particular genes to the transmission of the sex chromosome. The sex chromosome in eukaryotic organisms is the chromosome that is represented differently in the two sexes. In most organisms with sex chromosomes, the female has two X chromosomes, while the male has one X and one Y chromosome. The Y chromosome is structurally and genetically different from the X chromosome. The association of genes with the sex chromosomes of eukaryotes is called sex linkage. Such genes, and the phenotypes they control, are said to be sex-linked.

Sex Determination Mechanisms

In many cases, sex determination is related to the sex chromosomes. In humans, for example, the presence of a Y chromosome specifies maleness, while its absence results in femaleness. The recent identification and characterization of a gene on the Y chromosome—*SRY/Sry*, the product of which directs male sexual differentiation—opens the way to a detailed analysis of the genetic control of sex determination in mammals. Several other sex determination systems are known in eukaryotes, including X chromosome–autosome balance systems (in which sex is determined by the ratio of the number of X chromosomes to the number of sets of autosomes), genic systems (in which a simple allelic difference determines the sex of an individual), and environmental systems (in which sex is determined by environmental cues).

Analysis of Sex-Linked Traits in Humans

In Chapter 2, we discussed the fact that the inheritance patterns of traits in humans are usually studied by charting the family trees of individuals exhibiting the trait, a method called pedigree analysis. In this chapter, we considered examples of X- and Y-linked human traits to illustrate the features of those mechanisms of inheritance in pedigrees. Collecting reliable human pedigree data is a difficult task. In many cases, the accuracy of record keeping within the families involved is open to question. Also, particularly with small families, there may not be enough affected people to allow an unambiguous determination of the inheritance mechanism. Moreover, the degree to which a trait is expressed may vary, so some individuals may be erroneously classified as normal. It is also possible for the same mutant phenotype to be produced by mutations in different genes, and therefore different pedigrees may correctly indicate different mechanisms of inheritance of the "same" trait.

Analytical Approaches for Solving Genetics Problems

The concepts introduced in this chapter may be reinforced by solving genetics problems similar to those introduced in Chapter 2. When sex linkage is involved, remember that one sex has two kinds of sex chromosomes, whereas the other sex has only one; this feature alters the inheritance patterns slightly. Most of the problems presented in this section center on interpreting data and predicting the outcome of particular crosses.

Q3.1 A female from a true-breeding strain of *Drosophila* with vermilion-colored eyes is crossed with a male from a true-breeding wild-type, red-eyed strain. All the F_1 males have vermilion-colored eyes, and all the females have wild-type red eyes. What conclusions can you draw about the mechanism of inheritance of the vermilion trait, and how can you test them?

A3.1 The observation is the classic one that suggests that a sex-linked trait is involved. Since none of the F_1 daughters have the trait and all the F_1 males do, the trait is presumably X-linked recessive. The results fit this hypothesis, because the F_1 males receive the X chromosome with the *v* gene from their homozygous *v/v* mother. Furthermore, the F_1 females are *v⁺/v*, since they receive a *v⁺*-bearing X chromosome from the wild-type male parent and a *v*-bearing X chromosome from the female parent. If the trait were autosomal recessive, all the F_1 flies would have wild-type eyes. If it were autosomal dominant, both the F_1 males and females would have vermilion-colored eyes. If the trait were X-linked dominant, all the F_1 flies would have vermilion eyes.

The easiest way to verify this hypothesis is to let the F_1 flies interbreed. This cross is *v⁺/v* female × *v/Y* male, and the expectation is that there will be a 1:1 ratio of wild-type : vermilion eyes in both sexes in the F_2. That is, half the females are *v⁺/v* and half are *v/v*; half the males are *v⁺/Y* and half are *v/Y*. This ratio is certainly not the 3:1 ratio that would result from an $F_1 \times F_1$ cross for an autosomal gene.

Q3.2 In humans, hemophilia A or B is caused by an X-linked recessive gene. A woman who is a nonbleeder had a father who was a hemophiliac. She marries a nonbleeder,

and they plan to have children. Calculate the probability of hemophilia in the female and male offspring.

A3.2 Since hemophilia is an X-linked trait, and since her father was a hemophiliac, the woman must be heterozygous for this recessive gene. If we assign the symbol *h* to this recessive mutation and *h⁺* to the wild-type (nonbleeder) allele, she must be *h⁺/h*. The man she marries is normal with regard to blood clotting and hence must be hemizygous for *h⁺*—that is, *h⁺/Y*. All their daughters receive an X chromosome from the father, and so each must have an *h⁺* gene. In fact, half the daughters are *h⁺/h⁺* and the other half are *h⁺/h*. Since the wild-type allele is dominant, none of the daughters are hemophiliacs. However, all the sons of the marriage receive their X chromosome from their mother. Therefore, they have a probability of ½ that they will receive the chromosome carrying the *h* allele, which means they will be hemophiliacs. Thus, the probability of hemophilia among daughters of this marriage is 0; among sons, it is ½.

Q3.3 Tribbles are hypothetical animals that have an X-Y sex determination mechanism like that of humans. The trait blotchy (*b*), with pigment in spots, is X-linked and recessive to solid color (*b⁺*), and the trait light color (*l*) is autosomal and recessive to dark color (*l⁺*). If you make reciprocal crosses between true-breeding blotchy, light-colored tribbles and true-breeding solid, dark colored tribbles, do you expect a 9:3:3:1 ratio in the F_2 generation of either or both of these crosses? Explain your answer.

A3.3 This question focuses on the fundamentals of X chromosome and autosome segregation during a genetic cross, and it tests whether or not you have grasped the principles involved in gene segregation. Figure 3.A diagrams the two crosses involved, and we can discuss the answer by referring to it.

First, consider the cross of a wild-type female tribble (*b⁺/b⁺, l/l⁺*) with a male double-mutant tribble (*b/Y, l/l*). Part (a) of the figure diagrams this cross. These F_1 tribbles are all normal—that is, they are solid and light-colored—because for the autosomal character, both sexes are heterozygous, and for the X-linked character, the female is heterozygous and the male is hemizygous for the *b⁺* allele donated by the normal mother. For the production of the F_2 progeny, the best approach is to treat the X-linked and autosomal traits separately. For the X-linked trait, random combination of the gametes produced gives a 1:1:1:1 genotypic ratio of *b⁺/b⁺* (solid female) : *b⁺/b* (solid female) : *b⁺/Y* (solid male) : *b/Y* (blotchy male) progeny. Categorizing by phenotypes, ½ of the progeny are solid females, ¼ are solid males, and ¼ are blotchy males. For the autosomal leg trait, the $F_1 \times F_1$ is a cross of two heterozygotes, so we expect a 3:1 phenotypic ratio of dark : light tribbles in the F_2. Since auto-

Figure 3.A

a) Solid, dark (wild-type) ♀ × blotchy, light ♂

P generation

$b^+/b^+\ l^+/l^+$ ♀  × $b/Y\ l/l$ ♂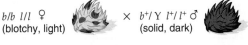
(solid, dark)   (blotchy, light)

F_1 generation

$b^+/b\ l^+/l$ ♀  × $b^+/Y\ l^+/l$ ♂
(solid, dark)   (solid, dark)

F_2 generation

Sex-linked phenotypes and genotypes	Autosomal phenotypes and genotypes

Genotypic results:

$\frac{1}{2}\ b^+\ (\frac{1}{2}b^+/b^+,\ \frac{1}{2}b^+/b\ ;$ solid) ♀ $\Big\langle \begin{array}{l} \frac{3}{4}\ l^+\ (l^+/l^+\ \text{and}\ l^+/l;\ \text{dark}) \\ \frac{1}{4}\ l\ (l/l\ ;\ \text{light}) \end{array}$

$\frac{1}{4}\ b^+\ (b^+/Y\ ;\ \text{solid})$ ♂ $\Big\langle \begin{array}{l} \frac{3}{4}\ l^+\ (\text{dark}) \\ \frac{1}{4}\ l\ (\text{light}) \end{array}$

$\frac{1}{4}\ b\ (b/Y\ ;\ \text{blotchy})$ ♂ $\Big\langle \begin{array}{l} \frac{3}{4}\ l^+\ (\text{dark}) \\ \frac{1}{4}\ l\ (\text{light}) \end{array}$

Phenotypic ratios:

	Solid-dark $b^+\ l^+$		Solid-light $b^+\ l$		Blotchy-dark $b\ l^+$		Blotchy-light $b\ l$
♀	6	:	2	:	0	:	0
♂	3	:	1	:	3	:	1
Total	9	:	3	:	3	:	1

b) Blotchy, light ♀ × solid, dark (wild-type) ♂

P generation

$b/b\ l/l$ ♀  × $b^+/Y\ l^+/l^+$ ♂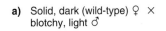
(blotchy, light)   (solid, dark)

F_1 generation

$b^+/b\ l^+/l$ ♀ × $b/Y\ l^+/l$ ♂
(solid, dark)   (blotchy, dark)

F_2 generation

Sex-linked phenotypes and genotypes	Autosomal phenotypes and genotypes

Genotypic results:

$\frac{1}{4}\ b^+\ (b^+/b^+;\ \text{solid})$ ♀ $\Big\langle \begin{array}{l} \frac{3}{4}\ l^+\ (l^+/l^+\text{and}\ l^+/l;\ \text{dark}) \\ \frac{1}{4}\ l\ (l\ /l\ ;\ \text{light}) \end{array}$

$\frac{1}{4}\ b\ (b/b\ ;\ \text{blotchy})$ ♀ $\Big\langle \begin{array}{l} \frac{3}{4}\ l^+\ (\text{dark}) \\ \frac{1}{4}\ l\ (\text{light}) \end{array}$

$\frac{1}{4}\ b^+\ (b^+/Y\ ;\ \text{solid})$ ♂ $\Big\langle \begin{array}{l} \frac{3}{4}\ l^+\ (\text{dark}) \\ \frac{1}{4}\ l\ (\text{light}) \end{array}$

$\frac{1}{4}\ b\ (b/Y\ ;\ \text{blotchy})$ ♂ $\Big\langle \begin{array}{l} \frac{3}{4}\ l^+\ (\text{dark}) \\ \frac{1}{4}\ l\ (\text{light}) \end{array}$

Phenotypic ratios:

	Solid-dark $b^+\ l^+$		Solid-light $b^+\ l$		Blotchy-dark $b\ l^+$		Blotchy-light $b\ l$
♀	3	:	1	:	3	:	1
♂	3	:	1	:	3	:	1
Total	6	:	2	:	6	:	2

some segregation is independent of the inheritance of the X chromosome, we can multiply the probabilities of the occurrence of the X-linked and autosomal traits to calculate their relative frequencies. The calculations are presented at the bottom of part (a) of the figure.

The first cross, then, has a 9:3:3:1 ratio of the four possible phenotypes in the F_2. However, note that the ratio in each sex is not 9:3:3:1, owing to the inheritance pattern of the X chromosome. This result contrasts markedly with the pattern of two autosomal genes segregating independently, where the 9:3:3:1 ratio is found for both sexes.

The second cross (a reciprocal cross) is diagrammed in part (b) of the figure. Since the parental female in this cross is homozygous for the sex-linked trait, all the F_1 males are blotchy. Genotypically, the F_1 males and females differ from those in the first cross with respect to the sex chromosome but are just the same with

respect to the autosome. Again, considering the X chromosome first as we go to the F_2, we find a 1:1:1:1 genotypic ratio of solid females : blotchy females : solid males : blotchy males. In this case, then, half of both males and females are solid, and half are blotchy, in contrast to the results of the first cross, in which no blotchy females were produced in the F_2. For the autosomal trait, we expect a 3:1 ratio of dark : light in the F_2, as before. Putting the two traits together, we get the calculations presented in part (b) of the figure. (Note: We use the total 6:2:6:2 here rather than 3:1:3:1 because the numbers add to 16, as does $9 + 3 + 3 + 1$.) So in this case, we do not get a 9:3:3:1 ratio; moreover, the ratio is the same in both sexes.

This question has forced us to think through the segregation of two types of chromosomes and has shown that we must be careful about predicting the outcomes of crosses in which sex chromosomes are involved. Nonetheless, the basic principles for this analysis are the same as those used before: Reduce the questions to their basic parts and then put the puzzle together step by step.

Questions and Problems

3.1 Depict each of the following crosses, first using Mendelian and then using *Drosophila* notation (see Box 3.1, p. 58). Give the genotype and phenotype of the F_1 progeny that can be produced.
a. In humans, a mating between two individuals, each heterozygous for the recessive trait phenylketonuria, whose locus is on chromosome 12
b. In humans, a mating between a female heterozygous for both phenylketonuria and X-linked color blindness and a male with normal color vision who is heterozygous for phenylketonuria
c. In *Drosophila*, a mating between a female with white eyes, curled wings, and normal long bristles and a male that has normal red eyes, normal straight wings, and short, stubble bristles. In these individuals, curled wings result from a heterozygous condition at a gene whose locus is on chromosome 2, while the short, stubble bristles result from a heterozygous condition at a gene whose locus is on chromosome 3.
d. In *Drosophila*, a mating between a female from a true-breeding line that has eyes of normal size that are white, black bodies (a recessive trait on chromosome 2), and tiny bristles (a recessive trait called *spineless* on chromosome 3) and a male from a true-breeding line that has normal red eyes, normal grey bodies, and normal long bristles but reduced eye size (a dominant trait called *eyeless* on chromosome 4)

3.2 In *Drosophila,* white eyes are a sex-linked character. The mutant allele for white eyes (w) is recessive to the wild-type allele for brick-red eye color (w^+). A white-eyed female is crossed with a red-eyed male. An F_1 female from this cross is mated with her father, and an F_1 male is mated with his mother. What will be the eye color of the offspring of these last two crosses?

*3.3 One form of color blindness in humans is caused by a sex-linked recessive mutant gene (c). A woman with normal color vision (c^+) and whose father was color-blind marries a man of normal vision whose father was also color-blind. What proportion of their offspring will be color-blind? (Give your answer separately for males and females.)

*3.4 In humans, red-green color blindness is recessive and X-linked, whereas albinism is recessive and autosomal. What types of children can be produced as the result of marriages between two homozygous parents: a normal-visioned albino woman and a color-blind, normally pigmented man?

*3.5 In *Drosophila*, vestigial (partially formed) wings (vg) are recessive to normal long wings (vg^+), and the gene for this trait is autosomal. The gene for the white-eye trait is on the X chromosome. Suppose a homozygous white-eyed, long-winged female fly is crossed with a homozygous red-eyed, vestigial-winged male.
a. What will be the appearance of the F_1 flies?
b. What will be the appearance of the F_2 flies?
c. What will be the appearance of the offspring of a cross of the F_1 back to each parent?

3.6 In *Drosophila*, two red-eyed, long-winged flies are bred together and produce the offspring given in the following table:

	Females	**Males**
red-eyed, long-winged	3/4	3/8
red-eyed, vestigial-winged	1/4	1/8
white-eyed, long-winged	—	3/8
white-eyed, vestigial-winged	—	1/8

What are the genotypes of the parents?

3.7 In chickens, a dominant sex-linked gene (B) produces barred feathers, and the recessive allele (b), when homozygous, produces nonbarred (solid color) feathers. Suppose a nonbarred cock is crossed with a barred hen.
a. What will be the appearance of the F_1 birds?
b. If an F_1 female is mated with her father, what will be the appearance of the offspring?
c. If an F_1 male is mated with his mother, what will be the appearance of the offspring?

*3.8 A man suffering from defective tooth enamel, which results in brown-colored teeth, marries a normal woman. All their daughters have brown teeth, but the sons are normal. The sons marry normal women, and all their children are normal. The daughters marry normal men, and 50 percent of their children have brown teeth. Explain these facts.

3.9 In humans, differences in the ability to taste phenylthiourea are due to a pair of autosomal alleles. Inability to taste is recessive to ability to taste. A child who is a nontaster is born to a couple who can both taste the substance. What is the probability that their next child will be a taster?

*3.10 Cystic fibrosis is inherited as an autosomal recessive. Two parents without cystic fibrosis have two children with cystic fibrosis and three children without. The parents come to you for genetic counseling.

a. What is the numerical probability that their next child will have cystic fibrosis?

b. Their unaffected children are concerned about being heterozygous. What is the numerical probability that a given unaffected child in the family is heterozygous?

3.11 Huntington disease is a human disease inherited as a Mendelian autosomal dominant. The disease results in choreic (uncontrolled) movements, progressive mental deterioration, and eventually death. The disease affects the carriers of the trait any time between 15 and 65 years of age. American folksinger Woody Guthrie died of Huntington disease, as did one of his parents. Marjorie Mazia, Woody's wife, had no history of this disease in her family. The Guthries had three children. What is the probability that a particular Guthrie child will die of Huntington disease?

*3.12 Suppose gene *A* is on the X chromosome, and genes *B*, *C*, and *D* are on three different autosomes. Thus, *A–* signifies the dominant phenotype in the male or female. An equivalent situation holds for *B–*, *C–*, and *D–*. The cross *AA BB CC DD* ♀ × *aY bb cc dd* ♂ is made.

a. What is the probability of obtaining an *A–* individual in the F₁?

b. What is the probability of obtaining an *a* male in the F₁?

c. What is the probability of obtaining an *A–B–C–D–* female in the F₁?

d. How many different F₁ genotypes will there be?

e. What proportion of F₂ individuals will be heterozygous for the four genes?

f. Determine the probabilities of obtaining each of the following types in the F₂: (1) *A– bb CC dd* (female); (2) *aY BB Cc Dd* (male); (3) *AY bb CC dd* (male); (4) *aa bb Cc Dd* (female).

3.13 A Turner syndrome individual would be expected to have how many Barr bodies in the majority of cells?

3.14 An XXY Klinefelter syndrome individual would be expected to have how many Barr bodies in the majority of cells?

*3.15 In human genetics, the pedigree is used for analysis of inheritance patterns. Females are represented by a circle, and males by a square. The following figure presents three 2-generation family pedigrees for a trait in humans. Normal individuals are represented by unshaded symbols, and people with the trait by shaded symbols. For each pedigree (A, B, and C), state, by answering yes or no in the appropriate blank space, whether transmission of the trait can be accounted for on the basis of each of the listed simple modes of inheritance.

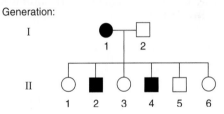

	Pedigree A	Pedigree B	Pedigree C
Autosomal recessive	_____	_____	_____
Autosomal dominant	_____	_____	_____
X-linked recessive	_____	_____	_____
X-linked dominant	_____	_____	_____

3.16 Shaded symbols in the following pedigree represent a trait. Which of the progeny eliminate X-linked recessiveness as a mode of inheritance for the trait?

a. I-1 and I-2

b. II-4

c. II-5

d. II-2 and II-4

Generation:

I   1   2

II   1   2   3   4   5   6

*3.17 When constructing human pedigrees, geneticists often refer to particular persons by a number. The generations are labeled by Roman numerals, and the individuals in each generation by Arabic numerals. For example, in the pedigree in the following figure, the female with the asterisk is I-2. Use this means to designate specific individuals in the pedigree. Determine the probable inheritance mode for the trait shown in the affected individuals (the shaded symbols) by answering the following questions. Assume that the condition is caused by a single gene.

Generation:

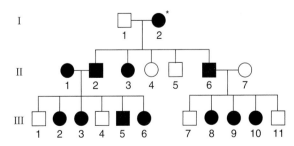

a. Y-linked inheritance can be excluded at a glance. What two other mechanisms of inheritance can be definitely excluded? Why can these be excluded?

b. Of the remaining mechanisms of inheritance, which is the most likely? Why?

3.18 A three-generation pedigree for a particular human trait is shown in the following figure.

Generation:

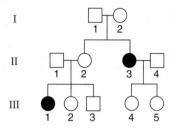

a. What is the mechanism of inheritance for the trait?
b. Which persons in the pedigree are known to be heterozygous for the trait?
c. What is the probability that III-2 is a carrier (heterozygous)?
d. If III-3 and III-4 marry, what is the probability that their first child will have the trait?

3.19 For each of the more complex pedigrees shown in Figure 3.B, determine the probable mechanism of inheritance: autosomal recessive, autosomal dominant, X-linked recessive, X-linked dominant, or Y-linked.

In problems 3.20–3.22, select the correct answer.

*3.20 If a genetic disease is inherited on the basis of an autosomal dominant gene, one would expect to find which of the following?
a. Affected fathers have only affected children.
b. Affected mothers never have affected sons.
c. If both parents are affected, all of their offspring have the disease.
d. If a child has the disease, one of his or her grandparents also had the disease.

*3.21 If a genetic disease is inherited as an autosomal recessive, one would expect to find which of the following?
a. Two affected individuals never have an unaffected child.
b. Two affected individuals have affected male offspring but no affected female children.
c. If a child has the disease, one of his or her grandparents will have had it.
d. In a marriage between an affected individual and an unaffected one, all the children are unaffected.

Figure 3.B

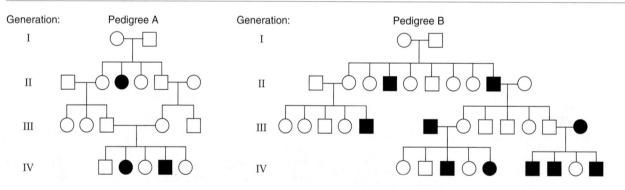

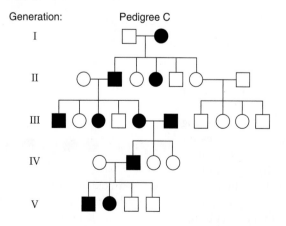

3.22 Which of the following statements is *not* true for a disease that is inherited as a rare X-linked dominant trait?

a. All daughters of an affected male will inherit the disease.

b. Sons will inherit the disease only if their mother has the disease.

c. Both affected males and affected females will pass the trait to half the children.

d. Daughters will inherit the disease only if their father has the disease.

*3.23 Women who were known to be carriers of the X-linked recessive hemophilia gene were studied to determine the amount of time required for the blood-clotting reaction. It was found that the time required for clotting was extremely variable from individual to individual. The values obtained ranged from normal clotting time at one extreme to clinical hemophilia at the other extreme. What is the most probable explanation for these findings?

Palomino horse.

4

Extensions of Mendelian Genetic Analysis

PRINCIPAL POINTS

More than two allelic forms of a gene can exist. However, any given diploid individual can possess only two different alleles of a given gene.

In complete dominance, the same phenotype results whether the dominant allele is heterozygous or homozygous. In incomplete dominance, the phenotype of the heterozygote is intermediate between those of the two homozygotes. In codominance, the heterozygote exhibits the phenotypes of both homozygotes.

In many cases, different genes interact to determine phenotypic characteristics. In epistasis, for example, modified Mendelian ratios occur because of gene interactions: The phenotypic expression of one gene depends on the genotype of another gene locus.

Alleles of certain genes may be fatal to the individual. The existence of such lethal alleles indicates that the product usually produced by the nonlethal allele is essential for the function of the organism; the gene is called an essential gene.

Penetrance is the frequency (in percent) at which an allele manifests itself phenotypically within a population. Expressivity is the kind or degree of phenotypic manifestation of a gene or genotype in a particular individual.

The zygote's genetic constitution only specifies the organism's potential to develop and function. As the organism develops and cells differentiate, many things can influence gene expression. One such influence is the organism's environment, both internal and external. Examples of the internal environment include age and sex; examples of the external environment include nutrition, light, chemicals, temperature, and infectious agents.

Variation in most of the genetic traits considered in the earlier discussion of Mendelian principles is determined predominantly by differences in genotype; that is, phenotypic differences result from genotypic differences. For many traits, however, phenotype is influenced by both genes and the environment.

iActivity

i A HALF-CENTURY BEFORE WATSON AND CRICK determined the structure of DNA, Karl Landsteiner discovered that different individuals have different blood types and that these blood types are inherited. However, the inheritance of blood type does not always follow the inheritance patterns predicted by Mendel's principles. As it turns out, blood types are one example of a trait that has an inheritance pattern more complex than Mendel described. In this chapter, you will learn about the inheritance of blood types and other traits that are exceptions to and extensions of Mendel's principles. Then, in the iActivity, you can use your understanding of these inheritance patterns to help solve a paternity suit involving the actor Charlie Chaplin.

Mendel's principles apply to all eukaryotic organisms and form the foundation for predicting the outcome of crosses in which segregation and independent assortment might occur. As more and more geneticists did experiments, though, they found exceptions to and extensions of Mendel's principles. Several of these cases are discussed in this chapter, with the goal of providing you with a broader knowledge of genetic analysis, particularly in terms of how genes relate to the phenotypes of an organism.

Multiple Alleles

So far in our genetic analyses, we have considered genes that have only two alleles, such as smooth versus wrinkled seeds in peas, red versus white eyes in *Drosophila*, and unattached versus attached earlobes in humans. The allele that predominates in populations of the organism in the wild is the wild-type allele, and the alternative allele is the mutant allele. In a population of individuals, however, a given gene may have several alleles (one wild type and the rest mutant), not just two. Such genes are said to have **multiple alleles,** and the alleles are said to constitute a *multiple allelic series* (Figure 4.1). Although a gene may have multiple alleles in a given population of individuals, *a single diploid individual can have only a*

maximum of two of these alleles, one on each of the two homologous chromosomes carrying the gene locus.

ABO Blood Groups

An example of multiple alleles of a gene is found in the human ABO blood group series. Since certain ABO blood groups are incompatible, these alleles are of particular importance during blood transfusions. (There are many blood group series other than ABO; these also can cause problems in blood transfusions and need to be checked through the process of cross-matching.)

There are four blood group phenotypes in the ABO system: O, A, B, and AB; Table 4.1 lists their possible genotypes. The six genotypes that give rise to the four phenotypes represent various combinations of three ABO blood group alleles: I^A, I^B, and i. Individuals homozygous for the recessive i allele are of blood group O. Both I^A and I^B are dominant to i. Individuals of blood group A are either I^A/I^A or I^A/i, and those of blood group B are either I^B/I^B or I^B/i. Heterozygous I^A/I^B individuals are of blood group AB—that is, essentially of both blood groups A and B (see the discussion of codominance on pp. 80–81).

The genetics of this system follows Mendelian principles. An individual who expresses blood group O, for example, must be i/i in genotype. The parents of this person could both be O ($i/i \times i/i$) or A ($I^A/i \times I^A/i$, to produce one-fourth i/i progeny) or B ($I^B/i \times I^B/i$), or one could be A and one could be B ($I^A/i \times I^B/i$). Each parent must be either homozygous i or heterozygous, with i as one of the two alleles.

Blood-typing (determining an individual's blood group) and analyzing blood group inheritance can help in cases of disputed paternity or maternity or when babies are inadvertently switched in a hospital. In such cases, genetic data cannot prove the identity of the parent. But genetic analysis on the basis of blood group can show that an individual is *not* the parent of a particular child;

Figure 4.1

Allelic forms of a gene.

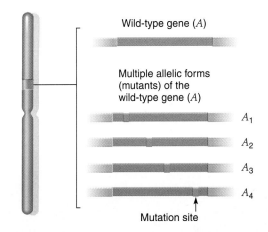

Wild-type gene (*A*)

Multiple allelic forms (mutants) of the wild-type gene (*A*)

A_1

A_2

A_3

A_4

Mutation site

Table 4.1	ABO Blood Groups in Humans, Determined by the Alleles I^A, I^B, and I
Phenotype (Blood Group)	**Genotype**
O	i/i
A	I^A/I^A or I^A/i
B	I^B/I^B or I^B/i
AB	I^A/I^B

antigens usually are not recognized as foreign by the organism expressing them (except in the case of auto-immune diseases).

The human ABO blood group series was discovered by Karl Landsteiner in the early 1900s. He received the 1930 Nobel Prize in Physiology or Medicine for this discovery. The I^A, I^B, and i alleles of the ABO locus specify the four blood types, O, A, B, and AB (see Table 4.1). The I^A allele specifies the A antigen. People of blood type A (genotypes I^A/I^A or I^A/i) have the A antigen on their red blood cells, and blood serum prepared from them contains naturally occurring antibodies against the B antigen (called anti-B antibodies) but none against the A antigen. Antibodies against the B antigen agglutinate, or clump, any red blood cells that have the B antigen on them. Since clumped cells cannot move through the fine capillaries, agglutination may lead to organ failure and possibly death. Conversely, people of blood type B (genotypes I^B/I^B or I^B/i) have the B antigen on their red blood cells, and blood serum prepared from them contains naturally occurring anti-A antibodies but no anti-B antibodies. People of blood type AB (genotype I^A/I^B) have both A and B antigens on their blood cells and neither anti-A nor anti-B antibodies in their blood serum. Lastly, in people with blood type O (i/i), the red blood cells have neither A nor B antigen, and their blood serum contains both anti-A and anti-B antibodies. The antigen-antibody relationships are summarized in Figure 4.2. Agglutination (clumping) of the red blood cells is seen in each case where an antibody interacts with the antigen for which it is specific.

What transfusions are safe, then, between people with different blood groups in the ABO system?

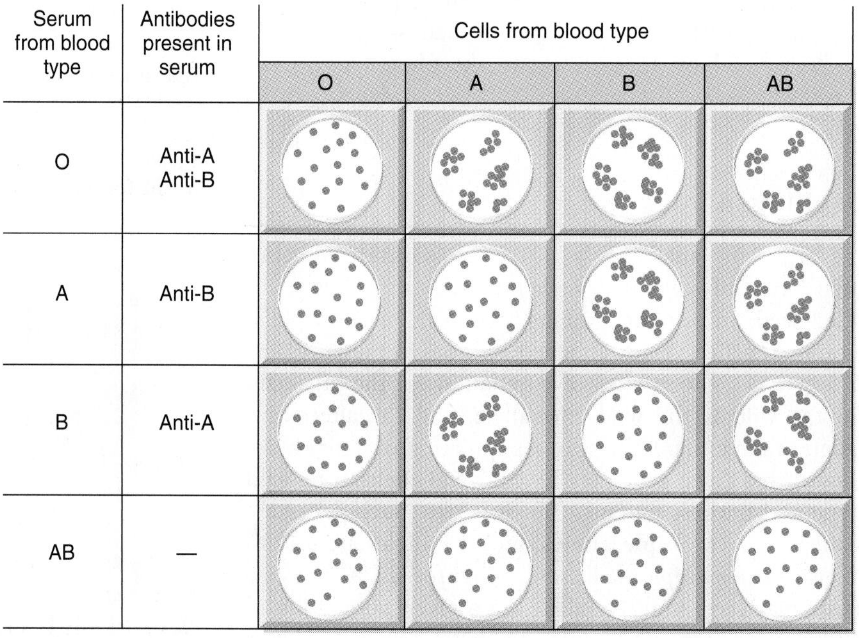

for example, a child of phenotype AB (genotype I^A/I^B) could not be the child of a parent of phenotype O (genotype i/i). (Note: In most states, blood type data alone usually are not sufficient for a legal decision about paternity or maternity. In addition, DNA fingerprinting typically is used; see Chapter 14, pp. 293–296.)

The blood types of transfusion donors and recipients must be carefully matched because the blood group alleles specify molecular groups, called *cellular antigens,* that are attached to the outside of the red blood cells. An antigen (*antibody-generating substance*) is any molecule that is recognized as foreign by an organism and that stimulates production of specific protein molecules called antibodies, which bind to the antigen. An antibody is a protein molecule that recognizes and binds to the foreign substance (antigen) introduced into the organism. A given individual has a large number of antigens on cells and tissues, many of which are foreign to another individual; hence the concern over blood type in blood transfusions and tissue type in organ transplants. However, the

Figure 4.2

Antigen-antibody reactions that characterize the human ABO blood groups. Blood serum from each of the four blood groups was mixed with blood cells from the four types, in all possible combinations. In some cases, such as a mix of B serum with A cells, the cells become clumped.

1. People with blood type A produce the A antigen, so their blood can be transfused only into recipients who do not have the anti-A antibody—that is, people of blood types A and AB.

2. Those with blood type B produce the B antigen, so their blood can be transfused only into recipients who do not have the anti-B antibody—that is, people of blood types B and AB.

3. People with blood type AB produce both the A and B antigens, so their blood can be transfused only into recipients who do not have either the anti-A antibody or the anti-B antibody—that is, people of blood type AB.

4. People with blood type O produce neither A nor B antigens, so their blood can be transfused into any recipient—that is, people of blood types A, B, AB, and O.

You will note from this presentation that people of blood type AB can receive transfusions of blood from people of any of the four blood types. Thus, people with blood type AB are called *universal recipients*. Similarly, blood type O blood can be used as donor blood for any recipient because it elicits no reaction. Thus, people with blood type O are called *universal donors*.

In the iActivity *Was She Charlie Chaplin's Child?* on the Web site, you will use your expertise to interpret the results of blood group tests that may prove whether silent movie great Charlie Chaplin was the father of Carol Ann Berry.

Drosophila Eye Color

Another example of multiple alleles concerns the white (*w*) locus of *Drosophila*. Recall from Chapter 3 (p. 56) that the w^+ allele results in wild-type brick-red eyes and that the recessive *w* allele, when homozygous or hemizygous, results in white eyes. There are actually more than 100 mutant alleles at the white locus. With w^+ symbolizing the wild-type (brick-red) allele of the white-eye gene, *w* is the recessive mutant white-eye allele, and, for example, w^e is the recessive mutant eosin (reddish-orange) allele.

The white locus alleles have different, characteristic eye colors ranging from white to near-wild-type when the alleles are homozygous or hemizygous. Eye color phenotype is related to the amount of pigment deposited in the eye cells. Table 4.2 lists the relative amounts of pigment present in wild-type females and in females homozygous for different mutant alleles of the white locus. As expected, the original mutant allele *w* has the least amount of pigment. Between white and wild, there is a wide range of pigment amounts. The phenotypic expressions of different alleles of the same gene reflect the vary-

Table 4.2	Eye Pigment Quantification for *Drosophila* White Alleles
Genotypes	**Relative Amount of Total Pigment**
w^+/w^+ (wild type)	1.0000
w/w (white)	0.0044
w^t/w^t (tinged)	0.0062
w^a/w^a (apricot)	0.0197
w^{bl}/w^{bl} (blood)	0.0310
w^e/w^e (eosin)	0.0324
w^{ch}/w^{ch} (cherry)	0.0410
w^{a3}/w^{a3} (apricot-3)	0.0632
w^w/w^w (wine)	0.0650
w^{co}/w^{co} (coral)	0.0798
w^{sat}/w^{sat} (satsuma)	0.1404
w^{col}/w^{col} (colored)	0.1636

ing extent to which the biological activity of the protein encoded by the white gene (its *gene product*) has been altered. Various combinations of wild-type and mutant alleles can occur in flies, depending on the parents involved, but a particular individual can have no more than two different alleles.

KEYNOTE

Many allelic forms of a gene can exist in a population. When they do, the gene is said to show multiple allelism, and the alleles involved constitute a multiple allelic series. However, any given diploid individual can possess only two different alleles of a given gene. Multiple alleles obey the same rule of transmission as alleles of which there are only two types.

Modifications of Dominance Relationships

Complete dominance is the phenomenon in which one allele is dominant to another, so that the phenotype of the heterozygote is the same as that of the homozygous dominant. With **complete recessiveness,** the recessive allele is phenotypically expressed only when the organism is homozygous. Complete dominance and complete recessiveness are the two extremes of a range of dominance relationships. Whereas all the allelic pairs Mendel studied showed complete dominance–complete recessiveness relationships, many allelic pairs do not.

Incomplete Dominance

When one allele is not completely dominant to another allele, it is said to show **incomplete,** or **partial, dominance.**

animation

a Incomplete Dominance and Codominance

With incomplete dominance, the heterozygote's phenotype is intermediate to those of individuals homozygous for either individual allele involved.

Plumage color in chickens is an example of incomplete dominance. Crosses between a true-breeding black strain (homozygous C^BC^B) and a true-breeding white strain (homozygous C^WC^W) give F_1 birds with bluish-grey plumage (Figure 4.3a), called Andalusian blues by chicken breeders. (Here the symbols are designed to give equal weight to the two alleles because neither dominates the phenotype. In this particular example, the C signifies color, and B and W indicate black and white, respectively.) An Andalusian blue does not breed true because it is heterozygous, so in Andalusian × Andalusian crosses (Figure 4.3b), the two alleles segregate in the offspring and pro-

duce black, Andalusian blue, and white fowl in a ratio of 1:2:1. The most efficient way to produce Andalusian blues is to cross black × white because all progeny of this mating are Andalusian blues. In very loose molecular terms, then, two doses of C^B are needed to give black, whereas in the heterozygote, the white dilutes the black to grey.

Codominance

Codominance is a modification of the dominance relationship that is related to incomplete dominance. In codominance, the heterozygote exhibits the phenotypes of *both* homozygotes. Codominance differs from incomplete dominance, in which the heterozygote exhibits a phenotype intermediate between the two homozygotes.

The ABO blood series discussed earlier in this chapter provides a good example of codominance. Heterozygous I^A/I^B individuals are of blood group AB because both

Figure 4.3

Incomplete dominance in chickens. (a) A cross between a white bird and a black bird produces F_1 birds of intermediate grey color, called Andalusian blues. **(b)** The F_2 generation shows the 1:2:1 phenotypic ratio characteristic of incomplete dominance.

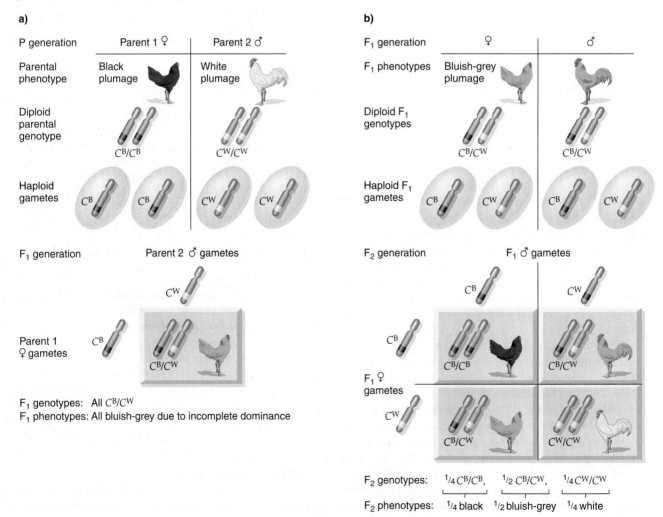

the A antigen (product of the I^A allele) and the B antigen (product of the I^B allele) are produced. Thus, the I^A and I^B alleles are codominant.

The human M-N blood group system is another example of codominance. For transfusion compatibility, this system is of less clinical importance than the ABO system. In the M-N system, three blood types occur: M, MN, and N. They are determined by the genotypes L^M/L^M, L^M/L^N, and L^N/L^N, respectively. As in the ABO system, the M-N alleles result in the formation of antigens on the red blood cell surface. The heterozygote in this case has both the M and the N antigens and shows the phenotypes of both homozygotes.

Molecular Explanations of Incomplete Dominance and Codominance

What explains incomplete dominance and codominance at the molecular level? A general interpretation is that in codominance, products result from both alleles in a heterozygote, so that often (but not always) both homozygote phenotypes are observed (e.g., L^M/L^N individuals express both M and N blood group antigens). In incomplete dominance, only one allele in a heterozygote is expressed to produce a product. A homozygote for the expressed allele, then, has two doses of the gene product, and full phenotypic expression results (for example, black chickens). In a homozygote for the allele that is not expressed, a phenotype characteristic of no gene expression results (white chickens). In a heterozygote, the single allele expressed results in only enough product for an intermediate phenotype (grey chickens).

K E Y N O T E

With complete dominance, the same phenotype results whether the dominant allele is heterozygous or homozygous. With complete recessiveness, the allele is phenotypically expressed only when the genotype is homozygous recessive; the recessive allele has no effect on the phenotype of the heterozygote. Complete dominance and complete recessiveness are two extremes between which all transitional degrees of dominance are possible. In incomplete dominance, the phenotype of the heterozygote is intermediate between those of the two homozygotes, whereas in codominance, the heterozygote exhibits the phenotypes of both homozygotes.

Gene Interactions and Modified Mendelian Ratios

No gene acts by itself in determining an individual's phenotype; the phenotype is the result of highly complex and integrated patterns of molecular reactions that are under direct gene control. All the genetic examples we have discussed and will discuss have discrete biochemical bases, and in a number of cases complex interactions between genes can be detected by genetic analysis. We discuss some examples in this section.

Consider two independently assorting gene pairs, each with two alleles: *A* and *a*, and *B* and *b*. The outcome of a cross between individuals that are doubly heterozygous (*A/a B/b* × *A/a B/b*) will be nine genotypes in the following proportions:

1/16 *A/A B/B*

2/16 *A/A B/b*

1/16 *A/A b/b*

2/16 *A/a B/B*

4/16 *A/a B/b*

2/16 *A/a b/b*

1/16 *a/a B/B*

2/16 *a/a B/b*

1/16 *a/a b/b*

If the phenotypes determined by the two allelic pairs are distinct—for example, smooth versus wrinkled peas, long versus short stems—and there is complete dominance, then we get the familiar dihybrid phenotypic ratio of 9:3:3:1 (see Figure 2.12b). Any alteration in this standard 9:3:3:1 ratio indicates that the phenotype is the product of the interaction of two or more genes.

As we introduced in Chapter 2, the 9:3:3:1 phenotypic ratio can be represented genotypically in a shorthand way as *A/– B/–*, *A/– b/b*, *a/a B/–*, and *a/a b/b*, respectively. The dash indicates that the phenotype is the same whether the gene is homozygous dominant or heterozygous (e.g., *A/–* means either *A/A* or *A/a*). This system cannot be used when incomplete dominance or codominance is involved because the *A/A* or *A/a* genotypes produce different phenotypes.

The following sections discuss the main processes that result in modified Mendelian ratios. The first section describes interactions between different genes that control the same general phenotypic attribute. In the second section, we discuss interactions of different genes in which an allele of one gene masks the expression of alleles of another gene. This second type of interaction is called *epistasis*. In both cases, the discussions are confined to dihybrid crosses in which the two pairs of alleles assort independently. In the real world, there are many more complex examples of gene interactions involving more than two pairs of alleles and/or genes that do not assort independently.

Gene Interactions That Produce New Phenotypes

If the two allelic pairs in a dihybrid cross affect the same phenotypic characteristic, there is a chance for interaction of their gene products to give novel phenotypes, and the result may or may not be modified phenotypic ratios, depending on the particular interaction between the products of the nonallelic genes.

Comb Shape in Chickens. In chickens, new comb shape phenotypes (Figure 4.4) result from interactions between the alleles of two gene loci. Each of the four comb types can be bred true if the alleles involved are homozygous.

Crosses made between true-breeding rose-combed and single-combed varieties show that rose is completely dominant over single. When the F_1 rose-combed birds are interbred, a ratio of 3 rose : 1 single results in the F_2. Similarly, pea comb is completely dominant over single, with a 3 pea : 1 single ratio in the F_2. When true-breeding rose and pea varieties are crossed, however, the result is different and interesting (Figure 4.5a). Instead of showing either rose or pea combs, all birds in the F_1 show yet another comb form, which is called walnut comb because it resembles half a walnut meat.

When the F_1 walnut-combed birds are interbred, not only do walnut-, rose-, and pea-combed birds appear, but so do single-combed birds (Figure 4.5b). These four comb

types occur in a ratio of 9 walnut : 3 rose : 3 pea : 1 single. Such a ratio is characteristic in F_2 progeny from two parents each heterozygous for two genes. The class in the F_2 generation with at least one copy of each dominant allele is walnut, and the proportion of the single-combed birds indicates that this class contains both recessive alleles.

The explanation of these results is as follows (see Figure 4.5b): The walnut comb depends on the presence of two dominant alleles, R and P, both located at two independently assorting gene loci. In $R/– p/p$ birds, a rose comb results; in $r/r P/–$ birds, a pea comb results; and in $r/r p/p$ birds, a single comb results.

Thus, it is the *interaction* of two dominant alleles, each of which individually produces a different phenotype, that produces a different phenotype. No modification of typical Mendelian ratios is involved. The molecular basis for the four comb types is not known.

Fruit Shape in Summer Squash (The 9:6:1 Ratio). Two of the many varieties of summer squash have long fruit and sphere-shaped fruit. This trait also shows complete dominance at two gene pairs, and interaction between both dominants results in a new phenotype. The molecular bases for the different shapes of squash fruit are not known.

The long-fruit varieties always breed true. However, in some crosses between different varieties of true-breeding sphere-shaped plants, the F_1 fruit is disk-shaped (Figure 4.6). In such instances, the F_2 fruit is approximately $9/16$ disk-shaped, $6/16$ sphere-shaped, and $1/16$ long-shaped; this is a modification of the typical Mendelian ratio, so two genes are involved. The explanation is as follows: A dominant allele of either gene and homozygosity for the recessive allele of the other gene ($A/– b/b$ or $a/a B/–$) results in the same phenotype: spherical fruit. Two dominant alleles, one of each gene ($A/– B/–$), interact together to produce a new phenotype: disk-shaped fruit. The doubly homozygous recessive ($a/a b/b$) gives a long fruit shape. Thus, the cross of two sphere-shaped squashes here is $A/A b/b$ and $a/a B/B$. The F_1 fruit is disk-shaped and genotypically $A/a B/b$. The F_2 disk-shaped fruit is $A/– B/–$, the spherical fruit is $A/– b/b$ or $a/a B/–$, and the long-shaped fruit is $a/a b/b$, giving the 9:6:1 ratio.

Epistasis

Epistasis is a form of gene interaction in which one gene masks the phenotypic expression of another. No new phenotypes are produced by this type of gene interaction. A gene that masks another gene's expression is said to be epistatic, and a gene whose expression is masked is said to be hypostatic. If we think about the F_2 genotypes $A/– B/–$, $A/– b/b$, $a/a B/–$, and $a/a b/b$, epistasis may be caused by the presence of homozygous recessives of one gene pair, so that a/a masks the effect of the B allele. Or epistasis may result from the presence of one domi-

Four distinct comb shape phenotypes in chickens, resulting from all possible combinations of a dominant and a recessive allele at each of two gene loci. (a) Rose comb ($R/– p/p$). (b) Walnut comb ($R/– P/–$). (c) Pea comb ($r/r P/–$). (d) Single comb ($r/r p/p$).

a) Rose comb

b) Walnut comb

c) Pea comb

d) Single comb

Figure 4.5

Complete dominance in chickens. The genetic crosses show the interaction of genes for comb shape. **(a)** The cross of a true-breeding rose-combed bird with a true-breeding pea-combed bird gives all walnut-combed offspring in the F$_1$ generation. **(b)** When the F$_1$ birds are interbred, a 9:3:3:1 ratio of walnut : rose : pea : single occurs in the F$_2$ generation.

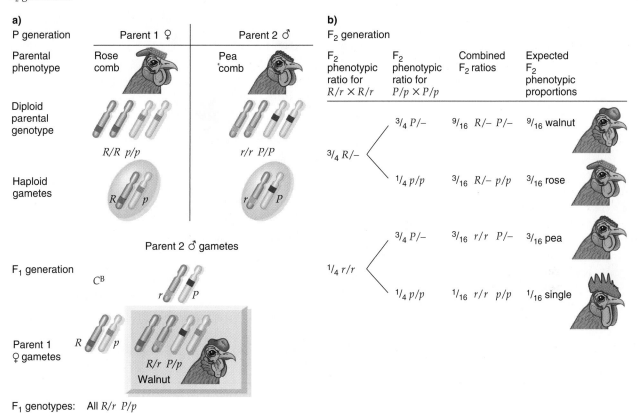

nant allele in a gene pair. For example, the *A* allele might mask the effect of the *B* allele. Epistasis can also occur in both directions between two gene pairs. All these possibilities produce quite a number of modifications of the 9:3:3:1 ratio in a dihybrid cross. Some examples of epistasis follow.

Recessive Epistasis: Coat Color in Rodents (The 9:3:4 Ratio). In *recessive epistasis*, *a/a B/–* and *a/a b/b* individuals have the same phenotype, so the phenotypic ratio in the F$_2$ is 9:3:4 rather than 9:3:3:1. An example is coat color in rodents. Wild mice have a greyish color because of the presence of black and yellow banded hairs in the fur. This coloration, the agouti pattern, aids in camouflage and is found in many wild rodents, including guinea pigs, grey squirrels, and wild mice.

Several other coat colors are seen in domesticated rodents. Albinos, for example, have no pigment in the fur or in the irises of the eyes, so they have a white coat and pink eyes. Albinos breed true, and this variation behaves

as a complete recessive to any other color. Another variant has black coat color as the result of the absence of the yellow pigment found in the agouti pattern. Black is recessive to agouti.

When true-breeding agouti mice are crossed with albinos, the F$_1$ progeny are all agouti, and when these F$_1$ agoutis are interbred, the F$_2$ progeny consist of approximately 9/16 agouti, 3/16 black, and 4/16 albino animals (Figure 4.7). This pattern occurs because the parents differ in whether they have a dominant allele, *A*, of a gene for the agouti pattern, which is a yellow banding of the black hairs (*A/–* are agouti and *a/a* are nonagouti). (Note: The symbols here are the actual ones used for the genes involved in rodent coat color. Do not confuse the *a* and *c* loci here with the *a* and *b* loci referred to in our continuing discussion of modified ratios that began on p. 81.) Phenotypically, *A/– C/–* are agouti, *a/a C/–* are black, and *A/– c/c* and *a/a c/c* are albino, giving a 9:3:4 phenotypic ratio of agouti : black : albino. Thus, this example demonstrates epistasis of *c/c* over *A/–*. In other words, white hairs

Figure 4.6

Generation of an F₂ 9:6:1 ratio for fruit shape in summer squash.

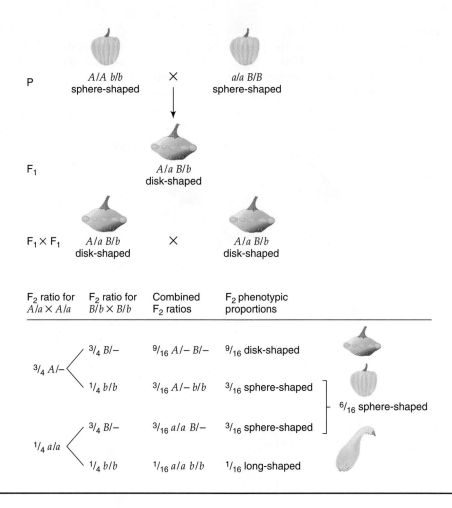

Figure 4.7

Recessive epistasis: generation of an F₂ 9:3:4 ratio for coat color in rodents.

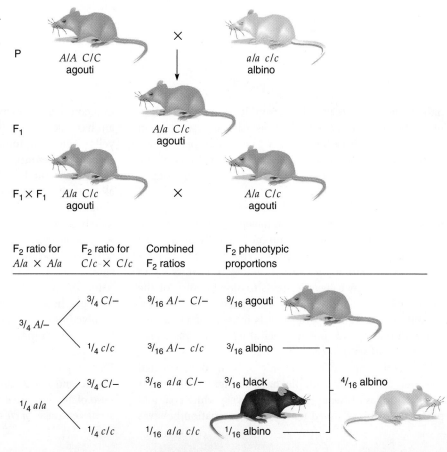

are produced in *c/c* mice, regardless of the genotype at the other locus.

Duplicate Recessive Epistasis: Flower Color in Sweet Peas (The 9:7 Ratio).

In the sweet pea, purple flower color is dominant to white and gives a typical 3:1 ratio in the F_2. White-flowered varieties of sweet peas breed true, and crosses between different white varieties usually produce white-flowered progeny. In some cases, however, crosses of two true-breeding white varieties give only purple-flowered F_1 plants. When these F_1 hybrids are self-fertilized, they produce F_2 plants consisting of about $9/16$ purple-flowered sweet peas and $7/16$ white-flowered sweet peas (Figure 4.8). The 9:7 ratio is a modification of the 9:3:3:1 ratio. Even though they are not all homozygous for the alleles in question, all the F_2 white-flowered plants breed true when self-fertilized. One-ninth of the purple-flowered F_2 plants—the *C/C P/P* genotypes—breed true.

These results can be explained by the interaction of two genes. The $9/16$ purple-flowered F_2 plants suggest that colored flowers appear only when two independent dominant alleles are present together and that the color purple results from some interaction between them. White flower color would then result from homozygosity for the recessive allele of one or both genes. Thus, gene pair *C/c* specifies whether or not the flower can be colored, and gene pair *P/p* specifies whether purple flower color will result. An interaction of two genes to give rise to a specific product is a form of epistasis called *duplicate recessive epistasis* or *complementary gene action*.

The following purely theoretical pathway—undoubtedly too simplistic for the actual phenotypes—can be envisioned for the production of the purple pigment:

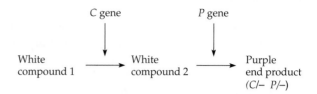

In this pathway, a colorless precursor compound (not shown here) is converted in a series of steps to a purple end product. Each step is controlled by the product of a gene. To explain the F_2 ratio, we can propose that gene *C* controls the conversion of white compound 1 to white compound 2 and that gene *P* controls the conversion of white compound 2 to the purple end product. Therefore, homozygosity for the recessive allele of either or both of the *C* and *P* genes will result in a block in the pathway, and only white pigment will accumulate. That is, *C/– p/p*, *c/c P/–*, and *c/c p/p* genotypes will all be white. The only plants that will produce purple flowers are those in which both steps of the pathway are completed so that the colored pigment is produced. This situation occurs only in *C/– P/–* plants.

In the cross described (see Figure 4.8), the two white parentals are *C/C p/p* and *c/c P/P*, and the F_1 plants are purple and doubly heterozygous. Interbreeding the F_1 plants gives a 9:7 ratio of purple : white in the F_2. Recessive

---

Figure 4.8

Duplicate recessive epistasis: generation of an F_2 9:7 ratio for flower color in sweet peas.

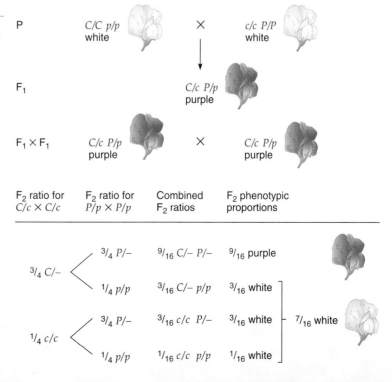

epistasis occurs in both directions between two gene pairs. The consequence of this gene interaction is that the same phenotype (white) is exhibited whenever one or the other gene pair is homozygous recessive.

In sum, many types of phenotypic modifications are possible as a result of interactions between the products of different gene pairs. Geneticists detect such interactions when they observe deviations from the expected phenotypic ratios in crosses. We have discussed some examples in which two genes assort independently and in which complete dominance is exhibited in each allelic pair. The ratios we discussed would necessarily be modified further if the genes did not assort independently or if incomplete dominance or codominance prevailed. Table 4.3 shows examples of epistatic F_2 phenotypic ratios from an $A/a\ B/b \times A/a\ B/b$ cross.

Epistasis plays a role in many human genetic diseases, further complicating their analysis. In these cases, complex interrelationships exist. For example, the majority of bipolar disorder (also called manic-depressive illness), a complex human genetic disorder involving pathological mood disturbances, involves epistasis between multiple genes and may include other, more complex genetic mechanisms.

KEYNOTE

In many instances, alleles of different genes interact to determine phenotypic characteristics. Sometimes the interaction between genes results in new phenotypes without modification of typical Mendelian ratios. In epistasis, interaction between genes causes modifications of Mendelian ratios because one gene interferes with the phenotypic expression of another gene (or genes). The phenotype is controlled largely by the former gene and not the latter when both genes occur together in the genotype. The analysis of epistasis is complicated further when one or both gene pairs involve incomplete dominance or codominance or when genes do not assort independently.

Essential Genes and Lethal Alleles

For a few years after the rediscovery of Mendel's principles, geneticists believed that mutations changed only the appearance of a living organism, but then they discovered that a mutant allele could cause death. In a sense, this

---

Table 4.3 Examples of Epistatic F_2 Phenotypic Ratios from an $A/a\ B/b \times A/a\ B/b$ in Which Complete Dominance is Shown for Each Gene Pair

		A/A B/B	A/A B/b	A/a B/B	A/a B/b	A/A b/b	A/a b/b	a/a B/B	a/a B/b	a/a b/b
More than four phenotypic classes	A and B both incompletely dominant	1	2	2	4	1	2	1	2	1
	A incompletely and B completely dominant	3		6		1	2	3		1
Four phenotypic classes	A and B both completely dominant (classic ratio)	9				3		3		1
Fewer than four phenotypic classes	a/a epistatic to B and b; recessive epistasis	9				3		4		
	A epistatic to B and b; dominant epistasis	12						3		1
	A epistatic to B and b; b/b epistatic to A and a; dominant and recessive epistasis	13[a]						3		
	a/a epistatic to B and b; b/b epistatic to A and a; duplicate recessive epistasis	9						7		
	A epistatic to B and b; B epistatic to A and a; duplicate dominant epistasis	15								1
	Duplicate interaction	9				6				1

[a]The 13 is composed of the 12 classes immediately above plus the one *a/a b/b* from the last column.

Source: Science of Genetics, 6th ed. by George W. Burns and Paul J. Bottino. Copyright © 1989. Reprinted by permission of Prentice Hall, Inc., Upper Saddle River, NJ.

mutation is still a change in phenotype, with the new phenotype being lethality. An allele that results in the death of an organism is called a **lethal allele,** and the gene involved is called an essential gene. **Essential genes** are genes that, when mutated, can result in a lethal phenotype. If the mutation is due to a **dominant lethal allele,** both homozygotes and heterozygotes for that allele will show the lethal phenotype. If the mutation is due to a **recessive lethal allele,** only homozygotes for that allele will have the lethal phenotype.

An example of an essential gene is the gene for yellow body color in mice. The yellow variety never breeds true. When yellows are bred with nonyellows, the progeny show an approximately 1:1 ratio of yellow : nonyellow mice. (The nonyellow color depends on other coat color genes.) This ratio is expected from the mating of a heterozygote with a recessive, suggesting that yellow mice are heterozygous. When the yellow heterozygotes are interbred, a phenotypic ratio of about 2 yellow : 1 nonyellow is observed instead of the predicted 3:1 ratio.

It turns out that yellow homozygotes are aborted in utero; in other words, the yellow allele has a *dominant* effect with regard to coat color, but it acts as a *recessive* allele with respect to the lethality phenotype because only homozygotes die. We now know that homozygotes for the yellow allele die at the embryo stage.

The yellow allele is an allele of the agouti locus (a) and has been given the symbol A^Y. The yellow × yellow cross is shown in Figure 4.9. Genotypically, the cross is $A^Y/A \times A^Y/A$. We expect a genotypic ratio of $1/4$ A^Y/A^Y : $2/4$ A^Y/A : $1/4$ A/A among the progeny. The $1/4$ A^Y/A^Y mice die before birth, giving a birth ratio of $2/3$ A^Y/A (yellow) : $1/3$ A/A (nonyellow). Since the A^Y allele causes lethality in the homozygous state, it is called a recessive lethal. Characteristically, when two heterozygotes are crossed, recessive lethal alleles are recognized by a 2:1 ratio of progeny types.

Essential genes are found in all diploid organisms. In humans there are many known recessive lethal alleles. One example is *Tay-Sachs disease* (OMIM 272800; see p. 213). The gene involved, *HEXA,* encodes the enzyme hexosaminidase A. Homozygotes appear normal at birth, but before about 1 year of age, they show symptoms of central nervous system deterioration. Progressive mental retardation, blindness, and loss of neuromuscular control follow. Afflicted children usually die at 3 to 4 years of age. The genetic defect in Tay-Sachs results in an enzyme deficiency that prevents proper nerve function.

There are X-linked lethal mutations as well as autosomal lethal mutations, dominant lethal mutations as well as recessive lethals. In humans, for example, the genetic disease hemophilia (OMIM 306700) is caused by an X-linked recessive allele. Untreated, hemophilia is lethal. Dominant lethals exert their effect in heterozygotes, resulting in the death of the organism usually at conception or at a fairly young age. Dominant lethals

Figure 4.9

Inheritance of a lethal gene in mice. A mating of two yellow mice gives $1/4$ nonyellow (black mice), $1/2$ yellow mice, and $1/4$ dead embryos. The viable yellow mice are heterozygous A^Y/A, and the dead individuals are homozygous A^Y/A^Y.

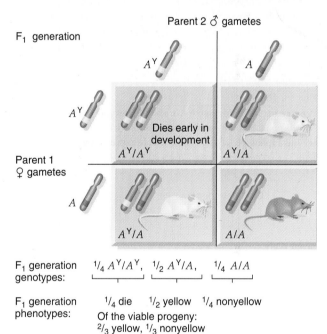

cannot be studied genetically unless death occurs after the organism has reached reproductive age. For example, the symptoms of the autosomal dominant trait *Huntington disease* (OMIM 143100)—involuntary movements and progressive central nervous system degeneration—may not begin until affected individuals reach their early 30s; as a result, parents may unknowingly pass on the gene to their offspring. Death usually occurs when the afflicted persons are in their 40s or 50s. The well-known American folksinger Woody Guthrie died from Huntington disease.

K E Y N O T E

A lethal allele is fatal to the individual. There are recessive lethal alleles and dominant lethal alleles, and they can be X-linked or autosomal. The existence of lethal alleles of a gene indicates that the gene's normal product is essential for the function of the organism; therefore, the gene is an essential gene.

Gene Expression and the Environment

The *development* of a multicellular organism from a zygote is a process of *regulated growth and differentiation* that results from the interaction of the organism's genome with the environment, both the internal cellular environment and the external environment. Development is a tightly controlled, programmed series of phenotypic changes that, under normal environmental conditions, is essentially irreversible. Four major processes interact to constitute the complex process of development: (1) replication of the genetic material, (2) growth, (3) differenti-

ation of the various cell types, and (4) arrangement of differentiated cells into defined tissues and organs.

Think of development as a series of intertwined, complex biochemical pathways. The internal or external environment may influence any of these pathways by affecting the products of the genes controlling the pathways. This phenomenon is most readily studied in experimental organisms where the genotype is unequivocally known. The extent to which the gene manifests its effects under varying environmental conditions can then be seen. We consider some examples in the following section.

Penetrance and Expressivity

In some cases, not all individuals with a particular genotype show the expected phenotype. The frequency with which a dominant or homozygous recessive gene manifests itself in individuals in a population is called the **penetrance** of the gene. Penetrance depends on both the genotype (e.g., the presence of epistatic or other genes) and the environment. Figure 4.10a illustrates the concept of penetrance. Penetrance is complete (100 per-

Figure 4.10

Penetrance and expressivity in the phenotypic expression of a genotype.
(**a**) Complete and incomplete penetrance. (**b**) Constant and variable expressivity.
(**c**) Incomplete penetrance and variable expressivity.

a)

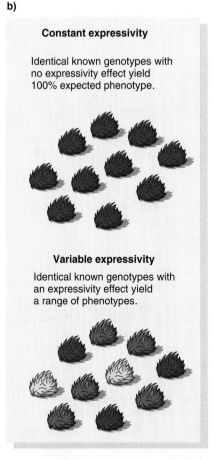

Complete penetrance

Identical known genotypes yield 100% expected phenotype.

Incomplete penetrance

Identical known genotypes yield <100% expected phenotype.

b)

Constant expressivity

Identical known genotypes with no expressivity effect yield 100% expected phenotype.

Variable expressivity

Identical known genotypes with an expressivity effect yield a range of phenotypes.

c)

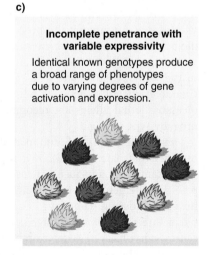

Incomplete penetrance with variable expressivity

Identical known genotypes produce a broad range of phenotypes due to varying degrees of gene activation and expression.

cent) when all the homozygous recessives show one phenotype, when all the homozygous dominants show another phenotype, and when all the heterozygotes are alike. For example, if all individuals carrying a dominant mutant allele show the mutant phenotype, the allele is completely penetrant. Many genes show complete penetrance; the seven gene pairs in Mendel's experiments and the alleles in the human ABO blood group system are examples.

If less than 100 percent of the individuals with a particular genotype exhibit the phenotype expected, penetrance is incomplete. For example, an organism may be genotypically $A/-$ or a/a but may not display the phenotype typically associated with that genotype. If, say, 80 percent of the individuals carrying a particular gene show the corresponding phenotype, we say that there is 80 percent penetrance.

In humans, many genes show reduced penetrance. For example, brachydactyly (OMIM 112500), an autosomal dominant trait that causes shortened and malformed fingers, shows 50 to 80 percent penetrance. It is also thought that a number of genes that confer predisposition to cancer exhibit low to moderate penetrance, making it more difficult to identify and characterize those genes.

Genes may influence a phenotype to different degrees. **Expressivity** is the degree to which a penetrant gene or genotype is phenotypically expressed in an individual. Figure 4.10b illustrates the concept of expressivity. Like penetrance, expressivity depends on both the genotype and the environment, and it may be constant or variable. Molecularly, we can think of expressivity at a simple level as the result of different degrees of mutational alteration of the protein encoded by the gene.

An example of variation in expressivity is found in the human condition called osteogenesis imperfecta (OMIM 166200). The three main features of this disease are blueness of the sclerae (the whites of the eyes), very fragile bones, and deafness. Osteogenesis imperfecta is inherited as an autosomal dominant with almost 100 percent penetrance. However, the trait shows variable expressivity: A person with the gene may have any one or any combination of the three traits. Moreover, the fragility of the bones for those who exhibit this condition is also quite variable.

Lastly, some genes exhibit both incomplete penetrance and variable expressivity. Figure 4.10c illustrates this concept. For example, individuals with neurofibromatosis (OMIM 162200), an autosomal dominant trait, develop tumorlike growths (neurofibromas) over the body. This genetic disease shows 50 to 80 percent penetrance, and it also shows variable expressivity (Figure 4.11). In its mildest form, the disease causes individuals to have only a few pigmented areas on the skin (called *café-au-lait spots* because they are the color of coffee with milk). In more severe cases, one or more other symptoms may be seen, including neurofibromas of var-

Figure 4.11

Variable expressivity in individuals with neurofibromatosis.
Top: Café-au-lait spot. Middle: Café-au-lait spot and freckling. Bottom: Large number of cutaneous neurofibromas (tumorlike growths).

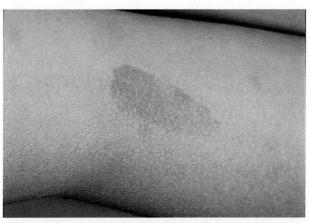

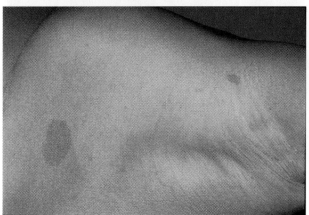

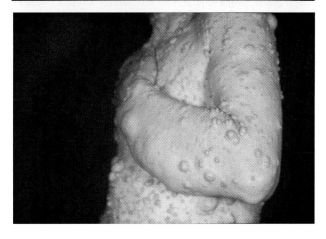

ious sizes; high blood pressure; speech impediments; headaches; large head; short stature; tumors of the eye, brain, or spinal cord; and curvature of the spine. Therefore, for medical genetics studies it is important to recognize that a gene may vary widely in its expression, and this makes the task of genetic counseling that much more difficult.

K E Y N O T E

Penetrance is the frequency with which a dominant or homozygous recessive allele manifests itself in the phenotype of individuals in a population. Expressivity is the type or degree of phenotypic manifestation of a penetrant allele or genotype in a particular individual.

Effects of the Environment

The phenotypic expression of a gene depends on a number of factors, including the influences of the environment. Next we consider some examples of environmental influences on gene expression.

Age of Onset. The age of the organism creates internal environmental changes that can affect gene function. All genes do not function continually; instead, over time, programmed activation and deactivation of genes occur as the organism develops and functions. Numerous age-dependent genetic traits occur in humans; pattern baldness (OMIM 109200) appears in males between 20 and 30 years of age, and Duchenne muscular dystrophy (DMD; OMIM 310200) appears in children between 2 and 5 years of age. In most cases, the nature of the age dependency is not understood.

Sex. The expression of particular genes may be influenced by the sex of the individual. In the case of sex-linked genes, as mentioned earlier, differences in the phenotypes of the two sexes are related to different complements of genes on the sex chromosomes. However, in some cases, genes that are on autosomes affect a particular character that appears in one sex but not the other. Traits of this kind are called **sex-limited traits.**

Examples of sex-limited traits in animals are milk production in dairy cattle (the genes involved obviously operate in females but not in males), the appearance of horns in certain species of sheep (males with genes for horns have horns, and females with genes for horns do not have horns), and the ability to produce eggs or sperm. An example in humans is the distribution of facial hair.

A slightly different situation is found in **sex-influenced traits,** which, like sex-limited traits, often are controlled by autosomal genes. Such traits appear in both sexes, but either the frequency of occurrence in the two sexes is different or the relationship between genotype and phenotype is different.

Pattern baldness (OMIM 109200) is an example of a sex-influenced trait in humans.[1] Pattern baldness is con-

trolled by an autosomal gene that acts as a dominant in males and as a recessive (or at least it is expressed at lower levels) in females. Pattern baldness begins in women much later in life than in men because of hormonal influences.

Other human examples of sex-influenced traits are cleft lip and palate (incomplete fusion of the upper lip and palate), in which there is a 2:1 ratio of the trait in males : females; clubfoot (2:1 ratio); gout (8:1 ratio); rheumatoid arthritis (1:3 ratio); osteoporosis (1:3 ratio); and systemic lupus erythematosus (an autoimmune disease; 1:9 ratio).

Temperature. Biochemical reactions in the cell are catalyzed by enzymes. Normally, enzymes are unaffected by temperature changes within a reasonable range. However, some alleles of an enzyme-coding gene may give rise to an enzyme that is temperature-sensitive; that is, it may function normally at one temperature but be nonfunctional at another temperature. An example of a temperature effect on gene expression is fur color in Himalayan rabbits. Certain genotypes of this white rabbit cause dark fur to develop at the ears, nose, and paws, where the local surface temperature is lower (Figure 4.12a). (A similar situation applies to Siamese cats.) Since all body cells develop from a single zygote, this distinct fur pattern cannot be the result of a genotypic difference of the cells in those areas. Rather, it can be hypothesized that the fur pattern results from environmental influences. This hypothesis can be tested by rearing Himalayan rabbits under different temperature conditions. When a rabbit is reared at a temperature above 30°C, all its fur, including that of the ears, nose, and paws, is white (Figure 4.12b). If a rabbit is raised at 25°C, the typical Himalayan phenotype results (Figure 4.12c). Finally, if a rabbit is raised at 25°C while part of its body is artificially cooled to a temperature below 25°C, the rabbit develops the Himalayan coat phenotype and exhibits an additional patch of dark fur on the cooled area (Figure 4.12d).

Chemicals. Certain chemicals can have significant effects on an organism, as the following two examples show.

Phenylketonuria. The human disease phenylketonuria (PKU; OMIM 261600) is an autosomal recessive trait that is defective in the biochemical pathway for the metabolism of the amino acid phenylalanine. In individuals homozygous for the recessive allele, various symptoms appear, most notably mental retardation at an early age. The diet determines how severe the symptoms of PKU will be. Problem foods include protein containing phenylalanine, such as the protein of mother's milk. PKU can be diagnosed soon after birth by a simple blood test and then treated by restricting the amount of phenylalanine in the diet.

Phenocopies Induced by Chemicals. Changes in the chemical composition of the environment can also influence the expression of one or more genes. The most

---

[1] Baldness is not a straightforward trait to study. One reason is that there is variable expressivity in the baldness phenotype: Baldness may appear first in the crown or on the forehead, and the degree of baldness varies from minimal to extreme. Moreover, a number of genes affect the presence of hair on the head, including the pattern baldness gene, and the final phenotype is the result of the interaction between the environment and the particular set of baldness genes present.

Figure 4.12

Effect of temperature on gene expression. (a) Himalayan rabbit. **(b)** White extremities result when a Himalayan rabbit is reared at above 30°C. **(c)** Normal Himalayan pattern when rabbit is reared at 25°C. **(d)** Normal Himalayan pattern when rabbit is reared at 25°C, with a dark patch on the side where the flank has been cooled to below 25°C.

a)

b)

White extremities, reared at >30°C

c)

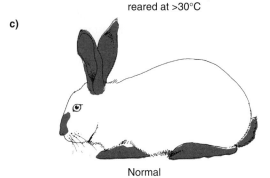

Normal Himalayan pattern, reared at 25°C

d)

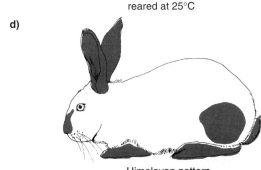

Himalayan pattern with dark patch on flank, reared at 25°C, flank cooled to below 25°C

sensitive time period is early development because small changes at that time can result in great changes later. When a developing embryo is exposed to certain drugs, chemicals, and viruses, these agents may produce a phenocopy (phenotypic copy). A **phenocopy** is defined as a nonhereditary phenotypic modification, caused by special environmental conditions, that mimics a similar phenotype caused by a known gene mutation. In other words, although the individual expresses a mutant phenotype, the genotype is *normal*. The agent that produces a phenocopy is called a *phenocopying agent*. There are many examples of phenocopies, and in some instances, studying the phenocopy has provided useful information about the actual molecular defects caused by the phenocopy's mutant counterpart.

In humans, cataracts, deafness, and heart defects sometimes are produced when an individual is homozygous for rare recessive alleles; these disorders may also result if the mother is infected with rubella (German measles) virus during the first 12 weeks of pregnancy. Another human trait for which there is a phenocopy is *phocomelia*, a suppression of the development of the long bones of the limbs, which is caused by a rare dominant allele with variable expressivity. Between 1959 and 1961, similar phenotypes were produced by the sedative thalidomide when it was taken by expectant mothers between the 35th and 50th days of gestation. The drug was removed from the market when its devastating effects were discovered.

KEYNOTE

The phenotypic expression of a gene depends on several factors, including its dominance relationships, the genetic constitution of the rest of the genome (e.g., the presence of epistatic or modifier genes), and the influences of the internal and external environments. In some cases, special environmental conditions can cause a phenocopy, a nonhereditary phenotypic modification that mimics a similar phenotype caused by a gene mutation.

Nature Versus Nurture

We are left with the nature-nurture question; that is, what are the relative contributions of genes and environment to the phenotype? (Nature-nurture is discussed more in Chapter 23.) Up to this point, variation in most of the traits we have examined has been determined largely by differences in genotype; that is, phenotypic differences have reflected genetic differences. However, we have already seen that the phenotypes of many traits are influenced by both genes and environment. Let us consider the nature-nurture issue in the context of some human examples.

Human height, or stature, is definitely influenced by genes. On the average, tall parents tend to have tall offspring, and short parents tend to have short offspring.

There are also a number of genetic forms of dwarfism in humans. Achondroplasia is a type of dwarfism in which the bones of the arms and legs are shortened but the trunk and head are of normal size; achondroplasia results from a single dominant allele. But the environment also plays a role in determining height. For example, human height has increased about 1 inch per generation over the past 100 years as a result of better diets and improved health care. Genes and the environment have interacted in determining human height.

For a trait such as height, genes set certain limits (or specify potential) for the phenotype. The phenotype an individual develops within these limits depends on the environment. The range of potential phenotypes that a single genotype could produce if exposed to a range of environmental conditions is called the **norm of reaction.** For some genotypes, the norm of reaction is small; that is, the phenotype produced by a genotype is nearly the same in different environments. For other genotypes, the norm of reaction is large, and the phenotype produced by the genotype varies greatly in different environments.

Many human behavioral traits are the result of interaction between genes and the external environment. One example is alcoholism, which is a major health problem in the United States. About 10 million Americans are problem drinkers, and 6 million are severely addicted to alcohol. Numerous studies have shown that alcoholism is influenced by genes. For instance, sons of alcoholic fathers who are separated from their biological parents at birth and adopted into a family with nonalcoholic parents are four times more likely to become alcoholic than sons adopted at birth whose biological fathers were not alcoholic. However, no gene forces a person to drink alcohol. That is, one cannot become an alcoholic unless one is exposed to an environment in which alcohol is available and drinking is encouraged. What genes do is make certain people more or less susceptible to alcohol abuse; they increase or decrease the risk of developing alcoholism. How genes influence our susceptibility to alcohol abuse is not yet clear. They may affect the way we metabolize alcohol, which might affect how much we drink. Or genes may influence certain of our personality traits that make us more or less likely to drink heavily. The important point is that a behavioral trait such as alcoholism may be influenced by genes, but the genes alone do not produce the phenotype.

Nowhere has the role of genes and environment been more controversial than in the study of human intelligence. In the past, people were divided on the issue, some insisting that human intelligence was genetically programmed, others arguing that it was produced entirely by the environment. The clash of these opposing views was called the nature-nurture controversy. Today, geneticists recognize that neither of these extreme views is correct; human intelligence is the product of both genes and environment.

That genes influence human intelligence is clearly evidenced by genetic conditions that produce mental retardation, such as PKU (OMIM 261600; see p. 90) and Down syndrome (trisomy 21:OMIM 190685; see Chapter 21, pp. 445–448). Numerous studies also indicate that genes influence differences in IQ among nonretarded people. (IQ, or intelligence quotient, is a standardized measure of mental age compared with chronological age; it is fairly stable over time.) For example, adoption studies show that the IQ of adopted children is closer to that of their biological parents than to the IQ of their adoptive parents.

However, IQ is also influenced by environment. Identical twins frequently differ in IQ, which can be explained only by environmental differences. Family size, diet, and culture are environmental factors known to affect IQ. Thus, IQ results from the interaction of genes and environment. Consequently, if two people (other than identical twins) differ in IQ, it is impossible to attribute that difference solely to either genes or environment, because both interact in determining the phenotype. So, although we cannot change our genes, we can alter the environment and thus affect a phenotypic trait such as intelligence.

KEYNOTE

Variation in most of the genetic traits considered in the discussion of Mendelian principles is determined predominantly by differences in genotype; that is, phenotypic differences resulted from genotypic differences. For many traits, however, the phenotypes are influenced by both genes and the environment. The debate over the relative contribution of genes and environment to the phenotype has been called the nature-nurture controversy.

Summary

A variety of exceptions to and extensions of Mendel's principles were discussed in this chapter. These include the following:

1. *Multiple alleles.* A gene may have many allelic forms in a population, and these alleles are called multiple alleles. A diploid individual can have only two different alleles of a given set of multiple alleles.

2. *Modified dominance relationships.* In complete dominance, the same phenotype results whether an allele is heterozygous or homozygous. In incomplete dominance, the phenotype of the heterozygote is intermediate between those of the two homozygotes. In codominance, the heterozygote shows the phenotypes of both homozygotes.

3. *Gene interactions and modified Mendelian ratios.* Many genes do not function independently in determining phenotypic characteristics. In epistasis, modified Mendelian ratios occur because of interactions of nonallelic genes. The phenotypic expression of one gene depends on the genotype at another gene locus. In other interactions, a new phenotype is produced.

4. *Essential genes and lethal alleles.* Alleles of certain genes result in the lack of production of a necessary functional gene product, and this gives rise to a lethal phenotype. Such lethal alleles may be recessive or dominant. The existence of lethal alleles of a gene indicates that the normal product of the gene is essential for the organism.

5. *Penetrance and expressivity.* Penetrance is the condition in which not all individuals who are known to have a particular allele show the phenotype specified by that allele. That is, penetrance is the frequency with which a dominant or homozygous recessive allele manifests itself in individuals in the population. The related phenomenon of expressivity describes the degree to which a penetrant gene or genotype is phenotypically expressed in an individual. Both penetrance and expressivity depend on the genotype and the external environment.

6. *Dual influence of genes and environment on phenotype.* An organism's potential to develop and function is specified by the zygote's genetic constitution. As an organism develops and differentiates, gene expression is influenced by a number of factors, including dominance relationships, the genetic constitution of the rest of the genome, and the influences of the internal and external environments. That is, the phenotypes of many traits are influenced by both genes and environment. The debate over the relative contributions of genes and environment to the phenotype has been called the nature-nurture controversy.

Analytical Approaches for Solving Genetics Problems

Q4.1 In snapdragons, red flower color (C^R) is incompletely dominant to white flower color (C^W); the heterozygote has pink flowers. Also, normal broad leaves (L^B) are incompletely dominant to narrow, grasslike leaves (L^N); the heterozygote has an intermediate leaf breadth. If a red-flowered, narrow-leaved snapdragon is crossed with a white-flowered, broad-leaved one, what will be the phenotypes of the F_1 and F_2 generations, and what will be the frequencies of the different classes?

A4.1 This basic question on gene segregation includes the issue of incomplete dominance. In the case of incomplete dominance, remember that the genotype can be directly determined from the phenotype. Therefore, we do not need to ask whether or not a strain breeds true because all phenotypes have a different (and therefore known) genotype.

The best approach here is to assign genotypes to the parental snapdragons. Let $C^R/C^R\ L^N/L^N$ represent the red, narrow-leaved plant, and let $C^W/C^W\ L^B/L^B$ represent the white, broad-leaved plant. The F_1 plants from this cross will all be double heterozygotes, $C^R/C^W\ L^B/L^N$. Because of the incomplete dominance, these plants are pink-flowered and have leaves of intermediate breadth. Interbreeding the F_1 plants gives the F_2 generation, but it does not have the usual 9:3:3:1 ratio. Instead, there is a different phenotype for each genotype. These genotypes and phenotypes and their relative frequencies are shown in Figure 4.A (see p. 94).

Q4.2 In snapdragons, red flower color is incompletely dominant to white, with the heterozygote being pink; normal flowers are completely dominant to peloric-shaped ones; and tallness is completely dominant to dwarfness. The three gene pairs segregate independently. If a homozygous red, tall, normal-flowered plant is crossed with a homozygous white, dwarf, peloric-flowered one, what proportion of the F_2 plants will resemble the F_1 plants in appearance?

A4.2 Let us assign symbols: C^R = red and C^W = white; N = normal flowers and n = peloric; T = tall and t = dwarf. The initial cross, then, becomes $C^R/C^R\ T/T\ N/N \times C^W/C^W\ t/t\ n/n$. From this cross we see that all the F_1 plants are triple heterozygotes with the genotype $C^R/C^W\ T/t\ N/n$ and with the phenotype pink, tall, normal-flowered. Interbreeding the F_1 generation will produce 27 different genotypes in the F_2; this answer follows from the rule that the number of genotypes is 3^n, where n is the number of heterozygous gene pairs involved in the $F_1 \times F_1$ cross (see Chapter 2).

Here we are asked specifically for the proportion of F_2 progeny that resemble the F_1 plants in appearance. We can calculate this proportion directly without needing to display all the possible genotypes and then grouping the progeny in classes according to phenotype. First, we calculate the frequency of pink-flowered plants in the F_2; then we determine the proportion of these plants that have the other two attributes. From a $C^R/C^W \times C^R/C^W$ cross, we calculate that half of the progeny will be heterozygous C^R/C^W and therefore pink. Next, we determine the proportion of F_2 plants that are phenotypically like the F_1 with respect to height (tall). Either T/T or T/t plants will be tall, so $\frac{3}{4}$ of the F_2 plants will be tall. Similarly, $\frac{3}{4}$ of the F_2 plants will be normal-flowered like the F_1 plants. To obtain the probability of all three of these phenotypes occurring together (pink, tall, normal), we must multiply the individual probabilities because the gene pairs segregate independently. The answer is $\frac{1}{2} \times \frac{3}{4} \times \frac{3}{4} = \frac{9}{32}$.

Figure 4.A

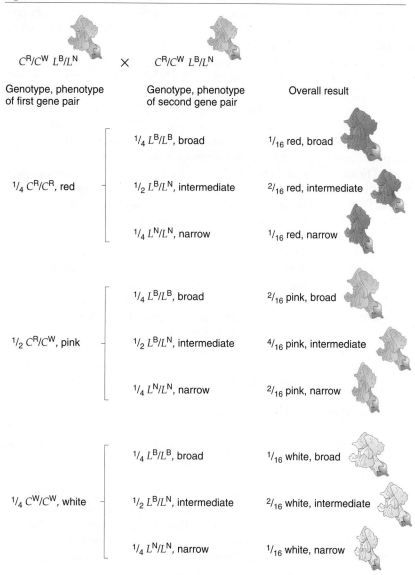

$C^R/C^W\ L^B/L^N$ × $C^R/C^W\ L^B/L^N$

Genotype, phenotype of first gene pair

Genotype, phenotype of second gene pair

Overall result

$1/4\ C^R/C^R$, red

$1/4\ L^B/L^B$, broad — $1/16$ red, broad

$1/2\ L^B/L^N$, intermediate — $2/16$ red, intermediate

$1/4\ L^N/L^N$, narrow — $1/16$ red, narrow

$1/2\ C^R/C^W$, pink

$1/4\ L^B/L^B$, broad — $2/16$ pink, broad

$1/2\ L^B/L^N$, intermediate — $4/16$ pink, intermediate

$1/4\ L^N/L^N$, narrow — $2/16$ pink, narrow

$1/4\ C^W/C^W$, white

$1/4\ L^B/L^B$, broad — $1/16$ white, broad

$1/2\ L^B/L^N$, intermediate — $2/16$ white, intermediate

$1/4\ L^N/L^N$, narrow — $1/16$ white, narrow

Q4.3

a. An $F_1 \times F_1$ self gives a 9:7 phenotypic ratio in the F_2. What phenotypic ratio would you expect if you test-crossed the F_1?

b. Answer the same question for an $F_1 \times F_1$ cross that gives a 9:3:4 ratio.

c. Answer the same question for a 15:1 ratio.

A4.3   This question deals with epistatic effects. In answering the question, we must consider the interaction between the different genotypes in order to proceed with the testcross. Let us set up the general genotypes that we will deal with throughout. The simplest are allelic pairs a^+ and a and b^+ and b, where the wild-type alleles are completely dominant to the other member of the pair.

a. A 9:7 ratio in the F_2 generation implies that both members of the F_1 cross are double heterozygotes and that epistasis is involved. Essentially, any genotype with a homozygous recessive condition has the same phenotype, so the 3, 3, and 1 parts of a 9:3:3:1 ratio are phenotypically combined into one class. In terms of genotype, $9/16$ are $a^+/-\ b^+/-$ types, and the other $7/16$ are $a^+/-\ b/b$, $a/a\ b^+/-$, and $a/a\ b/b$. (As always, the use of the dash after a wild-type allele signifies that the same phenotype results, whether the missing allele is a wild type or a mutant.) The testcross asked for is $a^+/a\ b^+/b \times a/a\ b/b$, and we can predict a 1:1:1:1 ratio of $a^+/a\ b^+/b : a^+/a\ b/b : a/a\ b^+/b : a/a\ b/b$. The first genotype will have the same phenotype as the $9/16$ class of the F_2 generation, but because of epistasis, the other three genotypes will have the same phenotype as the $\frac{7}{16}$ class of the F_2. In sum, the answer is a phenotypic ratio of 1:3 in the progeny of a testcross of the F_1.

b. We are asked to answer the same question for a 9:3:4 ratio in the F_2. Again, this question involves a modified dihybrid ratio, where two classes of the 9:3:3:1 have the same phenotype. Complete dominance for each of the two gene pairs occurs here also, so the F_1 individuals are $a^+/a\ b^+/b$. Perhaps both the $a^+/-\ b^+/b$ and $a/a\ b/b$ classes in the F_2 will have the same phenotype, whereas the $a^+/-\ b^+/-$ and $a/a\ b^+/-$ classes will have phenotypes distinct from each other and from the interaction class. The genotypic ratio of a testcross of the F_1 is the same as in part (a) of this question. Considering them in the same order as we did in part (a), the second and fourth classes would have the same phenotype because of epistasis. So there are only three possible phenotypic classes instead of the four found in the testcross of a dihybrid F_1, where there is complete dominance and no interaction. The phenotypic ratio here is 1:1:2.

c. This question is yet another example of epistasis. Since 15 + 1 = 16, this number gives the outcome of an F_1 self of a dihybrid where there is complete dominance for each gene pair and interaction between the dominant alleles. In this case, the $a^+/-\ b^+/-$, $a^+/-\ b/b$, and $a/a\ b^+/-$ classes have one phenotype and include $\frac{15}{16}$ of the F_2 progeny, and the $a/a\ b/b$ class has the other phenotype and $\frac{1}{16}$ of the F_2 progeny. The genotypic results of a testcross of the F_1 are the same as in parts (a) and (b) of this question; that is, the F_2 progeny exhibit a 1:1:1:1 ratio of $a^+/a\ b^+/b : a^+/a\ b/b : a/a\ b^+/b : a/a\ b/b$. The first three classes have the same phenotype, which is the same as that of the $\frac{15}{16}$ class of the F_2 generation, and the last class has the other phenotype. The answer, then, is a 3:1 phenotypic ratio.

Questions and Problems

4.1 In rabbits, C = agouti coat color, c^{ch} = chinchilla, c^h = Himalayan, and c = albino. The four alleles constitute a multiple allelic series. The agouti C is dominant to the three other alleles, c is recessive to all three other alleles, and chinchilla is dominant to Himalayan. Determine the phenotypes of progeny from the following crosses.
a. $C/C \times c/c$
b. $C/c^{ch} \times C/c$
c. $C/c \times C/c$
d. $C/c^h \times c^h/c$
e. $C/c^h \times c/c$

*4.2 If a given population of diploid organisms contains only three alleles of a particular gene (say w, $w1$, and $w2$), how many different diploid genotypes are possible in the populations? List all possible genotypes of diploids (consider *only* these three alleles).

4.3 The genetic basis of the ABO blood types seems most likely to be
a. multiple alleles
b. polyexpressive hemizygotes
c. allelically excluded alternates
d. three independently assorting genes

4.4 In humans, the three alleles I^A, I^B, and i constitute a multiple allelic series that determines the ABO blood group system, as described in this chapter. For the following problems, state whether the child mentioned can actually be produced from the marriage. Explain your answer.
a. an O child from the marriage of two A individuals
b. an O child from the marriage of an A to a B
c. an AB child from the marriage of an A to an O
d. an O child from the marriage of an AB to an A
e. an A child from the marriage of an AB to a B

4.5 A man is blood type O,M. A woman is blood type A,M, and her child is type A,MN. The man cannot be the father of this child because
a. O men cannot have type A children
b. O men cannot have MN children
c. an O man and an A woman cannot have an A child
d. an M man and an M woman cannot have an MN child

*4.6 A woman of blood group AB marries a man of blood group A whose father was group O. What is the probability that
a. their two children will both be group A?
b. one child will be group B and the other group O?
c. the first child will be a son of group AB and the second child a son of group B?

4.7 If a mother and her child belong to blood group O, what blood group could the father *not* belong to?

*4.8 A man of what blood group could *not* be the father of a child of blood type AB?

4.9 In snapdragons, red flower color (C^R) is incompletely dominant to white (C^W); the C^R/C^W heterozygotes are pink. A red-flowered snapdragon is crossed with a white-flowered one. Determine the flower color of
a. the F_1
b. the F_2
c. the progeny of a cross of the F_1 to the red parent
d. the progeny of a cross of the F_1 to the white parent

*4.10 In shorthorn cattle, the heterozygous condition of the alleles for red coat color (C^R) and white coat color (C^W) is roan coat color. If two roan cattle are mated, what proportion of the progeny will resemble their parents in coat color?

4.11 What progeny will a roan shorthorn have if bred to
a. a red?
b. a roan?
c. a white?

*4.12 In peaches, fuzzy skin (F) is completely dominant to smooth (nectarine) skin (f), and the heterozygous condition of oval glands at the base of the leaves (G^O) and no glands (G^N) gives round glands. A homozygous fuzzy, no-gland peach variety is bred to a smooth, oval-gland variety.
a. What will be the appearance of the F_1 peaches?
b. What will be the appearance of the F_2 peaches?
c. What will be the appearance of the offspring of a cross of the F_1 back to the smooth, oval-gland parent?

4.13 In guinea pigs, short hair (L) is dominant to long hair (l), and the heterozygous condition of yellow coat (C^Y) and white coat (C^W) gives cream coat. A short-haired cream guinea pig is bred to a long-haired white guinea pig, and a long-haired cream baby guinea pig is produced. When the baby grows up, it is bred back to the short-haired cream parent. What phenotypic classes and in what proportions are expected among the offspring?

4.14 The shape of radishes may be long (S^L/S^L), oval (S^L/S^S), or round (S^S/S^S), and the color of radishes may be red (C^R/C^R), purple (C^R/C^W), or white (C^W/C^W). If a long red radish plant is crossed with a round white plant, what will be the appearance of the F_1 and the F_2 plants?

4.15 In poultry, the dominant alleles for rose comb (R) and pea comb (P), if present together, give walnut comb. The recessive alleles of each gene, when present together in a homozygous state, give single comb. What will be the comb characters of the offspring of the following crosses?
a. $R/R\ P/p \times r/r\ P/p$
b. $r/r\ P/P \times R/r\ P/p$
c. $R/r\ p/p \times r/r\ P/p$

4.16 For the following crosses involving the comb character in poultry, determine the genotypes of the two parents.
a. A walnut crossed with a single produces offspring that are ¼ walnut, ¼ rose, ¼ pea, and ¼ single.
b. A rose crossed with a walnut produces offspring that are ⅜ walnut, ⅜ rose, ⅛ pea, and ⅛ single.
c. A rose crossed with a pea produces five walnut and six rose offspring.
d. A walnut crossed with a walnut produces one rose, two walnut, and one single offspring.

4.17 In poultry, feathered (F) shanks (part of the legs) are dominant to clean (f), and white plumage of white leghorns (I) is dominant to black (i). Comb phenotypes and genotypes are given in Figure 4.4.

a. A feathered-shanked, white, rose-combed bird crossed with a clean-shanked, white, walnut-combed bird produces these offspring: 2 feathered, white, rose; 4 clean, white, walnut; 3 feathered, black, pea; 1 clean, black, single; 1 feathered, white, single; 2 clean, white, rose. What are the genotypes of the parents?
b. A feathered-shanked, white, walnut-combed bird crossed with a clean-shanked, white, pea-combed bird produces a single offspring that is clean-shanked, black, and single-combed. In additional offspring from this cross, what proportion may be expected to resemble each parent?

*4.18 F_2 plants segregate ⁹/₁₆ colored : ⁷/₁₆ colorless. If a colored plant from the F_2 generation is chosen at random and selfed, what is the probability that there will be *no* segregation of the two phenotypes among its progeny?

*4.19 In peanuts, a plant may be either "bunch" or "runner." Two different strains of peanut, V4 and G2, in which "bunch" occurred were crossed, with the following results:

V4 bunch × V4 bunch
↓
all bunch
G2 bunch × G2 bunch
↓
all bunch

The two true-breeding strains of bunch were crossed in the following way:

V4 bunch × G2 bunch
↓
F_1 runner
F_1 × F_1
↓
F_2 9 runner : 7 bunch

What is the genetic basis of the inheritance pattern of runner and bunch in the F_2 peanuts?

*4.20 In rabbits, one enzyme (the product of a functional gene A) is needed to produce a substance required for hearing. Another enzyme (the product of a functional gene B) is needed to produce another substance required for hearing. The genes responsible for the two enzymes are not linked. Individuals homozygous for either one or both of the nonfunctional recessive alleles, a or $b,$ are deaf.
a. If a large number of matings were made between two double heterozygotes, what phenotypic ratio would be expected in the progeny?
b. This phenotypic ratio is a result of what well-known phenomenon?

c. What phenotypic ratio would be expected if rabbits homozygous recessive for trait A and heterozygous for trait B were mated to rabbits heterozygous for both traits?

4.21 In Doodlewags (hypothetical creatures), the dominant allele *S* causes solid coat color; the recessive allele *s* results in white spots on a colored background. The black coat color allele *B* is dominant to the brown allele *b*, but these genes are expressed only in the genotype *a/a*. Individuals that are *A/–* are yellow regardless of *B* alleles. Six pups are produced in a mating between a solid yellow male and a solid brown female. Their phenotypes are 2 solid black, 1 spotted yellow, 1 spotted black, and 2 solid brown.

a. What are the genotypes of the male and female parents?

b. What is the probability that the next pup will be spotted brown?

4.22 The allele *l* in *Drosophila* is recessive and sex-linked and lethal when homozygous or hemizygous (the condition in the male). If a female of genotype *L/l* is crossed with a normal male, what is the probability that the first two surviving progeny will be males?

*4.23 A locus in mice is involved with pigment production; when parents heterozygous at this locus are mated, $\frac{3}{4}$ of the progeny are colored and $\frac{1}{4}$ are albino. Another phenotype concerns coat color; when two yellow mice are mated, $\frac{2}{3}$ of the progeny are yellow and $\frac{1}{3}$ are agouti. The albino mice cannot express whatever alleles they may have at the independently assorting agouti locus.

a. When yellow mice are crossed with albinos, they produce F_1 mice consisting of $\frac{1}{2}$ albino, $\frac{1}{3}$ yellow, and $\frac{1}{6}$ agouti. What are the probable genotypes of the parents?

b. If yellow F_1 mice are crossed among themselves, what phenotypic ratio would you expect among the progeny? What proportion of the yellow progeny produced here would be expected to breed true?

4.24 In *Drosophila melanogaster*, a recessive autosomal allele, ebony (*e*), produces a black body color when homozygous, and an independently assorting autosomal allele, black (*b*), also produces a black body color when homozygous. Flies with genotypes *e/e b+/–*, *e+/– b/b*, and *e/e b/b* are phenotypically identical with respect to body color. Flies with genotype *e+/– b+/–* have a grey body color. True-breeding *e/e b+/b+* ebony flies are crossed with true-breeding *e+/e+ b/b* black flies.

a. What will be the phenotype of the F_1 flies?

b. What phenotypes and what proportions would occur in the F_2 generation?

c. What phenotypic ratios would you expect to find in the progeny of these backcrosses?
i. $F_1 \times$ true-breeding ebony
ii. $F_1 \times$ true-breeding black

*4.25 In four-o'clock plants, two genes, *Y* and *R*, affect flower color. Neither is completely dominant, and the two interact on each other to produce seven different flower colors:

Y/Y R/R = crimson	*Y/y R/R* = magenta
Y/Y R/r = orange-red	*Y/y R/r* = magenta-rose
Y/Y r/r = yellow	*Y/y r/r* = pale yellow
y/y R/R, y/y R/r, and *y/y r/r* = white	

a. In a cross of a crimson-flowered plant with a white one (*y/y r/r*), what will be the appearance of the F_1 plants, the F_2 plants, and the offspring of the F_1 backcrossed to the crimson parent?

b. What will be the flower colors in the offspring of a cross of orange-red × pale yellow?

c. What will be the flower colors in the offspring of a cross of a yellow with a *y/y R/r* white?

4.26 Two four-o'clock plants were crossed and gave the following offspring: $\frac{1}{8}$ crimson, $\frac{1}{8}$ orange-red, $\frac{1}{4}$ magenta, $\frac{1}{4}$ magenta-rose, and $\frac{1}{4}$ white. Unfortunately, the person who made the crosses was color-blind and could not record the flower colors of the parents. From the results of the cross, deduce the genotypes and flower colors of the two parents.

*4.27 Genes *A*, *B*, and *C* are independently assorting and control production of a black pigment.

a. Assume that *A*, *B*, and *C* act in a pathway as follows:

$$\text{colorless} \xrightarrow{A} \xrightarrow{B} \xrightarrow{C} \text{black}$$

The alternative alleles that give abnormal functioning of these genes are designated *a*, *b*, and *c*, respectively. A black *A/A B/B C/C* is crossed with a colorless *a/a b/b c/c* to give a black F_1. The F_1 is selfed. What proportion of the F_2 is colorless? (Assume that the products of each step except the last are colorless, so only colorless and black phenotypes are observed.)

b. Assume that *C* produces an inhibitor that prevents the formation of black by destroying the ability of *B* to carry out its function, as follows:

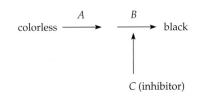

A colorless *A/A B/B C/C* individual is crossed with a colorless *a/a b/b c/c*, giving a colorless F$_1$. The F$_1$ is selfed to give an F$_2$. What is the ratio of colorless to black in the F$_2$? (Only colorless and black phenotypes are observed, as in part [a].)

4.28 In *Drosophila,* a mutant strain has plum-colored eyes. A cross between a plum-eyed male and a plum-eyed female gives ⅔ plum-eyed and ⅓ red-eyed (wild-type) progeny flies. A second mutant strain of *Drosophila,* called stubble, has short bristles instead of the normal long bristles. A cross between a stubble female and a stubble male gives ⅔ stubble and ⅓ normal-bristled flies in the offspring. Assuming that the plum gene assorts independently from the stubble gene, what will be the phenotypes and their relative proportions in the progeny of a cross between two plum-eyed, stubble-bristled flies? (Both genes are autosomal.)

*4.29 In sheep, white fleece (*W*) is dominant over black (*w*), and horned (*H*) is dominant over hornless (*h*) in males but recessive in females. If a homozygous horned white ram is bred to a homozygous hornless black ewe, what will be the appearances of the F$_1$ and F$_2$ sheep?

4.30 A horned black ram bred to a hornless white ewe produces the following offspring: Of the males, ¼ are horned and white; ¼ are horned and black; ¼ are hornless and white; and ¼ are hornless and black. Of the females, ½ are hornless and black, and ½ are hornless and white. What are the genotypes of the parents?

The fruit fly, *Drosophila melanogaster*, with the *vestigial* wing mutation.

Gene Mapping in Eukaryotes

PRINCIPAL POINTS

Genetic recombinants result from physical exchanges between homologous chromosomes in meiosis. A chiasma is the site of crossing-over, and crossing-over is the reciprocal exchange of chromosome parts at corresponding positions along homologous chromosomes by symmetrical breakage and rejoining.

Crossing-over in eukaryotes occurs at the four-chromatid stage in prophase I of meiosis.

The map distance between genes is measured in map units (mu); 1 mu is defined as the interval in which 1 percent crossing-over takes place. However, gene mapping crosses produce data in the form of recombination frequencies, which are used to estimate the map distance where, in this case, 1 mu is equivalent to a recombination frequency of 1 percent.

As the distance between genes increases, the incidence of multiple crossovers causes the recombination frequency to be an underestimate of the crossover frequency and hence of the true map distance. Mapping functions can be used to correct for this problem and thereby give a more accurate estimate of map distance.

The occurrence of a chiasma between two chromatids may physically impede the occurrence of a second chiasma nearby, a phenomenon called chiasma interference.

Tetrad analysis is a mapping technique that can be used to map the genes of certain haploid eukaryotic organisms in which the products of a single meiosis, the meiotic tetrad, are contained within a single structure. In these situations, map distance between genes is computed by analyzing the relative proportion of tetrad types rather than by analyzing individual progeny.

iActivity

i THE HUMAN GENOME PROJECT MAY BE THE BEST-known gene-mapping project in the world. The aim of the project is to determine both the location of all the genes in the human genome and the exact nucleotide sequence that comprises the 3 billion nucleotide pairs that make up the genome. But long before the development of the recombinant DNA technologies that allowed the Human Genome Project to come into being, scientists were creating genetic maps of eukaryotic organisms. What do these maps tell us? How are they constructed? How can they be used? After you have read this chapter, you can further explore the answer to these and other questions by trying the iActivity.

Genes on nonhomologous chromosomes assort independently during meiosis. In many instances, however, certain genes (and hence the phenotypes they control) are inherited together because they are located on the same chromosome. Genes on the same chromosome are said to be *syntenic*. Genes that do not appear to assort independently exhibit **linkage** and are called **linked genes**. Linked genes belong to a *linkage group*.

Genetic analysis is the dissection of the structure and function of the genetic material. In classic genetic analysis, progeny from crosses between parents with different genetic characters are analyzed to determine the frequency with which differing parental alleles are associated in new combinations. Progeny showing the parental combinations of alleles are called *parentals*, and progeny showing nonparental combinations of alleles are called *recombinants*. The process by which the recombinants are produced is called **genetic recombination.** Through testcrosses, we can determine which genes are linked to each other and can then construct a *linkage map*, or *genetic map*, of each chromosome.

Classic genetic mapping has provided information that is useful in many aspects of genetic analysis. For example, knowing the locations of genes on chromosomes has been useful in recombinant DNA research and in experiments directed toward understanding the DNA sequences in and around genes. These days, the focus of mapping studies is on constructing genetic maps of genomes using both gene markers and DNA markers. A *marker*, or *genetic marker*, is a mutation that gives a distinguishable phenotype; in other words, it is an allele that marks a chromosome or a gene. **Gene markers** are alleles of the kind we have discussed to this point in the text. **DNA markers** are molecular markers—that is, DNA regions in the genome that differ sufficiently between individuals so that they can be detected by molecular analysis of DNA. The goal of genome mapping studies is to generate high-resolution maps of chromosomes, which are useful for investigating genes and their functions. The ultimate genetic maps will be of the base pair sequence of organisms' genomes (see Chapter 15).

Your goal in this chapter is to learn how genetic linkage affects Mendelian gene segregation patterns and how genes are mapped in eukaryotes. You will also be introduced to tetrad analysis, a specialized mapping technique for studying gene linkage relationships in certain haploid eukaryotes in which the products of a single meiosis, the meiotic tetrad, are contained within a single structure that is readily analyzable.

Discovery of Genetic Linkage

Our modern understanding of genetic linkage comes from the work of Thomas Hunt Morgan and his colleagues with linkage in *Drosophila melanogaster*, done around 1911.

Morgan's Linkage Experiments with *Drosophila*

By 1911, Morgan had identified a number of X-linked genes, including *w* (white eye) and *m* (miniature wing; the wing is smaller than normal). Morgan crossed a female white miniature (*w m/w m*) fly with a wild-type male (*w+ m+*/Y) (Figure 5.1). For the former genotype, the slash signifies the pair of homologous chromosomes and indicates that the genes on either side of the slash are linked. For the latter genotype, because the genes are X-linked, a slash indicates the X chromosome and the Y indicates a Y chromosome. We will also use another special genetic symbolism for genes on the same chromosome: $\frac{a\,b}{a\,b}$ signifies that genes *a* and *b* are on the same chromosome, with the chromosome signified by the horizontal line. This particular genotype is homozygous for *a* and *b*. With this system, X-linked genes in a female are indicated by allele symbols separated by one or two continuous lines to indicate the homologous chromosomes: for example,

$$\frac{w\,m}{w\,m} \quad \text{or} \quad \frac{w\,m}{w\,m}$$

and X-linked genes in a male are shown, for example, as

$$\underrightarrow{w\,m}$$

where the straight line indicates the X chromosome and the bent line indicates the Y chromosome. This genotype representation is the same as *w m*/∧ (where the bent slash is the Y chromosome) or *w m*/Y. (Note: If a discontinuous line is used between a series of allele pairs, the extent of each segment signifies a different chromosome.)

In the cross, the F$_1$ males were white-eyed and had miniature wings (genotype *w m*/Y), whereas all females were heterozygous and wild-type for both eye color and wing size (genotype *w+ m+/w m*). The F$_1$ flies were interbred, and 2,441 F$_2$ flies were analyzed. In crosses of X-linked genes set up as in Figure 5.1, the F$_1$ × F$_1$ is

Figure 5.1

Morgan's experimental crosses of white-eye and miniature-wing variants of *Drosophila melanogaster*, **showing evidence of linkage and recombination in the X chromosome.**

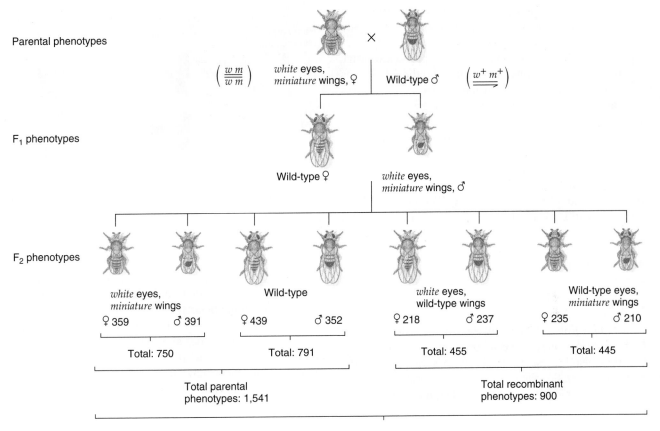

Parental phenotypes

$\left(\dfrac{w\ m}{w\ m}\right)$ *white* eyes, *miniature* wings, ♀   Wild-type ♂ $\left(\dfrac{w^+\ m^+}{}\right)$

F₁ phenotypes

Wild-type ♀   *white* eyes, *miniature* wings, ♂

F₂ phenotypes

white eyes, *miniature* wings
♀ 359   ♂ 391
Total: 750

Wild-type
♀ 439   ♂ 352
Total: 791

white eyes, wild-type wings
♀ 218   ♂ 237
Total: 455

Wild-type eyes, *miniature* wings
♀ 235   ♂ 210
Total: 445

Total parental phenotypes: 1,541

Total recombinant phenotypes: 900

Total progeny: 1,541 + 900 = 2,441

Percent recombinants: $^{900}/_{2,441} \times 100 = 36.9$

equivalent to doing a testcross because the F₁ males produce X-bearing gametes with recessive alleles of both genes and Y-bearing gametes that have no alleles for the genes being studied. In the F₂, the most frequent phenotypic classes in both sexes were the *grandparental phenotypes* of white eyes plus miniature wings or normal red eyes plus large wings. Conventionally, we call the original genotypes of the two chromosomes **parental genotypes, parental classes,** or, more simply, **parentals.** The term is also used to describe phenotypes, so the original white miniature females and wild-type males in these particular crosses are defined as the parentals.

Morgan observed that 900 of the 2,441 F₂ flies, or 36.9 percent, had nonparental phenotypic combinations of white eyes plus normal wings and red eyes plus miniature wings. Nonparental combinations of linked genes are called **recombinants.** Fifty percent recombinant phenotypes is expected if independent assortment is the case; thus, the lower percentage observed is evidence for linkage of the two genes. To explain the recombinants, Morgan proposed that in meiosis, exchanges of genes had occurred between the two X chromosomes of the F₁ females.

Morgan's group analyzed a large number of other crosses of this type. *In each case, the parental phenotypic classes were the most frequent, while the recombinant classes occurred much less frequently.* Approximately equal numbers of each of the two parental classes were obtained, and similar results were obtained for the recombinant classes. Morgan's general conclusion was that *during meiosis, alleles of some genes assort together because they lie near each other on the same chromosome.* To turn this around, the closer two genes are on the chromosome, the more likely they are to remain together during meiosis. This is because the recombinants are produced as a result of crossing-over between homologous chromosomes during meiosis and the closer two genes are together the less likely there will be a recombination event between them.

The terminology related to the physical exchange of homologous chromosome parts can be confusing. To clarify:

1. A chiasma (plural, *chiasmata*; see Figure 1.21) is the place on a homologous pair of chromosomes at which a physical exchange is occurring; it is the site of crossing-over.
2. Crossing-over is the process of reciprocal exchange of chromatid segments at corresponding positions along homologous chromosomes; the process involves symmetrical breakage of two chromatids and rejoining.
3. Crossing-over is also defined as the events leading to genetic recombination between linked genes in both prokaryotes and eukaryotes.

Figure 5.2 presents a very simplified diagram of the process of crossing-over. Crossing-over occurs at the four-chromatid stage in prophase I of meiosis. Each crossover involves two of the four chromatids, and along the length of a chromosome, all chromatids can be involved in crossing-over.

KEYNOTE

> The production of genetic recombinants results from physical exchanges between homologous chromosomes during meiotic prophase I. A chiasma is the site of crossing-over. Crossing-over is the reciprocal exchange of chromosome parts at corresponding positions along homologous chromosomes by breakage and rejoining. Crossing-over is also used to describe the events leading to genetic recombination between linked genes. Crossing-over in eukaryotes takes place at the four-chromatid stage in prophase I of meiosis.

Constructing Genetic Maps

We have learned that the number of genetic recombinants produced is characteristic of the two linked genes involved. We now examine how genetic experiments can be used in **genetic mapping,** the process of determining the relative position of genes on chromosomes.

Detecting Linkage Through Testcrosses

Before beginning experiments to construct a **genetic map** of the relative positions of genes on a chromosome (also called a **linkage map**), geneticists must show that the genes under consideration are linked. Unlinked genes assort independently. Therefore, a way to test for linkage is to analyze the results of crosses to see whether the data deviate significantly from those expected by independent assortment.

The best cross to use to test for linkage is the test-cross, a cross of an individual with another individual

Figure 5.2

Mechanism of crossing-over. A highly simplified diagram of a crossover between two nonsister chromatids during meiotic prophase, giving rise to recombinant (nonparental) combinations of linked genes.

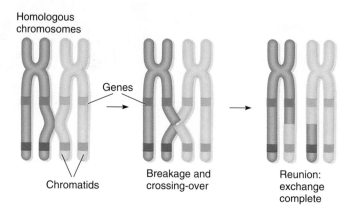

homozygous recessive for all genes involved. We saw in Chapter 2 that a testcross between $a^+/a\ b^+/b$ and $a/a\ b/b$, where genes a and b are unlinked, gives a 1:1:1:1 ratio of the four possible phenotypic classes $a^+\ b^+ : a^+\ b : a\ b^+ : a\ b$. A significant deviation from this ratio in the direction of too many parental types and too few recombinant types therefore suggests that the two genes are linked. It is important to know how large a deviation must be to be considered significant. The *chi-square test* can be used to make such a decision (see Chapter 2, pp. 43–44). Here we illustrate the use of the chi-square test for analyzing testcross data.

Consider data from a testcross involving fruit flies. In *Drosophila*, b is a recessive autosomal mutation that results in black body color, and vg is a recessive autosomal mutation that results in vestigial (short, crumpled) wings. Wild-type flies have grey bodies and long, uncrumpled (normal) wings. True-breeding black, normal ($b/b\ vg^+/vg^+$) flies were crossed with true-breeding grey, vestigial ($b^+/b^+\ vg/vg$) flies. F_1 grey, normal ($b^+/b\ vg^+/vg$) female flies were testcrossed to black, vestigial ($b/b\ vg/vg$) male flies. (The female is the heterozygote in this testcross because in *Drosophila*, no crossing-over occurs between *any* homologous pair of chromosomes in males.) The testcross progeny data were

	283 grey, normal
	1,294 grey, vestigial
	1,418 black, normal
	241 black, vestigial
Total	3,236 flies

We hypothesize that the two genes are unlinked (null hypothesis) and use the chi-square test to test the hypothesis, as shown in Table 5.1. We use this null hypothesis because the hypothesis must be testable; that is, we must be able to make meaningful predictions. A

Table 5.1	Chi-Square Test Used with Testcross Data to Test the Hypothesis That Two Genes Are Unlinked

(1) Phenotypes	(2) Observed Number (*o*)	(3) Expected Number (*e*)	(4) *d* (*o* − *e*)	(5) *d*²	(6) *d*²/*e*
Parentals: (black, normal and grey, vestigial)	2,712	1,618	1,094	1,196,836	739.7
Recombinants (black, vestigial and grey, normal)	524	1,618	−1,094	1,196,836	739.7
Total	3,236	3,236			1,479.4

(7) $\chi^2 = 1{,}479.4$ (8) df 1

hypothesis that two genes are linked is not testable because we cannot predict what progeny ratios would be.

If the two genes are unlinked, then a testcross should result in a 1:1 ratio of parentals : recombinants. Column 1 lists the parental and recombinant phenotypes expected in the progeny of the cross; column 2 lists the observed (*o*) numbers; and column 3 lists the expected (*e*) numbers for the parentals and recombinants, given the total number of progeny (3,236) and the hypothesis being tested (1:1 in this case). Column 4 lists the deviation value (*d*) calculated by subtracting the expected number (*e*) from the observed number (*o*) for each class. The sum of the *d* values is always zero.

Column 5 lists the deviation squared (d^2), and column 6 lists the deviation squared divided by the expected number (d^2/e). The chi-square value, χ^2 (item 7 in the table), is given by the formula

$$\chi^2 = \Sigma \frac{d^2}{e}$$

where $d^2 = (o - e)^2$ and Σ means "sum of."

In Table 5.1, χ^2 is the sum of the two values in column 6. In our example, $\chi^2 = 1{,}479.4$. The last value in the table, item 8, is the degrees of freedom (df) for the set of data; there are $n - 1 = 1$ degree of freedom in this case.

The chi-square value and the degrees of freedom are used with a table of chi-square probabilities (see Table 2.5) to determine the probability (*P*) that the deviation of the observed values from the expected values is due to chance. For $\chi^2 = 1{,}479.4$ with 1 degree of freedom, the *P* value is much lower than 0.001; in fact, it is not in the table. This is interpreted to mean that independent repetitions of this experiment would produce chance deviations from the expected as large as those observed in many fewer than 1 out of 1,000 trials. As a reminder, if the probability of obtaining the observed chi-square values is greater than 5 in 100 ($P > 0.05$), the deviation is considered not statistically significant and could have occurred by chance alone. If $P \le 0.05$, we consider the deviation

from the expected values statistically significant and not due to chance alone; the hypothesis may well be invalid. If $P \le 0.01$, the deviation is highly statistically significant; the data are not consistent with the hypothesis. Thus, in this case we would reject the independent assortment hypothesis. Genetically, the only alternative hypothesis that could logically apply is that the genes are linked.

The Concept of a Genetic Map.   In an individual doubly heterozygous for the *w* and *m* alleles, for example, the alleles can be arranged in two ways:

$$\frac{w^+ \, m^+}{w \, m} \quad \text{or} \quad \frac{w^+ \, m}{w \, m^+}$$

In the arrangement on the left, the two wild-type alleles are on one homologue and the two recessive mutant alleles are on the other homologue, an arrangement called **coupling** (or the *cis* configuration). Crossing-over between the two loci produces $w^+ \, m$ and $w \, m^+$ recombinants. In the arrangement on the right, each homologue carries the wild-type allele of one gene and the mutant allele of the other gene, an arrangement called **repulsion** (or the *trans* configuration). Crossing-over between the two genes produces $w^+ \, m^+$ and $w \, m$ recombinants.

The data obtained by Morgan from *Drosophila* crosses indicated that the frequency of crossing-over (and hence of recombinants) for linked genes is characteristic of the gene pairs involved: For the X-linked genes white (*w*) and miniature (*m*), the recombination frequency is 36.9 percent. Moreover, the recombination frequency for two linked genes is the same, regardless of whether the alleles of the two genes involved are in coupling or in repulsion. *Although the actual phenotypes of the recombinant classes are different for the two arrangements, the percentage of recombinants among the total progeny will be the same in each case (within experimental error).*

In 1913, a student of Morgan's, Alfred Sturtevant, suggested that the recombination frequencies could be used as a quantitative measure of the genetic distance between two genes on a genetic map. The genetic distance between genes is measured in **map units (mu)**, where 1 map unit is defined as the interval in which 1 percent crossing-over takes place. The map unit is sometimes called a **centimorgan (cM)** in honor of Morgan. It is important to note that for a pair of linked genes, crossover frequency is *not* the same as recombination frequency. The former refers to the frequency of physical exchanges between chromosomes in meiosis for the region between the genes, and the latter refers to the frequency of recombination of genetic markers in a cross as determined by analyzing the phenotypes of the progeny. Geneticists follow genetic markers in crosses, so the data obtained are in the form of recombination frequencies. In our discussions, then, we will use recombination frequencies as geneticists often do: as working estimates of map distances between genes, where a map unit is equivalent to a recombination frequency of 1 percent. Later we will discuss how such data relate to crossover frequencies and therefore to true map units.

The genes on a chromosome, then, can be represented by a one-dimensional genetic map that shows in linear order the genes belonging to the chromosome. Crossover and recombination values give the linear order of the genes on a chromosome and provide information about the genetic distance between any two genes. The farther apart two genes are, the greater will be the *crossover frequency*. Thus, in Figure 5.3, the probability of recombination occurring between genes *A* and *B* is much less than between genes *B* and *C*, because *A* and *B* are closer together than are *B* and *C*. The first genetic map ever constructed was based on *recombination frequencies* from *Drosophila* crosses involving the sex-linked genes *w*, *m*, and *y*, where *w* gives white eyes, *m* gives miniature wings, and *y* gives yellow body. From these mapping experiments, the recombination frequencies for the $w \times m$, $w \times y$, and $m \times y$ crosses were established as 32.6, 1.3, and 33.9 percent, respectively. (Note that in this independent experiment, the recombination frequency for *w* and *m* is a little lower than in the experiment discussed previously in this chapter on p. 101.) The percentages are quantitative measures of the distances between the genes involved.

We can construct a genetic map based on the recombination frequency data. The recombination frequencies show that *w* and *y* are closely linked and that *m* is quite far from the other two genes. Since the $w - m$ genetic distance is less than the $y - m$ distance (as shown by the smaller recombination frequency in the $w \times m$ cross), the order of genes must be *y w m* (or *m w y*); thus, the three genes are ordered and spaced with 1.3 mu between *y* and *w* and 32.6 mu between *w* and *m*:

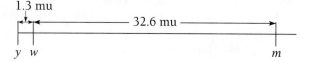

Figure 5.3

The relationship between recombination and map distance. The farther apart two genes are, the greater the number of possible sites for recombination. Thus, the probability of recombination occurring between genes *A* and *B* is much less than that between genes *B* and *C*. The percentage of recombinants can provide information about the relative genetic distance between two linked genes.

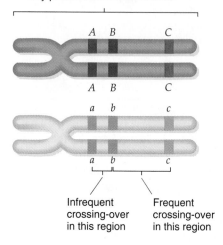

Crossing-over may occur at any point on chromosome arms

Infrequent crossing-over in this region

Frequent crossing-over in this region

Gene Mapping Using Two-Point Testcrosses

We have seen that recombination frequency may be used to obtain an estimate of the genetic distance between two linked genes. By carrying out two-point testcrosses such as those shown in Figure 5.4, we can determine the relative numbers of parental and recombinant classes in the progeny. For autosomal recessives (as in Figure 5.4), a double heterozygote is crossed with a doubly homozygous recessive mutant strain. When the double heterozygous $a^+ b^+/a\,b$ F_1 progeny from a cross of $a^+ b^+/a^+ b^+$ with $a\,b/a\,b$ are testcrossed with $a\,b/a\,b$, four phenotypic classes are found among the F_2 progeny. Two of these classes have the parental phenotypes $a^+ b^+$ and $a\,b$. Since both classes result from chromosomes that have not crossed over between these genes, approximately equal numbers of these two types are expected.

The other two F_2 phenotypic classes have recombinant phenotypes $a^+ b$ and $a\,b^+$, which come from diploids in which a single crossover occurred between the chromosomes. We expect approximately equal numbers of these two recombinant classes. Because a single crossover event occurs more rarely than no crossing-over, an excess of parental phenotypes over recombinant phenotypes in the progeny of a testcross indicates linkage between the genes involved.

Two-point testcrosses for the purposes of mapping are set up in similar ways for genes showing other mechanisms of inheritance. For autosomal dominants, the double heterozygous individuals are testcrossed with an

Figure 5.4

Testcross to show that two genes are linked. Genes *a* and *b* are recessive mutant alleles linked on the same autosome. A homozygous $a^+ b^+/a^+ b^+$ individual is crossed with a homozygous recessive $a b/a b$ individual, and the doubly heterozygous F$_1$ progeny ($a^+ b^+/a b$) are testcrossed with homozygous $a b/a b$ individuals.

Recessive linked autosomal mutant alleles

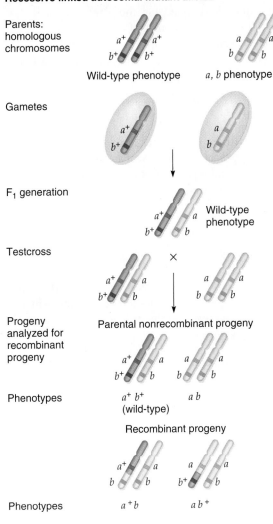

$$\frac{A\ B}{A^+ B^+} \times \xrightarrow{A^+ B^+}$$

For both X-linked cases, the females can be crossed with males of any genotype. If only male progeny are analyzed, the contribution of the male's X chromosome to the progeny is ignored.

In all cases, a two-point testcross should yield a pair of parental types that occur with about equal frequencies and a pair of recombinant types that also occur with about equal frequencies. Of course, the actual phenotypes depend on the relative arrangement of the two allelic pairs in the homologous chromosomes, that is, whether they are in coupling (*cis*) or repulsion (*trans*). To get a count of the representatives of each progeny class (this will give the recombination frequency), the following formula is used:

$$\frac{\text{number of recombinants}}{\text{total number of testcross progeny}} \times 100\% = \frac{\text{recombination}}{\text{frequency}}$$

The recombination frequency is used directly as an estimate of map units.

The two-point method of mapping is most accurate when the two genes examined are close together; when genes are far apart, there are inaccuracies, as we will see later. Large numbers of progeny must also be counted (scored) to ensure a high degree of accuracy. From mapping experiments carried out in all types of organisms, we know that genes are linearly arranged in linkage groups. There is a one-to-one correspondence of linkage groups and chromosomes, so the sequence of genes on the linkage group reflects the sequence of genes on the chromosome.

Generating a Genetic Map

We can now discover how a genetic map is generated from estimating the number of times crossing-over occurred in a particular segment of the chromosome out of all meioses examined. In many cases, the probability of a crossing-over event is not uniform along a chromosome, so we must be cautious about how far we extrapolate the genetic map (derived from data produced by genetic crosses) to the physical map of the chromosome (derived from determinations of the locations of genes along the chromosome itself—for example, from sequencing the DNA).

The recombination frequencies observed between genes may also be used to predict the outcome of genetic crosses. For example, a recombination frequency of 20 percent between genes indicates that for a doubly heterozygous genotype (such as $a^+ b^+/a b$), 20 percent of the gametes produced, on average, will be recombinants ($a^+ b$ and $a b^+$ for our example, 10 percent of each expected).

For any testcross, the recombination frequency in the progeny cannot exceed 50 percent. That is, if the genes are assorting independently, an equal number of

individual homozygous for the recessive alleles, which in this case are the wild-type alleles:

$$\frac{A\ B}{A^+ B^+} \times \frac{A^+ B^+}{A^+ B^+}$$

For X-linked recessives, a double heterozygous female is crossed with a hemizygous male carrying the recessive alleles:

$$\frac{a^+ b^+}{a\ b} \times \xrightarrow{a\ b}$$

And for X-linked dominants, a doubly heterozygous female is crossed with a male carrying wild-type alleles on the X chromosome:

recombinants and parentals are *expected* in the progeny, so the recombination frequency is 50 percent. If we get a recombination frequency of 50 percent from a cross, then, we state that the two genes are unlinked. Genes may be unlinked (i.e., show 50 percent recombination) in two ways. First, the genes may be on different chromosomes, a case we discussed before. Second, *the genes may be far apart on the same chromosome.*

The second case is illustrated in Figure 5.5, which shows the effects of single crossovers and double crossovers on the production of parental and recombinant chromosomes for two loci that are far apart on the same chromosome. (In reality, in such a situation, there would be multiple crossovers between the two loci in each meiosis.) Single crossovers between any pair of nonsister chromatids result in two parental and two recombinant chromosomes; that is, for two loci, 50 percent of the products are recombinant (see Figure 5.5a).

Double crossovers can involve two, three, or all four of the chromatids (see Figure 5.5b). For double crossovers involving the same two nonsister chromatids (a *two-strand double crossover*), all four resulting chromosomes are parental for the two loci of interest. For *three-strand double crossovers* (double crossovers involving three of the four chromatids), two parental and two recombinant chromosomes result. For a *four-strand double crossover,* all four resulting chromosomes are recombinant. Considering all possible double crossover patterns together, 50 percent of the products are recombinant for the two loci. Similarly, for any multiple number of crossovers between loci that are far apart, examination of a large number of meioses will show that 50 percent of the resulting chromosomes are recombinant. This is the reason for the recombination frequency limit of 50 percent exhibited by unlinked genes on the same chromosome.

When genes are more closely linked together, no crossovers occur between the two loci in some meioses, resulting in four parental chromosome products. In mapping linked genes, therefore, the map distance depends on the ratio of the meioses with no detectable crossovers to meioses with any number of crossovers between the loci.

The point is that if two genes show 50 percent recombination in a cross, they may not be on different chromosomes. More data would be needed to determine whether the genes are on the same chromosome or different chromosomes. One way to find out is to map a number of other genes in the linkage group. For example, if *a* and *m* show 50 percent recombination, perhaps we will find that *a* shows 27 percent recombination with *e*, and *e* shows 36 percent recombination with *m*. This result would indicate that *a* and *m* are in the same linkage group approximately 63 mu apart, as shown here:

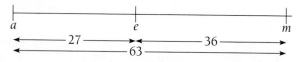

Gene Mapping Using Three-Point Testcrosses

Although genetic maps can be built using a series of two-point testcrosses, geneticists more typically have mapped several linked genes at a time in single testcrosses. Here we illustrate this more complex type of mapping analysis for three linked genes using a three-point testcross. In diploid organisms, the three-point testcross is a cross of a triple heterozygote with a triply homozygous recessive. If the mutant genes in the cross are all recessive, a typical three-point testcross might be

$$\frac{a^+\ b^+\ c^+}{a\ \ b\ \ c} \times \frac{a\ b\ c}{a\ b\ c}$$

If any of the mutant genes are dominant to the wild type, the *triply homozygous recessive parent in the testcross will carry the wild-type allele of these genes.* For example, a testcross of a strain heterozygous for two recessive mutations (*a* and *c*) and one dominant mutation (*B*) might be

$$\frac{a^+\ B^+\ c^+}{a\ \ B\ \ c} \times \frac{a\ B^+\ c}{a\ B^+\ c}$$

In a testcross involving sex-linked genes, the female is the heterozygous strain (assuming that the female is the homogametic sex) and the male is hemizygous for the recessive alleles.

Three-point mapping is also done in haploid eukaryotic organisms, but in this case a testcross is not needed. Thus, in yeast we might cross two haploid strains to generate the triply heterozygous diploid cell. A cross of an *a b c* strain with an *a⁺ b⁺ c⁺* strain, for instance, would give an *a⁺ b⁺ c⁺/a b c* diploid cell. That cell goes through meiosis to generate haploid spores from which progeny yeast cultures grow. Since the cells are haploid, the phenotypes of the progeny are determined directly by the genotypes.

Now let us work through a gene-mapping example. Suppose we have a hypothetical plant in which there are three linked genes, all of which control fruit phenotypes. A recessive allele *p* of the first gene determines purple fruit color versus yellow color of the wild type. A recessive allele *r* of the second gene results in a round fruit shape versus elongated fruit in the wild type. A recessive allele *j* of the third gene gives a juicy fruit versus the dry fruit of the wild type. The task before us is to determine the order of the genes on the chromosome and the map distances between the genes. To do so, we make the appropriate testcross of a triple heterozygote (*p⁺ r⁺ j⁺/p r j*) with a triply homozygous recessive (*p r j/p r j*) and then count the different phenotypic classes in the progeny (Figure 5.6).

For each gene in the cross, two different phenotypes occur in the progeny; therefore, for the three genes, a total of $(2)^3 = 8$ phenotypic classes appear in the progeny, representing all possible combinations of phenotypes. In an

Figure 5.5

Demonstration that the recombination frequency between two genes located far apart on the same chromosome cannot exceed 50 percent. (a) Single crossovers produce one-half parental and one-half recombinant chromatids. **(b)** Double crossovers (two-strand, three-strand, and four-strand) collectively produce one-half parental and one-half recombinant chromatids.

Parental genotypes

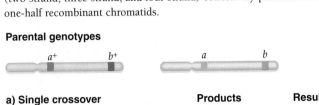

a) Single crossover   **Products**   **Resulting genotypes**   **Sum**

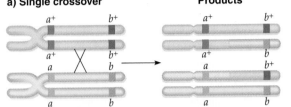

a^+ b^+	Parental	
a^+ b	Recombinant	Recombinants = 2
		Total = 4
a b^+	Recombinant	Therefore, 2/4 recombinants
a b	Parental	

b) Double crossovers

Two-strand double crossover

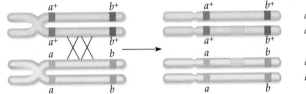

a^+ b^+	Parental	
a^+ b^+	Parental	
		Total: 0/4 recombinants
a b	Parental	
a b	Parental	

Three-strand double crossover (2 ways)

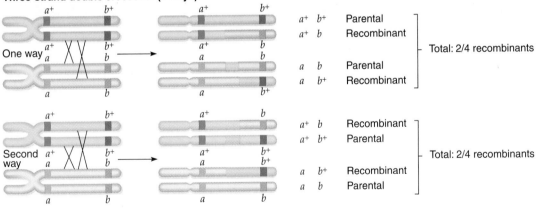

One way

a^+ b^+	Parental	
a^+ b	Recombinant	
		Total: 2/4 recombinants
a b	Parental	
a b^+	Recombinant	

Second way

a^+ b	Recombinant	
a^+ b^+	Parental	
		Total: 2/4 recombinants
a b^+	Recombinant	
a b	Parental	

Four-strand double crossover

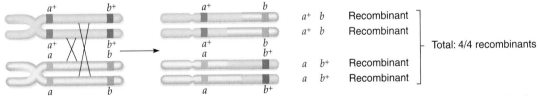

a^+ b	Recombinant	
a^+ b	Recombinant	
		Total: 4/4 recombinants
a b^+	Recombinant	
a b^+	Recombinant	

Sum: Recombinants = 0 + 2 + 2 + 4 = 8

Total = 4 + 4 + 4 + 4 = 16

Therefore, recombinants = 50%

Figure 5.6

Three-point mapping, showing the testcross used and the resultant progeny.

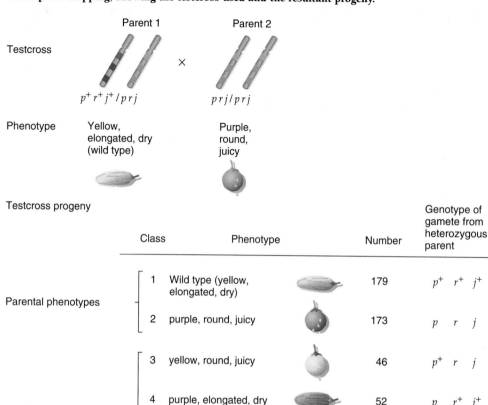

	Class	Phenotype		Number	Genotype of gamete from heterozygous parent		
Parental phenotypes	1	Wild type (yellow, elongated, dry)		179	p^+	r^+	j^+
	2	purple, round, juicy		173	p	r	j
Recombinant phenotypes	3	yellow, round, juicy		46	p^+	r	j
	4	purple, elongated, dry		52	p	r^+	j^+
	5	yellow, round, dry		22	p^+	r	j^+
	6	purple, elongated, juicy		22	p	r^+	j
	7	yellow, elongated, juicy		4	p^+	r^+	j
	8	purple, round, dry		2	p	r	j^+

Total = 500

actual experiment, not all the phenotypic classes may be generated. The absence of a phenotypic class is also important information, and the experimenter should enter a 0 in the class for which no progeny are found.

Establishing the Order of Genes. The first step in mapping the three genes is to determine the order of the genes on the chromosome. One parent carries the recessive alleles

for all three genes; the other is heterozygous for all three genes. Therefore, the phenotype of each of the progeny is determined by the alleles in the gamete from the triply heterozygous parent; the gamete from the other parent will carry only recessive alleles. We know from the genotypes of the original parents that all three genes are in coupling. Since the heterozygous parent in the testcross was $p^+ r^+ j^+/p\ r\ j$, classes 1 and 2 in Figure 5.6 are parental progeny: Class 1 is produced by the fusion of a ($p^+ r^+ j^+$ gamete with a $p\ r\ j$ gamete from the triply homozygous recessive parent. Class 2 is produced by the fusion of a $p\ r\ j$ gamete from the heterozygous parent and a $p\ r\ j$ gamete. These classes are generated from meioses in which no crossing-over occurs in the region of the chromosome in which the three genes are located.

Figure 5.7

Consequences of a double crossover in a triple heterozygote for three linked genes. In a double crossover, the middle allelic pair changes orientation relative to the outside alleleic pairs.

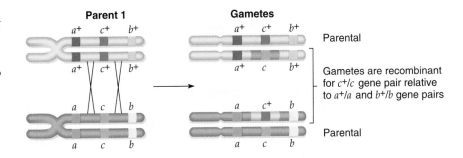

Gametes are recombinant for c^+/c gene pair relative to a^+/a and b^+/b gene pairs

The other six progeny classes result from crossovers within the region spanned by the three genes that gave rise to recombinant gametes. At the simplest level, there may have been a single crossover between a pair of linked genes or a double crossover—that is, two crossovers, one between *each* pair of linked genes. Statistically, the frequency of double crossovers in the region is less than the frequency of either single crossover, so *double-crossover gametes are the least frequent pair found.* Therefore, to identify the double-crossover progeny, we can examine the progeny to find the *pair* of classes that have the lowest number of representatives. In Figure 5.6, classes 7 and 8 are such a pair. The genotypes of the gametes from the heterozygous parent that give rise to these phenotypes are $p^+ r^+ j$ and $p r j^+$.

Figure 5.7 illustrates the consequences of a double crossover in a triple heterozygote for three linked genes, *a*, *b*, and *c,* where the alleles are in coupling and the *c* gene is in the middle. The key observation to make is that a double crossover changes the orientation of the allelic pair in the middle of the three genes (here, c^+/c) with respect to the two flanking allelic pairs. That is, after the double crossover, the *c* allele is now on the chromatid with the a^+ and b^+ alleles, and the c^+ allele is on the chromosome with the *a* and *b* alleles. Therefore, genes *p*, *r*, and *j* must be arranged in such a way that the center gene switches to give classes 7 and 8. To determine the arrangement, we first check the relative organization of the genes in the parental heterozygote to be sure which alleles are in coupling and which are in repulsion. In this

example, the parental (noncrossover) gametes are $p^+ r^+ j^+$ and $p r j$, so all are in coupling. The double-crossover gametes are $p^+ r^+ j$ and $p r j^+$, so the only possible gene order compatible with the data is $p j r$, with the genotype of the heterozygous parent being $p^+ j^+ r^+/p j r$. Figure 5.8 illustrates the generation of the double-crossover gametes from that parent.

Calculating the Recombination Frequencies for Genes. The cross data can be rewritten as shown in Figure 5.9 to reflect the newly determined gene order. For convenience in the analysis, the region between genes *p* and *j* is called region I and that between genes *j* and *r* is called region II.

Recombination frequency can now be calculated for two genes at a time. For the $p - j$ distance, all the crossovers that occurred in region I must be added together. Thus, we must consider the recombinant progeny resulting from a single crossover in that region (classes 3 and 4) *and* the recombinant progeny produced by a double crossover in which one crossover is between *p* and *j* and the other is between *j* and *r* (classes 7 and 8). The double crossovers must be included because each double crossover includes a single crossover in region I and therefore involves recombination between genes *p* and *j*. From Figure 5.9 there are 98 recombinant progeny in classes 3 and 4, and 6 in classes 7 and 8, giving a total of 104 progeny that result from recombination in region I. There are 500 progeny in all, so the percentage of progeny generated by crossing-over in region I is

Figure 5.8

Rearrangement of the three genes in Figure 5.6 to *p j r*. The evidence is that a double crossover involving the same two chromatids (as shown in this figure) generates the least frequent pair of recombinant phenotypes (in this case, class 7 with four progeny and class 8 with two progeny).

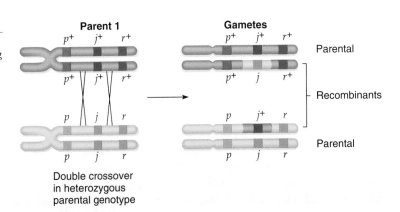

Double crossover in heterozygous parental genotype

Figure 5.9

Rewritten form of the testcross and testcross progeny in Figure 5.6, based on the actual gene order $p\ j\ r$.

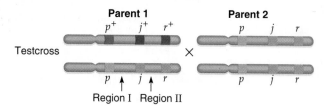

Testcross progeny

Class	Genotype of gamete from heterozygous parent			Number	Origin
1	p^+	j^+	r^+	179	Parentals, no crossover
2	p	j	r	173	
3	p^+	j	r	52	Recombinants, single crossover region I
4	p	j^+	r^+	46	
5	p^+	j^+	r	22	Recombinants, single crossover region II
6	p	j	r^+	22	
7	p^+	j	r^+	4	Recombinants, double crossover
8	p	j^+	r	2	

Total = 500

20.8 percent, determined as follows (sco = single crossovers; dco = double crossovers):

$$\frac{\text{sco in region I } (p-j) + \text{dco}}{\text{total progeny}} \times 100\%$$

$$= \frac{(52+46)+(4+2)}{500} \times 100\%$$

$$= \frac{98+6}{500} \times 100\%$$

$$= \frac{104}{500} \times 100\%$$

$$= 20.8\%$$

In other words, the recombination frequency for genes p and j is 20.8, which gives an estimated map distance of 20.8 mu. This map distance, which is quite large, is chosen mainly for illustration. We will see later that a recombination frequency of 20.8 in an actual cross would probably underestimate the true map distance.

The same method is used to calculate the recombination frequency for genes j and r. That is, we calculate the frequency of crossovers in the cross that gave rise to progeny recombinant for genes j and r and directly relate that frequency to map distance. In this case, all the crossovers that occurred in region II (see Figure 5.9) must be added (classes 5, 6, 7, and 8). The percentage of crossovers is calculated in the following manner:

$$\frac{\text{sco in region II } (j-r) + \text{dco}}{\text{total progeny}} \times 100\%$$

$$= \frac{(22+22)+(4+2)}{500} \times 100\%$$

$$= \frac{44+6}{500} \times 100\%$$

$$= \frac{50}{500} \times 100\%$$

$$= 10.0\%$$

Thus, the recombination frequency for genes j and r is 10.0, which gives an estimated map distance of 10.0 map units.

In summary, we have generated a genetic map of the three genes in our example (Figure 5.10). This example has illustrated that the three-point testcross is an effective way to establish the order of genes and to calculate map distances.

To compute the map distance between the two outside genes, we simply add the two map distances. In our example, the $p - r$ distance is 20.8 + 10.0 = 30.8 mu. This map distance also can be computed directly from the data by combining the two formulas discussed previously:

$$\text{distance} = \frac{(\text{sco in region I}) + (\text{dco}) + (\text{sco in region II}) + (\text{dco})}{\text{total progeny}} \times 100\%$$

$$= \frac{(\text{sco in region I}) + (\text{sco in region II}) + (2 \times \text{dco})}{\text{total progeny}} \times 100\%$$

$$= \frac{(52+46)+(22+22)+2(4+2)}{500} \times 100\%$$

$$= \frac{98+44+2(6)}{500} \times 100\%$$

$$= 30.8 \text{ mu}$$

KEYNOTE

The map distance between genes can be calculated from the results of testcrosses between strains carrying appropriate genetic markers. The unit of genetic distance is the map unit (mu), where 1 mu is defined as the interval in which 1 percent crossing-over takes place. Gene mapping crosses produce data in the form of recombination frequencies, which are used to estimate map distance, where 1 mu is equivalent to a recombination frequency of 1 percent. Recombination frequencies are not identical to crossover frequencies and typically underestimate the true map distance.

Interference and Coincidence. The recombination frequencies determined by three-point mapping are useful in elaborating the overall organization of genes on a chromosome and in telling us a little about the recombination mechanisms themselves. For example, we computed a recombination frequency of 20.8 between genes p and j and a recombination frequency of 10.0 between genes j and r in the previous three-point testcross example.

Figure 5.10

Genetic map of the *p-j-r* region of the chromosome computed from the recombination data in Figure 5.9.

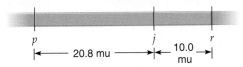

$$p \qquad j \qquad r$$
$$\longleftarrow 20.8 \text{ mu} \longrightarrow \longleftarrow 10.0 \text{ mu} \longrightarrow$$

KEYNOTE

The occurrence of a crossover may interfere with the occurrence of a second crossover nearby. The extent of interference is expressed by the coefficient of coincidence, which is calculated by dividing the number of observed double crossovers by the number of expected double crossovers. The coefficient of coincidence ranges from zero to 1, and the extent of interference is measured as 1 minus the coefficient of coincidence.

In the example shown in Figure 5.9, if crossing-over in region I is independent of crossing-over in region II, then the probability of a double crossover in the two regions is equal to the product of the probabilities of the two events occurring separately; that is,

$$\frac{\text{recombination frequency, region I}}{100} \times \frac{\text{recombination frequency, region II}}{100}$$

$$= 0.208 \times 0.100 = 0.0208$$

or 2.08 percent double crossovers are expected to occur. However, only 6/500 = 1.2 percent double crossovers occurred in this cross (classes 7 and 8).

It is characteristic of mapping crosses that double-crossover progeny typically do not appear as often as the map distances between the genes lead us to expect. That is, in some way, the presence of one crossover interferes with the formation of another crossover nearby; this phenomenon is called **interference.** The extent of interference is expressed as a **coefficient of coincidence;** that is,

$$\text{coefficient of coincidence} = \frac{\text{observed double-crossover frequency}}{\text{expected double-crossover frequency}}$$

and

$$\text{interference} = 1 - \text{coefficient of coincidence}$$

For the portion of the map in our example, the coefficient of coincidence is

$$0.012/0.0208 = 0.577$$

A coefficient of coincidence value of 1 means that in a given region, all double crossovers occurred that were expected on the basis of two independent events; there is no interference, so the interference value is zero. If the coefficient of coincidence is zero, none of the expected double crossovers occurred. Here there is total interference, with one crossover completely preventing a second crossover in the region under examination; the interference value is 1. These examples show that coincidence values and interference values are inversely related. In this example, the coefficient of coincidence of 0.577 means that the interference value is 0.423. Only 57.7 percent of the expected double crossovers took place in the cross.

Calculating Accurate Map Distances

Earlier we learned that map units between linked genes, strictly speaking, are defined in terms of crossover frequency, whereas operationally, geneticists quantify the frequency of recombinants in genetic crosses. We also learned that crossover frequency and recombination frequency are not synonymous and that recombination frequency often leads to an underestimation of true map distance. How, then, do we obtain accurate map distances for linked genes?

To answer this question, we need to focus on the consequences of crossovers between linked genes. Consider a hypothetical case of two allelic pairs (a^+/a and b^+/b) linked in coupling and separated by quite a distance on the same chromosome. Figure 5.11a shows that a single crossover results in recombination of the two allelic pairs, producing two parental and two recombinant gametes. The same result will occur for any odd number of crossovers in the region between the genes. Figure 5.11b shows that a double crossover involving two of the four chromatids does not result in recombination of the allelic pairs, so only parental gametes result. Parental gametes also result for any even number of crossovers between the two linked genes. However, the crossover frequency between genes is a measure of the distance between them. Therefore, because the double crossover in Figure 5.11b did not generate recombinant gametes, two crossover events will be uncounted, and the map distance based on recombination frequency between genes a and b will be underestimated.

In gene mapping, if no more than a single crossover occurs between linked genes, there is a direct linear relationship between genetic map distance and the observed recombination frequency because recombination frequency then equals crossover frequency. In practice, we see this relationship only when genetic map distances are small, that is, when genes are between 0 and approximately 7 mu apart. In other words, map distances based on recombination frequencies of 7 percent or less are very accurate. As the distance between genes increases beyond this point, the chance of multiple crossovers increases, and there is no longer an exact linear relationship between map distance and recombination frequency because some crossovers go uncounted. As a result, it is difficult to obtain an accurate measure of map distance.

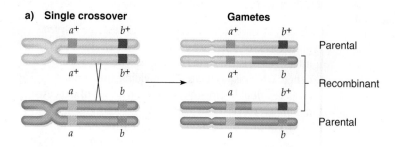

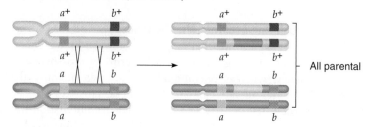

Figure 5.11

Progeny of single and double crossovers. (a) A single crossover between linked genes generates recombinant gametes. **(b)** A double crossover between linked genes gives parental gametes.

Fortunately, it has been possible to derive mathematical formulas, called **mapping functions,** to define the relationship between map distance and recombination frequency. A particular mapping function based on the assumption of no interference between crossovers is shown in Figure 5.12. You can see the direct relationship between map distance and recombination frequency at 7 mu or less, and the curve slowly approaches the limit recombination frequency of 50 percent. Just to pick a couple of points, when the recombination frequency is 20 percent, the true map distance is almost 30 mu, and when the recombination frequency is 30 percent, the true map distance is almost 50 mu. In general, mapping functions all require some basic assumptions about the frequency of crossovers compared with distance between genes. Therefore, the usefulness of applying the mapping functions depends on the validity of the assumptions.

KEYNOTE

At genetic distances greater than about 7 mu, the incidence of multiple crossovers causes the recombination frequency to be an underestimate of the crossover frequency and hence of the true map distance. Mapping functions can be used to correct for the effects of multiple crossovers and thereby give a more accurate map distance.

Tetrad Analysis in Certain Haploid Eukaryotes

Tetrad analysis is a special mapping technique that can be used to map the genes of haploid eukaryotic organisms in which the products of a single meiosis, the meiotic tetrad, are contained within a single structure. The

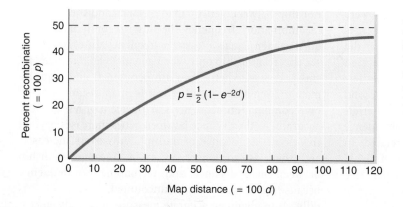

Figure 5.12

A mapping function for relating map distance and recombination frequencies. This particular mapping function was developed by J. B. S. Haldane and assumes no interference between crossovers; d is the crossover frequency and e is the base of the natural logarithms.

eukaryotic organisms in which this phenomenon occurs are either fungi or single-celled algae. Tetrad analysis often is used with the orange bread mold *Neurospora crassa* and the yeast *Saccharomyces cerevisiae* (both fungi) and *Chlamydomonas reinhardtii* (a single-celled alga). Tetrad analysis allows study of the details of events at meiosis that is not possible in any other system.

By analyzing the phenotypes of the meiotic tetrads, geneticists can directly infer the genotypes of each member of the tetrad. That is, because haploid organisms have only one copy of each gene, the phenotype is the direct result of the allele present. In other words, dominance and recessiveness do not come into play as they do in diploid organisms. We will outline this technique shortly.

Before we discuss the principles of tetrad analysis, let us learn a little about the life cycles of the organisms with which tetrad analysis can be done. For example, the life cycle of baker's yeast (also called the budding yeast), *Saccharomyces cerevisiae,* is diagrammed in Figure 5.13. Two mating types occur in yeast, **a** and α. The haploid cells of this organism reproduce mitotically (the vegetative life cycle), with the new cell arising from the parental cell by budding. Fusion of haploid **a** and α cells produces a diploid cell that is stable and that also reproduces by budding. Diploid **a**/α cells *sporulate;* that is, they go through meiosis. The four haploid meiotic products of the diploid cell, the ascospores, are contained within a roughly spherical ascus. Two of these ascospores are of mating type **a,** and two are of mating type α. When the ascus is ripe, the ascospores are released, and they germinate to produce haploid cells that, on solid medium, grow and divide to produce a colony. In yeast, the four ascospores are arranged randomly within the ascus; that is, they are so-called *unordered tetrads.* The green alga *Chlamydomonas reinhardtii* also produces four meiotic products that are arranged randomly in a sac.

The fungus *Neurospora crassa* has a somewhat similar life cycle, but in this organism, the ascospores are arranged in a linear ascus, an *ordered tetrad.* There are actually eight ascospores in this organism because each of the four meiotic products divides again by mitosis. The order of the four spore pairs within an ascus reflects exactly the orientation of the four chromatids of each tetrad at the metaphase plate in meiosis I. The spores can be isolated in the same order as they are in the ascus, or they can be isolated randomly from the ascus.

Using Tetrad Analysis to Map Two Linked Genes

Let us see how we can analyze gene linkage relationships by analyzing unordered tetrads.

By making an appropriate cross, a diploid is constructed that is heterozygous for both genes, and after meiosis, the resulting tetrads are analyzed. For yeast, a mechanical micromanipulator with a very fine needle is used to dissect each spore out of the ascus.

Consider the cross $a^+ b^+ \times a\, b$ in which a and b are linked genes. Figure 5.14 shows the three different tetrad types that result. If no crossing over occurs between the genes (Figure 5.14a), a **parental-ditype (PD)** tetrad

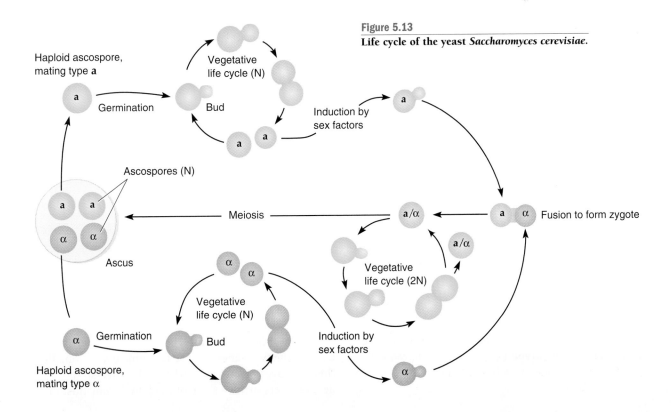

Figure 5.13

Life cycle of the yeast *Saccharomyces cerevisiae.*

Figure 5.14

Origin of tetrad types for a cross $a\ b \times a^+\ b^+$ in which both genes are located on the same chromosome. (a) No crossover. (b) Single crossover. (c–e) Three types of double crossovers.

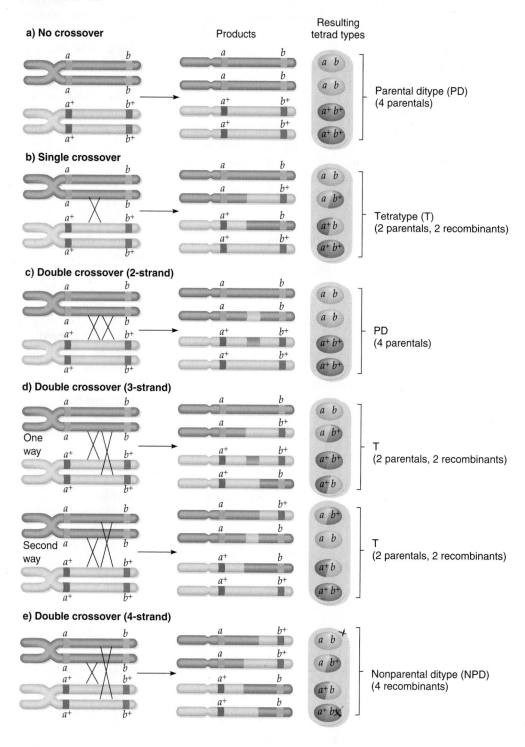

results. The PD tetrads contain only two types of meiotic products, both of which are of the parental type (hence the name parental ditype). A single crossover (Figure 5.14b) produces a **tetratype (T)** tetrad. A T tetrad contains two parentals (one of each type) and two recombinants (one of each type).

For double crossovers, we must take into account the chromatid strands involved. In a two-strand double crossover (Figure 5.14c), the two crossover events between the two genes involve the same two chromatids. This crossover results in a PD tetrad. Three-strand double crossovers involve three of the four chromatids; there are two possible ways in which this can occur (Figure 5.14d). In either case, the result is a T tetrad. Lastly, in four-strand double crossovers (Figure 5.14e), each crossover event involves two distinct chromatids, so all four chromatids of the tetrad are involved. This crossover results in a **nonparental-ditype** (**NPD**) tetrad. An NPD tetrad contains two types of meiotic products, both of which are the nonparental (recombinant) types.

From the relative numbers of each type of meiotic tetrad, the distance between the two genes can be computed by using a modification of the basic mapping formula

$$\frac{\text{number of recombinants}}{\text{total number of progeny}} \times 100\%$$

In tetrad analysis, we analyze types of tetrads, rather than individual progeny. To convert the basic mapping formula into tetrad terms, the recombination frequency between genes *a* and *b* becomes

$$\frac{1/2\,T + NPD}{\text{total tetrads}} \times 100\%$$

In essence, we are looking at tetrads with recombinants and determining the proportion of spores in those tetrads that are recombinant. So in the formula, the ½ T and the NPD represent the recombinants from the cross; the other ½ T and the PD represent the nonrecombinants (the parentals). Thus, the formula does indeed compute the percentage of recombinants. For instance, if there are 200 tetrads with 140 PD, 48 T, and 12 NPD, the recombination frequency between the genes is

$$\frac{1/2\,(48) + 12}{200} \times 100\% = 18\%$$

This conversion produces a formula for calculating the map distance between two linked genes by using the frequencies of the three possible types of tetrads rather than by analyzing individual progeny. If more than two genes are linked in a cross, the data can best be analyzed by considering two genes at a time and by classifying each tetrad into PD, NPD, and T for each pair.

KEYNOTE

In organisms in which all products of meiosis are contained within a single structure, the analysis of the relative proportion of tetrad types provides another way to compute the map distance between genes. The general formula when two linked genes are being mapped is

$$\frac{1/2\,T + NPD}{\text{total tetrads}} \times 100\%$$

Summary

In this chapter, we discussed linkage, crossing-over, and gene mapping in eukaryotes. The production of genetic recombinants results from physical exchanges between homologous chromosomes in meiosis. The exchange of parts of chromatids is called crossing-over, and the site of crossing-over is called a chiasma. Crossing-over is a reciprocal event that, in eukaryotes, occurs at the four-strand stage in prophase I of meiosis.

Gene mapping is the process of locating the position of genes in relation to one another on the chromosome. The first step is to show that genes are linked (located on the same chromosome), which is indicated by the fact that they do not assort independently in crosses. Then crosses are done to determine the map distance between the linked genes. The unit of genetic distance is the map unit, where 1 mu is the interval in which 1 percent crossing-over takes place. However, gene mapping crosses generate recombination frequency data; when the genes are not closely linked, multiple crossovers often occur between them, and this leads to underestimates of map distances. Nonetheless, recombination frequencies have been used traditionally by geneticists to give an approximation of map distances. Mathematically derived mapping functions can be used to calculate more accurate map distances based on recombination frequencies.

We also learned how to map genes in certain haploid microorganisms by using tetrad analysis. That is, in these microorganisms, the meiotic tetrads are kept together in structures, making it possible to isolate and analyze them. With this technique, distances between linked genes are calculated by analyzing the relative proportions of tetrad types, which are reflective of recombination events, rather than by analyzing individual progeny.

Analytical Approaches for Solving Genetics Problems

Q5.1 In corn, the gene for colored (*C*) seeds is completely dominant to the gene for colorless (*c*) seeds. Similarly, a single gene pair controls whether the endosperm (the part of the seed that contains the food stored for the embryo) is full or shrunken. Full (*S*) is dominant to shrunken (*s*). A true-breeding colored, full-seeded plant was crossed with a colorless, shrunken-seeded one. The F_1 colored, full plants were testcrossed to the doubly recessive type, that is, colorless and shrunken. The result was as follows:

colored, full	4,032
colored, shrunken	149
colorless, full	152
colorless, shrunken	4,032
Total	8,368

Is there evidence that the gene for color and the gene for endosperm shape are linked? If so, what is the map distance between the two loci?

A5.1 The best approach is to begin by diagramming the cross using gene symbols:

P   colored and full × colorless and shrunken
              CC SS   *cc ss*

F₁   colored and full
                              Cc Ss

Testcross: colored and full × colorless and shrunken
                    Cc Ss   *cc ss*

If the genes were unlinked, a 1:1:1:1 ratio of colored and full : colored and shrunken : colorless and full : colorless and shrunken would be the progeny of this test-cross. By inspection, we can see that the actual progeny deviate a great deal from this ratio, showing a 27:1:1:27 ratio. If we did a chi-square test (using the actual numbers, not the percentages or ratios), we would see immediately that the hypothesis that the genes are unlinked is invalid, and we must consider the two genes to be linked in coupling. More specifically, the parental combinations (colored, full and colorless, shrunken) are more numerous than expected, whereas the recombinant types (colorless, full and colored, shrunken) are correspondingly less numerous than expected. This result comes directly from the inequality of the four gamete types produced by meiosis in the colored and full F₁ parent.

Given that the two genes are linked, the crosses can be diagrammed to reflect their linkage as follows:

P $\frac{C\ S}{C\ S}$ × $\frac{c\ s}{c\ s}$

F₁ $\frac{C\ S}{c\ s}$

Testcross: $\frac{C\ S}{c\ s}$ × $\frac{c\ s}{c\ s}$

To calculate the map distance between the two genes, we need to compute the frequency of crossovers in that region of the chromosome during meiosis. We cannot do that directly, but we can compute the percentage of recombinant progeny that must have resulted from such crossovers:

Parental types: colored, full   4,032
                              colorless, shrunken   4,035
                                                                    8,067

Recombinant types: colored, shrunken   149
                              colorless, full   152
                                                                    301

This calculation gives about 3.6 percent recombinant types (301/8,368 × 100%) and about 96.4 percent parental types (8,067/8,368 × 100%). Since the recombi-

nation frequency can be used directly as an indication of map distance, especially when the distance is small, we can conclude that the distance between the two genes is 3.6 mu (3.6 cM).

We would get approximately the same result if the two genes were in repulsion rather than in coupling. That is, the crossovers are occurring between homologous chromosomes, regardless of whether there are genetic differences in the two homologues that we, as experimenters, use as markers in genetic crosses. This same cross in repulsion would be as follows:

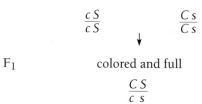

P   colorless and full × colored and shrunken
              $\frac{c\ S}{c\ S}$ $\frac{C\ s}{C\ s}$

F₁   colored and full
                              $\frac{C\ S}{c\ s}$

Data from an actual testcross of the F₁ with colorless and shrunken (*cc ss*) gave 638 colored and full (recombinant) : 21,379 colored and shrunken (parental) : 21,906 colorless and full (parental) : 672 colorless and shrunken (recombinant), with a total of 44,595 progeny. Thus, 2.94 percent were recombinants, for a map distance between the two genes of 2.94 mu, a figure reasonably close to the results of the cross made in coupling.

Q5.2 In the Chinese primrose, slate-colored flower (*s*) is recessive to blue flower (*S*); red stigma (*r*) is recessive to green stigma (*R*); and long style (*l*) is recessive to short style (*L*). All three genes involved are on the same chromosome. The F₁ of a cross between two true-breeding strains, when testcrossed, gave the following progeny:

Phenotype	Number of Progeny
slate flower, green stigma, short style	27
slate flower, red stigma, short style	85
blue flower, red stigma, short style	402
slate flower, red stigma, long style	977
slate flower, green stigma, long style	427
blue flower, green stigma, long style	95
blue flower, green stigma, short style	960
blue flower, red stigma, long style	27
Total	3,000

a. What were the genotypes of the parents in the cross of the two true-breeding strains?

b. Make a map of these genes, showing gene order and the distances between genes.

c. Derive the coefficient of coincidence for interference between these genes.

A5.2

a. With three gene pairs, eight phenotypic classes are expected, and eight are observed. The reciprocal pairs of classes with the most representatives are those resulting from no crossovers, and these pairs can tell us the genotypes of the original parents. The two classes are slate, red, long and blue, green, short. Thus, the F_1 triply heterozygous parent of this generation must have been $S R L/s r l$, so the true-breeding parents were $S R L/S R L$ (blue, green, short) and $s r l/s r l$ (slate, red, long).

b. The order of the genes can be determined by inspecting the reciprocal pairs of phenotypic classes that represent the results of double crossing-over. These classes have the least numerous representatives, so the double-crossover classes are slate, green, short ($s R L$) and blue, red, long ($S r l$). The gene pair that has changed its position relative to the other two pairs of alleles is the central gene, S/s in this case. Therefore, the order of genes is $R S L$ (or $L S R$). We can diagram the F_1 testcross as follows:

$$\frac{R S L}{r s l} \times \frac{r s l}{r s l}$$

A single crossover between the R and S genes gives the green, slate, long ($R s l$) and red, blue, short ($r S L$) classes, which have 427 and 402 members, respectively, for a total of 829. The double-crossover classes have already been defined, and they yield 54 progeny. The map distance between R and S is given by the crossover frequency in that region, which is the sum of the single crossovers and double crossovers divided by the total number of progeny, then multiplied by 100 percent. Thus,

$$\frac{829 + 54}{3,000} \times 100\% = \frac{883}{3,000} \times 100\%$$
$$= 29.43\% \text{ or } 29.43 \text{ mu}$$

With similar logic, the distance between S and L is given by the crossover frequency in that region, which is the sum of the single-crossover and double-crossover progeny classes divided by the total number of progeny. The single-crossover progeny classes are green, blue, long ($R S l$) and red, slate, short ($r s L$), which have 95 and 85 members, respectively, for a total of 180. The map distance is given by

$$\frac{180 + 54}{3,000} \times 100\% = \frac{234}{3,000} \times 100\%$$
$$= 7.8\% \text{ or } 7.8 \text{ mu}$$

The data we have derived give us the following map:

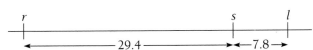

c. The coefficient of coincidence is given by

$$\frac{\text{frequency of observed double crossovers}}{\text{frequency of expected double crossovers}}$$

The frequency of observed double crossovers is $54/3,000 = 0.018$. The expected frequency of double crossovers is the product of the map distances between r and s and between s and l, that is, $0.294 \times 0.078 = 0.023$. The coefficient of coincidence, therefore, is $0.018/0.023 = 0.78$. In other words, 78 percent of the expected double crossovers did indeed take place; there was 22 percent interference.

Questions and Problems

5.1 A cross $a^+a^+ \ b^+b^+ \times aa \ bb$ results in an F_1 of phenotype $a^+ \ b^+$; the following numbers are obtained in the F_2 (phenotypes):

$a^+ \ b^+$	110
$a^+ \ b$	16
$a \ b^+$	19
$a \ b$	15
Total	160

Are genes at the a and b loci linked or independent? What F_2 numbers would otherwise be expected?

*5.2 In corn, a dihybrid for the recessives a and b is testcrossed. The distribution of the phenotypes is as follows:

$A \ B$	122
$A \ b$	118
$a \ B$	81
$a \ b$	79

Are the genes assorting independently? Test the hypothesis with a chi-square test. Explain tentatively any deviation from expectation, and tell how you would test your explanation.

5.3 In *Drosophila*, the mutant black (b) has a black body, and the wild type has a grey body; the mutant vestigial (vg) has wings that are much shorter and crumpled compared with the long wings of the wild type. In the following cross, the true-breeding parents are given together with the counts of offspring of F_1 females × black and vestigial males:

P	black and normal × grey and vestigial
F_1	females × black and vestigial males

Progeny:	grey, normal	283
	grey, vestigial	1,294
	black, normal	1,418
	black, vestigial	241

From these data, calculate the map distance between the black and vestigial genes.

*5.4 Use the following two-point recombination data to map the genes concerned. Show the order and the length of the shortest intervals.

Gene Loci	% Recombination	Gene Loci	% Recombination
a,b	50	b,d	13
a,c	15	b,e	50
a,d	38	c,d	50
a,e	8	c,e	7
b,c	50	d,e	45

5.5 A corn plant known to be heterozygous at three loci is testcrossed. The progeny phenotypes and frequencies are as follows:

+	+	+	455
a	b	c	470
+	b	c	35
a	+	+	33
+	+	c	37
a	b	+	35
+	b	+	460
a	+	c	475
	Total		2,000

Give the gene arrangement, linkage relations, and map distances.

*5.6 Genes a and b are linked, with 10 percent recombination. What would be the phenotypes, and the probability of each, among progeny of the following cross?

$$\frac{a\ b^+}{a^+\ b} \times \frac{a\ b}{a\ b}$$

*5.7 Genes a and b are sex-linked and are located 7 mu apart in the X chromosome of *Drosophila*. A female of genotype $a^+\ b/a\ b^+$ is mated with a wild-type fly ($a^+\ b^+$/Y).
a. What is the probability that one of her sons will be either $a^+\ b^+$ or $a\ b^+$ in phenotype?
b. What is the probability that one of her daughters will be $a^+\ b^+$ in phenotype?

*5.8 Genes a and b are on one chromosome, 20 mu apart; c and d are on another chromosome, 10 mu apart. Genes e and f are on yet another chromosome and are 30 mu apart. Cross a homozygous $A\ B\ C\ D\ E\ F$ individual with an $a\ b\ c\ d\ e\ f$ one, and cross the F$_1$ back to an $a\ b\ c\ d\ e\ f$ individual. What are the chances of getting individuals of the following phenotypes in the progeny?
a. $A\ B\ C\ D\ E\ F$
b. $A\ B\ C\ d\ e\ f$
c. $A\ b\ c\ D\ E\ f$
d. $a\ B\ C\ d\ e\ f$
e. $a\ b\ c\ D\ e\ F$

*5.9 Genes d and p occupy loci 5 map units apart in the same autosomal linkage group. Gene h is in a separate

autosomal linkage group and therefore segregates independently of the other two. What types of offspring are expected, and what is the probability of each, when individuals of the following genotypes are testcrossed?
a. $\dfrac{D\ P\ \ h}{d\ p\ \ h}$ **b.** $\dfrac{d\ P\ H}{D\ p\ \ h}$

5.10 A hairy-winged (h) *Drosophila* female is mated with a yellow-bodied (y), white-eyed (w) male. The F$_1$ are all wild type. The F$_1$ progeny are then crossed, and the F$_2$ that emerge are as follows:

Females:	wild type	757
	hairy	243
Males:	wild type	390
	hairy	130
	yellow	4
	white	3
	hairy, yellow	1
	hairy, white	2
	yellow, white	360
	hairy, yellow, white	110

Give genotypes of the parents and the F$_1$, and note the linkage relations and distances where possible.

5.11 For each of the following tabulations of testcross progeny phenotypes and numbers, state which locus is in the middle, and reconstruct the genotype of the tested triple heterozygotes.
a.

$A\ B\ C$	191
$a\ b\ c$	180
$A\ b\ c$	5
$a\ B\ C$	5
$A\ B\ c$	21
$a\ b\ C$	31
$A\ b\ C$	104
$a\ B\ c$	109

b.

$C\ D\ E$	9
$c\ d\ e$	11
$C\ d\ e$	35
$c\ D\ E$	27
$C\ D\ e$	78
$c\ d\ E$	81
$C\ d\ E$	275
$c\ D\ e$	256

c.

$F\ G\ H$	110
$f\ g\ h$	114
$F\ g\ h$	37
$f\ G\ H$	33
$F\ G\ h$	202
$f\ g\ H$	185
$F\ g\ H$	4
$f\ G\ h$	0

5.12 The following numbers were obtained for testcross progeny in *Drosophila* (phenotypes):

+ m +	218
w + f	236
+ + f	168
w m +	178
+ m f	95
w + +	101
+ + +	3
w m f	1
Total	1,000

Construct a genetic map.

*5.13 Three of the many recessive mutations in *Drosophila melanogaster* that affect body color, wing shape, or bristle morphology are black (*b*) body versus grey in the wild type; dumpy (*dp*), obliquely truncated wings versus long wings in the wild type; and hooked (*hk*) bristles at the tip versus not hooked in the wild type. From a cross of a dumpy female with a black, hooked male, all the F₁ were wild-type for all three characters. The testcross of an F₁ female with a dumpy, black, hooked male gave the following results:

wild type	169
black	19
black, hooked	301
dumpy, hooked	21
hooked	8
hooked, dumpy, black	172
dumpy, black	6
dumpy	304
Total	1,000

a. Construct a genetic map of the linkage group (or groups) these genes occupy. If applicable, show the order and give the map distances between the genes.

b. Determine the coefficient of coincidence for the portion of the chromosome involved in the cross. How much interference is there?

5.14 Two normal-looking *Drosophila* flies are crossed and yield the following phenotypes among the progeny:

Females:	+ + +	2,000

Males:	+ + +	3
	a b c	1
	+ b c	839
	a + +	825
	a b +	86
	+ + c	90
	a + c	81
	+ b +	75
	Total	4,000

Give parental genotypes, gene arrangement in the female parent, map distances, and the coefficient of coincidence.

5.15 The following questions make use of this genetic map:

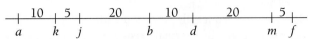

a. Calculate the frequency of *j b* gametes from a *J B/j b* genotype.

b. Calculate the frequency of *A M* gametes from an *a M/A m* genotype.

c. Calculate the frequency of *J B D* gametes from a *j B d/J b D* genotype.

d. Calculate the frequency of *J B d* gametes from a *j B d/J b D* genotype.

e. Calculate the frequency of *j b d/j b d* genotypes in a *j B d/J B D* × *j B d/J B D* mating.

f. Calculate the frequency of *A k F* gametes from an *A K F/a k f* genotype.

*5.16 A female *Drosophila* carries the recessive mutations *a* and *b* in repulsion on the X chromosome (she is heterozygous for both). She is also heterozygous for an X-linked recessive lethal allele, *l*. When she is mated to a true-breeding, normal male, she yields the following progeny:

Females:	1,000	+	+

Males:	405	a	+
	44	+	b
	48	+	+
	2	a	b

Draw a chromosome map of the three genes in the proper order and with map distances as nearly as you can calculate them.

5.17 A farmer who raises rabbits wants to break into the Easter market. He has stocks of two true-breeding lines. One is hollow and long-eared but not chocolate, and the second is solid, short-eared, and chocolate. Hollow (*h*), long ears (*le*), and chocolate (*ch*) are all recessive, autosomal, and linked as in the following map:

```
h                le                        ch
|                |                          |
|<---- 26 mu ---->|<------- 32 mu -------->|
```

The farmer can generate a trihybrid by crossing his two lines, and at great expense he is able to obtain the services of a male homozygous recessive at all three loci to cross with his F₁ females.

The farmer has buyers for both solid and hollow bunnies; however, all must be chocolate and long-eared. Assuming that interference is zero, if he needs 25 percent of the progeny of the desired phenotypes to be profitable, should he continue with his breeding? Calculate the percent of the total progeny that will be the desired phenotypes.

5.18 In *Drosophila*, many different mutations have been isolated that affect a normally brick-red eye color caused by the deposition of brown and bright-red pigments. Two

X-linked recessive mutations are *w* (white eyes, map position 1.5) and *cho* (chocolate-brown eyes, map position 13.0), with *w* epistatic to *cho*.

a. A white-eyed female is crossed with a chocolate-eyed male, and the normal, red-eyed F_1 females are crossed with either wild-type or white-eyed males. Determine the frequency of the progeny types produced in each cross.

b. The recessive mutation *st* causes scarlet (bright-red) eyes and maps to the third chromosome at position 44. Mutant flies with only *st* and *cho* alleles have white eyes, and *w* is epistatic to *st*. Suppose a true-breeding *w* male is crossed with a true-breeding *cho, st* female. Determine the frequency of the progeny types you would expect if the F_1 females are crossed with true-breeding scarlet-eyed males.

5.19 In *Saccharomyces*, *Neurospora*, and *Chlamydomonas*, what meiotic events give rise to PD, NPD, and T tetrads?

5.20 Double exchanges between two loci can be of several types, called two-strand, three-strand, and four-strand doubles.

a. Four recombination gametes would be produced from a tetrad in which the first of two exchanges is depicted in the figure below. Draw in the second exchange.

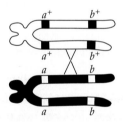

b. In the following figure, draw in the second exchange so that four nonrecombination gametes would result.

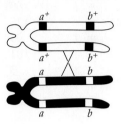

5.21 The following asci were obtained from the cross *leu* + × + *rib* in yeast. Draw the linkage map and determine the map distance.

110	45	6	39
leu +	*leu rib*	+ +	*leu* +
+ *rib*	*leu* +	*leu rib*	+ *rib*
leu +	+ +	*leu rib*	+ +
+ *rib*	+ *rib*	+ +	*leu rib*

*5.22 The genes *a*, *b*, and *c* are linked in *Neurospora crassa*. The following asci were obtained from the cross *a b* + × + + *c*.

45	5	146	1
a b +	*a b* +	*a b* +	*a b* +
+ *b c*	*a* + +	*a b* +	+ + +
a + +	+ *b c*	+ + *c*	*a b c*
+ + *c*	+ + *c*	+ + *c*	+ + *c*

10	20	15	58
a b +	*a b* +	*a b* +	*a b* +
a + *c*	+ + *c*	*a b c*	+ *b* +
+ *b* +	*a b* +	+ + +	*a* + *c*
+ + *c*	+ + *c*	+ + *c*	+ + *c*

Determine the correct gene order. Calculate all gene-gene distances.

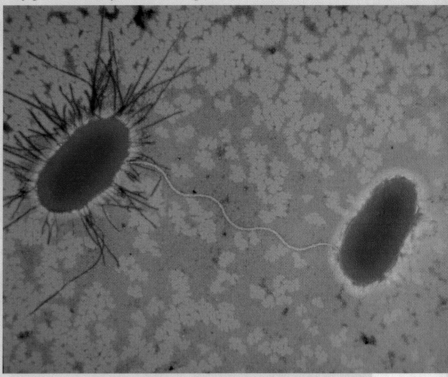

Conjugation between *Hfr* (left) and *F⁻* (right) strains of *E. coli*.

6

Gene Mapping in Bacteria and Bacteriophages

PRINCIPAL POINTS

Conjugation is a process in which there is a unidirectional transfer of genetic information through direct cellular contact between a donor and a recipient bacterial cell. The donor state is conferred by the presence of a plasmid called an *F* factor. Conjugation results in the unidirectional transfer of a copy of the *F* factor from donor to recipient.

The *F* factor can integrate into the bacterial chromosome. Strains in which this has occurred—*Hfr* strains—can conjugate with recipient strains, and transfer of the bacterial chromosome ensues. The order of genes and distances between genes can be determined by the time of acquisition of the genes by the recipient from the donor during conjugation.

Transformation is the transfer of genetic material between organisms by small extracellular pieces of DNA. By genetic recombination, part of the transforming DNA molecule can exchange with a portion of the recipient's chromosomal DNA. Transformation can be used experimentally to determine gene order and map distances between genes.

Transduction is a process whereby bacteriophages (phages) mediate the transfer of bacterial DNA from one bacterium (the donor) to another (the recipient). Transduction can be used experimentally to map bacterial genes.

The same principles used to map eukaryotic genes are used to map phage genes. That is, a bacterial host is simultaneously infected with two strains of phages differing from each other in one or more gene loci. The percentages of recombinants are determined and the gene arrangement and distances between genes are then postulated.

> The same principles of recombinational mapping in eukaryotes can be applied to mapping the distance between mutational sites in different genes (intergenic mapping) and to mapping mutational sites within the same gene (intragenic mapping).

> From fine-structure analysis of the *rII* region of bacteriophage T4, it was determined that the unit of mutation and of recombination is the DNA base pair.

> The number of genes that cause a particular mutant phenotype is determined by the complementation, or *cis-trans*, test. If two viral mutants, each carrying a mutation in a different gene, are combined in a single host cell, the mutations make up for each other's defect (complement), and a wild-type phenotype results. If two mutants, each carrying a mutation in the same gene, are combined, the mutations do not complement and the mutant phenotype is still expressed.

ⓘ BACTERIA AND BACTERIOPHAGES HAVE LONG played a key role in genetics. Frederick Griffith's and Oswald Avery's experiments with *Streptococcus* were key in the discovery of DNA as the genetic material. Herbert Boyer and Stanley Cohen first cloned a recombinant DNA molecule using bacteria. And F. Peyton Rous helped determine the nature of certain inherited cancers with his investigations of a virus that infects chickens.

In this chapter, you will learn about the genetics of bacteria and bacteriophages: how they reproduce, how new strains are produced, and the experimental techniques that geneticists use to map bacterial and viral genes. After you have read this chapter, you can apply what you've learned by trying the iActivity, in which you will create a genetic map of the *E. coli* chromosome.

In Chapter 5, we considered the principles of genetic mapping in eukaryotic organisms. To map genes in bacteria and bacteriophages, geneticists use essentially the same experimental strategies. Crosses are made between strains that differ in genetic markers, and recombinants, the products of the exchange of genetic material, are detected and counted. Analysis of data obtained from such crosses is the same as for eukaryotes: The frequency with which crossing-over occurs between two sets of genes relates to the map distance between the two gene loci. The major difference is the experimental techniques involved.

Recently, emphasis in genetic research has shifted from localizing individual genes on chromosomes by making crosses to determining the sequence of bases in the DNA. With DNA base sequence analysis, scientists can identify genes directly so that the ultimate genetic map of a species can be constructed. When all the genes of an organism are identified, at least at the nucleotide level, the door is opened to investigating the function of each gene. In the case of pathogenic microorganisms, the genomic sequence information is an extremely valuable resource for efforts to identify and understand the genes responsible for pathogenesis. Complete genomic sequences have been determined for many species of Bacteria and Archaea (see Chapter 15). Bacterial genomes sequenced include the 4.6 million-base-pair (4.6-megabase) genome of *E. coli*, the 1.44-megabase genome of the Lyme disease causative agent *Borrelia burgdorferi*, the 1.66-megabase genome of the stomach ulcer–causing *Helicobacter pylori*, and the 1.14-megabase genome of the syphilis bacterium *Treponema pallidum*. The archaeon genomes sequenced include the 1.66-megabase genome of *Methanococcus jannaschii*, a hyperthermophilic methanogen that grows optimally at 85°C and at pressures up to 200 atmospheres. With genome sequences, scientists can look for genes directly, thereby providing the ultimate in genetic maps—information at the nucleotide level about the organization of genes in the genome.

In this chapter, you will learn about the classic genetic studies of bacteria and bacteriophages. You will also learn about a series of classic genetic experiments that investigated the fine structure of the gene, that is, the detailed molecular organization of the gene as it relates to the mutational, recombinational, and functional events in which the gene is involved. A bacteriophage gene was the subject of these experiments.

Genetic Analysis of Bacteria

Genetic material can be transferred between bacteria by three main processes: conjugation, transformation, and transduction. It is possible to map bacterial genes by using any one of these methods. In each case, (1) transfer is unidirectional, and (2) no complete diploid zygote is formed (unlike in eukaryotes). However, not all methods can be used for all bacterial species, and the size of the region that can be mapped varies according to method.

Among bacteria, *E. coli* has been used extensively for genetic and molecular analysis. It is found in the large intestines of most animals, including humans. This bacterium is a good subject for study because it can be grown on a simple, defined medium and can be handled with simple microbiological techniques.

E. coli is a cylindrical organism about 1–3 μm long and 0.5 μm in diameter (see Figure 1.10); like most bacteria, it is small compared with eukaryotic cells. Its protoplasm is full of ribosomes, and a single circular DNA chromosome is in a region called the **nucleoid.** As in all prokaryotes, there is no membrane between the nucleoid region and the rest of the cell.

Like other bacteria, *E. coli* can be grown both in a liquid culture medium and on the surface of growth medium solidified with agar. Genetic analysis of bacteria typically is done by spreading (plating) cells on the surface of agar medium. Wherever a single bacterium lands on the agar surface, it will grow and divide repeatedly, resulting in the formation of a visible cluster of genetically identical cells called a *colony* (Figure 6.1). Each colony consists of a clone of cells that are genetically identical to the parental cell that initiated the colony. The concentration of bacterial cells in a liquid culture—the *titer*—can be determined by spreading known volumes of the culture or of a known dilution of the culture on the agar surface, incubating the plates at a constant temperature, and then counting the number of resulting colonies. The number you obtain is converted to colony-forming units (cfu) per milliliter (mL). For example, if 100 µL of a 1,000-fold dilution of a culture is spread on a plate and 165 colonies are produced, then this means there were 165 bacteria in 100 µL of the 1,000-fold dilution. Thus, in the original culture, there were 165 (colonies) × 1,000 (dilution factor) × 10 (because 0.1 mL were plated) = 1,650,000 cfu/mL = $1.65 × 10^6$ cfu/mL.

The composition of the culture medium used depends on the experiment and the genotypes of the strains being used. Each bacterial species (or any other microorganism, such as yeast) has a characteristic **minimal medium** on which it will grow. A minimal medium contains only the nutrients absolutely required for the growth of wild-type cells. The minimal medium for wild-type *E. coli*, for example, consists of a sugar (a carbon source) and some salts and trace elements. From the minimal medium, the organism can synthesize all the other components it needs for growth and reproduction, including amino acids, vitamins, DNA, and RNA. By contrast, the **complete medium** for a microorganism supplies vitamins and amino acids and all kinds of substances that might be expected to be essential metabolites and whose biosynthesis might be interfered with by mutation.

Genetic analysis of bacteria (and other microorganisms) typically involves mutants defective in their abilities to make one or more molecules essential for growth and perhaps also defective in genes affecting other metabolic processes. Strains that are unable to synthesize essential nutrients are called **auxotrophs** (also called *auxotrophic mutants, nutritional mutants,* or *biochemical mutants*). A strain that is wild-type and thus can synthesize all essential nutrients is called a **prototroph;** prototrophs need no nutritional supplements in the growth medium. By definition, the wild type, or prototroph, grows on the minimal medium for that organism, whereas an auxotroph grows on complete medium or on minimal medium plus the appropriate nutritional supplements.

For example, consider the *E. coli* strain with the genotype *trp ade thi*+. This strain will not grow on minimal medium because it has mutations for tryptophan and adenine biosynthesis. It will grow either on complete medium or on minimal medium supplemented with the amino acid tryptophan (because of the *trp* mutation) and the purine adenine (because of the *ade* mutation). It does not need the vitamin thiamine to grow because it carries the wild-type *thi* allele, as signified by the superscript +.

Some genes are involved not in biosynthetic pathways but in utilization pathways. For example, there are a number of genes for using various carbon sources such as lactose, arabinose, and maltose. In this case, the superscript + following the gene symbol means that the gene is wild-type and therefore that the bacterium can metabolize the substance. For example, a *lac*+ strain can metabolize lactose, whereas a *lac* mutant strain cannot. By varying the nutrient composition of the medium, it is possible to detect various genotypic classes in a genetic analysis.

In genetic experiments with microorganisms such as *E. coli*, crosses are made between strains differing in genotype (and therefore phenotype), and progeny are analyzed for parental and recombinant phenotypes. When auxotrophic mutations are involved, the determination of parental and progeny phenotypes (and therefore genotypes, because bacteria are haploid) involves testing colonies for their growth requirements. One convenient procedure for such testing is *replica plating* (see Figure 19.18). In replica plating, some of the bacteria of colonies on a plate of complete medium are transferred onto a sterile velveteen cloth mounted on a replica plater by stamping gently onto it. Replicas of the original colony pattern on the cloth are then made by gently pressing new plates onto the velveteen. If the new plate contains

Figure 6.1

Bacterial colonies growing on a nutrient medium in a Petri dish.

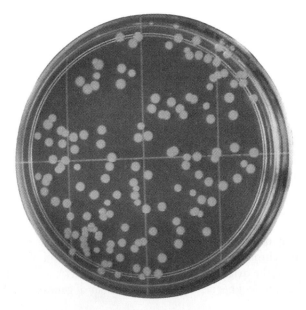

minimal medium, only prototrophic colonies can grow. Then, by comparing the patterns on the original master plate with those on the minimal-medium replica plate, researchers can readily identify auxotrophic colonies because they will be on the master plate but not on the minimal-medium plate. Using other plates containing minimal medium plus combinations of nutritional supplements appropriate for the strain or strains involved, one can determine the phenotypes and genotypes of all the auxotrophic colonies.

Genetic Mapping in Bacteria by Conjugation

Discovery of Conjugation in *E. coli*

Conjugation is a process in which there is a unidirectional transfer of genetic information through direct cellular contact between a donor bacterial cell and a recipient bacterial cell. The contact is followed by formation of a physical bridge between the cells. Then a segment (rarely all) of the donor's chromosome may be transferred into the recipient and may undergo genetic recombination with a homologous chromosome segment of the recipient cell. Recipients that have incorporated a piece of donor DNA into their chromosomes are called **transconjugants.**

Conjugation was discovered in 1946 by Joshua Lederberg and Edward Tatum. They studied two *E. coli* strains that differed in their nutritional requirements. Strain A had the genotype *met bio thr+ leu+ thi+*, and strain B had the genotype *met+ bio+ thr leu thi*. Strain A can grow only on a medium supplemented with the amino acid methionine (*met*) and the vitamin biotin (*bio*) but does not need the amino acids threonine (*thr*) or leucine (*leu*) or the vitamin thiamine (*thi*). Strain B can grow only on a medium supplemented with threonine, leucine, and thiamine, but does not require methionine or biotin.

Lederberg and Tatum mixed the two *E. coli* strains A and B together and plated them onto minimal medium (Figure 6.2). The mixed culture gave rise to some prototrophic colonies (*met+ bio+ thr+ leu+ thi+*) at a frequency of about 1 in 10 million cells. Since no colonies appeared when each strain was plated separately on minimal medium, mutation was ruled out as the cause of the prototrophic colonies. The mixing, then, is a genetic cross that produced recombinants.

In a separate experiment, Bernard Davis placed strains A and B in a liquid medium on either side of a U-tube apparatus separated by a filter with pores too small to allow bacteria to move through (Figure 6.3). The medium was moved between compartments by alternating suction and pressure, and then the cells were plated on minimal medium to check for the appearance of prototrophic

Figure 6.2

Lederberg and Tatum experiment showing that sexual recombination occurs between cells of *E. coli*. After the cells from strain A and strain B have been mixed and the mixture plated, a few colonies grow on the minimal medium, indicating that they can now make the essential constituents. These colonies are recombinants produced by an exchange of genetic material between the strains.

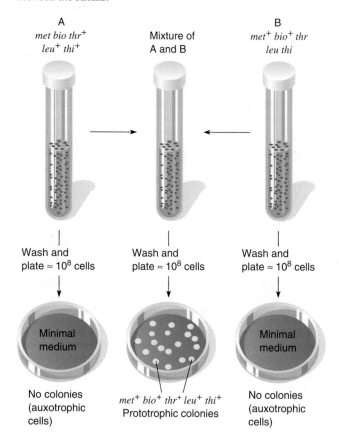

colonies. No prototrophic colonies appeared, meaning that cell-to-cell contact was required for the genetic exchange to occur. These experiments indicated that *E. coli* has the type of mating system called conjugation.

The Sex Factor F

In 1953, William Hayes showed that genetic exchange in *E. coli* occurs in only one direction, with one cell acting as a donor and the other cell acting as a recipient. Hayes proposed that the transfer of genetic material between the strains is mediated by a *sex factor* named *F* that the donor cell possesses (*F+*) and the recipient cell lacks (*F−*). The *F* factor found in *E. coli* is an example of a **plasmid,** a self-replicating, circular DNA found distinct from the main bacterial chromosome. About 1/40 the size of the host chromosome, the *F* factor contains a region of DNA called the **origin** (or O), the point where DNA transfer to the recipient begins, as well as a number of genes, includ-

Figure 6.3

Davis's U-tube experiment showing that physical contact between the two bacterial strains of the Lederberg and Tatum experiment was needed for genetic exchange to occur.

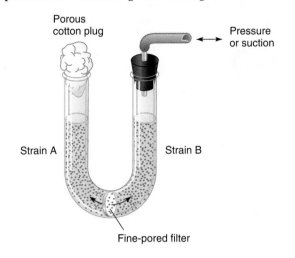

ing those that specify hairlike host cell surface components called **F-pili** (singular, *F-pilus*), or sex-pili, which allow the physical union of F^+ and F^- cells to take place.

When F^+ and F^- cells are mixed, they may conjugate ("mate"). Figure 6.4a1 diagrams the event. No conjugation can occur between two cells of the same mating type (i.e., two F^+ bacteria or two F^- bacteria). Genetic material is transferred from donor to recipient during conjugation beginning when one strand of the F factor is nicked at the origin, and DNA replication proceeds from that point (Figure 6.4a2). Beginning at the origin, a single strand of DNA is transferred to the F^- cell (Figure 6.4a3). Think of the process like unraveling a roll of paper towels. The origin is the first region of DNA unwound; as unwinding continues, replication maintains the remaining circular F factor in a double-stranded form. Once the F factor DNA enters the F^- recipient, the complementary strand is synthesized (Figure 6.4a4). When the complete F factor has been transferred, the F^- cell becomes an F^+ cell (Figure 6.4a5). In $F^+ \times F^-$ crosses, none of the bacterial chromosome is transferred; only the F factor is transferred.

K E Y N O T E

Some *E. coli* bacteria possess a plasmid, called the F factor, that is required for mating. *E. coli* cells containing the F factor are designated F^+, and those without it are F^-. The F^+ cells (donors) can mate with F^- cells (recipients) in a process called conjugation, which leads to the one-way transfer of a copy of the F factor from donor to recipient during replication of the F factor. As a result, both donor and recipient are F^+. None of the bacterial chromosome is transferred during $F^+ \times F^-$ conjugation.

High-Frequency Recombination Strains of *E. coli*

To get recombinants for *chromosomal* genes by conjugation involves special derivatives of F^+ strains called **Hfr (high-frequency recombination)** strains. Discovered separately by William Hayes and Luca Cavalli-Sforza, *Hfr* strains originate by a rare crossover event in which the F factor integrates into the bacterial chromosome (Figure 6.4b1–2). Plasmids such as F that are also capable of integrating into the bacterial chromosomes are called **episomes.** When the F factor is integrated, it no longer replicates independently, but is replicated as part of the host chromosome.

Because of the F factor genes, *Hfr* cells can conjugate with F^- cells (Figure 6.4b3 and the chapter opening photograph). When mating happens, events similar to those in the $F^+ \times F^-$ mating occur. The integrated F factor becomes nicked at the origin and replication begins (Figure 6.4b4). During replication, part of the F factor starting with the origin moves into the recipient cell, where the transferred strand is copied. In a short time, the donor bacterial chromosome begins to be transferred into the recipient. If there are allelic differences between donor genes and recipient genes, recombinants can be isolated (Figure 6.4b5). The recombinants are produced by double crossovers between the linear donor DNA and the circular recipient chromosome. In a double crossover, a segment of donor DNA is exchanged for the homologous segment of recipient DNA.

In $Hfr \times F^-$ matings, the F^- cell almost never acquires the *Hfr* phenotype. To become *Hfr,* the recipient cell must receive a complete copy of the F factor. However, only part of the F factor is transferred at the beginning of conjugation; the rest of the F factor is at the end of the donor chromosome. *All* of the donor chromosome would have to be transferred for a complete functional F factor to be found in the recipient, and that would require about 100 minutes at 37°C, the normal growth temperature for *E. coli.* This is an extremely rare event because all the while the bacteria are conjugating, they are "jiggling" around, so mating pairs typically break apart long before the second part of the F factor is transferred.

The low-frequency recombination of chromosomal gene markers in $F^+ \times F^-$ crosses can be understood when we consider that only about 1 in 10,000 F^+ cells in a population become *Hfr* cells by F factor integration. The reverse process, excision of the F factor, also occurs spontaneously and at low frequency, producing an F^+ cell from an *Hfr* cell. In excision, the F factor loops out of the *Hfr* chromosome, and by a single crossing-over event (just like the integration event), a circular host chromosome and a circular extrachromosomal F factor are generated.

F′ Factors

Hfr cells rarely become F^+ cells by excision of the F factor from the chromosome. Occasionally, excision of the F

Figure 6.4

Transfer of genetic material during conjugation in E. coli. (a) Transfer of the F factor from donor to recipient cell during $F^+ \times F^-$ matings. (b) Production of Hfr strain by integration of F factor and transfer of bacterial genes from donor to recipient cell during $Hfr \times F^-$ matings.

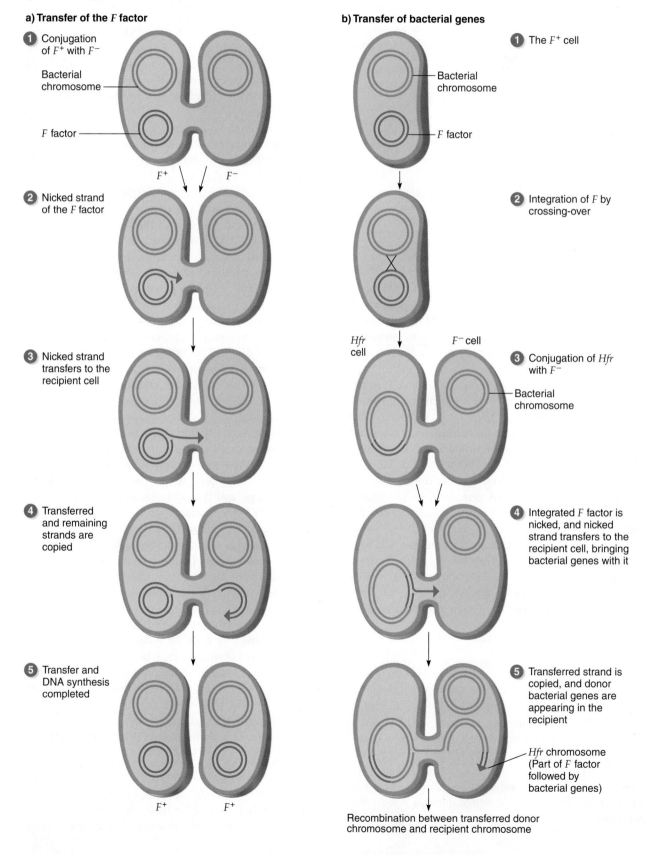

a) Transfer of the F factor

1 Conjugation of F^+ with F^-

Bacterial chromosome

F factor

F^+ F^-

2 Nicked strand of the F factor

3 Nicked strand transfers to the recipient cell

4 Transferred and remaining strands are copied

5 Transfer and DNA synthesis completed

F^+ F^+

b) Transfer of bacterial genes

1 The F^+ cell

Bacterial chromosome

F factor

2 Integration of F by crossing-over

Hfr cell F^- cell

3 Conjugation of Hfr with F^-

Bacterial chromosome

4 Integrated F factor is nicked, and nicked strand transfers to the recipient cell, bringing bacterial genes with it

5 Transferred strand is copied, and donor bacterial genes are appearing in the recipient

Hfr chromosome (Part of F factor followed by bacterial genes)

Recombination between transferred donor chromosome and recipient chromosome factor is not precise and an *F* factor is produced with a small section of the host chromosome that was adjacent to the integrated *F* factor. Since the *F* factor integrates at one of many sites on the chromosomes, many different host chromosome segments can be picked up in this way. Consider an *E. coli* strain in which the *F* factor has integrated next to the *lac*+ region, a set of genes required for the breakdown of lactose (Figure 6.5a). If the looping out

Figure 6.5

Production of an *F*′ factor. (a) Region of bacterial chromosome into which the *F* factor has integrated. **(b)** The *F* factor looping out incorrectly, so it includes a piece of bacterial chromosome, the *lac*+ genes. **(c)** Excision, in which a single crossover between the looped-out DNA segment and the rest of the bacterial chromosome results in an *F*′ factor, called *F*′ (*lac*).

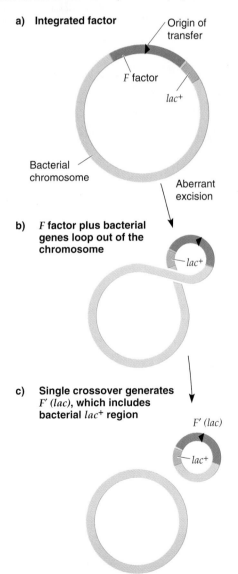

is not precise, then the adjacent *lac*+ host chromosomal genes can be included in the loop (Figure 6.5b). Then, by a single crossover, the looped-out DNA is separated from the host chromosome (Figure 6.5c) to produce an *F* factor carrying the host's *lac*+ genes. *F* factors containing bacterial genes are called *F*′ (*F* prime) factors, and they are named for the genes they have picked up. An *F*′ with the *lac* genes is called *F*′ (*lac*).

Cells with *F*′ factors can conjugate with F⁻ cells. As in *F*+ × F⁻ conjugation, a copy of the *F*′ factor is transferred to the F⁻ cell, which then becomes *F*′. The recipient also receives a copy of the bacterial gene(s) on the *F* factor (*lac* in our example). Since the recipient has its own copy of that DNA, the resulting cell line is partially diploid (*merodiploid*), having two copies of one or a few genes and only one copy of all the others. This particular type of conjugation is called **F-duction**, or *sexduction*, and it provides a way to study genes in a partly diploid state in *E. coli*.

Using Conjugation to Map Bacterial Genes

In genetic analysis using conjugation, the determining factor for whether a strain is a donor is the presence of an integrated *F* factor in the chromosome, making the strain an *Hfr* strain. In planning conjugation experiments to map genes, experimenters construct an appropriate donor strain by introducing an *F* factor through *F*+ × F⁻ mating and then selecting an *Hfr* derivative.

In the late 1950s, François Jacob and Elie Wollman studied the transfer of chromosomal genes from *Hfr* strains to F⁻ cells that had allelic differences for a number of genes. Their experimental design involved making an *Hfr* × F⁻ mating and, at various times after conjugation began, breaking apart the conjugating pairs using a kitchen blender and analyzing the transconjugants for which donor genes they had received. This is called an *interrupted mating experiment*.

The use of interrupted mating to map bacterial genes is illustrated by the following cross (Figure 6.6):

Donor:
*Hfr*H *thr*+ *leu*+ *azi*ᴿ *ton*ᴿ *lac*+ *gal*+ *str*ˢ

Recipient:
F⁻ *thr leu azi*ˢ *ton*ˢ *lac gal str*ᴿ

(The superscript S means "sensitive" and the superscript R means "resistant.")

The *Hfr*H strain (named for William Hayes) is prototrophic and is resistant to growth inhibition by the chemical sodium azide (*azi*ᴿ) and to infection by bacteriophage T1 (*ton*ᴿ) and is sensitive to the antibiotic streptomycin (*str*ˢ). The F⁻ strain is auxotrophic for threonine (*thr*) and leucine (*leu*), is sensitive to growth inhibition by the

Figure 6.6

Interrupted mating experiment involving the cross *HfrH thr⁺ leu⁺ azi^R ton^R lac⁺ gal⁺ str^S × F⁻ thr leu azi^S ton^S lac gal str^R*. The progressive transfer of donor genes with time is illustrated. Recombinants are generated by an exchange of a donor fragment with the homologous recipient fragment resulting from a double-crossover event. **(a)** At various times after mating commences, the conjugating pairs are broken apart and the transconjugant cells are plated on selective agar media to determine which genes have been transferred from the *Hfr* to the *F⁻*. **(b)** The graph shows the appearance of donor genetic markers in the *F⁻* cells as a function of time after the first genes that produced recombinants—*thr⁺* and *leu⁺*—have entered.

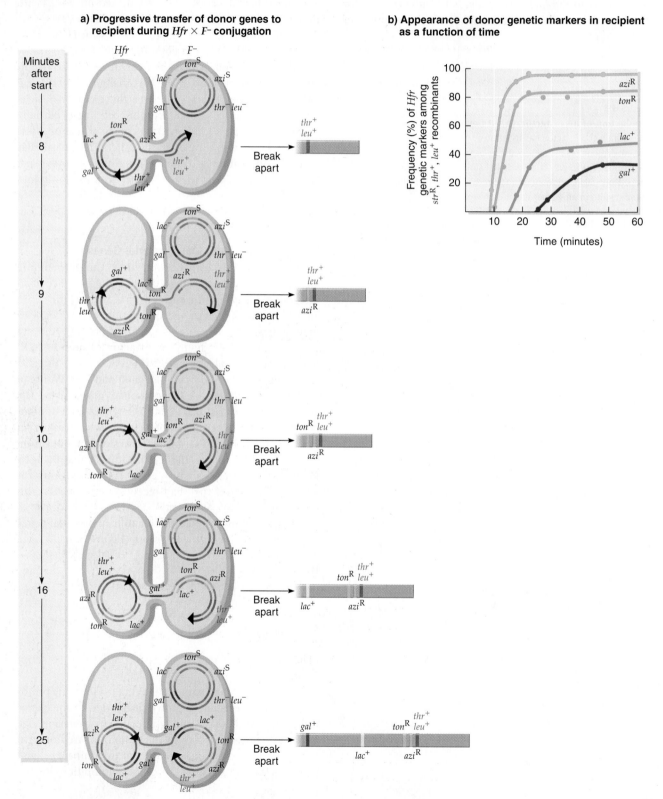

a) Progressive transfer of donor genes to recipient during *Hfr × F⁻* conjugation

b) Appearance of donor genetic markers in recipient as a function of time

chemical sodium azide (*azi*^{S}), is sensitive to infection by bacteriophage T1 (*ton*^{S}), is unable to ferment lactose (*lac*) or galactose (*gal*), and is resistant to growth inhibition by the antibiotic streptomycin (*str*^{R}).

In such a conjugation experiment, the two cell types are mixed together in a liquid medium at 37°C. Samples are removed from the mating mixture at various times and then agitated to break the pairs apart. By plating on selective agar media, recombinant recipients (the transconjugants) are then searched for and analyzed with respect to the time at which the first donor genes entered the recipient and produced recombinants.

For this particular cross, the medium contains streptomycin to kill the *HfrH* and lacks threonine and leucine so that the parental *F*⁻ cannot grow. In this cross, the threonine (*thr*⁺) and leucine (*leu*⁺) genes are the first donor genes to be transferred to the *F*⁻ to produce a merodiploid, so recombinants formed by the exchange of those genes with the *thr leu* genes of the *F*⁻ recipient grow on the selective medium. Appropriate media can be used to test for the appearance of other donor genes (*azi*^{R}, *ton*^{R}, *lac*⁺, and *gal*⁺) among the selected *thr*⁺ *leu*⁺ *str*^{R} transconjugants. For example, medium with sodium azide added can test for the presence of *azi*^{R} from the donor.

Figure 6.6b shows the results. The threonine (*thr*⁺) and leucine (*leu*⁺) genes are the first donor genes to be transferred to the *F*⁻, and that occurs at 8 minutes. (The two genes are inseparable timewise in a conjugation experiment because they are physically very close to one another.) The next gene to be transferred is *azi*^{R}, and recombinants for this gene are seen at about 9 minutes after the start of conjugation, that is, 1 minute after the *thr*⁺ and *leu*⁺ genes entered. Then *ton*^{R} recombinants are seen at 10 minutes, followed by *lac*⁺ recombinants at about 16 minutes and *gal*⁺ recombinants at about 25 minutes. Note that the maximum frequency of recombinants becomes smaller the later the gene enters the recipient, because with time there is an increasing chance that mating pairs will break apart.

In this experiment, each gene from the *Hfr* bacterium appears in recombinants at a different but reproducible time after mating begins. Thus, from the time intervals for the experiment described, the genetic map in Figure 6.7 may be constructed, with map units in minutes.

Figure 6.7

Genetic map of the genes in the experiment in Figure 6.6. The marker positions represent the time of entry of the genes into the recipient during the experiment.

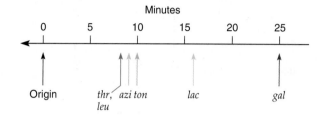

Hfr strains, *H, 1, 2,* and *3.* In each case, only one *Hfr* strain was used to cross with the recipient, and the order of gene transfer and the time between the appearance of each gene in the recipient was determined. The genetic distance in time units between a particular pair of genes is constant no matter which *Hfr* strain is used as donor; for example, the genetic distance between *thr* and *pro* is the same in *H, 1, 2,* and *3.* This validates the use of time units as a measure of genetic distance in *E. coli.*

From this sort of data, a genetic map of the chromosome is constructed by aligning the genes transferred by each *Hfr,* as shown in Figure 6.8b. In view of the overlap of the genes, the simplest map that can be drawn from these data is a circular one, as shown in Figure 6.9c. The map, then, is a composite of the results of the individual matings. The circularity of the map was itself a significant finding because all previous genetic maps—of eukaryotic chromosomes—were linear.

A complete genetic map of the *E. coli* chromosome eventually was constructed using conjugation experiments; it is 100 minutes long. Like genetic maps of other organisms, it provides information about the relative locations of *E. coli* genes on the circular chromosome. In 1997, the ultimate genetic map of *E. coli* was completed, that of the 4.6×10^6 (4.6-megabase) base pair sequence of the bacterium's genome. This map will be the key to unlocking the function of every gene in this model bacterium.

Circularity of the *E. coli* Map

Only one *F* factor is integrated in each *Hfr* strain. Different *Hfr* strains have the *F* factor integrated at different locations and in different orientations in the chromosome. Therefore, *Hfr* strains differ with respect to where the transfer of donor genes begins and the order of transfer of donor genes. Figure 6.8a shows the order of chromosomal gene transfer for four different

iActivity You are assisting Elie Wollman and François Jacob as they construct a genetic map of *E. coli* using the newly discovered interrupted mating procedure in the iActivity *Conjugation in E. coli* on the website.

Figure 6.8

Interrupted mating experiments with a variety of *Hfr* strains, showing that the *E. coli* linkage map is circular. **(a)** Orders of gene transfer for the *Hfr* strains H, 1, 2, and 3. **(b)** Alignment of gene transfer for the *Hfr* strains. **(c)** Circular *E. coli* chromosome map derived from the *Hfr* gene transfer data. The map is a composite showing various locations of integrated F factors. A given *Hfr* strain has only one integrated F factor.

a) Orders of gene transfer

Hfr strains:

H	origin–thr–pro–lac–pur–gal
1	origin–thr–thi–gly–his
2	origin–his–gly–thi–thr–pro–lac
3	origin–gly–his–gal–pur–lac–pro

b) Alignment of gene transfer for the *Hfr* strains

H	thr–pro–lac–pur–gal
1	his–gly–thi–thr
2	his–gly–thi–thr–pro–lac
3	pro–lac–pur–gal–his–gly

c) Circular *E. coli* chromosome map derived from *Hfr* gene transfer data

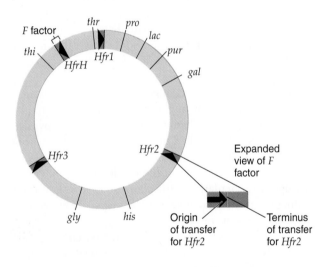

KEYNOTE

The circular *F* factor can integrate into the circular bacterial chromosome by a single crossover event. Strains in which this integration has happened can conjugate with *F⁻* strains, and transfer of the bacterial chromosome occurs. The strains containing the integrated *F* factor are called *Hfr* (high-frequency recombination) strains. In *Hfr* × *F⁻* matings, the chromosome is transferred in a one-way fashion from the *Hfr* cell to the *F⁻* cell, beginning at a specific site called the origin (O). The farther a gene is from O, the later it is transferred to *F⁻*, and this is the basis for mapping genes by their times of entry into the *F⁻* cell. Conjugation and interrupted mating allow mapping of the chromosome.

Genetic Mapping in Bacteria by Transformation

Transformation is the unidirectional transfer of extracellular DNA into cells, often resulting in a phenotypic change in the recipient. Bacterial transformation is used to map the genes of certain bacterial species in which mapping by other methods (conjugation or transduction) is not possible. In mapping experiments using transformation, DNA from a donor bacterial strain is extracted, purified, and broken into small fragments. This DNA is then added to recipient bacteria with a different genotype. If the donor DNA is taken up by a recipient cell and recombines with the homologous parts of the recipient's chromosome, a recombinant chromosome is produced. Recipients whose phenotypes are changed by transformation are called **transformants.** Only if there are genetic differences between donors and recipients will transformants be detected. In principle, any bacterial strain can serve as a donor strain, and any bacterial strain can serve as a recipient.

Bacterial species vary in their ability to take up DNA. To enhance the efficiency of transformation, cells typically are treated chemically or are exposed to a strong electric field in a process called *electroporation*, making the cell membrane more permeable to DNA. Cells prepared to take up DNA by transformation are called *competent cells.*

Only a small proportion of the cells involved in transformation will actually take up DNA. Consider an example of natural transformation of *Bacillus subtilis* (Figure 6.9). (Other systems may differ in the details of the process.) The donor double-stranded DNA fragment is wild-type (a^+) for a mutant allele a in the recipient cell (Figure 6.9a). During DNA uptake, one of the two DNA strands is degraded, so that only one intact linear DNA strand is left inside the cell (Figure 6.9b). This single, linear strand pairs with the homologous DNA of the recipient cell's circular chromosome to form a triple-stranded region (Figure 6.9c). Recombination then occurs by a double-crossover event involving the single-stranded DNA strand of the donor and the double-stranded DNA of the recipient (Figure 6.9d). The result is a recombinant recipient chromosome: In the region between the two crossovers, one DNA strand has the donor a^+ DNA segment, and the other strand has the recipient a DNA segment. In other words, in that region, *the two DNA strands are part donor, part recipient for the genetic information.* A region of DNA with different sequence information on the two strands is called **heteroduplex DNA.** (The other product of the double-crossover event is a single-stranded piece of DNA carrying an a DNA segment; that DNA fragment is degraded.)

After replication of the recipient chromosome, one progeny chromosome has donor genetic information on both DNA strands and is an a^+ transformant. The other progeny chromosome has recipient genetic information

Figure 6.9

Transformation in *Bacillus subtilis*. (**a**) Linear donor double-stranded bacterial DNA fragment carries the a^+ allele, and the recipient bacterium carries the a allele. (**b**) One donor DNA strand enters the recipient. (**c**) The single, linear DNA strand pairs with the homologous region of the recipient's chromosome, forming a triple-stranded structure. (**d**) A double crossover produces a recombinant a^+/a recipient chromosome and a linear a DNA fragment. The linear fragment is degraded, and by replication, one-half of the progeny are a^+ transformants and one-half are a nontransformants.

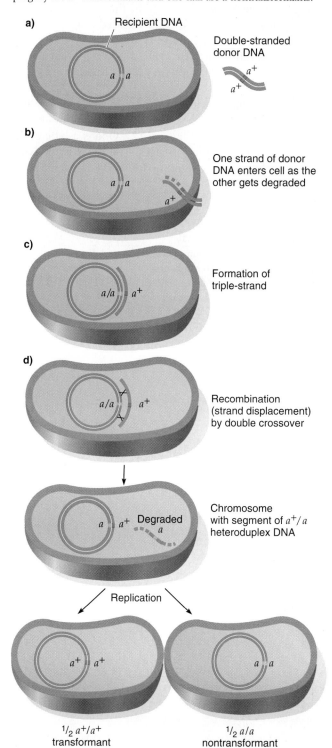

a) Recipient DNA

Double-stranded donor DNA

b) One strand of donor DNA enters cell as the other gets degraded

c) Formation of triple-strand

d) Recombination (strand displacement) by double crossover

Chromosome with segment of a^+/a heteroduplex DNA

Degraded

Replication

$^1\!/_2\ a^+/a^+$ transformant

$^1\!/_2\ a/a$ nontransformant

on both DNA strands and is an a nontransformant. Equal numbers of a^+ transformants and a nontransformants are produced. Given highly competent recipient cells, transformation of most genes occurs at a frequency of about 1 cell in every 10^3 cells.

Transformation can be used to determine whether genes are linked (in this case meaning physically close to one another on the single bacterial chromosome), to determine the order of genes on the genetic map, and to determine map distance between genes. The principles of determining whether two genes are linked are as follows. Efficient transformation of DNA involves fragments with a size sufficient to include only a few genes. If two genes, x^+ and y^+, are far apart on the donor chromosome, they will always be found on different DNA fragments. Thus, given an $x^+\ y^+$ donor and an $x\ y$ recipient, the probability of simultaneous transformation (*cotransformation*) of the recipient to $x^+\ y^+$ is (from the product rule) the product of the probability of transformation of each gene alone. If transformation occurred at a frequency of 1 in 10^3 cells per gene, $x^+\ y^+$ transformants would be expected to appear at a frequency of 1 in 10^6 recipient cells ($10^{-3} \times 10^{-3}$). However, if two genes are close enough that they often are carried on the same DNA fragment, the cotransformation frequency would be close to the frequency of transformation of a single gene. As determined experimentally, if the frequency of cotransformation of two genes is substantially higher than the products of the two individual transformation frequencies, the two genes must be close together.

KEYNOTE

Transformation is the transfer of small extracellular pieces of DNA between organisms. In bacterial transformation, DNA is extracted from a donor strain and added to recipient cells. A DNA fragment taken up by the recipient cell may associate with the homologous region of the recipient's chromosome. Part of the transforming DNA molecule can exchange with part of the recipient's chromosomal DNA. Frequent cotransformation of donor genes indicates close physical linkage of those genes. Transformation has been used to construct genetic maps for bacterial species for which conjugation or transduction analyses are not possible.

Genetic Mapping in Bacteria by Transduction

Transduction (literally, "leading across") is a process by which bacteriophages (bacterial viruses; phages, for short) transfer genes from one bacterium (the donor) to another (the recipient); such phages are called **phage vectors.** Since the amount of DNA a phage can carry is limited, the amount of genetic material that can be transferred is usually less than 1 percent of that in the bacterial chromosome. Once the donor genetic material has been introduced into the recipient, it may undergo genetic recombination with a homologous region of the recipient chromosome. The recombinant recipients are called **transductants.**

Bacteriophages

Most bacterial strains can be infected by specific phages. For example, *E. coli* can be infected by DNA-containing phages such as T2, T4, T5, T6, T7, and λ (lambda).

A phage contains its genetic material (either DNA or RNA) in a single chromosome surrounded by a coat of protein molecules. Figure 6.10 shows two phages of great genetic significance, T4 (Figure 6.10a) and λ (Figure 6.10b). Phage T4 is one of a series of arbitrarily numbered phages with similar properties named T2, T4, and T6 (the T-even phages). It has a number of distinct protein structures: a head (which contains DNA), a core, a

sheath, a base plate, and tail fibers. The last two structures enable the phage to attach itself to a bacterium.

Figure 6.11 shows the T4 life cycle. First (1), the phage particle attaches to the surface of the bacterial cell, and the phage chromosome is injected into the bacterium by a springlike contraction of the sheath. Then (2–5), the phage takes over the bacterium, breaking down the bacterial chromosome and using the bacterial molecules to produce progeny phages. Finally (6), the progeny phages are released from the bacterium as the cell is broken open (lysed). The released progeny phages in solution are called a **phage lysate.** This type of phage life cycle is called the **lytic cycle,** and phages that always follow the lytic cycle when they infect bacteria are called **virulent phages.**

Phage λ has a structure similar to that of T4. When its DNA is injected into *E. coli,* the phage follows one of two alternative paths (Figure 6.12). One is a lytic cycle, exactly like that of the T phages. The other is the **lysogenic pathway** (or *lysogenic cycle*). In the lysogenic path-

a)

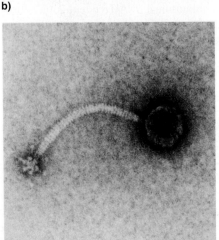
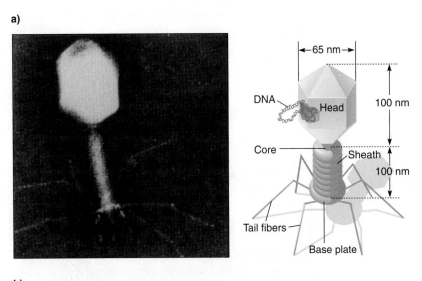

Figure 6.10

Electron micrographs and diagrams of two bacteriophages. (a) T4 phage, which is representative of T-even phages (1 nm = 10^{-9} m). **(b)** λ phage.

b)

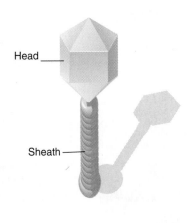

Figure 6.11

Lytic life cycle of a virulent phage, such as T4.

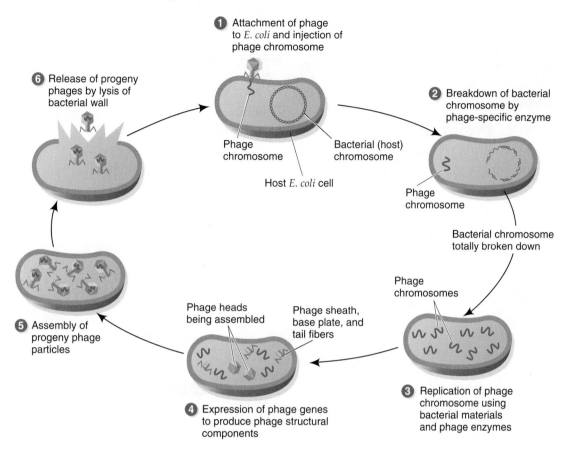

1 Attachment of phage to *E. coli* and injection of phage chromosome

Phage chromosome

Bacterial (host) chromosome

Host *E. coli* cell

6 Release of progeny phages by lysis of bacterial wall

2 Breakdown of bacterial chromosome by phage-specific enzyme

Phage chromosome

Bacterial chromosome totally broken down

Phage chromosomes

5 Assembly of progeny phage particles

Phage heads being assembled

Phage sheath, base plate, and tail fibers

3 Replication of phage chromosome using bacterial materials and phage enzymes

4 Expression of phage genes to produce phage structural components

way, the λ chromosome does not replicate; instead, it inserts (integrates) itself physically into a specific region of the host cell's chromosome, much like *F* factor integration. In this integrated state, the phage chromosome is called a **prophage.** Every time the host cell chromosome replicates, the integrated λ chromosome replicates as part of it. The bacterium that contains a phage in the prophage state is said to be **lysogenic** for that phage; the phenomenon of the insertion of a phage chromosome into a bacterial chromosome is called **lysogeny.** Phages that have a choice between lytic and lysogenic pathways are called *temperate phages.*

The prophage state is maintained by the action of a specific phage gene product (a repressor protein) that prevents expression of λ genes essential for the lytic cycle. When the repressor that maintains the prophage state is destroyed—for example, by environmental factors such as ultraviolet light irradiation—then the lytic cycle is induced. When induction occurs, the integrated λ chromosome is excised from the bacterial chromosome and the lytic cycle begins, resulting in the production and release from the cell of progeny λ phages.

Transduction Mapping of Bacterial Chromosomes

Transduction is a useful mechanism for mapping bacterial genes. Two types of transduction occur. In **generalized transduction,** any gene can be transferred between bacteria; in **specialized transduction,** only specific genes are transferred. In specialized transduction, by a process similar to the production of an *F′* factor (see pp. 125–127), an integrated phage genome excises imperfectly, picking up adjacent bacterial genes while leaving some phage genes behind. We will focus on generalized transduction here.

Joshua and Esther Lederberg and Norton Zinder discovered transduction in 1952. These researchers tested whether conjugation occurs in the bacterial species *Salmonella typhimurium.* Their experiment was similar to the one showing that conjugation exists in *E. coli.* They mixed together two multiple auxotrophic strains *phe⁺ trp⁺ tyr⁺ met his* (requiring methionine and histidine) and *phe trp tyr met⁺ his⁺* (requiring phenylalanine, tryptophan, and tyrosine) and found prototrophic recombinants—*phe⁺ trp⁺ tyr⁺ met⁺ his⁺*—at a low frequency. But unlike in the conjugation experiment, when they

Figure 6.12

Life cycle of a temperate phage, such as λ. When a temperate phage infects a cell, the phage may go through either the lytic or lysogenic cycle.

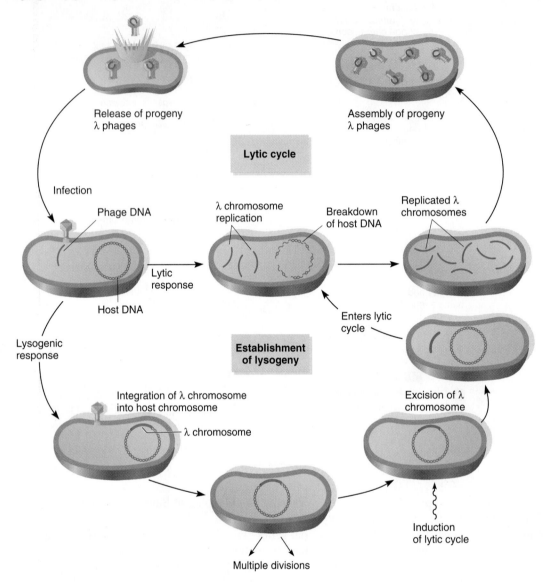

used the U-tube apparatus (which helped establish conjugation in *E. coli*), they still found prototrophs. This result indicated that recombinants were being produced by a mechanism that did not require cell-to-cell contact. The interpretation was that the agent responsible for the formation of recombinants was a *filterable agent* because it could pass through a filter with pores small enough to block bacteria. In this particular case, the filterable agent was identified as a temperate phage.

As an example, Figure 6.13 shows the mechanism for generalized transduction of *E. coli* by temperate phage P1. (1) Normally, the P1 phage enters the lysogenic state when it infects *E. coli*, existing in the prophage state in these cells. (2) If the lysogenic state is not maintained, the phage goes through the lytic cycle and produces progeny phages. (3) During the lytic cycle, the bacterial DNA is degraded and, rarely, a piece of bacterial DNA is packaged

into a phage head instead of phage DNA. These phages are called **transducing phages** because they are the vehicles by which genetic material is carried between bacteria. In Figure 6.13, the transducing phages are those carrying the donor bacterial genes a^+, b^+, or c^+. (4) The population of phages in the phage lysate, consisting mostly of normal phages but with about 1 in 10^5 transducing phages present, (5) can now be used to infect a new population of bacteria. The recipient bacteria are a in genotype. (6) If a transducing phage carrying the a^+ gene infects the recipient, genetic exchange of the donor a^+ gene with the recipient a gene can occur by double crossing-over. (7) The result is stable transduced bacteria called *transductants;* in this case, they are a^+ transductants.

Typically, a transduction experiment is designed so that the donor cell type and the recipient cell type have different genetic markers; then the transduction events

Figure 6.13

Generalized transduction between strains of *E. coli*. (**1**) Wild-type donor cell of *E. coli* infected with the temperate bacteriophage P1. (**2**) The host cell DNA is broken up during the lytic cycle. (**3**) During assembly of progeny phages, some pieces of the bacterial chromosome are incorporated into some of the progeny phages to produce transducing phages. (**4**) After cell lysis, a low frequency of transducing phages is found in the phage lysate. (**5**) The transducing phage infects an auxotrophic recipient bacterium. (**6**) A double-crossover event results in the exchange of the donor a^+ gene with the recipient mutant a gene. (**7**) The result is a stable a^+ transductant, with all descendants of that cell having the same genotype.

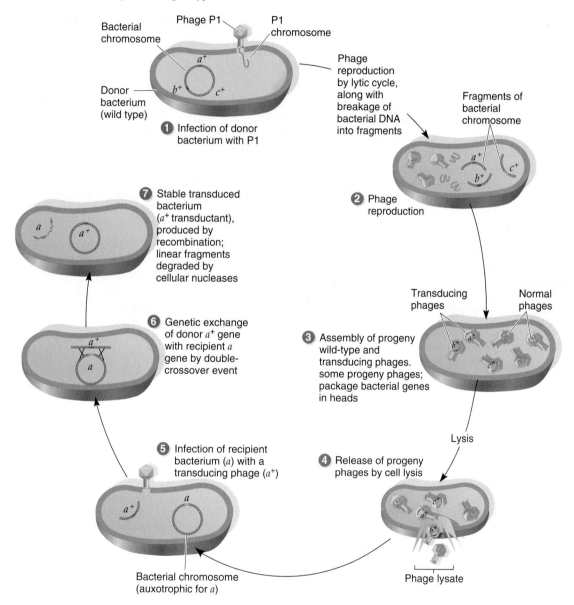

can be followed. For instance, if the donor cell is *thr*⁺ and the recipient cell is *thr*, prototrophic transductants can be detected because the cell no longer requires threonine to grow. In this way, researchers can pick out the extremely low number of transductants from among all the cells present by selecting for those cells that are able to do something the nontransduced cells cannot—namely, grow on a minimal medium. In this case, *thr*⁺ is called a *selected marker*. Other markers in the experiment are called *unselected markers*.

In generalized transduction, the process we have been describing, the piece of bacterial DNA that the phage erroneously picks up is a *random* piece of the fragmented bacterial chromosome. Thus, any genes can be transduced; only an appropriate phage and bacterial strains carrying different genetic markers are needed. Just as in transformation, generalized transduction can be used to determine gene linkage and gene order and to map the distance between genes. The logic is identical for the two processes. That is, the size of the DNA fragment that can be contained within a phage head is small, about 1.5 percent that of the *E. coli* chromosome. So only if two genes are physically close on the chromosome will they be found in the same transducing phage. Consider a donor bacterium with genotype a^+b^+ and a recipient bacterium with genotype $a\ b$. If the genes are close enough to be included in the transducing phage, then an a^+b^+ transductant—called a *cotransductant*—can be produced through infection by a single transducing phage. If the genes are not closely linked, then an a^+b^+ transductant can only be produced by the much rarer event of simultaneous infection of a recipient by two transducing phages, one carrying a^+ and the other carrying b^+. Therefore, if two genes are close enough that they often are packaged into the phage head on the same DNA fragment, the **cotransduction** frequency would be close to the frequency of transduction of a single gene; cotransduction of two or more genes is a good indication that the genes are closely linked.

K E Y N O T E

> Transduction is the process by which bacteriophages mediate the transfer of genetic information from one bacterium (the donor) to another (the recipient). The capacity of the phage particle is limited, so the amount of DNA transferred is usually less than 1 percent of that in the bacterial chromosome. In generalized transduction, any bacterial gene can be incorporated accidentally into the transducing phage during the phage life cycle and subsequently transferred to a recipient bacterium.

Mapping Genes of Bacteriophages

The same principles used to map eukaryotic genes are used to map phage genes. Crosses are made between phage strains that differ in genetic markers, and the proportion of recombinants among the total progeny is determined. The basic procedure of mapping phage genes in two-, three-, or four-gene crosses involves the mixed infection of bacteria with phages of different genotypes.

Before doing a mapping experiment involving phages, a basic experimental design must be chosen to ensure that we can accomplish two goals. First, we must be able to count the numbers of each phage type resulting from a phage cross. We do so by plating a mixture of phages and bacteria on a solid medium. The concentration of bacteria is chosen so that an entire lawn of bacteria grows. Phages are present in much lower concentrations. Each phage infects a bacterium and goes through the lytic cycle. The released progeny phages infect neighboring bacteria, and the cycle is repeated. The result is a clearing in the opaque lawn of bacteria. The clearing is called a **plaque,** and a single plaque derives from one of the original bacteriophages that was plated. Figure 6.14 shows plaques of the *E. coli* phage T4, which are uniformly clear (completely transparent).

The second goal is to distinguish phage phenotypes easily. Several mutations affect the phage life cycle, giving rise to differences in the appearance of plaques on a bacterial lawn. For example, there are strains of T2 that differ in either plaque morphology (the size of the plaque and shape of its borders) or host range (which bacterial strain the phage can lyse). Consider two phage strains. One has the genotype h^+r, meaning that it is wild-type for the host range gene (h^+, able to lyse the *B* strain but not the *B/2* strain of *E. coli*; i.e., strain *B* is the *permissive host* and strain *B/2* is the *nonpermissive host*

Figure 6.14

Plaques of the *E. coli* bacteriophage T4.

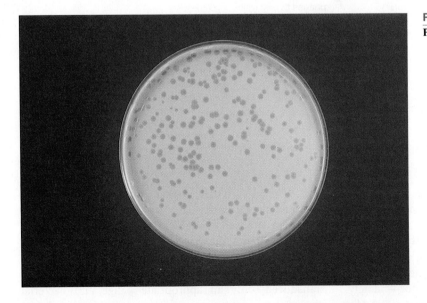

for h^+ phages) and mutant for the plaque morphology gene (r, producing large plaques with distinct borders). The other phage strain has the genotype $h\ r^+$, meaning that it is mutant for the host range gene (able to lyse both the B and $B/2$ strains of $E.\ coli$) and wild-type for the plaque morphology gene (r^+, producing small plaques with fuzzy borders). When plated on a lawn containing both the B and $B/2$ strains, any phage carrying the mutant host range allele h infects both B and $B/2$ and produces clear plaques, whereas phages carrying the wild-type h^+ allele produce cloudy plaques. The latter characteristic arises because phages bearing the h^+ allele can infect only the B bacteria, leaving a background cloudiness of uninfected $B/2$ bacteria.

To map these two genes, we make a genetic cross by infecting $E.\ coli$ strain B with the two (parental) phages, $h^+ r$ and $h\ r^+$ (Figure 6.15a). Once the two genomes are within the bacterial cell, each replicates (Figure 6.15b). If an $h^+ r$ and an $h\ r^+$ chromosome come together, a crossover can occur between the two gene loci to produce $h^+ r^+$ and $h\ r$ recombinant chromosomes (Figure 6.15c), which are assembled into progeny phages. When the bacterium lyses, the recombinant progeny are released into the medium, along with nonrecombinant (parental) phages (Figure 6.15d).

After the life cycle is completed, the progeny phages are plated onto a bacterial lawn containing a mixture of $E.\ coli$ strains B and $B/2$. Four plaque phenotypes are found from the experiment in Figure 6.15; two parental types and two recombinant types (Figure 6.16). The parental type $h\ r^+$ gives a small clear plaque with a fuzzy border; the other parental, $h^+ r$, gives a large cloudy plaque with a distinct border. The reciprocal recombinant types give recombined phenotypes: The $h^+ r^+$ plaques are cloudy and small with a fuzzy border, and the $h\ r$ plaques are clear and large with a distinct border.

Once the progeny plaques are counted, the recombination frequency between h and r is given by

$$\frac{(h^+\ r^+) + (h\ r)\ \text{plaques}}{\text{total plaques}} \times 100\%$$

As with eukaryotes, this recombination frequency reflects the relative genetic distance between the phage genes. When the genes are close enough together so that multiple crossovers are not likely to occur, the recombination frequency equals the crossover frequency, and in this case, the recombination frequency can be converted directly to map units.

KEYNOTE

The same principles used to map eukaryotic genes are used to map phage genes. That is, genetic material is exchanged between strains differing in genetic markers, and recombinants are detected and counted.

Figure 6.15

The principles of performing a genetic cross with bacteriophages. (a) Bacteria of $E.\ coli$ strain B are coinfected with the two parental bacteriophages, $h^+ r$ and $h\ r^+$. (b) Replication of both parental chromosomes occurs. (c) There is pairing of some chromosomes of each parental type, and crossing-over takes place between the two gene loci to produce $h^+ r^+$ and $h\ r$ recombinants. (d) Progeny phages are assembled and are released into the medium when the bacteria lyse; both parental and recombinant phages are found among the progeny.

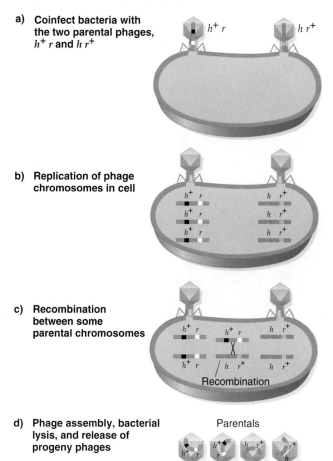

a) **Coinfect bacteria with the two parental phages, $h^+ r$ and $h\ r^+$**

b) **Replication of phage chromosomes in cell**

c) **Recombination between some parental chromosomes**

Recombination

d) **Phage assembly, bacterial lysis, and release of progeny phages**

Parentals

Recombinants

Fine-Structure Analysis of a Bacteriophage Gene

The recombinational mapping of the distance between genes, called *intergenic mapping* (*inter* means "between"), can be used to construct chromosome maps for both eukaryotic and prokaryotic organisms. The general principles of recombinational mapping can also be applied to mapping the distance between mutational sites within the same gene, a process called *intragenic mapping* (*intra* means "within").

Figure 6.16

Plaques produced by progeny of a cross of T2 strains
$h \, r^+$ × $h^+ r$. Four plaque phenotypes, representing both parental
types and the two recombinants, may be discerned. The parental
$h \, r^+$ phage produces a small clear plaque with a fuzzy border; the
other parental, $h^+ r$ phage, produces a large cloudy plaque with a
distinct border. The recombinant $h^+ r^+$ phage produces a small
cloudy plaque with a fuzzy border, and the recombinant $h \, r$ phage
produces a large clear plaque with a distinct border.

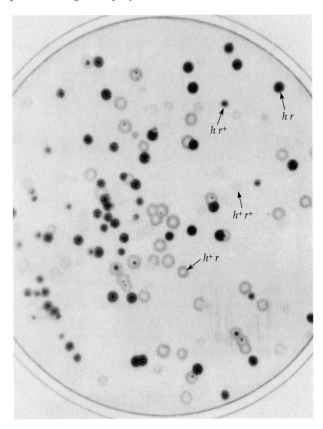

Intragenic mapping is possible because each gene
consists of many nucleotide pairs of DNA linearly
arranged along the chromosome. Much of the impetus
to analyze the fine details of gene structure came from
the elegantly detailed work of Seymour Benzer in the
1950s and 1960s on the *rII* region of bacteriophage T4.
His genetic experiments revealed much about the relationship between mapping and gene structure. In his
experiments, Benzer used phage T4 mainly because bacteriophages produce large numbers of progeny, which
facilitates the potential determination of very low
recombination frequencies. Benzer's initial experiments
involved **fine-structure mapping**, the detailed genetic
mapping of sites within a gene.

Benzer used strains of phage T4 carrying mutations
of the *rII* region. *rII* mutants have both a distinct *plaque
morphology* and distinct *host range properties*. Specifically,
when cells of *E. coli* growing on solid medium are
infected with wild-type (r^+) T4, small turbid plaques with

fuzzy edges are produced, whereas plaques produced by
rII mutants are large and clear. Regarding host range
properties, wild-type T4 can grow in and lyse cells of
either *E. coli* strain B or $K12(\lambda)$, whereas *rII* mutants can
grow in B but not $K12(\lambda)$. That is, strain B is the permissive host for *rII* mutants, and strain $K12(\lambda)$ is the nonpermissive host.

Recombination Analysis of *rII* Mutants

Benzer realized that the growth defect of *rII* mutants on
E. coli $K12(\lambda)$ could serve as a powerful selective tool for
detecting the presence of a very small proportion of r^+
phages within a large population of *rII* mutants. Initially,
Benzer set out to construct a fine-structure genetic map
of the *rII* region. He crossed 60 independently isolated *rII*
mutants in all possible combinations, using *E. coli* B as
the permissive host, and then collected the progeny
phages once the cells had lysed. For each cross—
rII1 × *rII2* in Figure 6.17—he plated a sample of the
phage progeny on *E. coli* B, the permissive host, in order
to count the total number of progeny phage per milliliter.
He plated another sample on *E. coli* $K12(\lambda)$, the nonpermissive host, to find the frequency of r^+ recombinants
resulting from genetic recombination between the pairs
of *rII* mutants used in each cross. In this way, Benzer calculated the percentage of very rare r^+ recombinants produced by crossing-over between closely linked alleles.
The recombination frequency for the two alleles is given
by the formula

$$\frac{2 \times \text{number of } r^+ \text{ recombinants}}{\text{total number of progeny}} \times 100\%$$

We multiply the number of r^+ recombinants by 2 to
account for the other class of recombinants that we cannot detect phenotypically, that is, the double mutants
(*rII1,2* in Figure 6.17).

An important control was set up for each cross. Each
rII parent alone was used to infect the permissive *E. coli*
B host and the progeny tested on plates of B and of
$K12(\lambda)$. Just as a mutation can occur to produce an *rII*
mutant from the r^+, a mutation can occur for an *rII*
mutant to change back (revert) to the r^+. Thus, it is
extremely important to calculate the reversion frequencies for the two *rII* mutations in a cross and subtract the
combined value from the computed recombination frequency. Fortunately, the reversion frequency for an *rII*
mutation is at least an order of magnitude lower than the
smallest recombination frequency that was found.

Benzer constructed a linear genetic map from the
recombination data obtained from all possible pairwise
crosses of the 60 *rII* mutants. Some pairs produced no r^+
recombinants when they were crossed, meaning that
those pairs carried mutations at exactly the same site.
Mutations that change the same nucleotide pair within a
gene are called *homoallelic*. However, most pairs of *rII*

Figure 6.17

Benzer's general procedure for determining the number of r^+ recombinants from a cross involving two *rII* mutants of T4.

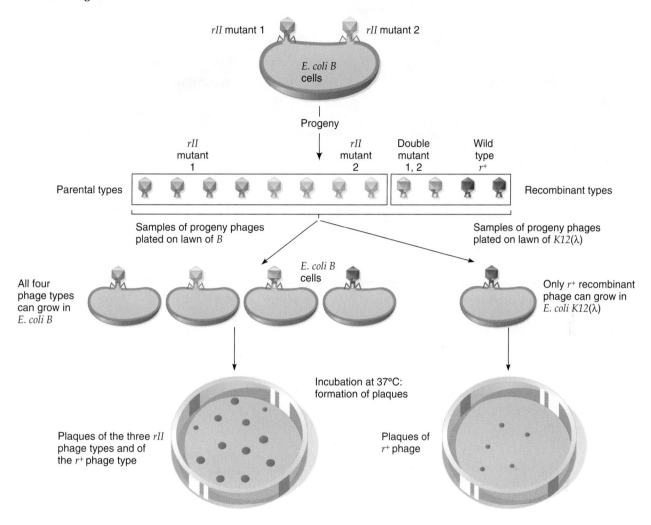

mutants did produce r^+ recombinants when crossed, indicating that they carried different altered nucleotide pairs in the DNA. Mutations that change different nucleotide pairs within a gene are called *heteroallelic*.

Benzer followed up these initial experiments with a more detailed mapping of more than 3,000 *rII* mutants. He showed from this that the *rII* region is subdivisible into more than 300 mutable sites that are separable by recombination (Figure 6.18). The distribution of mutants is not random; certain sites (called *hot spots*) are represented by a large number of independently isolated mutants.

The map shows that the lowest frequency with which r^+ recombinants were formed in any pairwise crosses of *rII* mutants involving heteroallelic mutations was 0.01 percent. The minimum map distance of 0.01 percent can be used to make a rough calculation of the molecular distance—the distance in nucleotide pairs—between mutant markers. The genetic map of phage T4

is about 1,500 map units. If two *rII* mutants produce 0.01 percent r^+ recombinants, this means that the mutations are separated by 0.02 map units, or by about $0.02/1,500 = 1.3 \times 10^{-5}$ of the total T4 genome. The total T4 genome contains about 2×10^5 nucleotide pairs (base pairs), so the smallest recombination distance that was observed was $(1.3 \times 10^{-5}) \times (2 \times 10^5)$, or about 3 base pairs. That means that Benzer's data showed that genetic recombination can occur within distances on the order of 3 base pairs. Later experiments by others showed conclusively that recombination can occur between mutations that affect adjacent base pairs in the DNA. That is, genetic experiments have shown that the base pair is both the *unit of mutation* and the *unit of recombination*. These definitions replaced the classical definitions that the gene was the unit of mutation and the unit of recombination—that is, that the gene was indivisible by the processes of mutation and recombination.

Figure 6.18

Fine-structure map of the *rII* region derived from Benzer's experiments. The number of independently isolated mutations that mapped to a given site is indicated by the number of blocks at the site. Hot spots are represented by a large number of blocks.

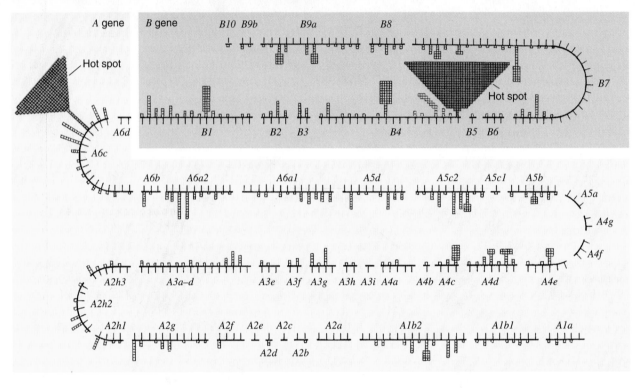

Defining Genes by Complementation (*Cis-Trans*) Tests

From the classic point of view, the gene is a *unit of function*; that is, each gene specifies one function. Benzer designed genetic experiments to determine whether this classic view was true of the *rII* region. To find out whether two different *rII* mutants belonged to the same gene (unit of function), Benzer used *cis-trans*, or **complementation, tests.** As you read the following discussion, it will help you to know that the complementation tests showed that the *rII* region actually consists of two genes (units of function), *rIIA* and *rIIB*. A mutation anywhere in either gene produces the *rII* plaque morphology phenotype and host range property. In

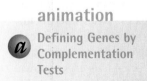

animation

a Defining Genes by Complementation Tests

other words, *rIIA* and *rIIB* each specify a different product needed for growth in *E. coli* K12(λ).

The complementation test is used to establish how many units of function (genes) are defined by a given set of mutations that express the same mutant phenotypes. In Benzer's work with the *rII* mutants, the nonpermissive strain *K12*(λ) was infected with a pair of *rII* mutant phages to see whether the two mutants, each unable by itself to grow in strain *K12*(λ), can work together to produce progeny phages. If the phages do produce progeny, the two mutants are said to complement each other, meaning that the two mutations must be in different genes (units of function) that encode different products. That is, those two products work together to allow progeny to be produced. If no progeny phages are produced, the mutants are not complementary, indicating that the mutations are in the same functional unit. Here, both mutants produce the same defective product, so the phage life cycle cannot proceed and no progeny phages result. (Note that genetic recombination is not necessary for complementation to occur; if genetic recombination does take place, a few plaques may occur on the lawn, but if complementation occurs, the entire lawn of bacteria will be lysed.)

These two situations are diagrammed in Figure 6.19. In the first case, the bacterium is infected with two phage genomes, one with a mutation in the *rIIA* gene and the other with a mutation in the *rIIB* gene (Figure 6.19a).

The *rIIA* mutant makes a nonfunctional *A* product and a functional *B* product, whereas the *rIIB* mutant makes a functional *A* product and a nonfunctional *B* product. Complementation occurs because the *rIIA* mutant still makes a functional *B* product and the *rIIB* mutant makes a functional *A* product, so phage propagation in *E. coli* K12(λ) can occur. In the second case, the bacterium is infected with two phage genomes, each with a different mutation in the same gene, *rIIA* (Figure 6.19b). No complementation occurs in this case because although both produce a functional *B* product, neither of the mutants makes functional *A* product, so the *A* function cannot take place and phage propagation in *E. coli* K12(λ) cannot occur.

On the basis of the results of such complementation tests, Benzer found that each *rII* mutant falls into one of two units of function, *rIIA* and *rIIB* (also called complementation groups, which directly correspond to genes). That is, all *rIIA* mutants complement all *rIIB* mutants, but *rIIA* mutants fail to complement other *rIIA* mutants, and *rIIB* mutants fail to complement other *rIIB* mutants. The dividing line between the *rIIA* and *rIIB* units of function is indicated in the fine-structure map in Figure 6.18.

For the complementation test examples shown in Figure 6.19, each phage that coinfects the nonpermissive *E. coli* strain K12(λ) carries one *rII* mutation, a configuration of mutations called the *trans* configuration. In this configuration, the two mutations are carried by different phages. As a control, it is usual to coinfect *E. coli* K12(λ) with an *r⁺* (wild-type) phage and an *rII* mutant phage carrying both mutations to see whether the expected wild-type function results. When both mutations under investigation are carried on the same chromosome, the configuration is called the *cis* configuration of mutations. (It is because of the *cis* and *trans* configurations of mutations that the complementation test is also called the *cis-trans* test.) In the *cis* test, the *r⁺* is expected to be dominant over the two muta-

tions carried by the *rII* mutant phage, so progeny phages are produced. Therefore, failure to produce progeny would not prove that the mutations are in different functional genes. Presumably, their two products act in common processes necessary for T4 propagation in strain K12(λ).

Benzer called the genetic unit of function revealed by the *cis-trans* test the *cistron*. A cistron may be defined as the smallest segment of DNA that encodes a piece of RNA. At present, *gene* is commonly used and *cistron* is used less often. Genetically, the *rIIA* cistron is about 6 mu and 800 base pairs long, and the *rIIB* cistron is about 4 mu and 500 base pairs long. The principles for performing a complementation test are the same in other organisms; only the practical details of performing the test are organism-specific. For example, in yeast, we could select two haploid cells that are of different mating types (**a** and α) and that carry different mutations conferring the same mutant phenotype. Mating these two types would produce a diploid, which would then be analyzed for complementation of the two mutations. In animal cells, two cells, each exhibiting the same mutant phenotype, could be fused together and analyzed; a wild-type phenotype would indicate that complementation occurred. Again, in neither of these cases is recombination necessary for complementation to occur.

KEYNOTE

The complementation, or *cis-trans*, test is used to determine how many units of function (genes) define a given set of mutations expressing the same mutant phenotypes. If two mutants, each carrying a mutation in a different gene, are combined, the mutations will complement and a wild-type function results. If two mutants, each carrying a mutation in the same gene, are combined, the mutations will not complement and the mutant phenotype is exhibited.

Figure 6.19

Complementation tests for determining the units of function in the *rII* region of phage T4; the nonpermissive host, *E. coli* K12(λ), is infected with two different *rII* mutants. (a) Complementation occurs. **(b)** Complementation does not occur.

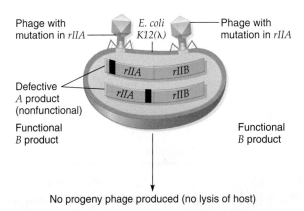

a) Complementation

Phage with mutation in *rIIA* — *E. coli* K12(λ) — Phage with mutation in *rIIB*

Defective *A* product (nonfunctional)

rIIA | *rIIB*

rIIA | *rIIB*

Defective *B* product (nonfunctional)

Functional *B* product

Functional *A* product

Progeny phage produced (lysis of host)

b) No complementation

Phage with mutation in *rIIA* — *E. coli* K12(λ) — Phage with mutation in *rIIA*

Defective *A* product (nonfunctional)

rIIA | *rIIB*

rIIA | *rIIB*

Functional *B* product

Functional *B* product

No progeny phage produced (no lysis of host)

Summary

In this chapter, we have seen how genes are mapped in prokaryotes such as bacteria and in bacteriophages. The same experimental strategy is used for all gene mapping; that is, genetic material is exchanged between strains differing in genetic markers and recombinants are detected and counted. In bacteria, the mechanism of gene transfer may be transformation, conjugation, or transduction. In each process, there is a donor strain and a recipient strain.

Conjugation is a plasmid-mediated process in which there is a unidirectional transfer of genetic information through direct cellular contact between a donor and a recipient. Transformation is the unidirectional transfer of extracellular DNA into cells, often resulting in a phenotypic change in the recipient. Transformation is the transfer of genetic material as small extracellular pieces of DNA between organisms. Transduction is a process whereby bacteriophages mediate the transfer of bacterial DNA from the donor bacterium to the recipient. Bacteriophage DNA can be mapped by infecting bacteria simultaneously with two phage strains and analyzing the resulting phage progeny for parental and recombinant phenotypes. Formally, this method of mapping is the same as that used for mapping genes in haploid eukaryotes.

Insights into the relationships between mapping and gene structure were obtained from a fine-structure analysis of the bacteriophage T4 *rII* region. The mutational sites within a gene were mapped through intragenic mapping. The resulting map indicated that the unit of mutation and the unit of recombination are the same: the base pair in DNA. These definitions replaced the classic definition of a gene as a unit of mutation, recombination, and function.

The number of units of function (genes) is determined by complementation tests. Given a set of mutations expressing the same mutant phenotype, two mutants are combined and the phenotype is determined. If the phenotype is wild-type, the two mutations have complemented and must be in different units of function. If the phenotype is mutant, the two mutations have not complemented and must be in the same unit of function.

Analytical Approaches for Solving Genetics Problems

Q6.1 In *E. coli*, the following *Hfr* strains donate the genes shown in the order given:

Hfr Strain	Order of Gene Transfer
1	G E B D N A
2	P Y L G E B
3	X T J F P Y
4	B E G L Y P

All the *Hfr* strains were derived from the same F⁺ strain. What is the order of genes in the original F⁺ chromosome?

A6.1 This question is an exercise in piecing together various segments of the circumference of a circle. The best approach is to draw a circle and label it with the genes transferred from one *Hfr* and then see which of the other *Hfr* strains transfers an overlapping set. For example, *Hfr* 1 transfers E, then B, then D, and so on, and *Hfr* 4 transfers B, then E, and so forth. Now we can juxtapose the two sets of genes transferred by the two *Hfr* strains and deduce that the polarities of transfer are opposite:

$$Hfr\ 1 \qquad G\ E\ B\ D\ N\ A$$
$$Hfr\ 2 \quad P\ Y\ L\ G\ E\ B$$

Extending this reasoning to the other *Hfr* strains, we can draw an unambiguous map (see the following figure), with the arrowheads indicating the order of transfer.

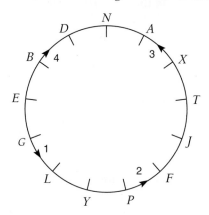

The same logic would be used if the question gave the relative time units of entry of each of the genes. In that case, we would expect that the time distance between any two genes would be approximately the same, regardless of the order of transfer or how far the genes were from the origin.

Questions and Problems

*6.1 In F⁺ × F⁻ crosses, the F⁻ recipient is converted to a donor with very high frequency. However, it is rare for a recipient to become a donor in *Hfr* × F⁻ crosses. Explain why.

*6.2 With the technique of interrupted mating, four *Hfr* strains were tested for the sequence in which they transmitted a number of different genes to an F⁻ strain. Each *Hfr* strain was found to transmit its genes in a unique order, as shown in the accompanying table (only the first six genes transmitted were scored for each strain).

Order of Transmission	Hfr Strain 1	2	3	4
First	O	R	E	O
	F	H	M	G
	B	M	H	X
	A	E	R	C
	E	A	C	R
Last	M	B	X	H

What is the gene sequence in the original strain from which these *Hfr* strains derive? Indicate on a diagram the origin and polarity of each of the four *Hfr* strains.

6.3 At time zero, an *Hfr* strain (strain 1) was mixed with an F⁻ strain, and at various times after mixing, samples were removed and agitated to separate conjugating cells. The cross may be written as

$$Hfr\ 1:\quad a^+\ b^+\ c^+\ d^+\ e^+\ f^+\ g^+\ h^+\ str^S$$
$$F^-:\quad a\ b\ c\ d\ e\ f\ g\ h\ str^R$$

(No order is implied in listing the markers.) The samples were then plated onto selective media to measure the frequency of $h^+\ str^R$ recombinants that had received certain genes from the *Hfr* cell. The graph of the number of recombinants against time is shown in the accompanying figure.

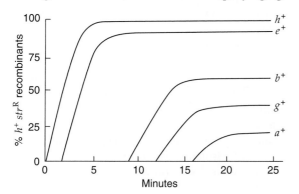

a. Indicate whether each of the following statements is true or false.
   i. All F⁻ cells that received a^+ from the *Hfr* in the chromosome transfer process must also have received b^+.
   ii. The order of gene transfer from *Hfr* to F⁻ was a^+ (first), then g^+, then b^+, then e^+, then h^+.
   iii. Most $e^+\ str^R$ recombinants are likely to be *Hfr* cells.
   iv. None of the $b^+\ str^R$ recombinants plated at 15 minutes are also a^+.
b. Draw a linear map of the *Hfr* chromosome indicating
   i. the point of nicking (the origin) and direction of DNA transfer
   ii. the order of the genes a^+, b^+, e^+, g^+, and h^+
   iii. the shortest distance between consecutive genes on the chromosomes

*6.4 If an *E. coli* auxotroph *A* could grow only on a medium containing thymine, and an auxotroph *B* could grow only on a medium containing leucine, how would you test whether DNA from *A* could transform *B*?

*6.5 Wild-type phage T4 grows on both *E. coli* B and *E. coli* K12(λ), producing turbid plaques. The *rII* mutants of T4 grow on *E. coli* B, producing clear plaques, but do not grow on *E. coli* K12(λ). This host range property permits the detection of a very low number of r^+ phages among a large number of *rII* phages. With this sensitive system, it is possible to determine the genetic distance between two mutations within the same gene, in this case the *rII* locus. Suppose *E. coli* B is mixedly infected with *rIIx* and *rIIy*, two separate mutants in the *rII* locus. Suitable dilutions of progeny phages are plated on *E. coli* B and *E. coli* K12(λ). A 0.1-mL sample of a thousandfold dilution plated on *E. coli* B produces 672 plaques. A 0.2-mL sample of undiluted phage plated on *E. coli* K12(λ) produces 470 turbid plaques. What is the genetic distance between the two *rII* mutations?

6.6 Construct a map from the following two-factor phage cross data (show map distance):

Cross	% Recombination
$r1 \times r2$	0.10
$r1 \times r3$	0.05
$r1 \times r4$	0.19
$r2 \times r3$	0.15
$r2 \times r4$	0.10
$r3 \times r4$	0.23

*6.7 The following two-factor crosses were made to analyze the genetic linkage between four genes in phage λ: *c*, *mi*, *s*, and *co*.

Parents	Progeny
$c + \times + mi$	1,213 $c +$, 1,205 $+ mi$, 84 $+ +$, 75 $c\ mi$
$c + \times + s$	566 $c +$, 808 $+ s$, 19 $+ +$, 20 $c\ s$
$co + \times + mi$	5,162 $co +$, 6,510 $+ mi$, 311 $+ +$, 341 $co\ mi$
$mi + \times + s$	502 $mi +$, 647 $+ s$, 65 $+ +$, 56 $mi\ s$

Construct a genetic map of the four genes.

6.8 Three gene loci in T4 that affect plaque morphology in easily distinguishable ways are *r* (rapid lysis), *m* (minute), and *tu* (turbid). A culture of *E. coli* is mixedly infected with two types of phage: $r\ m\ tu$ and $r^+\ m^+\ tu^+$. Progeny phages are collected, and the following genotype classes are found:

Class	Number
$r^+\ m^+\ tu^+$	3,729
$r^+\ m^+\ tu$	965
$r^+\ m\ tu^+$	520
$r\ m^+\ tu^+$	172
$r^+\ m\ tu$	162
$r\ m^+\ tu$	474
$r\ m\ tu^+$	853
$r\ m\ tu$	3,467
	10,342

Construct a map of the three genes. What is the coefficient of coincidence, and what does the value suggest?

*6.9 The *rII* mutants of bacteriophage T4 grow in *E. coli* B but not in *E. coli* K12(λ). The *E. coli* B strain is doubly infected with two *rII* mutants. A 6×10^7 dilution of the lysate is plated on *E. coli* B. A 2×10^5 dilution is plated on *E. coli* K12(λ). Twelve plaques appeared on strain K12(λ) and 16 on strain B. Calculate the amount of recombination between these two mutants.

6.10 Wild-type (r^+) strains of T4 produce turbid plaques, whereas *rII* mutant strains produce larger, clearer plaques. Five *rII* mutations (*a*–*e*) in the *A* cistron of the *rII* region of T4 give the following percentages of wild-type recombinants in two-point crosses:

Cross	% of Wild-type Recombinants	Cross	% of Wild-type Recombinants
$a \times b$	0.2	$e \times d$	0.7
$a \times c$	0.9	$e \times c$	1.2
$a \times d$	0.4	$e \times b$	0.5
$b \times c$	0.7	$b \times d$	0.2
$e \times a$	0.3	$d \times c$	0.5

What is the order of the mutational sites, and what are the map distances between the sites?

6.11 Some adenine-requiring mutants of yeast are pink because of the intracellular accumulation of a red pigment. Diploid strains were made by mating haploid mutant strains. The diploids exhibited the following phenotypes:

Cross	Diploid Phenotypes
1×2	pink, adenine-requiring
1×3	white, prototrophic
1×4	white, prototrophic
3×4	pink, adenine-requiring

How many genes are defined by the four different mutants? Explain.

*6.12 In *Drosophila* mutants *A, B, C, D, E, F,* and *G*, all have the same phenotype: the absence of red pigment in the eyes. In pairwise combinations in complementation tests, the following results were produced, where + = complementation and − = no complementation.

	A	B	C	D	E	F	G
G	+	−	+	+	+	+	−
F	−	+	+	−	+	−	
E	+	+	−	+	−		
D	−	+	+	−			
C	+	+	−				
B	+	−					
A	−						

a. How many genes are present?
b. Which mutants have defects in the same gene?

6.13 A homozygous white-eyed Martian fly (w_1/w_1) is crossed with a homozygous white-eyed fly from a different stock (w_2/w_2). It is well known that wild-type Martian flies have red eyes. This cross produces all white-eyed progeny. State whether each of the following is true or false. Explain your answer.
a. w_1 and w_2 are allelic genes.
b. w_1 and w_2 are nonallelic.
c. w_1 and w_2 affect the same function.
d. The cross was a complementation test.
e. The cross was a *cis-trans* test.
f. w_1 and w_2 are allelic by terms of the functional test.

The F_1 white-eyed flies are allowed to interbreed, and when you classify the F_2, you find 20,000 white-eyed flies and 10 red-eyed progeny. Concerned about contamination, you repeat the experiment and get exactly the same results. How can you best account for the presence of the red-eyed progeny? As part of your explanation, give the genotypes of the F_1 and F_2 generation flies.

6.14 A large number of mutations in *Drosophila* alter the normal deep-red eye color. As was discussed in Chapter 4 (p. 79), even alleles at one gene (*white*, *w*) can display a variety of phenotypes.

You are given a wild-type, deep-red strain and six independently isolated, true-breeding mutant strains that have varying shades of brown eyes, with the assurance that each mutant strain has only a single mutation. How would you proceed to determine
a. whether the mutation in each strain is dominant or recessive?
b. how many different genes are affected in the six mutant strains?
c. which mutants, if any, are allelic?
d. whether any of these mutants are alleles of genes already known to affect eye color?

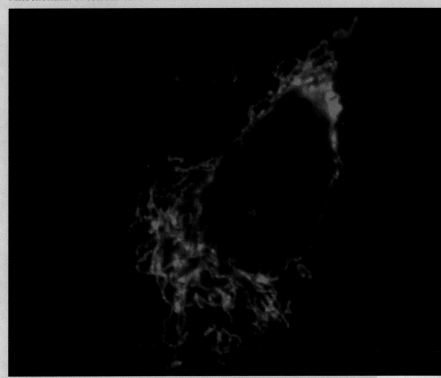

Mitochondria of osteosarcoma cells stained with MitoTracker Red.

7

Non-Mendelian Genetics

PRINCIPAL POINTS

Both mitochondria and chloroplasts contain their own DNA genomes, which contain a number of genes. The inheritance of mitochondrial and chloroplast genes follows rules different from those for nuclear genes: No meiotic segregation is involved, uniparental (usually maternal) inheritance is typically exhibited, extranuclear genes cannot be mapped to the known nuclear linkage groups, and a phenotype resulting from an extranuclear mutation persists after nuclear substitution.

Examples of gene mutations inherited in a non-Mendelian fashion, involving genes in mitochondrial DNA or chloroplast DNA, include shoot variegation in four o'clock plants, certain slow-growing mutants in fungi, and certain antibiotic resistance or dependence traits in *Chlamydomonas*.

Not all cases of non-Mendelian inheritance result from genes on mitochondrial DNA or chloroplast DNA. Many other examples in eukaryotes result from infectious heredity, in which symbiotic, cytoplasmically located bacteria or viruses are transmitted when cytoplasms mix.

Maternal effect is a phenotype in an individual that is established by the maternal nuclear genome as the result of mRNA or proteins that are deposited in the oocyte before fertilization. These inclusions direct early development of the embryo. Maternal effect is different from non-Mendelian inheritance. The maternal inheritance pattern of extranuclear genes occurs because the zygote obtains most of its cellular organelles (containing the extranuclear genes) from the female.

Although most genes are expressed independently of parental origin, the expression of certain genes is determined by whether the gene is inherited from the female or male parent. This phenomenon is called genomic imprinting.

iActivity

(i) IN 1909, A PLANT GENETICIST NAMED CARL CORRENS was investigating the inheritance of leaf color in a plant known as the four o'clock. Much to his surprise, he found that one trait did not conform to the expected pattern of Mendelian inheritance. Instead, the progeny phenotype was always the same as that of its female parent. In addition, the progeny phenotypes did not appear in Mendelian proportions. Years later, scientists discovered the reason for this unusual inheritance: The genes that controlled the color pattern were inherited via DNA from the plant's chloroplasts rather than from the nucleus! Since that time, scientists have identified other traits that are controlled by genes found in both chloroplasts and mitochondria.

In this chapter, you will learn about the structure and function of these extranuclear genes and the rules that govern non-Mendelian inheritance. Then, in the iActivity, you can use your understanding of both Mendelian and non-Mendelian inheritance patterns to study an unusual human disorder.

In our discussion of eukaryotic genetics up to now, we have analyzed the structure and expression of the genes located on chromosomes in the nucleus and have defined rules for the segregation of nuclear genes. Outside the nucleus, DNA is found in the mitochondrion (found in both animals and plants) and the chloroplast (found only in green plants and photosynthetic protists). **Mitochondria** (see Figure 1.8) contain the enzymes for carrying out most of the oxidative reactions that generate energy for cell functions. These enzymes include pyruvate dehydrogenase and the various enzymes needed to facilitate electron transport and oxidative phosphorylation, the citric acid cycle, and fatty acid oxidation. **Chloroplasts** are the site of photosynthesis in specialized cells in green plants and certain protists. Chloroplasts have a double membrane surrounding an internal, chlorophyll-containing lamellar structure embedded in a protein-rich stroma.

The genes in mitochondrial (mt) and chloroplast (cp) genomes are known as extrachromosomal genes, cytoplasmic genes, non-Mendelian genes, organellar genes, or extranuclear genes. The term *non-Mendelian* is particularly informative because extranuclear genes do not follow the rules of Mendelian inheritance, as do nuclear genes.

Cytoplasm is inherited from the mother in many organisms, so the inheritance of such cytoplasmic factors in these organisms is strictly maternal. In this chapter, we discuss the inheritance patterns of extranuclear genes.

Origin of Mitochondria and Chloroplasts

First, let us consider how mitochondria and chloroplasts might have arisen. The widely accepted **endosymbiont hypothesis** is that mitochondria and chloroplasts originated as free-living prokaryotes that invaded primitive eukaryotic cells and established a mutually beneficial (symbiotic) relationship. According to this hypothesis, eukaryotic cells started out as anaerobic organisms that lacked mitochondria and chloroplasts. At some point in their evolution, about a billion years ago, a eukaryotic cell established a symbiotic relationship with a purple nonsulfur photosynthetic bacterium. Over time, the oxidative phosphorylation activities of the bacterium became used for the benefits of the eukaryotic cell, and in the presence of atmospheric oxygen, photosynthetic activity (which is a generator of oxygen) was no longer needed and was lost. Eventually, the eukaryotic cell became dependent on the intracellular bacterium for survival, and the mitochondrion was formed. The chloroplasts of plants and algae are hypothesized to have occurred either simultaneously or later by the ingestion of an oxygen-producing photosynthetic bacterium (a cyanobacterium) by a eukaryotic cell.

Many proteins found in mitochondria and chloroplasts are encoded by nuclear genes. Thus, further evolution of mitochondria and chloroplasts probably involved the transfer of genes from the organelles to the nuclear DNA.

Rules of Non-Mendelian Inheritance

Since the pattern of inheritance shown by genes located in organelles other than the nucleus differs strikingly from the pattern shown by nuclear genes, the term **non-Mendelian inheritance** is appropriate when we are discussing extranuclear genes. In fact, if the results obtained from genetic crosses do not conform to predictions based on the inheritance of nuclear genes, there is a good reason to suspect extranuclear inheritance.

Here are the four main characteristics of non-Mendelian inheritance:

1. Ratios typical of Mendelian segregation are not found because meiosis-based Mendelian segregation is not involved.

2. In multicellular eukaryotes, the results of reciprocal crosses involving extranuclear genes are not the same as the results of reciprocal crosses involving nuclear genes because meiosis-based Mendelian segregation is not involved. (Recall from Chapters 2 and 3 that in

a reciprocal cross, the sexes of the parents are reversed in each case. For example, if A and B represent contrasting genotypes, A ♀ × B ♂ and A ♂ × B ♀ would represent a pair of reciprocal crosses.)

Extranuclear genes usually show **uniparental inheritance** from generation to generation. In uniparental inheritance, all progeny (both males and females) have the phenotype of only one parent. For multicellular eukaryotes, usually it is the mother's phenotype that is expressed exclusively, a phenomenon called **maternal inheritance.** Maternal inheritance occurs because the amount of cytoplasm in the female gamete usually greatly exceeds that in the male gamete. Therefore, the zygote receives most of its cytoplasm (containing the extranuclear genes in the mitochondria and, where applicable, the chloroplasts) from the female parent and a negligible amount from the male parent.

In contrast, the results of reciprocal crosses between a wild-type and a mutant strain are identical if the genes are located on nuclear chromosomes. One exception occurs when X-linked genes are involved (see Chapter 3), but even then the results are distinct from those for non-Mendelian inheritance.

3. Extranuclear genes cannot be mapped to the chromosomes in the nucleus.
4. Non-Mendelian inheritance is not affected by substituting a nucleus with a different genotype.

KEYNOTE

The inheritance of extranuclear genes follows rules different from those for nuclear genes. In particular, no meiotic segregation is involved, generally uniparental (and often maternal) inheritance is seen, extranuclear genes are not mappable to the known nuclear linkage groups, and the phenotype persists even after nuclear substitution.

Examples of Non-Mendelian Inheritance

In this section, we discuss the properties of a selected number of mutations in extranuclear chromosomes to illustrate the principles of non-Mendelian inheritance.

Shoot Variegation in the Four O'Clock

A variegated-shoot phenotype of the *albomaculata* strain of four o'clocks (*Mirabilis jalapa,* also called the marvel of Peru; Figure 7.1a) involves non-Mendelian inheritance. (A shoot of a plant consists of stem, leaves, and flowers.) This strain has mostly shoots with variegated leaves (leaves with yellowish-white patches) and occasional shoots that are wholly green or wholly yellowish (Figure 7.1b).

animation

a Shoot Variegation in the Four O'Clock

Figure 7.1

Variegation in the four o'clock, *Mirabilis jalapa.* (a) Photograph of the four o'clock. **(b)** Drawing showing variegation. Shoots that are all green, all white, and variegated are found on the same plant, and flowers may form on any of these shoots.

a)

b)

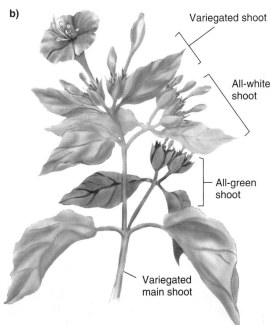

Table 7.1	Results of Crosses of Variegated Plants of *Mirabilis jalapa*	
Phenotype of Shoot-Bearing ♀ Parent (Egg)	**Phenotype of Shoot-Bearing ♂ Parent (Pollen)**	**Phenotype of Progeny**
White	White	White
	Green	White
	Variegated	White
Green	White	Green
	Green	Green
	Variegated	Green
Variegated	White	Variegated, green, or white
	Green	Variegated, green, or white
	Variegated	Variegated, green, or white

Table 7.1 summarizes the results of crosses between flowers on variegated, green, and white shoots. The seeds from flowers on green shoots give only green progeny, regardless of whether pollen is from green, white, or variegated shoots. Seeds from flowers on white shoots give only white progeny, regardless of the pollen source. (However, because the whiteness indicates the absence of chlorophyll and hence an inability to carry out photosynthesis, white progeny die soon after seed germination.) Finally, seeds from flowers on variegated shoots produce all three types of progeny—completely green, completely white, and variegated—regardless of the type of pollen. In subsequent generations, maternal inheritance is always seen in these same patterns. In sum, the *progeny phenotype in each case is the same as that of the maternal parent* (the color of the progeny shoots is the same as the color of the parental flower), which indicates maternal inheritance. Moreover, the results of reciprocal crosses differ from the expected patterns, and there is a lack of constant proportions of the different phenotypic classes in the segregating progeny. These last two properties are also characteristic of traits showing non-Mendelian inheritance and are not expected for traits showing Mendelian inheritance.

Flowering plants are green because of the presence of the green pigment chlorophyll in large numbers of chloroplasts. Green shoots in four o'clocks have a normal complement of chloroplasts. White shoots have abnormal, colorless chloroplasts called leukoplasts. Leukoplasts lack chlorophyll and therefore cannot carry out photosynthesis. The explanation for the inheritance of shoot color in the *albomaculata* strain of four o'clocks is that the abnormal chloroplasts are defective as a result of a mutant gene in the cpDNA. When a zygote receives both chloroplasts and leukoplasts from the maternal cytoplasm during plant growth and development, these organelles segregate so that a particular cell and its progeny cells may receive only chloroplasts (leading to green tissues), only leukoplasts (leading to white tissues), or a mixture of chloroplasts and leukoplasts (leading to variegation) (Figure 7.2). Let us take this model one step fur-

ther. The flowers on a green shoot have only chloroplasts; so through maternal inheritance, these chloroplasts form the basis of the phenotype of the next generation. Similar arguments can be made for the flowers on white or variegated shoots.

This simple model has three assumptions. One is that the pollen contributes essentially no cytoplasmic information—that is, no chloroplasts or leukoplasts—to the egg. This is a reasonable assumption because the egg is much larger than the pollen. Thus, we can assume that in the zygote, the extranuclear genetic determinants come from the egg. The second assumption is that the chloroplast genome replicates autonomously, and by growth and division of plastids (the general term for photosynthetic organelles, such as chloroplasts), the wild-type and mutant cpDNA molecules have the potential to segregate randomly to the new plastids, so that pure plastid lines can be generated from an original mixed line. The third assumption is that segregation of plastids to daughter cells is random, so that some daughters receive chloroplasts, some receive leukoplasts, and some receive mixtures.

KEYNOTE

The leaf color phenotypes in a variegated *albomaculata* strain of the four o'clock, *Mirabilis jalapa,* show maternal inheritance. The abnormal chloroplasts in white tissue are the result of a mutant gene in the cpDNA, and the observed inheritance patterns follow the segregation of cpDNA.

The [*poky*] Mutant of *Neurospora*

In the fungus *Neurospora crassa*, the sexual phase of the life cycle is initiated following a fusion of nuclei from mating type *A* and *a* parents. Even though this fungus is a simple eukaryote, the two gametes are not equal in cytoplasmic content, so we can think of a female gamete

Figure 7.2

Model for the inheritance of shoot color in the four o'clock, *Mirabilis jalapa*.

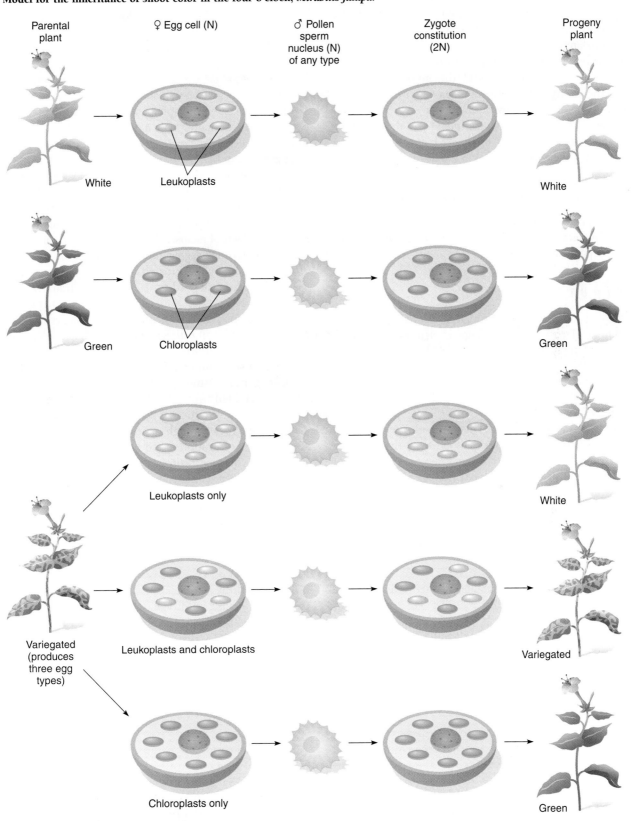

Parental plant · ♀ Egg cell (N) · ♂ Pollen sperm nucleus (N) of any type · Zygote constitution (2N) · Progeny plant

White · Leukoplasts · White

Green · Chloroplasts · Green

Leukoplasts only · White

Variegated (produces three egg types) · Leukoplasts and chloroplasts · Variegated

Chloroplasts only · Green

and a male gamete in much the same way as we can for animals and plants and make reciprocal crosses.

Neurospora crassa is an obligate aerobe; that is, it requires oxygen to grow and survive, so mitochondrial functions are essential for its growth. The [*poky*] mutant is partially defective in aerobic respiration as a result of changes in the cytochromes of the mitochondria that affect the ability of the mitochondria to generate enough ATP to support rapid growth; the result is slow growth.

Reciprocal crosses between [*poky*] and the wild type give the following results (the square brackets around the mutant symbols indicate an extranuclear gene):

[*poky*] ♀ × wild-type ♂ → all [*poky*] progeny
wild-type ♀ × [*poky*] ♂ → all wild-type progeny

As you can see, all the progeny show the same phenotype as the maternal parent, indicating maternal inheritance for the [*poky*] mutation.

KEYNOTE

The slow-growing [*poky*] mutant of *Neurospora crassa* shows maternal inheritance and deficiencies for some mitochondrial cytochromes.

Yeast *petite* Mutants

Yeast grows as single cells. Thus, on solid media, yeast forms discrete colonies consisting of many thousands of individual cells clustered together. Yeast can grow with or without oxygen. In the absence of oxygen, yeast obtains energy for growth and metabolism through fermentation, in which the mitochondria are not involved. In the presence of oxygen, mitochondria carry out aerobic respiration and facilitate a faster growth rate than is the case in the absence of oxygen.

Characteristics of *petite* Mutants. When yeast cells are spread onto solid medium, allowing growth to occur either by aerobic respiration or by fermentation, a very low proportion of the cells give rise to colonies that are much smaller than wild-type colonies. Since the discoverer of this phenomenon, Boris Ephrussi, was French, the small colonies are called *petites* (French for "small"), and the wild-type colonies sometimes are called *grandes* (French for "big"). *Petite* colonies are small because the growth rate of the mutant *petite* strain is significantly slower than that of the wild type, not because the cells are small. That is, there are fewer cells in the *petite* colonies.

Petites have cytochrome deficiencies and therefore are essentially incapable of carrying out aerobic respiration; they must obtain their energy primarily from fermentation, which is a less efficient process. The yeast *petite* system is particularly useful in studies of non-Mendelian inheritance because there is an abundance of *petite* mutants and because yeast cells that lack mito-

chondrial functions can still survive and grow. On a medium that can support fermentation and aerobic growth, the *petites* grow more slowly than the *grandes*, whereas on a medium that supports only aerobic respiration, *petites* are unable to grow.

Different Types of *petite* Mutants. Crosses between *petites* and wild-type yeast were made to determine how *petites* are inherited. Yeast are haploid, and crosses involve fusing two cells, one of mating type **a** and one of mating type α, to produce a diploid zygote (see p. 113 and Figure 5.13). That zygote can be grown into a colony to check its phenotype. When the zygote goes through meiosis (a process called sporulation), the four haploid meiotic products—the sexual spores (ascospores)—are contained within an ascus. Tetrad analysis can be done, that is, the analysis of all four products of meiosis (here, the four ascospores in an ascus) to determine segregation patterns.

Some *petites*, when crossed with wild-type yeast, give a 2:2 segregation of wild type : *petite* in tetrads (Figure 7.3a). This segregation pattern is that found for nuclear gene mutations, so these *petites* are *nuclear petites* (also called *segregational petites*) and symbolized *pet⁻*. The existence of *nuclear petites* is not surprising because some subunits of some mitochondrial proteins are encoded by nuclear genes. Using genetic symbols, when a *pet⁻* mutant is crossed with the wild type (*pet⁺*), the diploid is *pet⁺/pet⁻*, which produces wild-type colonies. When this *pet⁺/pet⁻* cell goes through meiosis, each resulting tetrad shows a 2:2 segregation of wild-type (*pet⁺*) : *petite* (*pet⁻*) phenotype. *Nuclear petites* occur much less frequently than extranuclear *petites*.

Two other classes of *petites*, the *neutral petites* and *suppressive petites*, show extranuclear inheritance. Figure 7.3b shows the inheritance pattern of *neutral petites* (symbolized [*rho⁻N*]). When a *neutral petite* is crossed with normal wild-type cells ([*rho⁺N*]), the [*rho⁺N*]/[*rho⁻N*] diploids all produce wild-type colonies. When these diploids go through meiosis, all resulting tetrads show a 0:4 ratio of *petite* : wild type, while nuclear genes segregate 2:2. The name *neutral*, then, refers to that fact that this class of *petites* does not affect the wild type. This result is a classic example of uniparental inheritance, in which all progeny have the phenotype of only one parent. *This phenomenon is not maternal inheritance, however, because the two haploid cells that fuse to produce the diploid are the same size and contribute equal amounts of cytoplasm.*

An examination of the mitochondrial genetic material in the *neutral petites* reveals a remarkable characteristic. Essentially 99 to 100 percent of the mtDNA is missing. Not surprisingly, then, the *neutral petites* cannot perform mitochondrial functions. They survive because of nuclear-encoded, cytoplasmically localized fermentation processes. In genetic crosses with the wild type, the normal mitochondria from the wild-type parent form a population from which new, normal mitochondria are

Figure 7.3

Inheritance of yeast *petite* mutants. **(a)** Mendelian inheritance of *nuclear petites* (*pet⁻*). **(b)** Non-Mendelian inheritance of *neutral petites* ([*rho⁻*N]). **(c)** Non-Mendelian inheritance of suppressive *petites* ([*rho⁻*S]).

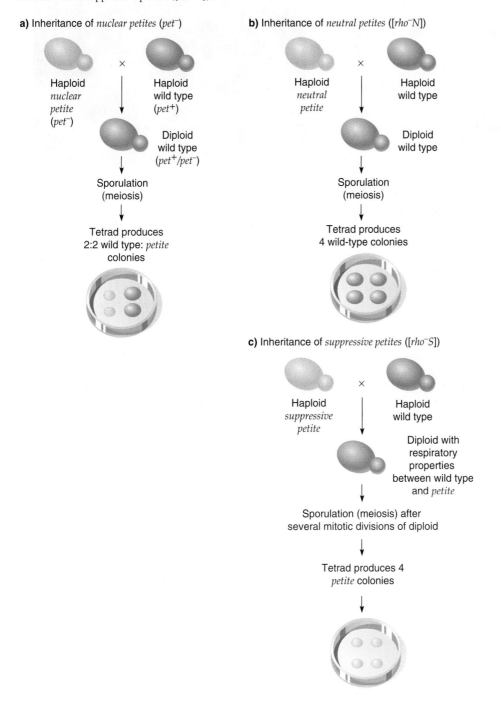

a) Inheritance of *nuclear petites* (*pet⁻*)

Haploid *nuclear petite* (*pet⁻*) × Haploid wild type (*pet⁺*)

Diploid wild type (*pet⁺*/*pet⁻*)

Sporulation (meiosis)

Tetrad produces 2:2 wild type: *petite* colonies

b) Inheritance of *neutral petites* ([*rho⁻*N])

Haploid *neutral petite* × Haploid wild type

Diploid wild type

Sporulation (meiosis)

Tetrad produces 4 wild-type colonies

c) Inheritance of *suppressive petites* ([*rho⁻*S])

Haploid *suppressive petite* × Haploid wild type

Diploid with respiratory properties between wild type and *petite*

Sporulation (meiosis) after several mitotic divisions of diploid

Tetrad produces 4 *petite* colonies

produced in all progeny, and therefore the *petite* trait is lost after one generation.

The second class of *petites* that shows extranuclear inheritance is the *suppressive petites* (symbolized [*rho⁻*S]). The *suppressive petites* are different from the *neutrals* because they *do* have an effect on the wild type. Most *petite* mutants are of the suppressive type.

The inheritance pattern of *suppressive petites* is different from that of *nuclear* and *neutral petites* (Figure 7.3c). When a [*rho⁺*]/[*rho⁻*S] diploid is formed, it has respiratory properties intermediate between those of normal and *petite* strains. If this diploid divides mitotically a number of times, the diploid population produced is up to 99 percent *petites*. The name *suppressive*, then, refers to that fact that this class of *petites* overwhelms the wild type so that a respiratory-deficient phenotype results. Sporulation of any of the *petites* in that population produces tetrads with a 4:0 ratio of *petite* : wild type (see Figure 7.3c). Sporulation of any of the few wild-type diploids in the population produces tetrads with a 0:4 ratio of *petite* : wild type.

The *suppressive petites* have changes in the mtDNA. The changes start out as deletions of part of the mtDNA, and then, by some correction mechanism, sequences that are not deleted become duplicated until the normal amount of mtDNA is restored. During the correction events, rearrangements of the mtDNA sometimes occur. These deletions and rearrangements disrupt genes or lead to losses of genes, thereby causing deficiencies in the enzymes involved in aerobic respiration, and so a *petite* colony results. *Suppressive petites* are proposed to have a suppressive effect over normal mitochondria either (1) by replicating faster than normal mitochondria and thereby outcompeting them or (2) by fusion with normal mitochondria followed by recombination between *suppressive* mtDNA and normal mtDNA, thereby altering the latter's gene organization.

KEYNOTE

> Yeast *petite* mutants grow slowly and have various deficiencies in mitochondrial functions as a result of alterations in mitochondrial DNA. Some *petite* mutants show Mendelian inheritance, while others show non-Mendelian inheritance. Particular patterns of inheritance are seen with different types of the latter *petite*.

Non-Mendelian Genetics of *Chlamydomonas*

The motile, haploid unicellular alga *Chlamydomonas reinhardtii* ("clammy-da-moan-us rhine-heart-E-eye"; Figure 7.4) has two flagella and a single chloroplast that contains many copies of cpDNA. There are two mating types, mt^+ and mt^-. In a process known as syngamous mating, a zygote is formed by fusion of two equal-sized cells (which therefore contribute an equal amount of cytoplasm), one of each mating type. A thick-walled cyst develops around the zygote. After meiosis, four haploid progeny cells are produced, and because mating type is determined by a nuclear gene, a 2:2 segregation of mt^+ : mt^- mating types results.

Chlamydomonas has proved to be an excellent system for studying chloroplast and mitochondrial inheritance. One chloroplast trait that is inherited in a non-Mendelian manner is erythromycin resistance ([ery^r]). Wild-type *Chlamydomonas* cells are erythromycin-sensitive ([ery^s]). From a cross of mt^+ [ery^r] × mt^- [ery^s], the offspring are all erythromycin-resistant (Figure 7.5a). This result is an example of uniparental inheritance. The reciprocal cross, mt^- [ery^r] × mt^+ [ery^s] (Figure 7.5b), also shows uni-

Figure 7.4

The unicellular green alga *Chlamydomonas*. This organism has a single chloroplast and a pair of flagella.

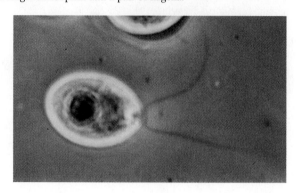

parental inheritance, although here the erythromycin-sensitive phenotype is inherited. Thus, even though both parents contribute equal amounts of cytoplasm to the zygote, the progeny resemble the mt^+ parent with regard to chloroplast-controlled phenotypes. Somehow, preferential segregation of one chloroplast type occurs in this organism, or perhaps one parental type is inactivated preferentially.

A number of *Chlamydomonas* chloroplast mutants in addition to ery^r show uniparental inheritance, with progeny *always* resembling the phenotype of the mt^+ parent. The molecular explanation for this is that the cpDNA from the mt^- parent always disappears from mt^+/mt^- zygotes. The reason is not known, however.

KEYNOTE

> *Chlamydomonas* chloroplast genes are inherited in a non-Mendelian manner, with progeny always resembling the phenotype of the mt^+ parent.

Human Genetic Diseases and Mitochondrial DNA Defects

A number of human genetic diseases result from mtDNA gene mutations. These diseases show maternal inheritance. The following are some brief examples.

Leber's hereditary optic neuropathy (*LHON*; OMIM 535000 at http://www.ncbi.nlm.nih.gov/Omim/) affects midlife adults and results in complete or partial blindness from optic nerve degeneration. Mutations in the mitochondrial genes for a number of electron transport chain proteins all lead to LHON. The electron transport chain drives cellular ATP production by oxidative phosphorylation. It appears that death of the optic nerve in LHON is a common result of oxidative phosphorylation defects, here brought about by inhibition of the electron transport chain.

In *Kearns-Sayre syndrome* (OMIM 530000), people exhibit three major types of neuromuscular defects: pro-

Individuals with Angelman syndrome (AS; OMIM 105830) have symptoms that include severe motor and intellectual retardation, smaller-than-normal head size, jerky limb movements, hyperactivity, and frequent unprovoked laughter. In about 50 percent of patients with AS, a deletion of region 15q11-q13 is seen. This is the same region affected in individuals with PWS. Indeed, it seems that AS can be caused in much the same way as PWS, except that in AS, maternally inherited alleles of the genes involved are needed for normal development. That is, the paternally inherited alleles are inactive because of methylation brought about by genomic imprinting, which causes AS to develop if the maternally inherited alleles are deleted or disrupted.

K E Y N O T E

> While most genes are expressed in a way that is totally independent of parental origin, the expression of certain genes is determined by whether the gene is inherited from the female or male parent. This phenomenon is called genomic imprinting.

Summary

Both mitochondria and chloroplasts contain a number of genes in their DNA genomes. There are many examples of mutations affecting those genes. The inheritance of these extranuclear genes follows rules different from those for nuclear genes, and this is how extranuclear genes were originally identified. For extranuclear genes, meiosis-based Mendelian segregation is not involved, the genes show uniparental (and usually maternal) inheritance, the trait involved persists after nuclear substitution, and the genes cannot be mapped to nuclear chromosomes. Not all cases of extranuclear inheritance result from genes on mtDNA or cpDNA. Other examples result from infectious heredity, in which cytoplasmically located bacteria or viruses are transmitted when cytoplasms mix.

Maternal effect is the phenotype in an individual that is established by the maternal nuclear genome as the result of mRNA and/or proteins that are deposited in the oocyte prior to fertilization. These inclusions direct early development of the embryo. Maternal effect is different from extranuclear inheritance. The maternal inheritance pattern of extranuclear genes occurs because the zygote receives most of its organelles (containing the extranuclear genes) from the female parent, whereas in maternal effect, the trait inherited is controlled by the maternal nuclear genotype before fertilization of the egg and does not involve extranuclear genes.

Finally, we discussed genomic imprinting, the phenomenon in which gene expression is determined by whether the gene is inherited from the female or male parent as opposed to being expressed independently of parental origin. This phenomenon does not involve extranuclear genes, but it does involve an influence on gene expression of one or the other parent. Methylation has been shown to be responsible for imprinting in some cases.

Analytical Approaches for Solving Genetics Problems

Q7.1 In *Neurospora*, strains of [*poky*] have been isolated that have reverted to nearly wild-type growth, although they still retain the abnormal metabolism and cytochrome pattern characteristic of [*poky*]. These strains are called *fast*-[*poky*]. A cross of a *fast*-[*poky*] female parent with a wild-type male parent gives a 1:1 segregation of [*poky*]–:–*fast*-[*poky*] ascospores in all asci. Interpret these results, and predict the results you would expect from the reciprocal of the stated cross.

A7.1 The [*poky*] mutation shows maternal inheritance, so when it is used as a female parent, all the progeny ascospores will carry [*poky*]. We can designate the cross as [*poky*] × [*N*], where *N* signifies normal cytoplasm. All progeny are [*poky*]. The asci show a 1:1 segregation for the [*poky*] and *fast*-[*poky*] phenotypes. This ratio is characteristic of a nuclear gene segregating in the cross. Therefore, the simplest explanation is that the factor that causes [*poky*] strains to be *fast*-[*poky*] is a nuclear gene mutation; we can call it *F* (its actual designated name). The *F* gene segregates in meiosis, as do all nuclear genes. The cross can be rewritten as [*poky*]*F* ♀ × [*N*] + ♂, which gives a 1:1 segregation of [*poky*]+ and [*poky*]*F*. The former is [*poky*] and the latter is *fast*-[*poky*].

With these results behind us, the reciprocal cross can be diagrammed as [*N*] + ♀ × [*poky*]*F* ♂. All the progeny spores from this cross are [*N*], half of them being *F* and half of them being +. If *F* has no effect on normal cytoplasm, then these two classes of spores would be phenotypically indistinguishable, which is the case.

Q7.2 Four slow-growing mutant strains of *Neurospora crassa*, coded *a*, *b*, *c*, and *d*, were isolated. All have an abnormal system of respiratory mitochondrial enzymes. The inheritance patterns of these mutants were tested in controlled crosses with the wild type, with the following results:

Female Parent		Male Parent	Progeny (Ascospores) Wild Type	Slow Growing
Wild type	×	*a*	847	0
a	×	Wild type	0	659
Wild type	×	*b*	1,113	0
b	×	Wild type	0	2,071
Wild type	×	*c*	596	590
Wild type	×	*d*	1,050	1,035

Give a genetic interpretation of these results.

A7.2 This question asks us to consider the expected transmission patterns for nuclear and extranuclear genes. The nuclear genes will have a 1:1 segregation in the offspring because this organism is a haploid organism and therefore should exhibit no differences in the segregation patterns, whichever strain is the maternal parent. On the other hand, a distinguishing characteristic of extranuclear genes is a difference in the results of reciprocal crosses. In *Neurospora*, this characteristic is usually manifested by all progeny having the phenotype of the maternal parent. With these ideas in mind, we can analyze each mutant in turn.

Mutant *a* shows a clear difference in its segregation in reciprocal crosses and is a classic case of maternal inheritance. The interpretation here is that the gene is extranuclear; therefore, the gene must be in the mitochondrion. The [*poky*] mutant described in this chapter shows this type of inheritance pattern.

By the same reasoning, the mutation in strain *b* must also be extranuclear.

Mutants *c* and *d* segregate 1:1, indicating that the mutations involved are in the nuclear genome. In these cases, we need not consider the reciprocal cross because there is no evidence of maternal inheritance. In fact, the actual mutations that are the basis for this question cause sterility, so the reciprocal cross cannot be done. We can confirm that the mutations are in the nuclear genome by doing mapping experiments, using known nuclear markers. Evidence of linkage to such markers would confirm that the mutations are not extranuclear.

Questions and Problems

*7.1 What features of non-Mendelian inheritance distinguish it from the inheritance of nuclear genes?

7.2 Distinguish between maternal effect and non-Mendelian inheritance.

7.3 Reciprocal crosses between two types of the evening primrose, *Oenothera hookeri* and *Oenothera muricata,* produce the following effects on the plastids:

O. hookeri female × *O. muricata* male → yellow plastids

O. muricata female × *O. hookeri* male → green plastids

Explain the difference in these results, noting that the chromosome constitution is the same in both types.

*7.4 A series of crosses are performed with a recessive mutation in *Drosophila* called *tudor*. Homozygous *tudor* animals appear normal and can be generated from the cross of two heterozygotes, but a true-breeding *tudor* strain cannot be maintained. When homozygous *tudor* males are crossed to homozygous *tudor* females, both of which appear to be phenotypically normal, a normal-appearing F$_1$

is produced. However, when F$_1$ males are crossed to wild-type females, or when F$_1$ females are crossed to wild-type males, no progeny are produced. The same results are seen in the F$_1$ progeny of homozygous *tudor* females crossed to wild-type males. The F$_1$ progeny of homozygous *tudor* males crossed to wild-type females appear normal, and they are capable of issuing progeny when mated either with each other or with wild-type animals.
a. How would you classify the *tudor* mutation? Why?
b. What might cause the *tudor* phenotype?

*7.5 A form of male sterility in corn is maternally inherited. Plants of a male-sterile line crossed with normal pollen give male-sterile plants. Some lines of corn carry a dominant, so-called restorer (*Rf*) gene that restores pollen fertility in male-sterile lines.
a. If a male-sterile plant is crossed with pollen from a plant homozygous for gene *Rf,* what will be the genotype and phenotype of the F$_1$?
b. If the F$_1$ plants of part (a) are used as females in a testcross with pollen from a normal plant (*rf/rf*), what would be the result? Give genotypes and phenotypes and designate the type of cytoplasm.

7.6 Distinguish between *nuclear* (segregational), *neutral,* and *suppressive petite* mutants of yeast.

*7.7 In yeast, a haploid *nuclear* (segregational) *petite* is crossed with a *neutral petite.* Assuming that both strains have no other abnormal phenotypes, what proportion of the progeny ascospores are expected to be *petite* in phenotype if the diploid zygote undergoes meiosis?

7.8 When grown on a medium containing acriflavin, a yeast culture produces a large number of very small (*tiny*) cells that grow very slowly. How would you determine whether the slow-growth phenotype was the result of a cytoplasmic factor or a nuclear gene?

7.9 *Drosophila melanogaster* has a sex-linked, recessive, mutant gene called *maroon-like (ma-l).* Homozygous *ma-l* females or hemizygous *ma-l* males have light-colored eyes because of the absence of the active enzyme xanthine dehydrogenase, which is involved in the synthesis of eye pigments. When heterozygous *ma-l+/ma-l* females are crossed with *ma-l* males, all the offspring are phenotypically wild-type. However, half the female offspring from this cross, when crossed back to *ma-l* males, give all *ma-l* progeny. The other half of the females, when crossed to *ma-l* males, give all phenotypically wild-type progeny. What is the explanation for these results?

*7.10 When females of a particular mutant strain of *Drosophila melanogaster* are crossed to wild-type males, all the viable progeny flies are females. Hypothetically, this result could be the consequence of either a sex-linked, male-specific lethal mutation or a maternally inherited factor that is lethal to males. What crosses would you perform to distinguish between these alternatives?

7.11 Reciprocal crosses between two *Drosophila* species, *D. melanogaster* and *D. simulans*, produce the following results:

melanogaster ♀ × *simulans* ♂ → females only
simulans ♀ × *melanogaster* ♂ → males, with few or no females

Propose a possible explanation for these results.

7.12 Some *Drosophila* are very sensitive to carbon dioxide; administering it to them anesthetizes them. The sensitive flies have a cytoplasmic particle called *sigma* that has many properties of a virus. Resistant flies lack *sigma*. The sensitivity to carbon dioxide shows strictly maternal inheritance. What would be the outcome of the following two crosses: (a) sensitive female × resistant male and (b) sensitive male × resistant female?

7.13 The pedigree in the following figure shows a family in which an inherited disease called Leber's hereditary optic atrophy is segregating. This condition causes blindness in adulthood. Studies have recently shown that the mutant gene causing Leber's hereditary optic atrophy is located in the mitochondrial genome.

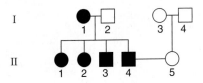

a. Assuming II-4 marries a normal person, what proportion of his offspring should inherit Leber's hereditary optic atrophy?

b. What proportion of the sons of II-2 should be affected?

c. What proportion of the daughters of II-2 should be affected?

*7.14 The inheritance of shell-coiling direction in the snail *Limnaea peregra* has been studied extensively. A snail produced by a cross between two individuals has a shell with a right-handed twist (dextral coiling). This snail produces only left-handed (sinistral) progeny on selfing. What are the genotypes of the F_1 snail and its parents?

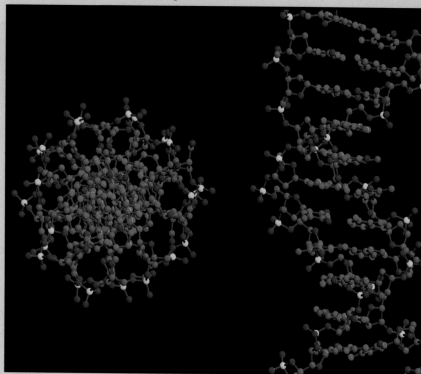

8

DNA: The Genetic Material

PRINCIPAL POINTS

Organisms contain genetic material that governs an individual's characteristics and that is transferred from parent to progeny.

DNA (deoxyribonucleic acid) is the genetic material of all living organisms and some viruses. RNA (ribonucleic acid) is the genetic material only of certain viruses.

DNA and RNA are macromolecules composed of smaller building blocks called nucleotides. Each nucleotide consists of a five-carbon sugar (deoxyribose in DNA, ribose in RNA) to which are attached a nitrogenous base and a phosphate group. In DNA, the four possible bases are adenine, guanine, cytosine, and thymine; in RNA, the four possible bases are adenine, guanine, cytosine, and uracil.

According to Watson and Crick's model, the DNA molecule consists of two polynucleotide chains (polymers of nucleotides) joined by hydrogen bonds between pairs of bases (adenine [A] and thymine [T], guanine [G] and cytosine [C]) in a double helix.

The genetic material of viruses may be double-stranded DNA, single-stranded DNA, double-stranded RNA, or single-stranded RNA, depending on the virus. The genomes of some viruses are organized into a single chromosome, whereas other viruses have a segmented genome.

The genetic material of prokaryotes is double-stranded DNA localized into one or a few chromosomes.

A bacterial chromosome is compacted by supercoiling of the DNA helix to produce looped domains.

The complete set of metaphase chromosomes in a eukaryotic cell is called its karyotype. The karyotype is species-specific.

The nuclear chromosomes of eukaryotes are complexes of DNA and histone and nonhistone chromosomal proteins. Each chromosome consists of one linear, unbroken, double-stranded DNA molecule running throughout its length; the DNA is variously coiled and folded. The histones are constant from cell to cell within an organism, whereas the nonhistones vary significantly between cell types.

The large amount of DNA present in the eukaryotic chromosome is compacted by its association with histones in nucleosomes and by higher levels of folding of the nucleosomes into chromatin fibers. Each chromosome contains a large number of looped domains of 30-nm chromatin fibers attached to a protein scaffold. The more condensed a part of a chromosome is, the less likely it is that the genes in that region will be active.

iActivity

(i) IMAGINE THAT YOU ARE HANDED A SEALED, BLACK box and are told it contains the secret of life. Determining the chemical composition, molecular structure, and function of the thing inside the box will allow you to save lives, feed the hungry, solve crimes, and even create new life forms. What's inside the box? What tools and techniques could you use to find out?

In this chapter, you will discover how scientists identified the contents of this "black box" and, in doing so, unraveled the secret of life. Later in this chapter, you can apply what you've learned by trying the iActivity, in which you use many of the same tools and techniques to determine the genetic nature of a virus that is ravaging rice plants in Asia.

In the previous chapters, we learned much about how traits are inherited, how recombination brings about variations in allele organization along chromosomes, and how genes are mapped. In all our discussions, however, we have taken as given the existence and function of the genetic material. In the next several chapters, we explore the molecular structure and function of genetic material—both **DNA (deoxyribonucleic acid)** and **RNA (ribonucleic acid)**—and examine the molecular mechanisms by which genetic information is transmitted from generation to generation. You will see exactly what the genetic message is, and you will learn how DNA stores and expresses that message. We begin by recounting how scientists discovered the nature and structure of the genetic material. These discoveries led to an explosion of knowledge about the molecular aspects of biology.

The Search for Genetic Material

Long before DNA and RNA were known to carry genetic information, scientists realized that living organisms contain some substance—a genetic material—that is responsible for the characteristics that are passed on from parent to child. Geneticists knew that the material responsible for hereditary information must have three principal characteristics:

1. It must contain, in a stable form, *the information* for an organism's cell structure, function, development, and reproduction.
2. It must *replicate accurately* so that progeny cells have the same genetic information as the parental cell.
3. It must be capable of *change*. Without change, organisms would be incapable of variation and adaptation, and evolution could not occur.

Around 1890, August Weismann proposed that there is a substance in cell nuclei that directs the development of cells and therefore controls the characteristics of the whole organism. In the early 1900s, experiments showed that **chromosomes**—the threadlike structures found in nuclei—are carriers of hereditary information. Chemical analysis over the next 40 years revealed that chromosomes are composed of protein and **nucleic acids,** a class of compounds that includes DNA and RNA. At first, many scientists believed that the genetic material must be protein. They reasoned that proteins have a great capacity for storing information because they are composed of 20 different amino acids. By contrast, DNA, with its four nucleotides, was thought to be too simple a molecule to account for the variation found in living organisms. However, beginning in the late 1920s, a series of experiments led to the clear identification of DNA as the genetic material.

Griffith's Transformation Experiment

One of the first studies was conducted in 1928 by Frederick Griffith, a British medical officer who was working with *Streptococcus pneumoniae* (also called pneumococcus), a bacterium that causes pneumonia (Figure 8.1). Griffith used two strains of the bacterium. One (the *S* strain) produces smooth, shiny colonies and is highly infectious (virulent); the other (the *R* strain) produces rough colonies and is harmless (avirulent). Although this was not known at the time, each cell of the *S* strain is surrounded by a polysaccharide (complex sugar) coat, or capsule, that results in the strain's infectious properties and is responsible for its smooth, shiny appearance. The *R* strain, which is a *mutation* of the *S* strain, lacks the polysaccharide coat.

Several variations of the *S* strain exist that have differences in the chemical composition of the polysaccharide coat. Griffith worked with two varieties, known as

Figure 8.1

Transmission electron micrograph of the bacterium *Streptococcus pneumoniae*. (Many of the cells are in the process of dividing.)

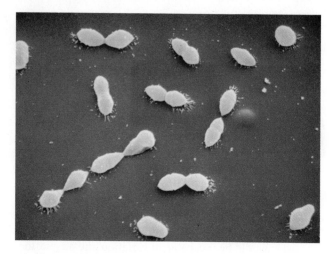

the *IIS* and *IIIS* strains. Occasionally, *S*-type cells can mutate, or change, into *R*-type cells, and vice versa. The mutations are type-specific; that is, if a *IIS* cell mutates into an *R* cell, then that *R* cell can mutate back only into a *IIS* cell, not a *IIIS* cell.

Griffith injected mice with different strains of the bacterium and observed whether the mice remained healthy or died (Figure 8.2). When mice were injected with *IIR* bacteria (*R* bacteria derived by mutation from *IIS* bacteria), the *IIR* bacteria did not affect the mice. When mice were injected with living *IIIS* bacteria, the mice died, and living *IIIS* bacteria could be isolated from their blood. However, if the *IIIS* bacteria were heat-killed before injection, the mice survived. These experiments showed that the bacteria had to be alive and had to have the polysaccharide coat to be infectious and kill the mice.

In his key experiment, Griffith injected mice with a mixture of living *IIR* bacteria and heat-killed *IIIS* bacteria. Surprisingly, the mice died, and living *S* bacteria were present in the blood. These bacteria were all of type *IIIS* and therefore could not have arisen by mutation of the *R* bacteria because mutation would have produced *IIS* colonies. Griffith concluded that some *IIR* bacteria had somehow been transformed into smooth, infectious *IIIS* cells by interaction with the dead *IIIS* cells. Griffith believed that the unknown agent responsible for the change in genetic material was a protein. He called the agent the **transforming principle.** (See Chapter 6 for a discussion of bacterial transformation.)

Figure 8.2

Griffith's transformation experiment. Mice injected with type *IIIS* pneumococcus died, whereas mice injected with either type *IIR* or heat-killed type *IIIS* bacteria survived. When injected with a mixture of living type *IIR* and heat-killed type *IIIS* bacteria, however, the mice died.

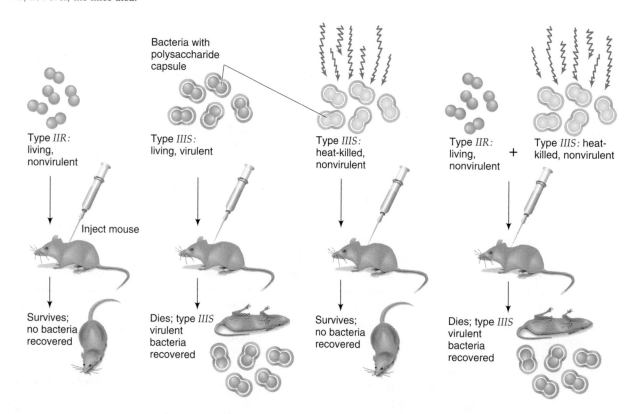

Avery's Transformation Experiment

Oswald T. Avery, along with his colleagues Colin M. MacLeod and Maclyn McCarty, followed up on Griffith's experiment in the 1930s and 1940s. The scientists tried to identify the transforming principle by studying the transformation of *R*-type bacteria to *S*-type bacteria in the test tube.

animation

a DNA as Genetic Material: The Avery Experiment

They lysed (broke open) *IIIS* cells and separated the cell extract into the various cellular macromolecular components, such as lipids, polysaccharides, proteins, and nucleic acids. They tested each component to see whether it contained the transforming principle by checking whether it could transform living *R* bacteria derived from *IIS* bacteria into *IIIS* bacteria. The nucleic acids (at this point not separated into DNA and RNA) were the only components from *IIIS* cells that could transform the *R* cells into *IIIS* cells.

Avery and his colleagues then used specific **nucleases**—enzymes that degrade nucleic acids—to determine whether DNA or RNA is the transforming principle (Figure 8.3). When they treated the nucleic acids with **ribonuclease (RNase),** which degrades RNA but not DNA, the transforming activity was still present. However, when they used **deoxyribonuclease (DNase),** which degrades only DNA, no transformation resulted. These results strongly suggested that DNA was the genetic material. Although Avery's work was important, it was criticized at the time because the nucleic acids isolated from the bacteria were contaminated by proteins.

The Hershey-Chase Bacteriophage Experiment

In 1953, Alfred D. Hershey and Martha Chase published a paper that provided more evidence that DNA is the genetic material. They were studying a bacteriophage called T2, which is a virtual twin of phage T4 (see EM in Figure 6.10a and the lytic life cycle in Figure 6.11).

animation

a DNA as Genetic Material: The Hershey-Chase Experiment

Hershey and Chase knew that T2 consisted of only DNA and protein and that the virus could somehow use its genetic material to reprogram its host cell to produce new phages. However, they did not know which component of the phage—DNA or protein—was responsible.

To determine the genetic material, Hershey and Chase grew cells of *Escherichia coli* in media containing either a radioactive isotope of phosphorus (^{32}P) or a radioactive isotope of sulfur (^{35}S) (Figure 8.4a). They used these isotopes because DNA contains phosphorus but no sulfur, and protein contains sulfur but no phosphorus. They infected the bacteria with T2 and collected the progeny phages. At this point, Hershey and Chase had two batches of T2; one had the proteins radioactively labeled with ^{35}S, and the other had the DNA labeled with ^{32}P.

They then infected *E. coli* with the two types of radioactively labeled T2 (Figure 8.4b). When the infecting phage was ^{32}P-labeled, most of the radioactivity was found within the bacteria soon after infection. Very little was found in protein parts of the phage (the *phage ghosts*) released from the cell surface after the cells were agitated in a kitchen blender. After lysis, some of the ^{32}P was found in the progeny phages. In contrast, after *E. coli*

Figure 8.3

Experiment that showed that DNA, not RNA, is the transforming principle. When the nucleic acid mixture of DNA and RNA was treated with ribonuclease (RNase), *S* transformants still resulted. However, when the DNA and RNA mixture was treated with deoxyribonuclease (DNase), no *S* transformants resulted. (Note: *R* colonies resulting from untransformed cells were present on both plates; they have been omitted from the drawings.)

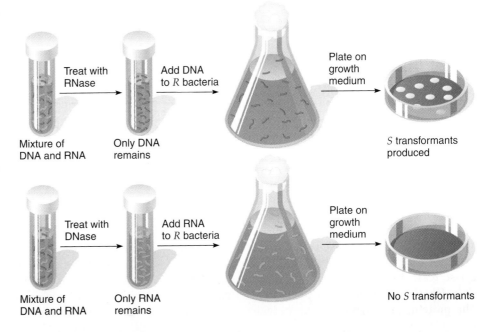

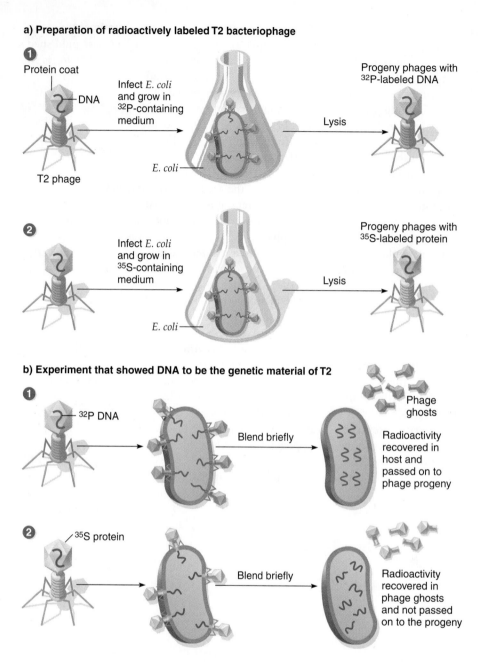

a) Preparation of radioactively labeled T2 bacteriophage

1 Protein coat — DNA — T2 phage — Infect *E. coli* and grow in ^{32}P-containing medium — *E. coli* — Lysis — Progeny phages with ^{32}P-labeled DNA

2 Infect *E. coli* and grow in ^{35}S-containing medium — *E. coli* — Lysis — Progeny phages with ^{35}S-labeled protein

b) Experiment that showed DNA to be the genetic material of T2

1 ^{32}P DNA — Blend briefly — Phage ghosts — Radioactivity recovered in host and passed on to phage progeny

2 ^{35}S protein — Blend briefly — Radioactivity recovered in phage ghosts and not passed on to the progeny

Figure 8.4

Hershey-Chase experiment. (a) The production of T2 phages either with (1) ^{32}P-labeled DNA or with (2) ^{35}S-labeled protein. **(b)** The experimental evidence showing that DNA is the genetic material in T2: (1) The ^{32}P is found within the bacteria and appears in progeny phages, whereas (2) the ^{35}S is not found within the bacteria and is released with the phage ghosts.

were infected with ^{35}S-labeled T2, almost none of the radioactivity appeared within the cell, and none was found in the progeny phage particles. Most of the radioactivity could be found in the phage ghosts released after the cultures were treated in the kitchen blender.

Since genes serve as the blueprint for making the progeny virus particles, it was also presumed that the blueprint must get into the bacterial cell for new phage particles to be built. Therefore, because *it was DNA and not protein that entered the cell,* as evidenced by the presence of ^{32}P and the absence of ^{35}S, Hershey and Chase reasoned that *DNA must be the material responsible for the function and reproduction of phage T2.* The protein, they hypothesized, provided a structural

framework that contains the DNA and the specialized structures required to inject the DNA into the bacterial cell.

Alfred Hershey shared the 1969 Nobel Prize in Physiology or Medicine for his "discoveries concerning the genetic structure of viruses."

Most of the organisms and viruses discussed in this book (such as humans, *Drosophila,* yeast, *E. coli,* and phage T2) have DNA as their genetic material. However, some bacterial viruses (e.g., Qβ), some animal viruses (e.g., poliovirus), and some plant viruses (e.g., tobacco mosaic virus) have RNA as their genetic material. No known prokaryotic or eukaryotic organism has RNA as its genetic material.

KEYNOTE

A series of experiments proved that the genetic material consists of one of two types of nucleic acids: DNA or RNA. Of the two, DNA is found in all living organisms and in some viruses, and RNA is the genetic material of the remaining viruses.

The Composition and Structure of DNA and RNA

Once DNA was shown to be the genetic material, scientists began trying to determine its molecular structure. It was already known that both DNA and RNA are *polymers*—large molecules that consist of many similar smaller molecules, called *monomers*, linked together. The monomers that make up DNA and RNA are called **nucleotides.** Each nucleotide consists of three distinct parts: a **pentose** (five-carbon) **sugar,** a **nitrogenous** (nitrogen-containing) **base,** and a **phosphate group.**

The pentose sugar in DNA is **deoxyribose,** and the sugar in RNA is **ribose** (Figure 8.5). The two sugars differ by the chemical groups attached to the 2′ carbon: a hydrogen atom (H) in deoxyribose and a hydroxyl group (OH) in ribose. (The carbon atoms in the pentose sugar are numbered 1′ to 5′ to distinguish them from the numbered carbon and nitrogen atoms in the rings of the bases.)

There are two classes of nitrogenous bases: the **purines,** which are nine-membered, double-ringed structures, and the **pyrimidines,** which are six-membered, single-ringed structures. There are two purines—**adenine (A)** and **guanine (G)**—and three different pyrimidines—**thymine (T), cytosine (C),** and **uracil (U).** The chemical structures of the five bases are

Figure 8.5

Structures of deoxyribose and ribose, the pentose sugars of DNA and RNA, respectively. The difference between the two sugars is highlighted.

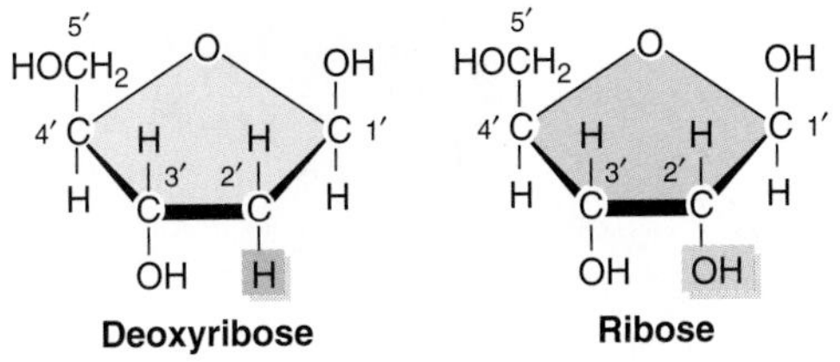

shown in Figure 8.6. (The carbons and nitrogens of the purine rings are numbered 1 to 9, and those of the pyrimidines are numbered 1 to 6.) Both DNA and RNA contain adenine, guanine, and cytosine; however, thymine is found only in DNA, and uracil is found only in RNA.

In DNA and RNA, bases are always attached to the 1′ carbon of the pentose sugar by a covalent bond. The purine bases are bonded at the 9 nitrogen, and the pyrimidines bond at the 1 nitrogen. The combination of a sugar and a base is called a **nucleoside.** When a phosphate is added to a nucleoside, the molecule is called a **nucleotide** or nucleoside phosphate. The phosphate group (PO_4^{2-}) is attached to the 5′ carbon of the sugar in both DNA and RNA. Examples of a DNA nucleotide (a **deoxyribonucleotide**) and an RNA nucleotide (a **ribonucleotide**) are shown in Figure 8.7a. A complete list of the names of the bases, nucleosides, and nucleotides is in Table 8.1.

To form **polynucleotides** of either DNA or RNA, nucleotides are linked together by a covalent bond between the phosphate group (which is attached to the 5′carbon of the sugar ring) of one nucleotide and the 3′ carbon of the

Figure 8.6

Structures of the nitrogenous bases in DNA and RNA. The parent compounds are purine (top left), and pyrimidine (bottom left). Differences between the bases are highlighted.

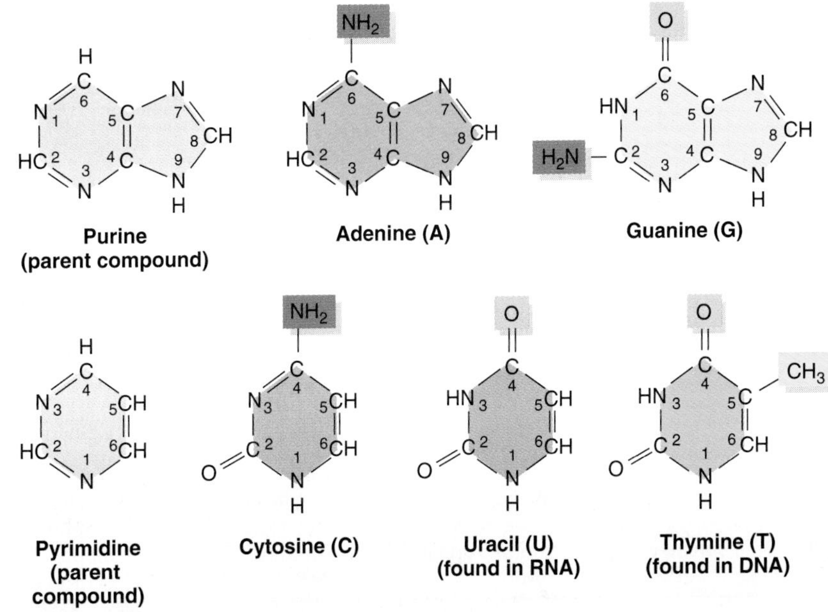

a) DNA and RNA nucleotides

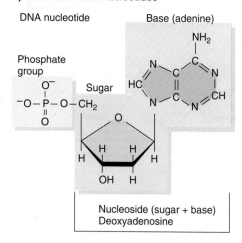

DNA nucleotide

Phosphate group

Sugar

Base (adenine)

Nucleoside (sugar + base)
Deoxyadenosine

Nucleotide (sugar + base + phosphate group)
Deoxyadenosine 5′ – monophosphate

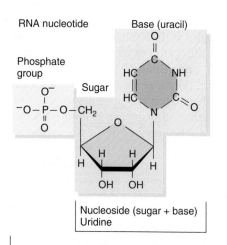

RNA nucleotide

Phosphate group

Sugar

Base (uracil)

Nucleoside (sugar + base)
Uridine

Nucleotide (sugar + base + phosphate group)
Uridine 5′– monophosphate or uridylic acid

b) DNA polynucleotide chain

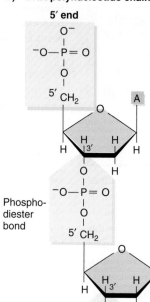

5′ end

Phospho-diester bond

Phospho-diester bond

3′ end

Figure 8.7

Chemical structures of DNA and RNA.
(a) Basic structures of DNA and RNA nucleosides (sugar plus base) and nucleotides (sugar plus base plus phosphate group), the basic building blocks of DNA and RNA molecules. Here the phosphate groups are orange, the sugars are red, and the bases are brown. **(b)** A segment of a polynucleotide chain, in this case a single strand of DNA. The deoxyribose sugars are linked by phosphodiester bonds (shaded) between the 3′ carbon of one sugar and the 5′ carbon of the next sugar.

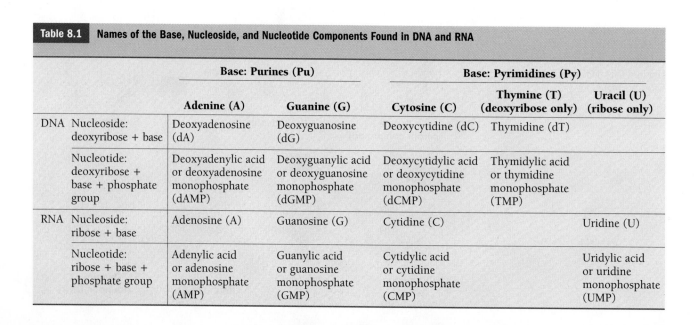

Table 8.1   **Names of the Base, Nucleoside, and Nucleotide Components Found in DNA and RNA**

| | | Base: Purines (Pu) | | Base: Pyrimidines (Py) | |
		Adenine (A)	**Guanine (G)**	**Cytosine (C)**	**Thymine (T) (deoxyribose only)**	**Uracil (U) (ribose only)**
DNA	Nucleoside: deoxyribose + base	Deoxyadenosine (dA)	Deoxyguanosine (dG)	Deoxycytidine (dC)	Thymidine (dT)	
	Nucleotide: deoxyribose + base + phosphate group	Deoxyadenylic acid or deoxyadenosine monophosphate (dAMP)	Deoxyguanylic acid or deoxyguanosine monophosphate (dGMP)	Deoxycytidylic acid or deoxycytidine monophosphate (dCMP)	Thymidylic acid or thymidine monophosphate (TMP)	
RNA	Nucleoside: ribose + base	Adenosine (A)	Guanosine (G)	Cytidine (C)		Uridine (U)
	Nucleotide: ribose + base + phosphate group	Adenylic acid or adenosine monophosphate (AMP)	Guanylic acid or guanosine monophosphate (GMP)	Cytidylic acid or cytidine monophosphate (CMP)		Uridylic acid or uridine monophosphate (UMP)

sugar of another nucleotide. These 5′-3′ phosphate linkages are called **phosphodiester bonds.** A short polynucleotide chain is diagrammed in Figure 8.7b. The phosphodiester bonds are relatively strong, and as a consequence, the repeated sugar-phosphate-sugar-phosphate backbone of DNA and RNA is a stable structure.

To understand how a polynucleotide chain is synthesized (which we will study in another chapter), we must be aware of one more feature of the chain: The two ends of the chain are not the same; that is, the chain has a 5′ carbon (with a phosphate group on it) at one end and a 3′ carbon (with a hydroxyl group on it) at the other end, as shown in Figure 8.7b. This asymmetry is called the *polarity* of the chain.

KEYNOTE

> DNA and RNA occur in nature as macromolecules composed of smaller building blocks called nucleotides. Each nucleotide consists of a five-carbon sugar (deoxyribose in DNA, ribose in RNA) to which is attached a phosphate group and one of four nitrogenous bases: adenine, guanine, cytosine, and thymine (in DNA) or adenine, guanine, cytosine, and uracil (in RNA).

The Discovery of the DNA Double Helix

In 1953, James D. Watson and Francis H. C. Crick (Figure 8.8) published a paper in which they proposed a model for the physical and chemical structure of the DNA molecule. The model they devised, which fit all the known data on the composition of the DNA molecule, is the now-famous double helix model for DNA. Unquestionably, the determination of the structure of DNA was a momentous occasion in biology, leading directly to the transformation in our understanding of all aspects of the life sciences.

At the time of Watson and Crick's discovery, DNA was already known to be composed of nucleotides. However, it was not known how the nucleotides formed the structure of DNA. The data Watson and Crick used to help generate their model came primarily from two sources: base composition studies conducted by Erwin Chargaff and X-ray diffraction studies conducted by Rosalind Franklin and Maurice H. F. Wilkins.

Base Composition Studies. By chemical treatment, Erwin Chargaff had hydrolyzed the DNA of a number of organisms and had quantified the purines and pyrimidines released. His studies showed that in all double-stranded DNAs, 50 percent of the bases were purines and 50 percent were pyrimidines. More important, the amount of adenine (A) was equal to that of thymine (T), and the amount of guanine (G) was equal to that of cytosine (C). These equivalencies have become known as Chargaff's rules. In comparisons of double-stranded DNAs from different organisms, the A/T ratio is 1, the G/C ratio is 1, but the $(A + T)/(G + C)$ ratio (typically denoted %GC) varies. Because the amount of purines equals the amount of pyrimidines, the $(A + G)/(C + T)$ ratio is 1.

X-Ray Diffraction Studies. Rosalind Franklin, working with Maurice H. F. Wilkins (Figure 8.9a), studied isolated fibers of DNA by using the X-ray diffraction technique, a procedure in which a beam of parallel X rays is aimed at molecules. The beam is diffracted (broken up) by the atoms in a pattern that is characteristic of the atomic weight and spatial arrangement of the molecules. The diffracted X rays are recorded on a photographic plate (Figure 8.9b). By analyzing the photographs, Franklin obtained information about the molecule's

Figure 8.8

James Watson (left in each photo) and Francis Crick (right) in 1993 at a fortieth anniversary celebration of their discovery of the structure of DNA and in 1953 with the model of DNA structure.

Figure 8.9

X-ray diffraction analysis of DNA. **(a)** Rosalind Franklin (left) and Maurice H. F. Wilkins (right, photographed in 1962, the year he received the Nobel Prize shared with Watson and Crick). **(b)** The X-ray diffraction pattern of DNA that Watson and Crick used in developing their double helix model. The dark areas that form an X shape in the center of the photograph indicate the helical nature of DNA. The dark crescents at the top and bottom of the photograph indicate the 0.34-nm distance between the base pairs.

a)

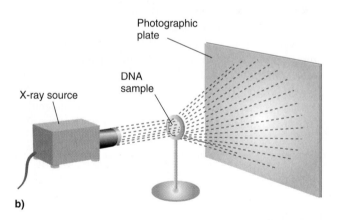

b)

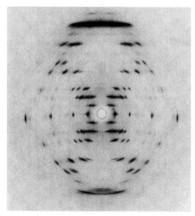

X-ray diffraction pattern

atomic structure. In particular, she concluded that DNA is a helical structure with two distinctive regularities of 0.34 nm and 3.4 nm along the axis of the molecule (1 nanometer [nm] = 10^{-9} meter = 10 angstrom units [Å]; 1 Å = 10^{-10} meter).

Watson and Crick's Model. Watson and Crick used these data to build three-dimensional models for the structure of DNA. Figure 8.10a shows a three-dimensional model of the DNA molecule, and Figure 8.10b is a diagram of the DNA molecule, showing the arrangement of the sugar-phosphate backbone and base pairs in a stylized way.

Watson and Crick's double helix model of DNA has the following main features:

1. The DNA molecule consists of two polynucleotide chains wound around each other in a right-handed double helix; that is, viewed on end (from either end), the two strands wind around each other in a clockwise (right-handed) fashion.

2. The two chains are *antiparallel* (show *opposite polarity*); that is, the two strands are oriented in opposite directions, with one strand oriented in the 5′ to 3′ way and the other strand is oriented 3′ to 5′. To put it in simpler terms, if the 5′ end is the "head" of the chain and the 3′ end is the "tail" of the chain, antiparallel means that the head of one chain is against the tail of the other chain and vice versa.

3. The sugar-phosphate backbones are on the outsides of the double helix, with the bases oriented toward

Figure 8.10

Molecular structure of DNA. (a) Three-dimensional molecular model of DNA as prepared by Watson and Crick. **(b)** Stylized representation of the DNA double helix.

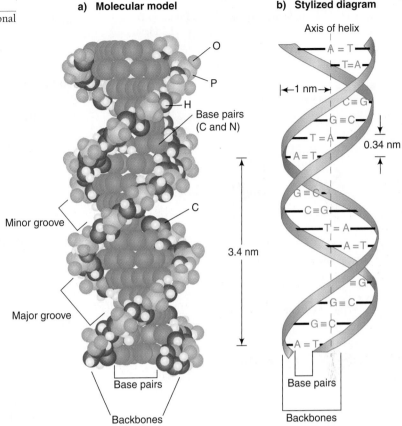

a) Molecular model

O
P
H
Base pairs (C and N)
Minor groove
C
Major groove
3.4 nm
Base pairs
Backbones

b) Stylized diagram

Axis of helix
A = T
T = A
←1 nm→
C ≡ G
G ≡ C
T = A
0.34 nm
A = T
G ≡ C
C ≡ G
T = A
A = T
≡ G
G ≡ C
G ≡ C
A = T
Base pairs
Backbones

the central axis (see Figure 8.10). The bases of both chains are flat structures oriented perpendicularly to the long axis of the DNA; that is, the bases are stacked like pennies on top of one another following the "twist" of the helix.

4. The bases of the opposite strands are bonded together by hydrogen bonds, which are relatively weak chemical bonds. The specific pairings observed are A with T (two hydrogen bonds; Figure 8.11a) and G with C (three hydrogen bonds; Figure 8.11b). The hydrogen bonds make it relatively easy to separate the two strands of the DNA—for example, by heating. Breaking the A-T base pair by heating is easier than breaking the G-C base pair because A-T has two hydrogen bonds and G-C has three hydrogen bonds. The A-T and G-C base pairs are the only ones that can fit the physical dimensions of the helical model, and they are in accord with Chargaff's rules. The specific A-T and G-C pairs are called **complementary base pairs,** so the nucleotide sequence in one strand dictates the nucleotide sequence of the other. For instance, if one chain has the sequence 5′-TATTCCGA-3′, the opposite, antiparallel chain must bear the sequence 3′-ATAAGGCT-5′.

5. The base pairs are 0.34 nm apart in the DNA helix. A complete (360°) turn of the helix takes 3.4 nm; therefore, there are 10 base pairs (bp) per turn. The external diameter of the helix is 2 nm.

6. Because of the way the bases bond with each other, the two sugar-phosphate backbones of the double helix are not equally spaced from one another along the helical axis. This results in grooves of unequal size between the backbones called the *major* (wider) *groove* and the *minor* (narrower) *groove* (see Figure 8.10a). Both of these grooves are large enough to allow protein molecules to make contact with the bases.

For their "discoveries concerning the molecular structure of nucleic acids and its significance for information transfer in living material," the 1962 Nobel Prize in Physiology or Medicine was awarded to Francis Crick, James Watson, and Maurice Wilkins. What about Rosalind Franklin and her contributions to the discovery? We will never know whether she would have shared the Nobel Prize. She had died by 1962, and Nobel Prizes are never awarded posthumously.

KEYNOTE

According to Watson and Crick's model, the DNA molecule consists of two polynucleotide chains joined by hydrogen bonds between pairs of bases (A and T, G and C) in a double helix. The diameter of the helix is 2 nm, and there are 10 base pairs in each complete turn (3.4 nm). The double helix model was proposed by Watson and Crick from chemical and physical analyses of DNA.

Figure 8.11

Structures of the complementary base pairs found in DNA. In both cases, a purine pairs with a pyrimidine. **(a)** The adenine-thymine bases, which pair through two hydrogen bonds. **(b)** The guanine-cytosine bases, which pair through three hydrogen bonds.

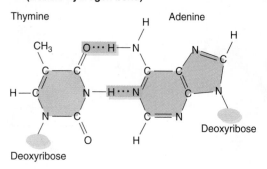

a) **Adenine-thymine base**
(Double hydrogen bond)

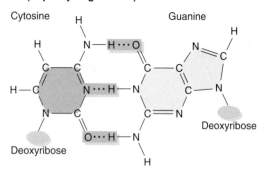

b) **Guanine-cytosine base**
(Triple hydrogen bond)

Different DNA Structures

In recent years, methods have been developed to synthesize short DNA molecules of defined sequences called **oligomers** (*oligo* means "few"). The pure DNA oligomers can be crystallized and analyzed by X-ray diffraction. These approaches have revealed that DNA can exist in several different forms, most notably the A-, B-, and Z-DNA forms (Figure 8.12). Watson and Crick's double helix model was made of B-DNA. Like B-DNA, A-DNA is a right-handed helix, but Z-DNA is a left-handed helix

with a zigzag sugar-phosphate backbone. (The latter property gave this DNA form its "Z" designation.) Some of the key structural features of each DNA type are presented in Table 8.2.

A-DNA and B-DNA. A-DNA and B-DNA are right-handed double helices with 10.9 and 10.0 base pairs per 360° turn of the helix, respectively. The A-DNA double helix is short and wide with a narrow, very deep major groove and a wide, shallow minor groove. (Think of these descriptions in terms of canyons: narrow or wide describes the distance from rim to rim, and shallow or deep describes the distance from the rim down to the bottom of the canyon.) The B-DNA double helix is thinner and longer for the same number of base pairs than A-DNA, with a wide major groove and a narrow minor groove; both grooves are of similar depths.

Z-DNA. Z-DNA is a left-handed helix with 12.0 base pairs per complete helical turn. The Z-DNA helix is thin and elongated with a deep minor groove. The major groove is very near the surface of the helix, so it is not distinct.

iActivity You must now determine the molecular composition and structure of a virus infecting the rice crops of Asia. Go to the iActivity *Cracking the Viral Code* on the website.

DNA in the Cell

In solution, DNA usually is found in the B form, so B-DNA is the most common form in cells. A-DNA is found only when the DNA is dehydrated, so it is unlikely that any lengthy sections of A-DNA exist within cells. Whether Z-DNA exists in living cells has long been a topic of debate among scientists. Some evidence for the existence of Z-DNA in cells includes the presence of Z-DNA-binding proteins in the nuclei of *Drosophila*, human, wheat, and bacterial cells. These proteins may stabilize the DNA in the Z-DNA form. The function of Z-DNA is not clearly understood, although roles in DNA replication,

Table 8.2	Properties of A-DNA, B-DNA, and Z-DNA		
Property	**A-DNA**	**B-DNA**	**Z-DNA**
Helix direction	Right-handed	Right-handed	Left-handed
Base pairs per helix turn	10.9	10.0	12.0
Overall morphology	Short and wide	Longer and thinner	Elongated and thin
Major groove	Extremely narrow and very deep	Wide and of intermediate depth	Flattened out on helix surface
Minor groove	Very wide and shallow	Narrow and of intermediate depth	Extremely narrow and very deep
Helix axis location	Major groove	Through base pairs	Minor groove
Helix diameter	2.2 nm	2.0 nm	1.8 nm

Figure 8.12

Space-filling models of different forms of DNA. (a) A-DNA. **(b)** B-DNA. **(c)** Z-DNA.

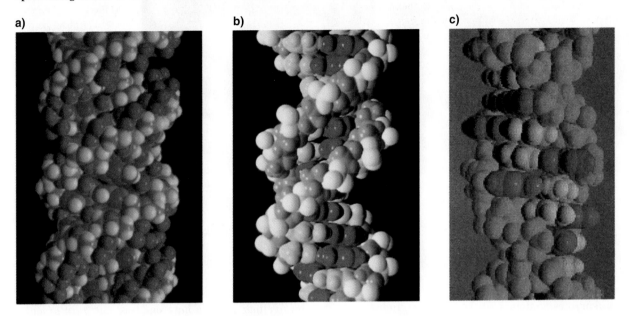

a)   b)   c)

genetic recombination, and transcription regulation have been proposed.

The Organization of DNA in Chromosomes

The DNA of the cell is organized into physical structures called chromosomes. The chromosome or set of chromosomes that contains all the DNA an organism possesses is called a **genome.** In prokaryotes, the genome is usually but not always a single circular chromosome. In eukaryotes, the genome is one complete haploid set of chromosomes contained in the cell nucleus. Eukaryotes also contain a mitochondrial genome and, in plants, a chloroplast genome. It is important to understand how DNA is organized in chromosomes to understand the process by which the information within a gene is accessed (Chapter 11). In the following sections, we discuss the organization of DNA molecules in chromosomes of viruses, prokaryotes, and eukaryotes.

Viral Chromosomes

Viruses are fragments of nucleic acid surrounded by proteins; they infect all types of living organisms. Depending on the virus, the genetic material may be double-stranded DNA, single-stranded DNA, double-stranded RNA, or single-stranded RNA, and it may be circular or linear. The genomes of some viruses are organized into a single chromosome, whereas other viruses have a segmented genome, where the genome is distributed among a number of DNA molecules. In this section, we look at the chromosome structure of three representative bacteriophages of genetic significance: T-even phages, ΦX174, and lambda (λ).

T2, T4, and T6, known collectively as the T-even bacteriophages, are DNA phages, meaning that their genomes consist of DNA. The T4 genome is 168,900 bp (168 kb, where 1 kb = 1 kilobase = 1,000 bp) in size. These virulent phages have similar structures (see Figure 6.10a) and contain a single, linear, double-stranded DNA chromosome surrounded by a protein coat. This chromosome is packaged within the head of the virus.

The virulent phage ΦX174 ("fi-X-one-seventy-four") is a DNA phage that infects *E. coli*. The ΦX174 phage is an icosahedron consisting of protein subunits surrounding the genetic material (Figure 8.13). The DNA of ΦX174 has a base ratio of 25A : 33T : 24G : 18C, a composition that does not fit the A = T and G = C complementary base-pairing rules. This is because the 5,386-nucleotide ΦX174 chromosome is single-stranded. In addition, the chromosome is resistant to digestion by an exonuclease, an enzyme that removes nucleotides from the ends of linear molecules. This is because the chromosome is circular rather than linear.

The λ (lambda) bacteriophage is similar in structure to the T-even phages (see Figure 6.10b). Unlike the T-even phages, however, the λ phage chromosome structure changes between linear and circular forms. Within the phage particle, the λ chromosome is linear, double-stranded DNA, and the two ends of the DNA molecule have 12-nucleotide, single-stranded segments that are complementary. When a λ phage infects a cell, the linear chromosome converts to a circular chromosome. The complementary ends of the molecule (called *sticky ends*) pair together to produce the circular form. When the phage reproduces, the chromosome is converted back to the linear form. Phage λ is a temperate phage; that is, when it infects a bacterial cell, it can go

a)

b)

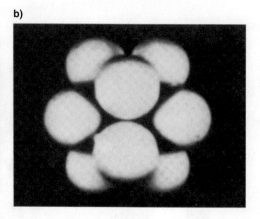

Figure 8.13

Bacteriophage ΦX174.
(**a**) Electron micrograph of ΦX174 phage particles. (**b**) Model of the ΦX174 phage particle.

through either the lytic or lysogenic cycle (see Figure 6.12 and pp. 132–133).

Prokaryotic Chromosomes

Most prokaryotes contain a single, double-stranded, circular DNA chromosome. The remaining prokaryotes have genomes that consist of one or more chromosomes that may be circular or linear. Typically in the latter cases, there is a main chromosome and one or more smaller chromosomes. When a minor chromosome is dispensable to the life of the cell, it is more correctly called a plasmid. For example, among the bacteria, *Borrelia burgdorferi*, the causative agent of Lyme disease, has a 0.91-Mb (1 Mb = 1 megabase = 1 million base pairs) linear chromosome and at least 17 small linear and circular chromosomes with a combined size of 0.53 Mb, and *Agrobacterium tumefaciens*, the causative agent of crown gall disease in some plants, has a 3.0-Mb circular chromosome and a 2.1-Mb linear chromosome. Among the archaea, chromosome organization also varies, although no linear chromosomes have yet been found. For example, *Methanococcus jannaschii* has 1.66-Mb, 58-kb, and 16-kb circular chromosomes, and *Archaeoglobus fulgidus* has a single 2.2-Mb circular chromosome.

In bacteria and archaeons, the chromosome is arranged in a dense clump in a region of the cell known as the **nucleoid.** Unlike eukaryotic nuclei, there is no membrane between the nucleoid region and the rest of the cell.

If an *E. coli* cell is lysed (broken open) gently, its DNA is released in a highly folded state (Figure 8.14). The double-stranded DNA is present as a single, 4.6-Mb circular chromosome, approximately 1,100 μm long. The length of DNA in the *E. coli* chromosome is approximately 1,000 times the length of the *E. coli* cell. So how does all that DNA fit into the nucleoid region of the cell? The DNA packs into the nucleoid area because it is **supercoiled**; that is, the double helix has been twisted in space about its own axis. Figure 8.15 pictures supercoiled circular DNA to show how much more compact a supercoiled molecule is compared with a nonsupercoiled (relaxed) molecule. There are two types of supercoiling: *negative supercoiling* and *positive supercoiling*. To visualize supercoiling, think of the DNA double helix as a spiral staircase that turns in a clockwise direction. If you untwist the spiral staircase by one complete turn, *you have the same number of stairs to climb, but you have one less 360° turn to make;* this is a negative supercoil. If, instead, you twist the spiral staircase by one more complete turn, *you have the same number of stairs to climb, but now there is one more 360° turn to make;* this is a positive supercoil. Either type of supercoiling causes the DNA to become more compact. The amount and type of DNA supercoiling is brought about by **topoisomerases,** enzymes found in all organisms.

Compacting of chromosomes also occurs because the DNA is organized into **looped domains.** In *E. coli*, each domain consists of a loop of about 40 kb of supercoiled DNA, so there are about 100 domains for the entire genome. The ends of each domain are held, presumably by proteins, such that the supercoiled DNA state in one domain is not affected by events that influence supercoiling of DNA in the other domains (Figure 8.16). The compaction achieved by organizing into looped domains is about 10-fold. Looped domains are found in other prokaryotes; the number of domains is species-specific and depends on the size of the genome.

animation

ⓐ DNA Supercoiling

KEYNOTE

Viral genomes may be double-stranded or single-stranded DNA, double-stranded or single-stranded RNA, and circular or linear. The genomes of some viruses are organized into a single chromosome, whereas other viruses have a segmented genome. The genetic material of bacteria and archaea is double-stranded DNA localized into one or a few chromosomes. The *E. coli* chromosome is circular and is organized into about 100 independent looped domains of supercoiled DNA.

Figure 8.14

Chromosome released from a lysed *E. coli* cell.

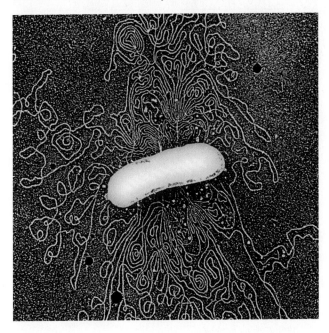

Eukaryotic Chromosomes

A fundamental difference between prokaryotes and eukaryotes is that most prokaryotes have a single type of chromosome (sometimes present in more than one copy in the cell), whereas most eukaryotes have a diploid number of chromosomes in almost all somatic cells. The number of human chromosomes is 46. Recall that this is because humans are diploid (2N) organisms, possessing one haploid (N) set of chromosomes (23 chromosomes) from the egg and another haploid set from the sperm.

The eukaryotic genome is the complete complement of genetic information in the haploid chromosome set. The total amount of DNA in the haploid genome of a species is known as its **C value.** Table 8.3 lists the C values for a selection of species. C value data show that the amount of DNA found among organisms varies widely, and there may or may not be significant variation in DNA amount between related organisms. For example, mammals, birds, and reptiles show little variation, whereas amphibians, insects, and plants vary over a wide range, often tenfold or more. There is also no direct relationship between the C value and the structural or organizational complexity of the organism, a situation called the *C value paradox.* At least one reason for this is variation in the amount of repetitive-sequence DNA in the genome (see pp. 178–179).

Each eukaryotic chromosome consists of one linear, double-stranded DNA molecule running throughout its length and complexed with about twice as much protein by weight as DNA. **Chromatin** is the complex of DNA and chromosomal proteins in the chromosome. The fundamental structure of chromatin is essentially identical in all eukaryotes.

Chromatin Structure.   Two major types of proteins are associated with DNA in chromatin: **histones** and **nonhistones.** Both types of proteins play an important role in determining the physical structure of the chromosome.

The histones are the most abundant proteins associated with chromosomes. They are small basic proteins, with a net positive charge that facilitates their binding to the negatively charged DNA. Five main types of histones are associated with eukaryotic DNA: H1, H2A, H2B, H3, and H4. Weight for weight, there is an equal amount of histone and DNA in chromatin.

The amounts and proportions of histones relative to DNA are constant from cell to cell in all eukaryotic organisms. The amino acid sequences of H2A, H2B, H3, and H4 histones are highly conserved (very similar) evolutionarily speaking, even between distantly related species. For example, there are only two amino acid differences in the H4 histones of cows and peas, and there is only one amino acid difference between sea urchin and calf thymus histone H3. Evolutionary conservation of these sequences is a strong indicator that histones perform the same basic role in organizing the DNA in the chromosomes of all eukaryotes.

Nonhistones are all the proteins associated with DNA apart from the histones. There are many different types of nonhistone chromosomal proteins. Some nonhistones play a structural role; others are only transiently associated with chromatin.

Figure 8.15

Electron micrographs of a circular DNA molecule showing relaxed and supercoiled states. (a) Relaxed (nonsupercoiled) DNA. **(b)** Supercoiled DNA. Both molecules are shown at the same magnification.

a)

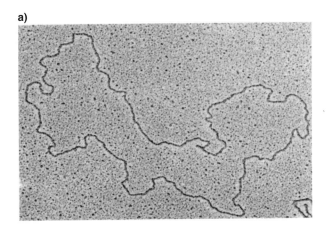

b)

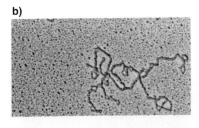

Figure 8.16

Model for the structure of a bacterial chromosome. The chromosome is organized into looped domains, the bases of which are anchored in an unknown way.

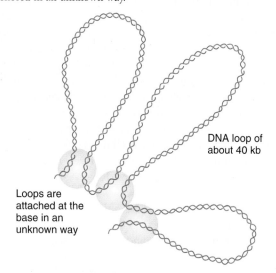

DNA loop of about 40 kb

Loops are attached at the base in an unknown way

The nonhistones are very different from the histones. Many nonhistones are acidic proteins—proteins with a net negative charge—and are likely to bind to positively charged histones in the chromatin. Each eukaryotic cell has many different nonhistones in the nucleus, but in contrast to the histones, the nonhistone proteins differ markedly in number and type from cell type to cell type within an organism, at different times in the same cell type, and from organism to organism. Relative to the mass of DNA in chromosomes, the nonhistone proteins also vary widely. Well-studied among the nonhistones are the high-mobility group (HMG) proteins, so named because of their rapid electrophoretic mobility in polyacrylamide gel electrophoresis. The HMG proteins are abundant and heterogeneous in chromatin and are thought to have a role in the formation of higher-order structures of chromatin (discussed later).

The histones play a crucial role in chromatin packing. A human cell, for example, has more than 700 times as much DNA as does *E. coli*. Without the compacting of the 3.4×10^9 base pairs of DNA in the diploid nucleus, the DNA of the chromosomes of a single human cell would be more than 2.3 meters long (about 7.5 feet) if the molecules were placed end to end. Several levels of packing enable chromosomes that would be several millimeters or even centimeters long to fit into a nucleus that is a few micrometers in diameter. The first level of packing involves the winding of DNA around histones in a structure called a **nucleosome.** In one model for the nucleosome, a short segment of DNA is wrapped around two molecules of each of four core histones: H2A, H2B, H3, and H4. This histone octamer is a flat-ended, cylindrical particle about 11 nm in diameter and 5.7 nm thick. The DNA winds around the outside of the nucleosome's protein core about 1¾ times, which condenses the DNA by a factor of about seven (Figure 8.17).

Table 8.3	Haploid DNA Content, or C Value, of Selected Species
Species	**C Value (bp)**
Viruses and Phages	
λ (bacteriophage)	48,502[a]
T4 (bacteriophage)	168,900
Feline leukemia virus (cat virus)	8,448[a]
Simian virus 40 (SV40)	5,243[a]
Human immunodeficiency virus-1 (HIV-1, causative agent of AIDS)	9,750
Measles virus (human virus)	15,894[a]
Bacteria	
Bacillus subtilis	4,214,814[a]
Borrelia burgdorferi (Lyme disease spirochete)	910,724[a]
Escherichia coli	4,639,221[a]
Helicobacter pylori (bacterium that causes stomach ulcers)	1,667,867[a]
Neisseria meningitis	2,272,351[a]
Archeae	
Methanococcus jannaschii	1,664,970[a]
Eukarya	
Saccharomyces cerevisiae (budding yeast; Brewer's yeast)	12,067,280[a]
Schizosaccharomyces pombe (fission yeast)	14,000,000
Lilium formosanum (lily)	36,000,000,000
Zea mays (maize, corn)	5,000,000,000
Amoeba proteus (amoeba)	290,000,000,000
Drosophila melanogaster (fruit fly)	180,000,000
Caenorhabditis elegans (nematode)	97,000,000
Danio rerio (zebrafish)	1,900,000,000
Xenopus laevis (African clawed frog)	3,100,000,000
Mus musculus (mouse)	3,454,200,000
Rattus rattus (rat)	3,093,900,000
Canis familiaris (dog)	3,355,500,000
Equus caballus (horse)	3,311,000,000
Homo sapiens (humans)	3,400,000,000

[a]These C values derive from the complete genome sequence; all others are estimates based on other measurements.

Figure 8.17

A possible nucleosome structure.

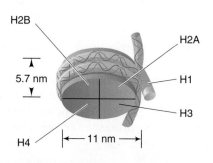

H2B

H2A

H1

5.7 nm

H3

H4

11 nm

Individual nucleosomes are connected by strands of linker DNA and molecules of H1 histone, known as *linker histone* (Figure 8.18). The resulting structures can be seen with an electron microscope as *10-nm chromatin fibers* (**10-nm nucleofilaments**), fibers with a diameter of about 10 nm. In its most unraveled state, the chromatin fiber of a nucleofilament has the appearance of beads on a string, where the beads are the nucleosomes (Figure 8.19) and the thinner "thread" connecting them is naked linker DNA.

In the cell, chromatin is more highly condensed than the "beads-on-a-string" nucleofilament. The nucleosomes associate with each other to form a more compact structure that is 30 nm in diameter. This structure is known as the *30-nm chromatin fiber*; an electron micrograph is shown in Figure 8.20a. Coiling of the 10-nm nucleofilament to produce the 30-nm fiber condenses the DNA to one-sixth its former size. The exact way in which the nucleofilament is coiled has not been determined, however. We know that histone H1 plays an important role in the formation of the 30-nm fiber because chromatin from which histone H1 has been removed can form 10-nm fibers but not 30-nm fibers. One possible model for producing the 30-nm fiber is shown in Figure 8.20b.

The next level of packing involves the formation of looped domains of DNA (Figure 8.21), similar to those found in supercoiled prokaryotic chromosomes. The amount of DNA in each loop ranges from tens to hundreds of kilobase pairs, so an average human chromosome has approximately 2,000 looped domains. The looped domains extend at an angle from the main chro-

Figure 8.20

The 30-nm chromatin fiber. (a) Electron micrograph. **(b)** A model for the packaging of nucleosomes into the 30-nm chromatin fiber.

a)

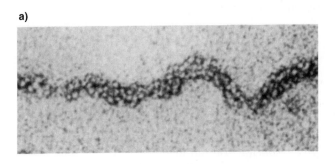

b)

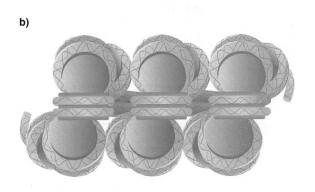

Figure 8.21

Schematic model for the organization of 30-nm chromatin fiber into looped domains that are anchored to a nonhistone protein chromosome scaffold.

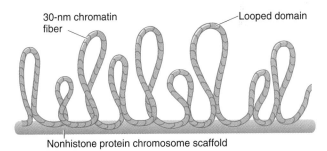

30-nm chromatin fiber   Looped domain

Nonhistone protein chromosome scaffold

Figure 8.18

Nucleosomes connected together by linker DNA and H1 histone to produce the "beads-on-a-string" extended form of chromatin.

"Beads-on-a-string" form of chromatin   11 nm

Figure 8.19

Electron micrograph of unraveled chromatin showing the nucleosomes in a "beads-on-a-string" morphology.

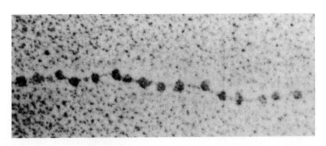

mosome axis and are anchored to a filamentous structural framework of protein inside the nuclear envelope called the *nuclear matrix*.

DNA sequences attached to the proteins in the matrix are called *matrix attachment regions* (MARs). Analysis of the sequences of the MARs has shown that a large proportion of them flank transcriptionally active genes and actively replicating regions. The organization of chromatin into loops appears to be important, then, both for condensing the chromatin fiber in the chromosome and for regulating gene expression. In this vein, it has been hypothesized that each loop is an independent unit of transcription and replication that functions independently of neighboring loops.

Euchromatin and Heterochromatin. The degree of DNA packing changes throughout the cell cycle. Metaphase chromatin of mitosis (M) and meiosis is the most highly condensed in the cell cycle. Throughout the G_1 phase of interphase, the chromatin gradually decondenses and is most dispersed in the S phase. In G_2, recondensation of chromatin begins, leading again to the highly condensed metaphase form before the process cycles again. Visually, in a metaphase chromosome, the looped domains are seen to have coiled and folded further to produce a compact chromatid 700 nm in diameter. Figure 8.22 shows the different orders of DNA packing that give rise to the highly condensed metaphase chromosome.

Figure 8.22

Schematic drawing of the many different orders of chromatin packing that are thought to give rise to the highly condensed metaphase chromosome.

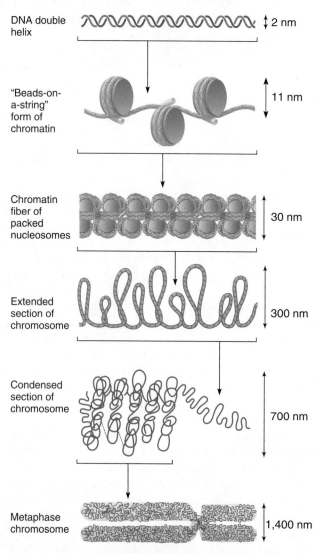

DNA double helix — 2 nm

"Beads-on-a-string" form of chromatin — 11 nm

Chromatin fiber of packed nucleosomes — 30 nm

Extended section of chromosome — 300 nm

Condensed section of chromosome — 700 nm

Metaphase chromosome — 1,400 nm

Historically, two forms of chromatin have been defined based on chromosome staining properties. **Euchromatin** consists of chromosomes or regions of chromosomes that show the normal cycle of chromosome condensation and decondensation in the cell cycle. Most of the genome of an active cell is euchromatic. Visually, euchromatin undergoes a change in intensity of staining ranging from the darkest in metaphase to the lightest in the S phase. Euchromatin typically is actively transcribed, meaning that the genes within it can be expressed.

Heterochromatin, by contrast, consists of chromosomes or chromosomal regions that usually remain condensed—more darkly staining than euchromatin—throughout the cell cycle, even in interphase. Heterochromatic DNA often replicates later than the rest of the DNA in the S phase. Genes within heterochromatic DNA are usually transcriptionally inactive. There are two types of heterochromatin. **Constitutive heterochromatin** is present in all cells at identical positions on both homologous chromosomes of a pair. This form of heterochromatin consists mostly of repetitive DNA and is exemplified by the centromere regions. **Facultative heterochromatin,** by contrast, varies in state in different cell types, or at different developmental stages, or sometimes from one homologous chromosome to another. This form of heterochromatin represents condensed, and therefore inactivated, segments of euchromatin. A well-known example of facultative heterochromatin is the *Barr body,* an inactivated X chromosome in somatic cells of XX mammalian females (see Chapter 3, pp. 61–63).

KEYNOTE

The nuclear chromosomes of eukaryotes are complexes of DNA, histone proteins, and nonhistone chromosomal proteins. Each chromosome consists of one linear, unbroken, double-stranded DNA molecule running throughout the length of the chromosome. There are five main types of histones (H1, H2A, H2B, H3, and H4), which are constant from cell to cell within an organism. Nonhistones, of which there are a large number, vary significantly between cell types, both within and among organisms as well as with time in the same cell type. The large amount of DNA present in the eukaryotic chromosome is compacted by its association with histones in nucleosomes and by higher levels of folding of the nucleosomes into chromatin fibers. Each chromosome contains a large number of looped domains of 30-nm chromatin fibers attached to a protein scaffold. The functional state of the chromosome is related to the extent of coiling: The more condensed a part of a chromosome is, the less likely it is that the genes in that region will be active.

Centromeric and Telomeric DNA. All eukaryotic chromosomes have two areas of special function: the centromere and the telomere. Recall from Chapter 1 that the behavior of chromosomes in mitosis and meiosis depends on the *kinetochores* that form on the centromeres. The telomeres also have a special role, facilitating the replication of the DNA in the chromosome. The DNA sequences that make up these regions are responsible for their function.

Centromeres are the DNA sequences found near the attachment point of mitotic or meiotic spindle fibers. The centromere region of each chromosome is responsible for the accurate segregation of the replicated chromosomes to the daughter cells during mitosis and meiosis. The error frequency for this process is low but significant and varies from organism to organism. In yeast, for example, segregation errors occur at a frequency of 1 in 10^5 or less.

The DNA sequences (called *CEN* sequences, after the *cen*tromere) of the yeast centromeres have been determined. Though each centromere has the same function, the *CEN* regions are very similar, but not identical, to one another in nucleotide sequence and organization. Centromere sequences have been determined for a number of other organisms; they are different both from those of yeast and from each other. Thus, while centromeres carry out the same function in all eukaryotes, there is no common sequence that is responsible for that function.

A telomere is required for replication and stability of the chromosome. Telomeres are characteristically heterochromatic. In most organisms that have been examined, the telomeres are positioned just inside the nuclear envelope and often are found associated with each other as well as with the nuclear envelope.

All telomeres in a given species share a common sequence. Most telomeric sequences may be divided into two types:

1. **Simple telomeric sequences** are at the extreme ends of the chromosomal DNA molecules. Simple telomeric sequences are the essential functional components of telomeric regions in that they are sufficient to supply a chromosomal end with stability. These sequences are species-specific and consist of a series of simple DNA sequences repeated one after the other (called *tandemly repeated DNA sequences*). In the ciliate *Tetrahymena*, for example, reading the sequence toward the end of one DNA strand, the repeated sequence is 5′-TTGGGG-3′ (Figure 8.23a), and in humans the repeated sequence is 5′-TTAGGG-3′. Note that the DNA is not double-stranded all the way out to the end. In a new model, the telomere DNA loops back on itself, forming a *t-loop* (Figure 8.23b). The single-stranded end invades the double-stranded telomeric sequences, causing a *displacement loop*, or *D-loop*, to form. You will see in the next chapter the role these telomere sequences have in the replication of eukaryotic chromosomes.

2. **Telomere-associated sequences** are regions internal to the simple telomeric sequences. These sequences often contain repeated but still complex DNA sequences extending many thousands of base pairs in from the chromosome end. The significance of these sequences is not yet known.

Whereas the telomeres of most eukaryotes contain short, simple, repeated sequences, the telomeres of *Drosophila* are quite different. *Drosophila* telomeres consist of *transposons,* DNA sequences that can move to other locations in the genome.

Chromosomes of Mitochondria and Chloroplasts

Many mitochondrial genomes are circular, double-stranded, supercoiled DNA molecules. Linear mitochondrial genomes are found in some protozoa and some fungi. No histones or similar proteins are associated with mtDNA. Multiple copies of the genomes are found within mitochondria located in multiple *nucleoid regions* (similar to those of bacterial cells).

The gene content is very similar in both number and function among mitochondrial genomes from different species. However, the size of the genome varies tremendously from organism to organism. In animals, the circular mitochondrial genome is less than 20 kb; for example, human mtDNA is 16,569 bp. In contrast, the mtDNA of yeast is about 80 kb (80,000 bp), and that of plants ranges

Figure 8.23

Telomeres. (a) Simple telomeric sequences at the ends of *Tetrahymena* chromosomes. (b) Model for telomere structure in which the telomere DNA loops back to form a t-loop. The single-stranded end invades the double-stranded telomeric sequences to produce a displacement loop (D-loop).

a) *Tetrahymena* **simple telomeric sequences**

b) t-loop model for telomeres

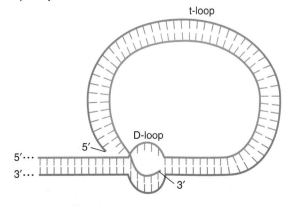

from 100,000 to 2 million bp. The main difference between animal, plant, and fungal mitochondria is that essentially the entire mitochondrial genomes of animals encode products, whereas the mitochondrial genomes of fungi and plants have extra DNA that does not code for products.

The structure of the chloroplast genome is similar to that of mitochondrial genomes. In all cases, the DNA is double-stranded, circular, devoid of structural proteins, and supercoiled. Chloroplast DNA is much larger than animal mtDNA—between 80 and 600 kb. The DNA sequences of the chloroplast genomes of a few organisms have been completely determined. For example, the tobacco genome is 155,844 bp, and the rice genome is 134,525 bp. All chloroplast genomes contain a significant proportion of noncoding DNA sequences.

The number of copies of cpDNA per chloroplast varies between species. In each case, there are multiple copies per chloroplast, typically distributed among a number of nucleoid regions. In the unicellular green alga *Chlamydomonas* (see Figure 7.4), for example, the one chloroplast in the cell contains 500 to 1,500 cpDNA molecules.

Unique-Sequence and Repetitive-Sequence DNA

Now that you know about the basic structure of DNA and its organization in chromosomes, we can discuss the distribution of certain sequences in the genomes of prokaryotes and eukaryotes. From molecular analyses, geneticists have found that some sequences are present only once in the genome, whereas other sequences are repeated. For convenience, these sequences are grouped into three categories: **unique-sequence (single-copy) DNA** (present in one to a few copies in the genome), **moderately repetitive DNA** (present in a few to about 10^5 copies in the genome), and **highly repetitive DNA** (present in about 10^5 to 10^7 copies in the genome). In prokaryotes, with the exception of the ribosomal RNA genes, transfer RNA genes, and a few other sequences, all of the genome is present as unique-sequence DNA. Eukaryotic nuclear genomes, on the other hand, consist of both unique-sequence and repetitive-sequence DNA, with the latter typically being quite complex in number of types, number of copies, and distribution. To date we have sketchy information about the distribution of the various classes of sequences in the genome. However, as the complete DNA sequences of more and more eukaryotic genomes are determined, we will develop a precise understanding of the molecular organization of genomes.

Unique-Sequence DNA. Unique sequences, sometimes called single-copy sequences, are defined as sequences present as single copies in the genome. (Thus, there are two copies per diploid cell.) Actually, in current usage, the term usually applies to sequences that have one to a few copies per genome. Most of the genes we know about—those that code for proteins in the cell—are in the

unique-sequence class of DNA. But not all unique-sequence material contains protein-coding sequences. In humans, unique sequences are estimated to make up roughly 65 percent of the genome.

Repetitive-Sequence DNA. Both moderately repetitive and highly repetitive DNA sequences are sequences that appear many times within a genome. These sequences can be arranged within the genome in one of two ways: distributed at irregular intervals (known as **dispersed repeated DNA** or **interspersed repeated DNA**) or clustered together so that the sequence repeats many times in a row (known as **tandemly repeated DNA**).

The distribution of dispersed repeated sequences has been studied in a variety of organisms. The picture that has emerged is one of families of repeated sequences interspersed through the genome with unique-sequence DNA. Each family consists of a set of related sequences; the sequence is characteristic of the family. Often small numbers of families have very high copy numbers and make up most of the dispersed repeated sequences in the genome. Two general interspersion patterns are encountered. In one, the families have sequences 100 to 500 bp long; these sequences are called **short interspersed repeated sequences (SINEs)**. In the other, the families have sequences about 5,000 bp or more long; these sequences are called **long interspersed repeated sequences (LINEs)**. All eukaryotic organisms have SINEs and LINEs, although the relative proportions vary widely. *Drosophila* and birds, for example, have mostly LINEs, whereas humans and frogs have mostly SINEs.

The SINE pattern of repeated sequences is found in a diverse array of eukaryotic species, including mammals, amphibians, and sea urchins. Each species with SINEs has its own characteristic array of SINE families. A well-studied SINE family is the *Alu* family of certain primates. This family is named for the cleavage site for the restriction enzyme *Alu*I ("Al-you-one") typically found in the repeated sequence. In humans, the *Alu* family is the most abundant SINE family in the genome, consisting of 200- to 300-bp sequences repeated as many as 9×10^5 times and making up about 9 percent of the total haploid DNA. One *Alu* repeat is located about every 5,000 bp in the genome.

Mammalian genomes have many copies of a particular LINE family of repeated sequence, the LINE-1 family, in addition to many SINE families. Other LINE families may be present also, but they are much less abundant than LINE-1. LINE-1 family members are up to 7,000 bp long, although a wide range of sizes is seen. Some of the longer LINE-1 family sequences are transposons.

Although we have a lot of information about the sequence organization and distribution of SINEs and LINEs, we have very little knowledge about the functions of those sequences. One hypothesis is that most of these sequences have no function at all. Another hypothesis is that some of the repeated sequences or their transcripts (or both) are somehow involved in mechanisms for regulating gene expression.

Unlike dispersed repeated DNA, tandemly repeated DNA is arranged one after the other in the genome in a head-to-tail organization. Tandemly repeated DNA is quite common in eukaryotic genomes. In some cases, the tandemly repeated DNA consists of short sequences 1 to 10 bp long; in other cases, the repeated sequences are associated with genes and are much longer. The simple telomeric sequences shown in Figure 8.23a exemplify the former type of tandemly repeated DNA. An example of the latter is seen in the genes for transfer RNA (tRNA), which are tandemly repeated in one or more clusters in most eukaryotes. However, the greatest amount of tandemly repeated DNA is associated with centromeres and telomeres. At each centromere, hundreds to thousands of copies of simple, short, tandemly repeated sequences (highly repetitive sequences) are found. Indeed, a significant proportion of the eukaryotic genome may consist of the highly repeated sequences found at centromeres: 8 percent in the mouse, about 50 percent in the kangaroo rat, and about 5 percent in humans. We describe nongenic tandemly repeated DNA in more detail in Chapter 15.

KEYNOTE

Prokaryotic genomes consist mostly of unique-sequence DNA, with only a few sequences and genes repeated. Eukaryotes have both unique and repetitive sequences in the genome, with an extensive spectrum of complexity of the repetitive sequences among species. Some of the repetitive sequences are genes, but most are not.

Summary

In this chapter you have learned that DNA or RNA can be the genetic material. All prokaryotic and eukaryotic organisms and most viruses have DNA as their genetic material, whereas some viruses have RNA as their genetic material. The form of the genetic material varies among organisms and their viruses. In living organisms, the DNA is always double-stranded; in viruses, the genetic material may be double- or single-stranded DNA or RNA, depending on the virus.

Chemical analysis has revealed that DNA and RNA are macromolecules composed of building blocks called nucleotides. Each nucleotide consists of a five-carbon sugar (deoxyribose in DNA, ribose in RNA) to which is attached one of four nitrogenous bases and a phosphate group. In DNA, the four nitrogenous bases are adenine, guanine, cytosine, and thymine; in RNA, the four bases are adenine, guanine, cytosine, and uracil. From chemical and physical analysis, it was determined that the DNA molecule consists of two polynucleotide chains joined by hydrogen bonds between pairs of bases (A and T, G and C) in a double helix. The diameter of the helix is 2 nm, and there are 10 base pairs in each complete turn of the helix (3.4 nm).

A number of different types of double-helical DNA have been identified by X-ray diffraction analysis. The three major types are the right-handed A- and B-DNAs and the left-handed Z-DNA. The common form found in cells is B-DNA (the form analyzed by Watson and Crick). A-DNA probably does not exist in cells. Z-DNA may exist in cells, but its function is unknown.

Examples of chromosome organization in bacteriophages were discussed, including the linear double-stranded DNA chromosomes in T-even phages, circular, single-stranded DNA in ΦX174, and the changing structure of the λ double-stranded DNA chromosome. The chromosomes of prokaryotic organisms consist of circular, double-stranded DNA molecules that are complexed with a number of proteins. The chromosome is condensed within the cell by supercoiling of the DNA helix and formation of looped domains of supercoiled DNA.

A distinguishing feature of eukaryotic chromosomes is that DNA is distributed among a number of chromosomes. The DNA in the genome specifies an organism's structure, function, and reproduction.

The amount of DNA in a genome (made up of a prokaryotic chromosome or of the haploid set of chromosomes in eukaryotes) is called the C value. There is no direct relationship between the C value and the structural or organizational complexity of an organism.

The organization of DNA in eukaryotic chromosomes was described in this chapter. The nuclear-located chromosomes of eukaryotes are complexes of DNA with histone and nonhistone chromosomal proteins. Each chromosome consists of one linear, double-stranded DNA molecule that runs through its length. As a class, the histones are constant from cell to cell within an organism and have been evolutionarily conserved. Nonhistone chromosomal proteins, on the other hand, vary significantly between cell types and between organisms. The DNA in a eukaryotic chromosome is highly condensed by its association with histones to form nucleosomes and by the several higher levels of folding of nucleosomes into chromatin fibers. Like prokaryotic chromosomes, eukaryotic chromosomes are organized into a large number of looped domains. These loops are attached to a protein scaffold. The most highly condensed structure is seen in metaphase chromosomes, and the least condensed structure is seen in interphase chromosomes. The factors controlling the transitions between different levels of chromosome folding are not known.

Both mitochondria and chloroplasts contain DNA, the length of which varies from organism to organism. The genomes of both organelles usually are circular, double-stranded DNAs without any associated histone-like proteins.

Two specialized eukaryotic chromosome structures are centromeres and telomeres. Significant progress has been made in defining the sequences of centromeres and telomeres; for example, the DNA sequences of centromeres are complex and species-specific, and the DNA sequences of telomeres are simple, short, tandemly repeated sequences that are also species-specific.

Molecular analysis has provided information about the distribution of DNA sequences in genomes. Such analysis has revealed that prokaryotic and viral genomes consist mostly of unique-sequence DNA, with only a few repeated sequences. In contrast, the genomes of eukaryotes contain both unique-sequence DNA and repeated-sequence DNA, and there is a wide spectrum of complexity of the repeated DNA sequences. Many genes are found in unique-sequence DNA, but not all unique-sequence DNA contains genes.

Repeated DNA sequences may be tandemly arranged or interspersed with unique-sequence DNA in the eukaryotic genome. Although some tandemly repeated DNA is composed of gene families, the greatest amounts of tandemly repeated DNA are not associated with genes but with centromeres and telomeres. At each centromere, for example, hundreds to thousands of copies of simple, short, tandemly repeated sequences may be found. Dispersed repeated sequences include both gene sequences and nongene sequences. Dispersed repeated gene sequences typically are present in low copy numbers (fewer than 50), whereas dispersed nongene sequences may be present in hundreds of thousands of copies.

Eukaryotic species have characteristic families of these repeated sequences interspersed through the genome with unique-sequence DNA. The two general interspersion families are SINEs (short interspersed repeats), in which the repeated sequences are 100 to 500 bp long, and LINEs (long interspersed repeats), in which the repeated sequences are about 5,000 bp or more long. SINEs and LINEs are found in all eukaryotes, although the relative proportions vary widely. Although much knowledge has been accumulated about SINEs and LINEs in a number of eukaryotic genomes, little is known about the functions of these sequences.

Analytical Approaches for Solving Genetics Problems

Q8.1 The linear chromosome of phage T2 is 52 μm long. The chromosome consists of double-stranded DNA, with 0.34 nm between each base pair. How many base pairs does a chromosome of T2 contain?

A8.1 This question involves the careful conversion of different units of measurement. The first step is to put the lengths in the same units: 52 μm is 52 millionths of a meter, or $52,000 \times 10^{-9}$ m, or 52,000 nm. One base occupies 0.34 nm in the double helix, so the number of base pairs in this chromosome is 52,000 divided by 0.34, or 152,941 bp.

The human genome contains 3×10^9 bp of DNA, for a total length of about 1 m distributed among 23 chromosomes. The average length of the double helix in a human chromosome is 3.8 cm, which is 3.8 hundredths of a meter, or 38 million nm—substantially longer than the T2 chromosome! There are more than 111.7 million base pairs in the average human chromosome.

Q8.2 The following table lists the relative percentages of bases of nucleic acids isolated from different species. For each one, what type of nucleic acid is involved? Is it double- or single-stranded? Explain your answer.

Species	Adenine	Guanine	Thymine	Cytosine	Uracil
(i)	21	29	21	29	0
(ii)	29	21	29	21	0
(iii)	21	21	29	29	0
(iv)	21	29	0	29	21
(v)	21	29	0	21	29

A8.2 This question focuses on the base-pairing rules and the difference between DNA and RNA. In analyzing the data, we should determine first whether the nucleic acid is RNA or DNA and then whether it is double- or single-stranded. If the nucleic acid has thymine, it is DNA; if it has uracil, it is RNA. Thus, species (i), (ii), and (iii) must have DNA as their genetic material, and species (iv) and (v) must have RNA as their genetic material. Next, we must analyze the data for strandedness. Double-stranded DNA must have equal percentages of A and T and of G and C. Similarly, double-stranded RNA must have equal percentages of A and U and of G and C. Therefore, species (i) and (ii) have double-stranded DNA, whereas species (iii) must have single-stranded DNA because the base-pairing rules are violated, with A = G and T = C but A ≠ T and G ≠ C. As for the RNA-containing species, (iv) contains double-stranded RNA because A = U and G = C, and (v) must contain single-stranded RNA.

Q8.3 Here are four characteristics of one 5′-3′strand of a particular long, double-stranded DNA molecule:

   i. 35 percent of the adenine-containing nucleotides (As) have guanine-containing nucleotides (Gs) on their 3′ sides.

   ii. 30 percent of the As have Ts as their 3′ neighbors.

   iii. 25 percent of the As have Cs as their 3′ neighbors.

   iv. 10 percent of the As have As as their 3′ neighbors.

Use this information to answer the following questions as completely as possible, explaining your reasoning in each case.

a. In the complementary DNA strand, what will be the frequencies of the various bases on the 3′ side of A?

b. In the complementary strand, what will be the frequencies of the various bases on the 3′ side of T?

c. In the complementary strand, what will be the frequency of each kind of base on the 5′ side of T?

d. Why is the percentage of A not equal to the percentage of T (and the percentage of C not equal to the percentage of G) among the 3′ neighbors of A in the 5′-3′ DNA strand described?

A8.3

a. This cannot be answered without more information. Although we know that the As neighbored by Ts in the original strand will correspond to As neighbored by Ts in the complementary strand, there will be additional As in the complementary strand about whose neighbors we know nothing.

b. This cannot be answered. All the As in the original strand correspond to Ts in the complementary strand, but we know only about the 5′ neighbors of these Ts, not the 3′ neighbors.

c. On the original strand, 35 percent were 5′-A G-3′, so on the complementary strand, 35 percent of the sequences will be 3′-T C-5′. So 35 percent of the bases on the 5′ side of T will be C. Similarly, on the original strand, 30 percent were 5′-A T-3′, 25 percent were 5′-A C-3′, and 10 percent were 5′-A A-3′, meaning that on the complementary strand, 30 percent of the sequences were 3′-T A-5′, 25 percent were 3′-T G-5′, and 10 percent were 3′-T T-5′. So 30 percent of the bases on the 5′ side of T will be A, 25 percent will be G, and 10 percent will be T.

d. The A = T and G = C rule applies only when considering both strands of a double-stranded DNA. Here we are considering only the original single strand of DNA.

Questions and Problems

8.1 Griffith's experiment injecting a mixture of dead and live bacteria into mice demonstrated that (choose the correct answer)

a. DNA is double-stranded

b. mRNA of eukaryotes differs from mRNA of prokaryotes

c. a factor was capable of transforming one bacterial cell type to another

d. bacteria can recover from heat treatment if live helper cells are present

*8.2 In the 1920s, while working with *Streptococcus pneumoniae,* the agent that causes pneumonia, Griffith injected mice with different types of bacteria. For each of the following bacteria types injected, indicate whether the mice lived or died.

a. type *IIR*

b. type *IIIS*

c. heat-killed *IIIS*

d. type *IIR* + heat-killed *IIIS*

8.3 Several years after Griffith described the transforming principle, Avery, MacLeod, and McCarty investigated the same phenomenon.

a. List the steps they used to show that DNA from dead *S. pneumoniae* cells was responsible for the change from a nonvirulent to a virulent state.

b. Did their work affirm Griffith's work or disaffirm it, and how?

c. What was the role of the enzymes used in their experiments?

*8.4 Hershey and Chase showed that when phages were labeled with ^{32}P and ^{35}S, the ^{35}S remained outside the cell and could be removed without affecting the course of infection, whereas the ^{32}P entered the cell and could be recovered in progeny phages. What distribution of isotope would you expect to see if parental phages were labeled with isotopes of

a. C?

b. N?

c. H?

Explain your answer.

8.5 The X-ray diffraction data obtained by Rosalind Franklin suggested that (choose the correct answer)

a. DNA is a helix with a pattern that repeats every 3.4 nm

b. purines are hydrogen bonded to pyrimidines

c. DNA is a left-handed helix

d. DNA is organized into nucleosomes

8.6 What evidence do we have that in the helical form of the DNA molecule the base pairs are composed of one purine and one pyrimidine?

8.7 What exactly is a deoxynucleotide made up of, and how many different deoxynucleotides are there in DNA? Describe the structure of DNA, and describe the bonding mechanism of the molecule (i.e., the kind of bonds on the sides of the "ladder" and the kind of bonds holding the two complementary strands together). Base pairing in DNA consists of purine-pyrimidine pairs, so why can't A-C and G-T pairs form?

*8.8 What is the base sequence of the DNA strand that would be complementary to the following single-stranded DNA molecules?

a. 5′-A G T T A C C T G A T C G T A-3′

b. 5′-T T C T C A A G A A T T C C A-3′

*8.9 Describe the bonding properties of G-C and T-A. Which would be the hardest to break? Why?

8.10 The double helix model of DNA, as suggested by Watson and Crick, was based on data gathered on DNA by other researchers. The facts fell into the following two general categories; give two examples of each.
a. chemical composition
b. physical structure

*8.11 For double-stranded DNA, which of the following base ratios always equals 1?
a. $(A + T)/(G + C)$
b. $(A + G)/(C + T)$
c. C/G
d. $(G + T)/(A + C)$
e. A/G

8.12 If the ratio of $(A + T)$ to $(G + C)$ in a particular DNA is 1, does this indicate that the DNA is probably constituted of two complementary strands of DNA or a single strand of DNA, or is more information necessary?

8.13 The percentage of cytosine in a double-stranded DNA is 17. What is the percentage of adenine in that DNA?

*8.14 A double-stranded DNA polynucleotide contains 80 thymidylic acid and 110 deoxyguanylic acid residues. What is the total nucleotide number in this DNA fragment?

*8.15 Analysis of DNA from a bacterial virus indicates that it contains 33 percent A, 26 percent T, 18 percent G, and 23 percent C. Interpret these data.

8.16 The genetic material of bacteriophage ΦX174 is single-stranded DNA. What base equalities or inequalities might we expect for single-stranded DNA?

8.17 Through X-ray diffraction analysis of crystallized DNA oligomers, different forms of DNA have been identified. These forms include A-DNA, B-DNA, and Z-DNA, and each has unique molecular attributes.
a. Which of these forms is the most common form in living cells?
b. Z-DNA has an unusual conformation resulting in more base pairs per helical turn than B-DNA. What is the conformation? Does this molecule have any function in living cells?
c. Which of these forms is never found in living cells?

8.18 If a virus particle contains double-stranded DNA with 200,000 bp, how many complete 360° turns occur in this molecule?

*8.19 A double-stranded DNA molecule is 100,000 bp (100 kb) long.
a. How many nucleotides does it contain?

b. How many complete turns are there in the molecule?
c. How long is the DNA molecule?

8.20 Different organisms have vastly different amounts of genetic material. *E. coli* has about 4.6 million bp of DNA in one circular chromosome; the haploid budding yeast (*S. cerevisiae*) has 12,067,280 bp of DNA in 16 chromosomes; and the gametes of humans have about 2.75 billion bp of DNA in 23 chromosomes.
a. If all of the DNA were B-DNA, what would be the average length of a chromosome in these cells?
b. On average, how many complete turns would be in each chromosome?
c. Would your answers to (a) and (b) be significantly different if the DNA were composed of, say, 20 percent Z-DNA and 80 percent B-DNA?
d. What implications do your answers to these questions have for the packaging of DNA in cells?

8.21 Define topoisomerases, and list the functions of these enzymes.

*8.22 In a particular eukaryotic chromosome (choose the best answer),
a. heterochromatin and euchromatin are regions where genes make functional gene products (where they are active)
b. heterochromatin is active, but euchromatin is inactive
c. heterochromatin is inactive, but euchromatin is active
d. both heterochromatin and euchromatin regions are inactive

*8.23 Compare and contrast eukaryotic chromosomes and bacterial chromosomes, considering the following features.
a. centromeres   f. nonhistone proteins
b. hexose sugars   g. DNA
c. amino acids   h. nucleosomes
d. supercoiling   i. circular chromosome
e. telomeres   j. looping

8.24 Discuss the structure of nucleosomes, the hierarchical packaging, and the composition of a core particle of a nucleosome. Also discuss the functions of nucleosomes.

*8.25 Answer the questions below after setting up the following "rope trick." Start with a belt (representing a DNA molecule; imagine the phosphodiester backbones lying along the top and bottom edges of the belt) and a soda can. Holding the belt buckle at the bottom of the can, wrap the belt flat against the side of the can, counterclockwise three times around the can. Now remove the "core" soda can and, holding the ends of the belt, pull the ends of the belt taut. After some reflection, answer the following questions.

Figure 9.2

The Meselson-Stahl experiment. The demonstration of semiconservative replication in *E. coli.* Cells were grown in ^{15}N-containing medium for several replication cycles and then transferred to ^{14}N-containing medium. At various times over several replication cycles, samples were taken; the DNA was extracted and analyzed by CsCl equilibrium density gradient centrifugation. Shown in the figure are a schematic interpretation of the DNA composition after various replication cycles, photographs of the DNA bands, and densitometric scans of the bands.

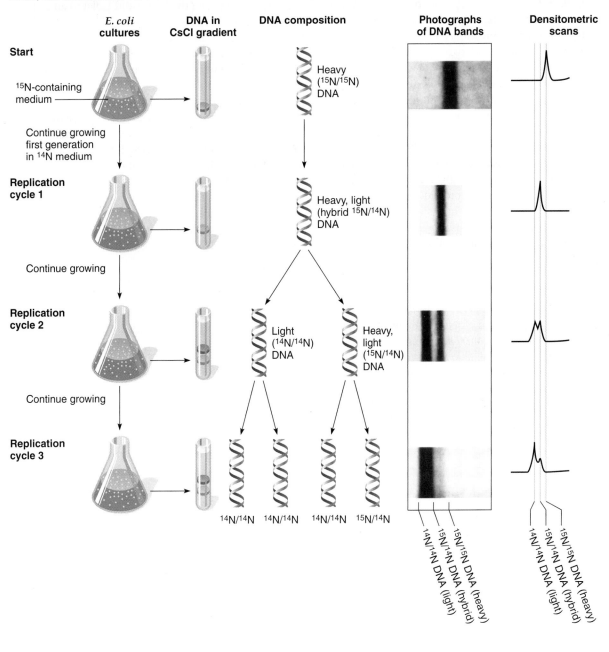

found in *one* band located halfway between the intermediate-density and light-density band positions. With subsequent replication cycles, there would continue to be one band, and it would become lighter in density with each replication cycle. The results of the Meselson-Stahl experiment did not match this prediction, so the dispersive model was ruled out.

Semiconservative DNA Replication in Eukaryotes

Replication of DNA in eukaryotic chromosomes can be visualized using a special staining procedure. The experimental system involves Chinese hamster ovary (CHO) cells growing in tissue culture. At the beginning of the experiment, the chemical 5-bromodeoxyuridine (BUdR) is added. BUdR is a *base analogue,* with a structure very similar to that of a normal base in DNA, in this case thymine. During replication, wherever a T is called for in the new strand, BUdR can be incorporated instead. After two cycles of replication in the presence of BUdR, mitotic chromosomes are stained with a fluorescent dye and Giemsa stain, with the result shown in Figure 9.3. Sister chromatids are visible, one darkly stained and one lightly stained. The dark-light staining pattern brings to mind the costumes of harlequins, so the chromosomes are called **harlequin chromosomes.** A chromatid with a DNA double helix consisting of two strands with BUdR stains less intensely than a chromatid with a DNA double helix consisting of one strand with BUdR and one strand with T. So if semiconservative replication occurs, after one replication cycle two progeny DNAs will be produced, each with one T strand and one BUdR strand. Then, after a second replication cycle in the presence of BUdR, we will get one sister chromatid consisting of a DNA with one T strand and one BUdR strand (darkly staining) and the other sister chromatid consisting of a DNA with two BUdR strands (lightly staining). This is the staining pattern that was observed, showing that semiconservative DNA replication also is the correct model in eukaryotes.

KEYNOTE

> DNA replication in *E. coli* and other prokaryotes and in eukaryotes occurs by a semiconservative mechanism in which the strands of a DNA double helix separate and a new complementary strand of DNA is synthesized on each of the two parental template strands. Semiconservative replication results in two double-stranded DNA molecules, each of which has one strand from the parent molecule and one newly synthesized strand.

Enzymes Involved in DNA Synthesis

In 1955, Arthur Kornberg and his colleagues set out to find the enzymes necessary for DNA replication so that they could analyze the reactions involved in detail. Their work focused on bacteria because it was assumed that bacterial replication machinery would be less complex than that of eukaryotes. Arthur Kornberg shared the 1959 Nobel Prize in Physiology or Medicine for his "discovery of the mechanisms in the biological synthesis of deoxyribonucleic acid."

DNA Polymerase I

Kornberg's approach was to identify all the ingredients required to synthesize *E. coli* DNA in vitro. The first successful DNA synthesis was accomplished in a reaction mixture containing DNA fragments, a mixture of four deoxyribonucleoside 5′-triphosphate precursors (dATP, dGTP, dTTP, and dCTP, collectively abbreviated dNTP for *deoxyribonucleoside triphosphate*), and an *E. coli* lysate (cells of the bacteria broken open to release their contents). To measure minute quantities of DNA expected to be synthesized in the reaction, Kornberg used radioactively labeled dNTPs.

Kornberg analyzed the lysate and isolated an enzyme that was capable of DNA synthesis. This enzyme was originally called the *Kornberg enzyme,* but it is now most commonly called **DNA polymerase I.** (By definition, enzymes that catalyze DNA synthesis are called **DNA polymerases.**)

With DNA polymerase I isolated, more detailed information could be obtained about DNA synthesis in vitro. Researchers found that four components were needed. If any one of the following four components was omitted, DNA synthesis would not occur:

1. All four dNTPs (If any one dNTP is missing, no synthesis occurs; these are the precursors for the nucleotide [phosphate-sugar-base] building blocks of DNA described in Chapter 8 [pp. 165–167].)
2. A fragment of DNA
3. DNA polymerase I
4. Magnesium ions (Mg^{2+}) (required for optimal DNA polymerase activity)

Figure 9.3

Visualization of semiconservative DNA replication in eukaryotes. Shown are harlequin chromosomes in Chinese hamster ovary cells that have been allowed to go through two rounds of DNA replication in the presence of the base analogue 5-bromodeoxyuridine (BUdR), followed by staining. Arrows indicate the sites where crossing-over has occurred.

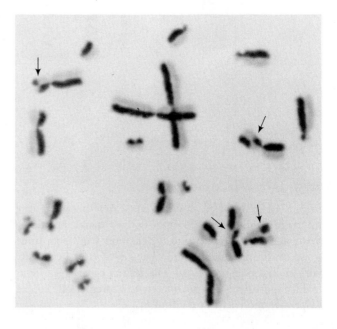

Subsequent experiments showed that the fragment of DNA acted as a template for synthesis of the new DNA; that is, the new DNA made in vitro was a faithful base pair–for–base pair copy of the original DNA.

Roles of DNA Polymerases

All DNA polymerases from prokaryotes and eukaryotes catalyze the polymerization of nucleotide precursors (dNTPs) into a DNA chain (Figure 9.4a). The same reaction in shorthand notation is shown in Figure 9.4b. The reaction has three main features:

1. At the growing end of the DNA chain, DNA polymerase catalyzes the formation of a phosphodiester bond between the 3′-OH group of the deoxyribose on the last nucleotide and the 5′-phosphate of the dNTP precursor. The energy for the formation of the phosphodiester bond comes from the release of two of three phosphates from the dNTP. The important concept here is that *the lengthening DNA chain acts as a primer in the reaction,* where a primer is a preexisting polynucleotide chain to which a new nucleotide can be added at the free 3′-OH.

2. At each step in lengthening the new DNA chain, DNA polymerase finds the correct precursor dNTP that can form a complementary base pair with the nucleotide on the template strand of DNA. Nucleotides can be added as rapidly as 800 per second. This does not occur with 100 percent accuracy, but the error frequency is extremely low.

3. The direction of synthesis of the new DNA chain is only from 5′ to 3′ because of the properties of DNA polymerase.

E. coli contains not just one but three DNA polymerases, the properties of which are summarized in Table 9.1. Both DNA polymerases I and III replicate DNA in the 5′→3′ direction in the cell, but the role of DNA polymerase II is unknown. All three *E. coli* DNA polymerases have 3′→5′ exonuclease activity. They can remove nucleotides from the 3′ end of a DNA chain. This enzyme activity is part of an error correction mechanism. If an incorrect base is inserted by DNA polymerase (which occurs at a frequency of about 10^{-6} for both DNA polymerase I and DNA polymerase III), in many cases the error is recognized immediately. Then the 3′→5′ exonuclease activity excises the erroneous nucleotide on the new strand. This process resembles that of the backspace key on a computer keyboard. The DNA polymerase then resumes the forward direction and inserts the correct character. Thus, in DNA replication, 3′→5′ exonuclease activity is a **proofreading** mechanism that helps keep the frequency of DNA replication errors very low. With proofreading, the frequency of replication errors by DNA polymerase I or III is reduced to lower than 10^{-9}.

DNA polymerase I has 5′→3′ exonuclease activity and can remove nucleotides from the 5′ end of a DNA strand or from an RNA primer strand. This activity, also important in DNA replication and repair, is examined later in this chapter.

KEYNOTE

The enzymes that catalyze the synthesis of DNA are called DNA polymerases. All known DNA polymerases synthesize DNA in the 5′→3′ direction. Some enzymes have proofreading activity.

Molecular Model of DNA Replication

Table 9.2 presents the functions of some of the *E. coli* DNA replication genes and key DNA sequences involved in replication. A number of the genes were identified by mutational analysis. In this section, we discuss a molecular model of DNA replication involving these genes and

Table 9.1 Comparison of the Structural and Functional Characteristics of the *E. coli* DNA Polymerases I, II, and III

DNA Polymerase	Polymerization: 5′→3′	Exonuclease: 3′→5′	Exonuclease: 5′→3′	Molecular Weight (daltons)	Molecules per Cell (approximately)	Structural Genes
I	Yes	Yes	Yes	103,000	400	*polA*
II	Yes	Yes	No	90,000	?	*polB*
III	Yes	Yes	No	Core of 130,000, 27,500, and 10,000 subunits; 7 other subunits[a]	10–20	*dnaE, dnaQ,* and *holE* for core

[a]Polymerase III consists of a catalytic core of α (alpha: 130,000 Da; *dnaE*), ε (epsilon: 27,500 Da; *dnaQ*; responsible for 3′→5′ exonuclease activity), and θ (theta: 10,000 Da; *holE*), and 7 other subunits: τ (tau: 71,000 Da; *dnaX*), γ (gamma: 47,500 Da; *dnaX*), δ (delta: 35,000 Da; *holA*), δ′ (delta prime: 33,000 Da; *holB*), χ (chi: 15,000 Da; *holC*), ψ (psi: 12,000 Da; *holD*), and β (beta: 40,600 Da; *dnaN*).

Figure 9.4

DNA chain elongation catalyzed by DNA polymerase. (**a**) Mechanism at molecular level. (**b**) The same mechanism, using a shorthand method to represent DNA.

a) Mechanism of DNA elongation

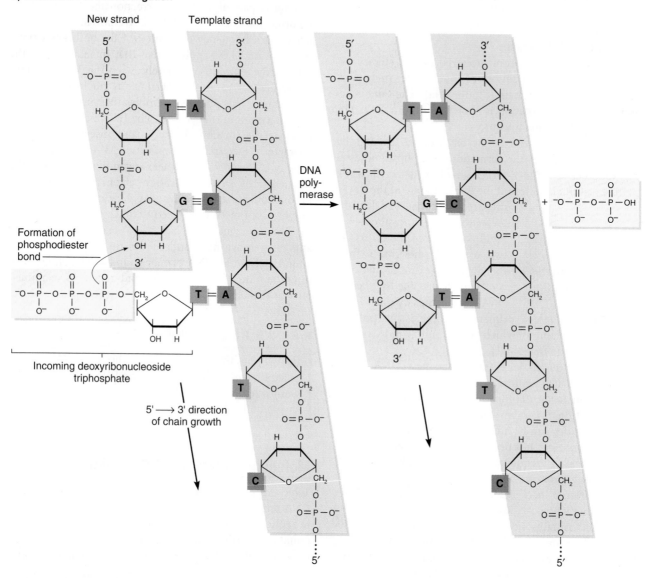

b) Shorthand notation

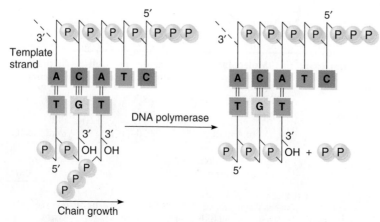

Table 9.2	Functions of Some of the Genes and DNA Sequences Involved in DNA Replication in *E. coli*

Gene Product and or Function	Gene
DNA polymerase I	*polA*
DNA polymerase II	*polB*
DNA polymerase III	*dnaE, dnaQ, dnaX, dnaN, dnaD holA→E*
Initiator protein; binds to *oriC*	*dnaA*
IHF protein (DNA-binding protein); binds to *oriC*	*himA*
FIS protein (DNA-binding protein); binds to *oriC*	*fis*
Helicase and activator of primase	*dnaB*
Complexes with *dnaB* protein and delivers it to DNA	*dnaC*
Primase; makes RNA primer for extension by DNA polymerase III	*dnaG*
Single-strand binding (SSB) proteins; bind to unwound single-stranded arms of replication forks	*ssb*
DNA ligase; seals single-stranded gaps	*lig*
Gyrase (type II topoisomerase); replication swivel to avoid tangling of DNA as replication fork advances	*gyrA, gyrB*
Origin of chromosomal replication	*oriC*
Terminus of chromosomal replication	*ter*
TBP (*ter* binding protein); stalls replication forks	*tus*

sequences. Note that for ease of representation, the drawings show the enzymes moving along a static DNA molecule, but Peter Cook has obtained evidence that the enzymes are actually stationary, with the DNA molecule passing by.

Initiation of Replication

Replication of prokaryotic and viral DNA usually starts at a specific site on the chromosome, the **origin of replication.** At that site, the double helix denatures into single strands, exposing the bases for the synthesis of new strands. The local denaturing of the DNA produces what is called a **replication bubble,** a region in the chromosome where the DNA is single-stranded and from which replication proceeds bidirectionally (in both directions). The segments of untwisted single strands on which the new strands are made (following complementary base-pairing rules) are called the **template strands.** In the circular *E. coli* chromosome, there is a single origin of replication, called *oriC*, from which replication proceeds bidirectionally.

Figure 9.5 shows a model for the formation of a replication bubble at the origin of replication in *E. coli* and the initiation of the new DNA strand. First, *gyrase* (a form of topoisomerase; see Chapter 8, p. 172) relaxes the supercoiled DNA. Then initiator proteins wrap the DNA around them at the origin of replication to form a large complex (Figure 9.5, step 1). **DNA helicase** (product of the *dnaB* gene) then binds to the initiator proteins (step 2) and is loaded onto the DNA (step 3). The DNA helicase untwists (denatures) the DNA, using the energy from hydrolysis of

adenosine triphosphate (ATP). Next, the enzyme **primase** (**DNA primase,** a form of RNA polymerase, product of the *dnaG* gene) binds to the helicase, forming a complex called the **primosome** (step 4).

Primase is important in DNA replication because no known DNA polymerases can initiate the synthesis of a DNA strand; they can only add nucleotides to a preexisting strand. Primase synthesizes a short **RNA primer** (about 11 nucleotides) to which new nucleotides can be added by DNA polymerase, and so the first new DNA chain is made (Figure 9.5, step 5). The RNA primer is removed later and replaced with DNA; we will return to discuss this event further.

We must be clear about the difference between *template* and *primer* with respect to DNA replication. A template strand is the one on which the new strand is synthesized according to complementary base-pairing rules. A primer is a short segment of nucleotides bound to the template strand. The primer acts as a substrate for DNA polymerase, which extends the primer as a new DNA strand, the sequence of which is complementary to the template strand.

KEYNOTE

The initiation of DNA synthesis first involves the denaturation of double-stranded DNA at an origin of replication, catalyzed by DNA helicase. Next, DNA primase binds to the helicase and the denatured DNA and synthesizes a short RNA primer. The RNA primer is extended by DNA polymerase as new DNA is made. The RNA primer is later removed.

Figure 9.5

Model for the formation of a replication bubble at a replication origin in *E. coli* and the initiation of the new DNA strand.

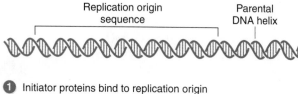

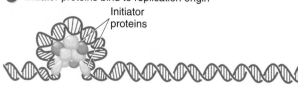

❶ Initiator proteins bind to replication origin

Initiator proteins

❷ DNA helicase binds to initiator proteins

DNA helicase

❸ Helicase loads onto DNA

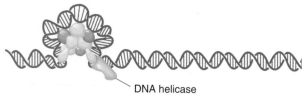

❹ Helicase denatures helix and binds with DNA primase to form primosome

Primosome

DNA primase

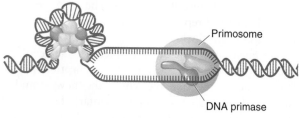

❺ Primase synthesizes RNA primer, which is extended as DNA chain by DNA polymerase

Replication bubble

DNA polymerase

RNA primer

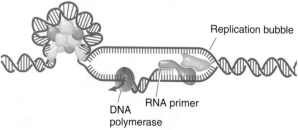

Semidiscontinuous DNA Replication

When DNA unwinds to expose the two single-stranded template strands for DNA replication, a Y-shaped structure called a **replication fork** is formed. A replication fork moves in one direction. When DNA unwinds in the middle of a DNA molecule, as in a circular chromosome, there are two replication forks; think of this as two Ys joined together at their tops. In many (but not all) cases, each replication fork is active, so that DNA replication proceeds bidirectionally. We will consider what happens at one replication fork.

animation

a Molecular Model of DNA Replication

As we just discussed (see Figure 9.5), helicase untwists the DNA to produce single-stranded template strands. **Single-strand DNA-binding (SSB) proteins** (encoded by the *ssb* gene) bind to the single-stranded DNA, stabilizing it (Figure 9.6) and preventing reannealing. Primase synthesizes an RNA primer on each template strand. These RNA primers are lengthened by DNA polymerase III, which synthesizes new DNA complementary to the template strands (see Figure 9.6a) while simultaneously displacing the bound SSB proteins. Recall that DNA polymerases can make DNA only in the 5′→3′ direction; however, the two DNA strands are of opposite polarity. To maintain the 5′→3′ polarity of DNA synthesis on each template and one overall direction of replication fork movement, DNA is made in opposite directions on the two template strands (see Figure 9.6a). The new strand being made 5′→3′ in the *same* direction as the movement of the replication fork is called the **leading strand,** and the new strand being made in the *opposite* direction as the movement of the replication fork is called the **lagging strand.** The leading strand, then, requires a single RNA primer for its synthesis, while the lagging strand requires a series of primers, as we shall see.

As the replication fork moves, helicase untwists more DNA (Figure 9.6b). Gyrase (a form of topoisomerase) relaxes the tension produced in the DNA ahead of the replication fork. This tension could be great because the replication fork rotates at about 3,000 rpm. On the leading-strand template (bottom strand in Figure 9.6), the leading strand is synthesized continuously toward the replication fork. Since DNA synthesis can proceed only in the 5′→3′ direction, however, lagging-strand synthesis has gone as far as it can. For DNA replication to continue on the lagging-strand template (top strand in Figure 9.6), a new initiation of DNA synthesis must occur. An RNA primer is made by the primase still bound to the helicase at the replication fork (Figure 9.6b). DNA polymerase III then adds DNA to the RNA primer to make another DNA fragment. Since the leading strand is being made continuously, whereas the lagging strand can be made only in pieces, or discontinuously, DNA replication as a whole occurs in a **semidiscontinuous** manner. The fragments

Figure 9.6

Model for the events occurring around a single replication fork of the *E. coli* chromosome. (a) Initiation. **(b)** Further untwisting and elongation of the new DNA strands. **(c)** Further untwisting and continued DNA synthesis. **(d)** Removal of the primer by DNA polymerase I. **(e)** Joining of adjacent DNA fragments by the action of DNA ligase. (Green = RNA; red = new DNA.)

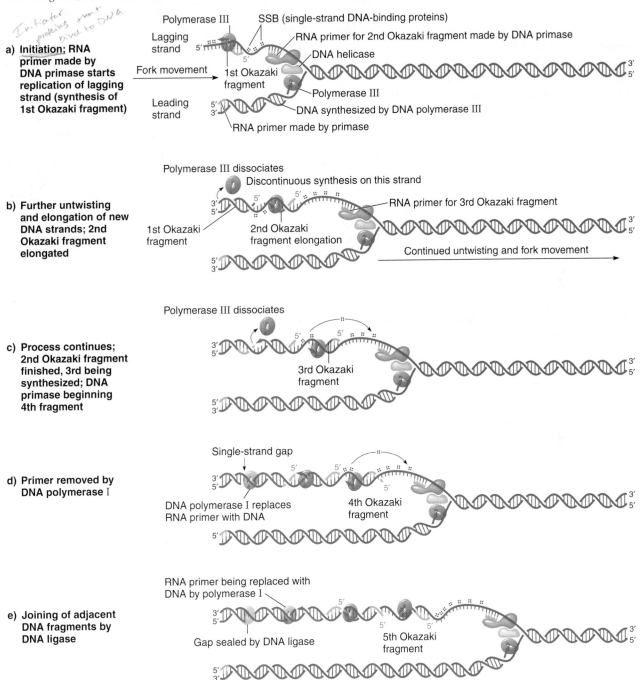

a) **Initiation; RNA primer made by DNA primase starts replication of lagging strand (synthesis of 1st Okazaki fragment)**

b) **Further untwisting and elongation of new DNA strands; 2nd Okazaki fragment elongated**

c) **Process continues; 2nd Okazaki fragment finished, 3rd being synthesized; DNA primase beginning 4th fragment**

d) **Primer removed by DNA polymerase I**

e) **Joining of adjacent DNA fragments by DNA ligase**

of lagging strand made in this process are called **Okazaki fragments** after their discoverers Reiji and Tuneko Okazaki and colleagues.

In Figure 9.6c, the process repeats itself: Helicase untwists the DNA, continuous DNA synthesis occurs on

the leading-strand template, and discontinuous DNA synthesis occurs on the lagging-strand template. Eventually, the unconnected Okazaki fragments on the lagging-strand template are joined into a continuous DNA strand. This requires the activities of DNA polymerase I and **DNA ligase.** Consider two adjacent Okazaki fragments. The 3′ end of the newer DNA fragment is adjacent to, but not joined to, the primer at the 5′ end of the previously made fragment. DNA polymerase III leaves the DNA, and DNA polymerase I continues the 5′→3′ synthesis of the DNA fragment, simultaneously removing the primer section of the older fragment by its 5′→3′ exonuclease activity (Figure 9.6d). When DNA polymerase I has completed replacement of RNA primer nucleotides with DNA nucleotides, a single-stranded gap is left between the two fragments. The two fragments are joined by DNA ligase to produce a longer DNA strand (Figure 9.6e). The whole sequence of events is repeated until all the DNA is replicated.

In sum, DNA replication in *E. coli* is a complicated process involving a number of enzymes and other proteins. We can relate the discontinuous nature of DNA synthesis to the replication of the circular DNA chromosome in *E. coli*. At *oriC* (see p. 191), the DNA is denatured to expose the two template strands. Denaturation produces a replication bubble that is actually *two* Y-shaped replication forks. Thus, DNA synthesis takes place in both directions (bidirectionally) away from the origin point. (Note that for some other circular chromosomes—certain plasmids, for example—replication is unidirectional.) The early stages of **bidirectional replication** are shown in Figure 9.7.

You can identify some of the specific elements and processes needed for DNA replication in the iActivity *Unraveling DNA Replication* on the Web site.

K E Y N O T E

During DNA replication, new DNA is made in the 5′→3′ direction, so chain growth is continuous on one strand and discontinuous (i.e., in segments that are later joined) on the other strand. This semidiscontinuous model is applicable to many other prokaryotic replication systems, each of which differs in the number and properties of the enzymes and other proteins needed.

DNA Replication in Eukaryotes

The biochemistry and molecular biology of DNA replication are very similar in prokaryotes and eukaryotes. However, an added complication in eukaryotes is that DNA is distributed among many chromosomes rather than just one. In this section, we summarize some of the important aspects of DNA replication in eukaryotes.

Figure 9.7

Diagram of the formation at a replication origin sequence of two replication forks that move in opposite directions. The top diagram is a DNA-only version of Figure 9.5, step 5.

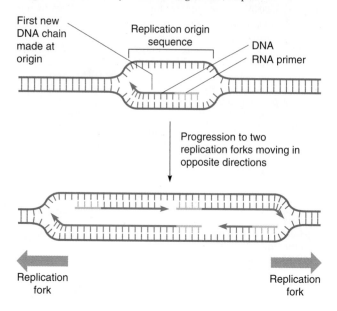

DNA Replication and the Cell Cycle

In every cell cycle, all chromosomes must be faithfully duplicated and a copy of each distributed to each of the two progeny cells. This means that both the DNA and the histones must be doubled with each cell cycle. Recall from Chapter 1 that the cell cycle in most somatic cells of higher eukaryotes is divided into four stages: gap 1 (G_1), synthesis (S), gap 2 (G_2), and mitosis (M) (see Figure 1.14). DNA replication and chromosomes duplication occur during the S phase, and the progeny chromosomes segregate into daughter cells during the M phase.

Genetic and molecular studies have elucidated a number of steps involved in controlling the progression of the cell through the cell cycle. A full discussion of this belongs more appropriately in a cell biology text, so only a simplified overview is given here. The focus is on yeasts, for which most details are known; mammalian systems work similarly, although they seem to involve more complex components.

Progression through the cell cycle is tightly controlled by the activities of many genes in an elaborate system of checks and balances (Figure 9.8). As a cell proceeds through G_1, it is gearing up for DNA replication and chromosome duplication in the S phase. A major checkpoint in G_1 called *START* in yeasts and *G_1 checkpoint* in mammalian cells determines whether the cell is able to or should continue into S. Unless a cell grows to a large enough size and the environment is favorable, it will stay in G_1. Another major checkpoint, called the G_2

Figure 9.8

Some of the molecular events that control progression through the cell cycle in yeasts.

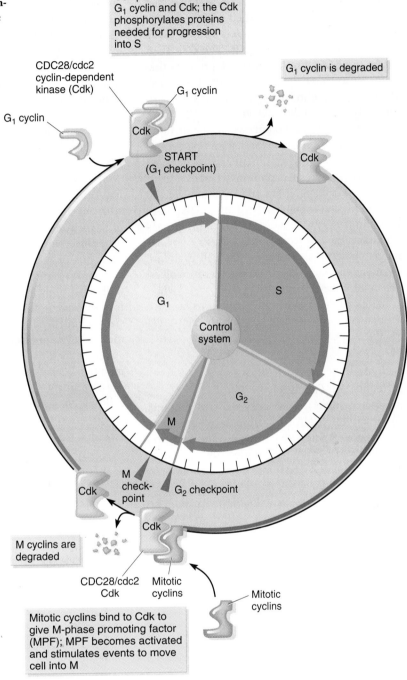

checkpoint both in yeasts and mammalian cells, occurs at the junction between G_2 and M. Unless all the DNA has replicated, the cell is big enough, and the environment is favorable, the cell cannot enter the mitotic phase of the cell cycle. A third checkpoint occurs during M. The chromosomes must be attached properly to the mitotic spindle to trigger the separation of chromatids and the completion of mitosis.

Key components in the regulatory events that occur at checkpoints are proteins known as *cyclins* (named because their concentration increases and decreases in a regular pattern through the cell cycle) and enzymes known as *cyclin-dependent kinases (Cdks)*. In yeasts, a single Cdk functions at both the G_1 and the G_2 checkpoints, while in mammalian cells at least two Cdks are involved at each checkpoint. In budding yeast

(*Saccharomyces cerevisiae*), the Cdk is called *CDC28* after the gene that encodes it, and in fission yeast (*Schizosaccharomyces pombe*) it is called *cdc2*.

At START, one or more G_1 *cyclins* bind to the CDC28/cdc2 Cdk kinase and activate it. The Cdk then phosphorylates (adds a phosphate to) key proteins that are needed for progression into S. Once the cyclin has activated the Cdk, the cyclin level decreases as a result of increased proteolysis (breakdown by enzymes). A similar process occurs at the G_2 checkpoint, when one or more *mitotic cyclins* bind to the Cdk to form the *M-phase promoting factor* (MPF). Then, when other enzymes dephosphorylate it, MPF is activated and in its active form catalyzes a series of phosphorylations that stimulate the cell to move into M. During mitosis, just after metaphase, the mitotic cyclin is degraded, which leads to MPF inactivation and allows the cell to complete mitosis.

The regulatory events controlling the cell cycle are complex. Many more steps and pathways are involved than have been described. We return to this subject when we discuss the genetics of cancer in Chapter 18 because some genes that normally control steps in the cell cycle often are mutated in cancer.

Eukaryotic Replication Enzymes

As we saw earlier, many of the enzymes and proteins involved in prokaryotic DNA replication have been identified. Less is known about the enzymes and proteins involved in eukaryotic DNA replication. However, it is clear that the steps described for DNA synthesis in prokaryotes also occur for DNA synthesis in eukaryotes: namely, denaturation of the DNA double helix and the semiconservative, semidiscontinuous replication of DNA. New DNA strands are initiated by RNA primers, which DNA polymerases extend as new DNA chains.

Six different DNA polymerases have been identified in eukaryotic cells, designated pol α (alpha), pol β (beta), pol δ (delta), pol γ (gamma), pol ε (epsilon), and pol ξ (xi). Pol α, pol δ, and pol ε polymerases are essential for DNA replication in eukaryotic nuclei; pol β and pol ξ are involved in DNA repair; and pol γ is required for the replication of mitochondrial DNA.

Replicons

Each eukaryotic chromosome consists of one linear DNA double helix. If there were only one origin of replication per chromosome, replicating each chromosome would take many, many hours. For example, there are about 3 billion base pairs of DNA in the haploid human genome (23 chromosomes), and the average chromosome is roughly 10^8 base pairs long. With a replication rate of 2 kb (2,000 bases) per minute, it would take approximately 830 hours (almost 35 days) to replicate one chromosome.

Actual measurements show that the chromosomes in eukaryotes replicate much faster than would be the case with only one origin of replication per chromosome. The diploid set of chromosomes in *Drosophila* embryos, for example, replicates in 3 minutes. This is six times faster than the replication of the *E. coli* chromosome, even though there is about 100 times more DNA in *Drosophila* than there is in *E. coli*.

Eukaryotic chromosomes duplicate rapidly because DNA replication initiates at many origins of replication throughout the genome. At each origin of replication, the DNA denatures (as in *E. coli*), and the replication proceeds bidirectionally. Eventually, each replication fork will run into an adjacent replication fork, initiated at an adjacent origin of replication. In eukaryotes, the stretch of DNA from the origin of replication to the two termini of replication (where adjacent replication forks fuse) on each side of the origin is called a **replicon** or a **replication unit** (Figure 9.9). Table 9.3 presents the number of replicons, their average size, and the rate of replication fork movement for six organisms. Note that the replicon size is much smaller, and the rate of fork movement much slower, in eukaryotic organisms than in bacteria. In general, each replicating eukaryotic chromosome has many replicons. Also, the number of origins and the average size of the replicon can vary with the nature of the cell (e.g., its growth rate and tissue of origin).

Figure 9.9

Replicating DNA of *Drosophila melanogaster*. (a) Electron micrograph showing replication units (replicons). (b) An interpretation of the electron micrograph shown in (a).

a)

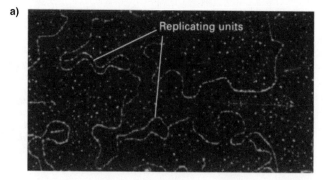

b)

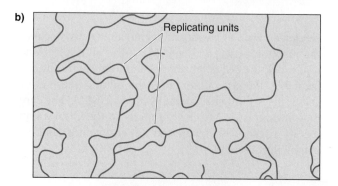

Table 9.3	Comparison of Bacterial and Eukaryotic Replicons		
Organism	**No. of Replicons**	**Average Length**	**Fork Movement**
Bacterium (*E. coli*)	1	4,200 kb	50,000 bp/min
Yeast (*Saccharomyces cerevisiae*)	500	40 kb	3,600 bp/min
Fruit fly (*Drosophila melanogaster*)	3,500	40 kb	2,600 bp/min
Toad (*Xenopus laevis*)	15,000	2,000 kb	500 bp/min
Mouse (*Mus musculus*)	25,000	150 kb	2,200 bp/min
Plant (bean: *Vicia faba*)	35,000	300 kb	n.a.[a]

[a] n.a.: value not available.

DNA replication does not occur simultaneously in all the replicons in an organism's genome. Instead, there is a cell-specific timing of initiation of replication at the various origins. Figure 9.10 shows one segment of one chromosome in which there are three replicons that always begin replicating at distinct times. When the replication forks fuse at the margins of adjacent replicons, the chromosome has replicated into two sister chromatids. In general, replication of a segment of chromosomal DNA occurs after the synchronous activation of a cluster of origins.

although progress is being made in characterizing some common sequence elements in the replicons of a number of eukaryotes.

KEYNOTE

In eukaryotes, DNA replication occurs in the S phase of the cell cycle and is similar molecularly to the replication process in prokaryotic cells. To enable the genome to be duplicated rapidly, DNA replication is initiated at a large number of sites throughout the chromosomes.

Origins of Replication

In the yeast *Saccharomyces cerevisiae*, specific chromosomal sequences have been identified that appear to act as origins of replication during chromosome duplication. These sequences are called **autonomously replicating sequences,** or **ARSs.** In mammals and other complex eukaryotes, these sequences are not as well defined,

Replicating the Ends of Chromosomes

Because DNA polymerases can synthesize new DNA only by extending a primer, there are special problems in replicating the ends, the telomeres, of eukaryotic chromosomes (Figure 9.11). A parental chromosome (Figure 9.11a) is replicated, resulting in two new DNA molecules,

Figure 9.10

Temporal ordering of DNA replication initiation events in replication units of eukaryotic chromosomes.

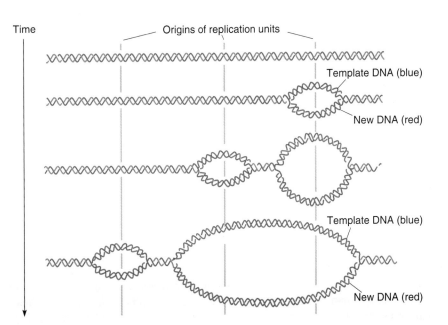

Time

Origins of replication units

Template DNA (blue)

New DNA (red)

Template DNA (blue)

New DNA (red)

Figure 9.11

The problem of replicating completely a linear chromosome in eukaryotes. (a) Diagram of a parent double-stranded DNA molecule representing the full length of a chromosome. **(b)** After semiconservative replication, new DNA segments hydrogen-bonded to the template strands have RNA primers at their 5′ ends. **(c)** The RNA primers are removed, DNA polymerase fills the resulting gaps, and DNA ligase joins the adjacent fragments. However, at the two telomeres, there are still gaps at the 5′ ends of the new DNA because RNA primers have been removed but no new DNA synthesis could fill them in.

a) Parent chromosome with multiple origins of replication

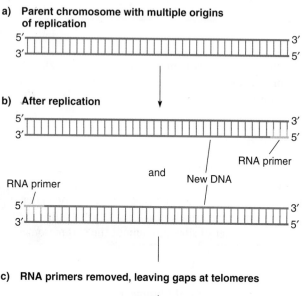

b) After replication

RNA primer

RNA primer   and   New DNA

c) RNA primers removed, leaving gaps at telomeres

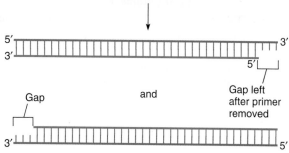

Gap   and   Gap left after primer removed

each of which has an RNA primer at the 5′ end of the newly synthesized strand in the telomere region (Figure 9.11b). The RNA primers are removed, leaving a single-stranded stretch of DNA—a gap—at the 5′ end of the new strand. The gap cannot be filled in by DNA polymerase because that enzyme cannot initiate new DNA synthesis. If nothing were done about these gaps, the chromosomes would get shorter and shorter with each replication cycle.

There is a special mechanism for replicating the ends of chromosomes. Most eukaryotic chromosomes have tandemly repeated, species-specific, simple sequences at their telomeres (see Chapter 8, pp. 176–177). The work of Elizabeth Blackburn and Carol W. Greider has shown that an enzyme called *telomerase* maintains chromosome lengths by adding telomere repeats to the chromosome ends. This mechanism does not involve the regular replication machinery.

Figure 9.12 shows the mechanism deduced for the protozoan *Tetrahymena* in a simplified way. The repeated sequence in *Tetrahymena* is 5′-TTGGGG-3′, reading toward the end of the DNA on the top strand in Figure 9.11. Telomerase acts at the stage shown in Figure 9.11c—that is, where a chromosome end has been produced with a gap at the end of the chromosome at the 5′ end of the new DNA (Figure 9.12a). Telomerase is an enzyme made up of both protein and RNA. The RNA component includes a base sequence that is complementary to the telomere repeat unit of the organism in which it is found. Therefore, the telomerase binds specifically to the overhanging telomere repeat at the end of the chromosome (Figure 9.12b). Next, the telomerase catalyzes the synthesis of three nucleotides of new DNA—TTG—using the telomerase RNA as a template (Figure 9.12c). The telomerase then slides toward the end of the chromosome so that its AAC at the 3′ end of the RNA template now pairs with the newly synthesized TTG on the DNA (Figure 9.12d). Telomerase then makes the rest of the TTGGGG telomere repeat (Figure 9.12e). The process is repeated to add more telomere repeats. In this way, the chromosome is lengthened by the addition of a number of telomere repeats. Then, by primer synthesis and DNA synthesis catalyzed by DNA polymerase in the conventional way, the former gap is filled in, and the new chromosomal DNA is lengthened (Figure 9.12f). After removal of the RNA primer, a new 5′ gap is left (Figure 9.12g), but any net shortening of the chromosome has been averted.

Telomere length, while not identical from chromosome end to chromosome end, is nonetheless regulated to an average length for the organism and cell type. In wild-type yeast, for example, the simple telomere sequences (TG$_{1-3}$, a repeating sequence of one T followed by one to three Gs) occupy an average of about 300 bp. Mutants that affect telomere length have been identified. For example, if the *TLC1* gene (which encodes the telomerase RNA) is deleted or the *EST1* (ever shorter telomeres) gene is mutated, telomeres shorten continuously until the cells die. This phenotype provides evidence that telomerase activity is necessary for long-term cell viability. The product of the *EST1* gene, the protein Est1p, is either a component of the telomere RNA-protein complex or a separate factor essential for telomerase function. Mutations of the *TEL1* and *TEL2* genes cause cells to maintain their telomeres at a new, shorter-than-wild-type length, making it clear that telomere length is regulated genetically.

Current evidence suggests many levels of regulation of telomere activity and telomere length. For example, attention is being given to the observation that telomerase activity in mammals is limited to immortal cells (such as tumor cells). The absence of telomerase activity in other cells results in progressive shortening of chromosome ends during successive divisions because of the failure to replicate those ends and in a limited number of cell divisions before the cell dies.

Figure 9.12

Synthesis of telomeric DNA by telomerase. The example is for *Tetrahymena* telomeres. **(a)** We start with a chromosome end with a 5′ gap left after primer removal. **(b)** Telomerase binds to the overhanging telomere repeat at the end of the chromosome. **(c)** A three-nucleotide DNA segment is synthesized at the chromosome end using the RNA template of telomerase. **(d)** The telomerase moves so that the RNA template can bind to the newly synthesized TTG in a different way. **(e)** Telomerase catalyzes the synthesis of a new telomere repeat using the RNA template. The process is repeated to add more telomere repeats. **(f)** After telomerase has left, new DNA is made on the template starting with an RNA primer. After the primer is removed **(g)**, the result is a longer chromosome than at the start, with a new 5′ gap.

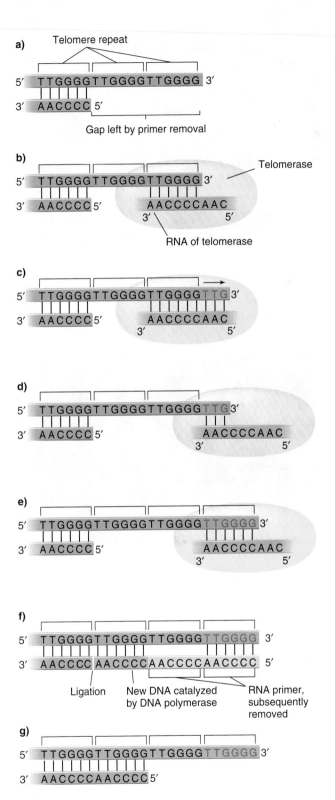

KEYNOTE

> Special enzymes—telomerases—are used to replicate the ends of chromosomes in eukaryotes. A telomerase is a complex of proteins and RNA. The RNA acts as a template for synthesizing the complementary telomere repeat of the chromosome, so telomerase is a type of reverse transcriptase enzyme.

Assembling New DNA into Nucleosomes

Eukaryotic DNA is complexed with histones in nucleosome structures, which are the basic units of chromosomes (see Chapter 8). Therefore, when the DNA is replicated, the histone complement must be doubled so that all nucleosomes are duplicated. This involves two processes: the synthesis of new histone proteins and the assembly of new nucleosomes.

Most histone synthesis is coordinated with DNA replication. The transcription of the genes for the five histones is initiated near the end of the G_1 phase, just before S. Translation of the histone mRNAs occurs throughout S, producing the histones to be assembled into nucleosomes as the chromosomes are duplicated.

Electron microscopy studies have shown that newly replicated DNA is assembled into nucleosomes almost immediately. Nonetheless, for replication to proceed, nucleosomes must disassemble during the short time when a replication fork passes. Measurements indicate that there is a nucleosome-free zone of about 200 to 300 base pairs around replication forks. Data from different kinds of experiments strongly suggest that new nucleosomes are made from an all-new set of histones while the old nucleosomes are conserved. Completed nucleosomes go randomly to the two DNA strands after the replication fork.

DNA Recombination

In the mid-1960s, Robin Holliday proposed a model for reciprocal recombination. Since then, the *Holliday model* has been refined and embellished by other geneticists, notably Matthew Meselson and Charles Radding, and T. Orr-Weaver and Jack Szostak. To give an idea of the recombination process at the molecular level, we present the Holliday model here.

The Holliday model is presented in Figure 9.13 for genetically distinguishable homologous chromosomes,

Figure 9.13

Holliday model for reciprocal genetic recombination.
Shown are two homologous DNA double helices that participate in the recombination process. Inset: Electron micrograph of a Holliday intermediate with some single-stranded DNA in the branch point region.

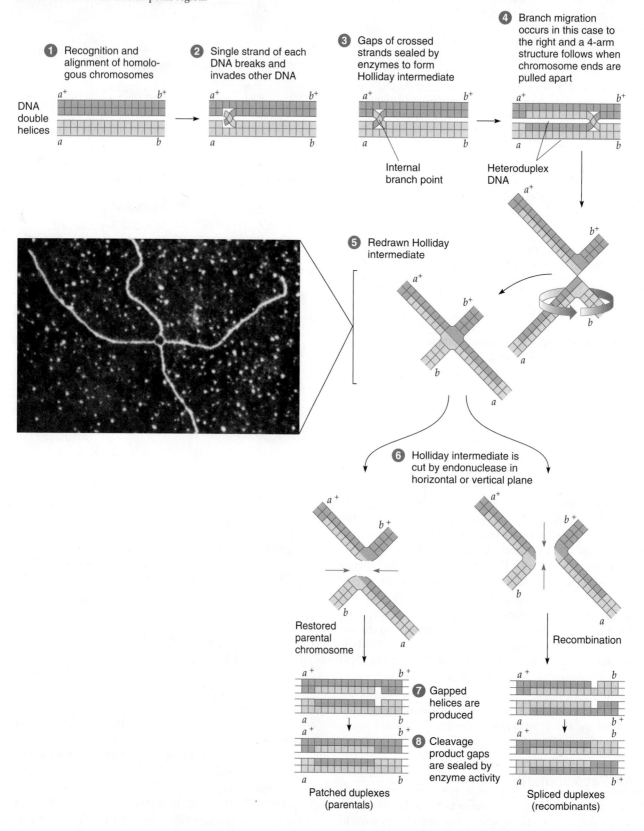

① Recognition and alignment of homologous chromosomes

② Single strand of each DNA breaks and invades other DNA

③ Gaps of crossed strands sealed by enzymes to form Holliday intermediate

④ Branch migration occurs in this case to the right and a 4-arm structure follows when chromosome ends are pulled apart

DNA double helices

Internal branch point

Heteroduplex DNA

⑤ Redrawn Holliday intermediate

⑥ Holliday intermediate is cut by endonuclease in horizontal or vertical plane

Restored parental chromosome

Recombination

⑦ Gapped helices are produced

⑧ Cleavage product gaps are sealed by enzyme activity

Patched duplexes (parentals)

Spliced duplexes (recombinants)

one with alleles a^+ and b^+ at opposite ends and the other with alleles a and b. The two DNA double helices in the figure participate in the recombination event. The first stage of the recombination process is *recognition and alignment* (Figure 9.13, step 1), in which two homologous DNA double helices become aligned precisely. In the second stage, one strand of each double helix breaks; each broken strand then invades the opposite double helix and base-pairs with the complementary nucleotides of the invaded helix (step 2). Enzymes are responsible for each of these steps. DNA polymerase and DNA ligase seal the gaps that are left, producing what is called a *Holliday intermediate*, with an internal branch point (step 3). The hybrid DNA molecules evident at this stage are called *heteroduplexes*; that is, the two strands of the double-stranded DNA molecules do not have completely complementary sequences. The two DNA double helices in the Holliday intermediate can rotate, causing the branch point to move to the right or to the left. Figure 9.13, step 4, shows a branch migration event that has occurred to the right. The four-armed structure for the DNA strands is produced simply by pulling the four chromosome ends apart. Branch migration generates complementary regions of hybrid DNA in both double helices (diagrammed as stretches of DNA helices with two different colors in steps 5–8).

The *cleavage and ligation* phase of recombination is best visualized if the Holliday intermediate is redrawn so that no DNA strand passes over or under another DNA strand. Thus, if the four-armed Holliday intermediate after Figure 9.13, step 4, is taken as a starting point and the lower two arms are rotated 180° relative to the upper arms, the structure shown in step 5 is produced.

Next, enzymes cut the Holliday intermediate at two points in the single-stranded DNA region of the branch point (step 6). The cuts can be in either the horizontal or vertical plane; both kinds of cuts occur with equal probability. Endonuclease cleavage in the horizontal plane (step 6, left) produces the two double helices shown in step 7, left. In each helix is a single-stranded gap. DNA ligase seals the gaps to produce the double helices shown in step 8, left. Since each of the resulting helices contains a segment of single-stranded DNA from the other helix flanked by nonrecombinant DNA, these double helices are called *patched duplexes*.

If the endonuclease cleavage in step 6 is in the vertical plane (step 6, right), the gapped double helices of step 7, right, are produced. In this case, there are segments of hybrid DNA in each duplex, but they are formed by what looks like a splicing together of two helices. The result is the double helices in step 8, right, which are called *spliced duplexes*.

In the example in Figure 9.13, the parental duplexes contain different genetic markers at the ends of the molecules. One parent is $a^+ b^+$ and the other is $a b$, as would be the case for a doubly heterozygous parent. However, the reciprocal recombination events shown in Figure 9.13 result in different products. In the patched duplexes (step 8, left) the markers are $a^+ b^+$ and $a b$, which is the parental configuration. However, in the spliced duplexes (step 8, right), the markers are recombinant, $a^+ b$ and $a b^+$. Since the enzymes that cut in the branch region during the final cleavage and ligation phase (step 6) cut randomly with respect to the plane of the cut (i.e., horizontal or vertical), the Holliday model predicts that a physical exchange between two gene loci on homologous chromosomes should result in the genetic exchange of the outside chromosome markers about half of the time.

Although the basic features of the Holliday model are generally accepted, a number of other models attempt to explain recombination at a more detailed level or in special systems. Discussion of these other models is beyond the scope of this text.

KEYNOTE

The Holliday model for genetic recombination involves a precise alignment of homologous DNA sequences of two parental double helices, endonucleolytic cleavage, invasion of the other helix by each broken end, ligation to produce the Holliday intermediate, branch migration, and finally, cleavage and ligation to resolve the Holliday intermediate into the recombinant double helices. Depending on the orientation of the cleavage events, the resulting double helices will be patched duplexes or spliced duplexes. If the recombination events occur between heterozygous loci, the patched duplexes produced are parental, whereas the spliced duplexes are recombinant for the two loci.

Summary

In this chapter, we discussed DNA replication and DNA recombination. Many aspects of DNA replication are similar in prokaryotes and eukaryotes, such as a semiconservative and semidiscontinuous mechanism, synthesis of new DNA in the $5' \rightarrow 3'$ direction, and the use of RNA primers to initiate DNA chains. The enzymes that catalyze the DNA synthesis are the DNA polymerases. In *E. coli,* there are three DNA polymerases, two of which are known to be involved in DNA replication along with several other enzymes and proteins. The DNA polymerases have $3' \rightarrow 5'$ exonuclease activity, which permits proofreading to take place during DNA synthesis if an incorrect nucleotide is inserted opposite the template strand. In eukaryotes, two DNA polymerases are involved in nuclear DNA replication. Neither has associated proofreading activity; that function presumably is the property of a separate protein.

In prokaryotes, DNA replication begins at specific chromosomal sites. Such sites are known also for yeast. A prokaryotic chromosome has one initiation site for DNA replication, whereas a eukaryotic chromosome has many initiation sites dividing the chromosome into replication units, or replicons. It is not clear whether the initiation sites used are the same sequences from cell generation to cell generation. The existence of replicons means that DNA replication of the entire set of chromosomes in a eukaryotic organism can proceed quickly, in some cases faster than with the single *E. coli* chromosome, despite the presence of orders of magnitude more DNA.

Since eukaryotic chromosomes are linear, there is a special problem of maintaining the lengths of chromosomes because removal of RNA primers results in a shorter new DNA strand. This problem is overcome by special enzymes called telomerases that maintain the length of chromosomes. Telomerases are a combination of proteins and RNA. The RNA component acts as a template to guide the synthesis of new telomere repeat units at the chromosome ends.

DNA replication in eukaryotes occurs in the S phase of the cell division cycle. Eukaryotic chromosomes are complexes of DNA with histones (and nonhistone chromosome proteins), so not only must DNA be replicated, but the nucleosome organization of chromosomes must be duplicated as the replication forks migrate. It is known that nucleosomes are "mature" soon after a replication fork passes, and evidence suggests that nucleosomes are conserved in chromosome duplication, with nucleosomes after the replication fork consisting of all new histones or all old histones.

A number of models have been proposed to describe the molecular events involved in the breakage and rejoining of DNA during crossing-over. One model, developed by Holliday, is described in this chapter. All molecular models for genetic recombination involve new DNA synthesis in small regions participating in crossing-over. In a later chapter (Chapter 19), we will see that DNA synthesis may also be involved in the repair of certain genetic damage. Thus, three major cellular processes involve DNA synthesis: DNA replication, genetic recombination, and DNA repair.

Analytical Approaches for Solving Genetics Problems

Q9.1 What would be the effect on chromosome replication in *E. coli* strains carrying deletions of the following genes?

a. *dnaE*
b. *polA*
c. *dnaG*

d. *lig*
e. *ssb*
f. *oriC*

A9.1 When genes are deleted, the function encoded by those genes is lost. All of the genes listed are involved in DNA replication in *E. coli*, and their functions are briefly described in Table 9.2 and in the text.

a. *dnaE* encodes a subunit of DNA polymerase III, the principal DNA polymerase in *E. coli* that is responsible for elongation of DNA chains. A deletion of the *dnaE* gene would undoubtedly lead to a nonfunctional DNA polymerase III. In the absence of DNA polymerase III activity, DNA strands could not be synthesized from RNA primers; hence, synthesis of new DNA strands could not occur, and there would be no chromosome replication.

b. *polA* encodes DNA polymerase I, which is used in DNA synthesis to extend DNA chains made by DNA polymerase III while simultaneously excising the RNA primer by 5'→3' exonuclease activity. As discussed in the text, in mutant strains lacking the originally studied DNA polymerase, DNA polymerase I, chromosome replication still occurred. Thus, chromosome replication would occur normally in an *E. coli* strain carrying a deletion of *polA*.

c. *dnaG* encodes DNA primase, the enzyme that synthesizes the RNA primer on the DNA template. Without the synthesis of the short RNA primer, DNA polymerase III cannot initiate DNA synthesis, so chromosome replication will not take place.

d. *lig* encodes DNA ligase, the enzyme that catalyzes the ligation of Okazaki fragments. In a strain carrying a deletion of *lig*, DNA synthesis would occur, but stable progeny chromosomes would not result because the Okazaki fragments could not be ligated together, so the lagging strand synthesized discontinuously on the lagging-strand template would be in fragments.

e. *ssb* encodes the single-strand binding proteins that bind to and stabilize the single-stranded DNA regions produced as the DNA is unwound at the replication fork. In the absence of single-strand binding proteins, impeded or absent DNA replication would result because the replication bubble could not be kept open.

f. *oriC* is the origin of replication region in *E. coli*, that is, the location at which chromosome replication initiates. Without the origin, the initiator protein cannot bind, no replication bubble can form, and so chromosome replication cannot take place.

Questions and Problems

9.1 Describe the Meselson-Stahl experiment, and explain how it showed that DNA replication is semiconservative.

*9.2 In the Meselson-Stahl experiment, ^{15}N-labeled cells were shifted to ^{14}N medium, at what we can designate as generation 0.

a. For the semiconservative model of replication, what proportion of ^{15}N-^{15}N, ^{15}N-^{14}N, and ^{14}N-^{14}N would you expect to find after 1, 2, 3, 4, 6, and 8 replication cycles?

b. Answer (a) in terms of the conservative model of DNA replication.

9.3 A spaceship lands on Earth and with it a sample of extraterrestrial bacteria. You are assigned the task of determining the mechanism of DNA replication in this organism.

You grow the bacteria in unlabeled medium for several generations, then grow it in the presence of ^{15}N for exactly one generation. You extract the DNA and subject it to CsCl centrifugation. The banding pattern you find is shown as follows:

It appears to you that this is evidence that DNA replicates in the semiconservative manner, but you are wrong. Why? What other experiment could you perform (using the same sample and technique of CsCl centrifugation) that would further distinguish between semiconservative and dispersive modes of replication?

*9.4 The elegant Meselson-Stahl experiment was among the first experiments to contribute to what is now a highly detailed understanding of DNA replication. Reconsider this experiment in light of current molecular models by answering the following questions.

a. Does the fact that DNA replication is semiconservative mean that it must be semidiscontinuous?

b. Does the fact that DNA replication is semidiscontinuous ensure that it is also semiconservative?

c. Do any properties of known DNA polymerases ensure that DNA is synthesized semiconservatively?

*9.5 List the components necessary to make DNA in vitro using the enzyme system isolated by Kornberg.

9.6 Kornberg isolated DNA polymerase I from *E. coli*. DNA polymerase I has an essential function in DNA replication. What are the functions of the enzyme in DNA replication?

9.7 Assume you have a DNA molecule with the base sequence TATCA going from the 5′ to the 3′ end of one of the polynucleotide chains. The building blocks of the DNA are drawn as in the following figure. Use this shorthand system to diagram the completed double-stranded DNA molecule, as proposed by Watson and Crick.

A G C T

PPP OH PPP OH PPP OH PPP OH

*9.8 Base analogues are compounds that resemble the natural bases found in DNA and RNA but are not normally found in those macromolecules. Base analogues can replace their normal counterparts in DNA during in vitro DNA synthesis. Four base analogues were studied for their effects on in vitro DNA synthesis using *E. coli* DNA polymerase. The results were as follows, with the amounts of DNA synthesized expressed as percentages of the DNA synthesized from normal bases only.

Analog	Normal Bases Substituted by the Analogue			
	A	**T**	**C**	**G**
A	0	0	0	25
B	0	54	0	0
C	0	0	100	0
D	0	97	0	0

Which bases are analogues of adenine? of thymine? of cytosine? of guanine?

9.9 a. Describe (draw) models of continuous, semidiscontinuous, and discontinuous DNA replication.

b. What was the contribution of Reiji and Tuneko Okazaki and colleagues with regard to these replication models?

9.10 How long would it take *E. coli* to replicate its entire genome (4.2×10^6 bp), assuming a replication rate of 1,000 nucleotides per second at each fork with no pauses?

*9.11 A diploid organism has 4.5×10^8 bp in its DNA. This DNA is replicated in 3 minutes. Assuming all replication forks move at a rate of 10^4 bp per minute, how many replicons (replication units) are present in this organism's genome?

9.12 The following events, steps, or reactions occur during *E. coli* DNA replication. For each entry in column A, select the appropriate entry in column B. Each entry in A may have more than one answer, and each entry in B can be used more than once.

Column A	Column B
_____ a. Unwinds the double helix	A. Polymerase I
_____ b. Prevents reassociation of complementary bases	B. Polymerase III
	C. Helicase
_____ c. Is an RNA polymerase	D. Primase
_____ d. Is a DNA polymerase	E. Ligase
_____ e. Is the "repair" enzyme	F. SSB protein
_____ f. Is the major elongation enzyme	G. Gyrase
	H. None of these
_____ g. A 5′→3′ polymerase	
_____ h. A 3′→5′ polymerase	
_____ i. Has 5′→3′ exonuclease function	
_____ j. Has 3′→5′ exonuclease function	
_____ k. Bonds free 3′-OH end of a polynucleotide to a free 5′-monophosphate end of a polynucleotide	
_____ l. Bonds 3′-OH end of a polynucleotide to a free 5′ nucleotide triphosphate	
_____ m. Separates daughter molecules and causes supercoiling	

*9.13 Describe the molecular action of the enzyme DNA ligase. What properties would you expect an *E. coli* cell to have if it had a temperature-sensitive mutation in the gene for DNA ligase?

*9.14 Chromosome replication in *E. coli* commences from a constant point, called the origin of replication. It is known that DNA replication is bidirectional. Devise a biochemical experiment to prove that the *E. coli* chromosome replicates bidirectionally. (Hint: Assume that the amount of gene product is directly proportional to the number of genes.)

9.15 Reiji Okazaki concluded that both strands could not replicate continuously. What evidence led him to this conclusion?

*9.16 A space probe returns from Jupiter and brings with it a new microorganism for study. It has double-stranded DNA as its genetic material. However, studies of replication of the alien DNA reveal that although the process is semiconservative, DNA synthesis is continuous on both the leading-strand and lagging-strand templates. What conclusions can you draw from that result?

*9.17 Compare and contrast eukaryotic and prokaryotic DNA polymerases.

9.18 In eukaryotes, cellular proliferation is tightly controlled. Explain why this is important, and outline how key regulatory proteins control the progression of the cell cycle.

*9.19 Autoradiography is a technique that allows radioactive areas of chromosomes to be observed under the microscope. The slide is covered with a photographic emulsion, which is exposed by radioactive decay. In regions of exposure, the emulsion forms silver grains on being developed. The tiny silver grains can be seen on top of the (much larger) chromosomes. Devise a method to find out which regions in the human karyotype replicate during the last 30 minutes of the S phase. (Assume a cell cycle in which the cell spends 10 hours in G_1, 9 hours in S, 4 hours in G_2, and 1 hour in M.)

*9.20 In Figure 9.3, semiconservative DNA replication is visualized in eukaryotic cells using the harlequin chromosome-staining technique.

a. Explain what the harlequin chromosome-staining technique is and how it provides evidence for semiconservative DNA replication in eukaryotes.

b. Propose a hypothesis to explain why, in Figure 9.3, some chromatids appear to contain segments of both DNA containing T and DNA containing BUdR, while others appear to consist entirely of DNA with T or DNA with BUdR.

*9.21 A mutant *Tetrahymena* has an altered repeated sequence in its telomeric DNA. What change in the telomerase enzyme would produce this phenotype?

The protein hemoglobin.

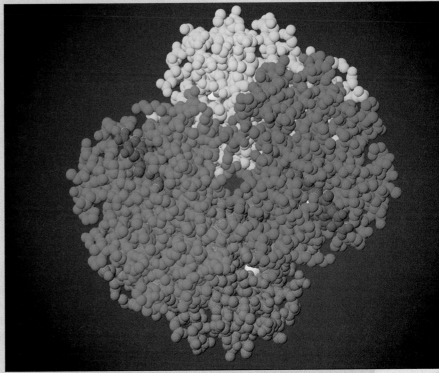

10

Gene Control of Proteins

PRINCIPAL POINTS

There is a specific relationship between genes and enzymes, embodied initially in the one gene–one enzyme hypothesis stating that each gene controls the synthesis or activity of a single enzyme. Since some enzymes consist of more than one polypeptide, and genes code for individual polypeptide chains, a more modern name for this hypothesis is one gene–one polypeptide.

Many human genetic diseases are caused by deficiencies in enzyme activities. Although some of these diseases are inherited as dominant traits, most are inherited as recessive traits.

From the study of alterations in proteins other than enzymes, convincing evidence was obtained that genes control the structure of all proteins.

Genetic counseling is the analysis of the risk that prospective parents may produce a child with a genetic defect and the presentation to appropriate family members of the available options for avoiding or minimizing those risks. Early detection of a genetic disease is done by carrier detection and fetal analysis.

i WITHIN THE FIRST FEW MINUTES OF LIFE, MOST newborns in the United States are subjected to a battery of tests: Reflexes are tested, respiration and skin color are assessed, and blood samples are collected and rushed to a lab. Assays of the blood samples help health practitioners determine whether the child has a debilitating or even lethal genetic disease. What are genetic diseases? What is the relationship between our genes, enzymes, and genetic disease? How can understanding gene function help prevent or minimize the risk of such diseases?

What do bread mold and certain human genetic disorders have in common? In the iActivity for this chapter, you will use Beadle and Tatum's experimental procedure to learn the answer to that question.

In this chapter, we examine gene function. We present some of the classic evidence that genes code for enzymes and other proteins. In particular, we examine the involvement of certain sets of genes in directing and controlling a particular biochemical pathway, the series of enzyme-catalyzed steps needed to break down or synthesize a particular chemical compound. Instead of thinking about the gene in isolation, we will see that the gene often must work in cooperation with other genes for cells to function properly. The experiments we will discuss represent the beginnings of molecular genetics, historically speaking, in that their goal was to gain a better understanding of the gene at the molecular level. In later chapters, we develop our modern understanding of gene structure and function further.

Gene Control of Enzyme Structure

Garrod's Hypothesis of Inborn Errors of Metabolism

In 1902, Archibald Garrod, an English physician, studied *alkaptonuria* (OMIM 203500), a human disease characterized by urine that turns black on exposure to air and by a tendency to develop arthritis later in life. Because of the urine phenotype, the disease is easily detected soon after birth.

Garrod and geneticist William Bateson concluded that alkaptonuria is a genetically controlled trait because several members of the same family often had alkaptonuria and because the disease was much more common among children of marriages involving first cousins than among children of marriages between unrelated partners. This finding was significant because first cousins have many alleles in common, so the chances are greater for recessive alleles to be homozygous in children of first-cousin marriages.

Garrod found that people with alkaptonuria excrete homogentisic acid (HA) in their urine, whereas normal people do not; it is the HA in urine that turns it black in air. This result indicated to Garrod that normal people can

metabolize HA but that people with alkaptonuria cannot. In Garrod's terms, this disease is an example of an *inborn error of metabolism;* that is, alkaptonuria is a genetic disease caused by the absence of a particular enzyme necessary for HA metabolism. Figure 10.1 shows part of the phenylalanine-tyrosine metabolic pathway, including the step from HA to maleylacetoacetic acid, which is not carried out in people with alkaptonuria. The mutation responsible for alkaptonuria is recessive, so only people homozygous for the mutant gene express the defect. Later analysis has pinpointed the location of this gene on chromosome 3. Garrod's work provided the first evidence of a specific relationship between genes and enzymes.

The One Gene–One Enzyme Hypothesis

George Beadle and Edward Tatum in 1942 heralded the beginnings of biochemical genetics, a branch of genetics that combines genetics and biochemistry to explain the nature of metabolic pathways. Results of their studies involving the haploid fungus *Neurospora crassa* (orange bread mold) showed a direct relationship between genes and enzymes and led to the *one gene–one enzyme hypothesis,* a landmark in the history of genetics. Beadle and Tatum shared one-half of the 1958 Nobel Prize in Physiology or Medicine for their "discovery that genes act by regulating definite chemical events."

animation

a The One Gene–One Enzyme Hypothesis

Isolation of Nutritional Mutants of *Neurospora*. To understand Beadle and Tatum's experiment, we must understand the life cycle of *Neurospora crassa,* the orange bread mold (Figure 10.2). *Neurospora crassa* is a mycelial-form fungus, meaning that it spreads over its growth medium in a weblike pattern. The mycelium produces asexual spores called conidia; their orange color gives the fungus its common name. *Neurospora* has important properties that make it useful for genetic and biochemical studies, including the fact that it is a haploid organism, so the effects of mutations may be seen directly, and the fact that it has a short life cycle, facilitating study of the segregation of genetic defects.

Neurospora is a haploid organism. It can be propagated vegetatively (asexually) by inoculating pieces of the mycelial growth or the asexual spores (conidia) on a suitable growth medium to give rise to a new mycelium. *Neurospora crassa* can also reproduce by sexual means. There are two mating types, called *A* and *a.* The sexual cycle is initiated by mixing *A* and *a* mating type strains on nitrogen-limiting medium. Under these conditions, cells of the two mating types fuse, followed by fusion of two haploid nuclei to produce a transient *A/a* diploid nucleus, which is the only diploid stage of the life cycle. The diploid nucleus immediately undergoes meiosis and produces four haploid nuclei (two *A* and two *a*) within an

Figure 10.1

Phenylalanine-tyrosine metabolic pathways. Individuals with alkaptonuria cannot metabolize homogentisic acid (HA) to maleylacetoacetic acid, and HA accumulates. Individuals with PKU cannot metabolize phenylalanine to tyrosine, and phenylpyruvic acid accumulates. Individuals with albinism cannot synthesize much melanin from tyrosine.

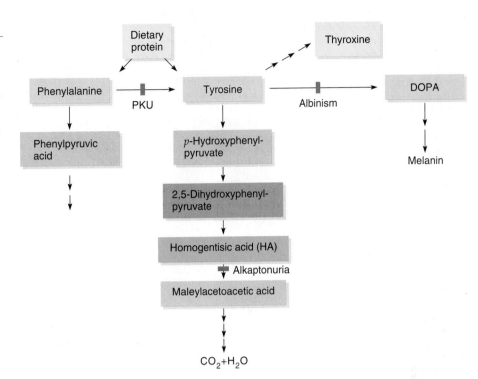

elongating sac called an *ascus*. A subsequent mitotic division results in a linear arrangement of eight haploid nuclei around which spore walls form to produce eight sexual ascospores (four *A* and four *a*). Each ascus, then, contains all the products of the initial, single meiosis. When the ascus is ripe, the ascospores (sexual spores) are shot out of the ascus and out of the fruiting body that encloses the ascus to be dispersed by wind currents. Germination of an ascospore begins the formation of a new haploid mycelium.

Important for Beadle and Tatum's experiments is the fact that *Neurospora* has simple growth requirements. Wild-type *Neurospora* can grow on a simple minimal medium containing only inorganic salts (including a source of nitrogen), an organic carbon source (such as glucose or sucrose), and the vitamin biotin. Beadle and Tatum reasoned that *Neurospora* synthesizes the other materials it needs for growth (e.g., amino acids, nucleotides, vitamins, nucleic acids, proteins) from the chemicals present in the minimal medium. They also realized that it should be possible to isolate **nutritional mutants** (also called **auxotrophic mutants** or **auxotrophs**) of *Neurospora* that require nutritional supplements to grow. Such auxotrophs could be isolated because they would not grow on the minimal medium.

Figure 10.3 shows how Beadle and Tatum isolated and characterized nutritional mutants. They treated conidia with X rays to induce genetic mutants and then crossed them with a wild-type strain (a strain that can grow on minimal medium, called a **prototrophic strain** or **prototroph**) of the opposite mating type. This cross was done because they wanted to study the genetics of

biochemical events, and they had to identify the nutritional mutations that were heritable. By crossing the mutagenized spores with the wild type, they ensured that any nutritional mutant they isolated had segregated in a cross and therefore had a genetic basis rather than a nongenetic basis for requiring the nutrient.

One progeny spore per ascus from the crosses was allowed to germinate in a nutritionally complete medium that contained all necessary amino acids, purines, pyrimidines, and vitamins, in addition to the sucrose, salts, and biotin found in minimal medium. In complete medium, any strain that could not make one or more necessary compounds from the basic ingredients in minimal medium could still grow by using the compounds supplied in the growth medium. Each culture grown on the complete medium was then tested for growth on minimal medium. The strains that did not grow were the auxotrophs. These mutants, in turn, were tested individually for their ability to grow on minimal medium plus amino acids and minimal medium plus vitamins. Theoretically, an amino acid auxotroph—a mutant strain that has lost the ability to synthesize a particular amino acid—would grow on minimal medium plus amino acids but not on minimal medium plus vitamins or on minimal medium alone. Similarly, vitamin auxotrophs would grow only on minimal medium plus vitamins.

Beadle and Tatum next did a second round of screening to determine which specific substance each auxotrophic strain needed for growth. Suppose an amino acid auxotroph was identified. To determine which of the 20 amino acids was required, the strain was inoculated into 20 tubes, each containing minimal medium

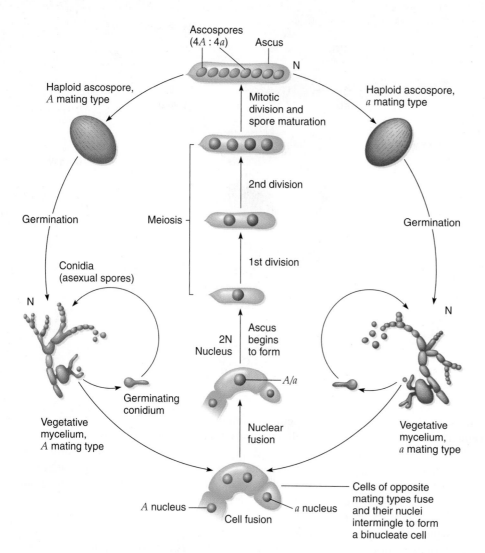

Figure 10.2
Life cycle of the haploid, mycelial-form fungus *Neurospora crassa*.
(Parts not to scale.)

(Figure labels, clockwise:)
Ascospores (4A : 4a) — Ascus — N

Haploid ascospore, *A* mating type

Mitotic division and spore maturation

2nd division

Germination — Meiosis

1st division

Conidia (asexual spores)

N

Ascus begins to form

2N Nucleus

Germinating conidium

Vegetative mycelium, *A* mating type

Nuclear fusion

A/a

A nucleus — *a* nucleus

Cell fusion

Haploid ascospore, *a* mating type

Germination

N

Vegetative mycelium, *a* mating type

Cells of opposite mating types fuse and their nuclei intermingle to form a binucleate cell

plus one of the 20 amino acids. In the example shown in Figure 10.3, a tryptophan auxotroph was identified because it grew only in the tube containing minimal medium plus tryptophan. Crosses between each mutant and wild type segregated in the 1:1 ratio of wild-type phenotype : tryptophan-requiring phenotype expected for a single gene defect.

Genetic Dissection of a Biochemical Pathway. Once Beadle and Tatum had isolated and identified nutritional mutants, they investigated the biochemical pathways affected by the mutations. They assumed that *Neurospora* cells, like all other cells, function through the interaction of the products of a very large number of genes. Furthermore, they reasoned that wild-type *Neurospora* converted the constituents of minimal medium into amino acids and other required compounds by a series of reactions that were organized into pathways. In this way, the synthesis of cellular components occurred through a series of small steps, each catalyzed by an enzyme. As an example of the analytical approach Beadle and Tatum used that led to an understanding of the relationship between genes

and enzymes, let us consider the genetic dissection of the pathway for the biosynthesis of the amino acid methionine in *Neurospora crassa*.

Starting with a set of methionine auxotrophs—mutants that require the addition of methionine to minimal medium to grow—genetic analysis identifies four genes. A mutation in any one of them gives rise to auxotrophy for methionine. These four genes in a wild-type cell are designated $met-2^+$, $met-3^+$, $met-5^+$, and $met-8^+$. Next, the growth pattern of the four mutant strains on media supplemented with chemicals thought to be intermediates involved in the methionine biosynthetic pathway—*O*-acetyl homoserine, cystathionine, and homocysteine—is determined, with the results shown in Table 10.1. By definition, all four mutant strains can grow on methionine, and none can grow on unsupplemented minimal medium.

The sequence of steps in a pathway can be deduced from the pattern of growth supplementation. The principles are as follows: The farther along in a pathway a mutant strain is blocked, the fewer intermediate compounds permit the strain to grow. If a mutant strain is blocked at early steps, a larger number of intermediates

Figure 10.3

Method devised by Beadle and Tatum to isolate auxotrophic mutations in *Neurospora*. Here, the mutant strain isolated is a tryptophan auxotroph.

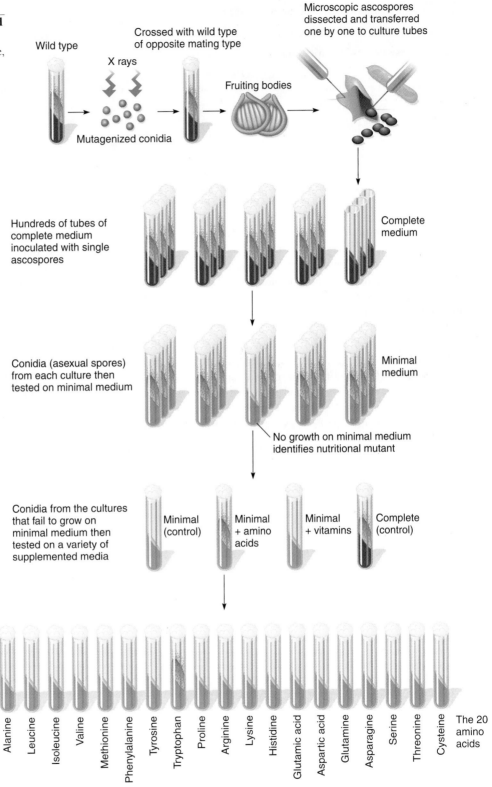

enable the strain to grow. Thus, in these analyses, not only is the pathway deduced, but the steps controlled by each gene are determined. In addition, a genetic block in a pathway may lead to an accumulation of the intermediate compound used in the step that is blocked.

Considering the results in Table 10.1 with this in mind, the *met-8* mutant strain grows when supplemented with methionine but not when supplemented with any of the intermediates. This means that the *met-8* gene must control the last step in the pathway, which leads to the

Table 10.1	Growth Responses of Methionine Auxotrophs				
Growth Response on Minimal Medium and . . .					
Mutant Strains	**Nothing**	**O-Acetyl Homoserine**	**Cystathionine**	**Homocysteine**	**Methionine**
Wild type	+	+	+	+	+
met-5	−	+	+	+	+
met-3	−	−	+	+	+
met-2	−	−	−	+	+
met-8	−	−	−	−	+

formation of methionine. The *met-2* mutant strain grows on media supplemented with methionine or homocysteine, so homocysteine must be immediately before methionine in the pathway, and the *met-2* gene must control the synthesis of homocysteine from another chemical. The *met-3* mutant strain grows on media supplemented with methionine, homocysteine, or cystathionine, so cystathionine must precede homocysteine in the pathway, and the *met-3* gene must control the synthesis of cystathionine from another compound. The *met-5* strain grows on media supplemented with methionine, homocysteine, cystathionine, or O-acetyl homoserine, so O-acetyl homoserine must precede cystathionine in the pathway, and the *met-5* gene must control the synthesis of O-acetyl homoserine from another compound. The methionine biosynthesis pathway involved here (which is part of a larger pathway) is shown in Figure 10.4. Gene *met-5*⁺ encodes the enzyme for converting homoserine to O-acetyl homoserine, so mutants for this gene can grow on minimal medium plus either O-acetyl homoserine, cystathionine, homocysteine, or methionine. Gene *met-3*⁺ codes for the enzyme that converts O-acetyl homoserine to cystathionine, so a *met-3* mutant strain can grow on minimal medium plus either cystathionine, homocysteine, or methionine. And so on.

From the results of experiments of this kind, Beadle and Tatum proposed that a specific gene encodes each enzyme. This proposed relationship between an organism's genes and the enzymes that catalyze the steps in a biochemical pathway is called the **one gene–one enzyme hypothesis.** Gene mutations that result in the loss of enzyme activity lead to the accumulation of precursors in the pathway (and to possible side reactions) and to the absence of the end product of the pathway. With the approach described, then, a biochemical pathway can be dissected genetically; that is, we can determine the sequence of steps in the pathway and relate each step to a specific gene.

However, more than one gene may control each step in a pathway. An enzyme[1] may have two or more different polypeptide chains, each of which is coded for by a specific gene. In this case, more than one gene specifies that enzyme and thus that step in the pathway. Therefore, the one gene–one enzyme hypothesis has been changed to the **one gene–one polypeptide hypothesis.** We should also be aware that some biochemical pathways are branched, making their analysis more complicated.

---

[1]We will learn later in the book that some enzymes are not proteins, but RNA.

Figure 10.4

Methionine biosynthetic pathway showing four genes in *Neurospora crassa* that code for the enzymes that catalyze each reaction. (The *met-5* and *met-2* genes are on the same chromosome; *met-3* and *met-8* are on two other chromosomes.)

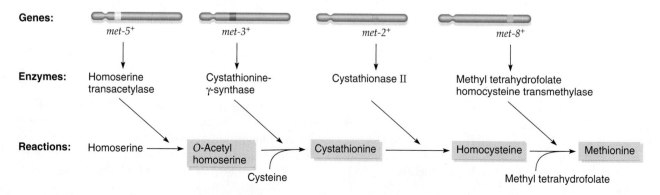

KEYNOTE

A specific relationship between genes and enzymes is embodied in Beadle and Tatum's one gene–one enzyme hypothesis, which states that each gene controls the synthesis or activity of a single enzyme. Because enzymes may consist of more than one polypeptide and genes code for individual polypeptide chains, a more modern version of this hypothesis is the one gene–one polypeptide hypothesis.

iActivity

Use the Beadle and Tatum experimental procedure to identify a nutritional mutant in the iActivity *Pathways to Inherited Enzyme Deficiencies* on the website.

Genetically Based Enzyme Deficiencies in Humans

Many human genetic diseases are caused by a single gene mutation that alters the function of an enzyme (Table 10.2). In general, an enzyme deficiency caused by a mutation may have either simple or **pleiotropic** (wide-reaching) consequences. Studies of these diseases have offered further evidence that many genes code for enzymes. Some genetic diseases are discussed in the following sections.

Phenylketonuria

Phenylketonuria (PKU; OMIM 261600) occurs in about 1 in 12,000 Caucasian births; it is most commonly caused by a recessive mutation of a gene on chromosome 12 (an autosome) at position 12q24.1, and people must therefore be homozygous for the mutation to exhibit the condition. The mutation is in the gene for phenylalanine hydroxylase. The absence of that enzyme activity prevents the conversion of the amino acid phenylalanine to the amino acid tyrosine (see Figure 10.1). Phenylalanine is one of the *essential amino acids*; that is, it is an amino acid that must be included in the diet because humans are unable to synthesize it. Phenylalanine is needed to make proteins, but excess amounts are harmful and are converted to tyrosine for further metabolism. Children born with PKU accumulate the phenylalanine they ingest, and that phenylalanine is converted to phenyl-pyruvic acid, which drastically affects the cells of the central nervous system and produces serious symptoms: severe mental retardation, a slow growth rate, and early death. (PKU children are unaffected before birth or at birth because excess phenylalanine that accumulates is metabolized by maternal enzymes.)

PKU has pleiotropic effects. Individuals with PKU cannot make tyrosine, an amino acid needed for protein synthesis, production of the hormones thyroxine and adrenaline, and production of the skin pigment melanin. This aspect of the phenotype is not very serious because tyrosine can be obtained from food. Yet food does not normally contain a lot of tyrosine. As a result, individuals with PKU make little melanin and therefore tend to have very fair skin and blue eyes (even if they have brown-eye genes). In addition, individuals with PKU have low adrenaline levels.

The symptoms of PKU depend on the amount of phenylalanine that accumulates, so the disease can be managed by controlling the dietary intake of phenylalanine. A mixture of individual amino acids with a very controlled amount of phenylalanine is used as a protein substitute in the PKU diet. The diet must maintain a level of phenylalanine in the blood that is high enough to facilitate normal development of the nervous system but low enough to prevent mental retardation. Treatment must begin in the first 1 to 2 months after birth, or the brain will be damaged and treatment will be ineffective. The diet is expensive, costing more than $5,000 per year. A difference of opinion exists as to whether the diet must be continued for life or whether it can be discontinued by about 10 years of age without subsequent defects developing in mental capacity or behavior. Additionally, women with PKU are advised either to maintain the restricted diet for life or to return to the PKU-restricted diet before becoming pregnant and maintain the diet through pregnancy. The reason is that children born to women with PKU living on normal diets are mentally retarded because high levels of phenylalanine in the maternal blood pass to the developing fetus across the placenta and adversely affect nervous system development independently of the fetus's genotype.

Given the very serious consequences of allowing PKU to go untreated, all U.S. states require that newborns be screened for PKU. The screen—the Guthrie test—is conducted by placing a drop of blood on a filter paper disk and placing the disk on solid culture medium containing the bacterium *Bacillus subtilis* and the chemical β-2-thienylalanine. The β-2-thienylalanine inhibits the growth of the bacterium. If phenylalanine is present, the inhibition is prevented; therefore, continued growth of the bacterium is evidence of the presence of high levels of phenylalanine in the blood and indicates the need for further tests to show that the infant has PKU.

Some foods and drinks containing the artificial sweetener aspartame (trade name NutraSweet) carry a warning that people with PKU should not use them. Aspartame is a dipeptide consisting of aspartic acid and phenylalanine. This combination signals to your taste receptors that the substance is sweet, yet it is not sugar and does not have the calories of sugar. Once ingested, aspartame is broken down to aspartic acid and phenylalanine, so it can have very serious effects on individuals with PKU.

Table 10.2	Selected Human Genetic Disorders with Demonstrated Enzyme Deficiencies		
Genetic Defect	**Locus**	**Enzyme Deficiency**	**OMIM Entry**
Acid phosphatase deficiency		Acid phosphatase	20090
Alkaptonuria	3q21–q23	Homogentisic acid oxidase	203500
Ataxia, intermittent[a]		Pyruvate decarboxylase	208800
Cystic fibrosis	7q31.2	Cystic fibrosis transmembrane conductance regulator (CFTR)	602421
Cataract	17q24	Galactokinase	230200
Citrullinemia[a]	9q34	Argininosuccinate synthetase	215700
Disaccharide intolerance I	3q25–q26	Invertase	222900
Fructose intolerance	9q22.3	Fructose-1-phosphate aldolase	229600
Galactosemia[a]	9p13	Galactose-1-phosphate uridyl transferase	230400
Gaucher disease[a]	1q21	Glucocerebrosidase	230800
G6PD deficiency (favism)[a]	Xq28	Glucose-6-phosphate dehydrogenase	305900
Glycogen storage disease I	17q21	Glucose-6-phosphatase	232200
Glycogen storage disease II[a]	17q25.2–q25.3	α-1,4-Glucosidase	232300
Glycogen storage disease III[a]	1p21	Amylo-1,β-glucosidase	232400
Glycogen storage disease IV[a]	3p12	Glycogen branching enzyme	232500
Hemolytic anemia[a]	3p21.1, 8p21.1, 20q11.2, 1q21	Glutathione peroxidase or glutathione reductase or glutathione synthetase or hexokinase or pyruvate kinase	138320, 138300, 231900, 266200
Hypoglycemia and acidosis	9q22.2–q22.3	Fructose-1,6-diphosphatase	229700
Immunodeficiency	1p32	Uridine monophosphate kinase	191710
Intestinal lactase deficiency (adult)		Lactase	223000
Ketoacidosis[a]	5p13	Succinyl CoA: 3-ketoacid CoA-transferase	245050
Kidney tubular acidosis with deafness	2cen–q13	Carbonic anhydrase B	267300
Leigh's necrotizing encephalopathy[a]	11q13.4–q13.5	Pyruvate carboxylase	266150
Lesch-Nyhan syndrome[a]	Xq26–q27.2	Hypoxanthine-guanine phosphoribosyltransferase	308000
Lysine intolerance		Lysine: NAD-oxidoreductase	247900
Male pseudohermaphroditism		Testicular 17,20-desmolase	309150
Maple sugar urine disease, type IA[a]	19q13.1–q13.2	Keto acid decarboxylase	248600
Muscular dystrophy, Duchenne and Becker types	Xp21.2	Dystrophin absent or defective; serum acetylcholinesterase or acetylcholine transferase or creatine phosphokinase elevated	310200
Niemann-Pick disease[a]	11p15.4–p15.1	Sphingomyelin hydrolase	257200
Orotic aciduria I[a]	3q13	Orotidylic decarboxylase and orotidylic pyrophosphorylase	258900
Phenylketonuria[a]	12q24.1	Phenylalanine hydroxylase	261600
Porphyria, acute intermittent[a]	11q23.3	Uroporphyrinogen III synthetase	176000
Porphyria, congenital erythropoietic[a]	10q25.2–q26.3	Uroporphyrinogen III synthase	263700
Pulmonary emphysema	14q32.1	α-l-Antitrypsin	107400
Pyridoxine dependency with seizures	2q31	Glutamic acid decarboxylase	266100
Ricketts, vitamin D–dependent		25-Hydroxycholciferol 1-hydroxylase	277420
Tay-Sachs disease[a]	15q23–q24	N-acetylhexosaminidase A	272800
Thyroid hormone synthesis, defect in	2p25	Iodide peroxidase or deiodinase	274500
Tyrosinemia, type III	12q24–qter	p-Hydroxyphenylpyruvate oxidase	276710

[a]Prenatal diagnosis possible.

Albinism

The classic form of *albinism* (see Figure 2.18; OMIM 203100) is caused by an autosomal recessive mutation. About 1 in 33,000 Caucasians and 1 in 28,000 African Americans in the United States have albinism. The mutation affects a gene for tyrosinase, an enzyme used in the conversion of tyrosine to DOPA, from which the brown pigment melanin derives (see Figure 10.1). Melanin absorbs light in the ultraviolet (UV) range and protects the skin against harmful UV radiation from the sun. People with albinism produce no melanin, so they have white skin, white hair, and red eyes (because of a lack of pigment in the iris) and visually are very light-sensitive. No apparent problems result from the accumulation of precursors in the pathway before the block.

There are at least two other kinds of albinism (see OMIM 203200 and OMIM 203290) because there are a number of biochemical steps in melanin biosynthesis from tyrosine. Thus, two parents with albinism that resulted from different enzyme deficiencies for two different steps of the pathway can produce normal children as a result of complementation of two mutations (see Chapter 6, pp. 140–141).

Lesch-Nyhan Syndrome

Lesch-Nyhan syndrome (OMIM 308000) is an ultimately fatal human trait caused by a recessive mutation in an X chromosome gene located at Xq26-q27.2. The gene spans 44 kb (44,000 bp) and encodes a polypeptide of 218 amino acids. An estimated 1 in 10,000 males exhibit Lesch-Nyhan syndrome. No Lesch-Nyhan females (who would be homozygous mutants) have been described. However, carrier females (heterozygotes who usually receive the mutant allele from their mothers) may have symptoms if the normal Lesch-Nyhan allele has undergone random X inactivation (lyonization; see Chapter 3, pp. 61–63).

Lesch-Nyhan syndrome is the result of a deficiency in the enzyme hypoxanthine-guanine phosphoribosyltransferase (HGPRT), an enzyme essential to purine use. When the biochemical pathway is highly impaired, as in this case, excess purines accumulate and are converted to uric acid. At birth, infants with Lesch-Nyhan syndrome are healthy and develop normally for several months. The uric acid excreted in the urine leads to the deposition of orange uric acid crystals in the diapers, an indication of disease. From 3 to 8 months, delays in motor development occur that lead to weak muscles. Later, the muscle tone changes radically, producing uncontrollable movements and involuntary spasms, seriously affecting feeding activities.

After 2 or 3 years, as a result of severe neurological complications, children with Lesch-Nyhan syndrome begin to show bizarre activity, such as compulsive biting of fingers, lips, and the inside of the mouth. This self-mutilation is painful and difficult to control. Typically, behavior toward others becomes aggressive. In intelligence tests, patients with Lesch-Nyhan syndrome score in the severely retarded region, although the difficulties they show in communicating with others may be a major contributing factor to the low scores. Most die before they reach their 20s, usually from infection, kidney failure, or uremia (uric acid in the blood).

Understandably, the elevated uric acid levels that result from an HGPRT deficiency might give rise to uremia, kidney failure, and mental deficiency. However, there is no clear explanation for how an HGPRT deficiency gives rise to self-mutilating activity, and there are no available therapies for these neurological complications.

Tay-Sachs Disease

Lysosomes are membrane-bounded organelles in the cell containing 40 or more different digestive enzymes that catalyze the breakdown of nucleic acids, proteins, polysaccharides, and lipids. A number of human diseases are caused by mutations in genes that code for lysosomal enzymes. Such diseases, collectively called *lysosomal storage diseases,* generally are caused by recessive mutations.

The best-known genetic disease of this type is *Tay-Sachs disease* (OMIM 272800), also called infantile amaurotic idiocy, which is caused by homozygosity for a rare recessive mutation of a gene on chromosome 15 at 15q23–q24. Although Tay-Sachs disease is rare in the population as a whole, it has a higher incidence in Ashkenazi Jews of central European origin, with about 1 in 3,600 children having the disease.

The gene in question (*HEXA*) codes for the enzyme N-acetylhexosaminidase A, which cleaves a particular chemical group from a brain ganglioside. (A ganglioside is one of a group of complex glycolipids found mainly in nerve membranes.) In infants with Tay-Sachs disease, the enzyme is nonfunctional, so the unprocessed ganglioside accumulates in the brain cells and causes a number of different clinical symptoms. Typically, the symptom first recognized is an unusually enhanced reaction to sharp sounds. A cherry-colored spot on the retina surrounded by a white halo also facilitates early diagnosis of the disease. About a year after birth, there is rapid neurological degeneration as the unprocessed ganglioside accumulates and the brain begins to lose control over normal function and activities. This degeneration involves generalized paralysis, blindness, a progressive loss of hearing, and serious feeding problems. By 2 years of age, the infant is essentially immobile, and death occurs at about 3 to 4 years of age, often from respiratory infection. There is no known cure for Tay-Sachs disease, but because carriers can be detected, the incidence can be controlled.

K E Y N O T E

> Many human genetic diseases are caused by deficiencies in enzyme activities. Most of these diseases are inherited as recessive traits.

Gene Control of Protein Structure

While most enzymes are proteins, not all proteins are enzymes. To understand completely how genes function, we next look at the experimental evidence that genes are responsible for the structure of nonenzymatic proteins, such as hemoglobin. Nonenzymatic proteins often are easier to study than enzymes. Enzymes usually are present in small amounts, whereas nonenzymatic proteins occur in large quantities in the cell, making them easier to isolate and purify.

Sickle-Cell Anemia

Sickle-cell anemia (SCA; OMIM 141900 and 603903) is a genetic disease affecting hemoglobin, the oxygen-transporting protein in red blood cells. Sickle-cell anemia was first described in 1910 by J. Herrick. He found that red blood cells from individuals with the disease lose their characteristic disk shape in conditions of low oxygen tension and assume the shape of a sickle (Figure 10.5). The

sickled red blood cells are more fragile than normal red blood cells and break easily; hence the anemia. Sickled cells also are not as flexible as normal cells and therefore tend to clog the capillaries rather than squeeze through them. As a result, blood circulation is impaired and tissues become deprived of oxygen. Although oxygen deprivation occurs particularly at the extremities, the heart, lungs, brain, kidneys, gastrointestinal tract, muscles, and joints can also become oxygen deprived and be damaged. An individual with SCA therefore may suffer from a pleiotropy of health problems, including heart failure, pneumonia, paralysis, kidney failure, abdominal pain, and rheumatism. Some individuals have a milder form of the disease called *sickle-cell trait*.

In 1949, E. A. Beet and J. V. Neel independently hypothesized that sickling was caused by a single mutant allele that was homozygous in sickle-cell anemia and heterozygous in sickle-cell trait. In the same year, Linus Pauling and coworkers showed that the hemoglobins of normal blood, sickle-cell anemia blood, and sickle-cell trait blood differ when they are subjected to electrophoresis, a technique used to separate molecules based on their electrical charges. They found that hemoglobin from normal individuals (called Hb-A) migrated in the opposite direction from the hemoglobin from individuals with sickle-cell anemia (called Hb-S; Figure 10.6). Hemoglobin from individuals with sickle-cell trait had a 1:1 mixture of Hb-A and Hb-S, indicating that heterozygous individuals make both types of hemoglobin. Pauling concluded that sickle-cell anemia results from a mutation that alters the chemical structure of the hemoglobin molecule. This experiment was one of the first rigorous proofs that protein structure is controlled by genes.

Hemoglobin, the molecule affected in sickle-cell anemia, consists of four polypeptide chains—two α polypep-

Figure 10.6

Electrophoresis of hemoglobin mutants. Hemoglobin found (*left*) in normal $\beta^A\beta^A$ individuals; (*center*) in $\beta^A\beta^S$ individuals, with sickle-cell trait; and (*right*) in $\beta^S\beta^S$ individuals, who exhibit sickle-cell anemia. Hemoglobins A and S migrate to different positions in an electric field and hence must differ in electrical charge.

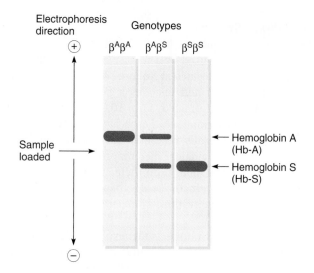

tides and two β polypeptides—each of which is associated with a heme group (involved in oxygen binding; Figure 10.7). In 1956, V. M. Ingram analyzed some of the amino acid sequences of Hb-A and Hb-S and found that the molecular defect in the Hb-S hemoglobin is a change from the acidic amino acid glutamic acid (negative electrical charge) at the sixth position from the N-terminal end of the β polypeptide to the neutral amino acid valine (no electrical charge) (Figure 10.8). We have learned that this particular substitution of amino acids causes the β polypeptide to fold up in a different way. In turn, this leads to sickling of the red blood cells in individuals with sickle-cell anemia and mild sickling of the red blood cells in individuals with sickle-cell trait.

Let us outline the genetics and the products of the genes involved. The β polypeptide sickle-cell mutant allele is β^S, and it is codominant with the wild-type allele β^A. Homozygous $\beta^A\beta^A$ individuals make normal Hb-A with two normal α chains and two normal β chains encoded by the wild-type α-globin gene and the wild-type β-globin gene (β^A). Homozygous $\beta^S\beta^S$ individuals make Hb-S, the defective hemoglobin, with two normal α chains specified by wild-type α-globin gene and two abnormal β chains specified by the mutant β-globin gene (β^S). These individuals have SCA. Heterozygous $\beta^A\beta^S$ individuals make both Hb-A and Hb-S and have sickle-cell trait. (Note: Only one type of β chain is found in any one hemoglobin molecule, so only two types of hemoglobin molecules are possible here; one has two normal β chains, and the other has two mutant β chains.) Under normal conditions, heterozygous individuals usually show few symptoms of the disease. However, after a sharp

Figure 10.5

Colorized scanning electron micrographs of (left) normal and (right) sickled red blood cells.

Figure 10.7

The hemoglobin molecule. The diagram shows the two α polypeptides and two β polypeptides, each of which is associated with a heme group.

Heme groups

β polypeptide   β polypeptide

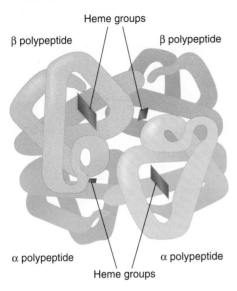

α polypeptide   α polypeptide

Heme groups

drop in oxygen tension (as in an unpressurized aircraft climbing into the atmosphere), sickling of red blood cells may occur, giving rise to symptoms similar to those found in people with severe anemia.

Other Hemoglobin Mutants

Over 200 hemoglobin mutants have been detected in general screening programs in which hemoglobin is isolated from red blood cells and analyzed by electrophoresis. Some mutations affect the α chain and others the β chain, and there is a wide variety in the types of amino acid substitutions that occur. From the changes in DNA that are assumed to be responsible for the substitutions, it is apparent that a single base change is involved in each case.

The identified hemoglobin mutants have various effects, depending on the amino acid substitution involved and its position in the polypeptide chains. Most have effects that are not as drastic as the SCA mutant. For example, in the Hb-C hemoglobin molecule, the same β polypeptide glutamic acid that is altered in SCA is changed to a lysine. Compared with the Hb-S change, this change is not as serious a defect because both amino acids are hydrophilic ("water loving"), so the conformation of the hemoglobin molecule is not as drastically altered. Individuals homozygous for the $β^C$ mutation experience only a mild form of anemia.

Biochemical Genetics of the Human ABO Blood Groups

The mechanism of inheritance of the human ABO blood groups was mentioned in Chapter 4. To summarize, the blood group depends on the presence of chemical substances called antigens on the red blood cell surface. When antigens are injected into a host organism, they may be recognized as foreign and removed from the circulation by the host's antibodies.

The I^A, I^B, and i alleles of the ABO locus specify the ABO blood groups. People of blood group A (genotypes I^A/I^A or I^A/i) have the A antigen on their red blood cells; people of blood group B (genotypes I^B/I^B or I^B/i) have the B antigen on their red blood cells; people of blood group AB (genotype I^A/I^B) have both the A and B antigens on their red blood cells; and people of blood group O (genotype i/i) have neither the A nor the B antigen on their red blood cells.

The relationship between the ABO alleles and the antigens on the red blood cells is as follows. The ABO locus encodes *glycosyltransferases,* which are enzymes that add sugar groups to a preexisting polysaccharide. The polysaccharides here are those that have combined with lipids to form glycolipids. These glycolipids then associate with red blood cell membranes to form the blood group antigens. Figure 10.9 shows the terminal sugar structures of the polysaccharide components of the glycolipids involved in the ABO blood type system. Most people produce a glycolipid called the H antigen, which has the terminal sugar sequence shown in the figure. The I^A allele produces a glycosyltransferase enzyme called α-N-acetylgalactosamyl transferase, which recognizes the H antigen and adds the sugar α-N-acetylgalactosamine to the end of the polysaccharide to produce the A antigen. The I^B allele, on the other hand, produces an α-D-galactosyltransferase, which also recognizes the H antigen but adds galactose to its polysaccharide to produce the B antigen. (Note that this small difference in the structure of the A and B antigens is sufficient to induce an antibody response.) In both cases, some H antigen remains unconverted.

For the I^A/I^B heterozygote, both enzymes are produced, so some H antigen is converted to the A antigen and some to the B antigen. The red blood cell has both antigens on the surface, so the individual is of blood group AB. It is the presence of both surface antigens that provides the molecular basis of the codominance of the I^A and I^B alleles (see Chapter 4, pp. 80–81).

People who are homozygous for the i allele produce no enzymes to convert the H antigen glycolipid. Therefore, their red blood cells carry only the H antigen. The H

Figure 10.8

The first seven N-terminal amino acids in normal and sickled hemoglobin β polypeptides. There is a single amino acid change from glutamic acid to valine at the sixth position in the sickled hemoglobin polypeptide.

Normal β polypeptide, Hb-A

H_3N^+

	1	2	3	4	5	6	7
	Val	His	Leu	Thr	Pro	Glu	Glu ...

Changes to

Sickle-cell β polypeptide, Hb-S

H_3N^+

	Val	His	Leu	Thr	Pro	Val	Glu ...

Figure 10.9

Human ABO blood type antigens. Conversion of (a) the H antigen to (b) the A antigen by the I^A allele-encoded glycosyltransferase and to (c) the B antigen by the I^B allele-encoded glycosyltransferase.

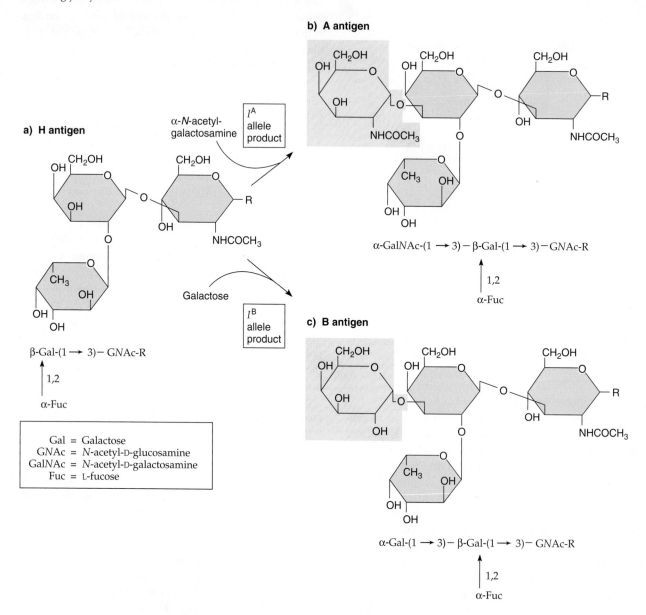

antigen does not elicit an antibody response in people of other blood groups because its polysaccharide component is the basic component of the A and the B antigens as well, so it is not detected as a foreign substance. People who are heterozygous for the *i* allele have the blood type of the other allele. For example, in I^B/i individuals, the I^B allele results in the conversion of some of the H antigen to the B antigen, determining the individual's blood group.

The H antigen is produced by the action of the dominant *H* allele at a locus distinct from the ABO locus. Individuals homozygous for the recessive mutant allele *h* do not make the H antigen, and therefore, regardless of the presence of I^A or I^B alleles at the ABO locus, no A or B

antigens can be produced. These very rare *h/h* individuals are like blood group O individuals in the sense that they lack A and B antigens; they are said to have the Bombay blood type. However, individuals with the Bombay blood groups produce anti-O antibodies (antibodies against the H antigen), while individuals with blood group O do not.

Cystic Fibrosis

Cystic fibrosis (CF; OMIM 219700) is a human disease that causes pancreatic, pulmonary, and digestive dysfunction in children and young adults. Typical of the disease is an abnormally high viscosity of secreted mucus. In some male patients, the vas deferens does not form properly, resulting in sterility. CF is managed by pounding the chest and back of a patient to help shake mucus free in different parts of the lungs and by giving antibiotics to treat any infections that develop. CF is a lethal disease; with present management procedures, life expectancy is about 40 years.

CF is caused by homozygosity for an autosomal recessive mutation located on the long arm of chromosome 7 at position 7q31.2-q31.3. CF is the most common lethal autosomal recessive disease among Caucasians, with about 1 in 2,000 newborns having the disease. About 1 in 23 Caucasians is estimated to be a heterozygous carrier. In the African American population, about 1 in 17,000 newborns have CF, and in Asians, the CF frequency is 1 in 90,000 newborns.

The defective gene product in patients with CF was not identified by biochemical analysis, as was the case for PKU and many other diseases, but by a combination of genetic and modern molecular biology techniques. The gene was localized to chromosome 7 and then was molecularly cloned from both a normal subject and from a patient with CF (described in detail in Chapter 14). Analysis of the DNA sequences of the cloned genes showed that in patients with a serious form of CF, the most common mutation—ΔF508—is the deletion of three consecutive base pairs in the ATP-binding nucleotide-binding fold (NBF) region toward the middle of the gene. Since each amino acid in a protein is specified by three base pairs in the DNA, this finding means that one amino acid is missing. But what does the CF protein do? From the DNA sequence of the gene, researchers deduced the amino acid sequence of the protein and then made some predictions about the type and three-dimensional structure of that pro-

tein. Their analysis indicated that the 1,480-amino acid CF protein is associated with membranes. The proposed structure for the CF protein—called cystic fibrosis transmembrane conductance regulator (CFTR)—is shown in Figure 10.10. By comparing the amino acid sequence of the CF protein with the amino acid sequences of other proteins in a computer database, researchers found CFTR protein to be homologous to a large family of proteins involved in active transport of materials across cell membranes. Functional analysis of this protein has shown that it is a chloride channel in certain cell membranes. In people with CF, the mutated CF gene results in an abnormal CFTR protein; as a result, ion transport across membranes is impaired. The symptoms of CF result, starting with abnormal mucus secretion and accumulation.

K E Y N O T E

From the study of alterations in proteins other than enzymes, such as those in hemoglobin responsible for sickle-cell anemia, convincing evidence was obtained that genes control the structures of all proteins.

Genetic Counseling

We have learned that many human genetic diseases are caused by enzyme or protein defects; those defects ultimately result from mutations at the DNA level. Many other genetic diseases arise from chromosome defects. We can now test for many enzyme or protein deficiencies and for many of the DNA changes associated with genetic diseases and thereby determine whether or not

Figure 10.10

Proposed structure for cystic fibrosis transmembrane conductance regulator (CFTR). The protein has two hydrophobic segments that span the plasma membrane, and after each segment is a nucleotide-binding fold (NBF) region that binds ATP. The site of the amino acid deletion resulting from the three-nucleotide-pair deletion in the CF gene most commonly seen in patients with severe cystic fibrosis is in the first (toward the amino end) NBF; this is the ΔF508 mutation. The central portion of the molecule contains sites that can be phosphorylated by the enzymes protein kinase A and protein kinase C.

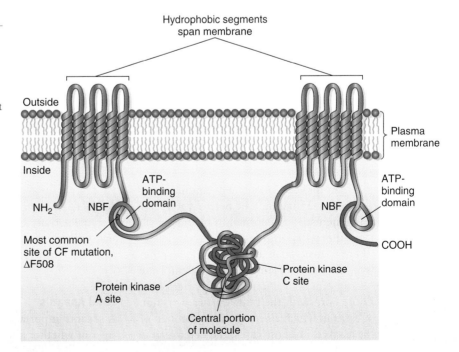

an individual has a genetic disease or is a carrier for that disease. It is also possible to determine whether individuals have any chromosomal abnormalities.

Genetic counseling is advice based on the analysis of the probability that patients have a genetic defect or of the risk that prospective parents may produce a child with a genetic defect. In the latter case, genetic counseling involves presenting the available options for avoiding or minimizing those possible risks. If a serious genetic defect is identified in a fetus, one option is abortion. Genetic counseling gives people an understanding of the genetic problems that are (or may be) in their family or prospective family.

Genetic counseling includes a wide range of information on human heredity. In many instances, the risk of having a child with a genetic condition may be stated in terms of precise probabilities; in other cases, where the role of heredity is not completely clear, the risk is estimated only generally. It is the responsibility of genetic counselors to give their clients clear, unemotional, and nonprescriptive statements based on the family history and on their knowledge of all relevant scientific information and the probable risks of giving birth to a child with a genetic defect.

Genetic counseling generally starts with pedigree analysis, a family tree study involving the careful compilation of phenotypic records of both families over a number of generations. (Pedigree analysis is described in Chapters 2 and 3.) Pedigree analysis is used to determine the likelihood that a particular allele is present in either family. Detection of a genetic condition occurs in one or both of two ways: carrier detection (identifying heterozygotes, or carriers, of recessive mutations) or fetal analysis. Assays for enzyme activities or protein amounts are limited to genetic diseases in which the biochemical condition is expressed in the parents or the developing fetus. Tests that measure actual changes in the DNA do not depend on expression of the gene in the parents or the fetus.

Although we can identify carriers of many mutant alleles and can determine whether fetuses have a genetic condition, in most cases there is no way to change the phenotype. Carrier detection and fetal analysis serve mainly to inform parents of the risks and probabilities of having a child with the mutation.

Carrier Detection

Carrier detection identifies individuals who are heterozygous for a recessive gene mutation. The heterozygous carrier of a mutant gene usually is normal in phenotype. If homozygosity for the mutation results in serious deleterious effects, there is great value in determining whether two people who are contemplating having a child are both carriers, because in that situation one-quarter of the children would be born with the trait. Carrier detection can be used in cases in which a gene product (protein or enzyme) can be assayed. In those cases, the heterozygote

Figure 10.11

Amniocentesis, a procedure used for prenatal diagnosis of genetic defects.

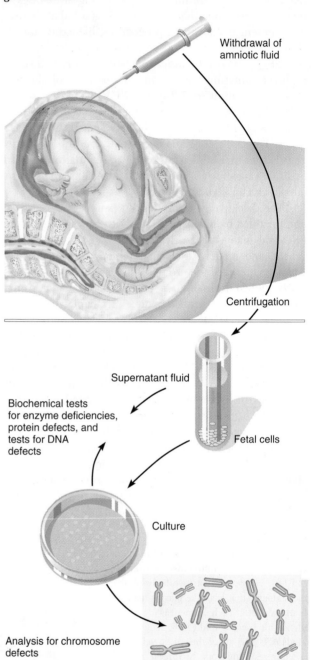

Withdrawal of amniotic fluid

Centrifugation

Supernatant fluid

Biochemical tests for enzyme deficiencies, protein defects, and tests for DNA defects

Fetal cells

Culture

Analysis for chromosome defects

(carrier) is expected to have approximately half the enzyme activity or protein amount as homozygous normal individuals. In Chapter 14, we will see how carriers can be identified by DNA tests.

Fetal Analysis

A second important aspect of genetic counseling is finding out whether a fetus is normal. **Amniocentesis** is one way in which this can be done (Figure 10.11). As a fetus develops in the amniotic sac, amniotic fluid surrounds it, serving as a cushion against shock. In amniocentesis, a syringe needle is carefully inserted through the mother's uterine wall and into the amniotic sac, and a sample of amniotic fluid is taken. The fluid contains cells that have sloughed off the fetus's skin; these cells can be cultured in the laboratory and then examined for protein or enzyme alterations or deficiencies, DNA changes, and chromosomal abnormalities. Amniocentesis is possible at any stage of pregnancy, but the small quantity of amniotic fluid available and the risk to the fetus makes it impractical to perform before the twelfth week of pregnancy. Because amniocentesis is complicated and costly, it is used primarily in high-risk cases, such as when a mother is over 35.

Another method for fetal analysis is **chorionic villus sampling** (Figure 10.12). The procedure can be done between the eighth and twelfth weeks of pregnancy, earlier than for amniocentesis. The chorion is a membrane layer surrounding the fetus consisting entirely of embryonic tissue. A chorionic villus tissue sample may be taken from the developing placenta through the abdomen (as in amniocentesis) or, preferably, via the vagina using biopsy forceps or a flexible catheter, aided by ultrasound. Once the tissue sample is obtained, the analysis is similar to that used in amniocentesis. Advantages of the technique are that the time of testing permits the parents to learn whether the fetus has a genetic defect earlier in the pregnancy than with amniocentesis and that it is not necessary to culture cells to obtain enough to do the biochemical assays. Fetal death and inaccurate diagnoses caused by the presence of maternal cells are more common in chorionic villus sampling than in amniocentesis, however.

KEYNOTE

Genetic counseling is advice based on analysis of the probability that patients have a genetic defect or of the risk that prospective parents may produce a child with a genetic defect. Early detection of a genetic disease is done by carrier detection and fetal analysis.

Summary

In this chapter, we discussed gene function, the evidence that genes code for enzymes and for nonenzymatic proteins. A number of the experiments described in this context were done before the unequivocal proof that DNA is the genetic material and before the elucidation of DNA structure.

As early as 1902, there was evidence of a specific relationship between genes and enzymes. This evidence was obtained by Archibald Garrod in his investigations of inborn errors of metabolism, human genetic diseases that result in enzyme deficiencies. The specific relationship between genes and enzymes is historically embodied in the one gene–one enzyme hypothesis, which states that each gene controls the synthesis or activity of a single enzyme. Since some protein enzymes and nonenzymatic proteins (such as hemoglobin) consist of more than one polypeptide, a more modern maxim is one gene–one polypeptide. Beadle and Tatum's experiments that led to the one gene–one enzyme hypothesis are considered landmarks in the development of molecular genetics.

In this chapter, we also discussed a number of examples of genetically based enzyme and protein deficiencies that give rise to genetic diseases in humans. Many are the result of homozygosity for recessive mutant genes. The severity of the disease depends on the effects of the loss of function of the particular enzyme involved. Thus, albinism is a mild genetic disease, whereas Tay-Sachs disease and cystic fibrosis ultimately are lethal. To make our understanding of gene function more complete, we examined experimental evidence that genes control the structure of nonenzymatic proteins.

Given the knowledge we currently have about a large number of genetically based enzyme and protein deficiencies in humans, it is possible to make some predictions about the existence of certain genetic diseases in individuals or families. Genetic counseling is advice based on analysis of the probability that patients have a genetic defect or of the risk that prospective parents may produce a child with a genetic defect. Techniques available to genetic counselors include amniocentesis and chorionic villus sampling.

Figure 10.12

Chorionic villus sampling, a procedure used for early prenatal diagnosis of genetic defects.

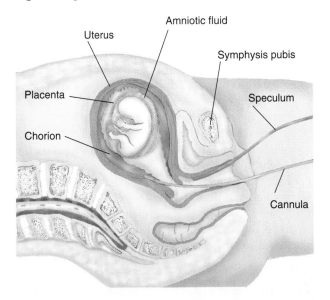

Analytical Approaches for Solving Genetics Problems

Q10.1 k^+, l^+, and m^+ are independently assorting genes that control the production of a red pigment. These three genes act in a biochemical pathway as follows:

$$\text{colorless 1} \xrightarrow{k^+} \text{colorless 2} \xrightarrow{l^+} \text{orange} \xrightarrow{m^+} \text{red}$$

The mutant alleles that produce abnormal functioning of these genes are k, l, and m; each is recessive to its wild-type counterpart. A red individual homozygous for all three wild-type alleles is crossed with a colorless individual that is homozygous for all three recessive mutant alleles. The F_1 is red. The F_1 is then selfed to produce the F_2 generation.

a. What proportion of the F_2 are colorless?
b. What proportion of the F_2 are orange?
c. What proportion of the F_2 are red?

A10.1

a. There are two ways to answer this question. One is to determine all of the genotypes that can specify the colorless phenotypes, and the other is to use subtractive logic, in which the proportions of orange and red are first calculated and these proportions are subtracted from 1 to give the proportion of colorless progeny in the F_2. We will first consider the second method.

To produce an orange phenotype, three things are needed: (1) At least one wild-type k^+ allele must be present so that the colorless 1 to colorless 2 step can occur; (2) at least one wild-type l^+ allele must be present so that the colorless 2 to orange step can occur; and (3) the individual must be m/m so that the orange to red step cannot proceed. Thus, an orange phenotype results from the genotype $k^+/-\ l^+/-\ m/m$. From an $F_1 \times F_1$ cross, the probability of getting an individual with that genotype is $\frac{3}{4} \times \frac{3}{4} \times \frac{1}{4} = \frac{9}{64}$.

To produce a red phenotype, three things are needed: (1) At least one wild-type k^+ allele must be present so that the colorless 1 to colorless 2 step can occur; (2) at least one wild-type l^+ allele must be present so that the colorless 2 to orange step can occur; and (3) at least one m^+ allele must be present so that the orange to red step can proceed. Thus, a red phenotype results from the genotype $k^+/-\ l^+/-\ m^+/-$. From an $F_1 \times F_1$ self, the probability of getting an individual with that genotype is $\frac{3}{4} \times \frac{3}{4} \times \frac{3}{4} = \frac{27}{64}$.

By default, all other F_2 individuals are colorless. The proportion of F_2 individuals that are colorless is therefore $1 - \frac{9}{64} - \frac{27}{64} = 1 - \frac{36}{64} = \frac{28}{64}$.

Using the first approach to calculate the proportion of F_2 colorless, we must determine all genotypes that will give a colorless phenotype and then add up the probabilities of each occurring. The genotypes and their probabilities are as follows:

Genotypes	Probabilities	
$k/k\ l/l\ m/m$	1/64	(cannot convert colorless 1)
$k/k\ l/l\ m^+/-$	3/64	(cannot convert colorless 1)
$k/k\ l^+/-\ m/m$	3/64	(cannot convert colorless 1)
$k/k\ l^+/-\ m^+/-$	9/64	(cannot convert colorless 1)
$k^+/-\ l/l\ m/m$	3/64	(cannot convert colorless 2)
$k^+/-\ l/l\ m^+/-$	9/64	(cannot convert colorless 2)
Total	28/64	

b. The proportion of the F_2 was calculated in part (a): $\frac{9}{64}$.

c. The proportion of the F_2 was calculated in part (a): $\frac{27}{64}$.

Q10.2 A number of auxotrophic mutant strains were isolated from wild-type, haploid yeast. These strains responded to the addition of certain nutritional supplements to minimal culture medium either by growth (+) or no growth (0). The following table gives the growth patterns for single gene mutant strains:

Mutant Strains	Supplements Added to Minimal Culture Medium				
	B	**A**	**R**	**T**	**S**
1	+	0	+	0	0
2	+	+	+	+	0
3	+	0	+	+	0
4	0	0	+	0	0

Diagram a biochemical pathway that is consistent with the data, indicating where in the pathway each mutant strain is blocked.

A10.2 The data to be analyzed are very similar to those discussed in the text for Beadle and Tatum's analysis of *Neurospora* auxotrophic mutants, from which they proposed the one gene–one enzyme hypothesis. Recall that the later in the pathway a mutant is blocked, the fewer nutritional supplements must be added to allow growth. In the data given, we must assume that the nutritional supplements are not necessarily listed in the order in which they appear in the pathway.

Analysis of the data indicates that all four strains will grow if given R, and that none will grow if given S. From this we can conclude that R is likely to be the end product of the pathway (all mutants should grow if given the end product) and that S is likely to be the first compound in the pathway (none of the mutants should grow given the first compound in the pathway). Thus, the pathway as deduced so far is

$$S \longrightarrow [B, A, T] \longrightarrow R$$

where the order of B, A, and T is as yet undetermined.

Now let us consider each of the mutant strains and see how their growth phenotypes can help define the biochemical pathway. Strain 1 will grow only if given B or R. Therefore, the defective enzyme in strain 1 must act somewhere before the formation of B and R and after the substances A, T, and S. Since we have deduced that R is the end product of the pathway, we can propose that B is the immediate precursor to R and that strain 1 cannot make B. The pathway so far is

$$S \longrightarrow [A,T] \xrightarrow{\ 1\ } B \longrightarrow R$$

Strain 2 will grow on all compounds except S, the first compound in the pathway. Thus, the defective enzyme in strain 2 must act to convert S to the next compound in the pathway, which is either A or T. We do not know yet whether A or T follows S in the pathway, but the growth data at least allow us to conclude where strain 2 is blocked in the pathway; that is,

$$S \xrightarrow{\ 2\ } [A,T] \xrightarrow{\ 1\ } B \longrightarrow R$$

Strain 3 will grow on B, R, and T, but not on A or S. We know that R is the end product and S is the first compound in the pathway. This mutant strain allows us to determine the order of A and T in the pathway. That is, since strain 3 grows on T but not on A, T must be later in the pathway than A, and the defective enzyme in 3 must be blocked in the yeast's ability to convert A to T. The pathway now is

$$S \xrightarrow{\ 2\ } A \xrightarrow{\ 3\ } T \xrightarrow{\ 1\ } B \longrightarrow R$$

Strain 4 will grow only if given the deduced end product R. Therefore, the defective enzyme produced by the mutant gene in strain 4 must act before the formation of R and after the formation of A, T, and B from the first compound S. The mutation in 4 must be blocked in the last step of the biochemical pathway in the conversion of B to R. The final deduced pathway and the positions of the mutant blocks are

$$S \xrightarrow{\ 2\ } A \xrightarrow{\ 3\ } T \xrightarrow{\ 1\ } B \xrightarrow{\ 4\ } R$$

Questions and Problems

10.1 Phenylketonuria (PKU) is an inheritable metabolic disease of humans; its symptoms include mental deficiency. This phenotypic effect results from
a. accumulation of phenylketones in the blood
b. absence of phenylalanine hydroxylase
c. deficiency of phenylketones in the blood
d. deficiency of phenylketones in the diet

*10.2 If a person were homozygous for both PKU and alkaptonuria (AKU), would you expect him or her to exhibit the symptoms of PKU, AKU, or both? Refer to the following pathway.

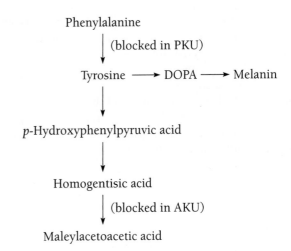

10.3 Refer to the pathway shown in problem 10.2. What effect, if any, would you expect PKU and AKU to have on pigment formation? Explain your answer.

*10.4 a^+, b^+, c^+, and d^+ are independently assorting Mendelian genes controlling the production of a black pigment. The alternate alleles that give abnormal functioning of these genes are a, b, c, and d. A black individual of genotype $a^+/a^+ \; b^+/b^+ \; c^+/c^+ \; d^+/d^+$ is crossed with a colorless individual of genotype $a/a \; b/b \; c/c \; d/d$ to produce a black F_1. $F_1 \times F_1$ crosses are then done. Assume that a^+, b^+, c^+, and d^+ act in a pathway as follows:

$$\begin{array}{ccccccc} & a^+ & & b^+ & & c^+ & & d^+ \\ \text{colorless} & \longrightarrow & \text{colorless} & \longrightarrow & \text{colorless} & \longrightarrow & \text{brown} & \longrightarrow & \text{black} \end{array}$$

a. What proportion of the F_2 are colorless?
b. What proportion of the F_2 are brown?

10.5 Using the genetic information given in problem 10.4, now assume that a^+, b^+, and c^+ act in a pathway as follows:

$$\begin{array}{c} \text{Colorless} \xrightarrow{\ a^+\ } \text{red} \searrow \\ \qquad\qquad\qquad\qquad \xrightarrow{\ c^+\ } \text{black} \\ \text{Colorless} \xrightarrow{\ b^+\ } \text{red} \nearrow \end{array}$$

Black can be produced only if both red pigments are present; that is, c^+ converts the two red pigments together into a black pigment.
a. What proportion of the F_2 are colorless?
b. What proportion of the F_2 are red?
c. What proportion of the F_2 are black?

*10.6
a. Three genes on different chromosomes are responsible for three enzymes that catalyze the same reaction in corn:

$$\text{colorless compound} \xrightarrow{\ a^+, b^+, c^+\ } \text{red compound}$$

The normal functioning of any one of these genes is sufficient to convert the colorless compound to the red compound. The abnormal functioning of these genes is designated by *a*, *b*, and *c*. A red a^+/a^+ b^+/b^+ c^+/c^+ is crossed with a colorless *a/a b/b c/c* to give a red F_1, a^+/a b^+/b c^+/c. The F_1 is selfed. What proportion of the F_2 are colorless?

b. It turns out that another step is involved in the pathway. It is controlled by gene d^+, which assorts independently of a^+, b^+, and c^+:

$$\text{colorless compound 1} \xrightarrow{d^+} \text{colorless compound 2} \xrightarrow{a^+, b^+, c^+} \text{red compound}$$

The inability to convert colorless 1 to colorless 2 is designated *d*. A red a^+/a^+ b^+/b^+ c^+/c^+ d^+/d^+ is crossed with a colorless *a/a b/b c/c d/d*. The F_1 are all red. The red F_1s are now selfed. What proportion of the F_2 are colorless?

10.7 In *Drosophila*, the recessive allele *bw* causes a brown eye, and the (unlinked) recessive allele *st* causes a scarlet eye. Flies homozygous for both recessives have white eyes. The genotypes and corresponding phenotypes, then, are as follows:

$bw^+/-$	$st^+/-$	red eye
bw/bw	$st^+/-$	brown
$bw^+/-$	*st/st*	scarlet
bw/bw	*st/st*	white

Outline a hypothetical biochemical pathway that would produce this type of gene interaction. Demonstrate why each genotype shows its specific phenotype.

*10.8 In J. R. R. Tolkien's *The Lord of the Rings*, the Black Riders of Mordor ride steeds with eyes of fire. As a geneticist, you are very interested in the inheritance of the fire-red eye color. You discover that the eyes contain two types of pigments, brown and red, that are usually bound to core granules in the eye. In wild-type steeds, precursors are converted by these granules to the above pigments, but in steeds homozygous for the recessive X-linked gene *w* (white eye), the granules remain unconverted and a white eye results. The metabolic pathways for the synthesis of the two pigments are shown in Figure 10.A. Each step of the pathway is controlled by a gene: A mutation *v* results in vermilion eyes; *cn* results in cinnabar eyes; *st* results in scarlet eyes; *bw* results in brown eyes; and *se* results in black eyes. All these mutations are recessive to their wild-type alleles and all are unlinked. For the following genotypes, show the proportions of steed phenotypes that would be obtained in the F_1 of the given matings.

a. *w/w* bw^+/bw^+ *st/st* × w^+/Y *bw/bw* st^+/st^+
b. w^+/w^+ *se/se* *bw/bw* × *w/Y* se^+/se^+ bw^+/bw^+
c. w^+/w^+ v^+/v^+ *bw/bw* × *w/Y* *v/v* *bw/bw*
d. w^+/w^+ bw^+/bw st^+/st × *w/Y* *bw/bw* *st/st*

*10.9 Upon infection of *E. coli* with bacteriophage T4, a series of biochemical pathways results in the formation of mature progeny phages. The phages are released after lysis of the bacterial host cells. Suppose that the following pathway exists:

$$A \xrightarrow{\text{enzyme}} B \xrightarrow{\text{enzyme}} \text{mature phage}$$

Also suppose that we have two temperature-sensitive mutants that involve the two enzymes catalyzing these sequential steps. One of the mutations is cold-sensitive (*cs*) in that no mature phages are produced at 17°C. The other is heat-sensitive (*hs*) in that no mature phages are produced at 42°C. Normal progeny phages are produced when phages carrying either of the mutations infect bacteria at 30°C. However, let us assume that we do not know the sequence of the two mutations. Two models are therefore possible:

(1) $A \xrightarrow{hs} B \xrightarrow{cs} \text{phage}$
(2) $A \xrightarrow{cs} B \xrightarrow{hs} \text{phage}$

Figure 10.A

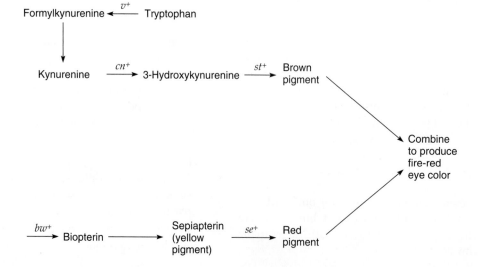

Outline how you would experimentally determine which model is the correct model without artificially lysing phage-infected bacteria.

10.10 Four mutant strains of *E. coli* (*a*, *b*, *c*, and *d*) all require substance X to grow. Four plates were prepared, as shown in the following figure. In each case, the medium was minimal, with just a trace amount of substance X to allow a small amount of growth of the mutant cells. On plate (a), cells of mutant strain *a* were spread over the entire surface of the agar and grew to form a thin lawn (continuous bacterial growth over the plate). On plate (b), the lawn is composed of mutant *b* cells. And so on. On each plate, cells of the four mutant types were inoculated over the lawn, as indicated by the circles. Dark circles indicate luxuriant growth. This experiment tests whether the bacterial strain spread on the plate can feed the four strains inoculated on the plate, allowing them to grow. What do these results show about the relationship of the four mutants to the metabolic pathway leading to substance X?

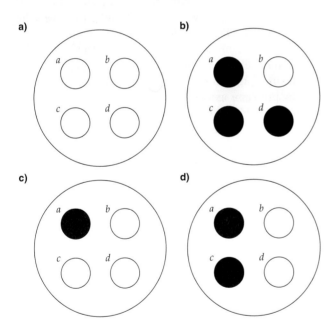

*10.11 The following growth responses (where + = growth and 0 = no growth) of mutants 1–4 were seen on the related biosynthetic intermediates A, B, C, D, and E. Assume that all intermediates are able to enter the cell, that each mutant carries only one mutation, and that all mutants affect steps after B in the pathway.

Mutant	A	B	C	D	E
1	+	0	0	0	0
2	0	0	0	+	0
3	0	0	+	0	0
4	0	0	0	+	+

Which of the schemes in the following figure best fits the data with regard to the biosynthetic pathway?

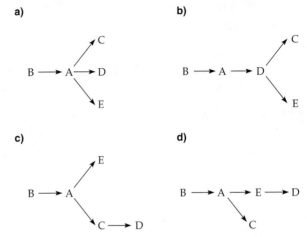

*10.12 Four strains of the haploid fungus *Neurospora*, all of which require arginine but have unknown genetic constitution, have the following nutrition and accumulation characteristics:

			Growth on		
Strain	Minimal Medium	Ornithine	Citrulline	Arginine	Accumulates
1	–	–	+	+	Ornithine
2	–	–	–	+	Citrulline
3	–	–	–	+	Citrulline
4	–	–	–	+	Ornithine

The pairwise complementation tests of the four strains gave the following results (+ = growth on minimal medium and 0 = no growth on minimal medium):

	4	3	2	1
1	0	+	+	0
2	0	0	0	
3	0	0		
4	0			

Crosses among mutants yielded prototrophs in the following percentages:

1 × 2:   25 percent
1 × 3:   25 percent
1 × 4:   none detected among 1 million progeny (ascospores)
2 × 3:   0.002 percent
2 × 4:   0.001 percent
3 × 4:   none detected among 1 million progeny (ascospores)

Analyze the data and answer the following questions.

a. How many distinct mutational sites are represented among these four strains?

b. In this collection of strains, how many types of polypeptide chains (normally found in the wild type) are affected by mutations?

c. Write the genotypes of the four strains, using a consistent and informative set of symbols.

d. Determine the map distances between all pairs of linked mutations.

e. Determine the percentage of prototrophs that would be expected among ascospores of the following types: (1) strain 1 × wild type; (2) strain 2 × wild type; (3) strain 3 × wild type; (4) strain 4 × wild type.

10.13 A breeder of Irish setters has a particularly valuable show dog that he knows is descended from the famous bitch Rheona Didona, who carried a recessive gene for atrophy of the retina. Before he puts the dog to stud, he must ensure that it is not a carrier for this allele. How should he proceed?

*10.14 Suppose you were on a jury to decide the following case:

The Jones family claims that Baby Jane, given to them at the hospital, belongs not to them but to the Smith family and that the Smiths' baby Joan really belongs to the Jones family. It is alleged that the two babies were accidentally exchanged soon after birth. The Smiths deny that such an exchange has been made. Blood group determinations show the following results:

Mrs. Jones, AB
Mr. Jones, O
Mrs. Smith, A
Mr. Smith, O
Baby Jane, A
Baby Joan, O

Which baby belongs to which family?

*10.15 Glutathione (GSH) is important for a number of biological functions, including prevention of oxidative damage in red blood cells, synthesis of deoxyribonucleotides from ribonucleotides, transport of some amino acids into cells, and maintenance of protein conformation. Mutations that have lowered levels of glutathione synthetase (GSS), a key enzyme in the synthesis of glutathione, result in one of two clinically distinguishable disorders. The severe form is characterized by massive urinary excretion of 5-oxoproline (a chemical derived from a synthetic precursor to glutathione), metabolic acidosis (an inability to regulate physiological pH appropriately), anemia, and central nervous system damage. The mild form is characterized solely by anemia. The characterization of GSS activity and the GSS protein in two affected patients, each with normal parents, is shown here:

Patient	Disease Form	GSS Activity in Fibroblasts (percentage of normal)	Effect of Mutation on GSS Protein
1	severe	9%	Arginine at position 267 replaced by tryptophan
2	mild	50%	Aspartate at position 219 replaced by glycine

a. What pattern of inheritance do you expect these disorders to exhibit?

b. Explain the relationship of the form of the disease to the level of GSS activity.

c. How can two different amino acid substitutions lead to dramatically different phenotypes?

d. Why is 5-oxoproline produced in significant amounts only in the severe form of the disorder?

e. Is there evidence that the mutations causing the severe and mild forms of the disease are allelic (in the same gene)?

f. How might you design a test to aid in prenatal diagnosis of this disease?

10.16 Some methods used to gather fetal material for prenatal diagnosis are invasive and therefore pose a small but very real risk to the fetus.

a. What specific risks and problems are associated with chorionic villus sampling and amniocentesis?

b. How are these risks balanced with the benefits of each procedure?

c. Fetal cells are reportedly present in maternal circulation after about 8 weeks of pregnancy. However, the number of cells is very low, perhaps no more than one in several million maternal cells. To date, it has not been possible to isolate fetal cells from maternal blood in sufficient quantities for routine genetic analysis. If the problems associated with isolating fetal cells from maternal blood were overcome, and sufficiently sensitive methods were developed to perform genetic tests on a small number of cells, what would be the benefits of performing genetic tests on these fetal cells?

10.17 In evaluating my teacher, my sincere opinion is that

a. he/she is a swell person whom I would be glad to have as a brother-in-law/sister-in-law

b. he/she is an excellent example of how tough it is when you do not have either genetics or environment going for you

c. he/she may have okay DNA to start with, but somehow all the important genes got turned off

d. he/she ought to be preserved in tissue culture for the benefit of other generations

*10.18 A *Neurospora* mutant has been isolated in the laboratory in which you are working. This mutant cannot make an amino acid we will call Y. Wild-type *Neurospora* cells make Y from a cellular product X through a biochemical pathway involving three intermediates called c, d, and e. How would you establish that your mutant contains a defective gene for the enzyme that catalyzes the d-e reaction?

10.19 You have been introduced to the functions and levels of proteins and their organization. List as many protein functions as you can, and give an example of each.

10.20 We have discovered that most enzymes are proteins, but not all proteins are enzymes. What are the functions of enzymes, and why are they essential for living organisms to carry out their biological functions?

10.21 We know that the function of any protein is tied to its structure. Give an example of how a disruption of structure by mutation can lead to a distinctive phenotypic effect.

*10.22 Define *autosomal recessive mutation,* and give some examples of diseases that are caused by autosomal recessive mutations. Explain how two parents who do not display any symptoms of a given disease (any of the diseases you have named) can have two or even three children who have the disease. How can these same parents have no children with the disease?

10.23 What can prospective parents do to reduce the risk of having offspring who have genetically based enzyme deficiencies?

11

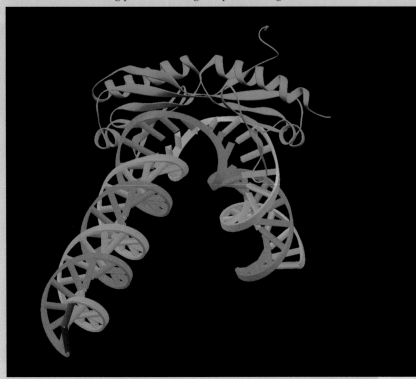

Gene Expression: Transcription

PRINCIPAL POINTS

Transcription is the process of copying genetic information in DNA into RNA base sequences. The DNA unwinds in a short region next to a gene, and an RNA polymerase catalyzes the synthesis of an RNA molecule. Only one strand of the double-stranded DNA is transcribed into an RNA molecule.

In *E. coli,* the initiation of transcription of protein-coding (structural) genes requires a complex of RNA polymerase and the sigma factor protein binding to the promoter. Once transcription has begun, the sigma factor dissociates and RNA synthesis is completed by the RNA polymerase core enzyme. Termination of transcription is signaled by specific sequences in the DNA.

In *E. coli,* a single RNA polymerase synthesizes mRNA, tRNA, and rRNA. Eukaryotes have three distinct nuclear RNA polymerases, each of which transcribes different gene types: RNA polymerase I transcribes the genes for the 28S, 18S, and 5.8S ribosomal RNAs; RNA polymerase II transcribes mRNA genes and some snRNA genes; and RNA polymerase III transcribes genes for the 5S rRNAs, the tRNAs, and the other snRNAs.

The transcripts of protein-coding genes are linear precursor mRNAs (pre-mRNAs). Prokaryotic mRNAs are modified little once they are transcribed, while eukaryotic mRNAs are modified by the addition of a 5′ cap and a 3′ poly(A) tail. Most eukaryotic pre-mRNAs contain sequences called introns (for *intervening* sequences) that do not code for amino acids. The introns are removed as the primary transcript is processed to produce the mature, functional mRNA molecule. The segments of RNA that remain in the mRNA are called exons (for *expressed* sequences).

Introns are removed from pre-mRNAs in a series of well-defined steps. Intron removal begins with the cleavage of the pre-mRNA at the 5′ splice junction. The free 5′ end of the intron loops back and bonds to a nucleotide in another region of the intron called the branch-point consensus sequence. Cleavage at the 3′ splice junction releases the intron, which is shaped like a lariat. The exons that flanked the intron are then spliced together. The removal of introns from eukaryotic pre-mRNA occurs in the nucleus in complexes called spliceosomes. The spliceosome consists of several small nuclear ribonucleoprotein particles (snRNPs) bound specifically to each intron.

Ribosomes are the cellular organelles on which protein synthesis takes place. In both prokaryotes and eukaryotes, ribosomes consist of two unequally sized subunits. Each subunit consists of a complex between one or more rRNA molecules and many ribosomal proteins. Eukaryotic ribosomes are larger and more complex than prokaryotic ribosomes.

In eukaryotes, three of the four rRNAs are encoded by tandem arrays of transcription units that can number in the thousands. Each transcription unit produces a single pre-rRNA molecule with the three rRNAs separated by spacer sequences. The individual rRNAs are generated by processing the pre-rRNA to remove the spacers. The fourth rRNA is encoded by separate genes.

In the precursor rRNAs of some organisms there are introns, the RNA sequences of which fold into a secondary structure that excises itself, a process called self-splicing. This process does not involve protein enzymes.

tRNA molecules bring amino acids to the ribosomes, where the amino acids are polymerized into a polypeptide chain. All tRNAs are about the same length, contain a number of modified bases, and have similar three-dimensional shapes.

For 5S rRNA genes and tRNA genes, the promoter for RNA polymerase III is located within the transcribed region of the gene. The internal promoter is called the internal control region. Transcription factors and RNA polymerase III bind to the internal control region for transcription of the genes.

iActivity

i DO YOU WANT TO MAKE A CLONE? MIX GENES TO create a new organism? Treat genetic disease with DNA? Investigate a murder? These biotechnology techniques, and many others, are made possible by an understanding of gene expression. And the first step in gene expression is transcription, during which information is transferred from the DNA molecule to a single-stranded RNA molecule. In this chapter, you will learn how DNA is transcribed into RNA and the structure and properties of different forms of RNA. Then, in the iActivity, you can investigate how mutations that affect the process of transcription can lead to an inherited disease.

The structure, function, development, and reproduction of an organism depend on the properties of the proteins present in each cell and tissue. A protein consists of one or more chains of amino acids. Each chain of amino acids is called a polypeptide, and the sequence of amino acids in a polypeptide chain is coded for by a gene. When a protein is needed in the cell, the genetic code for that protein's amino acid sequence must be read from the DNA and processed into the finished protein. Two major steps occur during protein synthesis: transcription and translation. **Transcription** is the synthesis of a single-stranded RNA copy of a segment of DNA. **Translation** (protein synthesis) is the conversion of the messenger RNA base sequence information into the amino acid sequence of a polypeptide. Unlike DNA replication, which typically occurs during only part of the cell cycle (at least in eukaryotes), transcription and translation generally occur throughout the cell cycle (although they are much reduced during the M phase of the cell cycle). In this chapter, you will learn about the transcription process. For ease of representation, in the diagrams of transcription we show the enzymes involved moving along a static DNA molecule as RNA is synthesized. However, a recent model by Peter Cook proposes that the enzymes are actually stationary, with the DNA molecule passing by. You should keep this in mind as you read about the transcription events.

Gene Expression: An Overview

In 1956, three years after Watson and Crick proposed their double helix model of DNA, Crick gave the name *Central Dogma* to the two-step process of DNA→RNA→protein (transcription followed by translation). Transcription is the synthesis of an RNA copy of a segment of DNA; only one of the two DNA strands is transcribed into RNA. The RNA is synthesized by RNA polymerase.

Not all genes encode proteins, so not all gene transcripts are the kind of RNA that is translated. In fact, there are four different types of RNA molecules, each encoded by its own type of gene:

1. **mRNA (messenger RNA)** encodes the amino acid sequence of a polypeptide. mRNAs are the transcripts of *protein-coding genes,* also called **structural genes.**
2. **tRNA (transfer RNA)** brings amino acids to ribosomes during translation.
3. **rRNA (ribosomal RNA),** with ribosomal proteins, makes up the ribosomes, the organelles on which mRNA is translated to produce a polypeptide.
4. **snRNA (small nuclear RNA),** with proteins, forms complexes used in eukaryotic RNA processing.

In the remainder of the chapter, we will learn about the transcription process and more about these four types of RNA.

The Transcription Process

In this section, we discuss how an RNA chain is synthesized.

RNA Synthesis

Associated with each gene are sequences called **gene regulatory elements,** which are involved in the regulation of transcription.

In both prokaryotes and eukaryotes, the enzyme **RNA polymerase** catalyzes the process of transcription (Figure 11.1). The DNA double helix unwinds for a short region next to the gene before transcription can begin. Only one of the two DNA strands is transcribed into an RNA.

In transcription, RNA is synthesized in the 5′→3′ direction. The 3′→5′ DNA strand that is read to make the RNA strand is called the *template strand.* The 5′→3′ DNA strand complementary to the template strand and which has the *same* polarity as the resulting RNA strand is called the *nontemplate strand.*

The RNA precursors for transcription are the ribonucleoside triphosphates ATP, GTP, CTP, and UTP, collectively called NTPs (*nucleoside triphosphates). RNA synthesis occurs by polymerization reactions that are very similar to the polymerization reactions involved in DNA synthesis (Figure 11.2; DNA polymerization is shown in Figure 9.4). RNA polymerase selects the next nucleotide to be added to the chain by its ability to pair with the exposed base on the DNA template strand. Unlike DNA polymerases, RNA polymerases can initiate new polynucleotide chains (no primer is needed), but they have no proofreading abilities.

Recall that RNA chains contain nucleotides with the base uracil instead of thymine and that uracil pairs with adenine. Therefore, where an A nucleotide occurs on the DNA template chain, a U nucleotide is placed in the RNA chain instead of a T.

For example, if the template DNA strand reads

$$3'\text{-ATACTGGAC-}5'$$

then the RNA chain will be synthesized in the 5′→3′ direction and will have the sequence

$$5'\text{-UAUGACCUG-}3'.$$

animation

a RNA Biosynthesis

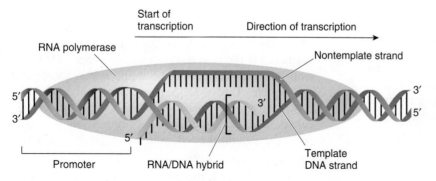

Start of transcription

Direction of transcription

RNA polymerase

Nontemplate strand

5′

3′

5′

3′

5′

Promoter

RNA/DNA hybrid

Template DNA strand

Figure 11.1

Transcription process. The DNA double helix is denatured by RNA polymerase in prokaryotes or by other proteins in eukaryotes. RNA polymerase then catalyzes the synthesis of a single-stranded RNA chain, beginning at the "start of transcription" point. The RNA chain is made in the 5′→3′ direction, using only one strand of the DNA as a template to determine the base sequence.

Figure 11.2

Chemical reaction involved in the RNA polymerase–catalyzed synthesis of RNA on a DNA template strand.

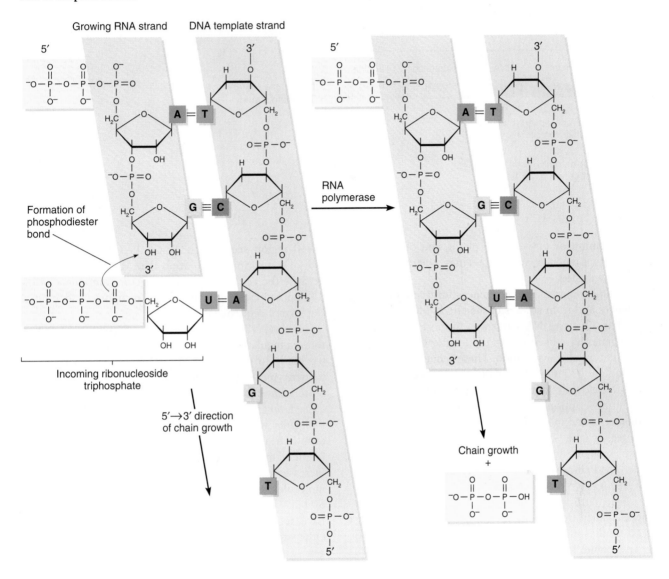

Transcription, the process of transcribing the genetic information in DNA into RNA base sequences, exhibits basic similarities in prokaryotes and eukaryotes. The DNA unwinds in a short region next to a gene, and an RNA polymerase catalyzes the synthesis of an RNA molecule in the 5'→3' direction along the 3'→5' template strand of the DNA. Only one strand of the double-stranded DNA is transcribed into an RNA molecule.

Initiation of Transcription at Promoters

In both prokaryotes and eukaryotes, the process of transcription occurs in three steps: initiation, elongation, and termination. Particular mechanisms are used to signal initiation and termination. In this section, we discuss the initiation of transcription in prokaryotes, focusing on *E. coli*. A prokaryotic gene may be divided into three sequences with respect to its transcription (Figure 11.3):

1. A sequence upstream of the start of the RNA-coding sequence called the **promoter,** with which the RNA polymerase interacts to begin transcription. The promoter ensures that initiation of every RNA occurs at the same site.

2. The RNA-coding sequence, the DNA sequence transcribed by RNA polymerase into the RNA transcript.

Figure 11.3

Promoter, RNA-coding sequence, and terminator regions of a gene. The promoter is "upstream" of the coding sequence, and the terminator is "downstream" of the coding sequence. The coding sequence begins at nucleotide +1.

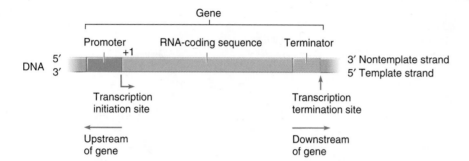

3. A **terminator** downstream of the end of the RNA-coding sequence, which specifies where transcription will stop.

From comparisons of sequences upstream of coding sequences and from studies of the effects of specific base pair changes, two DNA sequences in most promoters of *E. coli* genes have been shown to be critical for specifying the initiation of transcription. These sequences generally are found at −35 and −10; that is, they are centered at 35 and 10 base pairs upstream from +1, the base pair at which transcription starts. The **consensus sequence** (the sequence found most frequently at each position) for the −35 region (the −35 box) is 5′-TTGACA-3′. The consensus sequence for the −10 region (the *−10 box,* formerly called the **Pribnow box** after the researcher who first discovered it) is 5′-TATAAT-3′.

For transcription to begin, a form of RNA polymerase called the *holoenzyme* (or *complete enzyme*) must bind to the promoter. This holoenzyme consists of the **core enzyme** form of RNA polymerase (which has four polypeptides—two α, a β, and a β′) bound to another polypeptide called the *sigma factor* (σ). The sigma factor is essential for recognizing the −35 and −10 regions of the promoter.

The RNA polymerase holoenzyme binds to a promoter in two steps. First, it binds loosely to the −35 box while the DNA is still double-stranded (Figure 11.4a). Next, the RNA polymerase binds more tightly to the DNA as the DNA untwists for about 17 base pairs centered around the −10 box (Figure 11.4b). Once the RNA polymerase is bound at the −10 box, it is oriented properly to begin transcription at the correct nucleotide.

Promoters differ slightly in their actual sequence, so the binding efficiency of RNA polymerase varies. As a result, the rate at which transcription is initiated varies from gene to gene, which explains in part why different genes have different levels of expression. In other words, the relative strengths of promoters relate directly to how similar they are to the consensus sequence. For example,

a −10 region sequence of 5′-GATACT-3′ has a lower rate of transcription initiation than the 5′-TATAAT-3′ because the ability of the sigma factor component of the RNA polymerase holoenzyme to recognize and bind to the former sequence is lower.

Several different sigma factors in *E. coli* play important roles in regulating gene expression. Each type of sigma factor binds to the core RNA polymerase and permits the holoenzyme to recognize different promoters. Most promoters have the recognition sequences we have just discussed, and these are recognized by a sigma factor with a molecular weight of 70 kDa, called σ^{70}. Under conditions of high heat (heat shock) and other forms of stress, a different sigma factor called σ^{32} (molecular weight 32 kDa) increases in amount. This directs some RNA polymerase molecules to bind to the promoters of genes that encode proteins required to cope with the stress. Such promoters have recognition sequences specific for the σ^{32} factor. Other sigma factors control expression of yet other types of genes under various conditions. Multiple sigma factors are also found in other bacterial species.

Elongation and Termination of an RNA Chain

RNA synthesis takes place in a region of the DNA that has separated to form a transcription bubble. Once eight or nine RNA nucleotides have been linked together, the sigma factor dissociates from the RNA polymerase core enzyme (Figure 11.4c) and is used again in other transcription initiation reactions. The core enzyme completes the transcription of the gene.

As the core RNA polymerase moves along, it untwists the DNA double helix ahead of it (Figure 11.4d). Within the untwisted region, some bases of RNA are base-paired to the DNA in a temporary RNA-DNA hybrid, and the rest of the RNA is displaced away from the DNA. Transcription proceeds at a rate averaging 30 to 50 nucleotides per second.

Figure 11.4

Action of *E. coli* RNA polymerase in the initiation and elongation stages of transcription. (a) In initiation, the RNA polymerase holoenzyme first binds loosely to the promoter at the −35 region. **(b)** As initiation continues, RNA polymerase binds more tightly to the promoter at the −10 region, accompanied by a local untwisting of about 17 bp around the −10 region. At this point, the RNA polymerase is correctly oriented to begin transcription at +1. **(c)** After eight to nine nucleotides have been polymerized, the sigma factor dissociates from the core enzyme. **(d)** As the RNA polymerase elongates the new RNA chain, the enzyme untwists the DNA ahead of it; as the double helix re-forms behind the enzyme, the RNA is displaced away from the DNA.

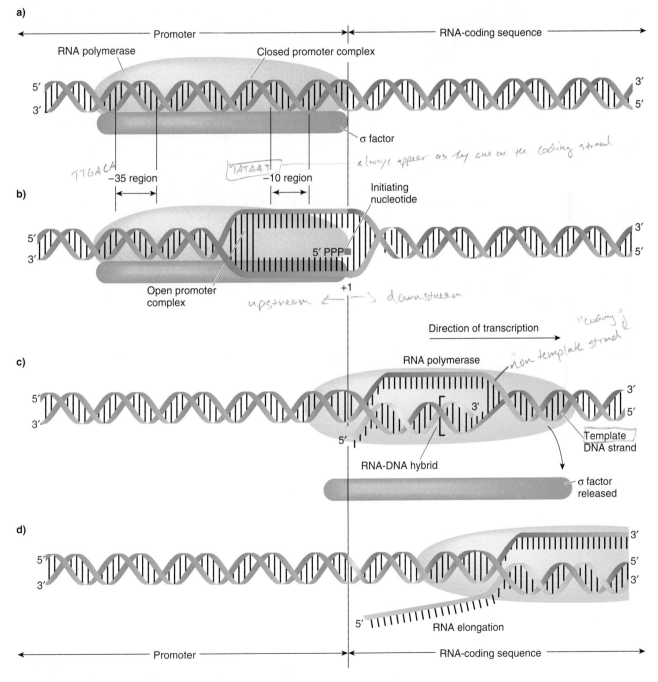

Termination of prokaryotic gene transcription is signaled by *terminators*. One important protein involved in the termination of transcription of some *E. coli* genes is *rho* (ρ). The terminators of such genes are called *rho-dependent terminators*. At many other terminators, the core RNA polymerase itself carries out the termination events. Those terminators are called *rho-independent terminators*.

Since only one type of RNA polymerase is found in prokaryotes, all classes of genes are transcribed by it, namely, protein-coding genes, tRNA genes, and rRNA genes.

KEYNOTE

In *E. coli,* the initiation and termination of transcription are signaled by specific sequences that flank the RNA-coding sequence of the gene. The promoter is recognized by the sigma factor component of the RNA polymerase–sigma factor complex. Two types of termination sequences are found, and a particular gene has one or the other. One type of terminator is recognized by the RNA polymerase alone, and the other type is recognized by the enzyme in association with the *rho* factor.

Transcription in Eukaryotes

Transcription is more complicated in eukaryotes than in prokaryotes because eukaryotes possess three different classes of RNA polymerases and because of the way transcripts are processed to their functional forms.

Eukaryotic RNA Polymerases

In eukaryotes, three different RNA polymerases transcribe the genes for the four types of RNAs. **RNA polymerase I,** located exclusively in the nucleolus, catalyzes the synthesis of three of the RNAs found in ribosomes: the 28S, 18S, and 5.8S rRNA molecules. The S values derive from the rate at which the rRNA molecules sediment during centrifugation in a sucrose gradient; they give a very rough indication of molecular size. **RNA polymerase II,** found only in the nucleoplasm of the nucleus, synthesizes messenger RNAs (mRNAs) and some small nuclear RNAs (snRNAs), some of which are involved in RNA-processing events. **RNA polymerase III,** found only in the nucleoplasm, synthesizes the transfer RNAs (tRNAs), which bring amino acids to the ribosome; 5S rRNA, a small rRNA molecule found in each ribosome; and the small nuclear RNAs (snRNA) not made by RNA polymerase II.

KEYNOTE

In *E. coli,* a single RNA polymerase synthesizes mRNA, tRNA, and rRNA. Eukaryotes have three distinct nuclear RNA polymerases, each of which transcribes different gene types: RNA polymerase I transcribes the genes for the 28S, 18S, and 5.8S ribosomal RNAs; RNA polymerase II transcribes mRNA genes and some snRNA genes; and RNA polymerase III transcribes genes for the 5S rRNAs, the tRNAs, and the remaining snRNAs.

Transcription of Protein-Coding Genes by RNA Polymerase II

In this section, we discuss the sequences and molecular events involved in transcribing a protein-coding gene. In eukaryotes, RNA polymerase II transcribes protein-coding genes. The product of transcription is a **precursor-mRNA (pre-mRNA)** molecule, a transcript that must be modified, processed, or both to produce the mature, functional mRNA molecule.

Promoters of protein-coding genes are analyzed in two principal ways. One way is to examine the effect of mutations that delete or alter base sequences upstream from the start point of transcription and to see whether those mutants affect transcription. Mutations that significantly affect transcription define important promoter elements. The second way is to compare the DNA sequences upstream of a number of protein-coding genes to see whether there are any regions with similar sequences. The results of these experiments show that the promoters of protein-coding genes contain *basal promoter elements* and *promoter proximal elements.*

The best-characterized basal promoter elements are the **TATA box** or **TATA element** (also called the **Goldberg-Hogness box** after its discoverers), located at about position −25, and a pyrimidine-rich sequence near the transcription start site called the *initiator element* or simply *Inr.* The TATA box has the seven-nucleotide consensus sequence TATAAAA. It is found in many genes, so it has only a general activity in the initiation of transcription. AT-rich DNA is easy to denature to single strands, so the TATA box probably facilitates strand separation for the initiation of transcription. The TATA box also determines the exact start point for transcription.

Promoter proximal elements are further upstream from the TATA box, about 50 to 200 nucleotides from the start of transcription. Examples of these elements are the **CAAT** ("cat") **box,** named for its consensus sequence and located at about −75, and the **GC box,** consensus sequence GGGCGG, located at about −90. Both the CAAT box and GC box work in either orientation, and like the TATA box, they have only general activities in transcription initiation.

Promoters contain various combinations of basal promoter elements and promoter proximal elements. In other words, a number of different elements can contribute to promoter function, and no one element is essential for transcription to take place.

Accurate initiation of transcription of a protein-coding gene involves the assembly of RNA polymerase II and a number of other proteins called **basal transcription factors (TFs)** on the basal promoter elements. All three eukaryotic RNA polymerases require basal transcription factors for transcription initiation. The basal transcription factors are numbered for the RNA polymerase with which they work and lettered to reflect their

Figure 11.5

Events that may occur during the initiation of transcription catalyzed by RNA polymerase II. (a) Assembly of the preinitiation complex of RNA polymerase II and transcription factors on the basal promoter elements. (b) Simplified model for the stimulation of transcription by the binding of an activator transcription factor at an enhancer. In actuality, many more proteins are involved than are shown.

a) Assembly of preinitiation complex

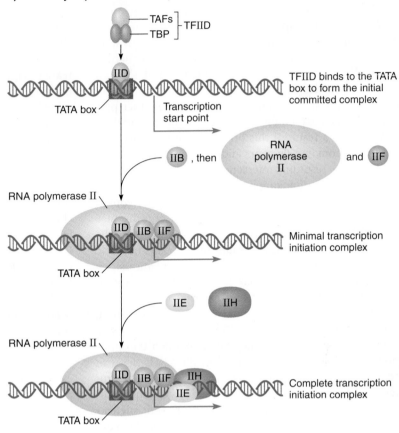

TFIID binds to the TATA box to form the initial committed complex

Minimal transcription initiation complex

Complete transcription initiation complex

b) Stimulation of transcription by activator binding to an enhancer

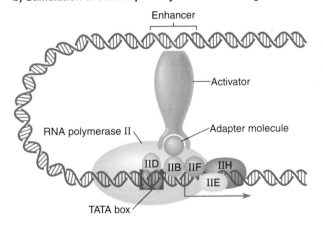

order of discovery. For example, TFIID is the fourth basal transcription factor (D) discovered that works with RNA polymerase II.

For protein-coding genes, the basal transcription factors and RNA polymerase II bind to promoter elements in a particular order (Figure 11.5a). First, TFIID binds to the T A T A box to form the *initial committed complex.* The multisubunit TFIID has one subunit called the T A T A-*binding protein* (TBP), which actually recognizes the T A T A box sequence, and a number of other proteins called *TBP-associated factors* (TAFs). The TFIID-T A T A box complex acts as a binding site for TFIIB, which then recruits RNA polymerase II and TFIIF to produce the *minimal transcription initiation complex.* (RNA polymerase II, like all eukaryotic RNA polymerases, cannot directly recognize and bind to promoter elements.) Next, TFIIE and TFIIH bind to produce the *complete transcription initiation complex,* also called the *preinitiation complex* (PIC) because it is ready to begin transcription.

The initiation complex is sufficient for only a low level of transcription. For a high level of transcription to occur, other transcription factors called **activators** control which promoters are transcribed actively. Activators bind to regulatory elements called **enhancers,** sequences required for maximal transcription. Figure 11.5b shows a simplified model in which an activator binds to an enhancer and, through an interaction with another protein called an adapter, forms a bridge to the preinitiation complex. As a result of this bridging, the DNA between the basal promoter elements and the enhancer becomes looped. The resulting interaction of activator proteins and the preinitiation complex stimulates transcription.

Enhancers are found singly or in multiple copies. They function in either orientation and at a large distance from the gene, either upstream, downstream, or in the gene itself. They may be thousands of base pairs from the gene they control. In most cases, the enhancers are upstream of the gene. Similar elements that have essentially the same properties as enhancer elements, except that they repress rather than activate gene transcription, are called **silencer elements.** Silencers are much less common than enhancers; they function when transcription factors called *repressors* bind to them. The level of transcription resulting from activities at the basal promoter elements is modulated by the effects of other transcription factors binding to enhancer and silencer elements. Because the activators and repressors are cell- and tissue-specific, the interactions of these transcription factors at enhancers and silencers are responsible for cell- and tissue-specific gene expression.

Investigate how mutations at different regions of the β-globin gene affect mRNA transcription and the production of β-globin in the iActivity *Investigating Transcription in Beta-Thalassemic Patients* on the website.

Eukaryotic mRNAs

Figure 11.6 shows the general structure of the mature, biologically active mRNA as it exists in both prokaryotic and eukaryotic cells. The mRNA molecule has three main parts. At the 5′ end is a **leader sequence,** or 5′ untranslated region (5′ UTR), which varies in length between mRNAs of different genes. Following the 5′ leader sequence is the actual **coding sequence** of the mRNA, the sequence that specifies the amino acid sequence of a protein during translation. The coding sequence varies in length, depending on the length of the protein for which it codes. Following the amino acid–coding sequence is a **trailer sequence,** or 3′ untranslated region (3′ UTR). The trailer sequence also varies in length from mRNA to mRNA.

mRNA production is different in prokaryotes and eukaryotes. In prokaryotes (Figure 11.7a), the RNA transcript functions directly as the mRNA molecule. That is, the base pairs of a prokaryotic gene are colinear with the bases of the translated mRNA. In eukaryotes (Figure 11.7b), the RNA transcript (the pre-mRNA) must be modified in the nucleus by a series of events known as *RNA processing* to produce the mature mRNA. In addition, because prokaryotes lack a nucleus, an mRNA begins to be translated on ribosomes before it has been transcribed completely; this process is called *coupled transcription and translation.* In eukaryotes, the mRNA must migrate from the nucleus to the cytoplasm (where the ribosomes are located) before it can be translated. Thus, a eukaryotic mRNA is always transcribed completely and processed before it is translated.

animation

α **mRNA Production in Eukaryotes**

Another fundamental difference between prokaryotic and eukaryotic mRNAs is that prokaryotic mRNAs often are *polycistronic,* meaning that they contain the amino acid–coding information from more than one gene, whereas eukaryotic mRNAs are always *monocistronic,* meaning that they contain the amino acid–coding information from just one gene. This concept is not shown in Figure 11.7.

Production of Mature mRNA in Eukaryotes. Unlike prokaryotic mRNA, eukaryotic mRNA is modified at both the 5′ and 3′ ends. In addition, most protein-coding genes have insertions of non–amino acid–coding sequences called **introns** between the other sequences present in mRNA—the **exons.**[1] The term *intron* is derived from *intervening sequence,* a sequence that is *not* translated

---

[1]Some bacteriophage genes also contain introns. For example, there is a single intron in the bacteriophage thymidylate kinase gene.

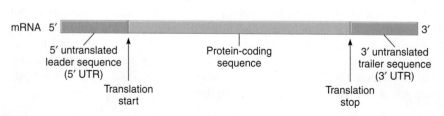

mRNA 5′ ⎯⎯⎯⎯⎯⎯⎯⎯⎯⎯⎯⎯⎯⎯⎯⎯⎯⎯⎯⎯ 3′

5′ untranslated leader sequence (5′ UTR)

Translation start

Protein-coding sequence

3′ untranslated trailer sequence (3′ UTR)

Translation stop

Figure 11.6

General structure of mRNA found in both prokaryotic and eukaryotic cells.

Figure 11.7

Processes for synthesis of functional mRNA in prokaryotes and eukaryotes.
(a) In prokaryotes, the mRNA synthesized by RNA polymerase does not have to be processed before it can be translated by ribosomes. Also, because there is no nuclear envelope, mRNA translation can begin while transcription continues, resulting in a coupling of transcription and translation. (b) In eukaryotes, the primary RNA transcript is a precursor-mRNA (pre-mRNA) molecule, which is processed in the nucleus by the addition of a 5′ cap and a 3′ poly(A) tail and removal of introns. Only when that mRNA is transported to the cytoplasm can translation occur.

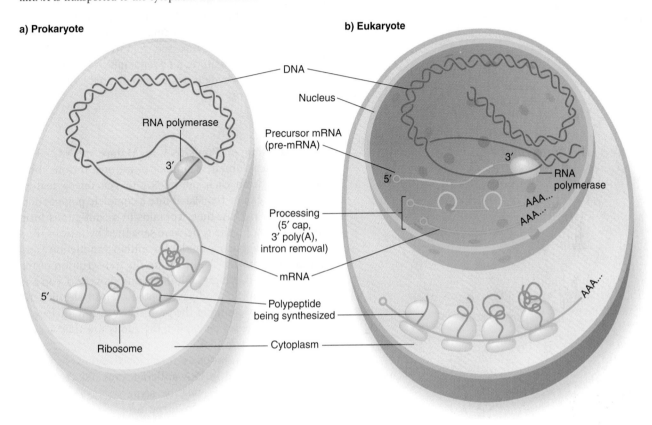

a) Prokaryote

b) Eukaryote

DNA

Nucleus

RNA polymerase

Precursor mRNA (pre-mRNA)

3′

3′

5′

RNA polymerase

AAA...

AAA...

Processing (5′ cap, 3′ poly(A), intron removal)

mRNA

5′

Polypeptide being synthesized

AAA...

Ribosome

Cytoplasm

into an amino acid sequence, and the term *exon* is derived from *expressed* sequence. Exons include the 5′ and 3′ UTRs as well as the amino acid–coding portions. The introns are removed in the processing of pre-mRNA to the mature mRNA molecule. The 1993 Nobel Prize in Physiology or Medicine was awarded to Richard Roberts and Philip Sharp for their independent discoveries of split genes.

5′ *and* 3′ *Modifications.* Once RNA polymerase II has made about 20 to 30 nucleotides of pre-mRNA, a *capping enzyme* adds a guanine nucleotide—most commonly 7-methyl guanosine (m^7G)—to the 5′ end by an unusual 5′-5′ linkage as opposed to the usual 5′-3′ linkage (Figure 11.8). This process is called **5′ capping.** The sugars of the next two nucleotides are also modified by methylation. The 5′ cap remains throughout processing and is present in the mature mRNA. The cap is essential for the ribo-

some to bind to the 5′ end of the mRNA, an initial step of translation.

Most eukaryotic pre-mRNAs become modified at their 3′ end by the addition of a sequence of about 50 to 250 adenine nucleotides called a **poly(A) tail.** There is no DNA template for the poly(A) tail, and the poly(A) tail remains while pre-mRNA is processed to mature mRNA. The poly(A) tail is important for determining the stability of the mRNA.

Addition of the poly(A) tail marks the 3′ end of the mRNA. That is, in many eukaryotes, there typically is no termination sequence in the DNA to signal the end of transcription of a protein-coding gene. Instead, mRNA transcription continues, in some cases for hundreds or thousands of nucleotides, past a site called the *poly(A) site.* The poly(A) site is about 10 to 30 nucleotides downstream of the poly(A) consensus sequence. In mammalian cells, poly(A) addition to the RNA occurs as

Figure 11.8

Cap structure at the 5′ end of a eukaryotic mRNA. The cap results from the addition of a guanine nucleotide and two methyl groups.

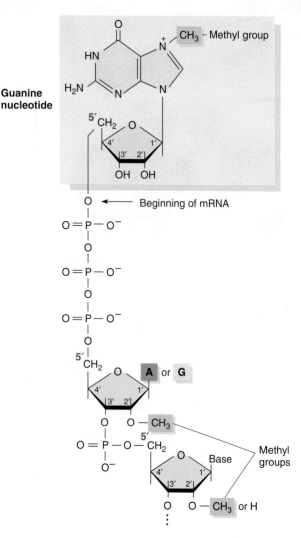

K E Y N O T E

The transcripts of protein-coding genes are messenger RNAs or their precursors. These molecules are linear and vary widely in length, depending on the size of the polypeptides they specify and whether they contain introns. Prokaryotic mRNAs are not modified once they are transcribed, whereas most eukaryotic mRNAs are modified by the addition of a cap at the 5′ end and a poly(A) tail at the 3′ end. Many eukaryotic pre-mRNAs contain non–amino acid–coding sequences called introns, which must be excised from the mRNA transcript to make a mature, functional mRNA molecule. The amino acid–coding segments separated by introns are called exons.

Processing of Pre-mRNA to Mature mRNA. Pre-mRNAs often contain a number of introns. Introns must be excised from each pre-mRNA to generate a mature mRNA that can be translated into a complete polypeptide. The mature mRNA, then, contains in a contiguous form the exons that in the gene were separated by introns. Figure 11.10 diagrams a model for mRNA production from genes with introns. The steps are the transcription of the gene by RNA polymerase II, the addition of the 5′ cap and the poly(A) tail to produce the pre-mRNA molecule, and finally the processing of the pre-mRNA in the nucleus to remove the introns and splice the exons together to produce the mature mRNA.

Introns in pre-mRNAs are removed and exons joined in a process called **mRNA splicing.** For mRNA splicing to take place, there must be some way for the machinery involved to determine what is an intron and what is an exon. Introns typically begin with 5′-GU and end with AG-3′, although more than just those nucleotides are needed to specify a junction between an intron and an exon; the 5′ splice junction probably involves at least seven nucleotides, and the 3′ splice junction involves at least ten nucleotides of intron sequence.

animation
a RNA Splicing

Splicing of pre-mRNA occurs in the nucleus. The splicing events occur in complexes called **spliceosomes,** which consist of the pre-mRNA bound to **small nuclear ribonucleoprotein particles** (**snRNPs;** also called *snurps*). These are snRNAs associated with proteins. There are five principal snRNAs (named U1, U2, U4, U5, and U6), and they are associated with six to ten proteins each to form the snRNPs.

Figure 11.11 shows a simplified model of splicing for two exons separated by an intron. The steps are as follows:

1. U1 snRNP binds to the 5′ splice junction of the intron. This binding is primarily the result of base pairing of the U1 snRNA in the snRNP to the 5′ splice site junction.

follows (Figure 11.9): A number of proteins, including CPSF (cleavage and polyadenylation specificity factor) protein, CstF (cleavage stimulation factor) protein, and two cleavage factor proteins (CFI and CFII) bind to and cleave the RNA. CPSF binds to the signal, and CstF binds to a GU-rich or U-rich sequence (GU/U in Figure 11.9) downstream of the poly(A) site. CPSF and CstF also bind to each other, producing a loop in the RNA. CFI and CFII are bound near the actual cleavage site. Once the RNA is cleaved, the enzyme **poly(A) polymerase (PAP)** uses ATP as a substrate and catalyzes the addition of A nucleotides to the 3′ end of the RNA to produce the poly(A) tail. During this process, PAP is bound to CPSF. As the poly(A) tail is synthesized, it becomes bound by poly(A)-binding protein II (PABII).

Figure 11.9

Schematic diagram of the 3′ end formation of mRNA and the addition of the poly(A) tail to that end in mammals. In eukaryotes, the formation of the 3′ end of an mRNA is produced by cleavage of the lengthening RNA chain. This process is signaled by specific nucleotide sequences in the RNA. See text for a discussion of the proteins involved.

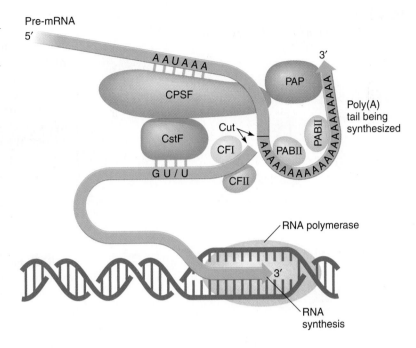

2. U2 snRNP binds to the **branch-point sequence** upstream of the 3′ splice junction.
3. A U4/U6 snRNP and a U5 snRNP interact, and the combination binds to the U1 and U2 snRNPs, causing the intron to loop, thereby bringing the two ends of the intron closer together.
4. U4 snRNP dissociates from the complex, and this results in the formation of the active spliceosome.
5. The snRNPs in the spliceosome cleave the intron

from exon 1 at the 5′ splice junction, and the now free 5′ end of the intron is bonded to a particular nucleotide (often an A nucleotide) in the branch-point sequence. The intron is now folded into an *RNA lariat structure.*

6. Next, the intron is excised (still in lariat shape) by cleavage at the 3′ splice junction, and then the exons 1 and 2 are ligated together. The snRNPs are released at this time. The process is repeated for each intron.

Figure 11.10

General sequence of steps in the formation of eukaryotic mRNA. Not all steps are necessary for all mRNAs.

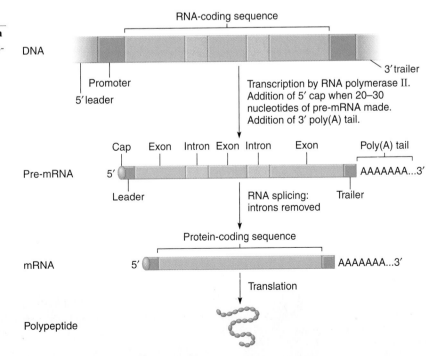

Figure 11.11

Model for intron removal by the spliceosome.

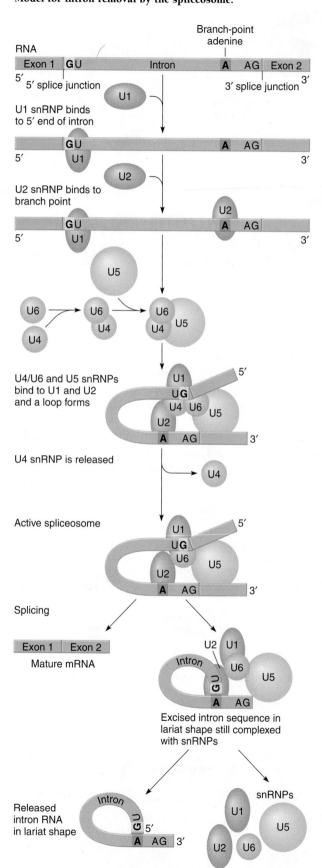

RNA

Exon 1 | GU | Intron | A AG | Exon 2

5′ splice junction
3′ splice junction

U1 snRNP binds to 5′ end of intron

U2 snRNP binds to branch point

U4/U6 and U5 snRNPs bind to U1 and U2 and a loop forms

U4 snRNP is released

Active spliceosome

Splicing

Exon 1 | Exon 2

Mature mRNA

Excised intron sequence in lariat shape still complexed with snRNPs

Released intron RNA in lariat shape

snRNPs

KEYNOTE

Introns are removed from pre-mRNAs in a series of well-defined steps. Intron removal begins with the cleavage of the pre-mRNA at the 5′ splice junction. The free 5′ end of the intron loops back and bonds to a site upstream of the 3′ splice junction. Cleavage at the 3′ splice junction releases the intron, which is shaped like a lariat. The exons that flanked the intron are spliced together once the intron is excised. The removal of introns from eukaryotic pre-mRNA occurs in the nucleus in complexes called spliceosomes, which consist of several snRNPs bound specifically to each intron.

RNA Editing. **RNA editing** involves the posttranscriptional insertion or deletion of nucleotides or the conversion of one base to another. As a result, the functional RNA molecule has a base sequence that does not match the DNA coding sequence.

RNA editing was discovered in trypanosomes, the protozoa that cause sleeping sickness. The phenomenon was shown to affect some mitochondrial mRNAs. The sequences of the mitochondrial *COIII* gene for subunit III of cytochrome oxidase and its mRNA transcripts were compared in the protozoans *Trypanosome brucei (Tb)*, *Crithridia fasiculata (Cf)*, and *Leishmania tarentolae (Lt)* (Figure 11.12). While the mRNA sequences are highly conserved among the three organisms, only the *Cf* and *Lt* mtDNA sequences are colinear with the mRNAs. Strikingly, the *Tb* gene has a sequence that cannot produce the mRNA it apparently encodes. The differences between the DNA and the mRNA can be accounted for by U nucleotides in the mRNA not encoded in the DNA and T nucleotides in the DNA not found in the transcript. The transcript of the *Tb COIII* gene is edited once it is made to add U nucleotides in the appropriate places and remove the U nucleotides encoded by the T nucleotides in the DNA. As Figure 11.12 shows, there are extensive insertions of U nucleotides and deletions of several templated T nucleotides. The magnitude of the changes is even more apparent when the whole sequence is examined: More than 50 percent of the mature mRNA consists of posttranscriptionally added U nucleotides. This RNA editing must be accurate to reconstitute the appropriate sequence for translation into the correct protein. A special RNA molecule, called a *guide RNA (gRNA)*, is involved in the process. The gRNA pairs with the mRNA transcript and is thought to be responsible for cleaving the transcript, templating the missing U nucleotides, and ligating the transcript back together again.

Since the discovery of RNA editing in *T. brucei*, other types of RNA editing have been described. In the slime mold *Physarum polycephalum*, single C nucleotides are added posttranscriptionally at many positions of several mitochondrial mRNA transcripts. In higher plants, the

Figure 11.12

Comparison of the DNA sequences of the cytochrome oxidase subunit III gene (COIII) in the protozoans *Trypanosome brucei (Tb)*, *Crithridia fasiculata (Cf)*, and *Leishmania tarentolae (Lt)*, aligned with the conserved mRNA for *Tb*. The lowercase *us* are the U nucleotides added to the transcript by RNA editing. The template Ts in *Tb* DNA that are not in the RNA transcript are shown in yellow.

Region of *COIII* gene transcript

Tb DNA	G GTTTTTGG AGG G G TTG G GA A GA GAG	
Tb RNA	uuGuGUUUUUGGuuuAGGuuuuuuuGuuG	UUGuuGuuuuGuAuuAuGAuuGAGu
Cf DNA	TTTTTATTTTGATTTCGTTTTTTTTTATG	TGTATTATTTGTGCTTTGATCCGCT
Lt DNA	TTTTTATTTTGATTTCGTTTTTTTTTATG	TGTTTTATTTATGTTATGAGTAGGA
Tb Protein	Leu Cys Phe Trp Phe Arg Phe Phe Cys Cys	Cys Cys Phe Val Leu Trp Leu Ser

sequences of many mitochondrial and chloroplast mRNAs are edited by C-to-U changes. C-to-U editing is also involved in producing an AUG initiation codon from an ACG codon in some chloroplast mRNAs in some higher plants. In mammals, C-to-U editing occurs in the nuclear gene-encoded mRNA for apolipoprotein B; this editing results in tissue-specific generation of a stop codon. Also in mammals, A-to-G editing has been shown to occur in the glutamate receptor mRNA, and pyrimidine editing occurs in a number of tRNAs.

Transcription of Other Genes

In this section, we discuss the transcription of non–protein-coding genes.

Ribosomal RNA and Ribosomes. Protein synthesis takes place on ribosomes. Each cell contains thousands of ribosomes. Ribosomes bind to mRNA and facilitate the binding of the tRNA to the mRNA so that a polypeptide chain can be synthesized.

Ribosome structure. In both prokaryotes and eukaryotes, the ribosomes consist of two unequally sized subunits—the large and small ribosomal subunits—each

of which consists of a complex between RNA molecules and proteins. Each subunit contains one or more ribosomal RNA (rRNA) molecules and a large number of **ribosomal proteins.**

We can use the *E. coli* ribosome as a model of a bacterial ribosome (Figure 11.13). It has a size of 70S, with subunit sizes of 50S (large subunit) and 30S (small subunit). The 50S subunit has 34 different proteins, a 23S rRNA (2,904 nucleotides), and a 5S rRNA (120 nucleotides). The small 30S ribosomal subunit has 20 different proteins and a 16S rRNA (1,542 nucleotides). Transcription of the ribosomal protein genes occurs by the mechanism we have already discussed for protein-coding genes of bacteria.

Eukaryotic ribosomes are larger and more complex than their prokaryotic counterparts, and they vary in size and composition among eukaryotic organisms. We will use the mammalian ribosome as a model for discussion. Mammalian ribosomes have a size of 80S and consist of a large 60S subunit and a small 40S subunit (Figure 11.14). The 40S subunit contains 18S rRNA (~1,900 nucleotides) and about 35 ribosomal proteins, and the 60S subunit contains 28S rRNA (~4,700 nucleotides), 5.8S rRNA (156 nucleotides), and 5S rRNA (120 nucleotides) and about 50 ribosomal proteins. Transcription of the genes that

Figure 11.13

Two views of a model of the complete (70S) ribosome of *E. coli*. The small (30S) ribosomal subunit is shown in yellow, and the large (50S) ribosomal subunit is shown in red.

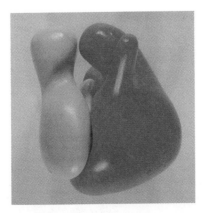

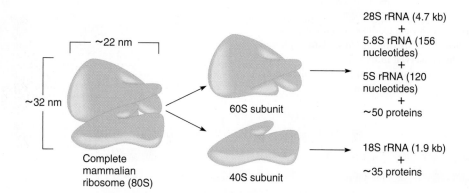

28S rRNA (4.7 kb)
+
5.8S rRNA (156 nucleotides)
+
5S rRNA (120 nucleotides)
+
~50 proteins

18S rRNA (1.9 kb)
+
~35 proteins

Figure 11.14

Composition of whole ribosomes and ribosomal subunits in mammalian cells.

code for ribosomal proteins occurs by the mechanism already discussed for protein-coding genes of eukaryotes.

Transcription of rRNA genes. In prokaryotes and eukaryotes, the regions of DNA that contain the genes for rRNA are called **ribosomal DNA (rDNA)** or *rRNA transcription units*. Figure 11.15a shows the general organiza-

tion of an *E. coli* rRNA transcription unit, named an *rrn* region. Seven *rrn* regions are scattered in the *E. coli* chromosome. Each *rrn* contains one copy each of the 16S, 23S, and 5S rRNA coding sequences arranged in the order 16S-23S-5S. There are one or two tRNA genes in the spacer region between the 16S and 23S rRNA sequences and another one or two tRNA genes in the 3′ spacer

Figure 11.15

rRNA genes and rRNA production in *E. coli*. (a) General organization of an *E. coli* rRNA transcription unit (an *rrn* region). (b) The scheme for the synthesis and processing of a precursor rRNA (p30S) to the mature 16S, 23S, and 5S rRNAs of *E. coli*. (The tRNAs have been omitted from this depiction of the processing scheme.)

a) *E. coli* **rRNA transcription unit**

b) **Processing of pre-rRNA**

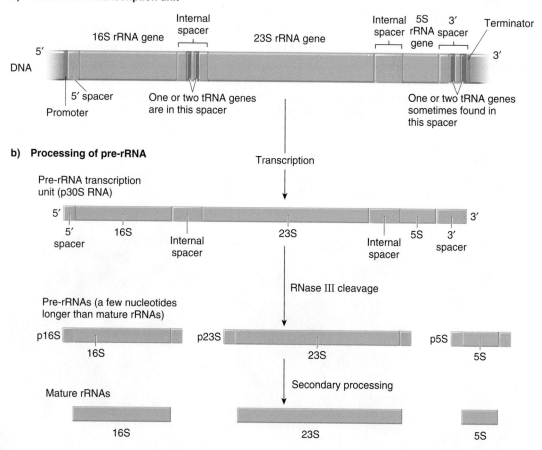

region between the end of the 5S rRNA sequence and the 3′ end of the transcription unit.

The rRNA transcription units are transcribed by RNA polymerase to produce a single **precursor rRNA (pre-rRNA)** molecule, the 30S pre-rRNA (p30S), which contains a 5′ leader sequence, the 16S, 23S, and 5S rRNA sequences (each separated by spacer sequences), and a 3′ trailer sequence (Figure 11.15b). RNase III cleaves p30S to produce the p16S, p23S, and p5S precursors. Other processing enzymes release the tRNAs from the spacers. As the rRNA genes are being transcribed, the pre-rRNA transcript rapidly becomes associated with ribosomal proteins. Cleavage of the transcript takes place within a complex formed between the rRNA transcript (as it is being transcribed) and ribosomal proteins. In this way, the functional ribosomal subunits are assembled.

Most eukaryotes have many copies of the genes for each of the four rRNA species 18S, 5.8S, 28S, and 5S. The genes for 18S, 5.8S, and 28S rRNAs are found adjacent to one another in the order 18S-5.8S-28S, with each set of three genes typically repeated 100 to 1,000 times (depending on the organism) to form tandem arrays called **rDNA repeat units** (Figure 11.16a). Human cells, for example, have 1,250 gene sets. The rDNA repeat units are organized into one or more clusters in the genome, and as a result of active transcription, a nucleolus forms around each cluster. Typically, the multiple nucleoli so formed fuse to form one nucleolus. Within the nucleolus, the 18S, 5.8S, and 28S rRNAs are synthesized, and they associate with 5S rRNA transcribed from multiple 5S rRNA genes located elsewhere in the genome and with ribosomal proteins to produce the ribosomal subunits.

Each eukaryotic rDNA repeat unit is transcribed by RNA polymerase I to produce a pre-rRNA molecule (45S pre-rRNA in human cells) (Figure 11.16b). The pre-rRNA contains the 18S, 5.8S, and 28S rRNA sequences

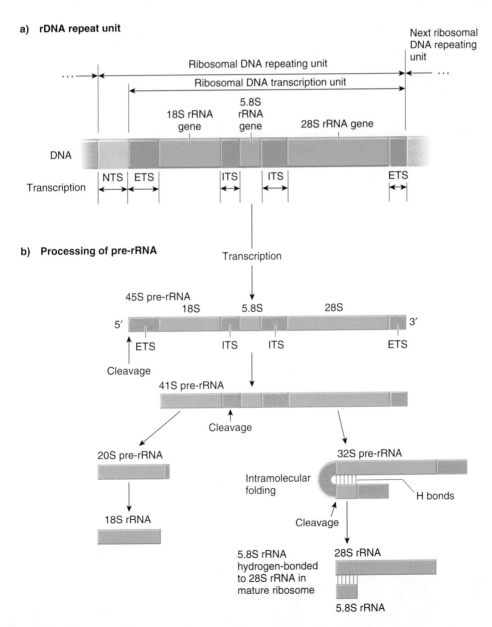

a) rDNA repeat unit

b) Processing of pre-rRNA

Figure 11.16

rRNA genes and rRNA production in eukaryotes. (a) Generalized diagram of a eukaryotic, ribosomal DNA repeat unit. The coding sequences for 18S, 5.8S, and 28S rRNAs are indicated in light brown. NTS = nontranscribed spacer; ETS = external transcribed spacer; ITS = internal transcribed spacer. **(b)** The transcription and processing of 45S pre-rRNA to produce the mature 18S, 5.8S, and 28S rRNAs.

and sequences between and flanking those three sequences: the **spacer sequences.** The external transcribed spacers (ETSs) are located upstream of the 18S sequence and downstream of the 28S sequence. The internal transcribed spacers (ITSs) are located on each side of the 5.8S sequence—that is, between the 18S sequence and the 5.8S sequence and between the 5.8S sequence and the 28S rRNA sequence. Between adjacent rDNA repeat units is a **nontranscribed spacer (NTS) sequence,** which is not transcribed (see Figure 11.16a).

The promoter for RNA polymerase I is upstream of the rRNA genes. Like other eukaryotic RNA polymerases, RNA polymerase I does not bind directly to the promoter; instead, specific transcription factors bind and form a complex to which RNA polymerase I binds to begin transcription. Termination of transcription of the pre-rRNA involves specific termination sites located downstream of the transcribed sequences of the rDNA.

To produce the 18S, 5.8S, and 28S rRNAs, the transcribed pre-rRNA is cleaved at specific sites to remove ITS and ETS sequences. As an example, Figure 11.16b shows how pre-rRNA is processed in human cells. First, the 5′ ETS sequence is removed, producing a precursor molecule containing all three rRNA sequences. The next cleavage produces the 20S precursor to 18S rRNA, and the 32S precursor is processed to 28S and 5.8S rRNAs. Removal of the ITS sequence generates the 18S rRNA from the 20S precursor.

All the pre-rRNA-processing events take place in complexes formed between the pre-rRNA, 5S rRNA, and ribosomal proteins. The 5S rRNA is produced by transcription of the 5S rRNA genes by RNA polymerase III (described later), and the ribosomal proteins are produced by transcription of the ribosomal protein genes by RNA polymerase II and the subsequent translation of the mRNAs. As pre-rRNA processing proceeds, the complexes undergo shape changes resulting in the formation of the 60S and 40S ribosomal subunits. The ribosomal subunits are transported to the cytoplasm, where they function in protein synthesis.

K E Y N O T E

Ribosomes consist of two unequally sized subunits. Each subunit contains one or more ribosomal RNA molecules and ribosomal proteins. The three prokaryotic rRNAs and three of the four eukaryotic rRNAs are encoded in rRNA transcription units. The fourth eukaryotic rRNA is encoded by separate genes. Transcription of rRNA transcription units by RNA polymerase produces pre-rRNA molecules that are processed to the mature rRNAs by removal of spacer sequences. The processing events occur in complexes of the pre-rRNAs with ribosomal proteins and other proteins and are part of the formation of the mature ribosomal subunits.

Self-Splicing of Introns in* Tetrahymena *Pre-rRNA. In some species of the protozoan *Tetrahymena,* the genes for the 28S rRNA are all interrupted by a 413-bp intron. Excision of this intron occurs in a particularly interesting way. The intron sequence is removed during processing of the pre-rRNA to produce the mature rRNAs. The excision of the intron—called a group I intron—unexpectedly was shown to occur by a *protein-independent reaction* in which the RNA intron folds into a secondary structure that promotes its own excision. This process is called **self-splicing** and was discovered in 1982 by Tom Cech and his research group. In 1989, Cech shared the Nobel Prize in Chemistry for his discovery. The self-splicing of the *Tetrahymena* pre-rRNA intron was the first example of what is now called **group I intron self-splicing.**

Figure 11.17 diagrams the self-splicing reaction for this group I intron in *Tetrahymena* pre-rRNA. The steps are as follows:

1. The pre-rRNA is cleaved at the 5′ splice junction as guanosine is added to the 5′ end of the intron.
2. The intron is cleaved at the 3′ splice junction.
3. The two exons are spliced together.
4. The excised intron circularizes to produce a lariat molecule, which is cleaved to produce a circular RNA and a short linear piece of RNA.

In the processing of *Tetrahymena* pre-rRNA, then, two separate events occur: (1) The pre-rRNA is cleaved to remove spacer sequences from the mature rRNAs, and (2) the intron in the 28S rRNA sequence is removed and the two 28S rRNA parts spliced together. The two events are not identical. *Removal of the spacer sequences releases rRNAs that remain separate. Intron removal, by contrast, results in the splicing together of the RNA sequences that flanked the intron.*

The self-splicing activity of the intron RNA sequence cannot be considered an enzyme activity. That is, while the RNA carries out the reaction, it is not regenerated in its original form at the end of the reaction, as is the case with protein enzymes. Modified forms of the *Tetrahymena* intron RNA and of other self-cleaving RNAs that function catalytically have been produced in the lab. These **RNA enzymes** are called **ribozymes;** they can be used experimentally to cleave RNA molecules at specific sequences.

The discovery that RNA can act like a protein was an important landmark in biology. The discovery has revolutionized theories about the origin of life. Previous theories proposed that proteins were required for replication of the first nucleic acid molecules. The new theories propose that the first nucleic acid was self-replicating through ribozyme-like activity of the molecule.

Figure 11.17

Self-splicing reaction for the group I intron in *Tetrahymena* pre-rRNA.

Tetrahymena **pre-rRNA**

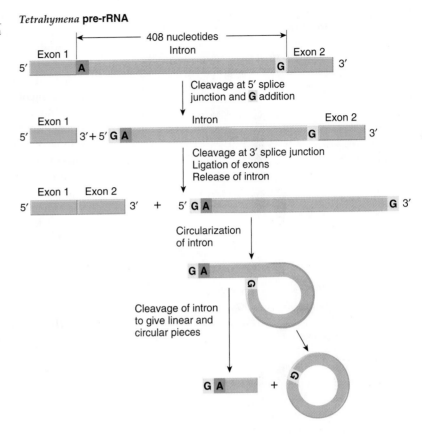

KEYNOTE

In some precursor rRNAs, there are introns, the RNA sequences of which fold into a secondary structure that excises itself, a process called self-splicing. The self-splicing reaction does not involve any proteins.

Transcription of Genes by RNA Polymerase III. RNA polymerase III transcribes eukaryotic 5S rRNA genes, tRNA genes, and some snRNA genes. 5S rRNA is a 120-nucleotide RNA found in the large subunit of the eukaryotic ribosome. tRNAs are 75- to 90-nucleotide RNAs that function in both prokaryotes and eukaryotes to bring amino acids to the ribosome-mRNA complex, where they are polymerized into protein chains in the translation (protein synthesis) process.

Prokaryotic 5S rRNA genes are part of the rDNA, whereas eukaryotic 5S rRNA genes are found in multiple copies in the genome, usually separate from the rDNA repeat units. Prokaryotic tRNA genes are found in one or at most a few copies in the genome, while eukaryotic tRNA genes are repeated many times in the genome. In the South African clawed toad, *Xenopus,* for example, there are about 200 copies of each tRNA gene. Each type of tRNA molecule has a different nucleotide sequence, although all tRNAs have the sequence CCA at their 3′ ends. This CCA sequence is added to the tRNA posttranscriptionally. The

differences in nucleotide sequences explain the ability of a particular tRNA molecule to bind a particular amino acid. All tRNA molecules are also extensively modified chemically by enzyme reactions after transcription.

The nucleotide sequences of all tRNAs can be arranged into what is called a *cloverleaf* (Figure 11.18). The cloverleaf results from complementary base pairing between different sections of the molecule, which results in four base-paired "stems" separated by four loops, I, II, III, and IV. Loop II contains the three-nucleotide **anticodon** sequence, which pairs with a codon (three-nucleotide sequence) in mRNA by complementary base pairing during translation. This codon-anticodon pairing is crucial for adding the amino acid specified by the mRNA to the growing polypeptide chain. Figure 11.19 shows the tertiary structure for tRNAs (the phenylalanine tRNA from yeast is shown). The cloverleaf is folded into a compact shape, like an upside-down L. In this L-shaped structure, the 3′ end of the tRNA—the end to which the amino acid attaches—is at the opposite end of the L from the anticodon loop.

How does RNA polymerase III transcribe the genes under its control? For 5S rRNA genes (5S rDNA) and tRNA genes (tDNA), the promoter for RNA polymerase III is *within the transcribed region.* This internal promoter is called the **internal control region (ICR).** snRNA genes transcribed by RNA polymerase III typically have promoters upstream of the genes.

Figure 11.18

Cloverleaf structure of yeast alanine tRNA. Py = pyrimidine. Modified bases: I = inosine; T = ribothymidine; ψ = pseudouridine; D = dihydrouridine; GMe = methylguanosine; GMe₂ = dimethylguanosine; IMe = methylinosine.

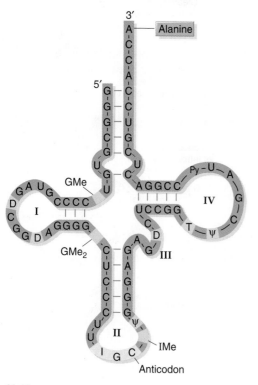

Like transcription initiation by RNA polymerases I and II, transcription initiation of 5S rDNA and tDNA involves the binding of transcription factors to the promoter, in this case TFIIIs to the ICR, to facilitate the binding of RNA polymerase III. The enzyme then initiates transcription at the beginning of the transcribed region of the gene.

Transcription of a 5S rRNA gene directly produces the mature 5S rRNA; there are no sequences that must be removed. Transcription of tRNA genes produces a **precursor tRNA (pre-tRNA)**, which has extra sequences at each end that are subsequently removed to produce the mature tRNA. In yeast, some tRNAs contain an intron in the anticodon loop; a tRNA-specific splicing mechanism is used to cut out the intron and to splice the adjacent tRNA segments together.

K E Y N O T E

RNA polymerase III transcribes 5S rRNA genes (5S rDNA), tRNA genes (tDNA), and some snRNA genes. For 5S rDNA and tDNA, the promoter for RNA polymerase III is located within the gene itself. The internal promoter is called the internal control region (ICR). Transcription factors and RNA polymerase III bind to the ICR for transcription of the genes.

Figure 11.19

tRNA structures. (a) Schematic of the three-dimensional structure of yeast phenylalanine tRNA as determined by X-ray diffraction of tRNA crystals. Note the characteristic L-shaped structure. **(b)** Photograph of a space-filling molecular model of yeast phenylalanine tRNA. The CCA end of the molecule is at the upper right, and the anticodon loop is at the bottom.

a)

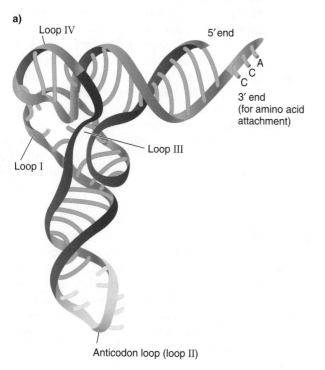

b)

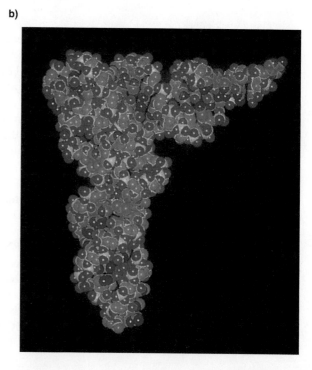

Summary

Transcription

When a gene is expressed, the DNA base pair sequence is transcribed into the base sequence of an RNA molecule. Transcription of four classes of genes produces messenger RNA (mRNA), transfer RNA (tRNA), ribosomal RNA (rRNA), and small nuclear RNA (snRNA). snRNA is found only in eukaryotes, and the other three classes are found in both prokaryotes and eukaryotes. Only mRNA is translated to produce a protein molecule.

When a gene is transcribed by RNA polymerase, only one of the two DNA strands is copied. The direction of RNA synthesis is 5′→3′. In addition to coding sequences, genes contain other sequences important for the regulation of transcription, including promoter sequences and terminator sequences. Promoter sequences specify where transcription of the gene is to begin, and terminator sequences specify where transcription is to stop.

Only one type of RNA polymerase is found in bacteria, so all classes of RNA are synthesized by the same enzyme. Consequently, the promoters for all three classes of genes are very similar. The promoter is recognized by a complex between the RNA polymerase core enzyme and a protein factor called sigma. Once transcription is initiated correctly, the sigma factor dissociates from the enzyme and is reused in other transcription initiation events. Termination is signaled by one of two possible termination sequences.

In eukaryotes, three different RNA polymerases are located in the nucleus. RNA polymerase I, in the nucleolus, transcribes the 18S, 5.8S, and 28S rRNA sequences. These rRNAs are part of ribosomes. RNA polymerase II, in the nucleoplasm, transcribes protein-coding genes (into mRNA precursors) and some snRNA genes. RNA polymerase III, in the nucleoplasm, transcribes tRNA genes, 5S rRNA genes, and the other snRNA genes. The promoters for the three types of RNA polymerase differ, but in each case, none of the three eukaryotic RNA polymerases directly binds to the promoter. Instead, the promoter sequences for the genes they transcribe are first recognized by transcription factors. The transcription factors bind to the DNA and facilitate binding of the polymerase to the transcription factor–DNA complex for correct initiation of transcription. For 5S rRNA genes and tRNA genes transcribed by RNA polymerase III, the promoter is located within the gene itself.

Although transcription is very similar in prokaryotes and eukaryotes, the molecular components of the process are very different. Even within eukaryotes, three different promoters have evolved along with three distinct RNA polymerases. Much remains to be learned about the associated transcription factors and regulatory proteins before we have a complete understanding of transcription and its regulation.

RNA Molecules and RNA Processing

We discussed the structure, synthesis, and function of mRNA, tRNA, and rRNA. Each mRNA encodes the amino acid sequence of a polypeptide chain. The nucleotide sequence of the mRNA is translated into the amino acid sequence of the polypeptide chain. Ribosomes consist of two unequal-sized subunits, each of which contains both rRNA and protein molecules. The amino acids that are assembled into proteins are brought to the ribosome bound to tRNA molecules.

mRNAs have three main parts: a 5′ untranslated leader sequence, the amino acid–coding sequence, and the 3′ untranslated trailer sequence. In prokaryotes, the gene transcript functions directly as the mRNA molecule, whereas in eukaryotes, the RNA transcript must be modified in the nucleus to produce mature mRNA. Modifications include the addition of a 5′ cap and a 3′ poly(A) tail and the removal of any introns. Intron removal is known as RNA splicing and involves specific interactions with snRNPs in structures called spliceosomes. Only when all processing events have been completed is the mRNA functional; at that point, it can be translated in the cytoplasm once it leaves the nucleus.

Ribosomal RNAs are important structural components of ribosomes. The smaller subunit contains one rRNA molecule, 16S in prokaryotes and 18S in eukaryotes. The prokaryotic larger subunit contains 23S and 5S rRNAs, and the eukaryotic larger subunit contains 28S, 5.8S, and 5S rRNAs.

In prokaryotes, the genes for the 16S, 23S, and 5S rRNAs are transcribed into a single pre-rRNA molecule. Some tRNA genes are found in the spacer regions of each transcription unit. The pre-rRNA molecules are processed to remove the noncoding sequences found at the ends of the molecules and spacers between the rRNA sequences. At the same time, the tRNAs are released. All the processing events occur while the pre-rRNA is associating with the ribosomal proteins, so that when processing is complete, the functional 50S and 30S subunits have been assembled.

In eukaryotes, the 18S, 5.8S, and 28S rRNA genes constitute a transcription unit, and many such transcription units are organized in a tandem array. The 5S rRNA genes are also present in many copies but are usually located elsewhere in the genome. The 18S, 5.8S, and 28S sequences are transcribed into pre-rRNA molecules that, in addition to the rRNA sequences, contain spacer sequences at the ends and between the rRNA sequences.

Transcription of tandem arrays of these rRNA genes results in the formation of a nucleolus around each cluster. The 60S and 40S ribosomal subunits are assembled in the nucleolus. That is, the spacers are removed by specific processing events that take place while the rRNA sequences are associated with ribosomal proteins and with 5S rRNA, which is transcribed elsewhere in the nucleus.

The completed ribosomal subunits exit the nucleus and participate in protein synthesis in the cytoplasm.

In the pre-rRNA of *Tetrahymena*, the 28S rRNA sequence is interrupted by an intron. The intron is removed during processing of the pre-rRNA in the nucleolus. The excision of this intron occurs by a protein-independent reaction in which the RNA sequence of the intron folds into a secondary structure that promotes its own excision. This process is called self-splicing.

All tRNAs bring amino acids to the ribosomes. Thus, all tRNAs are very similar in length (75 to 90 nucleotides) and in secondary and tertiary structure. Introns are present in some tRNA genes of yeast. Removal of these introns involves a different mechanism from the other RNA splicing mechanisms described.

In some organisms, RNA editing occurs to insert or delete nucleotides or convert one base to another in an RNA posttranscriptionally. As a result, the functional RNA molecule has a base sequence that does not match the DNA coding sequence. Many RNAs that are edited are encoded by the mitochondrial and chloroplast genomes.

Analytical Approaches for Solving Genetics Problems

Q11.1 If two RNA molecules have complementary base sequences, they can hybridize to form a double-stranded helical structure just as DNA can. Imagine that in a particular region of the genome of a certain bacterium, one DNA strand is transcribed to give rise to the mRNA for protein A, and the other DNA strand is transcribed to give rise to the mRNA for protein B.
a. Would there be any problem in expressing these genes?
b. What would you see in protein B if a mutation occurred that affected the structure of protein A?

A11.1
a. mRNA *A* and mRNA *B* would have complementary sequences, so they might hybridize with each other and not be available for translation.
b. Every mutation in gene *A* would also be a mutation in gene *B*, so protein B might also be abnormal.

Q11.2 Compare the following two events in terms of what their consequences would be. Event 1: An incorrect nucleotide is inserted into the new DNA strand during replication and not corrected by the proofreading or repair systems before the next replication. Event 2: An incorrect nucleotide is inserted into an mRNA during transcription.

A11.2 Event 1 would result in a mutation, assuming it occurred within a gene. The mistake would be inherited by future generations and would affect the structure of all mRNA molecules transcribed from the region; therefore, all molecules of the corresponding protein could be affected.

Event 2 would produce a single aberrant mRNA. This could produce a few aberrant protein molecules. Additional normal protein molecules would exist because other, normal mRNAs would have been transcribed. The abnormal mRNA would soon be degraded. The mRNA mistake would not be hereditary.

Questions and Problems

11.1 Compare DNA and RNA with regard to structure, function, location, and activity. Also, how do these molecules differ with regard to the polymerases used to synthesize them?

11.2 All base pairs in the genome are replicated during the DNA synthesis phase of the cell cycle, but only some of the base pairs are transcribed into RNA. How is it determined which base pairs of the genome are transcribed into RNA?

*11.3 Discuss the similarities and differences between the *E. coli* RNA polymerase and eukaryotic RNA polymerases.

11.4 What are the most significant differences between the organization and expression of prokaryotic genes and eukaryotic genes?

11.5 Discuss the molecular events involved in the termination of RNA transcription in prokaryotes. In what ways is this process fundamentally different in eukaryotes?

*11.6 Three different RNA polymerases are found in all eukaryotic cells, and each is responsible for synthesizing a different class of RNA molecules. How do the characteristics of the eukaryotic RNA polymerases differ in terms of their cellular location and products?

11.7 *E. coli* RNA polymerase can transcribe all the genes of *E. coli,* but the three eukaryotic RNA polymerases transcribe only specific, nonoverlapping subsets of eukaryotic genes. What mechanisms are used to restrict transcribing each of the three eukaryotic polymerases to a particular subset of eukaryotic genes?

11.8 More than 100 promoters in prokaryotes have been sequenced. One element of these promoters is sometimes called the Pribnow box, named after the investigator who compared several *E. coli* and phage promoters and discovered a region they held in common. Discuss the nature of this sequence (where is it located and why is it important?). Another consensus sequence appears a short distance from the Pribnow box. Diagram the positions of the two prokaryotic promoter elements relative to the start of transcription for a typical *E. coli* promoter.

*11.9 Eukaryotic promoters have two major types of regulatory regions, one of which consists of nucleotide sequences that serve as the recognition point for RNA polymerase binding. They function somewhat similarly in both eukaryotes and prokaryotes. What do you know about these sequences with regard to location, base makeup, and organization? A diagram with labels and a short explanation would suffice.

11.10 In addition to promoter regions, transcription of most eukaryotic genes is regulated by other DNA sequences called enhancers. How do these sequences differ from promoters?

11.11 How do the structures of mRNA, rRNA, and tRNA differ? Hypothesize a reason for the difference.

11.12 Many eukaryotic mRNAs, but not prokaryotic mRNAs, contain introns. Describe how these sequences are removed during the production of mature mRNA.

11.13 Distinguish between leader sequence, trailer sequence, coding sequence, intron, spacer sequence, nontranscribed spacer sequence, external transcribed spacer sequence, and internal transcribed sequence. Give examples of actual molecules in your answer.

*11.14 Discuss the posttranscriptional modifications and processing events that take place on the primary transcripts of eukaryotic rRNA and protein-coding genes.

11.15 Describe the organization of the ribosomal DNA repeating unit of a higher eukaryotic cell.

*11.16 Which of the following kinds of mutations would likely be recessive lethals in humans? Explain your reasoning.
a. deletion of the U1 genes
b. deletion within intron 2 of β-globin
c. deletion of four bases at the end of intron 2 and three bases at the beginning of exon 3 in β-globin

11.17 The figure below shows the transcribed region of a typical eukaryotic protein-coding gene. What is the size (in bases) of the fully processed, mature mRNA? Assume in your calculations a poly(A) tail of 200 As.

*11.18 Most human obesity does not follow Mendelian inheritance patterns because body fat content is determined by a number of interacting genes and environmental variables. Insights into how specific genes function to regulate body fat content have come from studies of mutant, obese mice. In one mutant strain, *tubby (tub)*, obesity is inherited as a recessive trait. Comparison of the DNA sequence of the *tub*⁺ and *tub* alleles has revealed a single base pair change: Within the transcribed region, a 5′ GC base pair has been mutated to a TA base pair. The mutation causes an alteration of the initial 5′ base of the first intron. Therefore, in the homozygous *tub/tub* mutant, a longer transcript is found. Propose a molecularly based explanation for how a single base change causes a nonfunctional gene product to be produced, why a longer transcript is found in *tub/tub* mutants, and why the *tub* mutant is recessive.

*11.19 Which of the following could occur in a single mutational event in a human? Explain.
a. deletion of 10 copies of the 5S ribosomal RNA genes only
b. deletion of 10 copies of the 18S rRNA genes only
c. simultaneous deletion of 10 copies of the 18S, 5.8S, and 28S rRNA genes only
d. simultaneous deletion of 10 copies each of the 18S, 5.8S, 28S, and 5S rRNA genes

11.20 During DNA replication in a mammalian cell, a mistake occurs: 10 wrong nucleotides are inserted into a 28S rRNA gene. This mistake is not corrected. What will probably be the effect on the cell?

*11.21 Choose the correct answers, noting that each blank may have more than one correct answer and that each answer (1–4) may be used more than once.
Answers:
1. eukaryotic mRNAs
2. prokaryotic mRNAs
3. transfer RNAs
4. ribosomal RNAs

a. _____ have a cloverleaf structure
b. _____ are synthesized by RNA polymerases
c. _____ display an anticodon each
d. _____ are the template of genetic information during protein synthesis
e. _____ contain exons and introns
f. _____ are of four types in eukaryotes and only three types in *E. coli*
g. _____ are capped on their 5′ end and polyadenylated on their 3′ end

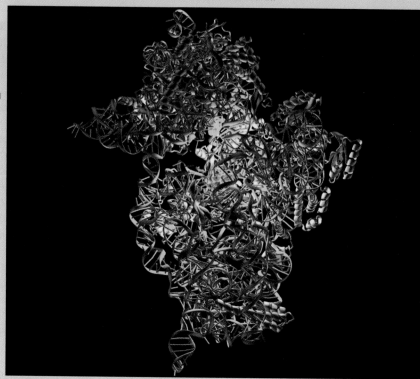

12

Gene Expression: Translation

PRINCIPAL POINTS

A protein consists of one or more subunits called polypeptides, which are composed of smaller building blocks called amino acids. The amino acids are linked together in the polypeptide by peptide bonds.

The amino acid sequence of a protein (its primary structure) determines its secondary, tertiary, and quaternary structures and hence its functional state.

The genetic code is a triplet code in which each three-nucleotide codon in an mRNA specifies one amino acid. Some amino acids are represented by more than one codon. The code is almost universal, and it is read without gaps in successive, nonoverlapping codons.

The mRNA is translated into a polypeptide chain on ribosomes. Amino acids are brought to the ribosome on tRNA molecules. The correct amino acid sequence is achieved by specific binding of each amino acid to its specific tRNA and by specific binding between the codon of the mRNA and the complementary anticodon of the tRNA.

In prokaryotes and eukaryotes, AUG (methionine) is the initiator codon for the start of translation. Elongation of the protein chain involves peptide bond formation between the amino acid on the tRNA in the A site of the ribosome and the growing polypeptide on the tRNA in the adjacent P site. Once the peptide bond has formed, the ribosome translocates one codon along the mRNA in preparation for the next tRNA. The incoming tRNA with its amino acid binds to the next codon occupying the A site.

Translation continues until a chain-terminating codon (UAG, UAA, or UGA) is reached in the mRNA. These codons are read by release factor proteins; then the polypeptide is released from the ribosome, and the other components of the protein synthesis machinery dissociate.

In eukaryotes, proteins are found free in the cytoplasm and in various cell compartments, such as the nucleus, mitochondria, chloroplasts, and secretory vesicles. Mechanisms exist to sort proteins to their appropriate cell compartments. For example, proteins to be secreted have N-terminal signal sequences that facilitate their entry into the endoplasmic reticulum for later sorting in the Golgi apparatus and beyond.

iActivity

i CHANGING A SINGLE LETTER IN A WORD CAN COMpletely change the meaning of the word. This, in turn, can change the meaning of the sentence containing that word. In living organisms, a sequence of three nucleotide "letters" produces an amino acid "word." The amino acids are strung together to form polypeptide "sentences." In this chapter, you will study the process by which nucleotide "letters" are translated into polypeptide "sentences."

One of the most important applications of human genome research is using sequence information to track down the causes of genetic diseases. In the iActivity for this chapter, you will investigate part of the gene responsible for cystic fibrosis, the most common fatal genetic disease in the United States, and try to identify possible causes of the disease.

The information for the proteins found in a cell is encoded in the structural genes of the cell's genome. Expression of a protein-coding gene occurs by transcription of the gene to produce an mRNA (discussed in Chapter 11), followed by **translation** of the mRNA: the conversion of the mRNA base sequence information into the amino acid sequence of a polypeptide. The nucleotide information that specifies the amino acid sequence of a polypeptide is called the **genetic code.** In this chapter, your goal is to learn how the nucleotide sequence of mRNA is translated into the amino acid sequence of a polypeptide.

Proteins

Chemical Structure of Proteins

A **protein** is a high-molecular-weight, nitrogen-containing organic compound of complex shape and composition.

Each cell type has a characteristic set of proteins that gives it its functional properties. A protein consists of one or more macromolecular subunits called **polypeptides,** which are composed of smaller building blocks, the **amino acids,** linked together in specific sequences in linear, unbranched chains. The sequence of amino acids gives the polypeptide its three-dimensional shape and its properties in the cell.

With the exception of proline, the amino acids have a common structure, shown in Figure 12.1. The structure consists of a central carbon atom (α-carbon) to which is bonded an amino group (NH$_2$), a carboxyl group (COOH), and a hydrogen atom. The figure shows an amino acid at the pH commonly found within cells, where the NH$_2$ and COOH groups are in a charged state, $-NH3^+$ and $-COO^-$, respectively. Also bound to the α-carbon is the *R group*. The R group varies from one amino acid to another and gives each amino acid its distinctive properties. Since different polypeptides have different sequences and proportions of amino acids, the arrangement and nature of the R groups gives a polypeptide its structural and functional properties.

Twenty amino acids are used to make proteins in living cells; their names, three-letter and one-letter abbreviations, and chemical structures are shown in Figure 12.2. The 20 amino acids are divided into subgroups based on whether the R group is acidic, basic, neutral and polar, or neutral and nonpolar.

Amino acids of a polypeptide are joined by a **peptide bond,** a covalent bond formed between the carboxyl group of one amino acid and the amino group of an adjacent amino acid (Figure 12.3). A polypeptide, then, is an unbranched molecule that consists of a sequence of many amino acids joined by peptide bonds. Every polypeptide has a free amino group at one end (called the N terminus, or the N-terminal end) and a free carboxyl group at the other end (called the C terminus, or the C-terminal end). The N-terminal end is defined as the beginning of a polypeptide chain.

Figure 12.1

General structural formula for an amino acid.

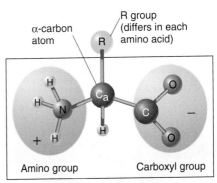

Structures common to all amino acids

Figure 12.2

Structures of the 20 naturally occurring amino acids organized according to chemical type. Below each amino acid name are its three-letter and one-letter abbreviations.

Acidic

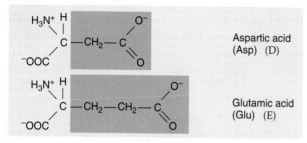

Aspartic acid
(Asp) (D)

Glutamic acid
(Glu) (E)

Neutral, nonpolar

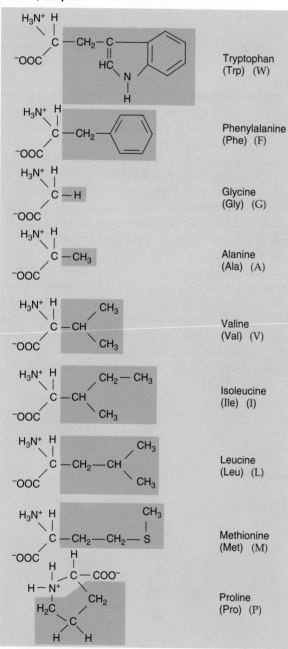

Tryptophan
(Trp) (W)

Phenylalanine
(Phe) (F)

Glycine
(Gly) (G)

Alanine
(Ala) (A)

Valine
(Val) (V)

Isoleucine
(Ile) (I)

Leucine
(Leu) (L)

Methionine
(Met) (M)

Proline
(Pro) (P)

Basic

Lysine
(Lys) (K)

Arginine
(Arg) (R)

Histidine
(His) (H)

Neutral, polar

Tyrosine
(Tyr) (Y)

Serine
(Ser) (S)

Threonine
(Thr) (T)

Asparagine
(Asn) (N)

Glutamine
(Gln) (Q)

Cysteine
(Cys) (C)

Figure 12.3

Mechanism for peptide bond formation between the carboxyl group of one amino acid and the amino group of another amino acid.

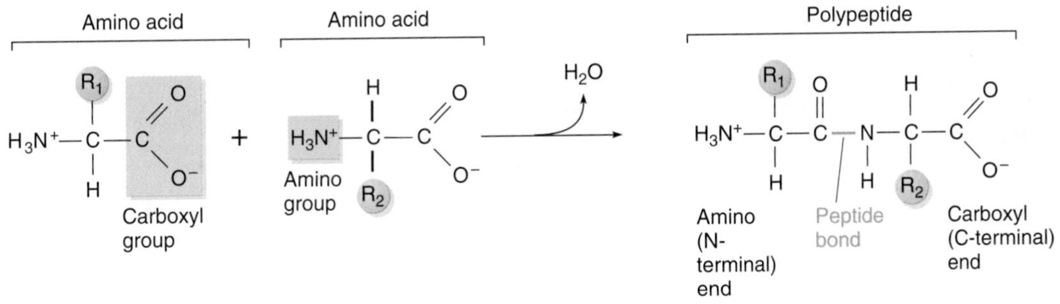

Molecular Structure of Proteins

Proteins can have four levels of structural organization, as shown in Figure 12.4:

1. The *primary structure* of a polypeptide chain is the amino acid sequence (Figure 12.4a). The amino acid sequence is directly determined by the base pair sequence of the gene that encodes it.

2. The *secondary structure* of a protein is the regular folding and twisting of a single polypeptide chain into a variety of shapes. A polypeptide's secondary structure is the result of weak bonds, such as electrostatic or hydrogen bonds, between NH and CO groups of amino acids that are near each other on the chain. One type of secondary structure found in regions of many polypeptides is the *α-helix* (Figure 12.4b).

   Another type of secondary structure is the β-pleated sheet (not illustrated). The β-pleated sheet involves a polypeptide chain (or chains) folded in a zigzag way with parallel regions or chains linked by hydrogen bonds. Many proteins contain a mixture of α-helical and β-pleated sheet regions.

3. A protein's *tertiary structure* (Figure 12.4c) is the three-dimensional structure of a single polypeptide chain. The three-dimensional shape of a polypeptide often is called its *conformation*. Tertiary folding is a direct property of the amino acid sequence of the chain and thus is related to the distribution of the R groups along the chain. Figure 12.4c shows the tertiary structure of the β polypeptide of hemoglobin. (The 1962 Nobel Prize in Chemistry was awarded to Max Perutz and Sir John Kendrew for their studies of the structures of globular proteins, and the 1972 Nobel Prize in Chemistry was awarded to Christian Anfinsen for his work on ribonuclease, especially concerning the connection between the amino acid sequence and the biologically active conformation.)

4. The *quaternary structure* is the complex of polypeptide chains in a multisubunit protein, so quaternary structure is found only in proteins having more than one polypeptide chain (Figure 12.4d). Shown in Figure 12.4d is the quaternary structure of a multimeric (many-subunit) protein, the oxygen-carrying protein hemoglobin, which consists of four polypeptide chains (two 141-amino-acid α polypeptides and two 146-amino-acid β polypeptides), each of which is associated with a heme group involved in the binding of oxygen. In the quaternary structure of hemoglobin, each α chain is in contact with each β chain, but there is little interaction between the two α chains or between the two β chains.

For many years, it was thought that the amino acid sequence alone was sufficient to specify how a protein folds into its functional state. We now know that for many proteins, folding depends on one or more of a family of proteins called *chaperones* (also called *molecular chaperones*). Chaperones act analogously to enzymes in that they interact with the proteins they help fold but do not become part of the functional protein produced. A detailed discussion of chaperones is beyond the scope of this book.

KEYNOTE

A protein consists of one or more molecular subunits called polypeptides, which are themselves composed of smaller building blocks, the amino acids, linked together by peptide bonds to form long chains. The primary amino acid sequence of a protein determines its secondary, tertiary, and quaternary structure and hence its functional state.

The Nature of the Genetic Code

How do nucleotides in the mRNA molecule specify the amino acid sequence in proteins? With four different nucleotides (A, C, G, U), a three-letter code generates 64 possible codons. If it were a one-letter code, only four amino acids could be encoded. If it were a two-letter

Figure 12.4

Four levels of protein structure. (a) Primary, the sequence of amino acids in a polypeptide chain. (b) Secondary, the folding and twisting of a single polypeptide chain into a variety of shapes. Shown is one type of secondary structure: the α-helix. The structure is stabilized by hydrogen bonds. (c) Tertiary, the specific three-dimensional folding of the polypeptide chain. Shown here is the β polypeptide chain of hemoglobin, a heme-containing polypeptide that carries oxygen in the blood. (d) Quaternary, the aggregate of polypeptide chains. Shown here is hemoglobin; it consists of two α chains, two β chains, and four heme groups.

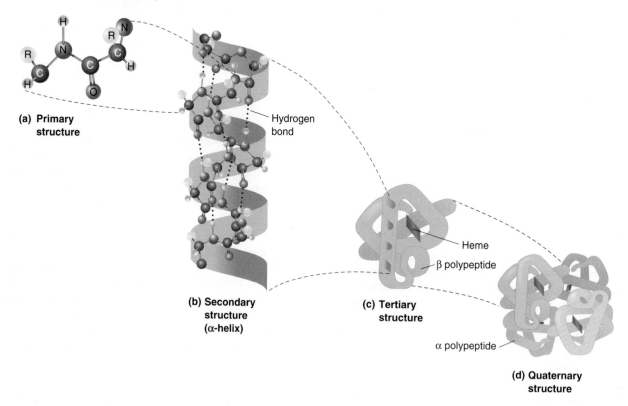

(a) Primary structure

Hydrogen bond

(b) Secondary structure (α-helix)

Heme

β polypeptide

(c) Tertiary structure

α polypeptide

(d) Quaternary structure

code, then only 16 (4 × 4) amino acids could be encoded. A three-letter code, however, generates 64 (4 × 4 × 4) possible codes, more than enough to code for the 20 amino acids found in living cells. Since there are only 20 different amino acids, a three-letter code suggests that some amino acids may be specified by more than one codon, which is in fact the case.

The Genetic Code Is a Triplet Code

The evidence that the genetic code is a triplet code—that is, that a set of three nucleotides (a codon) in mRNA code for one amino acid in a polypeptide chain—came from genetic experiments done by Francis Crick, Leslie Barnett, Sidney Brenner, and R. Watts-Tobin in the early 1960s. They studied mutations of a bacteriophage T4 gene that had additions or deletions of one or more base pairs. The consequence of a base pair addition or deletion in a gene is that the frame of reading during translation is shifted; hence, the mutations are called *frameshift mutations*. So when an mRNA transcribed from a gene with a frameshift is translated, the

amino acid sequence of the polypeptide made is incorrect after the point of the addition or deletion. Figure 12.5 illustrates this for a deletion in a hypothetical segment of DNA. For the purposes of discussion, we have assumed that the code is a triplet code. Thus, the mRNA transcript of the DNA would be read ACG ACG ACG, and so on, giving a polypeptide with a string of identical amino acids—threonine—each specified by ACG (Figure 12.5a). If the second AT base pair is deleted, the mRNA will now read ACG CGA CGA, and so on, giving a polypeptide starting with the amino acid specified by ACG (threonine) followed by a string of incorrect amino acids that are specified by CGA (arginine) (Figure 12.5b). That is, the reading frame of the message has been changed; the message is out of frame by one.

Crick and his colleagues combined more than one addition mutation and, in separate experiments, more than one deletion mutation to see how many additions or deletions were necessary to restore the reading frame. The answer in each case was three, indicating that the genetic code is a triplet code. Figure 12.6 is a hypothetical pre-

Figure 12.5

Frameshift mutation caused by deletion of a DNA base pair.
(a) Hypothetical segment of normal DNA, mRNA transcript, and polypeptide in the wild type. (b) Effect of a deletion mutation—here an AT base pair—on the amino acid sequence of a polypeptide. The reading frame is disrupted; all amino acids past the deletion are incorrect.

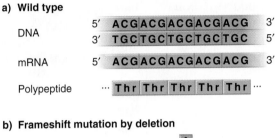

a) Wild type

DNA
5′ ACGACGACGACGACG 3′
3′ TGCTGCTGCTGCTGC 5′

mRNA
5′ ACGACGACGACGACG 3′

Polypeptide
··· Thr Thr Thr Thr Thr ···

b) Frameshift mutation by deletion

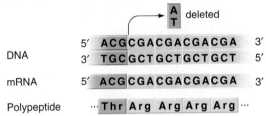

A
T deleted

DNA
5′ ACGCGACGACGACGA 3′
3′ TGCGCTGCTGCTGCT 5′

mRNA
5′ ACGCGACGACGACGA 3′

Polypeptide
··· Thr Arg Arg Arg Arg ···

sentation of the type of results they obtained, showing the effects of mutations just on the mRNA. The figure shows a 30-nucleotide segment of mRNA that codes for 10 different amino acids in the polypeptide. If we add three base pairs at nearby locations in the DNA coding for this mRNA segment, the result will be a 33-nucleotide segment that codes for 11 amino acids, one more than the original. However, the amino acids between the first and third insertions are not the same as in the wild-type mRNA. In essence, the reading frame is correct before the first insertion and again after the third insertion.

Deciphering the Genetic Code

The exact relationship of the 64 codons to the 20 amino acids was determined by experiments done mostly in the laboratories of Marshall Nirenberg and Ghobind Khorana. Nirenberg and Khorana shared the 1968 Nobel Prize in Physiology or Medicine with Robert Holley. Essential to these experiments was the use of **cell-free, protein-synthesizing systems** with components isolated and purified from *E. coli*. These systems contain ribosomes, tRNAs with amino acids attached, and all the necessary protein factors for polypeptide synthesis. Radioactively labeled amino acids were used to measure the incorporation of amino acids into new proteins.

In one approach to establish which codons specify which amino acids, synthetic mRNAs containing one, two, or three different types of bases were made and added to cell-free, protein-synthesizing systems. The polypeptides made in these systems were then analyzed. When the synthetic mRNA contained only one type of base, the results were unambiguous. Synthetic poly(U) mRNA, for example, directed the synthesis of a polypeptide consisting of a chain of phenylalanines, indicating that UUU is a codon for phenylalanine. Similarly, a synthetic poly(A) mRNA directed the synthesis of a lysine chain, and poly(C) directed the synthesis of a proline chain, indicating that AAA is a codon for lysine and CCC is a codon for proline.

Synthetic mRNAs made by the random incorporation of two different bases (called *random copolymers*) were also analyzed in the cell-free, protein-synthesizing systems. When mixed copolymers are made, the bases are incorporated into the synthetic molecule in a random way. Thus, poly(AC) molecules contain the eight different codons CCC, CCA, CAC, ACC, CAA, ACA, AAC, and AAA. In the cell-free, protein-synthesizing system, poly(AC) synthetic mRNAs caused the incorporation of asparagine, glutamine, histidine, and threonine into polypeptides, in addition to the lysine expected from AAA and the proline expected from CCC. The proportions of asparagine, glutamine, histidine, and threonine incorporated depended on the A : C ratio used to make the mRNA, and these observations were used to deduce information about the codons that specify the amino acids. For example, because an AC random copolymer containing

Figure 12.6

Hypothetical example showing how three nearby base pair addition mutations restore the reading frame. The mutations are shown here at the level of the mRNA.

Normal mRNA AUG ACA CAU AAC GGC UUC GUA UGG UGU GAA

Amino acids Met Thr His Asn Gly Phe Val Trp Cys Glu

3 + mutations

+U   +C   +A

mRNA AUG AUC ACA UAC ACG GCA UUC GUA UGG UGU GAA

Amino acids Met Ile Thr Tyr Thr Ala Phe Val Trp Cys Glu

Incorrect amino acids
in polypeptide

much more A than C resulted in the incorporation of many more asparagines than histidines, researchers concluded that asparagine is coded by two As and one C and histidine by two Cs and one A. With experiments of this kind, the base composition *(but not the base sequence)* of the codons for a number of amino acids was determined.

Another experimental approach also used copolymers, but these copolymers were synthesized so that they had a known sequence, not a random one. For example, a repeating copolymer of U and C gives a synthetic mRNA of UCUCUCUCUC. When this copolymer was tested in a cell-free, protein-synthesizing system, the resulting polypeptide had a repeating amino acid pattern of leucine-serine-leucine-serine. Therefore, UCU and CUC specify leucine and serine, although which coded for which could not be determined from the result.

Yet another approach used a *ribosome-binding assay,* developed in 1964 by Nirenberg and Philip Leder. This assay depends on the fact that in the absence of protein synthesis, specific tRNA molecules bind to ribosome-mRNA complexes. And, importantly, long mRNA molecules are not needed for the specific binding of the appropriate tRNA to the mRNA-ribosome complex; the binding of a trinucleotide (i.e., a codon-length piece of RNA) suffices. This trinucleotide-binding property made it possible to determine the specific relationships between many codons and the amino acids for which they code. Note that in this particular approach, the *specific nucleotide sequence of the codon is determined.* Using the ribosome-binding assay, many ambiguities that had arisen from other approaches were resolved. For example, UCU was found to be a codon for serine, and CUC was found to be a codon for leucine. All in all, about 50 codons were identified using this approach.

In sum, no single approach produced an unambiguous set of codon assignments, but information obtained through all the approaches enabled 61 codons to be assigned to amino acids; the other 3 codons do not specify amino acids (Figure 12.7). Each codon is written *as it appears in mRNA* and reads in a 5′→3′ direction.

Figure 12.7

The genetic code. Of the 64 codons, 61 specify one of the 20 amino acids. One of those 61 codons, AUG, is used in the initiation of protein synthesis. The other three codons are chain-terminating codons and do not specify any amino acid.

First letter	Second letter: U	Second letter: C	Second letter: A	Second letter: G	Third letter
U	UUU Phe (F)	UCU Ser (S)	UAU Tyr (Y)	UGU Cys (C)	U
	UUC Phe (F)	UCC Ser (S)	UAC Tyr (Y)	UGC Cys (C)	C
	UUA Leu (L)	UCA Ser (S)	UAA Stop	UGA Stop	A
	UUG Leu (L)	UCG Ser (S)	UAG Stop	UGG Trp (W)	G
C	CUU Leu (L)	CCU Pro (P)	CAU His (H)	CGU Arg (R)	U
	CUC Leu (L)	CCC Pro (P)	CAC His (H)	CGC Arg (R)	C
	CUA Leu (L)	CCA Pro (P)	CAA Gln (Q)	CGA Arg (R)	A
	CUG Leu (L)	CCG Pro (P)	CAG Gln (Q)	CGG Arg (R)	G
A	AUU Ile (I)	ACU Thr (T)	AAU Asn (N)	AGU Ser (S)	U
	AUC Ile (I)	ACC Thr (T)	AAC Asn (N)	AGC Ser (S)	C
	AUA Ile (I)	ACA Thr (T)	AAA Lys (K)	AGA Arg (R)	A
	AUG Met (M)	ACG Thr (T)	AAG Lys (K)	AGG Arg (R)	G
G	GUU Val (V)	GCU Ala (A)	GAU Asp (D)	GGU Gly (G)	U
	GUC Val (V)	GCC Ala (A)	GAC Asp (D)	GGC Gly (G)	C
	GUA Val (V)	GCA Ala (A)	GAA Glu (E)	GGA Gly (G)	A
	GUG Val (V)	GCG Ala (A)	GAG Glu (E)	GGG Gly (G)	G

= Chain termination codon (stop)

= Initiation codon

Characteristics of the Genetic Code

The characteristics of the genetic code are as follows:

1. *The code is a triplet code.* Each mRNA codon that specifies an amino acid in a polypeptide chain consists of three nucleotides.
2. *The code is comma-free; that is, it is continuous.* The mRNA is read continuously, three nucleotides at a time, without skipping any nucleotides of the message.
3. *The code is nonoverlapping.* The mRNA is read in successive groups of three nucleotides.
4. *The code is almost universal.* Almost all organisms share the same genetic language. Therefore, we can isolate an mRNA from one organism, translate it using the machinery from another organism, and produce the protein as if it had been translated in the original organism. The code is not completely universal, however. For example, the mitochondria of some organisms, such as mammals and yeast, have minor changes in the code, as does the nuclear genome of the protozoan *Tetrahymena.*
5. *The code is degenerate.* With two exceptions (AUG only codes for methionine, and UGG only codes for tryptophan), more than one codon occurs for each amino acid. This multiple coding is called the **degeneracy** of the code. There are particular patterns in this degeneracy (see Figure 12.7). When the first two nucleotides in a codon are identical and the third letter is U or C, the codon always codes for the same amino acid. For example, UUU and UUC specify phenylalanine, and CAU and CAC specify histidine. Also, when the first two nucleotides in a codon are identical and the third letter is A or G, the same amino acid is often specified. For example, UUA and UUG specify leucine, and AAA and AAG specify lysine. Finally, when the first two nucleotides in a codon are identical and the base in the third position is U, C, A, or G, the same amino acid is often specified. For example, CUU, CUC, CUA, and CUG all code for leucine.

6. *The code has start and stop signals.* Specific start and stop signals for protein synthesis are contained in the code. In both eukaryotes and prokaryotes, AUG (which codes for methionine) is almost always the start codon for protein synthesis.

Only 61 of the 64 codons specify amino acids; these codons are called **sense codons** (see Figure 12.7). The other three codons—UAG (amber), UAA (ochre), and UGA (opal)—do not specify amino acids, and no tRNAs in normal cells carry the appropriate anticodons. (The three-nucleotide anticodon pairs with the codon in the mRNA by complementary base pairing during translation.) These three codons are the **stop codons**, also called **nonsense codons** or **chain-terminating codons.** They are used singly or in tandem groups (UAG UAA, for example) to specify the end of translation of a polypeptide chain. Thus, when we read a particular mRNA sequence, we look for a stop codon located at a multiple of three nucleotides— *in the same reading frame*—from the AUG start codon to determine where the amino acid–coding sequence for the polypeptide ends.

7. *Wobble occurs in the anticodon.* Since 61 sense codons specify amino acids in mRNA, a total of 61 tRNA molecules could have the appropriate anticodons. According to the **wobble hypothesis** proposed by Francis Crick, the complete set of 61 sense codons can be read by fewer than 61 distinct tRNAs because of pairing properties of the bases in the anticodon (Table 12.1). Specifically, the base at the 5′ end of the anticodon complementary to the base at the 3′ end of the codon, or the third letter, is not as constrained three-dimensionally as the other two bases. This feature allows for less exact base pairing, so that the base at the 5′ end of the anticodon can pair with more than one type of base at the 3′ end of the codon; it can wobble. As Table 12.1 shows, no single tRNA molecule can recognize four different codons. But if the tRNA molecule contains the modified purine inosine at the 5′ end of the anticodon, then that tRNA can recognize three different codons. Figure 12.8 gives an example of how a single leucine tRNA can read two different leucine codons by base-pairing wobble.

iActivity Learn how to use sequencing information to track down part of the gene responsible for cystic fibrosis in the iActivity *Determining the Causes of Cystic Fibrosis* on the website.

K E Y N O T E

> The genetic code is a triplet code in which each codon (a set of three contiguous bases) in an mRNA specifies one amino acid. The code is degenerate: Some amino acids are specified by more than one codon. The genetic code is nonoverlapping and almost universal. Specific codons are used to signify the start and end of protein synthesis.

Translation: The Process of Protein Synthesis

Protein synthesis takes place on ribosomes, where the genetic message encoded in mRNA is translated. The mRNA is translated in the 5′→3′ direction, and the polypeptide is made in the N-terminal to C-terminal direction. Amino acids are brought to the ribosome bound to tRNA molecules. The correct amino acid sequence is achieved as a result of (1) the binding of each amino acid to its own specific tRNA and (2) the binding between the codon of the mRNA and the complementary anticodon in the tRNA.

Charging tRNA

The correct amino acid is attached to the tRNA by a type of enzyme called an **aminoacyl-tRNA synthetase.** The process is called aminoacylation, or **charging,** and it produces an **aminoacyl-tRNA** (or **charged tRNA**). Aminoacylation uses energy from ATP hydrolysis. Since there are 20 different amino acids, there are 20 different

Figure 12.8

Example of base-pairing wobble. Two different leucine codons (CUC, CUU) can be read by the same leucine tRNA molecule, contrary to regular base-pairing rules.

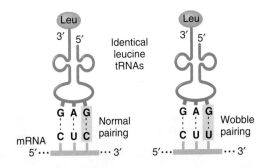

Table 12.1	Wobble in the Genetic Code	
Nucleotide at 5′ End of Anticodon		**Nucleotide at 3′ End of Codon**
G	can pair with	U or C
C	can pair with	G
A	can pair with	U
U	can pair with	A or G
I (inosine)	can pair with	A, U, or C

aminoacyl-tRNA synthetases. The recognition site for an aminoacyl-tRNA synthetase is enzyme-specific; in a number of cases, it is the anticodon region of the tRNA; but in other cases, sequences elsewhere in the tRNA are used.

Figure 12.9 shows charging of a tRNA molecule to produce valine-tRNA (Val-tRNA). First, the amino acid and ATP bind to the specific aminoacyl-tRNA synthetase enzyme. The enzyme then catalyzes a reaction in which the ATP loses two phosphates and is coupled to the amino acid as AMP to form aminoacyl-AMP. Next, the tRNA molecule binds to the enzyme, which transfers the amino acid from the aminoacyl-AMP to the tRNA and then displaces the AMP. The aminoacyl-tRNA molecule produced is then released from the enzyme. Chemically, the amino acid attaches at the 3′ end of the tRNA by a covalent linkage between the carboxyl group of the amino acid and the 3′-OH or 2′-OH group of the ribose of the adenine nucleotide found at the end of every tRNA.

Initiation of Translation

The three basic stages of protein synthesis—*initiation*, *elongation*, and *termination*—are similar in prokaryotes and eukaryotes. In the following sections, we discuss each of these stages in turn, concentrating on the processes in *E. coli*. In the discussions, we note where significant differences in translation occur between prokaryotes and eukaryotes.

The initiation of translation is shown in Figure 12.10. Initiation involves an mRNA molecule, a ribosome, a specific initiator tRNA, three different protein **initiation factors**, GTP (guanosine triphosphate), and magnesium ions.

In prokaryotes, the first step in translation is the binding of the 30S small ribosomal subunit to the region of the mRNA with the AUG initiation codon. The ribosomal subunit comes to the mRNA bound to three initiation factors (IF1, IF2, and IF3) and to a molecule of GTP and magnesium ions.

The AUG initiation codon is not sufficient to tell the ribosomal subunit where to bind to the mRNA; a sequence upstream (to the 5′ side in the leader of the mRNA) of the AUG is also required. This mRNA sequence—AGGAGG—is called the **Shine-Dalgarno sequence** after its discoverers. The model is that the for-

animation

a Initiation of Translation

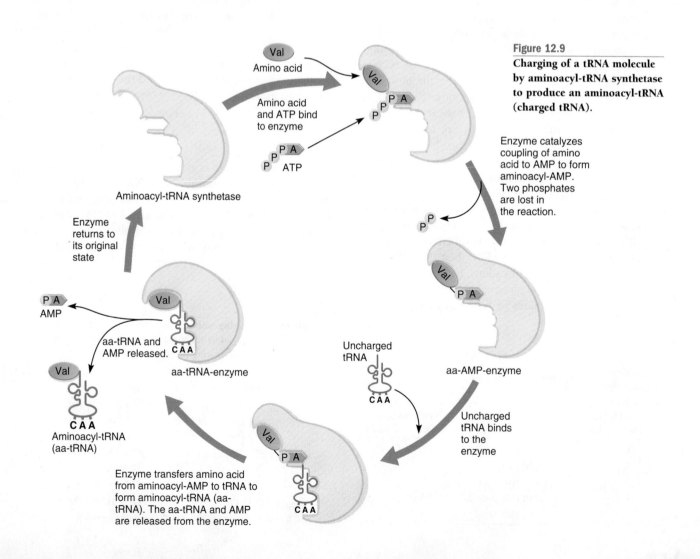

Figure 12.9

Charging of a tRNA molecule by aminoacyl-tRNA synthetase to produce an aminoacyl-tRNA (charged tRNA).

Val
Amino acid

Amino acid and ATP bind to enzyme

ATP

Aminoacyl-tRNA synthetase

Enzyme catalyzes coupling of amino acid to AMP to form aminoacyl-AMP. Two phosphates are lost in the reaction.

aa-AMP-enzyme

Uncharged tRNA
CAA

Uncharged tRNA binds to the enzyme

Enzyme returns to its original state

AMP

aa-tRNA and AMP released.
CAA
aa-tRNA-enzyme

Val
CAA
Aminoacyl-tRNA (aa-tRNA)

Enzyme transfers amino acid from aminoacyl-AMP to tRNA to form aminoacyl-tRNA (aa-tRNA). The aa-tRNA and AMP are released from the enzyme.
CAA

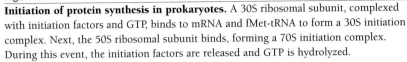

Figure 12.10

Initiation of protein synthesis in prokaryotes. A 30S ribosomal subunit, complexed with initiation factors and GTP, binds to mRNA and fMet-tRNA to form a 30S initiation complex. Next, the 50S ribosomal subunit binds, forming a 70S initiation complex. During this event, the initiation factors are released and GTP is hydrolyzed.

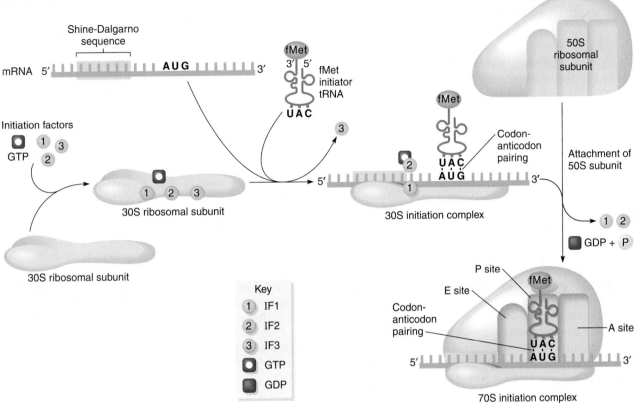

mation of complementary base pairs between this sequence in the mRNA and a sequence containing CCUCC at the 3′ end of 16S rRNA in the small subunit of the ribosome allows the ribosome to locate the start codon in the mRNA for the initiation of translation.

The next step in the initiation of translation is the binding of the initiator tRNA to the AUG start codon to which the 30S subunit is bound. In both prokaryotes and eukaryotes, the AUG initiator codon specifies methionine. As a result, newly made proteins in both types of organisms begin with methionine. In many cases, the methionine is removed later.

In prokaryotes, the initiator methionine is a modified form of methionine, called **formylmethionine** (abbreviated fMet), in which a formyl group has been added to the methionine's amino group. The initiator tRNA brings the fMet to the 30S-mRNA complex. When the fMet-tRNA binds to the 30S-mRNA complex, IF3 is released, and at this point, the *30S initiation complex*, consisting of the mRNA, 30S subunit, fMet-tRNA, IF1, and IF2, has been formed. Next, the 50S ribosomal subunit binds,

leading to GTP hydrolysis and release of IF1 and IF2. The final complex is called the *70S initiation complex*. The 70S ribosome has two binding sites for aminoacyl-tRNA, the peptidyl (P), and aminoacyl (A) sites. The fMet-tRNA is bound to the mRNA in the P site, and the A site is vacant (see Figure 12.10).

Translation initiation is very similar in eukaryotes. The main differences are that (1) the initiator methionine is unmodified, although a special initiator tRNA brings it to the ribosome, and (2) Shine-Dalgarno sequences are not found in eukaryotic mRNAs. Instead, the eukaryotic ribosome uses another way to find the AUG initiation codon. First, a *eukaryotic initiator factor,* eIF-4F, a multimer of several proteins including the *cap-binding protein* (CBP), binds to the cap at the 5′ end of the mRNA (see Chapter 11). Then a complex of the 40S ribosomal subunit with the initiator Met-tRNA, several eIF proteins, and GTP binds, along with other eIFs, and moves along the mRNA, scanning for the initiator AUG codon. The AUG codon is embedded in a short sequence—called the Kozak sequence—that indicates it is the initiator codon. This is called the *scanning model* for initiation. The AUG codon is almost always the first AUG codon from the 5′ end of the mRNA. Once it finds this AUG, the 40S subunit binds to it and then the 60S ribosomal subunit binds, displacing the eIFs, producing the *80S initiation complex* with the initiator Met-tRNA bound to the mRNA in the P site.

The poly(A) tail of the eukaryotic mRNA also plays a role in translation. Poly(A)-binding protein (PABP; see Figure 11.9) bound to the poly(A) tail also binds to one of the proteins of eIF-4F at the cap, thereby looping the 3′ end of the mRNA close to the 5′ end. In this way, poly(A) stimulates the initiation of translation.

Elongation of the Polypeptide Chain

animation

a Elongation of the Polypeptide Chain

Figure 12.11 depicts the elongation events—the addition of amino acids to the growing polypeptide chain one by one—as they take place in prokaryotes. This phase has three steps:

1. Aminoacyl-tRNA (charged tRNA) binds to the ribosome.
2. A peptide bond forms.
3. The ribosome moves (translocates) along the mRNA, one codon at a time.

Binding of Aminoacyl-tRNA. At the start of elongation, the anticodon of fMet-tRNA is hydrogen-bonded to the AUG initiation codon in the peptidyl (P) site of the ribosome (Figure 12.11, step 1). The next codon in the mRNA is in the aminoacyl (A) site. In Figure 12.11, this codon (UCC) specifies serine (Ser).

Next, the appropriate aminoacyl-tRNA (here, Ser-tRNA) binds to the codon in the A site (step 2). This aminoacyl-tRNA is brought to the ribosome bound to the protein elongation factor EF-Tu and a molecule of GTP. When the aminoacyl-tRNA binds to the codon in the A site, GTP hydrolysis releases EF-Tu–GDP. The EF-Tu is separated from the GDP and is recycled in other elongation events.

Peptide Bond Formation. The ribosome maintains the two aminoacyl-tRNAs in the P and A sites in the correct positions so that a peptide bond can form between the two amino acids (Figure 12.11, step 3). Two steps are involved in the formation of this peptide bond. First, the bond between the amino acid and the tRNA in the P site is cleaved. In this case, the breakage is between the fMet and its tRNA. Second, the peptide bond is formed between the now free fMet and the Ser attached to the tRNA in the A site. This reaction is catalyzed by **peptidyl transferase.** For many years, this enzyme activity was

thought to be a result of the interaction of a few ribosomal proteins of the 50S ribosomal subunit. However, newer data indicate that the 23S rRNA molecule of the large ribosomal subunit is the enzyme. This means that the rRNA is acting as a ribozyme (catalytic RNA; see Chapter 11, p. 240–241).

Once the peptide bond has formed (see step 3), a tRNA without an attached amino acid (an uncharged tRNA) is left in the P site. The tRNA in the A site, now called peptidyl-tRNA, has the first two amino acids of the polypeptide chain attached to it, in this case fMet-Ser.

Translocation. In the last step in the elongation cycle, **translocation** (Figure 12.11, step 4), the ribosome moves one codon along the mRNA toward the 3′ end. In prokaryotes, translocation requires the activity of another protein elongation factor, EF-G. An EF-G-GTP complex binds to the ribosome, GTP is hydrolyzed, and translocation of the ribosome occurs along with displacement of the uncharged tRNA away from the P site. Translocation is similar in eukaryotes; the elongation factor in this case is eEF-2, which functions similarly to EF-G.

A 50S subunit site called E (for *exit*) is involved in the release of the uncharged tRNA from the *E. coli* ribosome. The model is that the uncharged tRNA moves from the P site and then binds to the E site, effectively blocking the next aminoacyl-tRNA from binding to the A site until translocation is complete and the peptidyl-tRNA is bound correctly in the P site (Figure 12.11, step 5). After translocation, the EF-G is released in a reaction requiring GTP hydrolysis; EF-G is then reused, as shown in step 4. During the translocation step, the peptidyl-tRNA remains attached to its codon on the mRNA. And because the ribosome has moved, the peptidyl-tRNA is now located in the P site (hence the name *peptidyl site*).

After translocation is completed, the A site is vacant. An aminoacyl-tRNA with the correct anticodon binds to the newly exposed codon in the A site, repeating the process already described. The whole process is repeated until translation terminates at a stop codon (step 6).

In eukaryotes, the elongation and translocation steps are similar to those in prokaryotes, although there are differences in the number and properties of elongation factors and in the exact sequence of events.

In both prokaryotes and eukaryotes, once the ribosome moves away from the initiation site on the mRNA, another initiation event occurs. Thus, many ribosomes may simultaneously be translating each mRNA. The complex between an mRNA molecule and all the ribosomes that are translating it simultaneously is called a **polyribosome** or **polysome** (Figure 12.12). An average mRNA may have eight to ten ribosomes synthesizing protein from it. Simultaneous translation enables a large amount of protein to be produced from each mRNA molecule.

Figure 12.11
Elongation stage of translation in prokaryotes.

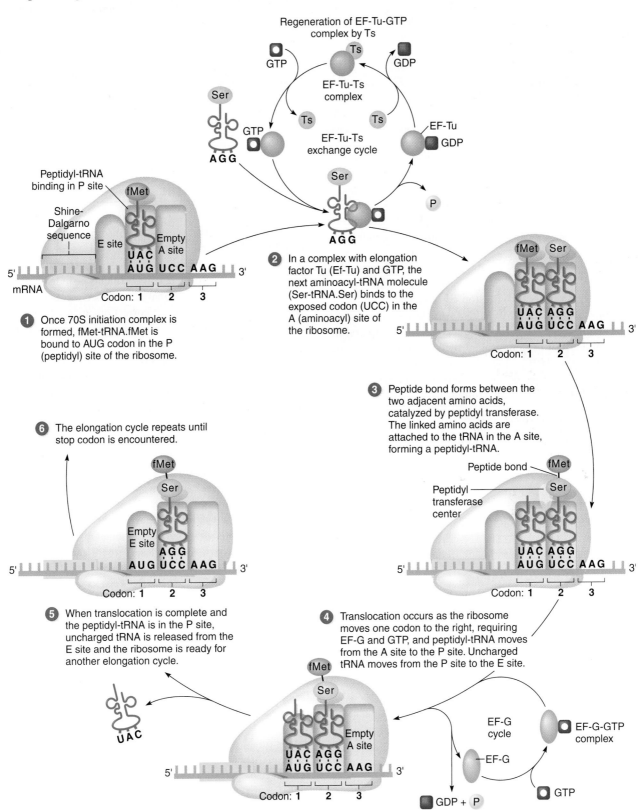

Regeneration of EF-Tu-GTP complex by Ts

EF-Tu-Ts complex

EF-Tu-Ts exchange cycle

EF-Tu

1 Once 70S initiation complex is formed, fMet-tRNA.fMet is bound to AUG codon in the P (peptidyl) site of the ribosome.

2 In a complex with elongation factor Tu (Ef-Tu) and GTP, the next aminoacyl-tRNA molecule (Ser-tRNA.Ser) binds to the exposed codon (UCC) in the A (aminoacyl) site of the ribosome.

3 Peptide bond forms between the two adjacent amino acids, catalyzed by peptidyl transferase. The linked amino acids are attached to the tRNA in the A site, forming a peptidyl-tRNA.

6 The elongation cycle repeats until stop codon is encountered.

5 When translocation is complete and the peptidyl-tRNA is in the P site, uncharged tRNA is released from the E site and the ribosome is ready for another elongation cycle.

4 Translocation occurs as the ribosome moves one codon to the right, requiring EF-G and GTP, and peptidyl-tRNA moves from the A site to the P site. Uncharged tRNA moves from the P site to the E site.

EF-G cycle

EF-G-GTP complex

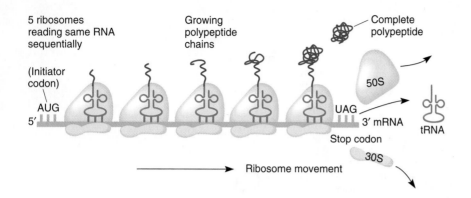

5 ribosomes reading same RNA sequentially

Growing polypeptide chains

Complete polypeptide

(Initiator codon)

AUG

5'

UAG

3' mRNA

Stop codon

Ribosome movement

50S

tRNA

30S

Figure 12.12
Diagram of a polysome, a number of ribosomes each translating the same mRNA sequentially.

Termination of Translation

Elongation continues until the polypeptide coded for in the mRNA is completed. The termination of translation is signaled by one of three stop codons (UAG, UAA, and UGA), which are the same in prokaryotes and eukaryotes (Figure 12.13, step 1). The stop codons do not code for any amino acid, and so no tRNAs in the cell have anticodons for them. The ribosome recognizes a stop codon with the help of proteins called **termination factors,** or **release factors (RF),** which read the codons (step 2) and then initiate a series of specific termination events. Different release factors are found in prokaryotes and eukaryotes.

animation

Translation
Termination

The specific termination events triggered by the release factors are (1) release of the polypeptide from the tRNA in the P site of the ribosome in a reaction catalyzed by peptidyl transferase (step 3), (2) release of the tRNA from the ribosome (step 4), and then (3) dissociation of the two ribosomal subunits and the RF from the mRNA (see step 4). In both prokaryotes and eukaryotes, the initiating amino acids—fMet and Met, respectively—usually are cleaved from the completed polypeptide.

KEYNOTE

Translation is a complicated process requiring many protein factors and energy. The AUG (methionine) initiator codon signals the start of translation in prokaryotes and eukaryotes. Elongation proceeds when a peptide bond forms between the amino acid attached to the tRNA in the A site of the ribosome and the growing polypeptide attached to the tRNA in the P site. Translocation occurs when the now uncharged tRNA in the P site is released from the ribosome and the ribosome moves one codon down the mRNA. Termination occurs as a result of the interaction of a release factor with a chain-terminating codon.

Protein Sorting in the Cell

In prokaryotes and eukaryotes, some proteins may be secreted, and in eukaryotes, some other proteins must be placed in different cell compartments, such as the nucleus, mitochondrion, chloroplast, and lysosome. The sorting of proteins to their appropriate compartments is under genetic control in that specific "signal" or "leader" sequences on the proteins direct them to the correct organelles. Similarly, in prokaryotes, certain proteins become localized in the membrane, while others are secreted.

Let us briefly describe protein distribution by the endoplasmic reticulum (ER) and Golgi apparatus in eukaryotes. Proteins synthesized on the rough ER are extruded into the space between the two membranes of the ER (the cisternal space). There they are modified by the attachment of sugar groups to produce glycoproteins and are then transferred to the Golgi apparatus in vesicles. The Golgi apparatus is the main distribution center where most of the sorting "decisions" occur. Proteins targeted for the plasma membrane are carried from the Golgi apparatus to the plasma membrane in transport vesicles. Proteins destined to be secreted are packaged into secretory storage vesicles. The secretory vesicles migrate to the cell surface, where they fuse with the plasma membrane and release their packaged proteins to the outside of the cell (Figure 12.14). Lysosomal proteins are collected into a specific class of Golgi export vesicles, which eventually become functional lysosomes.

In 1975, Gunther Blöbel, B. Dobberstein, and colleagues found that secreted proteins and other proteins sorted by the Golgi initially contain extra amino acids at the amino terminal end. Blöbel's work led to the **signal hypothesis,** which states that proteins sorted by the Golgi bind to the ER by a hydrophobic amino terminal extension (the **signal sequence**) to the membrane that is subsequently removed and degraded (Figure 12.15). Blöbel won the Nobel Prize in Physiology or Medicine in 1999 for this work.

In more detail, the signal sequence of a protein destined for the ER consists of about 15 to 30 N-terminal amino acids. Proteins without signal sequences remain in

Figure 13.4

The plasmid cloning vector pUC19. This plasmid has an origin of replication *(ori)*, an *amp*R selectable marker, and a polylinker located within part of the β-galactosidase gene, *lacZ*+.

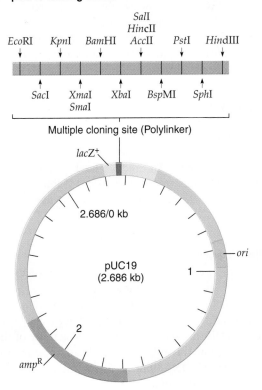

pUC19 cloning vector

Multiple cloning site (Polylinker)

2.686/0 kb

pUC19
(2.686 kb)

ori

*amp*R

ori = Origin of replication sequence
*amp*R = Ampicillin resistance gene
lacZ+ = Part of β-galactosidase gene

brane and letting the DNA enter. Resulting ampicillin-resistant colonies indicate cells that are transformed by plasmids, and blue-white colony screening distinguishes the recombinant DNA clones from plasmids without an inserted DNA fragment.

Plasmid cloning vectors have been developed for a large variety of prokaryotic and eukaryotic organisms, not just *E. coli*. Their general features are as we have discussed, although in some cases the sequences that are needed for replication in the organism of interest are not known, so the plasmids cannot replicate. Instead, they integrate into the genome.

Other types of cloning vectors are available for DNA cloning, including vectors derived from bacteriophage genomes; **shuttle vectors,** which permit DNA fragments to be cloned into two or more host organisms; **yeast artificial chromosomes (YACs,** pronounced "yaks"), which have a yeast telomere at each end, a yeast centromere sequence, a selectable marker, an origin of replication, and restriction sites for cloning very large pieces of DNA as essentially minichromosomes in yeast; and **bacterial artificial chromosomes (BACs,** pronounced "backs"), which are derived from the *E. coli F* factor (see Chapter 6, pp. 124–125) and which allow very large pieces of DNA to be cloned in *E. coli.*

Better beer through science? Go to the iActivity *Building a Better Beer* on the website and discover how genetically modified yeasts can improve your brew.

Figure 13.5

Insertion of a piece of DNA into the plasmid cloning vector pUC19 to produce a recombinant DNA molecule. pUC19 contains several unique restriction enzyme sites localized in a polylinker that are convenient for constructing recombinant DNA molecules. Insertion of a DNA fragment into the polylinker disrupts part of the β-galactosidase *(lacZ*+) gene, leading to nonfunctional β-galactosidase in *E. coli.* The blue-white color selection test described in the text can be used to select for vectors with or without inserts.

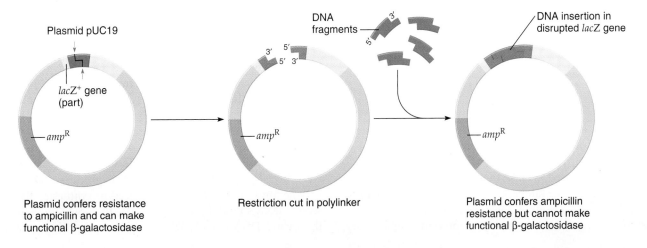

Plasmid pUC19

lacZ+ gene
(part)

*amp*R

DNA
fragments

*amp*R

DNA insertion in
disrupted *lacZ* gene

*amp*R

Plasmid confers resistance to ampicillin and can make functional β-galactosidase

Restriction cut in polylinker

Plasmid confers ampicillin resistance but cannot make functional β-galactosidase

KEYNOTE

Many different kinds of vectors have been developed to construct and clone recombinant DNA molecules. Most vectors replicate within their host organism. Cloning vectors also have restriction sites for inserting foreign DNA fragments, as well as one or more dominant selectable markers. The choice of the vector to use depends on the experimental goal.

Recombinant DNA Libraries

Typically, researchers want to study a particular gene or DNA fragment. When genomic DNA is isolated from an organism and cut with a restriction enzyme and the population of DNA fragments is cloned in a vector, we have a **genomic library,** a collection of clones containing at least one copy of every DNA sequence in the genome.

Genomic libraries have been made for many organisms, including humans (see the discussion of the Human Genome Project in Chapter 15) and many viruses. Related to genomic libraries are **complementary DNA (cDNA) libraries,** which are collections of clones of DNA copies of mRNAs isolated from cells. The following sections describe these two different types of libraries.

Genomic Libraries

A genomic library is a collection of clones that, when successfully made, contains at least one copy of every DNA sequence in the genome. The genomic library can be used to isolate and study a particular clone, such as that for a gene of interest. We will see how this can be done later. In this section, we focus on the construction of genomic libraries of eukaryotic DNA.

Genomic libraries are made using the cloning procedures already described. A restriction enzyme is used to cut up the genomic DNA, and a vector is chosen so that the entire genome is represented in a manageable number of clones. The number of clones needed to include all sequences in the genome depends on the size of the genome being cloned and the average size of the DNA fragments inserted into the vector. The probability of having at least one copy of any DNA sequence in the genomic library can be calculated from the formula

$$N = \frac{\ln(1-P)}{\ln(1-f)}$$

where N is the necessary number of recombinant DNA molecules, P is the probability desired, f is the fractional proportion of the genome in a single recombinant DNA molecule (i.e., f is the average size, in kilobase pairs, of the fragments used to make the library divided by the size of the genome, in kilobase pairs), and ln is the natural logarithm. For example, for a 99 percent chance that a particular yeast DNA fragment is represented in a library of 15-kb fragments, where the yeast genome size is about 12,000 kb, 3,682 recombinant DNA molecules would be needed. For the approximately 3,000,000-kb human genome, more than 920,000 clones would be needed, hence the use of large-capacity cloning vectors (such as YACs or BACs) for making libraries of large genomes. Whatever the genome or vector, to have confidence that all genomic sequences are represented, one must make a library with many times more than the calculated minimum number of clones.

KEYNOTE

A genomic library is a collection of clones that contains at least one copy of every DNA sequence in an organism's genome. Like regular book libraries, genomic libraries are great resources of information; in this case, the information is about the genome.

cDNA Libraries

DNA copies, called **complementary DNA (cDNA),** can be made from all mRNA molecules present in a population of eukaryotic cells at a particular time. These cDNA molecules can then be cloned to produce a cDNA library. Since a cDNA library reflects the gene activity of the cell type at the time the mRNAs are isolated, the construction and analysis of cDNA libraries are useful, for example, for comparing gene activities in different cell types of the same organism or in the same cell type at different times, as in cell differentiation during development.

The clones in the cDNA library represent the mature mRNAs found in the cell. In eukaryotes, mature mRNAs are processed molecules, so the sequences obtained are not equivalent to gene clones. In particular, intron sequences are present in gene clones but not in cDNA clones. For any mRNA, cDNA clones can be useful for subsequently isolating the gene that codes for that mRNA. The gene clone can provide more information than can the cDNA clone— for example, on the presence and arrangement of introns and on the regulatory sequences that control expression of the gene.

cDNA libraries are readily made from eukaryotic mRNAs because, uniquely among RNAs, these RNAs contain a poly(A) tail (see Chapter 11, pp. 234–235), which facilitates their isolation from other RNAs. Once the mRNAs have been isolated, cDNA copies are made as shown in Figure 13.6. The first step in cDNA synthesis is annealing a short oligo(dT) primer to the poly(A) tail. The primer is extended by **reverse transcriptase** (RNA-dependent DNA polymerase) to make a DNA copy of the mRNA strand. The result is a DNA-mRNA double-stranded molecule. Next, RNase H ("R-N-ase H," a type of ribonuclease), DNA polymerase I, and DNA ligase are used to synthesize the second DNA strand. RNase H degrades the RNA strand in the hybrid DNA-mRNA, DNA polymerase I makes new

Figure 13.6

The synthesis of double-stranded complementary DNA (cDNA) from a polyadenylated mRNA, using reverse transcriptase, RNase H, DNA polymerase I, and DNA ligase.

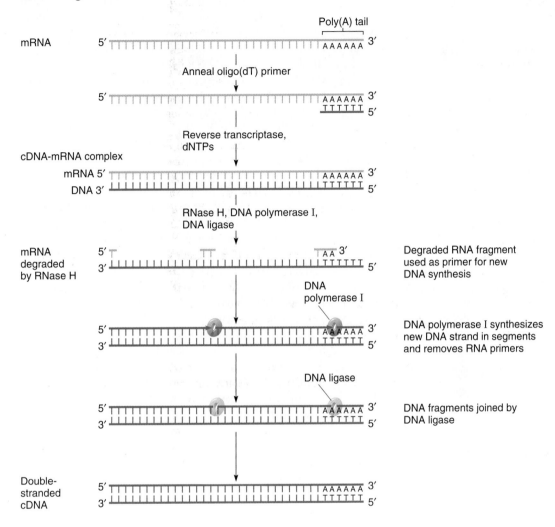

DNA fragments using the partially degraded RNA fragments as primers, and finally, DNA ligase ligates the new DNA fragments together to make a complete chain. The result is a double-stranded cDNA molecule that is a faithful DNA copy of the starting mRNA.

How do we clone cDNA molecules? Figure 13.7 illustrates the cloning of cDNA using a **restriction site linker**, or **linker**, which is a short, double-stranded piece of DNA (oligodeoxyribonucleotide) about 8 to 12 nucleotide pairs long that includes a restriction site, in this case *Bam*HI. Both the cDNA molecules and the linkers have blunt ends, and they can be ligated together at high concentrations of T4 DNA ligase. Sticky ends are produced in the cDNA molecule by cleaving the cDNA (with linkers now at each end) with *Bam*HI. The resulting DNA is inserted into a cloning vector that has also been cleaved with *Bam*HI, and the recombinant DNA molecule produced is transformed into an *E. coli* host cell for cloning.

KEYNOTE

DNA copies, called complementary DNA or cDNA, can be made of the population of mRNAs purified from a cell. First, a primer and the enzyme reverse transcriptase are used to make a single-stranded DNA copy of the mRNA; then RNase H, DNA polymerase I, and DNA ligase are used to make a double-stranded DNA copy called cDNA. This cDNA can be spliced into cloning vectors and cloned.

Finding a Specific Clone in a Library

Unlike libraries of books, clone libraries have no catalog, so they must be searched through—screened—to find a clone of interest. Fortunately, a number of screening procedures have been developed. For example, given the existence of a *probe*, such as a cloned cDNA, it is possible

Figure 13.7

The cloning of cDNA using *Bam*HI linkers.

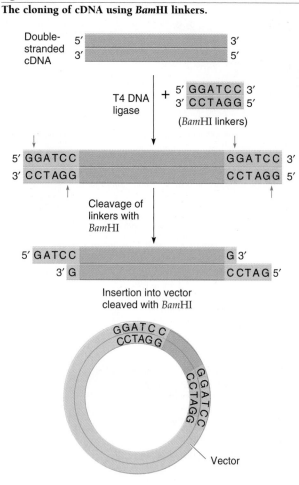

to identify in a genomic library the cloned gene that codes for the mRNA molecule from which the cDNA was made and then to isolate that gene for characterization. (A probe is a labeled sequence designed to home in on a sequence or gene of interest, as we will now see.) Here we discuss the screening of a genomic library made in a plasmid cloning vector.

The screening process is straightforward, as shown in Figure 13.8. First, *E. coli* cells are transformed with the genomic library (Figure 13.8, step 1), and the cells are plated on selective medium, where colonies are produced (step 2). Then the colonies are replica plated onto another plate of selective medium, this one with a membrane filter on its surface (step 3). (Replica plating involves pressing a pad of sterile velveteen onto the original master plate to pick up some of each colony in the pattern in which they grew on that plate and then pressing the velveteen gently onto the new plate, thereby "inoculating" it with cells from each colony in its original pattern.) Colonies grow on the membrane filter, which is then lifted off the plate and processed to lyse the bacterial cells, denature the DNA to single strands, and then

bind that DNA firmly to the filter (step 4). Next, the filter is placed in a heat-sealable plastic bag and incubated with the cDNA probe (step 5), which has been labeled radioactively or nonradioactively. To prepare the labeled DNA for use as a probe, the DNA is denatured to single strands by boiling. These labeled molecules are added to the membrane filters to which the denatured (single-stranded) DNA from each colony has been bound. Wherever the two sequences are complementary, DNA-DNA hybrids form between the probe and the colony DNA. If the cDNA probe is derived from the mRNA for β-globin, for example, that probe will hybridize with DNA bound to the filter that encodes the β-globin mRNA, that is, the β-globin gene. After sufficient time for hybridization has elapsed (approximately 24 hours), the filters are washed to remove unbound probe and subjected to the appropriate detection procedure, depending on whether the probe was radioactive or nonradioactive: *autoradiography* for a radioactive probe, chemiluminescent or colorimetric detection for a nonradioactive probe (step 6). Autoradiography involves placing the dried filter against X-ray film and leaving it in the dark for a period of time to produce an *autoradiogram*. When the film is developed, dark spots are seen wherever the radioactive probe is bound to the filter in the probe reaction. (The dark spots result from the decay of the radioactive atoms, which changes the silver grains in the film.) In chemiluminescent or colorimetric detection, substrates and reactions are used to visualize the probe-target DNA hybrids on the filter either by the generation of chemiluminescence, which is detected using an X-ray film, or by the generation of color, which is detected visually. From the positions of the spots on the film or filter, the locations of the bacterial colony or colonies on the original plate can be determined and the clones of interest isolated for further characterization.

Analysis of Genes and Gene Transcripts

Cloned DNA sequences are resources for experiments designed to answer many kinds of biological questions. The following experimental techniques are described in this section:

1. *The cloned DNA can be mapped with respect to the number and arrangement of restriction sites.* The resulting map, analogous to the arrangement of genes on a linkage map, is called a **restriction map.** This is commonly done for both cloned genes and cloned cDNAs. The restriction maps produced can be useful for making clones of subsections of the gene or cDNA or for comparing the gene with the cDNA.

2. *A cloned cDNA or a cloned gene can be used to analyze transcription of the corresponding gene in the cell.* For example, we can study the size of the initial transcript of the gene, the processing steps it goes

Figure 13.8

Using DNA probes to screen plasmid libraries for specific DNA sequences.

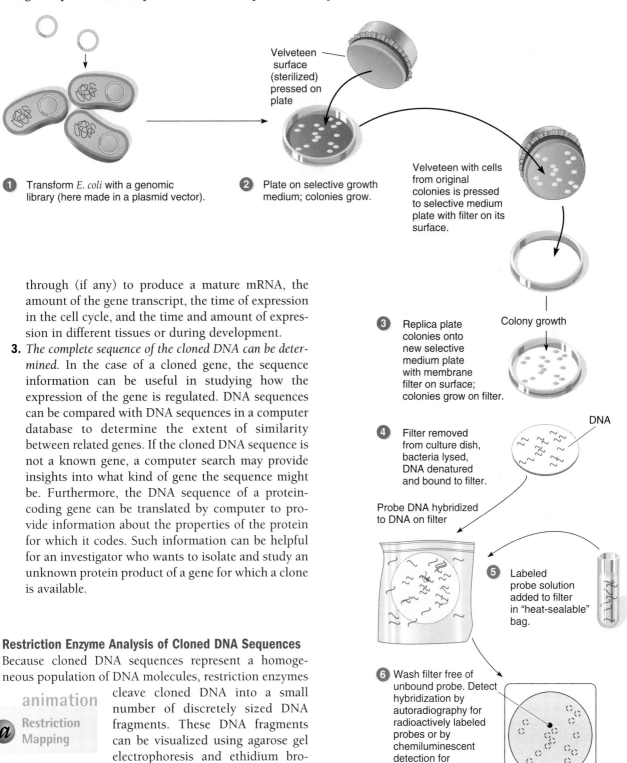

1. Transform *E. coli* with a genomic library (here made in a plasmid vector).

2. Plate on selective growth medium; colonies grow.

Velveteen surface (sterilized) pressed on plate

Velveteen with cells from original colonies is pressed to selective medium plate with filter on its surface.

3. Replica plate colonies onto new selective medium plate with membrane filter on surface; colonies grow on filter.

Colony growth

4. Filter removed from culture dish, bacteria lysed, DNA denatured and bound to filter.

DNA

Probe DNA hybridized to DNA on filter

5. Labeled probe solution added to filter in "heat-sealable" bag.

6. Wash filter free of unbound probe. Detect hybridization by autoradiography for radioactively labeled probes or by chemiluminescent detection for nonradioactively labeled probe. Dark spots indicate clones detected by probe.

through (if any) to produce a mature mRNA, the amount of the gene transcript, the time of expression in the cell cycle, and the time and amount of expression in different tissues or during development.

3. *The complete sequence of the cloned DNA can be determined.* In the case of a cloned gene, the sequence information can be useful in studying how the expression of the gene is regulated. DNA sequences can be compared with DNA sequences in a computer database to determine the extent of similarity between related genes. If the cloned DNA sequence is not a known gene, a computer search may provide insights into what kind of gene the sequence might be. Furthermore, the DNA sequence of a protein-coding gene can be translated by computer to provide information about the properties of the protein for which it codes. Such information can be helpful for an investigator who wants to isolate and study an unknown protein product of a gene for which a clone is available.

Restriction Enzyme Analysis of Cloned DNA Sequences

Because cloned DNA sequences represent a homogeneous population of DNA molecules, restriction enzymes cleave cloned DNA into a small number of discretely sized DNA fragments. These DNA fragments can be visualized using agarose gel electrophoresis and ethidium bromide staining. With such gels, restriction maps can be constructed without the need for hybridization with a labeled probe and subsequent detection.

animation

a Restriction Mapping

Let us assume that we have cloned a 5.0-kb piece of DNA and want to construct a restriction map of it (Figure 13.9, step 1). One sample of the DNA is digested with

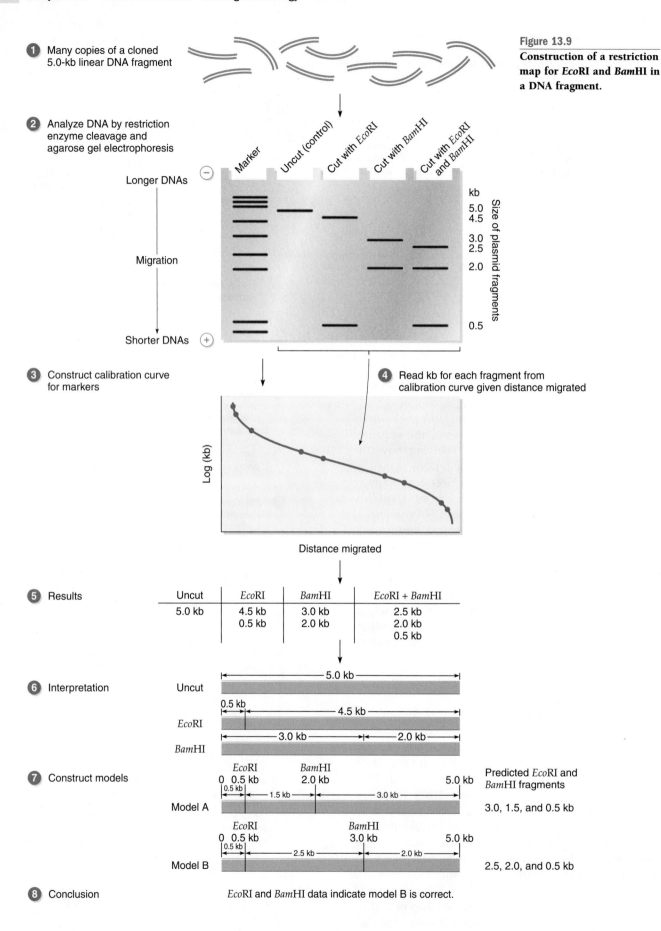

Figure 13.9

Construction of a restriction map for *Eco*RI and *Bam*HI in a DNA fragment.

1. Many copies of a cloned 5.0-kb linear DNA fragment

2. Analyze DNA by restriction enzyme cleavage and agarose gel electrophoresis

Marker | Uncut (control) | Cut with *Eco*RI | Cut with *Bam*HI | Cut with *Eco*RI and *Bam*HI

Longer DNAs (−)

Migration

Shorter DNAs (+)

kb
5.0
4.5
3.0
2.5
2.0
0.5

Size of plasmid fragments

3. Construct calibration curve for markers

4. Read kb for each fragment from calibration curve given distance migrated

Log (kb)

Distance migrated

5. Results

Uncut	*Eco*RI	*Bam*HI	*Eco*RI + *Bam*HI
5.0 kb	4.5 kb	3.0 kb	2.5 kb
	0.5 kb	2.0 kb	2.0 kb
			0.5 kb

6. Interpretation

Uncut — 5.0 kb

*Eco*RI — 0.5 kb | 4.5 kb

*Bam*HI — 3.0 kb | 2.0 kb

7. Construct models

*Eco*RI 0.5 kb | *Bam*HI 2.0 kb
0   5.0 kb
0.5 kb | 1.5 kb | 3.0 kb

Model A

Predicted *Eco*RI and *Bam*HI fragments

3.0, 1.5, and 0.5 kb

*Eco*RI 0.5 kb | *Bam*HI 3.0 kb
0   5.0 kb
0.5 kb | 2.5 kb | 2.0 kb

Model B

2.5, 2.0, and 0.5 kb

8. Conclusion

*Eco*RI and *Bam*HI data indicate model B is correct.

*Eco*RI, a second sample is digested with *Bam*HI, and a third sample is digested with both *Eco*RI and *Bam*HI. The DNA restriction fragments of each reaction are separated according to their molecular size by agarose gel electrophoresis; controls are a sample of the same DNA uncut with any enzyme and DNA fragments of known size—DNA fragment size markers, or simply size markers—so that the sizes of the unknown DNA fragments can be computed (step 2). The gel is a rectangular, horizontal slab of agarose (a firm, gelatinous material) that has a matrix of pores through which DNA passes in an electric field. Each gel is made by boiling a buffered agarose solution and allowing it to cool in a mold. A toothed comb is used to form discrete wells in the gels so that different samples can be analyzed simultaneously.

DNA is negatively charged because of its phosphates, so the DNA migrates toward the positive pole in an electric field. Migration is in a straight line from the well (called a lane). Because small DNA fragments can move more readily through the pores in the gel, small DNA fragments move through the gel more rapidly than large fragments.

After electrophoresis, the DNA is stained with ethidium bromide. The DNA complexed with ethidium bromide fluoresces under ultraviolet light, and the gel is photographed. (The chapter opener photograph shows such a gel with DNA bands illuminated with ultraviolet light.) From the photograph, the distance each DNA band migrated from the well can be measured. The molecular size of each DNA band in the size markers is known, so a calibration curve can be drawn of DNA size (in log kb) and migration distance (in mm) (Figure 13.9, step 3). The migration distances for the DNA bands from the uncut and cut DNA are then used with the calibration curve to determine the molecular sizes of the DNA fragments in the bands (step 4). For our example, the results are shown in Figure 13.9, step 5.

The results are analyzed as follows (Figure 13.9, steps 6–8):

1. When the 5.0-kb DNA is cut with *Eco*RI, 4.5-kb and 0.5-kb DNA fragments are obtained, indicating that there is one restriction site for *Eco*RI in the DNA located 0.5 kb from one end of the molecule (step 6).

2. When the same DNA is cut with *Bam*HI, 4.0-kb and 2.0-kb DNA fragments are produced. Using similar logic as in (1), there is one restriction site for *Bam*HI located 2 kb from one end of the molecule (step 6).

3. At this point, we know there is one restriction site for each enzyme, but we do not know the relationship between the two. However, we can make two models (step 7). In model A, the *Eco*RI site is 0.5 kb from one end and the *Bam*HI site is 2.0 kb from that same end. In model B, the *Eco*RI site is 0.5 kb from one end and the *Bam*HI site is 3.0 kb from that end (i.e., 2.0 kb from the other end). Model A predicts that

cutting with *Eco*RI and *Bam*HI will produce three fragments of 0.5, 1.5, and 3.0 kb (going from left to right along the DNA), and model B predicts that cutting with both enzymes will produce three fragments of 0.5, 2.5, and 2.0 kb. The actual data show three fragments with sizes 2.5, 2.0, and 0.5 kb, validating model B (step 8).

In real situations, restriction mapping involves data that are much more complicated (e.g., involving more restriction enzymes and a number of sites for each enzyme).

Restriction Enzyme Analysis of Genes in the Genome

As part of the analysis of genes, it can be helpful to determine the arrangement and specific locations of restriction sites. This information is useful, for example, for comparing homologous genes in different species, analyzing intron organization, or cloning parts of a gene, such as its promoter or controlling sequences, into a vector. The arrangement of restriction sites in a gene can be analyzed without actually cloning the gene if you have either a cDNA clone of the gene's mRNA or a clone of the same gene cloned from a closely related organism. Either one of those sequences is labeled radioactively or nonradioactively and is used as a probe to home in on the DNA fragments of the gene of interest among all the genomic fragments produced by restriction enzyme digestion of the genome. The process of analysis is as follows (Figure 13.10):

1. Samples of genomic DNA are cut with different restriction enzymes (Figure 13.10, steps 1 and 2), each of which produces DNA fragments of different lengths depending on the locations of the restriction sites.

2. The DNA restriction fragments are separated by size using agarose gel electrophoresis (step 3). After electrophoresis, the DNA is stained with ethidium bromide so that it can be seen under ultraviolet light. When genomic DNA is digested with a restriction enzyme, the result is a continuous smear of fluorescence down the length of the gel lane because the enzyme produces fragments of all sizes from large to small.

3. The DNA fragments are transferred to a membrane filter (step 4). The transfer to the membrane filter is done by the **Southern blot technique** (named after its inventor, Edward Southern). In brief, the gel is soaked in an alkaline solution to denature the double-stranded DNA into single strands. The gel is neutralized and placed on a piece of blotting paper that spans a glass plate. The ends of the paper are in a container of buffer and act as wicks. A piece of membrane filter is laid down so that it covers the gel. Sheets of blotting paper (or paper towels) and a weight are stacked on top of the membrane filter. The buffer solution in the bottom tray is wicked up by the

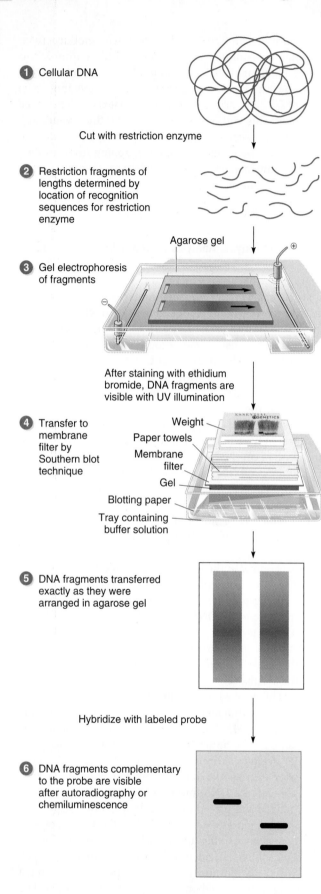

1 Cellular DNA

Cut with restriction enzyme

2 Restriction fragments of lengths determined by location of recognition sequences for restriction enzyme

Agarose gel

3 Gel electrophoresis of fragments

After staining with ethidium bromide, DNA fragments are visible with UV illumination

4 Transfer to membrane filter by Southern blot technique

Weight
Paper towels
Membrane filter
Gel
Blotting paper
Tray containing buffer solution

5 DNA fragments transferred exactly as they were arranged in agarose gel

Hybridize with labeled probe

6 DNA fragments complementary to the probe are visible after autoradiography or chemiluminescence

Figure 13.10

Southern blot procedure for analyzing cellular DNA for the presence of sequences complementary to a labeled probe, such as a cDNA molecule made from an isolated mRNA molecule. The hybrids, shown as three bands in this theoretical example, are visualized by autoradiography or chemiluminescence.

blotting paper, passing through the gel and the membrane filter and finally into the stack of blotting paper. During this process, the DNA fragments are picked up by the buffer and transferred from the gel to the membrane filter, to which they bind because of the membrane filter's chemical properties. The fragments on the filter are arranged in exactly the same way as they were in the gel (step 5).

4. A labeled single-stranded probe is added to the membrane filter, where it hybridizes to any complementary DNA fragment. Detection of the probe is carried out in a way appropriate for whether the probe is radioactive (by autoradiography) or nonradioactive (by chemiluminescence) to determine the positions of the hybrids (step 6). If DNA size markers are separated in a different lane in the agarose gel electrophoresis process, the sizes of the genomic restriction fragments that hybridized with the probe can be calculated. From the fragment sizes obtained, a restriction map can be generated to show the relative positions of the restriction sites. Suppose, for example, that using only *Bam*HI produces a DNA fragment of 3 kb that hybridizes with the radioactive probe. If a combination of *Bam*HI and *Pst*I is then used and produces two DNA fragments, one of 1 kb and the other of 2 kb, we would deduce that the 3-kb *Bam*HI fragment contains a *Pst*I restriction site 1 kb from one end and 2 kb from the other end. Further analysis with other enzymes, individually and combined, enables the researcher to construct a map of all the enzyme sites relative to all other sites.

Analysis of Gene Transcripts

A blotting technique related to the Southern blot technique—called **northern blot analysis**—has been developed to analyze RNA rather than DNA. (The name is not derived from a person but indicates that the technique is related to the Southern blot technique.) In northern blot analysis, RNA extracted from a cell is separated by size using gel electrophoresis, and the RNA molecules are transferred and bound to a filter in a procedure that is essentially identical to Southern blotting. After hybridization with a labeled probe and use of the appropriate detection system, bands show the locations of RNA fragments that were complementary to the probe. Given appropriate RNA size markers, the sizes of the RNA fragments identified with the probe can be determined.

Northern blot analysis is useful for revealing the size or sizes of the mRNA encoded by a gene. In some cases, a number of different mRNA species encoded by the same gene have been identified in this way, suggesting that different promoter sites or different terminator sites are used or that alternative mRNA processing can occur. Northern blot analysis can also be used to investigate whether an mRNA is present in a cell type or tissue and how much of it is present. This type of experiment is useful for determining levels of gene activity—for instance, during development, in different cell types of an organism, or in cells before and after they are subjected to various physiological stimuli.

K E Y N O T E

Cloned genes and other DNA sequences often are analyzed to determine the arrangement and specific locations of restriction sites. The analytical process involves cleavage of the DNA with restriction enzymes, followed by separation of the resulting DNA fragments by agarose gel electrophoresis. The sizes of the DNA fragments are calculated, enabling restriction maps to be constructed. DNA fragments produced by cleavage of genomic DNA show a wide range of sizes, resulting in a continuous smear of DNA fragments in the gel. In this case, specific gene fragments can be visualized only by transferring the DNA fragments to a membrane filter by the Southern blot technique, hybridizing a specific labeled probe with the DNA fragments, and detecting the hybrids. At that point, a restriction map can be made. A similar procedure—northern blot analysis—is used to analyze the sizes and quantities of RNAs isolated from a cell.

DNA Sequencing

Cloned DNA fragments can be analyzed to determine the nucleotide pair sequence of the DNA. This is the most detailed information one can obtain about a DNA fragment. The information is useful, for example, for identifying gene sequences and regulatory sequences within the fragment and for comparing the sequences of homologous genes from different organisms. Walter Gilbert and Frederick Sanger shared one-half of the 1980 Nobel Prize in Chemistry for their "contributions concerning the determination of base sequences in nucleic acids."

Dideoxy Sequencing. The most commonly used method of DNA sequencing, called **dideoxy sequencing** (developed by Fred Sanger in the 1970s), involves extension of a short primer by DNA polymerase.

Both linear DNA and circular DNA can be sequenced using the dideoxy DNA sequencing method. Linear DNA fragments can be generated, for example, by cutting plasmid DNA with one or more restriction enzymes.

The principle of dideoxy DNA sequencing is straightforward. The DNA is first denatured to single strands by heat treatment. Next, a short oligonucleotide primer is annealed to one of the two DNA strands (Figure 13.11). The oligonucleotide is designed so that its 3' end is next to the DNA sequence of interest. The oligonucleotide acts as a *primer* for DNA synthesis. Consider, for example, DNA fragments cloned into the plasmid cloning vector pUC19 (see Figure 13.4). With a pair of oligonucleotide primers complementary to the DNA flanking the

Figure 13.11

Dideoxy DNA sequencing of a theoretical DNA fragment.

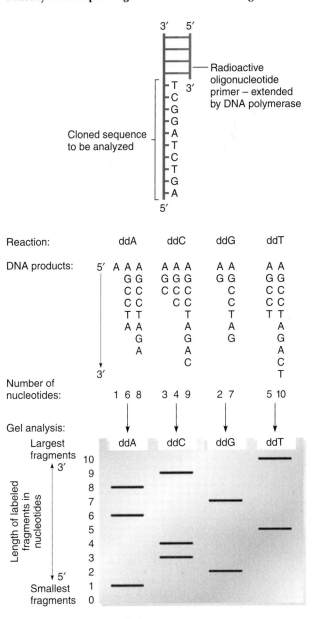

Sequence deduced from banding pattern of autoradiogram made from gel:

5'-A-G-C-C-T-A-G-A-C-T-3'

multiple cloning site, any DNA insert can be sequenced from each end. In fact, most plasmid cloning vectors have the same sequences flanking their multiple cloning sites, so that *universal sequencing primers* can be used to sequence any cloned insert in these vectors. Any other DNA fragment can be sequenced if some sequence information is available from which a primer can be made.

For each sequencing experiment, four separate reactions are set up. Each reaction contains the single-stranded DNA to be sequenced, the primer annealed to that DNA, DNA polymerase, the four normal deoxynucleotide precursors (dATP, dTTP, dCTP, and dGTP), and a small amount of one dideoxynucleotide precursor (ddATP, ddTTP, ddCTP, or ddGTP). A dideoxynucleotide (generically ddtNTP; Figure 13.12) is a modified nucleotide precursor that differs from a normal deoxynucleotide in that it has a 3′-H rather than a 3′-OH on the deoxyribose sugar. In Figure 13.11, the primer is labeled radioactively so that newly synthesized DNA can be detected easily.

As stated above, the four reactions differ according to which dideoxynucleotide is present (whether it has A, T, G, or C as the base). When the primer is extended, occasionally DNA polymerase inserts a dideoxynucleotide instead of the normal deoxynucleotide. Once that happens, no further DNA synthesis can then occur because the absence of a 3′-OH prevents the formation of a phosphodiester bond with an incoming DNA precursor.

For example, if an A is specified by the DNA template strand, a dideoxy A nucleotide (ddA) could be incorporated rather than the normal A nucleotide in the reaction mixture, and elongation of the chain would stop. In a population of molecules in the same DNA synthesis reaction, new DNA chains stop at *all* possible positions where the nucleotide is required because of the incorporation of the dideoxynucleotide. In the ddA reaction, the many different chains produced all end with ddA; all chains end with ddG in the ddG reaction; and so on (see Figure 13.11). It is important to realize that each DNA chain synthesized started from the same fixed point and ended at a particular base, the latter determined by the dideoxynucleotide incorporated.

The DNA chains in each reaction mixture are separated by polyacrylamide gel electrophoresis, and the locations of the DNA bands are revealed by autoradiography (in cases where radioactive DNA has been used). The DNA sequence of the newly synthesized strand is determined from the autoradiogram by reading up the *sequencing ladder* from the bottom to the top to give the sequence in 5′→3′ orientation. In the example, the band that moved the farthest ended with ddA, the band that moved the second farthest ended with ddG, and so on. The complete sequence determined is 5′-AGCCTAGACT-3′; this is *complementary* to the sequence on the template strand (see Figure 13.11).

Currently, automated procedures typically are used that enable DNA sequencing to proceed much more rapidly than with the manual dideoxy method. The automated procedures involve one reaction containing four dideoxynucleotides with different fluorescent labels, one for each of the four bases. The DNA fragments synthesized are separated by electrophoresis, and the results are scanned by a laser device that excites the fluorescent labels and determines which fluorescent label is present at each position. The output is a series of colored peaks corresponding to each nucleotide position in the sequence (Figure 13.13); the output is converted to a sequence of bases by computer. Automated procedures are of great utility for basic research as well as for genome-sequencing projects.

Figure 13.12

A dideoxynucleotide (ddNTP) DNA precursor.

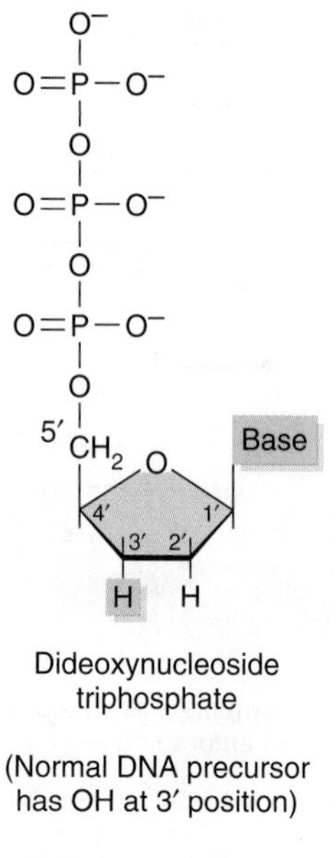

Dideoxynucleoside triphosphate

(Normal DNA precursor has OH at 3′ position)

K E Y N O T E

Methods have been developed for determining the sequence of a cloned piece of DNA. A commonly used method, the dideoxy procedure, uses enzymatic synthesis of a new DNA chain on a cloned template DNA strand. With this procedure, synthesis of new strands is stopped by the incorporation of a dideoxy analogue of the normal deoxyribonucleotide. With the incorporation of four different dideoxy analogues, the new strands stop at all possible nucleotide positions, thereby allowing the complete DNA sequence to be determined.

Analysis of DNA Sequences. Sequences determined by any sequencing method are entered into computer databases. Computer programs have been written to analyze DNA

Figure 13.13

Results of automated DNA sequence analysis using fluorescent dyes. The procedure is described in the text. The automated sequencer generates the curves shown in the figure from the fluorescing bands on the gel. The colors are generated by the machine and indicate the four bases: A is green, G is black, C is blue, and T is red. Where bands cannot be distinguished clearly, an N is listed.

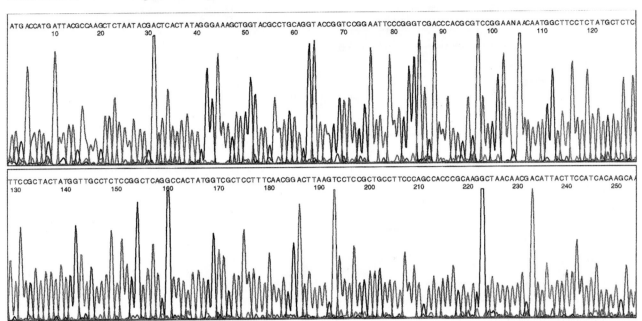

sequences for restriction site locations, for example, and to compare a variety of sequences, homologous regions, transcription regulatory sequences, and so on. Programs can search DNA sequences for possible protein-coding regions by looking for an initiator codon in a frame with a stop codon (called an open reading frame, or ORF). Other programs can be used to translate a cloned DNA sequence into a theoretical amino acid sequence and to make predictions about the structure and function of the protein. This is possible because the sequences of all sequenced proteins are in computer databases, enabling researchers to make detailed and rapid comparisons by computer. For example, many DNA-binding proteins have particular secondary structure motifs, and the detection of the sequence for such a motif in a gene sequence suggests that the gene encodes for a DNA-binding protein. Since many regulatory proteins are DNA-binding proteins, this conclusion would be of significant interest. Such a finding would then suggest the direction of future experimentation to define the actual function of the gene product. Increasingly, much of this computer analysis of sequences can be done using the Internet.

Polymerase Chain Reaction (PCR)

Generating large numbers of identical copies of DNA by the construction and cloning of recombinant DNA molecules was made possible in the 1970s. Recombinant DNA

techniques revolutionized molecular genetics by making it possible to analyze genes and their functions in new ways. However, cloning DNA is time consuming. In the mid-1980s, the **polymerase chain reaction (PCR)** was developed, and this has resulted in yet a new revolution in gene analysis. PCR produces an extremely large number of copies of a specific DNA sequence from a DNA mixture without having to clone it, a process called *amplification*. The amplified PCR products are called *amplimers*. PCR has become one of the most important tools in modern molecular biology. Kary Mullis, who developed PCR, shared the Nobel Prize in Chemistry in 1993. (The other recipient, M. Smith, received the prize for other work.)

The starting point for PCR is the double-stranded DNA containing the sequence to be amplified and a pair of oligonucleotide primers that flank that DNA. The primers usually are 20 or more nucleotides long and are made synthetically, so some information must be available about the sequence of interest. In brief, the PCR procedure is as follows (Figure 13.14):

1. Denature the double-stranded DNA to single strands by heating at 94–95°C (Figure 13.14, step 1). Cool, and anneal the primers (A and B in the figure) at 37–65°C, depending on how well the base sequences of the primers complement the base sequence of the DNA. The two primers are designed so that they anneal to the opposite strands of the template DNA at the boundaries of the sequence to be amplified. As a result, the 3′ ends of the primers face each other.

Figure 13.14

The polymerase chain reaction (PCR) for selective amplification of DNA sequences.

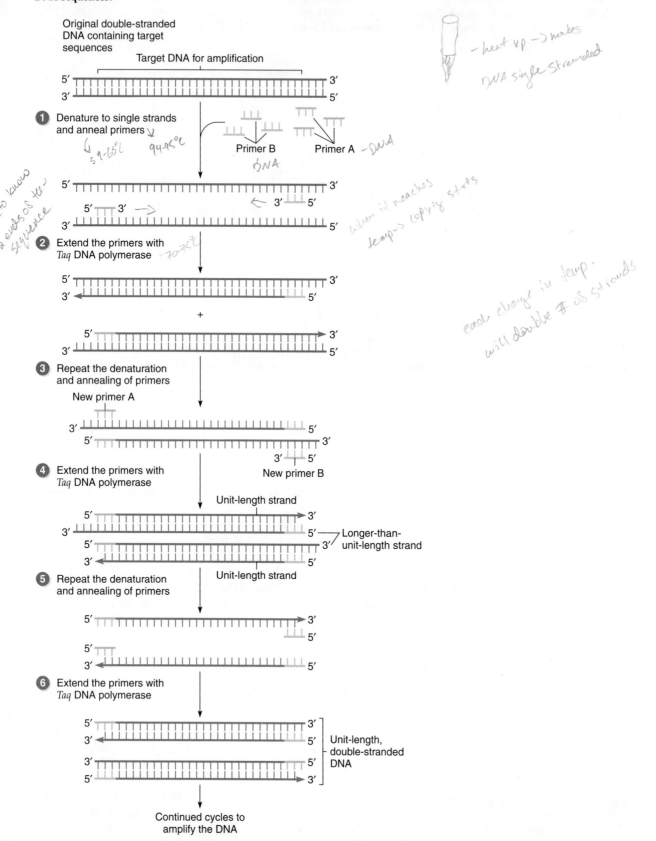

Handwritten annotations:

200–1000 bases

– heat up → makes DNA single stranded

Primer B   Primer A – DNA
DNA

have to know the 2 ends of the sequence

39–65℃   94–95℃

when it reaches temp → copying starts

70–75℃

each change in temp. will double # of strands

Original double-stranded DNA containing target sequences

Target DNA for amplification

5'   3'
3'   5'

1 Denature to single strands and anneal primers

2 Extend the primers with *Taq* DNA polymerase

3 Repeat the denaturation and annealing of primers

New primer A

New primer B

4 Extend the primers with *Taq* DNA polymerase

Unit-length strand

Longer-than-unit-length strand

Unit-length strand

5 Repeat the denaturation and annealing of primers

6 Extend the primers with *Taq* DNA polymerase

Unit-length, double-stranded DNA

Continued cycles to amplify the DNA

2. Extend the primers with DNA polymerase at 70–75°C (step 2). For this, a special heat-resistant DNA polymerase such as *Taq* ("tack") *polymerase* is used. This enzyme is the DNA polymerase of a thermophilic bacterium, *Thermus aquaticus.*

3. Repeat the heating cycle to denature the DNA to single strands and cool to anneal the primers again (step 3). (The further amplification of the original strands is omitted in the remainder of the figure.)

4. Repeat the primer extension with *Taq* DNA polymerase (step 4). In each of the two double-stranded molecules produced in the figure, one strand is of unit length; that is, it is the length of DNA between the 5′ end of primer A and the 5′ end of primer B—the length of the target DNA. The other strand in both molecules is longer than unit length.

5. Repeat the denaturation of DNA and annealing of new primers (step 5). (For simplification, the further amplification of the longer-than-unit-length strands continues with linear increase only and is omitted in the rest of the figure.)

6. Repeat the primer extension with *Taq* DNA polymerase (step 6). This produces unit-length, double-stranded DNA. Note that it took three cycles to produce the two molecules of target-length DNA. Repeated denaturation, annealing, and extension cycles result in the geometric increase in the amount of the unit-length DNA.

Using PCR, the amount of new DNA generated increases geometrically. Starting with one molecule of DNA, one cycle of PCR produces two molecules, two cycles produce four molecules, and three cycles produce eight molecules, two of which are the target DNA. A further 10 cycles produce 1,024 copies (2^{10}) of the target DNA, and in 20 cycles there will be 1,048,576 copies (2^{20}) of the target DNA. The procedure is rapid, each cycle taking only a few minutes using a *thermal cycler,* a machine that automatically cycles the reaction through the temperature changes in a programmed way.

There are many applications for PCR, including amplifying DNA for cloning, amplifying DNA from genomic DNA preparations for sequencing without cloning, mapping DNA segments, disease diagnosis, sex determination of embryos, forensics, and studies of molecular evolution. In disease diagnosis, for example, PCR can be used to detect bacterial pathogens or viral pathogens such as HIV (human immunodeficiency virus, the causative agent of AIDS) and hepatitis B virus. PCR can also be used in genetic disease diagnosis, which is discussed in Chapter 14.

In forensics, PCR can be used to amplify trace amounts of DNA in samples such as hair, blood, or semen collected from a crime scene. The amplified DNA can be analyzed and compared with DNA from a victim and a suspect, and the results can be used to implicate or exonerate suspects in the crime. This analysis, called *DNA typing,* is discussed in more detail in Chapter 14 (pp. 293–296).

Another interesting application of PCR is amplifying ancient DNA for analysis, using samples of tissues preserved hundreds or thousands of years ago, such as 440,000-year-old mammoths. PCR makes it possible to amplify selected DNA sequences and then analyze the sequences of those DNA molecules for comparison with contemporary DNA samples. These analyses enable us to make evolutionary comparisons between ancient forebears and present-day descendants.

K E Y N O T E

> The polymerase chain reaction (PCR) uses specific oligonucleotide primers to amplify a specific segment of DNA many thousandfold in an automated procedure. PCR has many applications in research and in the commercial arena, including generating specific DNA segments for cloning or sequencing and for amplifying DNA to detect specific genetic defects.

Summary

In this chapter, we discussed some of the procedures involved in recombinant DNA technology and DNA manipulation. Collectively, these procedures are also called genetic engineering. We have seen how it is possible to cut DNA at specific sites using restriction enzymes, how DNA can be cloned into specially constructed vectors, and how the cloned DNA can be analyzed in various ways. Through the construction of genomic libraries and cDNA libraries and the application of screening procedures to those libraries, a large number of genes from a wide variety of organisms have been cloned and identified.

Restriction mapping analysis has provided detailed molecular maps of genes and chromosomes analogous to the genetic maps constructed on the basis of recombination analysis. DNA-sequencing methods have given us an enormous amount of information about the DNA organization of genes, both coding sequences and regulatory sequences. The amount of DNA sequence information available is growing rapidly, and computer databases of such sequences are available for researchers to analyze. For example, when a new gene is sequenced, we can determine whether it has any sequences in common with genes already in the database.

In the mid-1980s, a new technique called polymerase chain reaction (PCR) was developed. Given some sequence information about a DNA fragment, synthetic oligonucleotide primers can be made and used to amplify the DNA fragment from the genome in a repeated cycle of DNA denaturation (strand separation), annealing of the primers, and extension of the primers with a special DNA polymerase. Numerous applications have been found for PCR, including cloning rare pieces of DNA, preparing DNA for sequencing without cloning, and diagnosing genetic diseases.

Analytical Approaches for Solving Genetics Problems

Q13.1 A piece of DNA 900 bp long is cloned and then cut out of the vector for analysis. Digestion of this linear piece of DNA with three different restriction enzymes singly and in all possible pairs gave the following restriction fragment size data:

Enzymes	Restriction Fragment Sizes
*Eco*RI	200 bp, 700 bp
*Hind*III	300 bp, 600 bp
*Bam*HI	50 bp, 350 bp, 500 bp
*Eco*RI + *Hind*III	100 bp, 200 bp, 600 bp
*Eco*RI + *Bam*HI	50 bp, 150 bp, 200 bp, 500 bp
*Hind*III + *Bam*HI	50 bp, 100 bp, 250 bp, 500 bp

Construct a restriction map from these data.

A13.1 The approach to this kind of problem is to consider a pair of enzymes and to analyze the data from the single and double digestions. First, consider the *Eco*RI and *Hind*III data. Cutting with *Eco*RI produces two fragments, one of 200 bp and the other of 700 bp, while cutting with *Hind*III also produces two fragments, one of 300 bp and the other of 600 bp. Thus, we know that both restriction sites are asymmetrically located along the linear DNA fragment with the *Eco*RI site 200 bp from an end and the *Hind*III site 300 bp from an end. When we consider the *Eco*RI + *Hind*III data, we can determine the positions of these two restriction sites relative to one another. For example, if the *Eco*RI site is 200 bp from the fragment end, and the *Hind*III site is 300 bp from that same end, then we would predict that cutting with both enzymes would produce three fragments of sizes 200 bp (end to *Eco*RI site), 100 bp (*Eco*RI site to *Hind*III site), and 600 bp (*Hind*III site to other end). On the other hand, if the *Eco*RI site is 200 bp from one fragment end and the *Hind*III site is 300 bp from the other fragment end, then cutting with both enzymes would produce three fragments of sizes 200 bp (end to *Eco*RI site), 400 bp (*Eco*RI site to *Hind*III site), and 300 bp (*Hind*III site to end). The actual data support the first model.

Now we pick another pair of enzymes, *Hind*III and *Bam*HI. (We could have picked *Eco*RI and *Bam*HI.) Cutting with *Hind*III produces fragments of 300 bp and 600 bp, as we have seen, and cutting with *Bam*HI produces three fragments of sizes 50 bp, 350 bp, and 500 bp, indicating that there are two *Bam*HI sites in the DNA fragment. Again, the double digestion products are useful in locating the sites. Double digestion with *Hind*III and *Bam*HI produces four fragments of 50 bp, 100 bp, 250 bp, and 500 bp. The simplest interpretation of the data is that the 300-bp *Hind*III fragment is cut into the 50-bp and 250-bp fragments by *Bam*HI and the 600-bp *Hind*III

fragment is cut into the 100-bp and 500-bp fragments by *Bam*HI. Thus, the restriction map shown in the accompanying figure can be drawn:

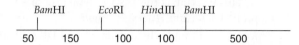

*Bam*HI		*Eco*RI	*Hind*III	*Bam*HI	
50	150	100	100		500

The *Bam*HI + *Eco*RI data are compatible with this model.

Q13.2 The recessive allele *bw*, when homozygous, results in brown eyes in *Drosophila*, in contrast to the wild-type bright red eye color. A restriction fragment length polymorphism (RFLP) for a particular DNA region results in either two restriction fragments (type I) or one restriction fragment (type II) when *Drosophila* DNA is cut with restriction enzyme C and the fragments are separated by DNA electrophoresis, blotted to a membrane filter, and probed with a particular DNA probe.

A true-breeding brown-eyed fly with type I DNA is crossed with a true-breeding, wild-type fly with type II DNA. The F_1 flies have wild-type eye color and exhibit both type I and type II DNA patterns. The F_1 flies are crossed with true-breeding brown-eyed flies with type I DNA, and the progeny are scored for eye color and RFLP type. The results are as follows:

Class	Phenotypes	Number
1	Red eyes, types I and II DNA	184
2	Red eyes, type I DNA	21
3	Brown eyes, type I DNA	168
4	Brown eyes, types I and II DNA	27
	Total progeny	400

Analyze these data.

A13.2 The eye color mutation is a familiar genetic marker. The restriction fragment length polymorphisms (RFLPs) are also genetic markers and can be analyzed just like any gene marker. If we symbolize the type I DNA as I and the type II DNA as II, the F_1 cross is

$$\frac{bw^+\ \text{II}}{bw\ \ \text{I}} \times \frac{bw\ \ \text{I}}{bw^+\ \text{II}}$$

This is a testcross with the exception that DNA markers do not exhibit dominance or recessiveness. The cross is drawn as if the markers are linked; of course, we have yet to show this. If the eye color and RFLP markers are unlinked, the result would be equal numbers of the four progeny classes. However, the data show a great excess of (a) red eyes, type I and II DNA and (b) brown eyes, type I DNA. Their origin is the pairing of F_1 bw^+ II and bw I gametes with bw I gametes to give bw^+ II/bw I and bw I/bw I progeny genotypes, respectively. Similarly, the progeny (a) red eyes, type I DNA and (b) brown eyes, type I and II DNA derive from pairing bw^+ I and bw II gametes with bw I gametes to give bw^+ I/bw I and bw II/bw I progeny, respectively. These latter two classes occur with about equal frequency, that is, a frequency much lower than for the other two classes. The simplest explanation is that the brown-eye gene and the RFLP marker are linked, so the F_1 cross is a mapping cross, much like those we analyzed in Chapter 5. Classes 1 and 3 are the parentals, and classes 2 and 4 are the recombinants. The map distribution between *bw* and the RFLP is therefore

$$\frac{21 + 27}{\text{total}} \times 100\%$$

$$\frac{48}{400} \times 100\%$$

$$= 12 \text{ map units}$$

Questions and Problems

13.1 There are many varieties of cloning vectors. One type of cloning vector used to clone recombinant DNA in *E. coli* is plasmids. What features do these vectors have that make them useful for constructing and cloning recombinant DNA molecules?

13.2 The ability to clone DNA fragments gives molecular biologists the power to understand the structure and function of our genes and to use fragments of DNA to produce proteins. What are the basic elements of research in recombinant DNA technology?

13.3 Much effort has been spent on developing cloning vectors that replicate in organisms other than *E. coli*.
a. Describe several different reasons one might want to clone DNA in an organism other than *E. coli*.
b. What is a shuttle vector, and why is it used?
c. Describe the salient features of a vector that could be used for cloning DNA in yeast.

*13.4 The human genome contains about 3×10^9 bp of DNA. How many 40-kb pieces would you have to clone into a library if you wanted to be 90 percent certain of including a particular sequence?

13.5 What is a cDNA library, and from what cellular genetic material is it derived?

13.6 A colleague has sent you a 4.5-kb DNA fragment excised from a plasmid cloning vector with the enzymes *Pst*I and *Bgl*II (see Table 13.1 for a description of these enzymes and the sites they recognize). Your colleague tells you that within the fragment there is an *Eco*RI site that lies 0.49 kb from the *Pst*I site. Suppose you want to clone this DNA fragment into the plasmid vector pUC19 (described in Figure 13.4).
a. How would you proceed?

b. How would you verify that you have cloned the correct fragment and determine the orientation of the fragment within the pUC19 cloning vector?

*13.7 Suppose you wanted to produce human insulin (a peptide hormone) by cloning. Assume that you could do this by inserting the human insulin gene into a bacterial host, where, given the appropriate conditions, the human gene would be transcribed and then translated into human insulin. Which would be better to use as your source of the gene: human genomic insulin DNA or a cDNA copy of this gene? Explain your choice.

13.8 You are given a genomic library of yeast prepared in a bacterial plasmid vector. You are also given a cloned cDNA for human actin, a protein that is conserved in protein sequences among eukaryotes. Outline how you would use these resources to attempt to identify the yeast actin gene.

*13.9 A cDNA library is made with mRNA isolated from liver tissue. When a cloned cDNA from that library is digested with the enzymes *Eco*RI (E), *Hind*III (H), and *Bam*HI (B), the restriction map shown in the figure below, part (a), is obtained. When this cDNA is used to screen a cDNA library made with mRNA from brain tissue, three identical cDNAs with the restriction map shown in part (b) are obtained. When either cDNA is used to synthesize a uniformly labeled ^{32}P-labeled probe and the probe is allowed to hybridize to a Southern blot prepared from genomic DNA digested singly with the enzymes *Eco*RI, *Hind*III, and *Bam*HI, an autoradiograph shows the pattern of bands seen in part (c). When either cDNA is used to synthesize a uniformly labeled ^{32}P-labeled probe and used to probe a northern blot prepared with poly(A)+ RNA isolated from liver and brain tissues, the pattern of bands in part (d) is seen. Fully analyze these data and then answer the following questions.

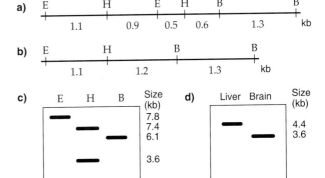

a. Do these cDNAs derive from the same gene?
b. Why are different-sized bands seen on the northern blot?

c. Why do the cDNAs have different restriction maps?

d. Why are some of the bands seen on the whole-genome Southern blot different sizes than some of the restriction fragments in the cDNAs?

13.10 Restriction endonucleases are used to construct restriction maps of linear or circular pieces of DNA. The DNA usually is produced in large amounts by recombinant DNA techniques. Generating restriction maps is like putting the pieces of a jigsaw puzzle together. Suppose we have a circular piece of double-stranded DNA that is 5,000 bp long. If this DNA is digested completely with restriction enzyme I, four DNA fragments are generated: fragment *a* is 2,000 bp long, *b* is 1,400 bp long, *c* is 900 bp long, and *d* is 700 bp long. If, instead, the DNA is incubated with the enzyme for a short time, the result is incomplete digestion of the DNA: Not every restriction enzyme site in every DNA molecule will be cut by the enzyme, and all possible combinations of adjacent fragments can be produced. From an incomplete digestion experiment of this type, fragments of DNA are produced from the circular piece of DNA that contains the following combinations of the above fragments: *a-d-b, d-a-c, c-b-d, a-c, d-a, d-b,* and *b-c.* Lastly, after digesting the original circular DNA to completion with restriction enzyme I, the DNA fragments are treated with restriction enzyme II under conditions conducive to complete digestion. The resulting fragments are 1,400, 1,200, 900, 800, 400, and 300 bp. Analyze all the data to locate the restriction enzyme sites as accurately as possible.

*13.11 A piece of DNA 5,000 bp long is digested with restriction enzymes A and B, singly and together. The DNA fragments produced are separated by DNA electrophoresis and their sizes are calculated, with the following results:

Digestion with		
A	**B**	**A + B**
2,100 bp	2,500 bp	1,900 bp
1,400 bp	1,300 bp	1,000 bp
1,000 bp	1,200 bp	800 bp
500 bp		600 bp
		500 bp
		200 bp

Each A fragment is extracted from the gel and digested with enzyme B, and each B fragment is extracted from the gel and digested with enzyme A. The sizes of the resulting DNA fragments are determined by gel electrophoresis, with the following results:

A Fragment	Fragments Produced by Digestion with B	B Fragment	Fragments Produced by Digestion with A
2,100 bp	→ 1,900, 200 bp	2,500 bp	→ 1,900, 600 bp
1,400 bp	→ 800, 600 bp	1,300 bp	→ 800, 500 bp
1,000 bp	→ 1,000 bp	1,200 bp	→ 1,000, 200 bp
500 bp	→ 500 bp		

Construct a restriction map of the 5,000-bp DNA fragment.

*13.12 Draw the banding pattern you would expect to see on a DNA-sequencing gel if you annealed the primer 5′-CTAGG-3′ to the following single-stranded DNA fragment and carried out a dideoxy sequencing experiment. Assume that the dNTP precursors were all labeled

3′-GATCCAAGTCTACGTATAGGCC-5′

13.13 PCR fragments can be separated by gel electrophoresis, and the size of a fragment can be estimated. Explain the process involved in determining the size of the PCR fragment in question.

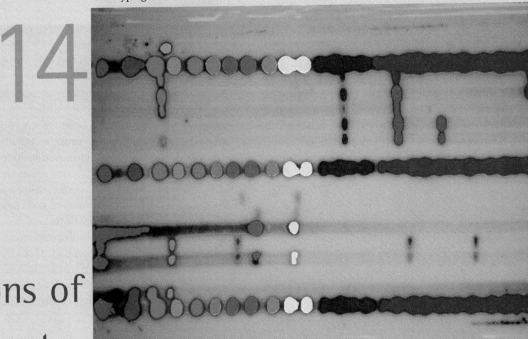
DNA typing (color enhanced).

14

Applications of Recombinant DNA Technology

PRINCIPAL POINTS

There are many applications of recombinant DNA technology and related procedures. Recombinant DNA techniques and PCR are widely used in analyzing basic biological processes, such as the arrangement of restriction sites in DNA, the sizes of RNA transcripts, the amount of RNA transcription, the steps in RNA processing, and protein-protein interactions in the cell.

Recombinant DNA and PCR techniques are used in DNA molecular testing for human genetic disease mutations. In general, human genetic testing is done for prenatal diagnosis, newborn screening, or carrier detection.

DNA typing, or DNA fingerprinting, distinguishes individuals and is based on the concept that no two individuals of the same species, save for identical twins, have the same genome sequence. DNA typing has many applications, including basic biological studies, forensics, detecting infectious species of bacteria, and analyzing old or ancient DNA.

Gene therapy is the treatment of a genetic disorder by introducing into the patient a normal gene to replace or overcome the effects of a mutant gene. For ethical reasons, only somatic gene therapy is being developed for humans. There are few examples of successful somatic gene therapy in humans, but there is great hope for treating many genetic diseases in this way in the future.

> Biotechnology and pharmaceutical companies develop products for the market by using the same kinds of recombinant DNA and PCR techniques employed in basic biological analysis, DNA molecular testing, gene cloning, DNA typing, and gene therapy.

> Genetic engineering of plants is possible using recombinant DNA or PCR techniques. Already, many genetically modified crops have been developed, and much of the processed food we buy contains modified genes. It is expected that many more types of improved crops will result from continued applications of this new technology.

i RECOMBINANT DNA TECHNOLOGY HAS BECOME SO prevalent in our society that on any given day, it is likely that you will hear or read a news report about a new application. Most common by far are stories about the use of recombinant DNA in the fields of medicine and agriculture; however, biotechnology has also revolutionized such fields as anthropology, conservation, industry, and forensics. In this chapter, you will learn about some of the specific uses of recombinant DNA technology. After you have read this chapter, you can apply what you've learned by trying the iActivity, in which you'll work with nonhuman DNA to help solve a murder.

In Chapter 13, we discussed a number of the techniques for constructing, cloning, and analyzing recombinant DNA molecules. Since the development of recombinant DNA technology and PCR, it has been possible to investigate many new questions in all areas of biology, and this has resulted in many exciting advances. The advances of the past few years include the ability to determine the sequences of a number of genomes, including the human genome. We discuss genome analysis in Chapter 15. In this chapter, we discuss some applications of these technologies. The applications are so wide-ranging that we can only scratch the surface. The examples chosen describe some of the applications as case studies. Thus, you can learn about the specific example while also seeing more generally the types of questions and hypotheses that can be investigated.

Analysis of Biological Processes

Recombinant DNA and PCR techniques have fundamental and widespread applications in basic biological research. Questions in most areas of biology are now being addressed with these techniques. In genetics, researchers are investigating such things as the functional organization of genes and the regulation of gene expression. In Chapter 13, for example, we described the principles of constructing a restriction map, using Southern blotting to focus on particular regions of the genome, using northern blotting to study RNA, and using DNA sequencing to obtain the most detailed information possible about genes. Similarly, in developmental genetics, key regulatory genes and target genes responsible for developmental events are being discovered and analyzed, and gene changes associated with aging and cancer are being investigated. In evolutionary biology, DNA sequence analysis is adding new information about the evolutionary relationships between organisms. In this section, we describe the use of recombinant DNA techniques to study a particular biological process to give a taste of the wide variety of questions that can be addressed.

Glucose Repression of Transcription of the Yeast *GAL1* Gene

This example illustrates how recombinant DNA technology can be used to study gene transcription.

In the yeast *Saccharomyces cerevisiae*, the expression of the *GAL* (galactose) genes is induced by the carbon source galactose in the growth medium. The products of the *GAL* genes are enzymes that catalyze the breakdown of galactose. However, when yeast is grown on glucose, the preferred carbon source, the *GAL* genes are not transcribed. (The genetics of transcriptional regulation of the *GAL* genes is described in Chapter 17, pp. 348–349.)

If glucose is added to a culture of yeast already growing in medium containing galactose, transcription of the *GAL* genes is stopped and the *GAL* mRNAs in the cell are rapidly degraded. The latter was shown in the following experiment, the results of which are illustrated in Figure 14.1. Yeast cells were grown so that their *GAL* genes were turned on. Then, at time zero, glucose was added and

Figure 14.1

Regulation of transcription of the yeast *GAL1* gene by glucose. Glucose was added at time zero, and the amount of *GAL1* transcripts was analyzed at various times thereafter by blotting and probing, as described in the text. (From Figure 5, Johnston, M., Flick, J. S., and Pexton, T., 1994. Multiple mechanisms provide rapid and stringent glucose repression of *GAL* gene expression in *Saccharomyces cerevisiae*. *Mol. Cell Biol.* 14:3834–3841.)

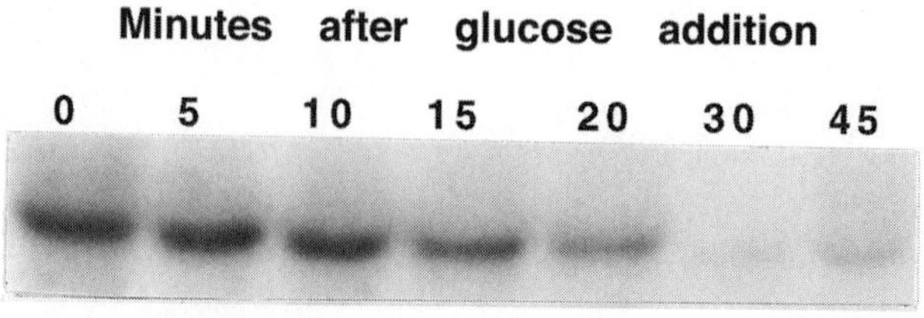

Minutes after glucose addition

| 0 | 5 | 10 | 15 | 20 | 30 | 45 |

samples were taken at various times thereafter. RNA was extracted from each of the samples, and the RNAs were separated by agarose gel electrophoresis. Northern blotting was then performed (see Chapter 13, pp. 278–279), and the blot was probed for the mRNA of one of the *GAL* genes, *GAL1*, using a radioactive probe. Visually, it is easy to see in Figure 14.1 that the amount of hybridization decreased rapidly in the 45-minute span of the sampling period. When these results were quantified and plotted on a graph, it was seen that there is a very rapid loss of mRNA in the first 10 minutes and a more gradual loss thereafter.

DNA Molecular Testing for Human Genetic Disease Mutations

Throughout this text, you will encounter many examples of human genetic diseases. These diseases are caused by enzyme or other protein defects that in turn are the result of mutations at the DNA level. For an increasingly large number of genetic diseases, including Huntington disease (OMIM 143100), hemophilia (OMIM 306700), cystic fibrosis (OMIM 219700), Tay-Sachs disease (OMIM 272800), and sickle-cell anemia (SCA; OMIM 141900), we can perform DNA molecular tests for the presence of mutations associated with the disease. In this section, practical issues of DNA molecular testing are discussed along with some examples of the testing approaches used.

Concept of DNA Molecular Testing

Genetic testing determines whether an individual who has symptoms, or is at a high risk of developing a genetic disease because of a family history of a heritable disease, actually has a particular gene mutation. **DNA molecular testing** is a type of genetic testing that focuses on the molecular nature of mutations associated with disease. Designing DNA molecular tests, then, depends on having knowledge about gene mutations that cause the disease of interest. This information comes from sequencing the gene involved.

A complication of genetic testing is that many different mutations of a gene can cause loss of function and therefore lead to the development of the disease. Often, no single molecular test can detect all mutations of the disease gene in question. For example, two genes, *breast cancer one* and *two* (*BRCA1* [OMIM 113705] and *BRCA2* [OMIM 600185]) are implicated in the development of breast and ovarian cancer. When functioning normally, the *BRCA1* and *BRCA2* gene products help control cell growth in breast and ovarian tissue. However, mutations that cause loss of or abnormal function of the genes' products can lead to the development of cancer. (See Chapter 18, p. 387, for more discussion of the *BRCA* genes and cancer.) Hundreds of mutations have been identified in *BRCA1* and *BRCA2*, but the risk of developing breast cancer varies widely among the mutations.

Obviously, this makes it impossible to develop a single DNA molecular test for *BRCA* gene mutations.

Some distinctions must now be made. A genetic test is done on a targeted population of people with symptoms of or a significant family history of the disease; the results inform an investigator whether an individual has a mutation known to be associated with a genetic disease. By contrast, screening for a disease is done on people without symptoms or a family history of the disease. For example, mammograms are clinical screening tests that detect breast lesions that might lead to cancer before there are clinical symptoms. Genetic testing for breast cancer, by contrast, reveals the presence or absence of mutations potentially associated with breast cancer development, although it cannot predict whether or when breast cancer will develop.

In the same vein, genetic tests are different from diagnostic tests for a disease. Diagnostic tests reveal whether a disease is present and to what extent the disease has developed. For example, a biopsy of a lump in the breast is a diagnostic test to determine whether the lesion is benign or cancerous.

Purposes of Human Genetic Testing

Genetic testing is done for three main purposes: prenatal diagnosis, newborn screening, and carrier (heterozygote) detection.

Prenatal diagnosis is done to assess whether a fetus is at risk for a genetic disorder. Amniocentesis or chorionic villus sampling can be performed and a sample analyzed for a specific gene mutation or for biochemical or chromosomal abnormalities (see Chapter 10, pp. 218–219). If both parents are asymptomatic carriers (heterozygotes) for a genetic disease, for example, there is a ¼ chance of the fetus being homozygous for the mutant allele, and the risk of developing the disease is likely to be very high. More recently, techniques have been developed to test embryos for genetic disorders before using them for in vitro fertilization. Embryos containing mutated genes that could lead to serious genetic disease can then be removed before implantation is performed.

Newborns can also be tested for specific mutations. For example, all newborns in the United States are tested for PKU (phenylketonuria; OMIM 261600). Other tests are available for groups at high risk for other genetic disorders, such as sickle-cell anemia (OMIM 141900) in African Americans and Tay-Sachs disease (OMIM 272800) in Ashkenazi Jews. These genetic tests, including the DNA molecular tests, typically are done using blood samples.

Testing individuals to see whether they are carriers (heterozygotes) for a recessive genetic disease is done to identify those who may pass on a deleterious gene to their offspring. Carriers can be detected now for a large number of genetic diseases, including Tay-Sachs disease, Duchenne muscular dystrophy (a progressive disease resulting in muscle atrophy and muscle dysfunction; OMIM 310200), and cystic fibrosis (OMIM 602421). DNA molecular tests can readily be done on blood samples.

Examples of DNA Molecular Testing

For DNA molecular testing, DNA samples typically are analyzed by restriction enzyme digestion and Southern blotting or by procedures involving PCR. In this section, we discuss examples of these testing approaches.

Testing by Restriction Fragment Length Polymorphism Analysis.

A mutation associated with a genetic disease may cause the loss or addition of a restriction site either within the gene or in a flanking region. In other words, in the human genome (and the genomes of other eukaryotes), different restriction maps can be found among individuals for the same region of a chromosome. The different patterns of restriction sites result in **restriction fragment length polymorphisms** (**RFLPs,** pronounced "riff-lips"), which are restriction enzyme–generated fragments of different lengths. (*Polymorphism* means the existence of many different forms, here a region of DNA that can have different lengths on different homologues or in different individuals.) A restriction map is independent of gene function, so a RFLP is detected whether or not the DNA sequence change responsible affects a detectable phenotype. A RFLP is a DNA marker that can be used in the same way as a conventional gene marker. In this case, we assay the DNA—that is, determine the genotype—directly in the form of a restriction map. Moreover, because we are looking directly at DNA, both parental types are seen in heterozygotes, so carriers can be identified easily.

RFLPs are useful in DNA molecular testing. In fact, many RFLPs are associated with genes known to cause disease, as the following example of sickle-cell anemia illustrates. In sickle-cell anemia, a single base pair change in the gene for hemoglobin's β-globin polypeptide results in an abnormal form of hemoglobin, Hb-S, instead of the normal Hb-A (see Chapter 10, pp. 213–215). Hb-S molecules associate abnormally, leading to sickling of the red blood cells, tissue damage, and possibly death.

The sickle-cell mutation changes an A T base pair to a T A base pair, so that the sixth codon for β-globin is changed from GAG to GUG. As a result, valine is inserted into the polypeptide instead of glutamic acid (Figure 14.2). The mutational change also generates a RFLP for the restriction enzyme *Dde*I (pronounced "D-D-E-one"). The *Dde*I restriction site is

$$5'\text{-CTNAG-}3'$$
$$3'\text{-GANTC-}5'$$

where the central base pair can be any of the four possible base pairs. The A T to T A mutation changes the fourth base pair in the restriction site. Thus, in the normal β-globin gene, β^A, there are three *Dde*I sites, one upstream of the start of the gene and the other two within the cod-

Figure 14.2

The beginning of the β-globin gene, mRNA, and polypeptide showing the normal Hb-A sequences and the mutant Hb-S sequences. The sequence differences between Hb-A and Hb-S are shown in red. The mutation alters a *Dde*I site (boxed in the Hb-A DNA).

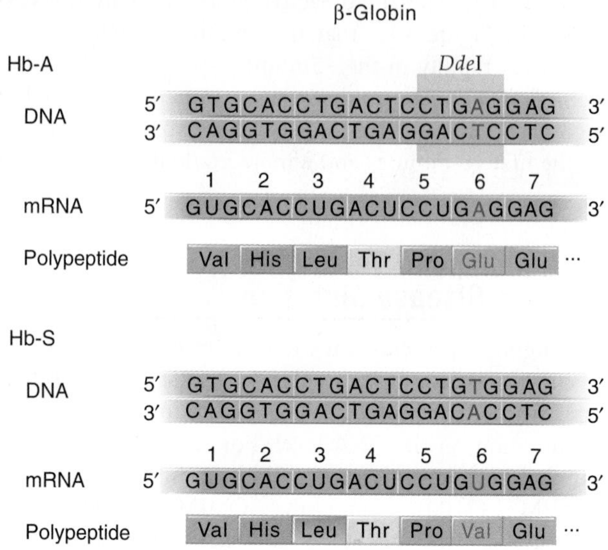

ing sequence (Figure 14.3a). In the sickle-cell mutant β-globin gene, β^S, the mutation has removed the middle *Dde*I site (see Figure 14.2), leaving only two *Dde*I sites (see Figure 14.3a). When DNA from normal individuals is cut with *Dde*I and the fragments separated by gel electrophoresis are transferred to a membrane filter by the Southern blot technique and then probed with the 5' end of a cloned β-globin gene, two fragments of 175 bp and 201 bp are seen (Figure 14.3b). DNA from individuals with SCA analyzed in the same way gives one fragment of 376 bp because of the loss of the *Dde*I site. Heterozygotes are detected by the presence of three bands of 376 bp, 201 bp, and 175 bp.

Not all RFLPs result from changes in restriction sites directly related to the gene mutations. Many result from changes to the DNA flanking the gene, sometimes a fair distance away. In these cases, detection of the mutation relies on the flanking RFLP segregating most of the time with the gene mutation. Recombination occurring between the RFLP and the gene of interest can occur, however, and this can cause some difficulty in interpreting the results.

Testing Using a PCR Approach.

DNA molecular tests using PCR can be developed only when sequence information is available; otherwise, the PCR primers cannot be designed. One common test using PCR is **allele-specific oligonucleotide (ASO) hybridization.** The principles are illustrated in this example of testing for mutations of the *GLC1A* gene (OMIM 137750), one of several genes that, when mutated, cause the most common form of glaucoma. This form of glaucoma has no symptoms ini-

Figure 14.3

Detection of sickle-cell gene by the *Dde*I restriction fragment length polymorphism. **(a)** DNA segments showing the *Dde*I restriction sites. **(b)** Results of analysis of DNA cut with *Dde*I, subjected to gel electrophoresis, blotted, and probed with a β-globin probe.

a) *Dde*I restriction sites

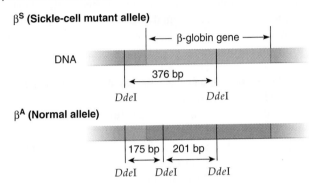

β^S (Sickle-cell mutant allele)

β^A (Normal allele)

b) *Dde*I fragments detected on a Southern blot by probing with beginning of β-globin gene

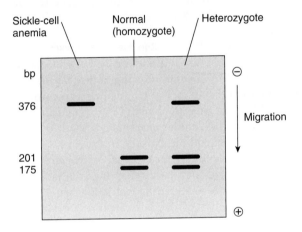

tially, but as the pressure in the eye builds, at some point peripheral vision is lost, and if the disease is not diagnosed and treated, total blindness can occur.

A number of glaucoma-causing mutations have been identified in the *GLC1A* gene. One of the mutations involves a change from a CG to a TA in the DNA, resulting in a codon change from CCG (Pro) to CUG (Leu). Figure 14.4a presents the sequence of part of the *GLC1A* gene to show the mutation; the DNA from a heterozygote was sequenced, so that both the wild-type C and the mutant T are seen at the mutation location.

animation

a **Polymerase Chain Reaction**

Based on the *GLC1A* gene sequence, primers were designed for PCR amplification of the region of the gene containing the mutation. The PCR product was separated by agarose gel electrophoresis, then extracted from the gel and dotted onto two membrane filters under conditions that denatured the DNA to single strands. Two ASOs were made, one for the wild-type allele and one for the mutant

allele (Figure 14.4b). In this case, each ASO was 19 nucleotides (nt) long, with the mutation position approximately in the middle. The ASOs were labeled radioactively and each hybridized with the DNA immobilized on one of the filters. The resulting autoradiograms for each DNA sample indicated whether the individual from whom the DNA was taken was homozygous normal, heterozygous, or homozygous mutant. As Figure 14.4c shows, for a homozygous normal individual, a signal is seen only for the wild-type ASO; for a heterozygous individual, a signal is seen for both ASOs; and for a homozygous mutant individual, a signal is seen only for the mutant ASO. This method has been used to analyze affected members of glaucoma families for the presence of particular mutations.

Figure 14.4

DNA molecular testing for mutations of the open-angle glaucoma gene, *GLC1A*, using PCR and allele-specific oligonucleotide (ASO) hybridization. **(a)** Sequence of part of the *GLC1A* gene from a heterozygote, showing a mutation from C to T, causing a Pro to Leu change in the polypeptide at amino acid 370. **(b)** Sequences of the two allele-specific oligonucleotides (ASOs), one for the wild-type allele and one for the mutant allele. **(c)** Results (theoretical) of hybridization with radioactive ASOs for homozygous normal, homozygous mutant, and heterozygous individuals.

a)

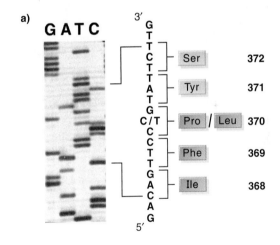

b) ASOs

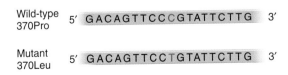

Wild-type 370Pro   5′ GACAGTTCC**C**GTATTCTTG 3′

Mutant 370Leu   5′ GACAGTTCC**T**GTATTCTTG 3′

c) ASO probe

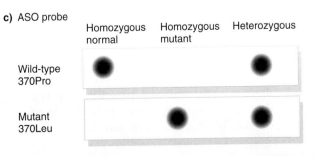

In general, ASO hybridization uses one radioactively labeled ASO as a probe for hybridization with a PCR product immobilized on a membrane filter. This approach allows one allele to be probed for on each filter and therefore is used to test individuals for the presence of a single particular mutation.

KEYNOTE

> Recombinant DNA and PCR techniques are used in DNA molecular testing for human genetic disease mutations. These tests have become possible as knowledge about the molecular nature of mutations associated with human genetic diseases has increased. In general, human genetic testing is done for prenatal diagnosis, newborn screening, or carrier detection. Many DNA molecular tests are based on restriction fragment length polymorphisms (RFLPs) or on PCR amplification followed by allele-specific oligonucleotide (ASO) hybridization.

Isolation of Human Genes

With a defined gene product isolated by biochemical procedures, a gene can be cloned. However, cloning a gene is difficult if the gene product that is altered is unknown. Fortunately, a number of approaches are available to solve this problem. One approach is to identify a RFLP marker that is genetically linked to the disease phenotype and then to home in on the gene, starting from the marker location on the chromosome. The isolation of a gene associated with a genetic disease on the basis of its approximate chromosomal position is called **positional cloning.** Some of the techniques used in positional cloning are illustrated in the story of how the cystic fibrosis gene was cloned. Fortunately, not all gene cloning is this lengthy or laborious.

Cloning the Cystic Fibrosis Gene

Cystic fibrosis (CF; OMIM 219700) is the most common lethal genetic disease in the United States today. CF is a disease inherited in an autosomal recessive fashion. Disease symptoms and genetic properties of the disease are described in Chapter 10 (pp. 216–217). The CF gene was the first human disease gene to be cloned solely by positional cloning. The effort took four years and the involvement of many researchers in many laboratories.

Identifying RFLP Markers Linked to the CF Gene. Many individuals in CF pedigrees were screened with a large number of RFLPs to determine whether any RFLPs were linked genetically to the CF gene. This was done by tracking the inheritance of the CF gene in the families and simultaneously analyzing their DNA by Southern blot analysis and hybridizing with probes to identify any RFLP marker (detected as characteristic DNA fragment sizes) that showed genetic linkage to the CF locus. One RFLP showed weak linkage to the CF locus.

Identifying the Chromosome on Which the CF Gene Is Located. The RFLP marker was used to identify the chromosome on which the CF gene is located. This was done by *in situ hybridization*, a technique in which chromosomes are spread on a microscope slide, hybridized with a labeled probe, and the results analyzed by autoradiography. By using a ^{3}H-labeled RFLP probe, the CF gene was shown to be on chromosome 7.

Identifying the Chromosome Region Where the CF Gene Is Located. Other known RFLPs on chromosome 7 were used to find those most closely linked to the CF gene. Two closely linked flanking markers (one marker on each side of the CF gene) were found that are 0.5 map unit apart, which corresponds to about 500,000 bp in the human genome. These two flanking markers were known to be located at region 7q31-q32 (7 = chromosome 7, q = the long arm [p is the short arm], 31–32 = subregions 31 and 32), so this localized the CF gene to that section of chromosome 7.

Cloning the CF Gene Between the Flanking Markers. An approach that can be used to find a gene between flanking markers is **chromosome walking,** a process used to identify adjacent clones in a genomic library. In chromosome walking, an initial cloned DNA fragment—for example, one of the flanking markers—is used to begin the walk. A labeled end piece of the initial clone is used to screen a genomic library for clones that hybridize with it, in particular the clones that overlap the original clone. These clones can be analyzed by restriction mapping to determine the extent of overlap. Then a new labeled probe can be made from the far end of a clone with minimal overlap, and the library is screened again. By repeating this process over and over, we can walk along the chromosome clone by clone.

A procedure related to chromosome walking is also used to move along chromosomes. Called *chromosome jumping*, it is a technique to span large amounts of DNA. Whereas in chromosome walking each step is an overlapping DNA clone, in chromosome jumping each jump is from one chromosome location to another without "touching down" on the intervening DNA.

For cloning the CF gene, a combination of chromosome jumps and chromosome walks were made. In the end, a large number of clones were isolated that spanned over 500 kb of DNA of the CF region.

Identifying the CF Gene in the Cloned DNA. The clones spanning the region between the RFLP markers were known to include the CF gene itself, but how does one identify the particular gene of interest in the set of clones?

One way to home in on genes in such clones is to use cloned DNA as probes to see whether they can hybridize with sequences in other species. The reasoning is that genes are conserved in sequence among related species, whereas nongene sequences are less likely to be conserved. The procedure is to digest genomic DNA from other organisms (such as mouse, hamster, or chicken) with restriction enzymes and analyze the fragments by Southern blotting and hybridization with a labeled probe. Since the blot contains DNA from a variety of organisms, it is often called a *zoo blot.* (By the way, a blot with DNA from males and females of a variety of organisms is called a *Noah's ark blot.*) In the CF project, five subcloned segments of the CF region used as probes cross-hybridized with DNA sequences from other organisms, identifying them as candidates for the CF gene. Two of the probes were ruled out based on linkage analysis, and a third was ruled out because it proved to be a *pseudogene,* that is, a sequence resembling a functional gene but lacking signals to permit it to be expressed.

Another property of protein-coding genes is that they produce mRNAs when they are expressed. Thus, a DNA probe made from a protein-coding gene or part of such a gene should hybridize with mRNAs on a northern blot (see Chapter 13, pp. 278–279). Northern blotting ruled out a fourth probe, leaving only one.

A probe made from this fifth clone was used to screen a cDNA library made from mRNA molecules isolated from cultured normal sweat gland cells. This tissue type was used as the source of mRNA because the symptoms of CF were believed to result from a defect in sodium and chloride transport, and such transport is an important function of sweat gland cells. The probe identified a single, positive clone on northern blots about 6,500 bp in length, indicating an mRNA of the same size. The cDNA clone was used to analyze the genomic clones in more detail: The *candidate CF gene* was shown to span approximately 250 kb of DNA and have 24 exons. Confirmation that the CF gene had been cloned came from comparing the DNA sequences of the candidate CF gene in a normal individual and in an individual with CF. The expectation was that mutational changes would be obvious if the correct gene had been identified. This proved to be the case: A 3-bp deletion was detected in the gene from the patient with CF. The product of the CF gene, cystic fibrosis transmembrane conductance regulator protein, was described in Chapter 10 (p. 217 and Figure 10.10).

K E Y N O T E

The isolation of human genes, particularly those associated with disease, is possible using an array of molecular techniques. Where the gene product is not known, the starting point for cloning is knowledge of the genetic linkage between the disease locus and one or more DNA markers. The isolation of a gene associated with a genetic disease on the basis of its approximate chromosomal position is called positional cloning.

DNA Typing

No two human individuals (except identical twins) have exactly the same genome, base pair for base pair, and this has led to techniques for **DNA typing** (also called **DNA fingerprinting** or **DNA profiling**) for use in forensic science, in paternity and maternity testing, and elsewhere. DNA typing relies on DNA analysis using molecular markers such as RFLPs detected by restriction enzyme digestion and blotting analysis, length polymorphisms detected by PCR amplification and agarose gel electrophoresis, and alleles detected by ASOs (allele-specific oligonucleotides). (These approaches are very similar to those described on pp. 289–292.)

The most useful kind of molecular markers here are those that are highly polymorphic, so that there are lots of differences among individuals in a population. For example, there are many regions in the genome where there are short, identical segments of DNA tandemly arranged head to tail. Individuals may show a great variation in the number of tandem repeats. When the repeating unit is two, three, or four base pairs long, it is called an **STR (short tandem repeat)** or **microsatellite.** When the repeating unit is about five to a few tens of base pairs long, it is called a **VNTR (variable number of tandem repeats)** or **minisatellite.**

The use of STRs or VNTRs as markers in RFLP-based DNA typing is illustrated in Figure 14.5. Individual A is heterozygous for two alleles at an STR or VNTR locus flanked by restriction sites. One allele has three copies of the repeat sequence, and the other allele has ten copies (in reality, the number of copies typically is larger). Individual B is homozygous for a six-repeat allele. When genomic DNA is digested with the restriction enzyme, restriction fragments of different sizes are produced because of the STRs or VNTRs. The results of the digests can be visualized by using a probe for the particular repeat sequence at the molecular marker locus. (A probe that is specific for the STR or VNTR sequences at one locus in the genome, as is the case here, is called a *monolocus, or single-locus, probe.* Probes that detect STR or VNTR sequences at a number of loci in the genome are known as *multilocus probes.*) In this case, two DNA bands would be detected for individual A, and one DNA band would be detected for individual B.

Such polymorphism can also be detected by PCR amplification using primers that flank the STRs or VNTRs. The only proviso here is that sequence information be available for designing the primers.

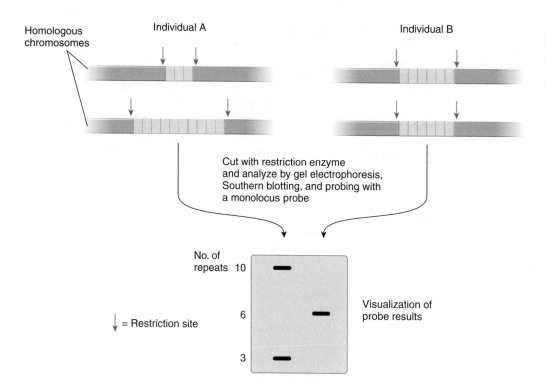

Figure 14.5
The concept involved in using STRs (microsatellites) or VNTRs (minisatellites) as DNA markers.

DNA Typing in a Paternity Case

Let us now consider how DNA typing is used in a paternity case. In this fictional scenario, a mother of a new baby has accused a particular man of being the father of her child, and the man denies it. The court will decide the case based on evidence from DNA typing. The DNA typing proceeds as follows (Figure 14.6): DNA samples are obtained from all three individuals involved (step 1). In a paternity case, the usual source of DNA is a blood sample. The DNA is cut with the restriction enzyme for the marker to be analyzed, and the resulting fragments are separated by electrophoresis (step 2), transferred to a membrane filter by Southern blotting (step 3), and probed with a labeled monolocus STR or VNTR probe (steps 4 and 5). The DNA banding pattern after autoradiography or chemiluminescence detection is then analyzed to compare the samples (step 6).

The data can be interpreted as follows: Two DNA fragments are detected for the mother, so she is heterozygous for one particular pair of alleles at the STR or VNTR locus under study. Likewise, two DNA fragments are detected for the baby, so the baby is also heterozygous. One of the fragments for the baby matches the larger, slow moving fragment for the mother, and the other fragment for the baby is much larger, indicating many more repeats in that allele. The baby receives one allele from its mother and one from its father. Does the paternal allele of the baby match an allele from the alleged father? Inspection of the paternal lane in the autoradiogram leads us to conclude that the answer is yes.

The data indicate that the man shares an allele with the baby, but they do not prove that he contributed that allele to the genome of the baby, although he obviously could have. If the man had no alleles in common with the baby, then the DNA typing data would have proved that he is not the father; this is the *exclusion* result. To establish positive identity—the *inclusion* result—through DNA typing is more difficult. It requires calculating the relative odds that the allele came from the accused or from another person. This calculation depends on knowing the frequencies of STR or VNTR alleles identified by the probe in the ethnic population from which the man comes. Most legal arguments focus on this matter because good estimates of STR or VNTR allele frequencies are known for only a limited array of ethnic groups, so that calculations of probability of paternity have questionable accuracy in many cases. To minimize possible inaccuracy, investigators use a number of different probes (often five or more) so that the combined probabilities calculated for the set of STRs or VNTRs can be high enough to convince a court that the accused is actually the parent (or is guilty in a criminal case), even allowing for problems knowing the true STR or VNTR allele frequencies for the population in question.

It is these combined probabilities that you hear or read about in the media with respect to DNA typing in court cases. In fact, DNA typing is not yet a generally accepted method for proving parenthood or guilt in U.S. courts. The scientific basis for the method is not in question; rather, DNA evidence is most commonly rejected for reasons such as possible errors in evidence collection or processing or weak population statistics. However, there are many cases in which DNA typing has excluded an accused individual because of mismatches, and such con-

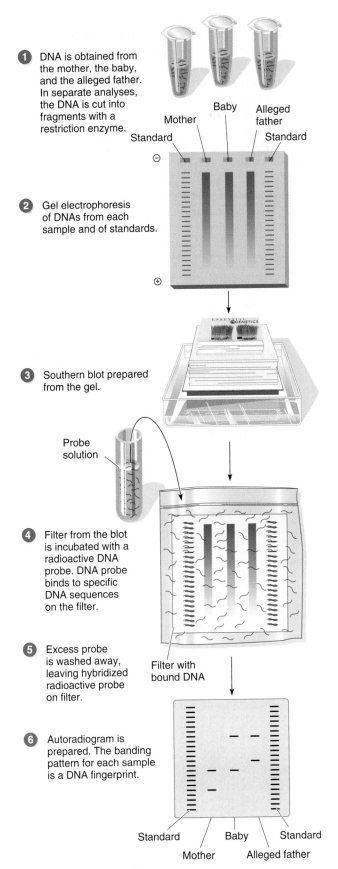

Figure 14.6
DNA typing as used in a paternity case.

① DNA is obtained from the mother, the baby, and the alleged father. In separate analyses, the DNA is cut into fragments with a restriction enzyme.

② Gel electrophoresis of DNAs from each sample and of standards.

③ Southern blot prepared from the gel.

④ Filter from the blot is incubated with a radioactive DNA probe. DNA probe binds to specific DNA sequences on the filter.

⑤ Excess probe is washed away, leaving hybridized radioactive probe on filter.

⑥ Autoradiogram is prepared. The banding pattern for each sample is a DNA fingerprint.

clusions are now widely accepted by U.S. courts. In our paternity case, we would probably be more persuaded that the accused was the father of the child if the data for each of five different monomorphic probes indicated that he contributed a particular allele to the child.

Other Applications of DNA Typing

There are many uses of DNA typing. Here we list just a few examples to illustrate the scope of usefulness of DNA typing for human testing and for tests involving other organisms.

1. *Forensic analysis in murder, rape, and other violent crimes.* Typically, DNA samples are taken during criminal investigations and compared with those of victims and suspects. In a murder case, DNA can be isolated from blood or hair at the scene; in a rape case, DNA can be taken from a semen sample. If only minuscule amounts of DNA can be collected, PCR is used to amplify the DNA for typing experiments.

2. *Conservation biology studies.* DNA typing can be used in studies of endangered species to determine genetic variability in those species.

3. *Forensic analysis in wildlife crimes.* Wild animals sometimes are killed illegally, and DNA typing is increasingly helping to solve the crimes. For example, a set of six STR markers was used in a poaching investigation in Wyoming involving pronghorn antelope. Six headless pronghorn antelope carcasses were discovered and reported to authorities. An investigation turned up a suspect who had a skull with horns. DNA samples were taken from the skull and compared with DNA samples from carcass samples, and a match was found. At the trial, the suspect was convicted on six counts of wanton destruction of big or trophy game.

4. *Testing for pathogens in food.* PCR using strain-specific primers can test for the presence of pathogenic *E. coli* strains in foods such as hamburger meat.

5. *Detection of genetically modified organisms (GMOs).* GMOs have been introduced widely into agriculture in the United States. Genetically modified crops typically contain genes that were introduced in the development of the new crop. PCR primers based on those sequences or the promoters that control them can be used to test for the presence of those genes. We can do these tests with the plants themselves or with processed foods. One current estimate is that between 50 and 75 percent of produce and processed foods in a supermarket are genetically modified or contain GMOs.

There are also an increasing number of interesting applications of DNA typing to help resolve historical controversies and mysteries. For example, in 1795, a 10-year-old boy died of tuberculosis in the tower of the Temple Prison in France. The great mystery was whether the boy was the dauphin, the sole surviving son of Louis XVI and Marie Antoinette, who were executed by republicans on the guillotine, or whether he was a stand-in while the true heir to the throne escaped. The dead boy's heart was saved after the autopsy, and despite some very rough handling and storage conditions since his death, in December 1999 two small tissue samples were taken from the heart; remarkably, DNA could be extracted from them. This DNA was typed against DNA extracted from locks of the dauphin's hair kept by Marie Antoinette, from two of the queen's sisters, and from present-day descendants. The results showed that the dead boy was the dauphin.

KEYNOTE

> DNA typing, or DNA fingerprinting, is done to distinguish individuals based on the concept that no two individuals of a species, save for identical twins, have the same genome sequence. The variations are manifested in restriction fragment length polymorphisms and length variations resulting from different numbers of short, tandemly repeated sequences. DNA typing has many applications, including basic biological studies, forensics, detecting infectious species of bacteria, and analysis of old or ancient DNA.

Gene Therapy

Theoretically, two types of gene therapy are possible: somatic cell therapy, in which somatic cells are modified to correct a genetic defect, and germ-line cell therapy, in which germ-line cells are modified to prevent a genetic defect in the offspring. Somatic cell therapy results in a treatment for the genetic disease in the individual, but progeny can still inherit the mutant gene. Germ-line cell therapy, however, could prevent the disease because the mutant gene would be replaced by the normal gene, and that normal gene would be inherited by the offspring. Both somatic cell therapy and germ-line cell therapy have been used successfully in nonhuman organisms, but only somatic cell therapy has been used in humans because of ethical issues raised by germ-line cell therapy.

The most promising candidates for somatic cell therapy are genetic disorders that result from a simple defect of a single gene and for which the cloned normal gene is available. Gene therapy involving somatic cells proceeds as follows: A sample of the individual's mutant cells is taken. Then normal, wild-type copies of the mutant gene

are introduced into the cells, and the cells are reintroduced into the individual. There, it is hoped, the cells will produce a normal gene product and the symptoms of the genetic disease will be completely or partially reversed. A cell that has had a gene introduced into it by artificial means such as this is called a **transgenic cell**, and the gene involved is called a **transgene.**

The source of the mutant cells varies with the genetic disease. For example, blood disorders, such as thalassemia or sickle-cell anemia, require modification of blood-line cells isolated from the bone marrow. For genetic diseases affecting circulating proteins, a promising approach is the gene therapy of skin fibroblasts, cells that are constituents of the dermis (the lower layer of the skin). Modified fibroblasts can easily be implanted back into the dermis, where blood vessels invade the tissue, allowing gene products to be distributed.

Successful somatic gene therapy has been demonstrated repeatedly in experimental animals such as mice, rats, and rabbits. In recent years, a few gene therapy trials have been done with humans. For example, in 1990, a 4-year-old girl suffering from severe combined immunodeficiency (SCID; OMIM 102700) caused by a deficiency in adenosine deaminase (ADA), an enzyme needed for normal function of the immune system, was treated by somatic gene therapy. T cells (cells involved in the immune system) were isolated from the girl and grown in the laboratory, and the normal ADA gene was introduced using a viral vector. The "engineered" cells were then reintroduced into the patient. Since T cells have a finite life in the body, continued infusions of engineered cells have been necessary. The introduced ADA gene is expressed, probably throughout the life of the T cell. As a result, the patient's immune system is functioning more normally, and she now gets no more than the average number of infections. The gene therapy treatment has enabled her to live a more normal life.

Successful somatic gene therapy has also been achieved for sickle-cell anemia (OMIM 141900). In December 1998, 13-year-old Keone Penn's bone marrow cells were replaced with stem cells from the umbilical cord of an unrelated infant, with the hope that the new cells would produce healthy bone marrow, the source of blood cells. After one year, there were no signs of sickled cells, and the patient was declared cured of the disease.

With time, many other genetic diseases are expected to be treatable with somatic gene therapy, including thalassemias, phenylketonuria, Lesch-Nyhan syndrome, cancer, Duchenne muscular dystrophy, and cystic fibrosis. For example, after successful experiments with rats, human clinical trials are under way for transferring the normal CF gene to patients with cystic fibrosis. However, many scientific, ethical, and legal questions must be addressed before gene therapy is implemented routinely.

KEYNOTE

> Gene therapy is the treating of a genetic disorder by introducing into the individual a normal gene to replace or overcome the effects of a mutant gene. For ethical reasons, only somatic gene therapy is being developed for humans. There are few examples of successful somatic gene therapy in humans, but there is great hope for treating many genetic diseases in this way in the future.

Commercial Products

The development of cloning and other DNA manipulation techniques has spawned the formation of many *biotechnology* companies, some of which focus on using DNA manipulations for making a wide array of commercial products. Although the details vary, the general approach to making a product is to express a cloned gene or cDNA in an organism that will transcribe the cloned sequence and translate the mRNA. For example, the production of recombinant protein products in transgenic mammals (in this case, sheep) is illustrated in Figure 14.7. Here the gene of interest (*GOI* in the figure) is placed in an expression vector so that it is adjacent to a promoter that is active only in mammary tissue, such as the β-lactoglobulin promoter. The recombinant DNA molecules are microinjected into sheep ova, and each ovum is then implanted into a foster mother. Transgenic offspring are identified using PCR to detect the recombinant DNA sequences. When these transgenic animals mature, the β-lactoglobulin promoter begins to express the associated gene in the mammary tissue, the milk is collected, and the protein of interest is obtained by biochemical separation techniques.

Here are some of the many other products produced by biotechnology companies:

- Tissue plasminogen activator (TPA), used to prevent or dissolve blood clots, thereby preventing strokes, heart attacks, or pulmonary embolisms
- Human growth hormone, used to treat pituitary dwarfism
- Human insulin ("humulin"), used to treat insulin-dependent diabetes
- DNase, used to treat cystic fibrosis
- Recombinant vaccines, used to treat human and animal viral diseases (such as hepatitis B in humans)
- Genetically engineered bacteria and other microorganisms used to improve production of, for example, industrial enzymes (such as amylases to break down starch to glucose), citric acid (for flavoring), and ethanol
- Genetically engineered bacteria used to accelerate the degradation of oil pollutants or certain chemicals (such as dioxin) in toxic wastes

Figure 14.7

Production of a recombinant protein product (here the protein encoded by the gene of interest, *GOI*) in a transgenic mammal, in this case a sheep.

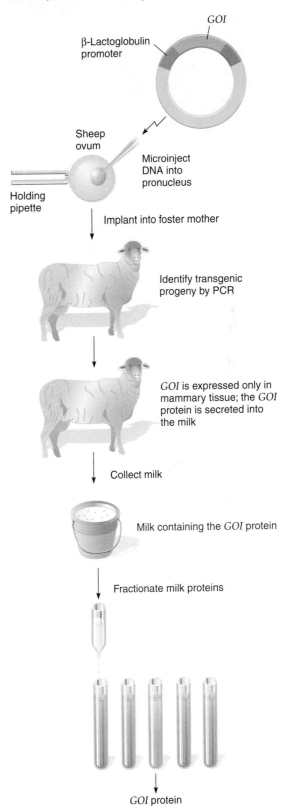

KEYNOTE

> With the same kinds of recombinant DNA and PCR techniques used in basic biological analysis, DNA molecular testing, gene cloning, DNA typing, and gene therapy, biotechnology and pharmaceutical companies develop useful products. Many types of products are now available or are in development, including pharmaceuticals and vaccines for humans and animals and genetically engineered organisms for improved production of important compounds in the food industry or for cleaning up toxic wastes.

Genetic Engineering of Plants

For many centuries, the traditional genetic engineering of plants involved selective breeding experiments in which plants with desirable traits were used to produce offspring with those traits. As a result, humans have produced hardy varieties of plants (e.g., corn, wheat, and oats) and increased yields, all using long-established plant-breeding techniques. (Similar techniques have also been used with animals, such as dogs, cattle, and horses, to produce desired breeds.) Now, vectors developed by recombinant DNA technology are available for transforming cells of crop plants; this has made possible the genetic engineering of plants for agricultural use.

We mentioned in the section on DNA typing that a very large number of genetically modified crops already have been developed, and a lot of the processed food we buy contains them. Let us briefly consider approaches to generating transgenic plants that are tolerant to the broad-spectrum herbicide Roundup to illustrate the types of approaches that are possible. Roundup contains the active ingredient glyphosate, which kills plants by inhibiting EPSPS, a chloroplast enzyme required for the biosynthesis of essential aromatic amino acids. Roundup is used widely to kill weeds because it is active in low doses and is degraded rapidly in the environment by microbes in the soil. If a crop plant is resistant to Roundup, a field can be sprayed with the herbicide to kill weeds without affecting the crop plant. Approaches for making transgenic, Roundup-tolerant plants include introducing a modified bacterial form of EPSPS that is resistant to the herbicide, so that the aromatic amino acids can still be synthesized even when the chloroplast enzyme is inhibited (Figure 14.8), and introducing genes that encode enzymes for converting the herbicide to an inactive form. Monsanto brought Roundup Ready soybeans to market in 1996, although their use has been controversial because of opposition by groups questioning the safety of genetically engineered plants for human consumption.

In the near future, we can expect many more genetically engineered plants to be developed. Of particular value will be crop plants that have increased yield, insect pest resistance, and herbicide tolerance. Such plants could be very helpful in alleviating world hunger. However, there is significant public resistance to genetically modified plants in many countries, including a growing resistance in the United States.

KEYNOTE

> Genetic engineering of plants is also possible using recombinant DNA and/or PCR techniques. It is expected that many more types of improved crops will result from future applications of this new technology.

Figure 14.8

Making a transgenic, Roundup-tolerant tobacco plant by introducing a modified form of the bacterial gene for the enzyme EPSPS that is resistant to the herbicide. The gene encoding the bacterial EPSPS was spliced to a petunia sequence encoding a transit peptide for directing polypeptides into the chloroplast, and the modified gene was introduced into tobacco. Both the native and the modified bacterial EPSPS are transported into the chloroplast. When plants are sprayed with Roundup, wild-type plants die because only the native chloroplast EPSPS is present and it is sensitive to the herbicide, but the transgenic plants live because they contain the bacterial EPSPS that is resistant to the herbicide.

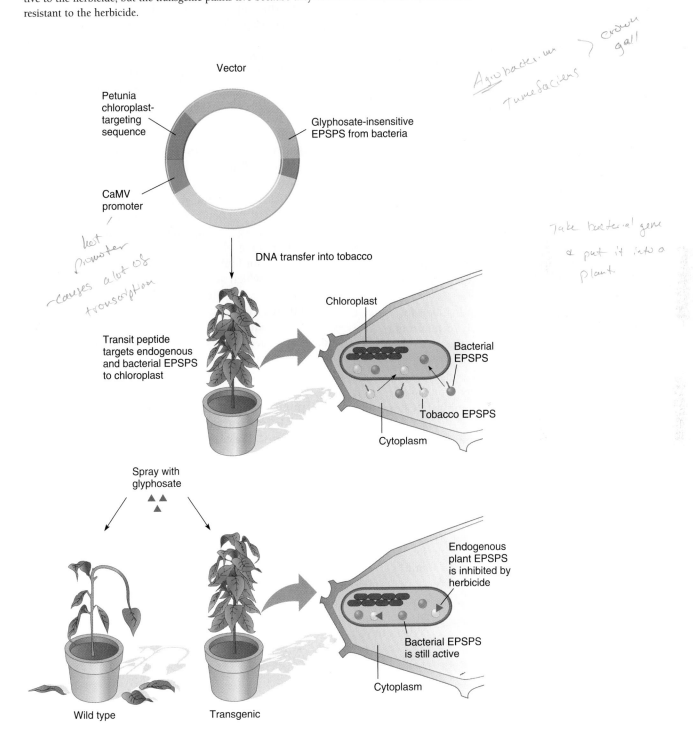

Summary

Recombinant DNA technology and PCR are being widely applied in both basic research and commerce. All areas of modern biology have been revolutionized by these new molecular techniques. Perhaps the most ambitious project that would not have been possible without these techniques is the Human Genome Project (HGP), which has the mandate to generate a complete map of all the genes in the human genome and to obtain the complete sequence of the human genome. We discuss the HGP in detail in Chapter 15.

Recombinant DNA and PCR techniques are also being used to develop new pharmaceuticals (including drugs, other therapeutics, and vaccines) and to develop new tools for diagnosing infectious and genetic diseases, as well as for use in human gene therapy, forensic analysis, and agriculture.

Analytical Approaches for Solving Genetics Problems

Q.14.1 ROC is a hypothetical polymorphic STR (microsatellite) locus in humans with a repeating unit of CAGA. The locus is shown in Figure 14.A as a box with 25 bp of flanking DNA sequences.

a. You plan to use PCR to type individuals for the ROC locus. If PCR primers must be 18 nucleotides long, what are the sequences of the pair of primers required to amplify the ROC locus?

b. Consider ROC alleles with 10 and 7 copies of the repeating unit. Using the primers you have designed, what will be the sizes of the amplified PCR products for each allele?

c. There are four known alleles of ROC with 15, 12, 10, and 7 copies of the repeating unit. How many possible human genotypes are there for these alleles, and what are they?

d. If one parent is heterozygous for the 15 and 10 alleles of the ROC locus and the other parent is heterozygous for the 10 and 7 alleles, what are the possible genotypes of their offspring for this locus, and in what proportion will they be found?

e. Growing up in the house with the two parents mentioned in part (d) are three children. When you type them for the ROC locus, you find that their genotypes are (10,10), (15,10), and (12,7). What can you conclude?

A.14.1

a. You need PCR primers that immediately flank the ROC locus. The primers must be of the correct polarity to amplify the DNA between them. Thus, the left primer is 5′-TTGATCTCCTTTAGCTTC, the rightmost 18 nucleotides of the flanking sequence to the left of ROC (reading left to right on the top strand), and the right primer is 5′-TCACATAATGAATTATAC, the leftmost 18 nucleotides of the flanking sequence to the right of ROC (reading right to left on the bottom strand).

b. PCR amplifies the DNA between the two primers used in the reaction. The size of a PCR product is the length of the DNA between the primers plus the lengths of the two primers. So, for a 10-copy allele of the ROC locus, with a repeating unit length of 4 nucleotides, the PCR product is $18 + (10 \times 4) + 18 = 76$ bp. For a 7-copy allele of the ROC locus, the PCR product is $18 + (7 \times 4) + 18 = 64$ bp.

c. Humans are diploid, so there are two copies of each locus in the genome. Each locus can be homozygous or heterozygous. Figuring out the genotypes involves determining all possible pairwise combinations of alleles. For four STR alleles, there are ten genotypes, four of which are homozygous and six of which are heterozygous. The genotypes are (15,15), (12,12), (10,10), (7,7), (15,12), (15,10), (15,7), (12,10), (12,7), and (10,7).

d. This question concerns the segregation of alleles. Each diploid parent produces haploid gametes, and the gametes from each parent pair randomly to produce the diploid progeny. Thus, a (15,10) parent produces equal numbers of 15 and 10 gametes, and a (10,7) parent produces equal numbers of 10 and 7 gametes. They will fuse randomly, as in the following figure.

	(10,7) parent gametes	
	10	7
(15,10) parent gametes — 15	(15,10)	(15,7)
(15,10) parent gametes — 10	(10,10)	(10,7)

Figure 14.A

```
5′-CTGATTCTTGATCTCCTTTAGCTTC          GTATAATTCATTATGTGATAATGCC-3′
3′-GACTAAGAACTAGAGGAAATCGAAG   ROC    CATATTAAGTAATACACTATTACGG-5′
```

ings. The other approach is the *direct shotgun approach,* in which the entire genome is broken up into random, overlapping fragments that are then sequenced. The genome sequence is then assembled by computer based on the sequence overlaps between fragments. In this approach, there is no prior knowledge of the location of the fragment. This is analogous to taking ten books that have been torn randomly into smaller leaflets of a few pages each and, by matching overlapping pages of the leaflets, assembling a complete copy of each book with the pages in the correct order. We will now consider these two approaches briefly.

Genome Sequencing Using a Mapping Approach. In this approach, genetic maps and physical maps are made first, with the goal of having markers mapped relatively closely throughout the genome. One of the initial goals of the HGP was a genetic map with a density of at least one genetic marker per 1 Mb of the genome.

Genetic maps of genomes are constructed using genetic crosses and, for humans, pedigree analysis. The principles used in genetic mapping are those described in detail in Chapter 5. Genetic crosses are used to establish the locations of genetic markers on chromosomes and to determine the genetic distance between them. Genetic distance is given by the frequency of crossing-over between the markers.

The genetic maps are made using genes, the traditional genetic markers for genetic mapping experiments, as well as **DNA markers.** DNA markers are genetic markers that are detected using molecular tools that focus on the DNA itself. With genes and DNA markers, map distances can be calculated between genes, between DNA markers, or between a gene and a DNA marker.

Four major types of DNA markers are used in human genome mapping:

1. **Restriction fragment length polymorphisms** (RFLPs, pronounced "riff-lips"), introduced in Chapter 14 (p. 290 and Figures 14.2 and 14.3). A RFLP can result from a base pair mutation that causes the loss or addition of a restriction site or from the insertion or deletion of DNA between the restriction sites being studied.

2. **Variable number of tandem repeats** (VNTRs), also called minisatellites, introduced in Chapter 14 (p. 293 and Figure 14.5). VNTR alleles differ in the number of tandem repeats, and the size of the repeat is about five to a few tens of base pairs long. There are tens to thousands of tandem copies in a particular allele. Only about 100 VNTR loci are currently known in the human genome.

3. **Short tandem repeats** (STRs), also called microsatellites, introduced in Chapter 14 (p. 293 and Figure 14.5). Like VNTRs, STRs contain a variable number of tandemly repeated sequences, and a specific locus tends to be highly polymorphic. In this case, the size

of the repeat typically is one to four base pairs. There are about 100,000 STR loci in the human genome, with an average of 30,000 bp between them.

4. **Single nucleotide polymorphisms** (SNPs, pronounced "snips"). The DNA sequences of two individuals are mostly identical, with base pair differences approximately every 500 to 1,000 bp. One base pair at a position showing a difference is the common one, present in most individuals, and another base pair at that position is a less common variant. If that less common base pair is present in at least 1 percent of the human population, that genomic base pair location is defined as the site of a SNP. There are about 3 million SNPs in the human genome, many more than the other types of DNA markers described. This type of polymorphism is very important in human genetics because it represents about 98 percent of all DNA polymorphisms.

K E Y N O T E

Genetic maps of genomes are constructed using recombination data from genetic crosses in the case of experimental organisms or from pedigree analysis in the case of humans. Both gene markers and DNA markers are used in genetic mapping analysis, the latter including restriction fragment length polymorphisms (RFLPs), variable number of tandem repeats (VNTRs), short tandem repeats (STRs), and single nucleotide polymorphisms (SNPs).

For organisms with small genomes, genetic maps have sufficient resolution to provide landmarks for a genome-sequencing effort. When the *E. coli* genome-sequencing project began, for example, there were 1,400 genetic markers, with an average of one per 3.3 kb. In humans, however, the genetic map did not have sufficient resolution to begin sequencing. Researchers had to supplement the genetic map with a detailed **physical map,** a map of genetic markers made by analyzing genomic DNA directly rather than by analyzing recombinants from pedigree analysis (for humans) or genetic crosses (for experimental organisms), as is the case with genetic maps.

There are several types of physical maps, including, in order of increasing resolution, cytogenetic maps of chromosomal banding patterns (see Chapter 1, p. 13), restriction maps, and **clone contig maps.** A clone contig map is a set of ordered, partially overlapping clones comprising all the DNA of an entire chromosome, or part of a chromosome, without any gaps. (The word *contig* is a shortened form of *contiguous.*) The clones are made using special vectors capable of incorporating very large DNA inserts. High-resolution clone contig maps provided the framework for sequencing the entire genome.

With a high-resolution map completed, sequencing is done using automated DNA sequencing as described in Chapter 13 (p. 280). Current technology is limited to sequencing about 500 nucleotides in a single reaction. To obtain an accurate sequence, both strands must be sequenced, and the sequencing must be repeated several times; hence the enormous task of sequencing the 3 billion base pairs of the human genome. Fortunately, improving sequencing and sequence analysis technologies is an important part of genome projects. In recent years, automated sequencing has become much quicker through the development of more efficient machines and the increasing use of robotics for preparing samples for sequencing.

For the human genome, clone contig maps were used for sequencing following the mapping approach to genome sequencing. However, the clone inserts were too large to be sequenced in a single sequencing reaction. Instead, each insert was sequenced using a *shotgun approach*. That is, each insert was cut out, sheared mechanically into a partially overlapping set of fragments of sizes amenable to sequencing, and cloned into a plasmid vector. Each of these subclones was then sequenced, and the overlapping sequences were assembled into a contiguous sequence by computer. Because each clone is mapped onto the chromosome, a complete sequence for a chromosome was assembled by integrating the sequences for the individual clone inserts into one contiguous sequence. Doing this for each chromosome gives the sequence for the complete genome. In its June 2000 announcement, the HGP reported that it has decoded 97 percent of the genome. The remaining 3 percent generally is considered to be very difficult or impossible to sequence. However, the sequence has been only partially assembled. That is, the actual order of only 53 percent of the sequences was known at that time.

In sum, genome sequencing using a mapping approach involves generating genetic and physical maps of increasing resolution, integrating the two maps, and then sequencing a set of contiguous clones that contain partially overlapping DNA fragments. This sequencing approach has been used mostly for eukaryotic genomes.

KEYNOTE

Except for organisms with small genomes, a genetic map has insufficient resolution to provide the landmarks necessary for genome sequencing. So a high-resolution physical map—a map of genetic markers made by analyzing genomic DNA directly by molecular means—is made to supplement the genetic map. A variety of different types of physical maps can be constructed, including cytogenetic maps, restriction maps, and clone contig maps. The genome sequencing itself is done with automated sequencers.

Genome Sequencing Using a Direct Shotgun Approach.   The first genomic sequence of a living cell, that of the bacterium *Haemophilus influenzae,* was achieved using a *direct shotgun approach* to sequencing. In this approach, the whole genome is broken into partially overlapping fragments, each fragment is cloned and sequenced, and the genome sequence is assembled using a computer. Once thought to be of limited usefulness for sequencing whole genomes greater than 100 kb, sophisticated computer algorithms for assembling sequences from hundreds to thousands of 300- to 500-bp sequences and robotic procedures for preparing DNA for sequencing have opened the door for sequencing large genomes using this shotgun approach. The human genome was sequenced in this way by J. Craig Venter's Celera Genomics. Celera's human genome sequence was also reported to be about 97 percent of the genome, but in contrast to the HGP's, it was completely assembled with the exception of gaps resulting from the missing 3 percent of the sequence.

Figure 15.1 outlines the direct shotgun approach to sequencing. First, random, partially overlapping fragments of genomic DNA are generated by mechanical shearing, and the fragments are cloned to form a library. In contrast to the mapping approach, the insert size for each clone is small—about 2 kb—enabling the clones to be made using simple plasmid vectors. Five hundred base pairs are sequenced from each end of each insert, and the sequence data are entered into the computer. Because of the partial overlapping of the clones, the sequence of the central approximately 1 kb of DNA is obtained when the overlapping clones are sequenced. For example, if a second clone overlapped the first clone by 500 bp, then sequencing the second clone would generate 500 bp of sequence from the middle unsequenced section of the first clone. The result of sequencing this library is a number of contig sequences covering most of the genome.

In sum, the direct shotgun approach has proven to be a successful alternative method for obtaining the sequence of most of the human genome. With both sequencing approaches, the raw sequences obtained must be *assembled* into sequence contigs; that is, the bases must be pieced together in their correct order as they are found in the genome. This is done by computer. The next step is *finishing*, in which sequence errors are fixed and gaps are filled in.

KEYNOTE

Sequencing a genome by the shotgun approach involves constructing a partially overlapping library of genomic DNA fragments, sequencing each clone, and assembling the genomic sequence by computer based on the sequence overlaps.

Figure 15.1

The direct shotgun approach to obtaining the genomic DNA sequence of an organism.

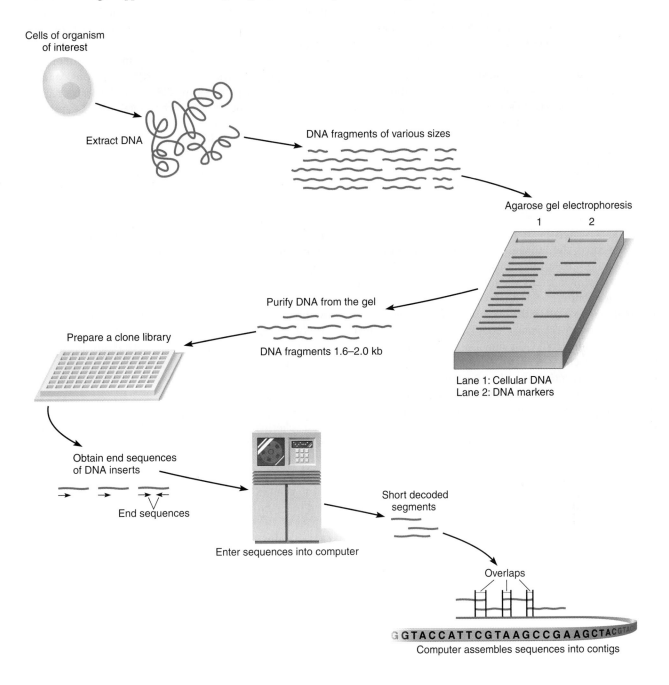

Overview of Genomes Sequenced

We now discuss some of the genomes that have been sequenced and what we have learned from those sequences in general. We can only scratch the surface of what is known here; for more information, go to the Internet sites for the National Center for Biotechnology Information (http://www.ncbi.nlm.nih.gov/Genomes/index.html) and the Institute for Genomic Research (http://www.tigr.org/).

The first "genome" completely sequenced was the circular chromosome of the human mitochondrion in 1981. Since then, a large number of viral genomes have been sequenced completely, along with an increasing number of genomes from living organisms. We will concentrate on the latter here.

At this writing, the genomes of 60 bacteria, 9 archaea, and 9 eukaryotes have been completely sequenced. Genome-sequencing projects are in progress for many more bacteria, archaeons, and microbial eukaryotes (yeasts, fungi, and protozoa) and a number of multicellular eukaryotes. In this section, the features of some selected sequenced genomes determined at the time that their genome sequences were reported are presented. Note that when genome sequences are published, it is generally the case that the sequence is mostly complete rather than finished. Sequencing continues after publication to fill in gaps and handle ambiguities, so the information presented in these sections is evolving.

Bacterial Genomes.

Haemophilus influenzae. The first cellular organism to have its genome sequenced was the eubacterium *H. influenzae.* This task was completed by the Institute for Genomic Research in 1995. The only natural host for *H. influenzae* is the human; in some cases, it causes ear and respiratory tract infections. The genome of this bacterium was sequenced by the direct shotgun approach to test the feasibility of the method because no genetic or physical map existed at the beginning of the project.

The genome of *H. influenzae* is 1.83 Mb (1,830,137 bp) in size, with an overall GC content of 38 percent. With all genome projects, the next step after obtaining the sequence is *annotation*, the identification and description of putative genes and other important sequences. There are usually no introns in protein-coding genes of bacteria and archaeons, so putative protein-coding genes are identified by a computer that searches both DNA strands for open reading frames (ORFs), potential amino acid–coding regions (i.e., potential protein-coding genes). An ORF begins with a start codon (AUG; ATG in the DNA sequence obtained) and ends a multiple of three nucleotides downstream with a stop codon (TAG, TAA, or TGA in the DNA sequence). The identified ORFs for the entire genome are searched against nucleotide and amino acid databases to attempt to identify genes specifically. With the current state of computer searching algorithms and the amount of defined information in sequence databases, a complete microbial genome sequence can be annotated for essentially all coding regions and other elements, such as repeated sequences, operons, and transposable elements (DNA sequences that are mobile in the genome; see Chapter 20).

Figure 15.2 shows the results of annotating the *H. influenzae* genome. The figure is a representation of the circular chromosome, indicating the location of predicted protein-coding genes, tRNA genes, rRNA genes, and repeated sequences, and illustrates the types of genome representations that derive from genome projects with small genomes. Genome analysis predicted 1,743 protein-coding genes comprising 85 percent of the genome. Of these predicted genes, 736 have no role assignment, meaning either that they did not match any protein in the databases or that they matched only proteins designated hypothetical. The remaining 1,007 predicted ORFs were

assigned functions in different categories based on database matches to genes of known function. For example, 68 are genes for amino acid metabolism, 30 are genes for central intermediary metabolism, 64 are genes for regulatory functions, 87 are genes for replication, 27 are genes for transcription, and 141 are genes for translation. Among the other sequences, there are six rRNA operons, and as described in Chapter 11 (pp. 238–239), the spacer regions between rRNA-coding sequences contain tRNA genes.

Mycoplasma genitalium. Also in 1995, the annotated genome sequence of the bacterium *M. genitalium* was reported by scientists at the Institute for Genomic Research. *Mycoplasmas* are members of a large group of bacteria that lack a cell wall and have a characteristically low GC content. They are of interest for genomics because their genome size is much smaller than average among prokaryotes. *Mycoplasmas* are parasites of a wide range of organisms, including humans, and they have pathogenic potential.

The circular genome of *M. genitalium* was sequenced by the direct shotgun approach. The genome is 580,070 bp long, the smallest known genome of any cellular organism. What was most remarkable about this particular genome project was the finding that only 470 genes are needed for cellular life. These genes make up 88 percent of the genome.

Escherichia coli. In 1997, the annotated genome sequence of the bacterium *E. coli* K12, was reported by researchers at the *E. coli* Genome Center at the University of Wisconsin, Madison. An unannotated sequence of the *E. coli* genome made up of sequence segments from more than one strain was reported at the same time by Takashi Horiuchi, of Japan. The *E. coli* genome-sequencing project took almost six years to complete.

E. coli (see Figure 1.10) is an extremely important organism. It is found in the lower intestines of animals, including humans, and survives well when introduced into the environment. Pathogenic *E. coli* strains make the news all too frequently as humans develop sometimes deadly enteric and other infections after contacting the bacterium at restaurants (e.g., in tainted meat) or in the environment (e.g., in contaminated lakes). In the laboratory, *E. coli* has been an extremely important model system for molecular biology, genetics, and biotechnology. Thus, the complete genome sequence of this bacterium was eagerly awaited. It was the sixth genome of a cellular organism to be reported.

The circular genome was sequenced using the direct shotgun approach. The genome of *E. coli* is 4,639,221 bp in size. ORFs make up 87.8 percent of the genome. Another 0.8 percent of the genome encodes rRNAs (seven operons) and tRNAs (86 genes), 0.7 percent is composed of repeated sequences, and the remaining approximately 11 percent includes regulatory and other sequences. There are 4,288 ORFs, 38 percent of which have no known function. Because of the extensive genetic mapping that has

Figure 15.2

The annotated genome of *H. influenzae*. The figure shows the location of each predicted ORF containing a database match as well as selected global features of the genome. Outer perimeter: key restriction sites. Outer concentric circle: coding regions for which a gene identification was made. Each coding region location is color coded with respect to its function. Second concentric circle: Regions of high GC content are shown in red (>42 percent) and blue (>40 percent), and regions of high AT content are shown in black (>66 percent) and green (>64 percent). Third concentric circle: locations of the six ribosomal RNA gene clusters (green), the tRNAs (black), and the cryptic mu-like prophage (blue). Fourth concentric circle: simple tandem repeats. The origin of replication is illustrated by the outward-pointing arrows (green) originating near base 603,000. Two possible replication termination sequences are shown near the opposite midpoint of the circle (red).

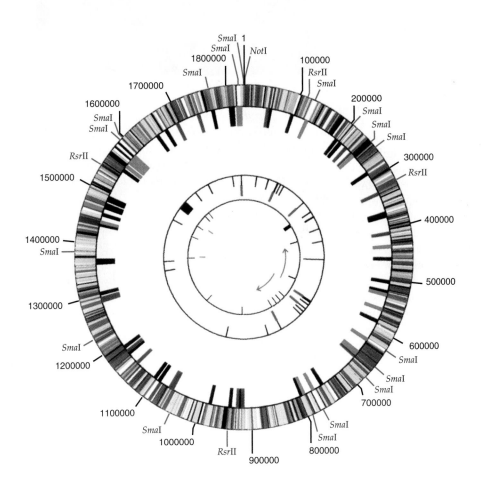

been done with *E. coli*, the sequence could be correlated with the genetic map, something that could not be done with the other two bacterial genomes discussed.

Archaeon Genomes. The *Methanococcus jannaschii* genome was the first genome of an archaeon to be completely sequenced. *M. jannaschii* is a hyperthermophilic methanogen that grows optimally at 85°C and at pressures up to 200 atmospheres. It is a strict anaerobe, and it derives its energy from the reduction of carbon dioxide to methane. Sequencing was by the direct shotgun approach and was a collaborative effort funded by the Department of Energy's Microbial Genome Initiative. The sequence was reported in 1996. The complete genome consists of three parts: a large, main circular chromosome of 1,664,976 bp; a circular, extrachromosomal element (ECE) of 58,407 bp; and a smaller, circular ECE of 16,550 bp. The main chromosome has 1,682 ORFs, the larger ECE has 44 ORFs, and the smaller ECE has 12 ORFs. Of the total 1,738 ORFs, only 38 percent have been assigned functions based on matches in databases. Most of the genes involved in energy production, cell division, and metabolism are similar to their counterparts in domain Bacteria, and most of the genes involved in DNA replication, transcription, and translation are similar to their counterparts in domain Eukarya. The genome sequence of this organism therefore affirmed the existence of a third major branch of life on Earth.

Eukaryotic Genomes.

Yeast. For decades, the budding yeast *Saccharomyces cerevisiae* (Figure 15.3) has been a model eukaryote for many kinds of research. Some of the reasons for this are that it can be cultured on simpler media, it is highly amenable to genetic analysis, and it is highly tractable for sophisticated molecular manipulations. Moreover, functionally it resembles mammals in many ways. Therefore, it was logical that its genome would be a target for early genome-sequencing efforts. In fact, the *S. cerevisiae* (yeast) genome was the first eukaryotic genome to be sequenced completely; the sequence was reported in 1996. The yeast genome effort was an international collaborative effort of more than 100 laboratories and involved more than 600 researchers. Sequencing was done using the mapping approach. The 16-chromosome genome was reported to be 12,067,280 bp, with chromosome sizes ranging from 230,195 bp in chromosome I, the smallest chromosome, to 1,522,191 bp in chromosome IV, the largest chromosome. Approximately 969,000 bp of repeated sequences were estimated not to be included in the published sequence. The sequence revealed 6,183 ORFs, 120 to 150 genes for rRNAs, 37 genes for snRNAs, and 262 genes for tRNAs; 233 of the ORFs have introns, and 80 of the tRNA genes have introns. Overall, about 70 percent of the total genome consists of ORFs, a lower percentage than is the case with the bacterial genomes sequenced. One protein-coding gene is found for every approximately 2 kb of the genome. To show the power of genome-sequencing analysis, at the outset of the yeast genome project only about 1,000 genes had been defined by genetic analysis.

The Nematode Worm Caenorhabditis elegans.

The genome of the nematode *C. elegans* (Figure 15.4), also called the "worm," was the first multicellular eukaryotic genome to be sequenced. Nematodes are smooth, nonsegmented worms with long, cylindrical

Figure 15.3

Scanning electron micrograph of the yeast *Saccharomyces cerevisiae*.

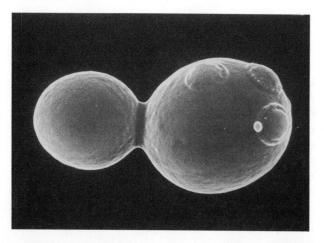

Figure 15.4

The nematode worm *Caenorhabditis elegans*.

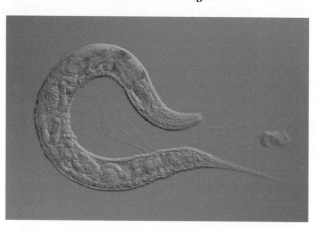

bodies. *C. elegans* is about 1 mm long; it lives in the soil, where it feeds on microbes. *C. elegans* has become an important model organism for studying the genetic and molecular aspects of embryogenesis, morphogenesis, development, nerve development and function, aging, and behavior.

The *C. elegans* genome project was initiated by Sydney Brenner and carried out by an international consortium. An almost complete genome sequence of 97 Mb was reported in December 1998. There are 19,099 ORFs, with an average density of one per 5 kb. Each gene has an average of five introns, and 27 percent of the genome sequenced consists of exons. The number of protein-coding genes is approximately three times that of yeast and about one-half the estimated number of human genes. There are also several hundred genes for non-protein-coding RNAs, including 659 tRNA genes, a single tandem array of rRNA genes, a single tandem array of 5S rRNA genes, and a number of noncoding RNA genes in introns of protein-coding genes.

The Fruit Fly Drosophila melanogaster.

The genome sequence of an organism of particular historical importance in genetics, the fruit fly *D. melanogaster* (see Figure 1.7b), was reported in March 2000 by a collaborative group including Gerald Rubin's *Drosophila* Genome Project at the University of California, Berkeley, and J. Craig Venter's company Celera Genomics. The fruit fly has been the subject of much genetics research and has contributed to our understanding of the molecular genetics of development.

The *Drosophila* genome is estimated to be about 180 Mb in size, about a third of which is heterochromatin mostly located around centromeres. These heterochromatic regions are unclonable, making the complete genome sequence of organisms with them (e.g., fruit flies, humans) impossible to obtain. As a consequence, the genome sequence reported constitutes 97 to 98 percent of the approximately 120 Mb euchromatic part of the genome and includes more than 99 percent of the genes.

The fruit fly genome has 13,600 genes. Surprisingly, the number of fruit fly genes is just over twice that found in yeast, yet the fruit fly is a much more complex organism. We must conclude that higher complexity in animals such as flies and humans does not require a correspondingly larger repertoire of gene products. The fruit fly's value as a model system for studying human biology and disease was affirmed by the finding that *D. melanogaster* has homologues for 177 of 289 genes known to be involved in human disease, including cancer.

The *Flowering Plant* Arabidopsis thaliana. The genome of *A. thaliana* (see Figure 1.7d) was the first genome sequenced for a flowering plant. *Arabidopsis* has been an important model organism for studying the genetic and molecular aspects of plant development. The *Arabidopsis* genome was sequenced by a multinational consortium called the *Arabidopsis* Genome Initiative (AGI), involving scientists and funding from eight countries. A 115-Mb sequence out of the estimated 125-Mb total genome sequenced was reported in December 2000. About 25,900 genes are distributed over the plant's five chromosomes, but only about 1,000 of those genes have a known function based on direct experimental evidence. About 100 *Arabidopsis* genes are similar to disease-causing genes in humans, including the genes for breast cancer and cystic fibrosis. The next step is to fill in the gaps in the sequence and explore the structure and function of the genome in detail. Toward this end, an initiative called the 2010 Project has been set up. It has an ambitious set of goals, including defining the function of every gene, determining where and when every gene is expressed, showing where the encoded protein ends up in the plant, and defining any proteins with which the protein interacts.

Homo sapiens. Lastly, a working draft of the genome of most interest to us, the genome of *Homo sapiens*, was announced in a joint press conference in June 2000 involving Francis Collins, of the National Human Genome Research Institute (NHGRI, representing the HGP), and J. Craig Venter, of Celera Genomics. In both cases, the human genome sequence being generated is an amalgamation of sequences and will not be an exact match for any one person's genome in the human population.

The sequencing part of the HGP was carried out by an international Human Genome Sequencing Project Consortium of scientists at 16 institutions in the United States, Great Britain, France, Germany, and China. Their sequencing approach was built on a foundation of genetic and physical maps. Approximately 50 percent of this genome sequence was considered in a near-finished form, and 24 percent was completely finished. Work is continuing to achieve the consortium's final goal of a finished sequence of the euchromatic part of the genome, defined as one with 99.9 accuracy and no gaps.

Celera Genomics used the direct shotgun approach for sequencing and assembled their sequence data with the help of human genome physical map data in public databases. According to Celera, they have achieved the first complete assembly of the human genome sequence. With an assembled sequence in hand, the next step is annotation: determining the genes it contains and analyzing other features of the genome.

The draft genome sequences and initial interpretations of assembled sequences were published by the Human Genome Project Sequencing Consortium in the 15 February 2001 issue of *Nature* and by Celera Genomics in the 16 February 2001 issue of *Science*. The genome scientists estimated that there are 32,000 genes in the human genome, a number increased to 42,000 with more recent analysis. This low number is making scientists drastically change their thinking about organism complexity and development. In fact, the two sequencing groups estimate that only 1.1 or 1.5 percent of the genome codes for proteins, so in between genes there is a great deal of non-protein-coding DNA. Almost one-half of this DNA consists of repeated sequences of various types. Interestingly, the human genome shares 223 genes with bacteria, but those genes are not found in yeast, the worm, or the fruit fly. This is just one of the puzzles to ponder for future analysis. All in all, the two human genome sequences are proving a great resource for scientists to learn about our species, and *data mining*, searching through genome sequences for information, will continue for many years. Undoubtedly there will be a strong focus on human disease genes, with an eye toward treatment and therapy.

K E Y N O T E

> Many genomes of both viruses and living organisms have now been sequenced. Analysis of the sequences has affirmed the division of living organisms into Bacteria, Archaea, and Eukarya. In general, completed genome sequences are analyzed by computer to identify ORFs and the array of gene functions they encode. Each organism has an array of genes needed for basic metabolic processes and genes whose products determine the specialized form and function of the organism.

Functional Genomics

Obtaining the complete genome sequence for an organism heralds the beginning of the post-genome-sequencing era, in which the sequence is analyzed in great detail. One important research direction is to describe the functions of all the genes in the genomes, including studying gene expression and its control, and this defines the field of *functional genomics*. The difficulty in assigning gene function is that going from gene sequence to gene function is the opposite direction of that classically taken in genetic analysis, in which researchers start with a phenotype and set out to identify and study the genes responsible. Present-day functional genomics relies on laboratory experiments by molecular biologists and sophisticated computer analysis by researchers in the rapidly growing field of **bioinformatics.** Bioinformatics fuses biology with mathematics and computer science. It is used for many things, including finding genes within a genomic sequence, aligning sequences in databases to determine the degree of matching, predicting the structure and function of gene products, describing the interactions between genes and gene products at a global level within the cell, between cells, and between organisms, and postulating phylogenetic relationships for sequences.

Identifying Genes in DNA Sequences

The next step after obtaining the complete sequence of a genome is *annotation,* the identification and description of putative genes and other important sequences. Annotation begins the process of assigning the functions of all genes of an organism. Of particular interest are the protein-coding genes, and we focus our attention on them here.

Procedurally, annotation involves using computer algorithms to search both DNA strands of the sequence for protein-coding genes. Putative protein-coding genes are found by searching for ORFs, that is, start codons in frame (separated by a multiple of three nucleotides) with a stop codon. This process is straightforward with prokaryotic genomes because there are no introns. However, the presence of introns in many eukaryotic protein-coding genes necessitates the use of more sophisticated algorithms designed to include the identification of junctions between exons and introns in scanning for ORFs.

ORFs of all sizes are found in the computer scan, so a size must be set below which it is deemed unlikely that the ORF encodes a protein in vivo, and it is not analyzed further. For the yeast genome, for instance, the lower limit was set to 100 codons. However, a few genes may be below this limit, and not all ORFs above 100 codons encode proteins.

Homology Searches to Assign Gene Function

The function of an ORF identified in genome scans may be assigned by searching databases for a sequence match with a gene whose function has been defined. Such searches are called *homology searches* and involve computer-based comparisons of an input sequence with all sequences in the database. Such similarity searches can be done using an Internet browser to access the computer programs. For example, the BLAST program at the National Center for Biotechnology Information (http://www.ncbi.nlm.nih.gov/) enables a user to paste the sequence to be studied into a window, either in nucleotide form or in amino acid form, and to get results indicating the degree to which the sequence of interest is similar to sequences in the database.

The principle of homology searches is as follows: If a newly sequenced gene (e.g., from a genome sequence project) is similar to a previously sequenced gene, the two genes are related in an evolutionary sense, so the function of the new gene probably is the same as or at least similar to the function of the previously sequenced gene. Homology searching with an amino acid sequence is preferred over that with a DNA sequence because with 20 different amino acids and only four different nucleotides, unrelated genes appear more different from one another at the amino acid level than at the nucleotide level. Given the information in current databases, there is a less than 50 percent chance that a new gene sequence will match a gene sequence in the database with a high enough degree of similarity to infer function.

A homology search can indicate a match for either the whole protein sequence or for parts of it. In the latter case, this means that a domain of the new gene product matches a domain of a previously identified gene product, so at least part of the new protein's function can be inferred. Evolutionarily speaking, such a result means that the domains have a common ancestor, but the genes as a whole may not.

Homology searching plays an important part in assigning gene function for sequences obtained from genome projects. For example, Figure 15.5 shows the

Figure 15.5

The distribution of predicted ORFs in the genome of yeast.
(From B. Dijon. 1996. *Trends Genet.* 12:263–270.)

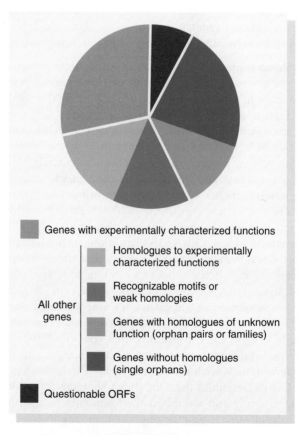

- Genes with experimentally characterized functions
- Homologues to experimentally characterized functions
- Recognizable motifs or weak homologies
- Genes with homologues of unknown function (orphan pairs or families)
- Genes without homologues (single orphans)

All other genes

- Questionable ORFs

distribution of ORFs in the yeast genome. About 30 percent of the genes were known as a result of standard genetic analysis before genome sequencing, including direct assays for function. The remaining 70 percent of the ORFs are genes whose functions have not been directly determined. These latter ORFs break down as follows. Thirty percent encode a protein that is related to functionally characterized proteins or has a domain related to domains in functionally characterized proteins. Another 10 percent have homologues in databases, but the functions of those homologues are unknown. Such yeast ORFs are called *FUN* (*function unknown*) genes, and those genes and their homologues are called *orphan families*. The remaining 30 percent of ORFs have no homologues in the databases. Within this class are the 6 to 7 percent of ORFs that are questionable in terms of being real genes. The remainder probably are real genes but at present are unique to yeast; they are called *single orphans*.

The problem of "function unknown" genes applies to the genomes of other organisms, both prokaryotic and eukaryotic. As more and more genes with defined functions are added to the databases, the percentage of ORFs with no matches to database sequences is decreasing.

Assigning Gene Function Experimentally

One key approach to assigning gene function experimentally is to knock out the function of a gene and determine what phenotypic changes occur. This is done by deleting the gene, or making a *gene knockout*. Figure 15.6 shows how this can be done in yeast using a PCR-based strategy.

Using PCR primers based on the known genome sequence, an artificial linear DNA deletion module is constructed and amplified. This module consists of part of the gene sequence upstream of and including the start codon and part of the gene sequence downstream of and including the stop codon, flanking a DNA fragment containing the *kan*^R (kanamycin) selectable marker that confers resistance to the inhibitory chemical G418. This linear DNA is transformed into yeast, and G418-resistant colonies are selected. The desired transformants are those generated when the fragment replaces the target ORF by homologous recombination. This event completely inactivates—knocks out—the gene because most of it is replaced. In genetic terms, a *loss-of-function mutation* is produced.

Using this approach, a yeast knockout (YKO) project is under way in which each yeast gene is being systematically deleted and the resulting strains are being studied under various conditions for changes in phenotype to assign function. The work involved in this is substantial because of the many areas of cell function that must be screened for a change in phenotype, including cell cycle events, meiosis, DNA synthesis, RNA synthesis and processing, protein synthesis, DNA repair, energy metabolism, and molecular transport mechanisms. Of the genes that have been knocked out to date, about one-third are essential genes (the functions of these genes are needed for yeast growth), and the remainder are not essential. Of the latter, approximately one-half result in no significant changes in phenotype, while the other half do. Similar knockout projects are also under way with a number of other organisms.

Figure 15.6

Creating a gene knockout in yeast. Schematic of a PCR-based gene deletion strategy involving a DNA fragment constructed by PCR from gene sequences flanking the *kan*^R selectable marker that is transformed into yeast and replaces the chromosomal ORF by homologous recombination.

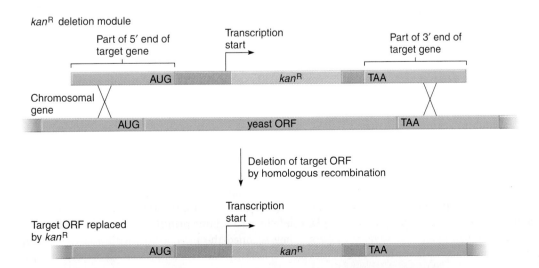

Describing Patterns of Gene Expression

In classic genetic analysis, research begins with a phenotype and leads to the gene or genes responsible. Once the gene is found and isolated, experiments can be done to study the expression of the gene in normal and mutant organisms as a way to understand the role of the gene in determining the phenotype. When the complete genome sequence is obtained for an organism, exciting new lines of research are possible: the analysis of expression of *all* genes in a cell at the transcriptional and translational levels and the analysis of all protein-protein interactions. Measuring the levels of RNA transcripts (usually focusing on mRNA transcripts), for example, gives us insight into the global gene expression state of the cell. To go along with this new research, a new term has been coined for the set of mRNA transcripts in a cell: the **transcriptome.** Since the mRNAs specify the proteins that are responsible for cellular function, the transcriptome is a major indicator of cellular phenotype and function. By extension, the complete set of proteins in a cell is called the **proteome.** Studies of the transcriptome and proteome are described in this section.

The Transcriptome. The transcriptome is not stable in a cell. Rather, the types and levels of mRNA transcripts change significantly as a cell responds to its environment and as it proceeds through DNA replication and cell division. By defining exactly what genes are expressed, when they are expressed, and their levels of expression, we can begin to understand cellular function at a global level.

DNA microarrays, also known as **DNA chips, GeneChip® arrays** (trademark of Affymetrix, Inc.), **oligonucleotide arrays,** or **probe arrays,** are among the most powerful tools for global gene expression experiments. A DNA microarray is an ordered grid of DNA molecules of known sequence fixed at known positions on a solid substrate, either a silicon chip, glass, or, less commonly, a nylon membrane.

There are two major types of DNA microarray technology. In one method, developed at Stanford University, premade DNA molecules (e.g., genomic DNA fragments, PCR products, cDNAs, or oligonucleotides) are spotted onto glass using mechanical microspotting (Figure 15.7a). The DNA is loaded into a spotting pin by capillary action, and a small volume of DNA is released onto the glass when the pin touches the surface. The pin is washed, and the next DNA is loaded and spotted on the adjacent spot. Rapid production of the microarray is made possible by having a multipin print head under robotic control (Figure 15.7b). Microarrays made in this way have the capacity of 10,000 or more DNA molecules in an area of 3.6 cm².

In another method, oligonucleotides are synthesized in situ on the substrate at defined positions. That is, Affymetrix, Inc. has adapted the process for putting thousands of transistors onto a single integrated circuit to making oligonucleotide arrays—GeneChips®—that are about

1.6 cm², the size of a small postage stamp, with a density of about 1 million oligonucleotides per square centimeter. First, the sequences of a set of DNA oligonucleotides (up to about 25 nucleotides long) are defined based on the experiment, and the probe arrays then are made by a proprietary, light-directed chemical synthesis process using photolithographic masks (see http://www.affymetrix.com for a movie of the process).

As with Southern and northern hybridization, experiments involving DNA microarrays or GeneChips® involve probes and target nucleic acids. With microarrays, though, the probes are the DNA molecules fixed to the glass or silicon chip, and they are unlabeled, and the targets are labeled free DNA molecules whose identities or quantities are being analyzed. For this reason, a useful generic term for a microarray or chip is *probe array.* The label usually is a fluorescent tag, such as a fluorescein or cyanine dye (commonly, a green Cy3 dye or a red Cy5 dye). After hybridization, the fluorescence pattern is recorded by laser scanning and the data are analyzed according to the goals of the experiment. If both green and red fluorescent dyes are used simultaneously, the laser detector can distinguish the two because they produce different wavelengths of fluorescence emission.

Let us consider as an example the use of probe arrays to study global gene expression during yeast sporulation, the process of producing haploid spores by meiosis (Figure 15.8a). Yeast sporulation involves four major stages: DNA replication and recombination, meiosis I, meiosis II, and spore maturation. The sequential transcription of at least four classes of genes—early, middle, mid–late, and late—correlates with these stages. When these probe array experiments began, about 150 genes had been identified that are differentially expressed during sporulation. In the new research, the researchers induced diploid yeast cells to sporulate, and at seven timed intervals, they took cell samples and used DNA microarrays containing 97 percent of the known or predicted yeast genes to analyze the temporal program of gene expression during meiosis and spore formation. Light and electron microscopy were used to correlate the sampling time with the exact stage of sporulation. For quantifying gene expression, they isolated mRNAs from the cell samples and synthesized fluorescently labeled cDNAs by reverse transcription in the presence of Cy5 (red)-labeled dUTP (Figure 15.8b). For a nonsporulating cell control, they isolated mRNAs from cells at a time point immediately before inducing sporulation and again synthesized fluorescently labeled cDNAs, this time using Cy3 (green)-labeled dUTP. For each time point, they hybridized a mixture of reference green-labeled cDNAs and experimental red-labeled cDNAs to DNA microarrays. The DNA microarrays were made by using PCR to amplify each ORF (using primers based on the genome sequence) and printing the sequences onto a glass slide using a robotic printing device. After hybridization, they scanned the microarrays with a laser detector device to quantify the red and green fluorescence loca-

Figure 15.7

Preparing a DNA microarray using robot-driven, mechanical microspotting of premade DNA molecules onto glass. (**a**) Schematic of the microspotting process. (**b**) Photograph of a robotic device for making the DNA microarray.

a)

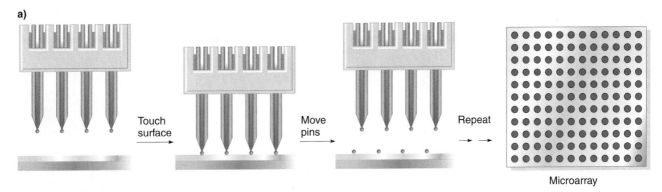

Touch surface → Move pins → Repeat →

Microarray

b)

tions and intensities. Figure 15.8c shows an example of the results obtained. The relative abundance of transcripts from each gene in sporulating versus nonsporulating yeast cells is seen by the ratio of red to green fluorescence. If an mRNA is more abundant in sporulating cells than in nonsporulating cells, as is the case for the *TEP1* gene (see Figure 15.8c), this results in a higher ratio of red-labeled to green-labeled cDNAs prepared from the two types of cells and therefore the same higher ratio of red to green fluorescence detected on the array. In general, a gene whose expression is induced by sporulation is seen as a red spot, and a gene whose expression is repressed by sporulation is seen as a green spot. Genes that are expressed at approximately equal levels in nonsporulating cells and during sporulation are seen as yellow spots.

With this approach, the researchers found that more than 1,000 yeast genes showed significant changes in mRNA levels during sporulation. About one-half of the genes are repressed during sporulation, and one-half are not repressed. At least seven distinct temporal patterns of

gene induction are seen, and this is providing some insights into the functions of many orphan genes.

DNA microarrays are becoming widespread, despite their high cost. For example, DNA microarrays are being used to study the expression profiles of 40,000 human genes during cell division. DNA microarrays are also being used to study cancer. For example, by analyzing global gene expression in normal individuals and in individuals with a particular type of cancer, it is theoretically possible to identify characteristic gene expression patterns that could be used to classify the cancer and to screen for its onset. The former has already proved possible for diffuse large B cell lymphoma, where it has been shown based on gene expression profiles (*transcriptional "fingerprints"*) that there are previously unknown distinct types of the cancer.

DNA microarrays are also useful for screening for genetic diseases. Of particular interest are genetic diseases characterized by a large number of possible mutations, making simple DNA typing methods (see Chapter

Figure 15.8

Global gene expression analysis of yeast sporulation using a DNA microarray. **(a)** The stages of sporulation in yeast, correlated with the sequential transcription of at least four classes of genes. (Adapted from Chu et al. 1998. *Science* 282:699–705.) **(b)** Outline of the DNA microarray experiment. **(c)** Example of results of a global gene expression analysis in yeast using a DNA microarray. The entire yeast genome is represented on the DNA chip, and the colored dots represent levels of gene expression, as described in the text.

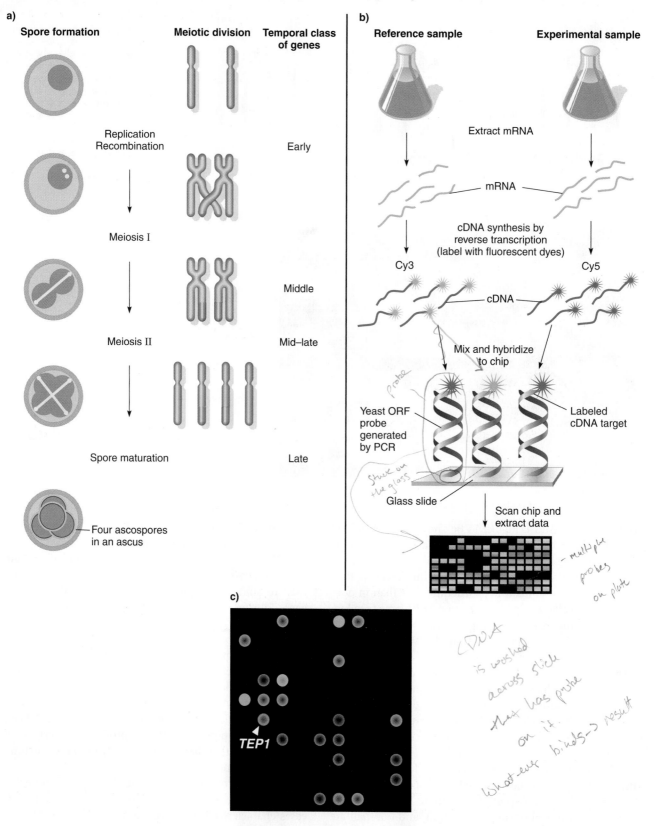

14, pp. 293–296) inefficient. For example, the genes *BRCA1* (breast cancer 1) and *BRCA2* cause approximately 60 percent of all cases of hereditary breast and ovarian cancers. However, at least 500 different mutations have been discovered in *BRCA1* that can lead to the development of cancer. Assaying for that many different mutations is well within the scope of DNA chip technology, and such chips are being developed for this and for many other diseases of similar genetic complexity. The principles for use are very similar to those for the yeast sporulation example. Blood is taken from a patient, and green-labeled DNA is produced by PCR and mixed with red-labeled DNA from a normal individual. The chip in this case consists of oligonucleotides that collectively represent the entirety of the *BRCA1* and *BRCA2* genes. If the patient has a mutation in one or other of the genes, the red (normal) DNA will hybridize to the DNA on the chip, but the green (patient's) DNA will not hybridize in the region where the mutation is located. Normal hybridization is seen as a yellow (red + green) spot, and a mutation is seen as a red spot. Because the position of the spot on the array is known and because the oligonucleotides for each spot are known, the mutation is localized within a very narrow region of the *BRCA1* or *BRCA2* gene and can be analyzed in more detail.

 In the iActivity *Personalized Prescriptions for Cancer* on the Web site, you are a researcher at the Russellville clinic trying to determine the gene expression profile for a patient with cancer.

The Proteome. The proteome is the complete set of expressed proteins in a cell at a particular time. **Proteomics** is the cataloging and analysis of those proteins to determine when a protein is expressed, how much is made, and with what other proteins the protein can interact. The approaches in proteomics are mostly biochemical and molecular.

The goals of proteomics are (1) to identify every protein in the proteome (this involves making a comprehensive two-dimensional acrylamide gel electrophoresis map of the proteins, isolating each protein, and analyzing the protein by mass spectrometry); (2) to determine the sequences of each protein and enter the data into databases; and (3) to analyze globally protein levels in different cell types and at different stages in development.

Identifying and sequencing all the proteins from a cell is much more complex than mapping and sequencing a genome. Companies working in this area are striving to speed up dramatically the identification and sequencing of proteins and the computer analysis of the data. And, coinciding with the publication of the human genome sequences, a global Human Proteome Organisa-

tion (HUPO) was launched. HUPO is intended to be the postgenomic analogue of HUGO, with a mission to increase awareness of and support for proteomics research at scientific, political, and financial levels.

Proteomics is an extremely important field because it focuses on the functional products of genes, which determine the phenotypes of a cell. Of particular human interest are diseases, and proteins and peptides are closer to the actual disease process than are the genes that encode them. However, the challenges for proteomics are much greater than those for genomics.

KEYNOTE

The goal of functional genomics is to define the functions of all the genes in a genome of a particular species, including the patterns and control of gene expression. Gene expression is analyzed at two levels: the mRNA transcripts, called the transcriptome, and proteins, called the proteome. Functional genomics relies on laboratory experiments and computer analysis. DNA microarrays are used, for example, to generate a gene expression profile for a cell in which the mRNA transcripts are described qualitatively and quantitatively.

Comparative Genomics

Comparative genomics involves comparing entire genomes of different species, with the goal of enhancing our understanding of their functions and evolutionary relationships. Comparative genomics is rooted in the tenet that all present-day genomes have evolved from common ancestral genomes. Therefore, studying a gene in one organism can provide meaningful information about the homologous gene in another organism, and, more globally, comparing the overall arrangements of genes and nongene sequences of different organisms can tell us about the evolution of genomes. Since direct experimentation with humans is unethical, comparative genomics provides a valuable way to determine the functions of human genes by studying homologous genes in nonhuman organisms. Identifying and studying homologues to human disease genes in another organism is potentially valuable for developing an understanding of the biochemical function and malfunction of the human gene.

Like functional genomics, comparative genomics focuses on the genome level. Comparative genomics involves analyzing genomes from two or more species, with the goal of defining the extent and specifics of similarities and differences between sequences, either gene sequences or nongene sequences. An obvious question that comparative genomics can address is the evolutionary relationships between two or more genomes. For example, as we discussed earlier, complete genome sequence analysis affirmed the evolutionary relationships and distinctions among Bacteria, Archaea, and Eukarya.

KEYNOTE

> Comparative genomics is the comparison of complete genomes of different species, with the goal of increasing our understanding of the genes and nongene sequences of each genome and their evolutionary relationships.

Ethics and the Human Genome Project

The Human Genome Project is raising ethical issues. With the entire human genome sequence in hand, we will be able to identify and isolate all human genes that cause diseases. This will lead to the development of tests for many gene defects that we cannot test for now, including those that may lead to the development of a disease later in life, such as cancer. However, these tests will become available before we have developed a cure for those diseases, as is already the case for most testable genetic diseases. A number of ethical questions arise from this scenario. Should a patient be told if a test for an incurable genetic disease is positive? (This issue applies now for Huntington disease, a dominant lethal disease in which the symptoms typically do not appear until later in life.) Should employers be able to ask for results of a genetic test if the employee does not want to know? Should health insurance companies or employers have access to genetic testing data, and, if so, how can the patient protect his or her insurability and employability? Should states be able to collect genetic data on their populace? The last three questions raise fundamental privacy issues.

Fortunately, these issues are not being ignored. The federal agencies funding the HGP are devoting 3 to 5 percent of their annual budgets to study the ethical, legal, and social issues (ELSIs) related to the availability of genetic information. This amounts to the world's largest bioethics program. Four areas are being emphasized by the ELSI program: (1) privacy of genetic information; (2) safe and effective introduction of genetic information in the clinical setting; (3) fairness in the use of genetic information; and (4) professional and public education. Appropriate laws and regulations are expected to be developed as a result of the activities of the ELSI program and of continuing dialogues among scientists, physicians, lawmakers, and members of the public.

Summary

With the development of recombinant DNA techniques, it became possible to clone individual genes and therefore to broaden the questions one could ask about phenotypes. With genomic analysis, the scope of questions becomes much broader. That is, one can now ask global questions about gene expression and begin to understand in detail how a cell or an organism functions rather than how a gene functions.

The analysis of the entire genomes of species of organisms defines the field of genomics. Structural genomics, a subfield of genomics, involves the genetic mapping, physical mapping, and sequencing of entire genomes. Two general approaches have been taken to sequence genomes. The mapping approach, conceived originally for large genomes, involves creating genetic maps and physical maps of increasing levels of resolution and then sequencing the segments. The direct shotgun approach, originally shown to be effective with small genomes, involves making a partially overlapping library of genomic DNA fragments, sequencing each clone, and assembling the genomic sequence by computer based on the sequence overlaps. With the increased power of computers to handle large amounts of sequence data, the direct shotgun approach can now be used with large genomes, including that of humans.

Although the original focus of the Human Genome Project was on the genome of humans and those of a few model organisms, it seems that everyone working with a particular organism is interested in sequencing its genome. To date, the genomes of many viruses and living organisms have been sequenced completely. Analysis of the genomes of the latter have affirmed the division of living organisms into Bacteria, Archaea, and Eukarya. In June 2000, a working draft of the human genome was reported, with the complete sequence expected in one to two years.

Through functional genomics, the genes of each species' genome are identified and their patterns of expression described. Functional genomics involves both laboratory analysis and computer analysis (bioinformatics) to identify genes in DNA sequences (annotation), determine function by homology searches and by experimental means such as gene knockouts, and to describe the patterns of gene expression at the mRNA and protein levels. The set of mRNA transcripts in a cell is called the transcriptome, and the complete set of proteins in a cell is called the proteome. The transcriptome is specific to the metabolic state of the cell, so by defining the transcriptome qualitatively and quantitatively, we can begin to understand cellular function at a global level. An extremely valuable tool for studying the transcriptome is the DNA microarray.

Since proteins govern the phenotypes of a cell, the study of the proteome—proteomics—will provide much more information about cellular function at a global level. At the moment, though, the tools for analyzing the transcriptome are much more sophisticated than those for analyzing the proteome, so we must await technological improvements before we will see a really rich set of data.

Comparative genomics, another subfield of genomics, involves the comparison of entire genomes of different species, either related or not. The goal of comparative genomics is to enhance our understanding of the functions of each genome for both gene and nongene sequences and to develop an understanding of evolutionary relationships. We can make conclusions about homologous genes in species of different organisms because all present-day genomes have evolved from common ancestral genomes. Comparative genomics will be particularly important for studies of the human genome because direct human experimentation is unethical. Information about a gene in closely related organisms will inform us about the function of that gene in humans.

Finally, the future of human genomics raises ethical issues. Conceivably, we can all have our genome completely sequenced and analyzed and deposit that information in a database or even on a chip we carry. Our genome sequences will reveal, among other things, whether we have genetic diseases, whether we have the potential to develop a genetic disease or cancer, and whether we have some mental or physical aberration that might affect our life or work. Such information could affect an individual's ability to obtain life insurance or health insurance or affect his or her job potential. Thus, fundamental privacy issues will have to be addressed as this brave new world of genomics marches forward.

Analytical Approaches for Solving Genetics Problems

Q15.1 Halushka and colleagues used specially designed DNA chips to search for SNPs in 75 protein-coding genes in 74 individuals. They scanned about 189 kb of genomic sequence consisting of 87 kb of coding, 25 kb of intron, and 77 kb of untranslated but transcribed (i.e., 5′-UTR and 3′-UTR) sequences. They identified a total of 874 possible SNPs, of which 387 were within coding sequences; these are designated cSNPs. Of the cSNPs, 209 would change the amino acid sequence in one of 62 predicted proteins.

a. In their sample, what is the frequency of SNPs (number of base pairs per SNP)?

b. Are the SNPs evenly distributed in coding and noncoding sequences? Is this an expected result? What implications does this result have?

c. Another type of marker used in genome mapping is the expressed sequence tag (EST). An EST is a marker produced by PCR using primers based on the sequence of a cDNA. Since a cDNA is a DNA copy of an mRNA, an EST marker corresponds to a functional protein-coding gene. At least 40,000 human ESTs have already been identified, and a reasonable estimate of the human gene number is about 75,000. Use the data given to estimate the number and distribution of SNPs in human genes.

i. About how many SNPs exist in human genes?

ii. How many are estimated to be in noncoding regions?

iii. How many are in coding regions but do not affect protein structure?

iv. How many are in coding regions and could affect protein structure?

A15.1 SNPs are *single nucleotide polymorphism:* differences of just one base pair in the DNA of different individuals. These alterations in DNA sequence are not necessarily detrimental to the organism. Rather, they are initially identified simply as differences, or polymorphisms, in DNA sequence. This problem asks you to analyze their frequency and distribution in humans and consider the implications of your analysis.

a. In 189,000 bp there are 874 SNPs, so on average, there are 189,000/874 = 216 bp of DNA sequence per SNP. Note that this sampling assessed the number of SNPs *in genes* and does not estimate the number of SNPs in genomic regions in between genes.

b. 387/874 = 44 percent of the SNPs lie in coding sequences, and 487/874 = 56 percent of the SNPs lie in noncoding sequences. The observation that there are fewer SNPs in coding sequences suggests that there is less sequence variation in coding sequences. This is expected because coding sequences specify amino acids that confer a protein's function. A SNP within a coding sequence might result in the insertion of an amino acid that alters the normal function of the protein. This could be disadvantageous and be selected against. Indeed, only 209/874 = 24 percent of the SNPs alter amino acid sequences, and SNPs that alter amino acid sequences are not found in all 75 genes examined. This indicates that although some sequence constraints may be present in noncoding sequences (e.g., if they bind a regulatory protein), more sequence variation is tolerated in noncoding regions.

c.

i. If there are 75,000 genes, one expects to find about 8.74×10^5 SNPs in the human genome: (874 SNP/75 genes) $\times$ 75,000 genes = 874,000.

ii. About 487/874 = 56 percent, or 4.9×10^5, will be in noncoding regions.

iii. About 387/874 = 44 percent, or 3.8×10^5, will be in coding regions. About (387 − 209)/874 = 20 percent, or 1.7×10^5, will not affect protein structure because they do not change the amino acid sequence in a protein.

iv. About 209/874 = 24 percent, or 2.1×10^5, could affect protein structure because they change the amino acid sequence in a protein. However, not all of these significantly affect protein structure. If a SNP results in the substitution of a similar (conserved) amino acid, it may not significantly alter the structure (or function) of the protein. For example, a SNP might result in aspartate being replaced by glutamate. Both are acidic amino acids, so this substitution may not significantly alter the protein's structure.

Questions and Problems

15.1 RFLP, VNTR, STR, and SNP are abbreviations for DNA markers used to map eukaryotic genomes. Explain what each abbreviation stands for and what each marker is.

*15.2 The frequency of individuals in a population with two different alleles at a DNA marker is called the marker's heterozygosity. Why would a STR DNA marker with nine known alleles and a heterozygosity of 0.79 be more useful for mapping studies than a nearby STR having three alleles and a heterozygosity of 0.20?

*15.3 Abbreviations used in genomics typically facilitate the quick and easy representation of longer tongue-twisting terms. Explore the nuances associated with some abbreviations by answering the following questions: Could a RFLP, VNTR, or STR be identified as a SNP? Why or why not?

15.4 Genomes can be sequenced using a direct shotgun approach or a mapping approach.
a. What is the difference between these approaches?
b. What limitations have traditionally been associated with the direct shotgun approach, and how have they been overcome?

*15.5 How has genomic analysis provided evidence that Archaea is a branch of life distinct from Bacteria and Eukarya?

15.6 How do the following compare between *S. cerevisiae* and *C. elegans*?
a. the percentage of genes with introns
b. the number of tRNA genes
c. the number of ORFs
d. the (approximate) percentage of the genome devoted to exons

15.7 What is bioinformatics, and what is its role in structural, functional, and comparative genomics?

15.8 What is the difference between a gene and an ORF? How might you identify the functions of ORFs whose functions are not yet known?

*15.9 Once a genomic region is sequenced, computerized algorithms can be used to scan the sequence to identify potential ORFs.
a. Devise a strategy to identify potential prokaryotic ORFs by listing features assessable by an algorithm checking for ORFs.
b. Why does the presence of introns within transcribed eukaryotic sequences preclude direct application of this strategy to eukaryotic sequences?
c. How might you modify the strategy to overcome the problems posed by the presence of introns in transcribed eukaryotic sequences?

15.10 Annotation of genomic sequences makes them much more useful to researchers. What features should be included in an annotation, and in what different ways can they be depicted? For some examples of current annotations in databases, see the following websites:
http://genome-www.stanford.edu/Saccharomyces/ maps.html (*S. cerevisiae*)
http://www.flybase.org and http://flybase.bio.indiana. edu:82/annot/ (*Drosophila*)
http://www.tigr.org/tdb/ath1/htmls/ath1.html (*Arabidopsis*)
http://www.ncbi.nlm.nih.gov/genome/guide/ (humans)
http://www.cybergenome.com/ (compilation of links to many organisms)

*15.11 One of the central themes in genetics is that an organism's phenotype results from an interaction between its genotype and the environment. Because some diseases have strong environmental components, researchers have begun to assess how disease phenotypes arise from the interactions of genes with their environments, including the genetic background in which the genes are expressed. (See http://pga.tigr.org/desc.shtml for additional discussion.) How might DNA microarrays be useful in a functional genomic approach to understanding human diseases that have environmental components, such as some cancers?

15.12 How does a cell's transcriptome compare with its proteome?
a. For a specific eukaryotic cell, can you predict which has more total members? Can you predict which has more unique members?
b. Suppose you are interested in characterizing changes in the pattern of gene expression in the mouse nervous system during development. Describe how you would efficiently assess changes in the transcriptome from the time the nervous system forms during embryogenesis to its maturation in the adult.
c. How would your analyses differ if you were studying the proteome?

*15.13 Distinguish between structural, functional, and comparative genomics by completing the following exercise. The following list describes specific activities and goals associated with genome analysis. Indicate the area associated with the activity or goal by placing a letter (S, structural; F, functional; C, comparative) next to each item. Some items will have more than one letter associated with them.
_____ Aligning DNA sequences within databases to determine the degree of matching
_____ Annotation of sequences within a sequenced genome
_____ Characterizing the transcriptome and proteome present in a cell at a specific developmental stage or in a particular disease state

_____ Comparing the overall arrangements of genes and nongene sequences in different organisms to understand how genomes evolve

_____ Describing the function of all genes in a genome

_____ Determining the functions of human genes by studying their homologues in nonhuman organisms

_____ Developing a comprehensive two-dimensional polyacrylamide gel electrophoresis map of all proteins in a cell

_____ Developing a physical map of a genome

_____ Developing DNA microarrays (DNA chips)

_____ Identifying homologues to human disease genes in organisms suitable for experimentation

_____ Identifying a large collection of simple tandem repeat or microsatellite sequences to use as DNA markers within one organism

_____ Making gene knockouts and observing the phenotypic changes associated with them

15.14 Genomic analyses of *Mycoplasma genitalium* have not only identified its 517 genes but also allowed identification of the genes that this organism needs for life. The analyses are presented in detail at the website of the Institute for Genomic Research (http://www.tigr.org/minimal/). Explore this website and then address the following questions.

a. How many protein-coding genes have unknown functions?

b. How do we know which genes are required for life?

c. Are any of the protein-coding genes with unknown functions required for life?

d. What functional genomics approaches might be helpful to discern their function?

e. What practical value might there be in identifying the minimal gene set needed for life?

f. Is it possible to synthesize DNA sequences containing all the genes identified as essential for life and assemble these on one chromosome? Would this be enough to have a template for a new life form? If not, what additional sequence information would be needed to generate a new life form?

g. What is the difference between the life form generated in (f) and *Mycoplasma genitalium*?

h. What ethical issues arise from these analyses? Who should decide how to address them?

16

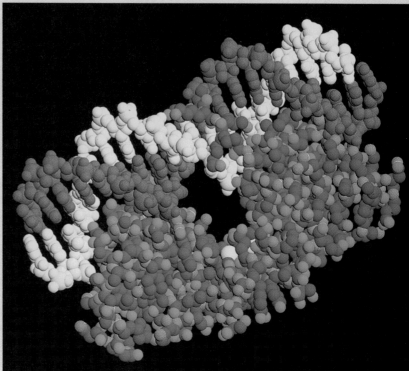

Regulation of Gene Expression in Bacteria and Bacteriophages

PRINCIPAL POINTS

In the lactose system of *E. coli,* the addition of lactose to cells brings about a rapid synthesis of three enzymes. In the absence of lactose, the synthesis of the three enzymes is turned off. The genes for the enzymes are contiguous on the *E. coli* chromosome and are adjacent to a controlling site (an operator) and a single promoter. The genes, the operator, and the promoter constitute an operon. Transcription of the genes results in a single polygenic mRNA. A regulatory gene is associated with an operon. To turn on gene expression in the lactose system, a lactose metabolite binds with a repressor protein (the product of the regulatory gene), inactivating it and preventing it from binding to the operator. As a result, RNA polymerase can bind to the promoter and transcribe the three genes as a single polycistronic (polygenic) mRNA. Operons are commonly involved in the regulation of gene expression in a large number of prokaryotic and bacteriophage systems.

Expression of a number of bacterial amino acid synthesis operons is controlled by a repressor-operator system and through attenuation at a second controlling site, called an attenuator. The repressor-operator system functions essentially like that for the *lac* operon, except that the addition of amino acid to the cell activates the repressor, thereby turning the operon off. An attenuator is located between the operator region and the first structural gene; it is a transcription termination site that allows only a fraction of RNA polymerases to transcribe the rest of the operon. Attenuation requires a coupling between transcription and translation and formation of particular RNA secondary structures that signal whether transcription can continue.

Bacteriophages such as lambda are especially adapted for undergoing reproduction within a bacterial host. Many genes related to the production of progeny phages or to the establishment or reversal of lysogeny in temperate phages are organized into operons. These operons, like bacterial operons, are controlled through the interaction of regulatory proteins with operators that are adjacent to clusters of structural genes. Phage lambda has been an excellent model for studying the genetic switch that controls the choice between lytic and lysogenic pathways in a lysogenic phage.

ONE OF THE BEST STRATEGIES FOR AN ORGANISM'S survival is the ability to adapt quickly to changes in its environment. In fact, a fundamental property of living cells is their ability to turn their genes on and off in response to extracellular signals. This control of gene expression makes it possible for cells to produce specific kinds of proteins when and where they are needed. In this chapter, you will learn some of the ways in which gene expression in microorganisms is regulated. Then, in the iActivity, you can investigate how mutations affect the process of regulation in *E. coli.*

Bacteria are free-living organisms that grow by increasing in mass and then divide by binary fission. Growth and division are controlled by genes, the expression of which must be regulated appropriately. Genes whose activity is controlled in response to the needs of a cell or organism are called **regulated genes.** An organism also has a large number of genes whose products are essential to the normal functioning of a growing and dividing cell. These genes are always active in growing cells and are known as **constitutive genes;** examples include genes that code for the enzymes needed for protein synthesis and glucose metabolism. Note that all genes are regulated at some level, so the distinction between regulated and constitutive genes is somewhat arbitrary.

The goal of this chapter is to learn about some of the mechanisms by which gene expression is regulated in bacteria and bacteriophages. Significantly, genes that encode proteins that work together in the cell typically are organized into operons; that is, the genes are adjacent to each other and are transcribed together onto a **polygenic (polycistronic) mRNA.** Regulation of synthesis of this mRNA depends on interactions between a regulatory protein and a regulatory site called an operator that is next to the gene array. Such studies of bacterial and bacteriophage gene regulation also have provided important insights into how genes are regulated in higher organ-

isms, including humans. Of course, much remains to be done to understand completely the regulation of gene expression in bacteria. The 4.6-mb (4.6 × 10⁶-bp) genome of *E. coli,* for example, has 4,288 protein-coding genes according to the genomic sequence, and more than 35 percent of those genes have no attributed function.

The *lac* Operon of *E. coli*

When gene expression is turned on in a bacterium by adding a substance (such as lactose) to the medium, the genes involved are said to be *inducible.* The regulatory substance that brings about this gene induction is called an **inducer,** and the phenomenon of producing a gene product in response to an inducer is called **induction.** The inducer is an example of a class of small molecules, called **effectors** or **effector molecules,** that help control expression of many regulated genes. Transcription of an inducible gene occurs in response to a regulatory event occurring at a specific DNA sequence—a **controlling site**—adjacent to or near the protein-coding sequence (Figure 16.1). The regulatory event typically involves an inducer and a regulatory protein. When the regulatory event occurs, RNA polymerase initiates transcription at the promoter (usually upstream of the controlling site). The gene is turned on, mRNA is made, and the protein coded for by the gene is produced. The controlling site itself does not code for any product. As an example of such gene regulation, we now discuss regulation of the inducible *lac* genes of the *E. coli lac* operon.

Lactose as a Carbon Source for *E. coli*

E. coli can grow in a simple medium containing salts (including a nitrogen source) and a carbon source such as glucose. The energy for biochemical reactions in the cell comes from glucose metabolism. The enzymes required for glucose metabolism are coded for by constitutive genes. If lactose is provided to *E. coli* as a carbon source

Figure 16.1

General organization of an inducible gene.

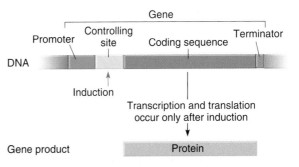

Inducible genes are expressed only when induced.

instead of glucose, a number of enzymes are rapidly synthesized that are required to metabolize lactose. (The same series of events, each involving a sugar-specific set of enzymes, is triggered by other sugars as well.) The enzymes are synthesized because the genes that code for them become actively transcribed in the presence of the sugar; the same genes are inactive if the sugar is absent. In other words, the genes are regulated genes whose products are needed only at certain times.

Lactose is a disaccharide consisting of the monosaccharides glucose and galactose. When lactose is present as the sole carbon source in the growth medium, three proteins are synthesized:

1. *β-Galactosidase*. This enzyme breaks down lactose into glucose and galactose, and it catalyzes the isomerization (conversion to a different form) of lactose to *allolactose*, a compound important in regulating expression of the *lac* operon (Figure 16.2). (In the cell the galactose is converted to glucose by enzymes encoded by a gene system specific for galactose catabolism. The glucose is then utilized by constitutively produced enzymes.)

2. *Lactose permease* (also called *M protein*). This protein, found in the *E. coli* cytoplasmic membrane, actively transports lactose into the cell.

3. *Transacetylase*. The function of this enzyme is poorly understood.

In wild-type *E. coli* growing in a medium containing glucose, only a low concentration of each of these three proteins is produced. In the presence of lactose but the absence of glucose, the amount of each enzyme increases coordinately (simultaneously) about a thousandfold. This occurs because the three essentially inactive genes are now being actively transcribed. This process is called **coordinate induction.** Allolactose, not lactose, is the inducer molecule directly responsible for the increased production of the three enzymes (see Figure 16.2). Furthermore, the mRNAs for the enzymes have a short half-life, so the transcripts must be made continually for the enzymes to be produced. When lactose is no longer present, transcription of the three genes is stopped and any mRNAs already present are broken down, so no more of these proteins are made. Existing proteins are diluted out by cell growth and division.

Experimental Evidence for the Regulation of *lac* Genes

Our basic understanding of the organization and regulation of the *lac* genes of *E. coli* as well as the controlling sites involved in lactose utilization came largely from the

Figure 16.2

Reactions catalyzed by the enzyme β-galactosidase. Lactose brought into the cell by the permease is converted to glucose and galactose (*top*) or to allolactose (*bottom*), the true inducer for the lactose operon of *E. coli*.

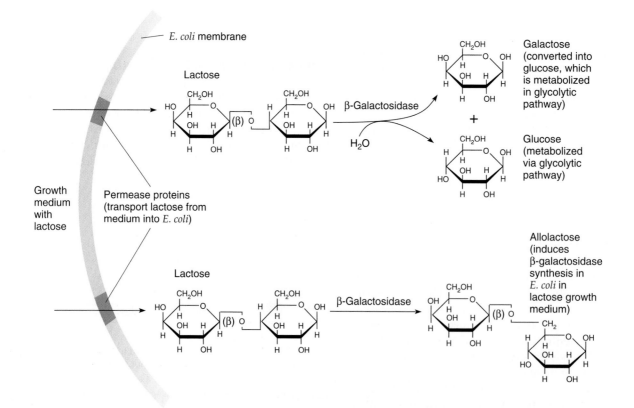

genetic experiments of François Jacob and Jacques Monod, for which they shared (along with Andre Lwoff for his work on the genetic control of virus synthesis) the 1965 Nobel Prize in Physiology or Medicine. We now summarize their experiments.

Mutations in the Protein-Coding Genes.

Mutations in the protein-coding (structural) genes obtained after treatment of cells with mutagens (chemicals that induce mutations)

animation

a Regulation of Expression of the *lac* Operon Genes

were used to study the organization and functions of the genes. The β-galactosidase gene was named *lacZ*, the permease gene *lacY*, and the transacetylase gene *lacA*. Standard genetic mapping experiments showed that the three genes are tightly linked in the order *lacZ-lacY-lacA*. These genes are transcribed onto a single polygenic mRNA molecule rather than onto three separate mRNAs. That is, RNA polymerase initiates transcription at a single promoter and a polygenic mRNA is synthesized with the gene transcripts in the order $5'-lacZ^+-lacY^+-lacA^+-3'$. In translation, a ribosome loads onto the polycistronic mRNA at the 5′ end, synthesizes β-galactosidase, then reinitiates translation at the permease sequence, synthesizes permease, then reinitiates at the transacetylase sequence, synthesizes transacetylase, and finally dissociates from the mRNA (Figure 16.3).

Mutations Affecting the Regulation of Gene Expression.

In wild-type *E. coli*, the three gene products are induced coordinately when lactose is present. Jacob and Monod isolated mutants in which all gene products of the operon were synthesized *constitutively*; that is, they were synthesized whether or not the inducer was present. They hypothesized that the mutations were regulatory mutations that affected the normal mechanisms controlling the expression of the structural genes for the enzymes. Jacob and Monod identified two classes of constitutive mutations: One class mapped to a small region next to the *lacZ* gene they called the **operator** (*lacO*). The other class mapped to a gene upstream of the operator that they called the *lacI* gene or lac **repressor gene.** Figure 16.4 depicts the organization of the *lac* structural gene cluster and the associated regulatory elements. This complex is the *lac* operon.

Operator Mutations.

The mutations of the operator were called operator-constitutive, or *lacO*ᶜ, mutations. Through the use of partial diploid strains (F′ strains in

---

Figure 16.3

Translation of the polygenic mRNA encoded by *lac* utilization genes in wild-type *E. coli*.

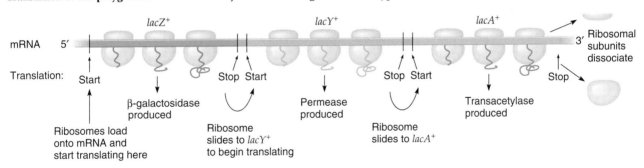

Figure 16.4

Organization of the *lac* genes of *E. coli* and the associated regulatory elements, the operator, promoter, and regulatory gene (*lacI*). The complex is called the *lac* operon.

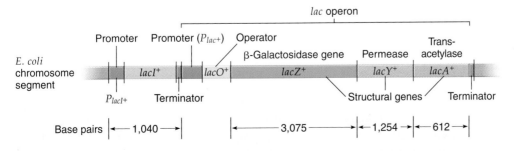

which a few chromosomal genes on an extrachromosomal genetic element called the *F* factor are introduced into a bacterial cell; see Figure 6.5), Jacob and Monod were able to define better the role of the operator in regulating expression of the *lac* genes. One such partial diploid was $\frac{F'\ lacO^+\ lacZ^-\ lacY^+}{lacO^c\ lacZ^+\ lacY^-}$ (both gene sets have a normal promoter, and the *lacA* gene is omitted because it is not important to our discussions).

One *lac* region in the partial diploid has a normal operator (*lacO⁺*), a mutated β-galactosidase gene (*lacZ⁻*), and a normal permease gene (*lacY⁺*). The other *lac* region has an operator-constitutive mutation (*lacO^c*), a normal β-galactosidase gene (*lacZ⁺*), and a mutated permease gene (*lacY⁻*). This partial diploid was tested for production of β-galactosidase (from the *lacZ⁺* gene) and of permease (from the *lacY⁺* gene), both in the presence and the absence of the inducer.

Jacob and Monod found that active β-galactosidase is synthesized in the absence of inducer and that permease is synthesized but is inactive because of the mutation. Only when lactose is added to the culture and the allolactose inducer is produced does synthesis of active permease occur. That is, the *lacZ⁺* gene (which is on the same DNA molecule as *lacO^c*) is *constitutively expressed* (meaning that the gene is active in the presence or absence of inducer), whereas the *lacY⁺* gene is under normal inducible control: It is inactive in the absence of inducer and active in the presence of inducer. In other words, a *lacO^c mutation affects only the genes downstream from it on the same DNA molecule.* Similarly, the *lacO⁺* region controls only *lac* structural genes adjacent to it and has no effect on the genes on the other DNA molecule. This phenomenon of a gene or DNA sequence controlling only genes that are on the same, contiguous piece of DNA is called **cis-dominance.** The *lacO^c* mutation is cis-dominant because the defect affects the adjacent genes only and cannot be overcome by a normal *lacO⁺* region elsewhere in the cell. That is, the operator must not encode a diffusible product. If it did, then in the *lacO⁺/lacO^c* diploid state, one or the other of the regions would have controlled all the lactose utilization genes regardless of their location.

lacI *Gene Regulatory Mutations.* The second class of *lac* constitutive mutants defined the *lacI* gene. Again, the use of partial diploid strains illuminated the normal function of the *lacI* gene.

The partial diploid here is $\frac{lacI^+\ lacO^+\ lacZ^-\ lacY^+}{lacI^-\ lacO^+\ lacZ^+\ lacY^-}$; both gene sets have normal operators and normal promoters. In the absence of the inducer, no β-galactosidase or permease was produced; both were synthesized in the presence of inducer. In other words, the expression of both operons was inducible. This means that the *lacI⁺* gene in the cell can overcome the defect of the *lacI⁻* mutation. Since the two *lacI* genes are located on different DNA molecules (i.e., they are in a trans configuration), the *lacI⁺* gene is said to be *trans-dominant* to the *lacI⁻* gene.

Since the *lacI⁺* gene controls the genes on the other DNA molecule, Jacob and Monod proposed that the *lacI⁻* gene is a repressor gene that produces a **repressor molecule.** No functional repressor molecules are produced in *lacI⁻* mutants. Thus, in a haploid *lacI⁻* bacterial strain, the *lac* operon is constitutive. In a partial diploid with both a *lacI⁺* and a *lacI⁻*, however, the functional repressor molecules produced by the *lacI⁺* gene control the expression of both *lac* operons in the cell, making both operons inducible.

Promoter Mutations. The promoter for the structural genes (located at the *lacZ* end of the cluster of *lac* genes; see Figure 16.4) is also affected by mutations. Most of the known promoter mutants (P_{lac}–) affect all three structural genes. Even in the presence of inducer, the lactose utilization enzymes are not made or are made only at very low rates. Since the promoter is the recognition sequence for RNA polymerase and does not code for any product, the effect of a *P* mutation is confined to the genes that it controls on the same DNA strand. The P_{lac}– mutations are another example of cis-dominant mutations.

Jacob and Monod's Operon Model for the Regulation of *lac* Genes

Based on the results of their studies, Jacob and Monod proposed their now classic *operon model.* By definition, an **operon** is a *cluster of genes, the expressions of which are regulated together by operator–repressor protein interactions plus the operator region itself and the promoter.* The promoter was not part of Jacob and Monod's original model; its existence was demonstrated in later studies. The order of the controlling elements and genes in the *lac* operon is promoter-operator-*lacZ-lacY-lacA*, and the regulatory gene *lacI* is located close to the structural genes, just upstream of the promoter (see Figure 16.4). The *lacI* gene has its own constitutive promoter and terminator.

The following description of the Jacob-Monod model for regulation of the *lac* operon has been embellished with up-to-date molecular information. Figure 16.5 depicts the state of the *lac* operon in wild-type *E. coli* growing in the absence of lactose. The repressor gene (*lacI⁺*) is transcribed constitutively, and translation of its mRNA produces a 360-amino-acid polypeptide. Four of these polypeptides associate together to form a tetramer, the functional repressor protein (Figure 16.6).

The repressor binds to the operator (*lacO⁺*). When the repressor is bound to the operator, RNA polymerase can bind to the operon's promoter but is blocked from initiating transcription of the protein-coding genes; the *lac* operon is said to be under *negative control.* The low level of transcription of the genes that produces a few molecules of the enzymes, even in the absence of inducer, occurs because repressors do not just bind and stay; they bind and unbind. In the split second after one repressor unbinds and before another binds, an RNA polymerase

Figure 16.9 (continued)

b) Partial diploid in the absence of inducer

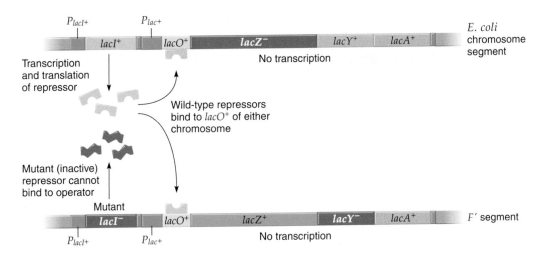

c) Partial diploid in the presence of inducer

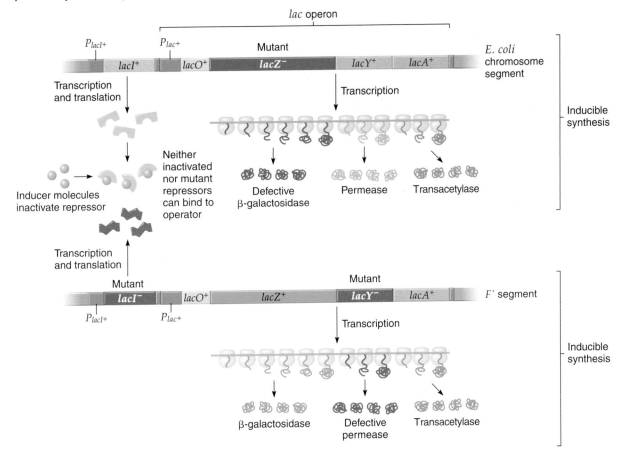

Regulation of the *trp* Operon

Two regulatory mechanisms are involved in controlling the expression of the *trp* operon. One mechanism uses a repressor-operator interaction, and the other determines whether initiated transcripts include the structural genes or are terminated before those genes are reached.

Expression of the *trp* Operon in the Presence of Tryptophan.

The regulatory gene for the *trp* operon is *trpR*; it is located some distance from the operon (and therefore does not appear in Figure 16.12). The product of *trpR* is an *apore-pressor protein*, which is basically an inactive repressor that alone cannot bind to the operator. When tryptophan

Figure 16.10

Dominant effect of *lacI*ˢ mutation over wild-type *lacI*⁺ in a *lacI*⁺ *lacO*⁺ *lacZ*⁺ *lacY*⁺ *lacA*⁺/ *lacI*ˢ *lacO*⁺ *lacZ*⁺ *lacY*⁺ *lacA*⁺ partial diploid cell growing in the presence of lactose.

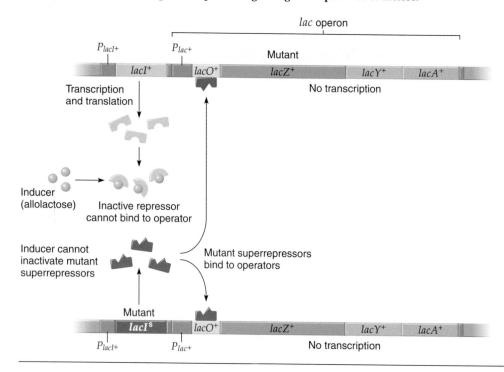

Figure 16.11

Role of cyclic AMP (cAMP) in the functioning of glucose-sensitive operons such as the *lac* operon of *E. coli*.

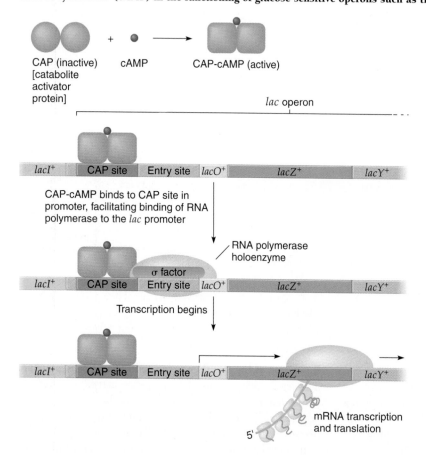

Figure 16.12

Organization of controlling sites and structural genes of the *E. coli* trp operon.

Also shown are the steps catalyzed by the products of the structural genes *trpA, trpB, trpC, trpD,* and *trpE.*

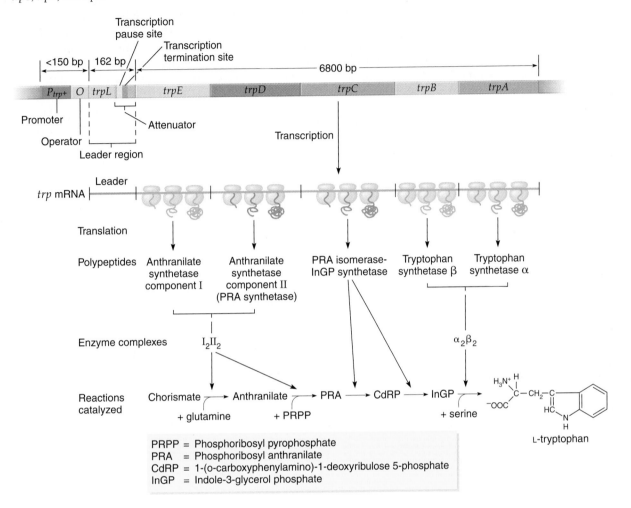

PRPP = Phosphoribosyl pyrophosphate
PRA = Phosphoribosyl anthranilate
CdRP = 1-(o-carboxyphenylamino)-1-deoxyribulose 5-phosphate
InGP = Indole-3-glycerol phosphate

is abundant within the cell, it interacts with the aporepressor and converts it to an active repressor. (Tryptophan is an example of an *effector* molecule, just as allolactose is the effector molecule for the *lac* operon.) The active repressor binds to the operator and prevents the initiation of transcription of the *trp* operon protein-coding genes by RNA polymerase. As a result, the tryptophan biosynthesis enzymes are not produced. By repression, transcription of the *trp* operon can be reduced about 70-fold.

Expression of the *trp* Operon in the Presence of Low Concentrations of Tryptophan. The second regulatory mechanism is involved in the expression of the *trp* operon under conditions of tryptophan starvation or tryptophan limitation. Under severe tryptophan starvation, the *trp* genes are expressed maximally, and under less severe starvation conditions, the *trp* genes are expressed at less than maximal levels. This is accomplished by a mecha-

nism that controls the ratio of full-length transcripts that include the five *trp* structural genes to short, 140-bp transcripts that have terminated at the attenuator site within the *trpL* region (see Figure 16.12). The short transcripts are terminated by a process called **attenuation.** The proportion of the transcripts that include the structural genes is inversely related to the amount of tryptophan in the cell; the more tryptophan there is, the greater the proportion of short transcripts. Attenuation can reduce transcription of the *trp* operon by a factor of 8 to 10. Thus, repression and attenuation together can regulate the transcription of the *trp* operon by a factor of about 560 to 700.

Molecular Model for Attenuation. The mRNA transcript of the leader region includes a sequence that can be translated to produce a short polypeptide. Just before the stop codon of the transcript are two adjacent codons for tryptophan that play an important role in attenuation.

There are four regions of the leader peptide mRNA that can fold and form secondary structures by complementary base pairing (Figure 16.13). Pairing of regions 1 and 2 results in a transcription *pause signal*; pairing of 3 and 4 results in a *termination of transcription signal*; and pairing of 2 and 3 results in an *antitermination signal*, signaling transcription to continue.

Crucial to the attenuation model is the fact that transcription and translation are tightly coupled in prokaryotes, made possible by the absence of a nuclear envelope. In the *trp* regulatory system, this coupling of transcription and translation is brought about by a pause of the RNA polymerase caused by the pairing of RNA regions 1 and 2 just after they have been synthesized (see Figure 16.13). The pause lasts long enough for the ribosome to load onto the mRNA and to begin translating the leader peptide, so that translation of the leader mRNA transcript occurs just behind the RNA polymerase.

As coupled transcription and translation continues, the position of the ribosome on the leader transcript plays an important role in the regulation of transcription termination at the attenuator. If the cells are starved for tryptophan, the amount of Trp-tRNA molecules (charged tryptophanyl-tRNA) drops dramatically because very few tryptophan molecules are available for the aminoacylation of the tRNA. A ribosome translating the leader transcript stalls at the tandem Trp codons in region 1 because the next specified amino acid in the peptide is in short supply; the leader peptide cannot be completed (Figure 16.14a). Since the ribosome now "covers" region 1 of the attenuator region, the 1-2 pairing cannot happen because region 1 is no longer available. However, RNA region 2 will pair with RNA region 3 once region 3 is synthesized. Because region 3 is paired with region 2, region 3 cannot pair with region 4 when it is synthesized. The 2-3 pairing is an antitermination signal because the termination signal of 3 paired with 4 does not form, and this allows RNA polymerase to continue past the attenuator and transcribe the structural genes.

If, instead, enough tryptophan is present that the ribosome can translate the Trp codons (Figure 16.14b), then the ribosome continues to the stop codon for the leader peptide. Since the ribosome is then covering part of RNA region 2, region 2 is unable to pair with region 3, and region 3 is then able to pair with region 4 when it is transcribed. The bonding of region 3 with region 4 is a termination signal. The 3-4 structure is called the attenuator. The key signal for attenuation is the concentration of Trp-tRNA in the cell because that determines how far the ribosome gets on the leader transcript, either to the Trp codons or to the stop codon.

Attenuation is involved in the genetic regulation of a number of other amino acid biosynthetic operons of *E. coli* and *Salmonella typhimurium*. In every case, there is a leader sequence with two or more codons for the particular amino acid, the synthesis of which is controlled by the enzymes encoded by the operon. Attenuation has also been shown to regulate a number of genes not involved with amino acid biosynthesis, such as the rRNA operons (*rrn*) and the *ampC* gene of *E. coli* (for ampicillin resistance).

Figure 16.13

Four regions of the *trp* operon leader mRNA and the alternate secondary structures they can form by complementary base pairing.

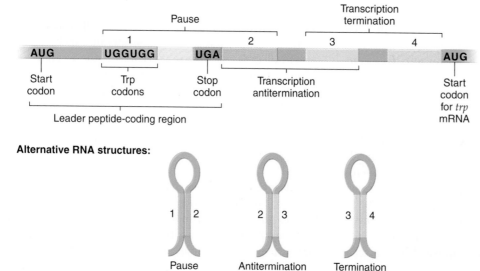

Figure 16.14

Models for attenuation in the *trp* operon of *E. coli*. The light blue structures are ribosomes that are translating the leader transcript. (**a**) Tryptophan-starved cells. (**b**) Non–tryptophan-starved cells.

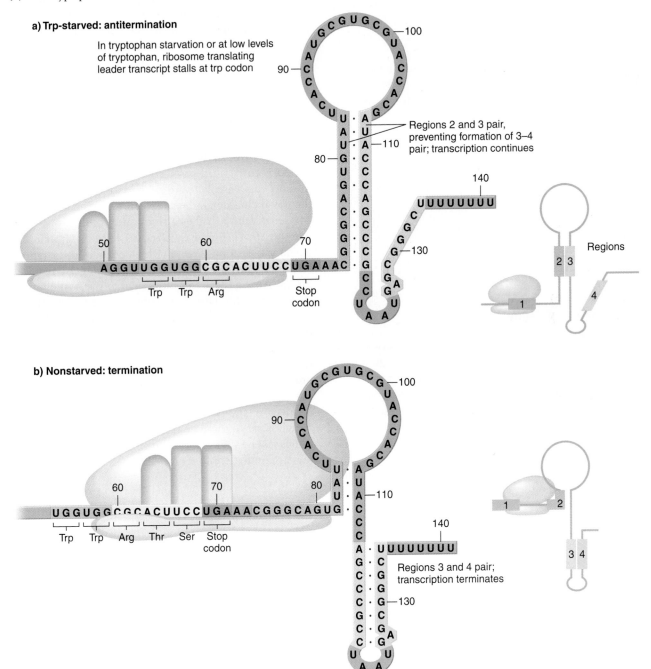

a) Trp-starved: antitermination

In tryptophan starvation or at low levels of tryptophan, ribosome translating leader transcript stalls at trp codon

Regions 2 and 3 pair, preventing formation of 3–4 pair; transcription continues

Trp   Trp   Arg

Stop codon

Regions

b) Nonstarved: termination

Trp   Trp   Arg   Thr   Ser   Stop codon

Regions 3 and 4 pair; transcription terminates

KEYNOTE

Regulation of the tryptophan (*trp*) operon of *E. coli* is at the level of initiating or completing a transcript of the operon. This is accomplished through a repressor-operator system, which responds to free tryptophan levels, and through attenuation at a second controlling site called an attenuator, which responds to Trp-tRNA levels. The attenuator is located in the leader region between the operator region and the first *trp* structural gene. The attenuator acts to terminate transcription in a way that is dependent on the concentration of tryptophan. In the presence of large amounts of tryptophan, attenuation is very effective; that is, enough Trp-tRNA is present so that the ribosome can move past the attenuator and allow the leader transcript to form a secondary structure that causes transcription to be blocked. In the absence of or at low levels of tryptophan, the ribosomes stall at the attenuator, and the leader transcript forms a secondary structure that permits continued transcription activity.

Regulation of Gene Expression in Phage Lambda

Bacteriophages exist by invading and manipulating bacterial cells. Many or all the essential components for phage reproduction are provided by the bacterial host cell, and the use of those components is controlled by the products of phage genes. Most genes of a phage, then, code for products that control the life cycle and the production of progeny phage particles. Much is known about gene regulation in a number of bacteriophages. In this section, we discuss the regulation of gene expression as it relates to the lytic cycle and lysogeny in bacteriophage lambda (λ). (Recall from Chapter 6, pp. 132–133, that the lytic cycle is a type of phage life cycle in which the phage takes over the bacterium and directs its growth and reproductive abilities to express the phage's genes and to produce progeny phages. Lysogeny is the insertion of a temperate phage chromosome into a bacterial chromosome, where it replicates when the bacterial chromosome replicates. In this state, the phage genome is repressed and is said to be in the prophage state.)

Early Transcription Events

Figure 16.15 shows the genetic map of λ. The mature λ chromosome is linear and has complementary "sticky" ends. Once free in the host cell, the λ chromosome circularizes, so we show the genetic map in a circular form. Recall that λ is a temperate phage (see Figure 6.12), so when it infects a bacterial cell and its genome circularizes, the phage has a choice of whether to enter the lytic pathway (in which progeny phages are assembled and

released from the cell) or the lysogenic pathway (in which the λ chromosome integrates into the host chromosome and no progeny phages are produced). This choice involves a sophisticated *genetic switch*. First, transcription begins at promoters P_L and P_R (Figure 16.16, step 1). Promoter P_L is for leftward transcription of the left early operon, and promoter P_R is for rightward transcription of the right early operon.

The first gene to be transcribed from P_R is *cro* (control of repressor and other), the product of which is the Cro protein. This protein plays an important role in setting the genetic switch to the lytic pathway. The first gene to be transcribed from P_L is *N*. The resulting N protein is a transcription *antiterminator* that allows RNA synthesis to proceed past certain transcription terminators, in this case leftward of *N* and rightward of *cro*, thereby including all the early genes (Figure 16.16, step 2). Genes transcribed as a result of the action of N protein are *cII*, *O*, *P*, and *Q*. Gene *cII* encodes protein cII, which can turn on gene *cI* (which encodes the λ repressor) and gene *int* (which encodes the integrase required for integrating the λ chromosome into the host chromosome during the lysogenic pathway). However, cII protein performs this function only when the phage follows the lysogenic pathway. Genes *O* and *P* encode two DNA replication proteins, and gene *Q* encodes a protein needed to turn on late genes for lysis and phage particle proteins. The Q protein is another antiterminator, permitting transcription to continue into the late genes involved in the lytic pathway. However, only when the switch is set to the lytic pathway and transcription continues from P_R for a sufficient time does enough Q protein accumulate to function effectively.

The Lysogenic Pathway

After the early transcription events, either the lysogenic or lytic pathway is followed (Figure 16.16, step 3). The switch is set for the lysogenic pathway as follows.

The establishment of lysogeny requires the protein products of the *cII* (right early operon) and *cIII* (left early operon; see Figure 16.15) genes. The cII protein (stabilized by cIII protein) activates transcription of the *cI* gene (located between the P_L and P_R promoters; Figure 16.16, step 4a) leftward from a promoter called P_{RE} (promoter for repressor establishment). The product of the *cI* gene, the λ repressor, binds to two operator regions, O_L and O_R (see Figure 16.15 and Figure 16.16, step 5a), whose sequences overlap the P_L and P_R promoters, respectively. The binding of the λ repressor prevents the further transcription by RNA polymerase of the early operons controlled by P_L and P_R. As a result, transcription of the *N* and *cro* genes is blocked, and because the two proteins specified by these genes are unstable, the concentrations of these two proteins in the cell drop dramatically. Furthermore, a repressor bound to O_R stimulates synthesis of

Figure 16.15

A map of phage λ, showing the major genes. (Promoters discussed in text: P_L = promoter for leftward transcription of the left early operon, P_R = promoter for rightward transcription of the right early operon, P_{RE} = promoter for repressor establishment, and P_{RM} = promoter for repressor maintenance.)

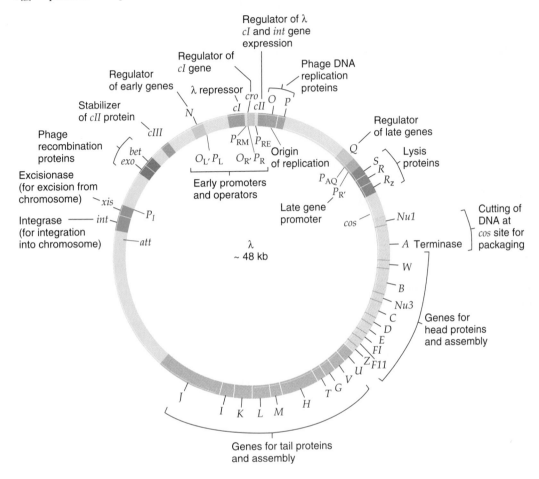

more repressor mRNA from a different promoter, P_{RM} (promoter for repressor maintenance), thereby maintaining repressor concentrations in the cell (see Figure 16.16, step 5a). Thus, if enough λ repressors are present, lysogeny is established by the binding of the repressor to operators O_L and O_R, followed by integration of λ DNA catalyzed by integrase. Integrase is the product of the cII-regulated promoter P_I. As the concentration of cII drops, P_I transcription shuts off, leaving P_{RM} as the only active promoter.

In sum, the lysogenic pathway is favored when enough λ repressor is made so that early promoters are turned off, thereby repressing all the genes needed for the lytic pathway. One important lytic pathway gene that is repressed is Q; the Q protein is a positive regulatory protein required for the production of phage coat proteins and lysis proteins (see the next section).

The Lytic Pathway

Let us consider the induction of the lytic pathway caused by ultraviolet light irradiation. Inducers such as ultravio-

let light typically damage DNA, and this somehow causes a change in the function of the bacterial protein RecA (product of the recA gene). Normally, RecA functions in DNA recombination, but when DNA is damaged, RecA stimulates the λ repressor polypeptides to cleave themselves in two and therefore become inactivated. The resulting absence of repressor at O_R allows RNA polymerase to bind at P_R, and the cro gene is then further transcribed. The Cro protein produced then acts to decrease RNA synthesis from P_L and P_R, and this reduces synthesis of cII protein, the regulator of λ repressor synthesis, and blocks synthesis of λ repressor mRNA from P_{RM} (Figure 16.16, step 4b). At the same time, transcription of the right early operon genes from P_R is decreased, but enough Q proteins are accumulated to set the genetic switch for transcription of the late genes for starting the lytic pathway (Figure 16.16, step 5b).

In summary, phage λ uses complex regulatory systems to choose either the lytic or the lysogenic pathway. This decision depends on a sophisticated genetic switch that involves competition between the products of the cI

Figure 16.16

Expression of λ genes after infecting *E. coli* and the transcriptional events that occur when either the lysogenic or lytic pathways are followed.

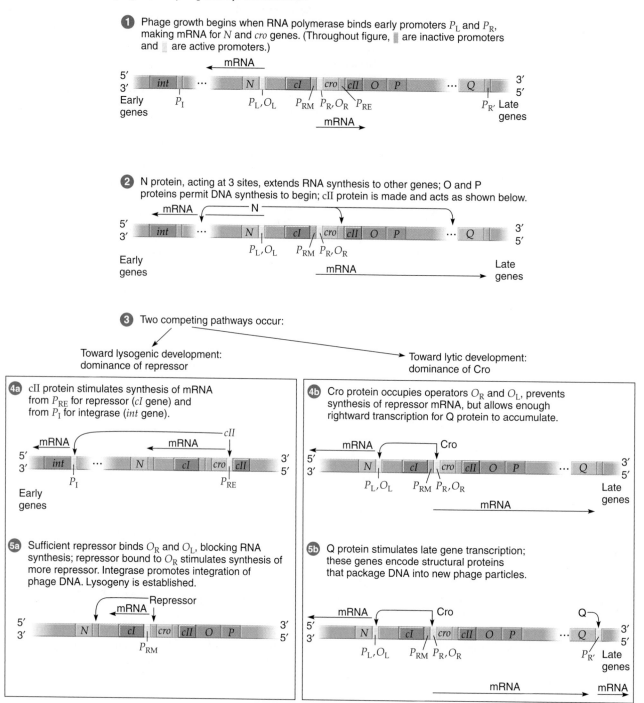

1 Phage growth begins when RNA polymerase binds early promoters P_L and P_R, making mRNA for N and *cro* genes. (Throughout figure, ▌ are inactive promoters and ░ are active promoters.)

2 N protein, acting at 3 sites, extends RNA synthesis to other genes; O and P proteins permit DNA synthesis to begin; cII protein is made and acts as shown below.

3 Two competing pathways occur:

Toward lysogenic development: dominance of repressor

Toward lytic development: dominance of Cro

4a cII protein stimulates synthesis of mRNA from P_{RE} for repressor (*cI* gene) and from P_I for integrase (*int* gene).

4b Cro protein occupies operators O_R and O_L, prevents synthesis of repressor mRNA, but allows enough rightward transcription for Q protein to accumulate.

5a Sufficient repressor binds O_R and O_L, blocking RNA synthesis; repressor bound to O_R stimulates synthesis of more repressor. Integrase promotes integration of phage DNA. Lysogeny is established.

5b Q protein stimulates late gene transcription; these genes encode structural proteins that package DNA into new phage particles.

("c-one") gene (the repressor) and the *cro* gene (the Cro protein). If the repressor dominates, the lysogenic pathway is followed; if the Cro protein dominates, the lytic pathway is followed.

Summary

From studies of the regulation of expression of the three lactose utilization genes in *E. coli*, Jacob and Monod developed a model that is the basis for the regulation of gene expression in a large number of bacterial and bacteriophage systems. The genes for the enzymes are contiguous in the chromosome and are adjacent to a controlling site (an operator) and a single promoter. This complex constitutes a transcriptional regulatory unit called an operon. A regulator gene, which may or may not be nearby, is associated with the operon. The addition of an appropriate substrate to the cell results in the coordinate induction of the operon's structural genes. Induction of the *lac* operon occurs as follows: Lactose binds with a repressor protein that is encoded by the regulator gene, inactivating the repressor protein and preventing it from binding to the operator. As a result, RNA polymerase can transcribe the three genes from a single promoter onto a single polygenic mRNA. As long as lactose is present, mRNA continues to be produced and the enzymes are made. When lactose is no longer present, the repressor protein is no longer inactivated, and it binds to the operator, thereby preventing RNA polymerase from transcribing the *lac* genes.

If both glucose and lactose are present in the medium, the lactose operon is not induced because glucose (which requires less energy to metabolize than does lactose) is the preferred energy source. This phenomenon is called catabolite repression and involves cellular levels of cyclic AMP. That is, in the presence of lactose and the absence of glucose, cAMP complexes with CAP to form a positive regulator needed for RNA polymerase to bind to the promoter. Addition of glucose results in a lowering of cAMP concentration, so no CAP-cAMP complex is produced; therefore, RNA polymerase cannot bind efficiently to the promoter and transcribe the *lac* genes.

The genes for a number of bacterial amino acid biosynthesis pathways are also arranged in operons. Expression of these operons is accomplished by a repressor-operator system or, in some cases, through attenuation at a second controlling site called an attenuator, or by both mechanisms. The *trp* operon is an example of an operon with both types of transcription regulation systems. The *trp* repressor-operator system functions essentially like that of the *lac* operon, except that the addition of tryptophan to the cell activates the repressor, turning the operon off. The attenuator, located downstream from the operator in a leader region that is translated, is a transcription termination site that allows only a fraction of RNA polymerases that initiate at the *trp* promoter to transcribe the rest of the operon. Attenuation requires a coupling between transcription and translation and the formation of particular RNA secondary structures that signal whether transcription can continue.

Key to the attenuation phenomenon is the presence of multiple copies of codons for the amino acid synthesized by the enzymes encoded by the operon. When enough of the amino acid is present in the cell, enough charged tRNAs are produced so that the ribosome can quickly translate the key codons in the leader region, and this causes the RNA being made by the RNA polymerase ahead of the ribosome to assume a secondary structure that signals transcription to stop. However, when the cell is starved for that amino acid, there are insufficient charged tRNAs to be used at the key codons, so the ribosome stalls at that point. As a result, the RNA ahead of that point assumes a secondary structure that permits continued transcription into the structural genes. The combination of repressor-operator regulation and attenuation control permits a fine degree of control of operon transcription.

Operons are also extensively used by bacteriophages to coordinate the synthesis of proteins with related functions. Many genes related to the production of progeny phages or to the establishment or reversal of lysogeny of temperate phages are organized into operons. These operons, like bacterial operons, are controlled through the interaction of regulatory proteins with operators that are adjacent to clusters of structural genes. Bacteriophage lambda has been an excellent model for studying the elaborate genetic switch that controls the choice between lytic and lysogenic pathways in a lysogenic phage. The switch involves two regulatory proteins, the λ repressor and the Cro protein, which have opposite affinities for three binding sites within each of the two major operators that control the λ life cycle.

In sum, operons are commonly encountered in bacteria and their viruses. They provide a simple way to coordinate the expression of genes with related function. The following generalizations can be made about the regulation of operons. First, for most operons, a regulator protein (e.g., the *lac* repressor) plays a key role in the regulation process because it is able to bind to a controlling site in the operon, the operator. Second, the transcription of a set of clustered structural genes is controlled by an adjacent operator through interaction with a regulator protein, which can exert positive or negative control, depending on the operon. Third, the trigger for changing the state of an operon from off to on and vice versa is an effector molecule (e.g., the inducer allolactose in the case of the *lac* operon), which controls the conditions under which the regulator protein will or will not bind to the operator.

Analytical Approaches for Solving Genetics Problems

Q16.1 In the laboratory, you are given ten strains of *E. coli* with the following *lac* operon genotypes, where $I = lacI$ (the repressor gene), $P = P_{lac}$ (the promoter), $O = lacO$ (the operator), and $Z = lacZ$ (the β-galactosidase gene).

For each strain, predict whether β-galactosidase will be produced (a) if lactose is absent from the growth medium and (b) if lactose is present in the growth medium.

1. $I^+ P^+ O^+ Z^+$

2. $I^- P^+ O^+ Z^+$

3. $I^+ P^+ O^c Z^+$

4. $I^- P^+ O^c Z^+$

5. $I^+ P^+ O^c Z^-$

6. $\dfrac{F' \; I^+ \; P^+ \; O^c \; Z^-}{I^+ \; P^+ \; O^+ \; Z^+}$

7. $\dfrac{F' \; I^+ \; P^+ \; O^+ \; Z^-}{I^+ \; P^+ \; O^c \; Z^+}$

8. $\dfrac{F' \; I^+ \; P^+ \; O^+ \; Z^+}{I^- \; P^+ \; O^+ \; Z^-}$

9. $\dfrac{F' \; I^+ \; P^+ \; O^c \; Z^-}{I^- \; P^+ \; O^+ \; Z^-}$

10. $\dfrac{F' \; I^- \; P^+ \; O^+ \; Z^-}{I^- \; P^+ \; O^c \; Z^+}$

(Note: In the partial diploid strains (6–10), one copy of the *lac* operon is in the host chromosome and the other copy is in the extrachromosomal F factor.)

A16.1 The answers are as follows, where + means that β-galactosidase is produced and − means that it is not produced:

Genotype	Noninduced: Lactose Absent	Induced: Lactose Present
(1)	−	+
(2)	+	+
(3)	+	+
(4)	+	+
(5)	−	−
(6)	−	+
(7)	+	+
(8)	−	+
(9)	−	+
(10)	+	+

To answer this question completely, we need a good understanding of how the lactose operon is regulated in the wild type and of the consequences of particular mutations on the regulation of the operon.

Strain (1) is the standard wild-type operon. No enzyme is produced in the absence of lactose because the repressor produced by the I^+ gene binds to the operator (O^+) and blocks the initiation of transcription. When lactose is added, it binds to the repressor, changing its conformation so that it no longer can bind to the O^+ region, thereby facilitating transcription of the structural genes for RNA polymerase.

Strain (2) is a haploid strain with a mutation in the lacI gene (I^-). The consequence is that the repressor protein cannot bind to the (normal) operator region, so there is no inhibition of transcription, even in the absence of lactose. This strain, then, is constitutive, meaning that β-galactosidase is produced by the $lacZ^+$ gene in the presence or absence of lactose.

Strain (3) is another constitutive mutant. In this case, the repressor gene is a wild type and the β-galactosidase gene $lacZ^+$ is a wild type, but there is a mutation in the operator region (O^c). Therefore, repressor protein cannot bind to the operator, and transcription occurs in the presence or absence of lactose.

Strain (4) carries both regulatory mutations of the previous two strains. Functional repressor is not produced, but even if it were, the operator is changed so that it cannot bind. The consequence is the same: constitutive enzyme production.

Strain (5) produces functional repressor, but the operator (O^c) is mutated. Therefore, transcription cannot be blocked, and lac polygenic mRNA is produced in the presence or absence of lactose. However, because there is also a mutation in the β-galactosidase gene (Z^-), no functional enzyme is generated.

In the partial diploid strain (6), one lac operon is completely wild-type and the other carries a constitutive operator mutation and a mutant β-galactosidase gene. In the absence of lactose, no functional enzyme is produced. For the wild-type operon, the repressor binds to the operator and blocks transcription. For the operon with the two mutations, the operator region is mutated and cannot bind repressor, so the mRNA for the mutated operon is produced; however, the lacZ gene is also mutated, so that functional enzyme cannot be produced. In the presence of lactose, functional enzyme is produced because repression of the wild-type operon is relieved so that the Z^+ gene can be transcribed. This type of strain provided one of the pieces of evidence that the operator region does not produce a diffusible substance.

In partial diploid (7), functional enzyme is produced in the presence or absence of lactose because one of the operons has an O^c mutation that does not respond to a repressor and that is linked to a wild-type Z^+ gene. That operon is transcribed constitutively. The other operon is inducible, but because there is a Z^- mutation, no functional enzyme is produced.

Partial diploid (8) has a wild-type operon and an operon with an I^- regulatory mutation and a Z^- mutation. The I^+ gene product is diffusible, so it can bind to the O^+ region of both operons, thereby putting both operons under inducer control. This strain demonstrates that the I^+ gene is trans-dominant to an I^- mutation. In this case, the particular location of the one Z^- mutation is irrelevant: The same result would have been obtained had the Z^+ and Z^- been switched between the two operons. In this partial diploid, β-galactosidase is not produced unless lactose is present.

In strain (9), β-galactosidase is produced only when the lactose is present because the O^c region controls only the genes that are adjacent to it on the same chromosome (cis-dominance), and in this case one of the adjacent genes is Z^-, which codes for a nonfunctional enzyme. The partial diploid is heterozygous I^+/I^-, but I^+ is trans-dominant, as discussed for strain (8). Thus, the only normal Z^+ gene is under inducer control.

Partial diploid (10) has defective repressor protein as well as an O^c mutation adjacent to a Z^+ gene. On the latter ground alone, this partial diploid is constitutive. The other operon is also constitutively transcribed, but because there is a Z^- mutation, no functional enzyme is generated from it.

Questions and Problems

16.1 How does lactose bring about the induction of synthesis of β-galactosidase, permease, and transacetylase? Why does this event not occur when glucose is also in the medium?

16.2 Operons produce polygenic mRNA when they are active. What is a polygenic mRNA? What advantages, if any, does it confer in terms of the cell's function?

*16.3 If an *E. coli* mutant strain synthesizes β-galactosidase whether or not the inducer is present, what genetic defect(s) might be responsible for this phenotype?

16.4 Elucidation of the regulatory mechanisms associated with the enzymes of lactose utilization in *E. coli* was a landmark in our understanding of regulatory processes in microorganisms. In formulating the operon hypothesis as applied to the lactose system, Jacob and Monod found that results from particular partial diploid strains were invaluable. Specifically, in terms of the operon hypothesis, what information did the partial diploids provide that the haploids could not?

*16.5 For the *E. coli lac* operon, write the partial diploid genotype for a strain that will produce β-galactosidase constitutively and permease by induction.

16.6 Mutants were instrumental in elaborating the model for regulation of the *lac* operon.
a. Discuss why *lacO^c* mutants are cis-dominant but not trans-dominant.

b. Explain why *lacI^s* mutants are trans-dominant to the wild-type *lacI^+* allele but *lacI^-* mutants are recessive.
c. Discuss the consequences of mutations in the repressor gene promoter as compared with mutations in the structural gene promoter.

*16.7 This question involves the *lac* operon of *E. coli*, where *I* = *lacI* (the repressor gene), *P* = *P_lac* (the promoter), *O* = *lacO* (the operator), *Z* = *lacZ* (the β-galactosidase gene), and *Y* = *lacY* (the permease gene). Complete Table 16.A, using + to indicate that the enzyme in question will be synthesized and − to indicate that the enzyme will not be synthesized.

16.8 A new sugar, sugarose, induces synthesis of two enzymes from the *sug* operon of *E. coli*. Some properties of mutations affecting the appearance of these enzymes are as follows (here, + = enzyme induced normally, i.e., synthesized only in the presence of the inducer; C = enzyme synthesized constitutively; 0 = enzyme cannot be detected, p. 342):

Table 16.A

	Genotype	Inducer Absent: β-galactosidase	Permease	Inducer Present: β-galactosidase	Permease
a.	$I^+ P^+ O^+ Z^+ Y^+$				
b.	$I^+ P^+ O^+ Z^- Y^+$				
c.	$I^+ P^+ O^+ Z^+ Y^-$				
d.	$I^- P^+ O^+ Z^+ Y^+$				
e.	$I^s P^+ O^+ Z^+ Y^+$				
f.	$I^+ P^+ O^c Z^+ Y^+$				
g.	$I^s P^+ O^c Z^+ Y^+$				
h.	$I^+ P^+ O^c Z^+ Y^-$				
i.	$I^- P^+ O^+ Z^+ Y^+$ / $I^+ P^+ O^+ Z^- Y^-$				
j.	$I^- P^+ O^+ Z^+ Y^-$ / $I^+ P^+ O^+ Z^- Y^+$				
k.	$I^s P^+ O^+ Z^+ Y^-$ / $I^+ P^+ O^+ Z^- Y^+$				
l.	$I^+ P^+ O^c Z^- Y^+$ / $I^+ P^+ O^+ Z^+ Y^-$				
m.	$I^- P^+ O^c Z^+ Y^-$ / $I^+ P^+ O^+ Z^- Y^+$				
n.	$I^s P^+ O^+ Z^+ Y^+$ / $I^+ P^+ O^c Z^+ Y^+$				
o.	$I^+ P^- O^c Z^+ Y^-$ / $I^+ P^+ O^+ Z^- Y^+$				
p.	$I^+ P^- O^+ Z^+ Y^-$ / $I^+ P^+ O^c Z^- Y^-$				
q.	$I^- P^- O^+ Z^+ Y^+$ / $I^+ P^+ O^+ Z^- Y^-$				
r.	$I^- P^+ O^+ Z^+ Y^-$ / $I^+ P^- O^+ Z^- Y^+$				

Mutation of	Enzyme 1	Enzyme 2
Gene A	+	0
Gene B	0	+
Gene C	0	0
Gene D	C	C

a. The genes are adjacent, in the order A B C D. Which gene is most likely to be the structural gene for enzyme 1?

b. Complementation studies using partial diploid (F′) strains were made. The extrachromosomal element (F′) and chromosome each carried one set of *sug* genes. The results were as follows:

Genotype of F′	Chromosome	Enzyme 1	Enzyme 2
$A^+ B^- C^+ D^+$	$A^- B^+ C^+ D^+$	+	+
$A^+ B^- C^- D^+$	$A^- B^+ C^+ D^+$	+	0
$A^- B^+ C^- D^+$	$A^+ B^- C^+ D^+$	0	+
$A^- B^+ C^+ D^+$	$A^+ B^- C^+ D^-$	+	+

From all the evidence given, determine whether the following statements are true or false:

1. It is possible that gene D is a structural gene for one of the two enzymes.

2. It is possible that gene D produces a repressor.

3. It is possible that gene D produces a cytoplasmic product required to induce genes A and B.

4. It is possible that gene D is an operator locus for the *sug* operon.

5. The evidence is also consistent with the possibility that gene C could be a gene that produces a cytoplasmic product required to induce genes A and B.

6. The evidence is also consistent with the possibility that gene C could be the controlling end of the *sug* operon (the end from which mRNA synthesis presumably commences).

*16.9 What consequences would a mutation in the catabolite activator protein (*CAP*) gene of E. coli have for the expression of a wild-type *lac* operon?

16.10 The *lac* operon is an inducible operon, whereas the *trp* operon is a repressible operon. Discuss the differences between these two types of operons.

*16.11 In the presence of high intracellular concentrations of tryptophan, only short transcripts of the *trp* operon are synthesized because of attenuation of transcription at a point 5′ to the structural genes. This is mediated by the recognition of two Trp codons in the leader sequence. If these codons were mutated to UAG stop codons, what effect would this have on the regulation of the operon in the presence or absence of tryptophan? Explain.

16.12 In the bacterium *Salmonella typhimurium*, seven of the genes coding for histidine biosynthetic enzymes

are located adjacent to one another in the chromosome. If excess histidine is present in the medium, the synthesis of all seven enzymes is coordinately repressed, whereas in the absence of histidine, all seven genes are coordinately expressed. Most mutations in this region of the chromosome result in the loss of activity of only one of the enzymes. However, mutations mapping to one end of the gene cluster result in the loss of all seven enzymes, even though none of the structural genes have been lost. What is the counterpart of these mutations in the *lac* operon system?

16.13 On infecting an E. coli cell, bacteriophage λ has a choice between the lytic and lysogenic pathways. Discuss the molecular events that determine which pathway is taken.

16.14 How do the λ repressor protein and the Cro protein regulate their own synthesis?

*16.15 If a mutation in the phage λ *cI* gene results in a nonfunctional *cI* gene product, what phenotype would you expect the phage to exhibit?

16.16 Bacteriophage λ can form a stable association with the bacterial chromosome because the virus manufactures a repressor. This repressor prevents the virus from replicating its DNA and making lysozyme and all the other tools used to destroy the bacterium. When you induce the virus with UV light, you destroy the repressor, and the virus goes through its normal lytic cycle. The repressor is the product of a gene called the *cI* gene and is a part of the wild-type viral genome. A bacterium that is lysogenic for λ⁺ is full of repressor protein, which confers immunity against any λ virus added to these bacteria. Added viruses can inject their DNA, but the repressor from the resident virus prevents replication, presumably by binding to an operator on the incoming virus. Thus, this system has many elements analogous to the *lac* operon. We could diagram a virus as shown in the following figure. Several mutations of the *cI* gene are known. The c_i mutation results in an inactive repressor.

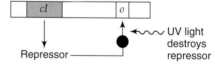

a. If you infect E. coli with λ containing a c_i mutation, can it lysogenize (form a stable association with the bacterial chromosome)? Why or why not?

b. If you infect a bacterium simultaneously with a wild-type c^+ and a c_i mutant of λ, can you obtain stable lysogeny? Why or why not?

c. Another class of mutants called c^{IN} makes a repressor that is insensitive to UV destruction. Will you be able to induce a bacterium lysogenic for c^{IN} with UV light? Why or why not?

17

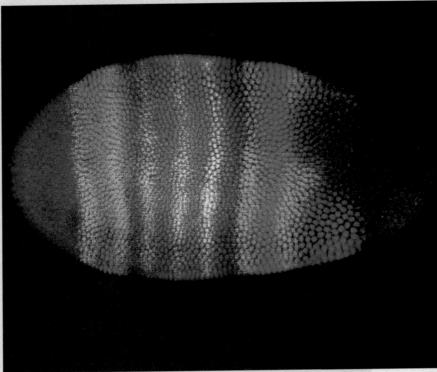

Differential expression of three genes—shown by a red, blue, and yellow immunofluorescence reaction to the proteins they produce—in a developing *Drosophila* embryo.

Regulation of Gene Expression in Eukaryotes

PRINCIPAL POINTS

There are no operons in eukaryotes. However, genes for related functions often are regulated coordinately.

In eukaryotes, gene expression is regulated at a number of distinct levels. That is, there are regulatory systems for the control of transcription, precursor RNA processing, transport of the mature RNA out of the nucleus, translation of the mRNAs, degradation of the mature RNAs, and degradation of the protein products.

Eukaryotic protein-coding genes contain both promoter and enhancer elements that interact with a number of regulatory proteins (transcription factors) that activate or repress transcription.

Chromatin configuration poses a barrier to transcription. High nucleosome density is associated with transcriptionally inactive chromatin, whereas transcriptionally active chromatin exhibits a lower nucleosome density. DNA methylation may also be involved in the regulation of gene transcription.

In complex, multicellular eukaryotes, steroid hormones regulate the expression of particular sets of genes. To function in this short-term regulatory system, a steroid hormone enters a cell. If that cell contains a receptor molecule specific for the hormone, the hormone binds to the receptor and activates it, and the hormone-receptor complex binds to hormone regulatory elements next to genes in the nucleus, thereby regulating the expression of those genes. Other hormones (polypeptide hormones) act at the cell surface, activating a system to produce cyclic AMP, a substance that acts as a second messenger to control gene activity. The specificity of hormone action results from the presence of hormone receptors in only certain cell types and from interactions of steroid-receptor complexes with cell type–specific regulatory proteins.

Regulation at the level of RNA includes processing, nuclear transport, translation, and stability of the mRNA.

Long-term regulation is involved in controlling events that activate and repress genes during development and differentiation. Those two processes result from differential gene activity of a genome that contains a constant amount of DNA rather than from a programmed loss of genetic information.

Antibodies are specialized proteins called immunoglobulins, which bind specifically to antigens (*anti*body *gene*rators: chemicals that are recognized as foreign by an organism and that induce an immune response). Antibody molecules consist of two light chains and two heavy chains. In germ-line DNA, the coding regions for immunoglobulin chains are scattered in tandem arrays of gene segments. During development, somatic recombination occurs to bring particular gene segments together to form functional antibody chain genes. A large number of different antibody chain genes result from the many possible ways in which the gene segments can recombine.

Drosophila has become an important model system in which to study the genetic control of development. *Drosophila* body structures result from specific gradients in the egg and the subsequent determination of embryo segments that directly correspond to adult body segments. Both processes are under genetic control, as shown by mutations that disrupt the developmental events. Studies of the mutations indicate that *Drosophila* development is directed by a temporal regulatory cascade.

Once the basic segmentation pattern has been laid down in *Drosophila,* homeotic genes determine the developmental identity of the segments. Homeotic genes share common DNA sequences called homeoboxes. Homeoboxes have been found in developmental genes in other organisms, and homeodomains—the regions of the proteins the homeoboxes encode—probably play a role in regulating transcription by binding to specific DNA sequences.

iActivity

ⓘ HOW DOES THE SINGLE CELL CREATED BY THE FUSION of sperm and egg transform itself into a complex organism? How do the cells of a developing human "know" how to arrange themselves in the shape of a human? What makes dividing cells form a leg rather than an eye, a heart, or a hand? As you will learn in this chapter, the development of humans and other eukaryotic organisms requires the precise regulation of groups of genes. After you have read this chapter, you can apply what you've learned by trying the iActivity, in which you will attempt to identify some of the genes responsible for changing embryonic stem cells into different forms of tissue.

As we learned in Chapter 16, gene expression in prokaryotes is commonly regulated in a unit of protein-coding sequences and adjacent controlling sites (promoter and operator), collectively called an operon. The molecular mechanisms in operon function provide a simple means of controlling the coordinated synthesis of proteins with related functions. In eukaryotes, gene expression is also regulated in a unit of protein-coding sequences and adjacent controlling sites. However, eukaryotic genes have nothing like the prokaryotic operator region, so eukaryotes are not considered to have operons. In this chapter, we look at some of the well-studied regulatory systems in eukaryotes.

Eukaryotic gene regulation falls into two categories. Short-term regulation involves regulatory events in which gene sets are quickly turned on or off in response to changes in environmental or physiological conditions in the cell's or organism's environment. Long-term gene regulation involves regulatory events other than those required for rapid adjustment to local environmental or physiological changes; that is, it involves the events required for an organism to develop and differentiate. Both of these categories of gene regulation are addressed in this chapter.

Levels of Control of Gene Expression in Eukaryotes

Most prokaryotic organisms are unicellular. They respond quickly to the environment by making changes in gene regulation, primarily at the transcriptional level, with some translational control (see Chapter 16). Transcriptional response is accomplished through the interaction of regulatory proteins with upstream regulatory DNA sequences. Rapid changes in levels of protein synthesis are achieved by switching off gene transcription and by rapid degradation of the mRNA molecules.

In eukaryotes, both unicellular and multicellular, the control of gene expression is more complicated. This results in part from the compartmentalization of eukaryotic cells and the demands imposed by the need for multicellular eukaryotes to generate large numbers and types of cells. Figure 17.1 diagrams some of the levels at which the expression of protein-coding genes can be regulated in eukaryotes: mRNA transcription, processing, transport, translation, and degradation and protein processing and degradation. We consider each of these in turn.

Transcriptional Control

General Aspects of Transcriptional Control. Transcriptional control regulates whether a gene is transcribed and the rate at which transcripts are produced.

Protein-coding eukaryotic genes contain both promoter elements and enhancers (see Chapter 11). The promoter elements are located just upstream of the site at which transcription begins. The enhancers usually are some distance away, either upstream or downstream. We can think of the different promoter elements as modules that function in the regulation of expression of the gene. Certain promoter elements, such as the TATA element, are required to specify where transcription by RNA polymerase II is to begin. Other promoter elements control whether transcription of the gene occurs; specific regulatory proteins (transcription factors) bind to these elements (Figure 17.2). If transcription is activated, the promoter element is a *positive regulatory element*, and the protein is a *positive regulatory protein*. If transcription is turned off, the promoter element is a *negative regulatory element*, and the protein is a *negative regulatory protein*.

A regulatory promoter element is specialized with respect to the gene (or genes) it controls because it binds a signaling molecule that is involved in the regulation of that gene's expression. Depending on the particular gene, there can be one, a few, or many regulatory promoter elements because under various conditions there may be one, a few, or many regulatory proteins that control the gene's expression. The remarkable specificity of regulatory proteins in binding to their specific regulatory element in the DNA and to no others ensures careful control of which genes are turned on and which are turned off.

Whereas promoter elements are crucial for determining whether transcription can occur, enhancer elements determine that maximal transcription of the gene occurs. Regulatory proteins bind to the enhancer elements, and which regulatory proteins bind depends on the DNA sequence of the enhancer element. A model for how the enhancers affect transcription from a distance is as follows: Specific transcription factors bind to the enhancer element; the DNA then forms a loop, so that the enhancer-bound regulatory proteins interact with the regulatory proteins and transcription factors bound to the promoter elements. Through the interactions of the proteins, transcription is activated or repressed (see Figure 11.5).

Both promoters and enhancers are important in regulating transcription of a gene. Each regulatory promoter element and enhancer element binds a special regulatory protein. Some regulatory proteins are found in most or all cell types, whereas others are found in only a limited number of cell types. Because some of the regulatory proteins activate transcription when they bind to the enhancer or promoter element, whereas others repress transcription, the net effect of a regulatory element on transcription depends on the combination of different proteins bound. If positive

Figure 17.1

Levels at which gene expression can be controlled in eukaryotes.

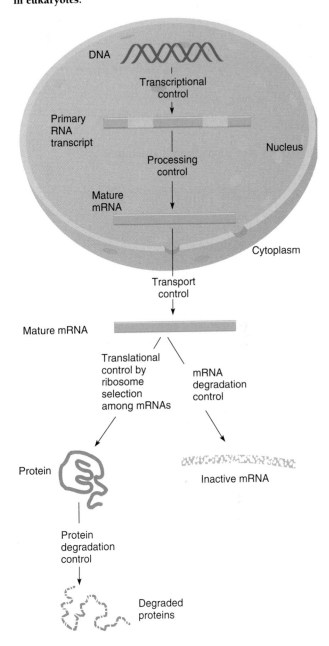

regulatory proteins are bound at both the enhancer and promoter elements, the result is activation of transcription. However, if a negative regulatory protein binds to the enhancer and a positive regulatory protein binds to the promoter element, the result depends on the interaction between the two regulatory proteins. If the negative regulatory protein has a strong effect, the gene is repressed. In this case, the enhancer is called a silencer element.

Enhancer and promoter elements appear to bind many of the same proteins. This implies that both types of regulatory elements affect transcription by a similar mechanism, probably involving interactions of the regulatory proteins, as described earlier. Interestingly, there appear to be a small number of regulatory proteins that control

Figure 17.2

Schematics of (a) positive regulation of gene transcription and (b) negative regulation of gene transcription.

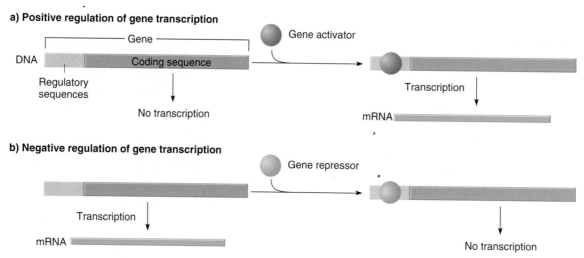

a) Positive regulation of gene transcription

b) Negative regulation of gene transcription

transcription. Therefore, by combining a few regulatory proteins in particular ways, the transcription of different arrays of genes is regulated and a large number of cell types is specified. This is called **combinatorial gene regulation.**

Many of the transcription factors and regulatory proteins involved in the process of transcription initiation in eukaryotes (and prokaryotes) are DNA-binding proteins. They interact with specific DNA sequences and exert an effect on transcription initiation. For a number of these proteins, but by no means all, certain well-defined structural motifs of the protein are responsible for binding of the protein to the DNA. Examples of some of the structural motifs (called *DNA-binding domains*) are the zinc finger, leucine zipper, and helix-turn-helix.

K E Y N O T E

Eukaryotic protein-coding genes contain both promoter elements and enhancer elements. Some promoter elements are required for transcription to begin. Other promoter elements have a regulatory function; these are specialized for the gene they control, binding specific regulatory proteins that control expression of the gene. Specific regulatory proteins bind also to the enhancer elements and activate transcription through their interaction with proteins bound to the promoter elements. Enhancer elements and promoter elements appear to bind many of the same proteins. This implies that both types of regulatory elements affect transcription by a similar mechanism, probably involving interactions of the regulatory proteins.

Chromosome Changes and Transcriptional Control. As we learned in Chapter 8, the eukaryotic chromosome consists of DNA complexed with histones (to form nucleosomes) and nonhistone chromosomal proteins. In this section, you will learn about the relationship between chromosome structure and transcriptional control.

DNase I Sensitivity and Gene Expression. Compared with inactive genes, almost all transcriptionally active genes have an increased sensitivity to the DNA-degrading enzyme, DNase I. The interpretation is that transcriptionally active genes have a looser chromosome structure than transcriptionally inactive genes. The extent of the DNase I–sensitive region varies with the particular gene, ranging from a few kilobase pairs around the gene to as much as 20 kb of flanking DNA. However, increased DNase I sensitivity does not mean that the DNA is not organized into nucleosomes, only that the chromosome is less highly coiled in those regions. In fact, recent experiments show that the histone octamer can "step around" the transcribing RNA polymerase, indicating that there is no reason for the nucleosome to disassemble for transcription to occur.

DNase-Hypersensitive Sites. Certain sites around transcriptionally active genes, called **hypersensitive sites** or **hypersensitive regions,** are even more highly sensitive to digestion by DNase I. These sites, or regions, typically are the first to be cut with DNase I. Most DNase-hypersensitive sites are in the regions upstream from the start of transcription including the promoter region, probably corresponding to the DNA sequences where RNA polymerase and other gene regulatory proteins bind.

K E Y N O T E

Chromosome regions that are transcriptionally active have looser DNA-protein structures than chromosome regions that are transcriptionally inactive, resulting in sensitivity of the DNA to digestion by DNase I. The promoter regions of active genes typically have an even looser DNA-protein structure, resulting in hypersensitivity to DNase I.

Histones and Gene Regulation. In the chromosome, DNA is wrapped around histones to form nucleosomes. Histones are extremely conserved evolutionarily speaking; their amino acid sequence, structure, and function are all very similar among all eukaryotic organisms. Furthermore, the five different types of histones are arranged in an essentially constant way along the DNA in a cell. This uniform distribution of histones along the DNA gives nowhere near the specificity needed to control the expression of thousands of genes. Histones do become modified, however, primarily through acetylation and phosphorylation, thereby altering their ability to bind to the DNA. If the gene they cover is to be expressed, the histones must be modified to loosen their grip on the DNA or be displaced from the DNA so that the DNA strands can interact with transcription factors or regulatory proteins. In essence, the histones act as general repressors of transcription. That is, when promoter elements (particularly the TATA box, which is important for transcription initiation; see Chapter 11) are associated with the histones of a nucleosome core, they cannot be found by the regulatory proteins and transcription factors that must bind to them to initiate transcription.

Some experiments have given important insights into how histones influence gene expression. These experiments investigated what happens to transcription when histones, promoter-binding proteins, and enhancer-binding proteins interact with DNA. In brief, the results were as follows:

1. If DNA is mixed simultaneously with histones and promoter-binding proteins, the histones compete more strongly for promoters and form nucleosomes at TATA boxes. As a result, the promoter-binding proteins cannot bind and transcription cannot occur.

2. If DNA is mixed first with promoter-binding proteins, the proteins assemble on TATA boxes and other promoter elements and block nucleosome assembly on those sites when histones are added. As a result, transcription occurs.

3. If DNA is mixed *simultaneously* with histones, promoter-binding proteins, and enhancer-binding proteins, the enhancer-binding proteins bind to enhancers and help promoter-binding proteins bind to TATA boxes by blocking access by the histones. As a result, transcription occurs.

These results indicate that histones are effective repressors of transcription but that other proteins can overcome that repression.

How does gene activation occur in the cell? One model is as follows: Histones block transcription by forming nucleosomes at TATA boxes. Promoter-binding proteins cannot disrupt the nucleosomes on the TATA boxes. However, enhancer-binding proteins bind to the enhancers (displacing any histones that are bound there) and interact with the histones in the nucleosome-bound TATA boxes. This causes the nucleosome core particle to break up, and promoter-binding proteins can now bind to the freed DNA.

Other research has established a correlation between the level of histone acetylation and transcriptional activity. Specifically, histones in transcriptionally active chromatin are hyperacetylated, whereas histones in transcriptionally inactive chromatin are hypoacetylated. When the histones become acetylated, nucleosome conformation is altered and higher-order chromatin structure is destabilized. As a result, the DNA associated with the nucleosomes becomes more accessible to transcription factors, and the generally repressive action of histones on transcription is overcome.

KEYNOTE

> Histones repress transcription by assembling nucleosomes on TATA boxes associated with genes. Gene activation occurs when proteins become bound to enhancers and disrupt the nucleosomes on the TATA boxes, thereby allowing the appropriate proteins to bind to the promoter elements to initiate transcription.

DNA Methylation and Transcriptional Control. Once DNA has been replicated within the eukaryotic cell, a small proportion of bases become chemically modified through the action of enzymes. A particular modified base has received significant attention because it has been correlated with gene activity in a number of cases. This base, 5-methylcytosine (5^mC), is generated by the modification of cytosine shortly after replication by the action of the enzyme DNA methylase (Figure 17.3).

The relationship between gene methylation and transcriptional activity of the gene has been studied for a number of genes. For at least 30 genes examined, a negative correlation exists between DNA methylation and transcription; that is, there is a lower level of methylated DNA (the DNA is hypomethylated) in transcriptionally active genes than in transcriptionally inactive genes. We must exercise caution in taking this broad generalization too far because not all methylated C nucleotides in a gene region become demethylated when the gene is expressed.

Figure 17.3

Production of 5-methylcytosine from cytosine in DNA by the action of the enzyme DNA methylase.

Cytosine (in DNA) → 5-Methylcytosine (5^mC)

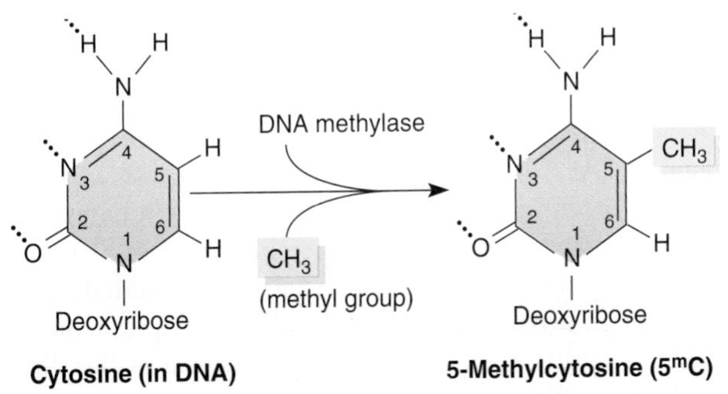

And methylation may not be a universal phenomenon in eukaryotes, especially since some eukaryotes have very little or no detectable methylation. Furthermore, the fact that a correlation exists between the level of DNA methylation and the transcriptional state of a gene provides no information about cause and effect. In other words, we do not know whether methylation changes are necessary for the onset of transcriptional activity (in organisms in which there is significant DNA methylation) or whether the changes are the result of transcriptional activity initiated by other means.

Nonetheless, there are now some observations indicating the importance of methylation in gene expression. For example, a gene encoding an enzyme that adds methyl groups to DNA has been shown to be essential for development in mice; that is, mice homozygous for mutations in that gene die at an early stage. Also, methylation is involved in the development of fragile X syndrome (OMIM 309550), which is the leading cause of inherited mental retardation. The syndrome develops after expansion (significant increase in copy number) of a triplet repeat (a repeated three-base-pair sequence) in the *FMR-1* gene and abnormal methylation of the gene to the point that transcription of the *FMR-1* gene is silenced. Fragile X syndrome is discussed in Chapter 21, pp. 443–444.

KEYNOTE

> The DNA of many eukaryotes has been shown to be methylated at a certain proportion of the bases, with most methylations occurring on cytosine residues. Transcriptionally active genes exhibit lower levels of DNA methylation than transcriptionally inactive genes.

Examples of Transcriptional Control of Gene Expression in Eukaryotes.

In this section, we look at some examples of the short-term regulation of gene expression in eukaryotes. This regulation involves rapid responses to changes in environmental or physiological conditions. The responses occur at the transcriptional level or at the level of posttranslational modulation of protein activity. We concentrate on transcriptional regulation here.

Galactose-Utilizing Genes of Yeast.

In the galactose-utilizing system of yeast, three genes, *GAL1*, *GAL7*, and *GAL10*, encode enzymes that function in a pathway to metabolize the sugar galactose (Figure 17.4a). The enzymes are galactokinase (*GAL1* gene), galactose transferase (*GAL7* gene), and galactose epimerase (*GAL10* gene). Once D-glucose 6-phosphate is made, it enters the glycolytic pathway. In the absence of galactose, the *GAL* genes are not transcribed. When galactose is added, there is a rapid, coordinate induction of transcription of the *GAL* genes and therefore a rapid production of the three galactose-utilizing enzymes.

Figure 17.4

The galactose-metabolizing pathway of yeast. **(a)** The steps catalyzed by the enzymes encoded by the *GAL* genes. **(b)** The organization of the *GAL* protein-coding genes. The arrows indicate the direction of transcription. (UDP = uridine diphosphate.)

a) Pathway

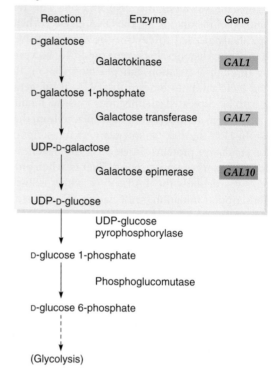

b) Gene organization

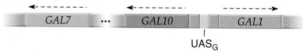

The *GAL1*, *GAL7*, and *GAL10* genes are located near each other (Figure 17.4b). An unlinked regulatory gene, *GAL4*, encodes the dimeric protein Gal4p, which has a zinc finger DNA-binding domain and a domain for activating transcription. Gal4p binds to a specific sequence in the DNA called *upstream activator sequence-galactose* (UAS_G); this sequence is a promoter element. The UAS_G consists of four similar 17-bp sequences, each of which binds a Gal4p dimer. Transcription occurs in both directions from the UAS_G (the genes are said to be *divergently transcribed*) (see Figure 17.4b).

A model for the regulation of transcription of the *GAL* genes is presented in Figure 17.5. In the absence of galactose, Gal4p is bound to UAS_G. Another protein, Gal80p (encoded by the *GAL80* gene), is bound to the Gal4p activation domain that is needed to turn on transcription. No transcription can occur under these circumstances. When galactose is added, a metabolite of galactose binds to Gal80p, and this causes Gal4p to become phosphorylated at certain amino acid residues. This changes the shape of the Gal4p-Gal80p complex so that Gal4p is changed to

Figure 17.5

Model for the activation of the *GAL* genes of yeast.

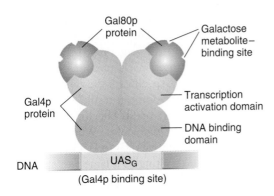

No galactose present

Gal80p protein

Galactose metabolite–binding site

Gal4p protein

Transcription activation domain

DNA binding domain

DNA

UAS$_G$

(Gal4p binding site)

Gal4p dimer is bound to UAS$_G$
Gal80p binds to Gal4p

In this form Gal4p cannot activate transcription of nearby *GAL* gene(s)

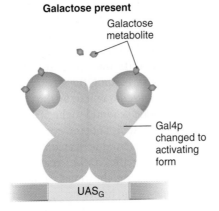

Galactose present

Galactose metabolite

Gal4p changed to activating form

UAS$_G$

Gal4p dimer is bound to UAS$_G$, Gal80p remains bound to Gal4p and binds galactose metabolite; this causes Gal4p to change shape

Altered Gal4p activates transcription of nearby *GAL* gene(s)

the activating form needed to turn on transcription of the *GAL* enzyme-coding genes. In this system, then, the Gal4p protein acts as a positive regulator (activator), and galactose is an effector molecule.

Steroid Hormone Regulation of Gene Expression in Animals. The cells of higher eukaryotes are not exposed to rapid changes in environment, as are cells of bacteria and of microbial eukaryotes. This is because most cells of higher eukaryotes are exposed to the intercellular fluid, which is nearly constant in the nutrients, ions, and other important molecules it supplies. The constancy of the cell's environment is maintained in part through the action of chemicals called *hormones*, which are secreted by various cells in response to signals and circulate in the blood until they stimulate their target cells. Elaborate feedback loops control the amount of hormone secreted, controlling the response so that appropriate levels of chemicals in the blood and tissues are maintained.

A hormone, then, is an effector molecule that is produced by one cell and that causes a physiological response in another cell. Some hormones—for example, steroid hormones—act by binding to a specific cytoplasmic receptor called a steroid hormone receptor (SHR), and then the complex binds directly to the cell's genome to regulate gene expression. Other hormones—for example, polypeptide hormones—may act at the cell surface to activate a membrane-bound enzyme, adenylate cyclase, which produces cyclic AMP (cAMP) from ATP. The cAMP acts as an intracellular signaling compound (called a *second messenger*) in a process called *signal transduction* to activate the cellular events associated with the hormone.

animation

Regulation of Gene Expression by Steroid Hormones

A key to hormone action is that hormones act on specific target cells that have receptors capable of recognizing and binding that particular hormone. For most of the polypeptide hormones (e.g., insulin, ACTH, vasopressin), the receptors are on the cell surface, whereas the receptors for steroid hormones are inside the cell.

Since the action of steroid hormones on gene expression has been well studied, we concentrate on steroid hormones here. Steroid hormones have been shown to be important in the development and physiological regulation of organisms ranging from fungi to humans. Steroid hormones have tissue-specific effects. For example, estrogen induces the synthesis of the protein prolactin in the rat pituitary gland, the protein vitellogenin in frog liver, and the proteins conalbumin, lysozyme, ovalbumin, and ovomucoid in the hen oviduct. Glucocorticoids induce the synthesis of growth hormone in the rat pituitary gland and the enzyme phosphoenolpyruvate carboxykinase in the rat kidney. The specificity of the response to steroid hormones is controlled by the hormone receptors. With the exception of receptors for the steroid hormone glucocorticoid, which are widely distributed among tissue types, steroid receptors are found in a limited number of target tissues. Steroid hormones have well-substantiated effects on transcription, and they also can affect the stability of mRNAs and possibly the processing of mRNA precursors.

Mammalian cells contain between 10,000 and 100,000 steroid hormone receptor molecules. The SHRs are proteins with a very high affinity for their respective steroid hormones. All steroid hormones work in the same general way. For example, in the absence of a particular steroid hormone, the appropriate SHR is found in the cell associated with a large complex of proteins called *chaperones,* one of which is Hsp90. In this association, the SHR is inactive. When a steroid hormone such as glucocorticoid enters a cell, it binds to its specific SHR molecule, displacing Hsp90 and forming a glucocorticoid-SHR complex (Figure 17.6). In this diagram, the SHR is located in the nucleus, although there is controversy over the exact location of SHRs in the cell. That is, SHRs are found in both the cytoplasm and the nucleus, and it is not clear whether both are active in binding the steroid hormone, although nuclear SHRs are thought to be the active ones. When the steroid hormone is bound to the SHR, the receptor protein becomes activated, and the complex is found only in the nucleus. The steroid-activated receptor complex now binds to specific DNA regulatory sequences, activating or repressing the transcription of the specific genes controlled by the hormone.

All genes regulated by a specific steroid hormone have in common a DNA sequence to which the steroid-receptor complex binds. The binding regions are called **steroid hormone response elements (HREs).** The H in the acronym is replaced with another letter to indicate the specific steroid involved. Thus, GRE is the glucocorticoid response element, and ERE is the estrogen response element. The HREs are located, often in multiple copies, in the enhancer regions of genes. The GRE, for example, is located about 250 bp upstream from the transcription start point. The consensus sequence for GRE is AGAACANNNTGTTCT, where N is any nucleotide.

How the hormone-receptor complexes, once bound to the correct HREs, regulate transcriptional levels is not completely known. Potentially, the hormone-receptor complexes interact with transcription factors in the transcription initiation complex and facilitate the initiation of transcription by RNA polymerase II. It is presumed that each steroid hormone regulates its specific transcriptional activation by the same general mechanism. The unique action of each type of steroid results from the different receptor proteins and HREs involved.

In different types of cells, the same steroid hormone may activate different sets of genes, even though the various cells have the same SHR. This is because a steroid-receptor complex can activate a gene only if the correct array of other regulatory proteins is present. Since the other regulatory proteins are specific for the cell type, different patterns of gene expression can result.

In sum, steroid hormones act as effector molecules, and SHRs act as regulatory molecules. When the two combine, the resulting complex binds to DNA and regulates gene transcription, resulting in a large and specific increase or decrease in cellular mRNA levels.

KEYNOTE

In multicellular eukaryotes, one of the well-studied systems of short-term gene regulation is the control of protein synthesis by hormones. Polypeptide hormones act at the cell surface, activating a system to produce cAMP, which acts as a second messenger to control gene activity. Steroid hormones exert their action by forming a complex with a specific receptor protein, thereby activating the receptor; the complex then binds directly to specific sequences on the cell's genome to regulate the expression of specific genes. The specificity of hormone action is caused by the presence of hormone receptors in only certain cell types and by interactions of steroid-receptor complexes with cell type–specific regulatory proteins.

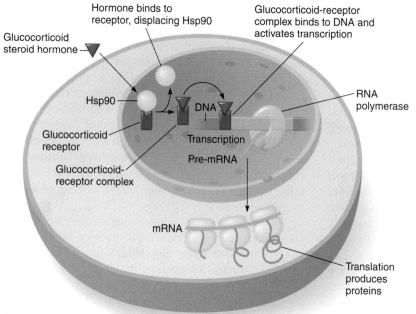

Figure 17.6

Model for the action of the steroid hormone glucocorticoid in mammalian cells.

Glucocorticoid steroid hormone

Hormone binds to receptor, displacing Hsp90

Glucocorticoid-receptor complex binds to DNA and activates transcription

Hsp90

DNA

RNA polymerase

Glucocorticoid receptor

Transcription

Pre-mRNA

Glucocorticoid-receptor complex

mRNA

Translation produces proteins

RNA Processing Control

RNA processing control regulates the production of mature RNA molecules from precursor RNA molecules. In Chapter 11, we discussed the synthesis of pre-mRNA and its processing to mature mRNA. When we examine living systems, we do not always see the "textbook" processing steps. For example, there are many cases in which *alternative polyadenylation* sites may be used to produce different pre-mRNA molecules, and *alternative splicing* (also called *differential splicing*) may be used to produce different functional mRNAs. Which product is generated depends on regulatory signals. The products of alternative polyadenylation or alternative splicing are proteins that are encoded by the same gene but differ structurally and functionally. Such proteins are called *protein isoforms,* and their synthesis may be tissue-specific. Alternative polyadenylation is independent of alternative splicing.

Figure 17.7 shows an example of how alternative polyadenylation and alternative splicing result in tissue-specific products of the human calcitonin gene (*CALC*). *CALC* consists of five exons and four introns. This gene is transcribed in certain cells of the thyroid gland and in certain neurons of the brain. Alternative polyadenylation occurs such that the polyadenylation site next to exon 4, pA_1, is used in thyroid cells, and the polyadenylation site next to exon 5, pA_2, is used in the neuronal cells (Figure 17.7, step 1).

Alternative splicing occurs at the next stage of intron removal (Figure 17.7, step 2). The pre-mRNA in the

animation

a RNA Processing Control

Figure 17.7

Alternative polyadenylation and alternative splicing resulting in tissue-specific products of the human calcitonin gene, *CALC*. In the thyroid gland, calcitonin is produced, whereas in certain neurons, CGRP (calcitonin gene–related peptide) is produced.

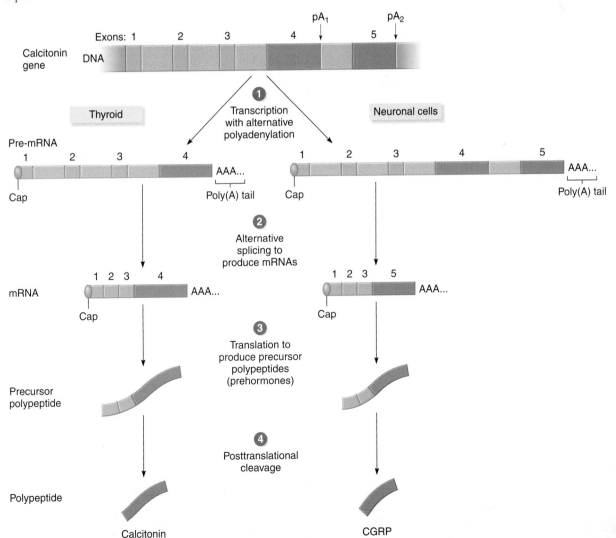

thyroid is spliced to remove the three introns and bring together exons 1, 2, 3, and 4. The pre-mRNA in the neuronal cells is spliced to remove introns and to bring together introns 1, 2, 3, and 5; exon 4 is excised and discarded. The mRNAs produced are translated to produce precursor polypeptides (prehormones; step 3) from which the functional hormones are generated posttranslationally by protease cleavage (step 4). The two products are calcitonin in the thyroid, with its amino acid sequence encoded by exon 4, and CGRP (calcitonin gene–related peptide), with its amino acid sequence encoded by part of exon 5. (The remainder of exon 5 is the 3′ trailer part of the mRNA.) The mechanisms by which this alternative polyadenylation and alternative splicing occur are not known. The outcome is two different polypeptides encoded by the same gene synthesized in two different tissues. The thyroid hormone calcitonin is a circulating calcium ion homeostatic hormone that aids the kidney in retaining calcium. CGRP is found in the hypothalamus and appears to have neuromodulatory and trophic (growth-promoting) activities.

KEYNOTE

> Gene expression in eukaryotes can be regulated at the level of RNA processing. This type of regulation operates to direct the production of mature RNA molecules from precursor RNA molecules. Two regulatory events that exemplify this level of control are choice of poly(A) site and choice of splice site. In both cases, different types of mRNAs are produced, depending on the choices made.

mRNA Translation Control

Messenger RNA molecules are subject to **translational control** by ribosome selection among mRNAs. Differential translation can greatly affect gene expression. For example, mRNAs are stored in many unfertilized vertebrate and invertebrate eggs. In the unfertilized egg, the rate of protein synthesis is very slow; however, protein synthesis increases significantly after fertilization. Since this increase occurs without new mRNA synthesis, it is likely that translational control is responsible. The mechanisms involved are not completely understood, but typically, stored mRNAs are associated with proteins that both protect the mRNAs and inhibit their translation. Furthermore, the poly(A) tail is known to promote the initiation of translation. It has been found that, in general, stored, inactive mRNAs have shorter poly(A) tails (15–90 As) than active mRNAs (100–300 As). In growing oocytes, mRNAs destined for storage and later translation have short poly(A) tails.

In principle, an mRNA molecule can have a short poly(A) tail either because only a short string of A nucleotides was added at the time of polyadenylation or

because a normal-length poly(A) tail was added that was subsequently trimmed. At least for some messages stored in growing oocytes of mouse and frog, the latter mechanism is involved. In one example, examination of one particular mRNA in this class has shown that the pre-mRNA still in the process of intron removal has a long poly(A) tail (300–400 As), whereas the mature, stored message has a short poly(A) tail (40–60 As). The decrease in length of the poly(A) tail for this message class occurs rapidly in the cytoplasm by a deadenylation enzyme. The rapid deadenylation is signaled by the *adenylate/uridylate (AU)-rich element (ARE)* in the 3′ untranslated region (3′ UTR) of the mRNA upstream of the AAUAAA polyadenylation sequence. Interestingly, to activate a stored mRNA in this class, a cytoplasmic polyadenylation enzyme recognizes the ARE and adds 150 A nucleotides or so. Thus, the same sequence element is used to control poly(A) tail length and mRNA translatability at different times and in opposite ways.

mRNA Degradation Control

Once in the cytoplasm, all RNA species are subjected to **degradation control,** in which the rate of RNA breakdown (also called RNA turnover) is regulated. Both rRNA (in ribosomes) and tRNA are very stable species, whereas mRNA molecules exhibit a range of stability from minutes to months. The stability of particular mRNA molecules may change in response to regulatory signals. For example, Table 17.1 presents examples of changes in mRNA stability for a number of cell types that occur in the presence and absence of specific effector molecules.

mRNA degradation is a major control point in the regulation of gene expression in eukaryotes. Various sequences or structures have been shown to affect the half-lives of mRNAs, including the AU-rich elements (AREs) discussed earlier and various secondary structures. Two major mRNA decay pathways are the deadenylation-dependent decay and deadenylation-independent decay pathways. In the deadenylation-dependent decay pathway, the poly(A) tails are deadenylated until the tails are too short (5–15 As) to bind PAB, poly(A)-binding protein. Once the tail is almost removed, the 5′ cap structure is removed by an enzyme in a step called decapping. After an mRNA molecule is decapped, it is degraded from the 5′ end by a 5′→3′ exonuclease.

Our understanding of mRNA degradation is increasing through our study of mutants that affect the process and through our study of the molecular components involved. For example, in yeast, the decapping enzyme is encoded by the *DCP1* gene. Yeast strains with a mutant *DCP1* gene are viable, and mRNA degradation still occurs but by a distinct deadenylation-independent decay pathway. In one such pathway, mRNAs are decapped without being deadenylated, thereby exposing them to rapid degradation by 5′→3′ exonucleases.

Figure 17.16

Locations of imaginal disks in a mature *Drosophila* larva and the adult structures derived from each disk.

Imaginal disks in larva   **Adult structures**

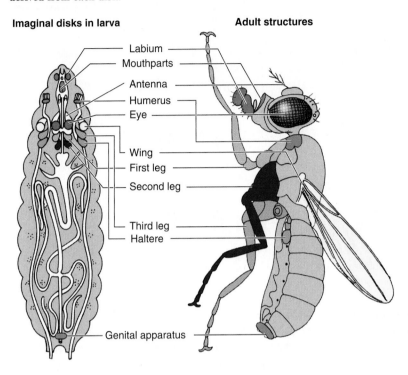

Labium
Mouthparts
Antenna
Humerus
Eye
Wing
First leg
Second leg
Third leg
Haltere
Genital apparatus

localized to the posterior pole of the egg cytoplasm by products of other posterior group maternal effect genes. These mRNAs are translated after fertilization to produce the NANOS protein, which forms a posterior-to-anterior gradient and acts as a morphogen that directs abdomen formation. The NANOS protein is a translational repressor, repressing the translation particularly of mRNA transcripts of the *hunchback (hb)* gene. These transcripts are deposited in the egg during oogenesis and are distributed evenly. However, for development to proceed correctly, HUNCHBACK proteins must be present in a gradient, decreasing in amount from anterior to posterior. The NANOS protein is present in a high posterior–to–low anterior gradient, and its translational repression activity creates the necessary HUNCHBACK protein gradient.

Segmentation Genes.   Next, the embryo is subdivided into regions through the action of **segmentation genes,** which determine the segments of the embryo and adult. Mutations in segmentation genes alter the number of segments or their internal organization but do not affect the overall organizational polarity of the egg. The segmentation genes are subclassified on the basis of their mutant phenotypes into gap genes, pair rule genes, and segment polarity genes (Figure 17.17). Mutations in *gap genes* (e.g., *Krüppel, hunchback, giant, tailless*) result in the deletion of regions consisting of several adjacent segments; mutations in *pair rule genes* (e.g., *hairy, even-skipped, runt, fushi tarazu*) result in the deletion of the same part

of the pattern in every other segment; and mutations in *segment polarity genes* (e.g., *engrailed, hedgehog, armadillo, gooseberry*) have portions of segments replaced by mirror images of adjacent half segments.

Segmentation genes have specific roles in specifying regions of the embryo. Gap genes are activated or repressed by maternal effect genes. For example, many gap genes are activated by the BICOID protein. Gap gene transcription leads to an organization of the embryo into broad regions, each of which covers areas that will later develop into several distinct segments. Critical to this broad definition of regions is expression of the *hunchback* gene.

Next, through the transcription-regulating action of the gap genes, the pair rule genes are expressed, leading to a division of the embryo into a number of regions, each of which contains a pair of parasegments (see Figure 17.15). The transcription factors encoded by the pair rule genes regulate the expression of the segment polarity genes, which determines regions that will become the segments seen in larvae and adults.

Homeotic Genes.   Once the segmentation pattern has been determined, a major class of genes called **homeotic** (structure-determining) **genes** (also called *selector genes*) specifies the identity of each segment with respect to the body part that will develop at metamorphosis. **Homeotic mutations** alter the identity of particular segments, transforming them into copies of other segments.

Figure 17.17

Functions of segmentation genes as defined by mutations.

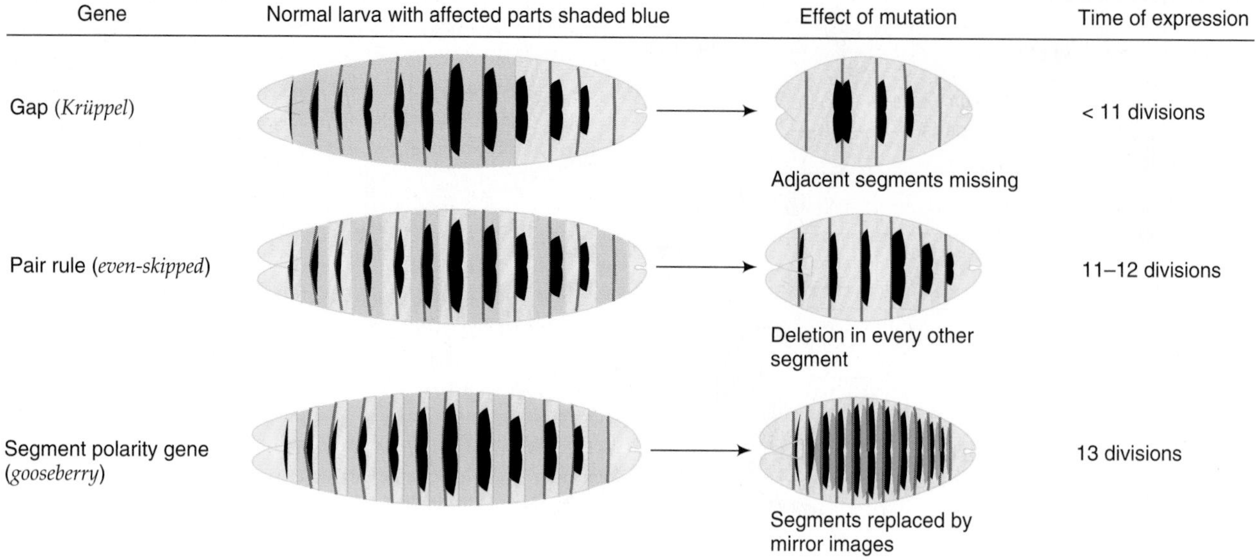

Gene	Normal larva with affected parts shaded blue	Effect of mutation	Time of expression

Gap (*Krüppel*) — Adjacent segments missing — < 11 divisions

Pair rule (*even-skipped*) — Deletion in every other segment — 11–12 divisions

Segment polarity gene (*gooseberry*) — Segments replaced by mirror images — 13 divisions

Pioneering studies by Edward Lewis were done on a cluster of homeotic genes called the *bithorax* complex (*BX-C*). *BX-C* determines the posterior identity of the fly, namely thoracic segment T3 and abdominal segments A1–A8. *BX-C* contains three genes called *Ultrabithorax* (*Ubx*), *abdominal-A* (*abd-A*), and *Abdominal-B* (*Abd-B*). Mutations in these homeotic genes often are lethal, and the fly typically does not survive past embryogenesis. Some nonlethal mutations have been characterized, however, that allow an adult fly to develop. Figure 17.18 shows the abnormal adult structures that can result from *bithorax* mutations. A diagram showing the segments of a normal adult fly is in Figure 17.18a; the wings are located on thoracic segment Thorax 2 (T2), and the pair of halteres (rudimentary wings used as balancers in flight) are on segment T3. A photograph of a normal adult fly clearly showing the wings and halteres is presented in Figure 17.18b. Figure 17.18c shows one type of developmental abnormality that can result from nonlethal homeotic mutations in *BX-C*; shown is a fly homozygous for three separate mutations in the *Ubx* gene: *abx*, *bx3*, and *pbx*. Collectively these mutations transform segment T3 into an adult structure similar to T2. The transformed segment has a fully developed set of wings. The fly lacks halteres, however, because no normal T3 segment is present.

Another well-studied group of mutations defines another large cluster of homeotic genes called the *Antennapedia* complex (*ANT-C*). *ANT-C* determines the anterior identity of the fly, namely the head and thoracic segments T1 and T2. *ANT-C* contains five genes, called *labial* (*lab*), *proboscipedia* (*pb*), *Deformed* (*Dfd*), *Sex combs reduced* (*Scr*), and *Antennapedia* (*Antp*). Most *ANT-C* mutations are lethal. Among the nonlethal mutations is a group of mutations in *Antp* that result in leg parts instead

of an antenna growing out of the cells near the eye during the development of the eye disk (Figure 17.19a and b). The leg has a normal structure but is positioned in an abnormal location. A different mutation in *Antp*, called *Aristapedia*, has a different effect: Only the distal part of the antenna, the arista, is transformed into a leg (Figure 17.19c). Therefore, the homeotic genes *ANT-C* and *BX-C* encode products that are involved in controlling the normal development of the relevant adult fly structures.

The *Antennapedia* complex (*ANT-C*) and the *bithorax* complex (*BX-C*) have been cloned. Both complexes are very large. In *ANT-C*, for example, the *Antp* gene is 103 kb long, with many introns; this gene encodes a mature mRNA of only a few kilobases. *BX-C* covers more than 300 kb of DNA and contains only three protein-coding regions amounting to about 50 kb of that DNA: *Ubx*, *abdA*, and *AbdB* (Figure 17.20). The other 250 kb of DNA in the complex is not transcribed and is believed to consist of regulatory regions of significant size and complexity. The functions of these regulatory regions are to control the expression of the protein-coding genes.

The *ANT-C* and *BX-C* protein-coding genes have similar functions, but are located in different places in the genome. Analysis of the DNA sequences for the genes reveals the presence of similar sequences of about 180 bp that has been named the **homeobox.** The homeobox is part of the protein-coding sequence of each gene, and the corresponding 60-amino-acid part of each protein is called the **homeodomain.**

Homeoboxes have been found in more than 20 *Drosophila* genes, most of which regulate development. All homeodomain-containing proteins are DNA-binding proteins. The homeodomain of such proteins binds strongly to an 8-bp consensus recognition sequence upstream of all

Figure 17.18

Adult structures that result from *bithorax* mutations.
(a) Drawing of a normal fly. The haltere (rudimentary wing) is on thoracic segment T3 (see Figure 17.20). (b) Photograph of a normal fly with a single set of wings. (c) Photograph of a fly homozygous for three mutant alleles (*bx3, abx, pbx*) that result in the transformation of segment T3 into a structure like segment T2: a segment with a pair of wings. These flies therefore have two sets of wings but no halteres.

a)

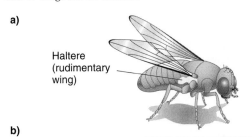

Haltere
(rudimentary
wing)

b)

c)

genes controlled by the proteins. Thus, homeodomain-containing proteins are transcriptional regulators.

The complete set of homeotic genes and complexes in *Drosophila*—generically the *Hox* genes—consists of *lab, pb, Dfd, Scr, Antp, Ubx, abdA,* and *AbdB*. Most interestingly, these complexes are arranged in the same order along the chromosome as they are expressed along the anterior-posterior body axis; this is known as the colinearity rule. Homeotic gene complexes are found also in all major animal phyla with the exception of sponges and coelenterates. The homeobox sequences in the *Hox* genes are highly conserved, indicating common function in the wide range

of organisms involved. As in *Drosophila*, the homeotic genes of vertebrates—the *Hox* genes—follow the colinearity rule. In mammals, for example, there are four clusters of homeotic genes designated *HoxA–D*. Each cluster is thought to have originated by duplication of a primordial gene cluster followed by evolutionary divergence. The patterns of *Hox* gene expression, the effects of mutations, and embryological analyses all indicate that the vertebrate genes have homeotic effects similar to those of *Drosophila* homeotic genes. Furthermore, the studies indicate that the *Hox* genes specify the vertebrate body plan.

Homeotic genes are also found in plants. For example, many homeotic mutations that affect flower development have been identified in *Arabidopsis*. Studies of homeotic mutants have led to models of flower development and, more generally, of plant development. In parallel with *Drosophila* homeotic genes, plant homeotic genes appear to be part of a sequential array of genes that regulate development.

KEYNOTE

Development of *Drosophila* body structures results from gradients along the anterior-posterior and dorsal-ventral axes of the egg and from the subsequent determination of regions in the embryo that directly correspond to adult body segments. As defined by mutations, genes control *Drosophila* development in a temporal regulatory cascade. First, maternal effect genes specify the gradients in the egg; then segmentation genes (gap genes, pair rule genes, and segment polarity genes) determine the segments of the embryo and adult; and homeotic genes next specify the identity of the segments.

Microarray Analysis of *Drosophila* Development

With the *Drosophila* genome sequence completed, studies are now under way to use that information to enrich our molecular understanding of *Drosophila* development. For example, DNA microarrays (see Chapter 15, pp. 314–317) are being used to study changes in gene expression patterns in various developmental transitions. One such study has examined metamorphosis brought about by the hormone ecdysone from 18 hours before pupal formation (BFP) to 12 hours after pupal formation (APF). The expression of 6,240 cDNA clones (about 40 percent of the estimated genes of *Drosophila*) was analyzed. The results indicated that 534 genes were differentially expressed during the metamorphosis, some being repressed and some being induced. At a more specific level, the study cataloged the ecdysone-caused induction of a number of genes involved in the dramatic differentiation of the central nervous system during early metamorphosis at 4 hours BPF. Similarly,

Figure 17.19

Effects of some mutations in the *Antennapedia* complex. (**a**) Scanning electron micrograph of the antennal area of a wild-type fly. (**b**) Scanning electron micrograph of the antennal area of a homeotic mutant of *Drosophila, Antennapedia,* in which the antenna is transformed into a leg. (**c**) Scanning electron micrograph of the homeotic mutant of *Drosophila, Aristapedia,* in which the arista is transformed into a leg.

a) **Normal**

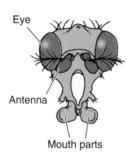

b) **Antennapedia**

c) **Aristapedia**

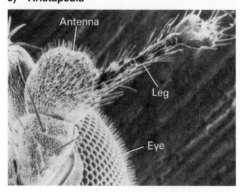

the repression of a number of genes encoding proteins required for muscle formation was shown to occur at 4 hours BPF; this prepares the metamorphosing *Drosophila* for breakdown of larval muscle tissues beginning at 2 hours APF.

Undoubtedly, more of these kinds of studies will be done in the future. Thus, we can expect to see continued rapid advances in our knowledge of *Drosophila* development and of the development of other important model organisms.

Figure 17.20

Organization of the *bithorax* complex (*BX-C*). The DNA spanned by this complex is 300 kb long. The transcription units for *Ubx*, *abdA*, and *AbdB* are shown below the DNA; the exons are shown by colored blocks and the introns by bent, jointed lines. All three genes are transcribed from right to left. Shown above the DNA are regulatory mutants that affect the development of different fly segments.

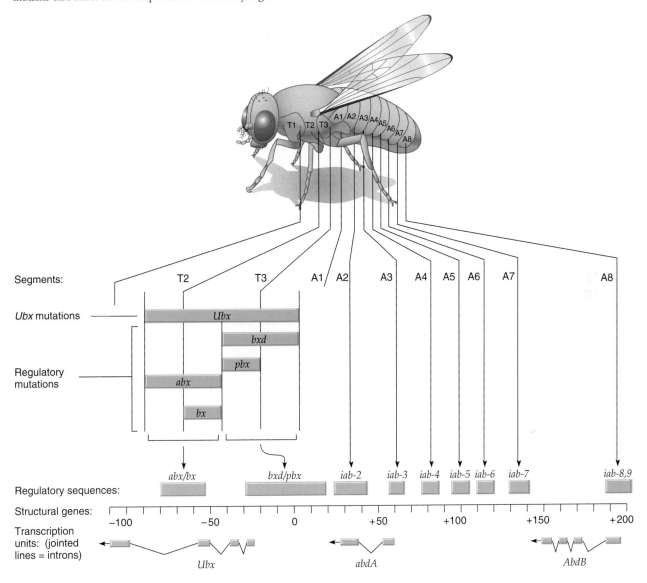

Summary

In this chapter, we have considered a number of examples of gene regulation in eukaryotes. The general picture is one of much greater complexity than in prokaryotes. Genes are not organized into operons because genes of related function often are scattered around the genome; nonetheless, genes are regulated coordinately. The chromosome organization in eukaryotes also makes for greater complexity in regulating gene expression.

Three main topics were discussed in the chapter: the levels of control of gene expression, gene regulation dur-

ing development and differentiation, and genetic regulation of development in *Drosophila*.

Levels of Control of Gene Expression

Gene expression in eukaryotes is regulated at a number of distinct levels, some of which we discussed in the chapter. Regulatory systems have been found for (1) the control of transcription, (2) the control of precursor RNA processing, (3) the transport of the mature RNA out of the nucleus, (4) the translation of the mRNAs,

(5) the degradation of the mature RNAs, and (6) the degradation of the protein products.

The control of transcription itself involves a number of elements. At the DNA level, transcription of protein-coding genes involves the interaction of transcription factors with promoter elements and of regulatory proteins with both promoter elements and enhancer elements. Depending on the element and the protein that binds to it, the effect on transcription may be positive or negative. Although unique regulatory proteins certainly bind to promoter elements and to enhancer elements, some regulatory proteins are shared by the two, indicating that both types of regulatory elements affect transcription by a similar mechanism.

At the chromosome level, the regulation of transcription must deal with the interaction of histones and non-histones with the DNA. Generally, the eukaryote chromosome is repressed for transcription by this interaction, so activation of transcription is a derepression phenomenon that results from a loosening of the DNA-protein structure in the region of a gene being activated. Relevant to this, transcriptionally active chromatin is more sensitive to digestion with DNase I than is transcriptionally inactive chromatin. Moreover, upstream of active genes are DNase-hypersensitive sites, which are highly sensitive to DNase I digestion. These sites probably correspond to the binding sites for RNA polymerase and regulatory proteins.

Histones play an important role in gene repression by assembling nucleosomes on TATA boxes in promoter regions. Gene activation involves regulatory proteins binding to enhancers and then disrupting the nucleosomes on the TATA boxes. This allows regulatory proteins and transcription factors to bind to promoters to initiate transcription.

Gene activation in eukaryotes is accompanied in many instances by a decrease in the level of DNA methylation, although the relationship, if any, to chromosome organization changes is not clear.

Gene expression can be regulated at the level of RNA processing. This type of regulation operates to determine the production of mature RNA molecules from precursor RNA molecules. Two regulatory events that exemplify this level of control are choice of poly(A) site and choice of splice site. In both cases, different types of mRNAs are produced, depending on the choice made. Examples are known of both types of regulation in developmental systems. For example, alternative polyadenylation and alternative splicing result in tissue-specific products of the human calcitonin gene.

Gene expression is also regulated by mRNA translation control and by mRNA degradation control. The latter is believed to be a major control point in the regulation of gene expression, as evidenced by the wide range of mRNA stabilities found within organisms. Nucleases are ultimately responsible for the degradation of the RNAs, and the signals for the differential mRNA stabilities are a property of the structural features of individual mRNAs. For example, a group of AU-rich sequences in the 3′ untranslated regions of some short-lived mRNAs is responsible for their instability.

We discussed the immune response, the structure of antibody molecules, and generation of antibody diversity in mammals. Here chromosomal rearrangements occur during cell development to bring together parts of the coding regions for the light and heavy polypeptide chains of antibody molecules into the genes that are transcribed. The various permutations of segments that are possible in this process are the basis for the huge number of different types of antibodies that can be produced in mammals.

Gene Regulation During Development and Differentiation

Classic and recent experiments have shown that development and differentiation result from differential gene activity of a genome that contains a constant amount of DNA rather than from a programmed loss of genetic information. Development and differentiation, then, must involve regulation of gene expression, using the levels of control we have just discussed. Adding to the complexity is the fact that we must consider the regulation of a large number of genes for each developmental process and communication between differentiating tissues, as well as systems for timing the activation and repression of genes during those important events. For example, the structure and function of a cell often are determined early, even though the manifestations of this determination process are not seen until later in development. Such early determination events may involve some preprogramming of genes that will be turned on later. Neither the nature of these determination events nor the mechanism of timing in developmental processes is well understood in vertebrates, although progress is being made in understanding the genetic regulation of development in certain organisms such as *Drosophila*.

In sum, a great deal of information has been learned in the past decade or so about gene regulation in eukaryotes. We have merely scratched the surface in this chapter. Thousands of researchers are working to elaborate the molecular details of gene regulation in model systems. Much of our advancing knowledge has been made possible by the application of recombinant DNA and related technologies, and we can look forward to sustained and rapid increases in our understanding of eukaryotic gene regulation.

Analytical Approaches for Solving Genetics Problems

Q17.1 We learned in this chapter that in humans, there are several distinct genes that code for α- and β-like globin polypeptides. These α-, β-, γ-, δ-, ε-, and ζ-globin genes are transcriptionally active at specific stages of development, resulting in the synthesis of polypeptides that are assembled in specific combinations to form different types of hemoglobin (see pp. 355–356 and Figure 17.9). Fill in the following table, indicating whether the globin gene in question is sensitive (S) or resistant (R) to DNase I digestion at each of the developmental stages listed.

Globin Gene	Tissue		
	Embryonic Yolk Sac	Fetal Spleen	Adult Bone Marrow
α			
β			
γ			
δ			
ζ			
ε			

A17.1 The correctly filled in table is as follows:

Globin Gene	Tissue		
	Embryonic Yolk Sac	Fetal Spleen	Adult Bone Marrow
α	R	S	S
β	R	R	S
γ	R	S	R
δ	R	R	S
ζ	S	R	R
ε	S	R	R

The explanation for the answers is as follows: DNase I typically digests regions of DNA that are transcriptionally active, while not digesting regions of DNA that are transcriptionally inactive. This is because transcriptionally inactive DNA is more highly coiled than transcriptionally active DNA. R means, then, that the gene was transcriptionally inactive, while S means that the gene was transcriptionally active.

To consider each globin gene in turn, the α gene is transcriptionally inactive in the embryonic yolk sac, but active in fetal spleen and adult bone marrow. That is, in the spleen fetal hemoglobin (Hb-F) is made; Hb-F contains two α polypeptides and two γ polypeptides. In the bone marrow, Hb-A is made, which contains two α and two β polypeptides.

The β-globin gene is inactive in yolk sac and spleen and is active in bone marrow, making one of the two polypeptides found in Hb-A, the main adult form of hemoglobin. The β-like γ polypeptide is found in Hb-F, which is made only in the liver and the spleen; thus, the γ gene is active in spleen and inactive in yolk sac and bone marrow. The β-like δ polypeptide is found in $α_2δ_2$ hemoglobin, which is a minor class of hemoglobin found in adults; thus, the δ gene is active only in adult bone marrow.

The ζ gene makes an α-like polypeptide found only in the hemoglobin of the embryo, so the ζ gene is active in the yolk sac but inactive in spleen and bone marrow.

Finally, the ε gene encodes the β-like polypeptide of the embryo's hemoglobin, so this gene is also active in the yolk sac but inactive in spleen and bone marrow.

Questions and Problems

17.1 Promoters, enhancers, transcription factors, and regulatory proteins that are active for one gene typically share structural similarities with these elements in other genes. Nonetheless, the transcriptional control of a gene can be exquisitely specific: It will be specifically transcribed in some tissues at very defined times. Explore how this specificity arises by addressing the following questions.

a. Distinguish between the functions of promoters and enhancers in transcriptional regulation.

b. What structural features are found in proteins that bind these DNA elements?

c. Can an enhancer bound by a regulatory protein stimulate as well as suppress transcription? If so, how?

d. Given that several different genes may contain the same types of promoter and enhancer elements, and a number of transcription factors contain the same structural features, how is transcriptional specificity generated?

17.2 A cloned DNA sequence was used to probe a Southern blot. There were two DNA samples on the blot, one from white blood cells and the other from a liver biopsy of the same individual. Both samples had been digested with *Hpa*II. The probe bound to a single 2.2-kb band in the white blood cell DNA but bound to two bands (1.5 and 0.7 kb) in the liver DNA.

a. Is this difference likely to result from a somatic mutation in a *Hpa*II site? Explain.

b. How would it affect your answer if you knew that white blood cell and liver DNA from this individual both showed the two-band pattern when digested with *Msp*I?

*17.3 Both fragile X syndrome and Huntington disease are caused by trinucleotide repeat expansion. Individuals with fragile X syndrome have at least 200 CGG repeats at the 5′ end of the *FMR-1* gene. Individuals with Huntington disease have at least 36 CAG repeats within the protein-coding region of the huntingtin gene.

a. How is gene expression affected by these repeat expansions?

b. Based on your answer to (a), why is the fragile X syndrome recessive, whereas Huntington disease is dominant?

c. Why is the number of trinucleotide repeats needed to cause the phenotype different for each disease?

17.4 Both peptide and steroid hormones can affect gene regulation of a targeted population of cells.

a. What is a hormone?

b. Distinguish between the mechanisms by which a peptide hormone and a steroid hormone affect gene expression.

c. What role do each of the following have in a physiological response to a peptide hormone or a steroid hormone?

   i. steroid hormone receptor (SHR)

   ii. steroid hormone response element (HRE)

   iii. second messenger

   iv. cAMP and adenylate cyclase

*17.5 The following figure shows the effect of the hormone estrogen on ovalbumin synthesis in the oviduct of 4-day-old chicks. Chicks were given daily injections of estrogen ("Primary stimulation") for 10 days and then the injections were stopped. Two weeks after withdrawal (25 days), the injections were resumed ("Secondary stimulation").

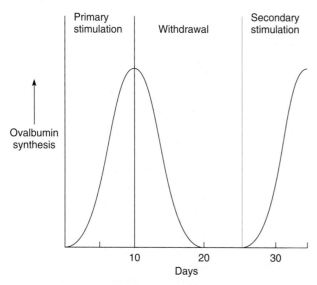

Provide possible explanations of these data.

17.6 Although the primary transcript of a gene may be identical in two different cell types, the translated mRNAs can be quite different. Consequently, in different tissues, distinct protein products can be produced from the same gene. Discuss two different mechanisms by which the production of mature mRNAs can be regulated to this end; give a specific example for each mechanism.

*17.7 Although eukaryotes lack operons like those found in prokaryotes, the exceptional conserved organization of the *ChAT/VAChT* locus is reminiscent of a prokaryotic operon. *ChAT* is the gene for the enzyme choline acetyltransferase, which synthesizes acetylcholine, a neurotransmitter released by one neuron to signal another neuron. *VAChT* is the gene for the vesicular acetylcholine transporter protein, which packages acetylcholine into vesicles before its release from a neuron. Both *ChAT* and *VAChT* are expressed in the same neuron.

Part of the *VAChT* gene is nested within the first intron of the *ChAT* gene, and the two genes share a common regulatory region and a first exon. The structure of a primary mRNA and two processed mRNA transcripts produced by this locus are diagrammed in the following figure. The common regulatory region important for transcription of the locus in neurons is shown in the DNA, black rectangles in RNA represent exons, lines connecting the exons represent spliced intronic regions, and AUG indicates the translation start codons within the *ChAT* and *VAChT* mRNAs. Polyadenylation sites are not shown.

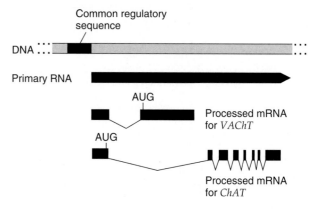

a. In what ways is the organization of the *VAChT/ChAT* locus reminiscent of a bacterial operon?

b. Why is the organization of this locus not structurally equivalent to a bacterial operon?

c. Based on the transcript structures shown, what modes of regulation might be used to obtain two different protein products from the single primary mRNA?

17.8 Distinguish between the terms *development* and *differentiation*.

17.9 What is totipotency? Give an example of the evidence for the existence of this phenomenon.

*17.10 In Woody Allen's 1973 film *Sleeper,* the aging leader of a futuristic totalitarian society has been dismembered in a bomb attack. The government wants to clone the leader from his only remaining intact body part, a nose. The characters Miles and Luna thwart the cloning by abducting the nose and flattening it under a steam roller.

a. In light of the 1996 cloning of the sheep Dolly, how should the cloning have proceeded if Miles and Luna had not intervened?

b. If methods like those used for Dolly had been successful, in what genetic ways would the cloned leader be unlike the original?

c. Suppose that instead of a nose, only mature B cells (B lymphocytes of the immune system) were available. What genetic deficits would you expect in the "new leader"?

d. In the set of experiments used to clone Dolly, six additional live lambs were obtained. Why is the production of Dolly more significant than the production of the other lambs?

e. If the cloning of the leader had succeeded, can you make any prediction about whether the "cloned leader" would be interested in perpetuating the totalitarian state?

17.11 In humans, β-thalassemia is a disease caused by failure to produce sufficient β-globin chains. In many cases, the mutation causing the disease is a deletion of all or part of the β-globin structural gene. Individuals homozygous for certain of the β-thalassemia mutations are able to survive because their bone marrow cells produce γ-globin chains. The γ-globin chains combine with α-globin chains to produce fetal hemoglobin. In these people, fetal hemoglobin is produced by the bone marrow cells throughout life, whereas normally it is produced in the fetal liver. Use your knowledge about gene regulation during development to suggest a mechanism by which this expression of γ-globin might occur in β-thalassemia.

*17.12 The following figure shows the percentage of ribosomes found in polysomes in unfertilized sea urchin oocytes (0 hours) and at various times after fertilization.

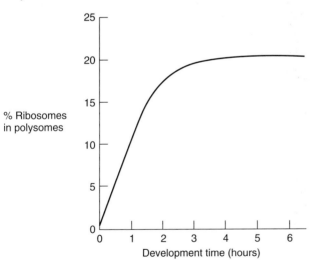

In the unfertilized egg, less than 1 percent of ribosomes are present in polysomes, and at 2 hours postfertilization, about 20 percent of ribosomes are present in polysomes. It is known that no new mRNA is made during the time period shown. How can the data be interpreted?

17.13 The mammalian genome contains about 10^5 genes. Mammals can produce about 10^6 to 10^8 different antibodies. Explain how it is possible for both of these statements to be true.

17.14 Antibody molecules (Ig) are composed of four polypeptide chains (two of one light chain type and two

of one heavy chain type) held together by disulfide bonds.

a. If for the light chain there were 300 different V_κ segments and 4 J_κ segments, how many different light chain combinations would be possible?

b. If for the heavy chain there were 200 V_H segments, 12 D segments, and 4 J_H segments, how many heavy chain combinations would be possible?

c. Given the information in parts (a) and (b), what would be the number of possible types of IgG molecules (L + H chain combinations)?

17.15 Define imaginal disk, homeotic mutant, and transdetermination.

*17.16 Imagine that you observed the following mutants (*a–e*) in *Drosophila*. Based on the characteristics given, assign each of the mutants to one of the following categories: maternal effect gene, segmentation gene, or homeotic gene.

a. Mutant *a*: In homozygotes, phenotype is normal, except wings are oriented backward.

b. Mutant *b*: Homozygous females are normal but produce larvae that have a head at each end and no distal ends. Homozygous males produce normal offspring (assuming the mate is not a homozygous female).

c. Mutant *c*: Homozygotes have very short abdomens, which are missing segments A2 through A4.

d. Mutant *d*: Affected flies have wings growing out of their heads in place of eyes.

e. Mutant *e*: Homozygotes have shortened thoracic regions and lack the second and third pair of legs.

*17.17 If actinomycin D, an antibiotic that inhibits RNA synthesis, is added to newly fertilized frog eggs, there is no significant effect on protein synthesis in the eggs. Similar experiments have shown that actinomycin D has little effect on protein synthesis in embryos up until the gastrula stage. After the gastrula stage, however, protein synthesis is significantly inhibited by actinomycin D, and the embryo does not develop further. Interpret these results.

*17.18 It is possible to excise small pieces of early embryos of the frog, transplant them to older embryos, and follow the course of development of the transplanted material as the older embryo develops. A piece of tissue is excised from a region of the late blastula or early gastrula that would later develop into an eye and is transplanted to three different regions of an older embryo host (see part a of Figure 17.A). If the tissue is transplanted to the head region of the host, it will form eye, brain, and other material characteristic of the head region. If the tissue is transplanted to other regions of the host, it will form organs and tissues characteristic of those regions in normal development (e.g., ear, kidney).

In contrast, if tissue destined to be an eye is excised from a neurula and transplanted into an older embryo host to exactly the same places as used for the blastula or gastrula transplants, in every case the transplanted tissue differentiates into an eye (see part b of Figure 17.A). Explain these results.

Figure 17.A

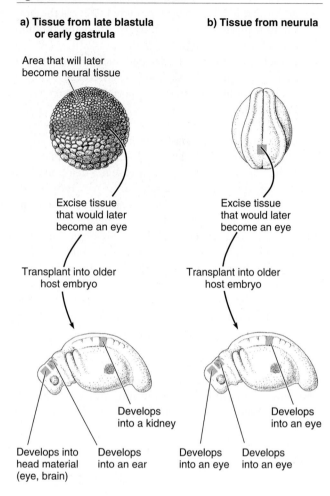

a) **Tissue from late blastula or early gastrula**

Area that will later become neural tissue

Excise tissue that would later become an eye

Transplant into older host embryo

Develops into head material (eye, brain)

Develops into an ear

Develops into a kidney

b) **Tissue from neurula**

Excise tissue that would later become an eye

Transplant into older host embryo

Develops into an eye

Develops into an eye

Develops into an eye

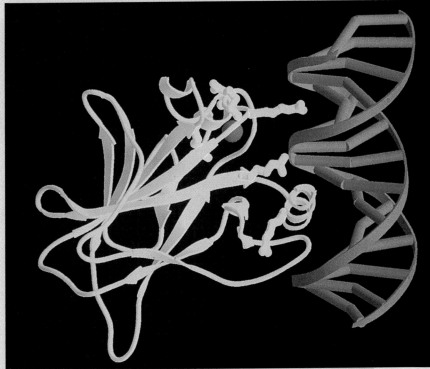
p53 protein binding to DNA.

18

Genetics of Cancer

PRINCIPAL POINTS

Progression of a normal eukaryotic cell through the cell cycle is tightly controlled by a number of molecular factors. Healthy cells grow and divide only when the balance of stimulatory and inhibitory signals received from outside the cell favors cell proliferation. A cancerous cell does not respond to the usual signals and reproduces without constraints.

Mutant forms of three classes of genes—proto-oncogenes, tumor suppressor genes, and mutator genes—have the potential to contribute to the transformation of a cell to a cancerous state. The products of proto-oncogenes normally stimulate cell proliferation; the products of tumor suppressor genes normally inhibit cell proliferation; and the products of mutator genes are involved in DNA replication and repair.

The two-hit mutation model for cancer states that two mutational events are necessary for cancer to develop, one in each allele of a cancer-causing gene. In familial (hereditary) cancers, one mutation is inherited, predisposing the person to cancer; the other mutation occurs later in somatic cells. In sporadic (nonhereditary) cancers, both mutations occur in somatic cells. This simple two-hit model applies to very few cancers; other cancers involve mutations in many genes.

Some DNA viruses and RNA viruses cause cancers. All RNA tumor-causing viruses are retroviruses—viruses that replicate via a DNA intermediate—but not all retroviruses cause cancer. When a retrovirus infects a cell, the RNA genome is released from the virus particle, and through the action of reverse transcriptase, a cDNA copy of the genome, called the proviral DNA, is synthesized. The proviral DNA integrates into the genome of the host cell. Then, using host transcriptional machinery, viral genes are transcribed, and full-length viral RNAs are produced. Progeny viruses are assembled and exit the cell, where they can infect other cells.

When tumor induction occurs after retrovirus infection, it is because of the activity of a viral oncogene (v-*onc*) in that retroviral genome. Retroviruses carrying an oncogene are known as transducing retroviruses.

Normal animal cells contain genes with DNA sequences that are similar to those of the viral oncogenes. These cellular genes are proto-oncogenes. When a proto-oncogene is mutated to produce a cellular oncogene (c-*onc*), it induces tumor formation.

The normal products of tumor suppressor genes have inhibitory roles in cell growth and division. Therefore, when both alleles of a tumor suppressor gene are inactivated or lost, the inhibitory activity is lost, and uncontrolled cell proliferation can occur.

Mutator genes are genes that, when mutant, increase the spontaneous mutation frequencies of other genes. In the cell, the normal (unmutated) forms of mutator genes are involved in key activities, such as DNA replication and DNA repair.

The development of most cancers involves the accumulation of mutations in a number of genes over a significant period of a person's life. This multistep path typically involves mutational events that change proto-oncogenes to oncogenes and inactivate tumor suppressor genes and mutator genes, thereby breaking down the multiple mechanisms that regulate growth and differentiation.

Various types of radiation and many chemicals increase the frequency with which cells become cancerous. These agents are known as carcinogens. Practically all carcinogens act by causing changes in the genome of the cell. In the case of chemical carcinogens, a few act directly on the genome, but most act indirectly by being converted to active derivatives by cellular enzymes. All carcinogenic forms of radiation act directly.

iActivity

i AT CURRENT RATES, OVER A THIRD OF THE PEOPLE who read this will die of cancer. Cancers are diseases characterized by the uncontrolled and abnormal division of eukaryotic cells. When cells divide unchecked within the body, they can give rise to tissue masses known as tumors. Some of these are not life threatening, but others can invade and disrupt surrounding tissues. What are the mechanisms that regulate cell growth and division? What causes uncontrolled growth in a cell? What genes are involved in the development of cancer? Is cancer inherited? In this chapter, you will learn the answer to these and other questions. Then, in the iActivity, you can apply what you learned as you investigate the origins of a form of bladder cancer.

In Chapter 17, we learned about some of the genetically controlled processes involved in development and differentiation. The picture we have is that during development, specific tissues and organs arise by genetically programmed cell division and differentiation. Occasionally, dividing and differentiating cells deviate from their normal genetic program and give rise to tissue masses called *tumors*, or *neoplasms* ("new growth"). Figure 18.1 shows a mammogram indicating the presence of a tumor. The process by which a cell loses its ability to remain constrained in its growth properties is called **transformation** (not to be confused with the transformation of a cell by uptake of exogenous DNA). If the transformed cells stay together in a single mass, the tumor is said to be *benign*. Benign tumors usually are not life threatening, and their surgical removal generally results in a complete cure.

Figure 18.1

A mammogram showing a tumor.

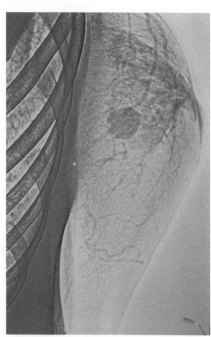

Exceptions include many brain tumors, which are life threatening because they impinge on essential cells. If the cells of a tumor can invade and disrupt surrounding tissues, the tumor is said to be *malignant* and is identified as a **cancer.** Cells from malignant tumors can also break off and move through the blood system or lymphatic system, forming new tumors at other locations in the body. The spreading of malignant tumor cells throughout the body is called *metastasis*. Malignancy can result in death because of damage to critical organs, secondary infection, metabolic problems, second malignancies, or hemorrhage.

The initiation of tumors in an organism is called **oncogenesis** (from the Greek *onkos,* meaning "mass" or "bulk," and *genesis,* meaning "birth"). There are many genetic causes of cancer, such as spontaneous genetic changes (spontaneous gene mutations or chromosome mutations, for instance), exposure to mutagens or radiation, or the action of genes in cancer-inducing viruses (*tumor viruses*). There is also hereditary predisposition to cancer. In this chapter, we focus on the genetic basis of tumors and cancers.

Relationship of the Cell Cycle to Cancer

During development, a tissue is produced by cell proliferation. During a series of divisions, progeny cells begin to express genes that are specific for the tissue, a process called cell differentiation. Cell differentiation is also associated with the progressive loss of the ability of cells to proliferate: The most highly differentiated cell, the one that is fully functional in the tissue, can no longer divide. Such cells are known as *terminally differentiated cells.* They have a finite life span in the tissue and are replaced with younger cells produced by division of *stem cells,* a small fraction of cells in the tissue that are capable of *self-renewal*. To understand neoplastic diseases, both benign and malignant ones, we must realize that the linkage of growth with differentiation of any tissue is not necessary. That is, cells *can* divide without undergoing terminal differentiation.

Eukaryotic cells proliferate by going through the cell cycle—the cycle of cell growth, mitosis, and cell division (see Chapter 1, p. 14, and Chapter 9, pp. 194–196). Recall that the cell cycle consists of the mitotic phase (M) and an interphase between divisions consisting of three stages: G_1, S, and G_2. Several molecular factors control the progression of a normal cell through the cell cycle, the most important being the proteins that operate through interaction with receptors embedded in the plasma membrane. When these factors bind to the cell surface receptor, a *signal transduction pathway* is induced whereby the signal is transmitted into the cytoplasm, and the regulatory effect on cell division occurs. *Growth factors* turn on stimulatory pathways for cell division (Figure 18.2a), and *growth-inhibiting factors* turn on inhibitory pathways for cell division (Figure 18.2b). Growth factors cause genes that encode proteins needed for the cell division process to be turned on; growth-inhibiting factors cause genes that encode proteins with inhibitory effects on cell division to be turned on. Normal healthy cells give rise to progeny cells only when the balance of stimulatory and inhibitory signals from outside the cell favors cell division. A neoplastic cell, on the other hand, has lost control of cell division and reproduces without constraints (although not at a faster rate). This can occur when genes that encode inhibitory factors mutate or when genes that encode stimulatory factors mutate.

animation
a Regulation of
Cell Division in
Normal Cells

K E Y N O T E

Progression of a normal eukaryotic cell through the cell cycle is tightly controlled by a number of molecular factors. Healthy cells grow and divide only when the balance of stimulatory and inhibitory signals received from outside the cell favor cell proliferation. A cancerous cell does not respond to the usual signals and reproduces without constraints.

The Two-Hit Mutation Model for Cancer

With the exception of cancers caused by viruses, cancers are genetic disorders in that they are caused by changes in DNA that are stably inherited by progeny cells. That is, an accumulation of genetic mutations in particular classes of genes in a cell over a period of time causes cancer. The escalating genetic damage causes a progressive loss in the ability of the cell to respond properly to growth regulatory signals, so that eventually the cell will divide uncontrollably, thereby giving rise to a tumor. Research over the past 20 years or so has led to the identification of a number of the particular genes related to the onset of cancer.

Anecdotal evidence that genes have a role in cancer came from the observation that there was a high incidence of particular cancers in some human families. Cancers that run in families are known as *familial (hereditary) cancers;* cancers that do not appear to be inherited are known as *sporadic* (or *nonhereditary*) *cancers.* Sporadic cancers are more frequent than familial cancers.

One model for the relationship of mutations to cancer came from the study of the onset of retinoblastoma (OMIM 180200), a childhood cancer of the eye (Figure 18.3). Retinoblastoma occurs from birth to age 4 years and is the most common eye tumor in children. If discovered early enough, more than 90 percent of the eye tumors can be permanently destroyed, usually by gamma radiation. There are two forms of retinoblastoma. In *sporadic retinoblastoma* (60 percent of cases), an eye tumor develops spontaneously in a patient from a family with no history of the

Figure 18.2

General events for regulation of cell division in normal cells. (a) When a growth factor binds to its cell membrane receptor, it acts as a signal to stimulate cell growth. To do that, the signal is transduced into the cell and relayed to the nucleus, activating the expression of one or more genes that encode one or more proteins required for the stimulation of cell division. (b) When a growth-inhibiting factor binds to its cell membrane receptor, it acts as a signal to inhibit cell growth. In this case, the signal is transduced into the cell and relayed to the nucleus, activating the expression of one or more genes that encode one or more proteins required for the inhibition of cell division.

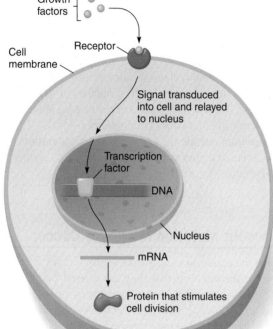
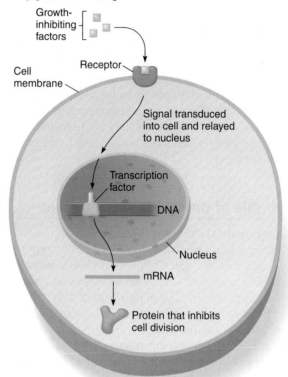

a) Stimulation of cell division induced by growth factor

b) Inhibition of cell division induced by growth-inhibiting factor

disease. In these cases, a *unilateral tumor* develops (a tumor in one eye only). In *hereditary retinoblastoma* (40 percent of cases), the susceptibility to develop eye tumors is inherited. Patients with this form of retinoblastoma typically develop multiple eye tumors involving both eyes *(bilateral tumors)*, and they usually develop the disease at an earlier age than patients with sporadic retinoblastoma. A single gene is responsible for retinoblastoma.

In 1971, Alfred Knudson developed a two-hit mutational model for retinoblastoma (Figure 18.4). In sporadic retinoblastoma (Figure 18.4a), a child is born with two wild-type copies of the retinoblastoma gene (genotype RB^+/RB^+), and mutation of each to a mutant RB allele must then occur in the same eye cell. Since the chance of having two independent mutational events in the same cell is very low, sporadic retinoblastoma patients would be expected to develop mostly unilateral

tumors, as is the case. Furthermore, the rarity of the mutation event means that the two gene copies are mutated at different times, the first mutation producing an RB/RB^+ cell, and the second mutation in that cell giving rise to the RB/RB genotype that results in eye tumor development. In hereditary retinoblastoma, patients inherit one copy of the mutated retinoblastoma gene through the germ line; that is, they are RB/RB^+ heterozygotes (Figure 18.4b). Only a single additional mutation of the retinoblastoma gene in an eye cell is needed to produce an RB/RB homozygote that would result in tumor formation. Given the number of cells in a developing retina and the rate of mutation per cell, *loss of heterozygosity* (LOH; here, a mutation in the RB^+ allele) is very likely for at least a few cells. Furthermore, because only a single mutation is needed to produce homozygosity for RB, hereditary retinoblastoma is characterized on

Figure 18.3
An eye tumor in a patient with retinoblastoma.

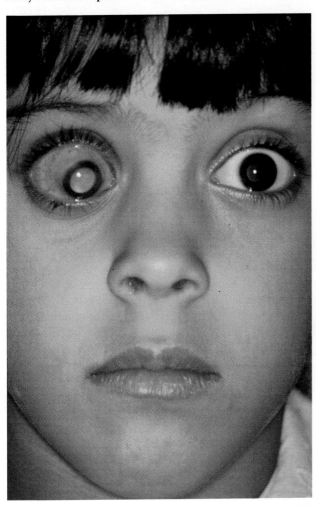

Figure 18.4

Knudson's two-hit mutation model. This model was proposed to explain **(a)** sporadic retinoblastoma by two independent mutations of the retinoblastoma (*RB*) gene and **(b)** hereditary retinoblastoma by a single mutation of the wild-type retinoblastoma gene in retinal cells in which a mutant *RB* was inherited through the germ line.

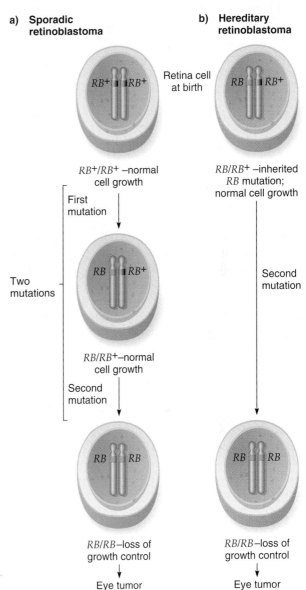

a) **Sporadic retinoblastoma**   b) **Hereditary retinoblastoma**

the average by earlier onset than sporadic retinoblastoma and by multiple bilateral tumors.

According to Knudson's model, the *retinoblastoma mutation is recessive* because cancer develops only if both alleles are mutant. However, if one mutation is inherited through the germ line, tumor formation requires only a mutational event in the remaining wild-type allele in any one of the cells in that particular tissue. Because of the high likelihood of such an event, the *disease appears dominant* in pedigrees. So for hereditary retinoblastoma and in hereditary neoplasms in general, we say that inheritance of just one gene mutation predisposes a person to cancer but does not cause it directly; a second mutation is required for loss of heterozygosity. Commonly, therefore, we talk about there being a hereditary disposition for cancer in such families.

Support for Knudson's hypothesis came in the 1980s from the analysis of the chromosomes of tumor cells and normal tissues in patients with retinoblastoma. Many patients carried deletions of a region of chromosome 13,

and through genetic analysis, the *RB* gene was mapped to chromosome location 13q14.1-q14.2. Retinoblastoma is among a very few cancers for which only one gene is critical for its development, in this case a mutation in a gene for a growth inhibitory factor, that is, a tumor suppressor gene (see pp. 384–385). In most cases, cancer develops as a multistep process involving mutations in several different key genes related to cell growth and division.

KEYNOTE

The two-hit mutation model for cancer explains the difference between familial (hereditary) cancers and sporadic (nonhereditary) cancers. In familial cancers, one mutation is inherited, thereby predisposing the person to cancer. When the second mutation occurs later in somatic cells, cancer may then develop. In sporadic cancers, both mutations occur in the somatic cells, so such cancers typically occur later in life than familial cancers because the probability of two mutations is lower than the probability of one mutation.

Genes and Cancer

Three classes of genes are mutated frequently in cancer. These are *proto-oncogenes, tumor suppressor genes,* and *mutator genes.* The products of proto-oncogenes normally stimulate cell proliferation. Mutant proto-oncogenes—now called oncogenes—either are more active than normal or are active at inappropriate times. The products of unmutated tumor suppressor genes normally inhibit cell proliferation. Mutant tumor suppressor genes have lost their inhibitory function. The products of wild-type mutator genes are needed to ensure fidelity of replication and maintenance of genome integrity. Mutant mutator genes have lost their normal function, and this makes the cell prone to accumulate mutational errors.

Oncogenes

Transformation of cells to the neoplastic state can result from infection with **tumor viruses,** which induce the cells they infect to proliferate in an uncontrolled fashion and produce a tumor. Tumor viruses, which may have RNA or DNA genomes, are widely found in animals. *RNA tumor viruses* and *DNA tumor viruses* cause cancer by entirely different mechanisms, as we will see. RNA tumor viruses transform a cell because of the property of one or more genes in the viral genome called *viral oncogenes.* By definition, an **oncogene** is a gene whose action stimulates unregulated cell proliferation.

Retroviruses and Oncogenes. RNA tumor viruses are all **retroviruses,** and the oncogenes carried by RNA tumor viruses are altered forms of normal host cell genes.

Structure of Retroviruses. Examples of retroviruses are Rous sarcoma virus, feline leukemia virus, mouse mammary tumor virus, and human immunodeficiency virus (HIV-1, the causative agent of *acquired immunodeficiency syndrome*—AIDS). A retrovirus particle is shown in Figure 18.5. Within a protein core, which often is icosahedral in shape, are two copies of the 7-kb to 10-kb single-stranded RNA genome. The core is surrounded by an envelope derived from host membranes with virus-encoded glycoproteins inserted into it. When the virus infects a cell, the envelope glycoproteins interact with a host cell surface receptor to begin the process by which the virus enters the cell.

Life Cycle of Retroviruses. A well-studied retrovirus is *Rous sarcoma virus (RSV).* RSV causes sarcomas—cancers of the connective tissue or muscle cells—in chickens. The RNA genome organization of RSV is shown in Figure 18.6a. When RSV infects a cell, the RNA genome is released from the virus particle, and a double-stranded DNA copy of the genome (*proviral DNA*) is made by reverse transcriptase, an enzyme brought into the cell as part of the virus particle and encoded by the *pol* gene (Figure 18.6b). This RNA-to-DNA copying process is called reverse transcription.

The proviral DNA next integrates into the host chromosome as follows: The ends of all retroviral RNA genomes consist of the sequences R and U_5 (at the left in Figure 18.6a) and U_3 and R (at the right in Figure 18.6a). During proviral DNA synthesis by reverse transcriptase, the end sequences of the genome are duplicated to produce long terminal repeats (LTRs) of U_3-R-U_5 (see Figure 18.6b). The two ends of the proviral DNA are ligated to produce a circular, double-stranded molecule (Figure 18.6c). This brings the two LTRs next to each other. Staggered nicks are made in both viral and cellular DNAs, and integration of the viral DNA begins (Figure 18.6d). By recombination, the viral ends become joined to the ends of the cellular DNA (Figure 18.6e); at this point, integration has occurred (Figure 18.6f). Finally, the single-stranded gaps are filled in. The integration of retrovirus proviral DNA results in a duplication of DNA at the target site, producing short, direct repeats in the host cell DNA flanking the provirus.

Once integrated, the proviral DNA is transcribed by the host RNA polymerase II to produce by alternative splicing (see Chapter 17) the viral mRNAs that encode the individual viral proteins. Typical retroviruses have three protein-coding genes for the viral life cycle: *gag, pol,* and *env.* The *gag* gene encodes a precursor protein that, when cleaved, produces virus particle proteins. The *pol* gene encodes a precursor protein that is cleaved to produce reverse transcriptase and an enzyme needed for the integration of the proviral DNA into the host cell chromosome. The *env* gene encodes the precursor to the envelope glycoprotein. Progeny RNA genomes are produced by transcription of the entire integrated viral DNA and packaged into new virus particles that exit the cell and can infect other cells.

Some retroviruses also carry an oncogene that gives them the ability to transform the cells they infect; these are the *oncogenic retroviruses.* In the case of RSV, the oncogene is called *src* (see Figure 18.6a), and like other retroviral oncogenes, it is not involved in the viral life cycle. Different retroviruses carry different oncogenes. Most oncogenic retroviruses (RSV is an exception) cannot replicate because they do not have a full set of life cycle genes. Retroviruses without oncogenes direct their own life cycle but do not change the growth properties of the cells they infect; these are *nononcogenic retroviruses.*

Figure 18.5

Stylized drawing of a retrovirus.

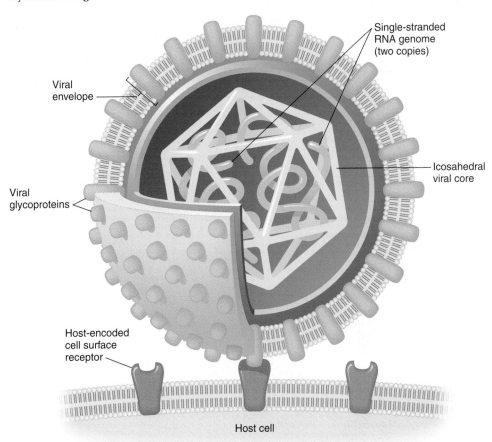

Concerning retroviruses, it is appropriate to discuss briefly HIV-1, the causative agent of AIDS, even though this virus does not cause cancer. The retrovirus HIV-1 has a bullet-shaped capsid and is surrounded by a viral envelope in which are embedded virus-encoded gp120 glycoproteins. The HIV-1 genome contains complete *gag*, *pol*, and *env* genes, so HIV can self-propagate. In addition, HIV contains several other genes that are not oncogenes but that help control gene expression. For example, one of these genes, *tat*, encodes a protein that regulates transcription of the *gag* and *pol* genes and the translation of the resulting mRNA.

The gp120 glycoprotein of the HIV-1 envelope is the main basis for the infection of cells. This glycoprotein is recognized by the CD4 receptor found on the surface of immune system cells called *helper T cells* (a type of T lymphocyte), as well as on the surface of certain other cell types. Once the virus has bound to a receptor on the cell surface, the virus particles enter the cell. Next, the viral protein coat is lost, and the viral life cycle is initiated, starting with reverse transcription of the viral RNA into proviral DNA, which integrates into the cell's genome.

Through normal viral replication, HIV-1 causes the death of the cell it infects. Thus, by repeated viral replication and the infection of more cells, a steady destruc-

tion of helper T cells and other infected cells by the virus takes place. The decrease in the population of helper T cells leads to a progressive decrease in function of the immune system. As a result, a person infected with HIV-1 is unable to combat infections by pathogens such as bacteria, viruses, and fungi and also becomes susceptible to numerous types of cancers. AIDS patients die most frequently from infections. To date, at least 19 million people worldwide have been infected with HIV-1.

KEYNOTE

Retroviruses are RNA viruses that replicate via a DNA intermediate. All RNA tumor viruses are retroviruses, but not all retroviruses cause cancer. When a retrovirus infects a cell, the RNA genome is released from the virus particle, and through the action of reverse transcriptase, a cDNA copy of the genome—called the proviral DNA—is synthesized. The proviral DNA integrates into the genome of the host cell. Then, using host transcriptional machinery, viral genes are transcribed, and full-length viral RNAs are produced. Progeny viruses assembled within the cell exit the cell and can infect other cells.

Figure 18.6

The Rous sarcoma virus (RSV) RNA genome and a suggested mechanism for the integration of the proviral DNA into the host (chicken) chromosome. (a) RSV genome RNA. (b) RSV proviral DNA produced by reverse transcriptase. (c) Circularization of the proviral DNA. (d) Staggered nicks are made in viral and cellular DNAs. (e) By recombination, the viral ends become joined to the ends of the cell's DNA. (f) The single-stranded gaps are filled in, and a complete, double-stranded, integrated RSV provirus results.

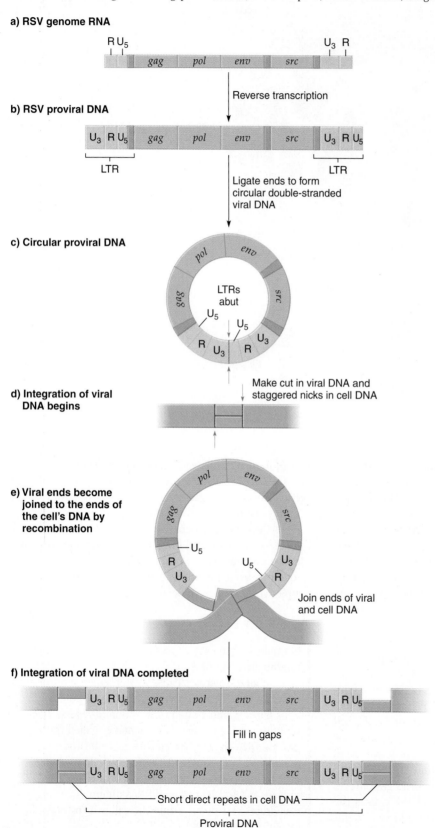

a) RSV genome RNA

b) RSV proviral DNA

c) Circular proviral DNA

d) Integration of viral DNA begins

e) Viral ends become joined to the ends of the cell's DNA by recombination

f) Integration of viral DNA completed

Viral Oncogenes. In the case of RSV, tumor induction is caused by a particular **viral oncogene** in the retroviral genome. Viral oncogenes (generically called v-*onc*s) are responsible for many different cancers. Only retroviruses that contain a v-*onc* gene are tumor viruses. The v-*onc* genes are named for the tumor that the virus causes, with the v indicating that the gene is of viral origin. Thus, the v-*onc* gene of RSV is v-*src*. Bacteriophages that have picked up cellular genes are said to transduce the genes to other cells, so such retroviruses are called **transducing retroviruses** because they have picked up an oncogene from the genome of the cell. (We learn how this happens later.) Table 18.1 lists some transducing retroviruses and their viral oncogenes. Retroviruses that do not carry oncogenes are called *nontransducing retroviruses.*

Cells infected by RSV rapidly transform into the cancerous state because of the activity of the v-*src* gene. RSV contains all the genes necessary for viral replication (*gag, env,* and *pol*), so an RSV-transformed cell produces progeny RSV particles. In this ability, RSV is an exception; all other transducing retroviruses are defective in some of their viral replication genes (Figure 18.7): They can transform cells but are unable to produce progeny viruses because they lack one or more genes needed for virus reproduction. These defective retroviruses can produce progeny virus particles if cells containing them are also infected with a normal virus (a *helper virus*) that can supply the missing gene products.

Cellular Proto-Oncogenes. In the mid-1970s, J. Michael Bishop, Harold Varmus, and others demonstrated that normal animal cells contain genes with DNA sequences very closely related to the viral oncogenes. These genes are called proto-oncogenes. (Bishop and Varmus received the 1989 Nobel Prize in Physiology or Medicine for their "discovery of the cellular origin of retroviral oncogenes.") In the early 1980s, R. A. Weinberg and M. Wigler showed independently that a variety of human tumor cells contain oncogenes. These genes, when introduced into other cells growing in culture, transformed those cells into cancer cells. The human oncogenes were found to be very similar to viral oncogenes that had been characterized earlier, even though viruses did not induce the human cancers involved. These human oncogenes also were shown to be closely related to proto-oncogenes found in normally growing cells.

In short, most human and other animal oncogenes are mutant forms of normal cellular genes. Such genes in their normal state are called **proto-oncogenes.** Proto-oncogenes have important roles in regulating cell division and differentiation. When proto-oncogenes become mutated or translocated such that they induce tumor formation, they are called oncogenes (*onc*s). If they are carried by a virus, oncogenes are known as v-*onc*s. If they reside in the host chromosome, oncogenes are called **cellular oncogenes,** or **c-*onc*s.** A transducing retrovirus, then, carries a significantly altered form of a cellular proto-oncogene (now a v-*onc*). When the transducing retrovirus infects a normal cell, the hitchhiking oncogene transforms the cell into a cancer cell. As we will see, normal cells can also be transformed into cancer cells, even if a tumor virus does not infect them, if the proto-oncogene is converted into a cellular oncogene.

One significant difference between a cellular proto-oncogene and its viral oncogene counterpart is that most proto-oncogenes contain introns that are not present in the corresponding v-*onc*. This is the result of splicing that occurs in transcription of the genomic viral RNA from proviral DNA.

Formation of Transducing Retroviruses. The location at which retroviral DNA (the provirus) integrates

Table 18.1	Some Transducing Retroviruses and Their Viral Oncogenes			
Oncogene	**Retrovirus Isolate**	**v-*onc* Origin**	**v-*onc* Protein**	**Type of Cancer**
src	Rous sarcoma virus	Chicken	pp60^{src}	Sarcoma
abl	Abelson murine leukemia virus	Mouse	P90-P160$^{gag\text{-}abl}$	Pre–B cell leukemia
erbA	Avian erythroblastosis virus	Chicken	P75$^{gag\text{-}erbA}$	Erythroblastosis and sarcoma
erbB	Avian erythroblastosis virus	Chicken	gp65^{erbB}	Erythroblastosis and sarcoma
fms	McDonough (SM)-FeSV	Cat	gp180$^{gag\text{-}fms}$	Sarcoma
fos	FBJ (Finkel-Biskis-inkins)-MSV	Mouse	pp55^{fos}	Osteosarcoma
myc	MC29	Chicken	P100$^{gag\text{-}myc}$	Sarcoma, carcinoma, and myelocytoma
myb	Avian myeloblastosis virus (AMV)	Chicken	p45^{myb}	Myeloblastosis
	AMV-E26	Chicken	P135$^{gag\text{-}myb\text{-}ets}$	Myeloblastosis and erythroblastosis
raf	3611-MSV	Mouse	P75$^{gag\text{-}raf}$	Sarcoma
H-*ras*	Harvey MSV	Rat	pp21^{ras}	Sarcoma and erythroleukemia
	RaSV	Rat	P29$^{gag\text{-}ras}$	Sarcoma?
K-*ras*	Kirsten MSV	Rat	pp21^{ras}	Sarcoma and erythroleukemia

Figure 18.7

Structures of four defective transducing viruses (not to scale).
(a) Avian myeloblastosis virus (AMV) contains the v-*myb* onco-gene, which replaces the 3′ end of *pol* and most of *env*. **(b)** Avian defective leukemia virus (DLV) contains the v-*myc* oncogene, which replaces the 3′ end of *gag*, all of *pol*, and the 5′ end of *env*. **(c)** Feline sarcoma virus (FeSV) contains the v-*fes* oncogene, which replaces the 3′ end of *gag* and all of *pol* and *env*. **(d)** Abelson murine leukemia virus (AbMLV) contains the v-*abl* oncogene, which replaces the 3′ end of *gag* and all of *pol* and *env*.

a) Avian myeloblastosis virus (AMV) genomic RNA

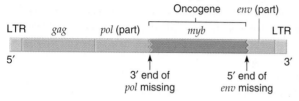

b) Avian defective leukemia virus (DLV) genomic RNA

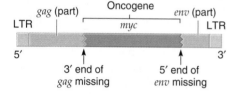

c) Feline sarcoma virus (FeSV) genomic RNA

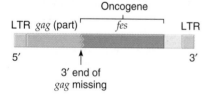

d) Abelson murine leukemia virus (AbMLV) genomic RNA

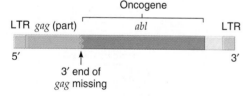

into cellular DNA is random. Sometimes there occurs a genetic rearrangement by which the transcriptional unit of the provirus connects to nearby cellular genes, often by a deletion event involving the loss of some or all of the *gag*, *pol*, and *env* genes. In this way, viral RNA contains all or parts of a cellular gene. All viral progeny then carry the cellular gene and, under the influence of viral promoters in the LTR, express the cellular protein in infected cells. If the cellular gene picked up was an oncogene, the modified retrovirus is oncogenic (see Figure 18.7). If the cellular gene picked up is a proto-oncogene, the modified retrovirus may still be oncogenic if the increased expression of the proto-oncogene causes oncogenesis.

Protein Products of Proto-Oncogenes. About 100 onco-genes have been identified. Based on DNA sequence similarities and similarities in amino acid sequences of the protein products, proto-oncogenes fall into several distinct classes, each with a characteristic type of protein product, as outlined in Table 18.2.

In the following sections, we illustrate the stimulatory function of protein products of proto-oncogenes on cell growth and division by considering just two examples: growth factors and protein kinases.

Growth Factors. The effect of oncogenes on cell growth and division led to an early hypothesis that proto-oncogenes might be regulatory genes involved with the control of cell multiplication during differentiation. There is a lot of evidence supporting this hypothesis.

We can generalize and say that some cancer cells can result from the excessive or untimely synthesis of growth factors in cells that do not normally produce the factors.

Table 18.2	Classes of Oncogene Products
Growth factors	
sis	PDGF B-chain growth factor
int-2	FGF-related growth factor
Receptor and nonreceptor protein-tyrosine and protein-serine/threonine kinases	
src	Membrane-associated nonreceptor protein-tyrosine kinase
fgr	Membrane-associated nonreceptor protein-tyrosine kinase
fps/fes	Nonreceptor protein-tyrosine kinase
kit	Truncated stem cell receptor protein-tyrosine kinase
pim-1	Cytoplasmic protein-serine kinase
mos	Cytoplasmic protein-serine kinase (cytostatic factor)
Receptors lacking protein kinase activity	
mas	Angiotensin receptor
Membrane-associated G proteins activated by surface receptors	
H-*ras*	Membrane-associated GTP-binding GTPase
K-*ras*	Membrane-associated GTP-binding GTPase
gsp	Mutant-activated form of G α
Cytoplasmic regulators	
crk	SH-2/3 protein that binds to (and regulates?) phosphotyrosine-containing proteins
Nuclear transcription factors (gene regulators)	
myc	Sequence-specific DNA-binding protein
fos	Combines with c-*jun* product to form AP-1 transcription factor
jun	Sequence-specific DNA-binding protein; part of AP-1
erbA	Dominant negative mutant thyroxine (T3) receptor
ski	Transcription factor?

Introduction of an altered growth factor gene such as a v-*onc* or mutation of a c-*onc* can cause tumor development.

Protein Kinases. Many proto-oncogenes encode protein kinases, enzymes that add phosphate groups to target proteins, thereby modifying the proteins' function. Protein kinases are integral members of signal transduction pathways. The *src* gene product, for example, is a nonreceptor protein kinase called pp60*src*. The viral protein, pp60v-*src*, and the protein encoded by the cellular oncogene, pp60c-*src*, differ in only a few amino acids, and both proteins bind to the inner surface of the plasma membrane. A large class of proteins, including the receptors for growth factors, uses protein phosphorylation to transmit signals through the membrane. Thus, the action of protein kinases appears also to be linked to growth factors and their activities through their role in signal transduction, and this explains how *src* can transform a normal cell into a metabolically different cancer cell.

Changing Cellular Proto-Oncogenes into Oncogenes. In normal cells, expression of proto-oncogenes is tightly controlled, so that cell growth and division occur only as appropriate for the cell type involved. However, when proto-oncogenes are changed into oncogenes, the tight control can be lost, and unregulated cell proliferation can take place.

Here are three examples of the types of changes that have been found:

1. *Point mutations* (base pair substitutions). Point mutations in the coding region of a gene or in the controlling sequences (promoter, regulatory elements, enhancers) can change a proto-oncogene into an oncogene by causing an increase in either the activity of the gene product or the expression of the gene, leading in turn to an increase in the amount of gene product.
2. *Deletions*. Deletions of part of the coding region or of part of the controlling sequences of a proto-oncogene have been found frequently in oncogenes. The deletions cause changes in the amount or activity of the encoded growth stimulatory protein, causing unprogrammed activation of some cell proliferation genes.
3. *Gene amplification* (increased number of copies of the gene). Some tumors have multiple (sometimes hundreds of) copies of proto-oncogenes. These probably result from a random overreplication of small segments of the genomic DNA. In general, extra copies of the proto-oncogene in the cell result in an increased amount of gene product, thereby inducing or contributing to unscheduled cell proliferation. For example, multiple copies of *ras* are found in mouse adrenocortical tumors.

Cancer Induction by Retroviruses. Retroviruses are common causes of cancer in animals, although only one type of cancer is known for humans. A retrovirus can cause cancer if it is a transducing retrovirus and the v-*onc* it carries is expressed. In this case, transcription of the v-*onc* takes place under the control of retroviral promoters. Another way in which a retrovirus can cause cancer is if the proviral DNA integrates near a proto-oncogene. In this situation, expression of the proto-oncogene can come under control of retroviral promoter and enhancer sequences in the retroviral LTR. These retroviral sequences do not respond to the environmental signals that normally regulate proto-oncogene expression, so overexpression of the proto-oncogene occurs, transforming the cell to the tumorous state. This process of proto-oncogene activation is called *insertional mutagenesis*. It is rare in animals and is not known to occur in humans.

KEYNOTE

After retrovirus infection, tumor induction can occur as a result of the activity of a viral oncogene (v-*onc*) in the retroviral genome. Retroviruses carrying an oncogene are known as transducing retroviruses. Normal cellular genes, called proto-oncogenes, have DNA sequences that are similar to those of the viral oncogenes. Proto-oncogenes encode proteins that stimulate cell growth and division. In their mutated state, proto-oncogenes are called cellular oncogenes (c-*onc*s), and they may induce tumors. Retroviral oncogenes are modified copies of the cellular proto-oncogenes that have been picked up by the retrovirus.

DNA Tumor Viruses. DNA tumor viruses are oncogenic—they induce cell proliferation—but they do not carry oncogenes like those in RNA tumor viruses. As mentioned previously (p. 378), their mechanism for transforming cells is completely different. DNA tumor viruses transform cells to the cancerous state through the action of one or more genes in the viral genome. Examples of DNA tumor viruses are found among five of six major families of DNA viruses: papovaviruses, hepatitis B viruses, herpes viruses, adenoviruses, and pox viruses.

DNA tumor viruses normally progress through their life cycles without transforming the cell to a cancerous state. Typically, the virus produces a viral protein that activates DNA replication in the host cell. Then, through the use of host proteins, the viral genome is replicated and transcribed, ultimately producing a large number of progeny viruses, which results in lysis and death of the cell. The released viruses can then infect other cells. Rarely, the viral DNA is not replicated and becomes integrated into the host cell genome. If the viral protein that activates DNA replication of the host cell is now synthesized, this protein transforms the cell to the cancerous state by stimulating the host cell to proliferate.

The papovavirus family includes examples of DNA tumor viruses. In this family are the many known papillomaviruses, some of which cause benign tumors such as skin and venereal warts in humans. Other human papillomaviruses (*HPV-16, HPV-18,* or both) cause cervical cancer, which is a leading cause of cancer deaths among women worldwide.

Tumor Suppressor Genes

In the late 1960s, Henry Harris fused normal rodent cells with cancer cells and observed that some of the resultant hybrid cells did not form tumors but grew normally. Harris hypothesized that the normal cells contained gene products that could suppress the uncontrolled cell proliferation characteristic of cancer cells. The genes involved were called **tumor suppressor genes.**

The normal products of tumor suppressor genes have an inhibitory role in cell growth and division. Thus, when tumor suppressor genes are inactivated, the inhibitory activity is lost, and unprogrammed cell proliferation can begin. Inactivation of tumor suppressor genes has been linked to the development of a wide variety of human cancers, including breast, colon, and lung cancer. In essence, tumor suppressor genes are the opposites of proto-oncogenes. Two mutations are needed to inactivate a tumor suppressor gene and thereby cause a potential loss of control over cell growth and division (Figure

18.8a), whereas only one mutation is needed to change a proto-oncogene to an oncogene and thereby stimulate cell growth and division (Figure 18.8b). Table 18.3 lists some of the known tumor suppressor genes in humans. The products of tumor suppressor genes are found throughout the cell.

The Retinoblastoma Tumor Suppressor Gene, *RB.* Retinoblastoma was introduced earlier in this chapter in the context of Knudson's two-hit mutation model for cancer.

Genetics of the Human RB *Tumor Suppressor Gene.* The human *RB* tumor suppressor gene (OMIM 180200) has been mapped to 13q14.1-q14.2. The *RB* gene spans 180 kb of DNA and encodes a 4.7 kb of mRNA that is translated to produce a 928-amino-acid nuclear phosphoprotein (a phosphorylated protein), pRB. pRB is expressed in every tissue type examined and is involved in regulating the cell cycle and all major cellular processes.

Cell Biology. The protein pRB plays a significant role in the cell cycle by regulating the passage of cells from G_1 to S, a transition that commits the cell to progressing through the rest of the cell cycle. In G_1 in a normal cell or in a cell heterozygous for an *RB* mutation, unphosphorylated pRB binds to a complex of two tran-

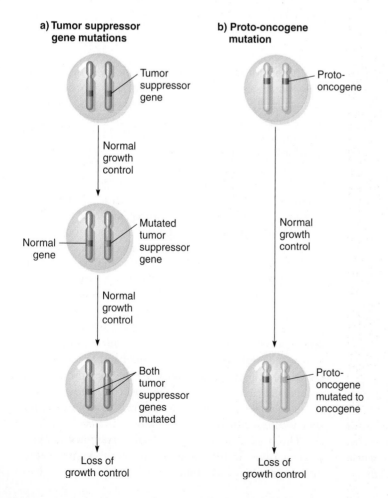

a) Tumor suppressor gene mutations

Tumor suppressor gene

↓ Normal growth control

Normal gene — Mutated tumor suppressor gene

↓ Normal growth control

Both tumor suppressor genes mutated

↓

Loss of growth control

b) Proto-oncogene mutation

Proto-oncogene

↓ Normal growth control

Proto-oncogene mutated to oncogene

↓

Loss of growth control

Figure 18.8

Comparison of the effects of tumor suppressor gene and proto-oncogene mutations. **(a)** Mutations in both alleles of a tumor suppressor gene are needed for the cell to lose growth control. **(b)** A mutation in only one allele of a proto-oncogene, converting it to an oncogene, is needed for the cell to lose growth control.

Table 18.3   Some Known or Candidate Tumor Suppressor Genes

Gene	Cancer Type	Product Location	Mode of Action	Hereditary Syndrome	Chromosome Location
APC	Colon carcinoma	Cytoplasm?	Cell adhesion molecule	Hereditary adenomatous polyposis	5q21-q22
BRCA1	Breast cancer	Nucleus	Transcription factor	Breast cancer and ovarian cancer	17q21
BRCA2	Breast cancer	Nucleus	Transcription factor?	Breast cancer	13q12-q13
DCC	Colon carcinoma	Membrane	Cell adhesion molecule	Involved in colorectal cancer	18q21.3
NF1	Neurofibromas	Cytoplasm	GTPase activator	Neurofibromatosis type 1	17q11.2
NF2	Schwannomas and meningiomas	Inner membrane?	Links membrane to skeleton?	Neurofibromatosis type 2	22q12.2
p16	Melanoma	Nucleus	Transcription factor	Melanoma	9p21
p53	Colon cancer; many others	Nucleus	Transcription factor	Li-Fraumeni syndrome	17p13.1
RB	Retinoblastoma	Nucleus	Transcription factor	Retinoblastoma	13q14.1-q14.2
VHL	Kidney carcinoma	Membrane?	Transcription elongation factor	von Hippel-Lindau disease	3p26-p25
WT1	Nephroblastoma	Nucleus	Transcription factor	Wilms tumor	11p13

Source: Adapted with permission from J. Marx, *Science* 261 (1993): 1385–1387. Copyright © 1993 American Association for the Advancement of Science.

scription factors called E2F and DP1 (Figure 18.9a). As long as this status is maintained, the cells remain in G_1 or enter the quiescent state (the G_0 phase). If progression through the cell cycle is signaled, phosphorylation of pRB by a cyclin/cyclin-dependent kinase (Cdk) complex occurs (cyclins are proteins involved in control of the cell cycle), and pRB is no longer able to bind to E2F. The released E2F molecules bind to genes with binding sites for this transcription factor, and those genes, whose activities are required for entry into S phase, are turned on. Progression of the cell into S is then ensured. Once a cell has completed mitosis, pRB is dephosphorylated.

In a cell with two mutant *RB* alleles, pRB is nonfunctional and does not bind to E2F/DP1, allowing the E2F to activate genes for transition to S phase (Figure 18.9b). As a result, unprogrammed cell division takes place.

Despite our present knowledge of the cellular activities of pRB, we do not yet know how children with *RB* mutations develop retinoblastoma. It may be through the role of pRB in development rather than its role in the cell cycle specifically. That is, cells programmed for terminal differentiation may depend critically on pRB to establish a state of permanent nonproliferation.

The *p53* Tumor Suppressor Gene.   The tumor suppressor gene *p53* (OMIM 191170) is so named because it encodes a protein of molecular weight 53 kDa called p53. When both alleles are mutated, *p53* may be involved in the development of perhaps 50 percent of all human cancers,

including breast, brain, liver, lung, colorectal, bladder, and blood cancer. This does not mean that *p53* causes 50 percent of human cancers but that mutations in *p53* are among the several genetic changes usually found in those cancers.

Genetics of the p53 Tumor Suppressor Gene.   The human *p53* gene, isolated by positional cloning (see Chapter 14, pp. 292–293), is at chromosome location 17p13.1. Individuals who inherit one mutant copy of *p53* develop Li-Fraumeni syndrome, a rare form of cancer that is an autosomal dominant trait because the cancer develops when the second copy of *p53* becomes mutated. Individuals with this syndrome develop cancers in a number of tissues, including breast and blood.

Cell Biology.   The 393-amino-acid p53 tumor suppressor protein is involved in many important cellular processes, including transcription, cell cycle control, DNA repair, and programmed cell death (*apoptosis*). Wild-type p53 binds to DNA and acts as a transcription factor. p53 binds to several genes, one of which, *WAF1*, is specifically activated by wild-type p53. *WAF1* encodes a 21-kDa protein called p21, which, when its synthesis is activated by p53, causes cells to arrest in G_1. The arrest occurs because p21 binds to cyclin/Cdk (see previous retinoblastoma discussion) and blocks the kinase activity required to activate the genes needed for the cell to make the transition from G_1 to S.

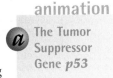

animation

a The Tumor Suppressor Gene *p53*

Figure 18.9

Role of pRB in regulating the passage of cells from G_1 to S. (a) In a normal cell, unphosphorylated pRB is in a complex with the E2F/DP1 transcription factors. When pRB becomes phosphorylated, that binding is blocked, and the transcription factors activate specific genes involved in the G_1-to-S transition. (b) A cell with two mutant *RB* alleles produces a shortened or unstable pRB that does not bind to E2F/DP1, allowing the transcription factors to activate genes for G_1-to-S transition. This leads to unprogrammed cell division.

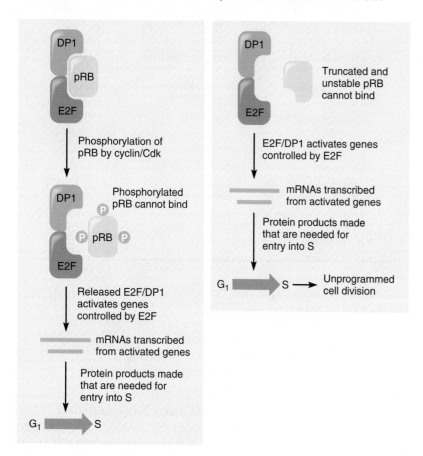

a) **Normal cell**

b) **Cell with two mutant *RB* alleles**

In a normal cell, damage to the cellular DNA, for example by irradiating the cell, is one of the most effective ways of causing p53 to initiate the cascade of events leading to arrest in G_1. DNA damage results in stabilization of p53, and the cascade of events in Figure 18.10 occurs. The G_1 arrest gives the cell time to induce the necessary pathways to repair the DNA damage, after which the cell cycle can resume. If the extent of DNA damage is too great and all lesions cannot be repaired, the cell cycle will not resume, and the damaged cell will die by apoptosis. The induction of apoptosis is an important function of p53.

If both alleles of *p53* are inactivated in the cell, active p53 is not present. Thus, *WAF1* cannot be activated, no p21 is available to block Cdk activity, and the cell is unable to arrest in G_1. Therefore, the cell cycle may proceed to S. Cells with mutated *p53* alleles do not go into growth arrest after DNA damage because of the lack of functional p53 protein, and apoptosis does not occur. The progression of unrepaired cells through the cell cycle can cause the cells to accumulate further genetic damage and thereby increase the probability of cancer.

Finally, experiments have been done in which transgenic mice carried deletions of both *p53* alleles. That these so-called *p53⁻/p53⁻* knockout mice actually developed and were fully viable indicated that the *p53* gene is not essential for the processes of cell growth, cell division, or cell differentiation, at least in the mouse. The knockout mice showed only one major phenotype, that of a very high frequency of cancers from the sixth month (in 75 percent of the mice) to the tenth month (in 100 percent of the mice). These results support the role of p53 in tumor suppression and in maintaining the genetic stability of cells.

Figure 18.10

Cascade of events by which DNA lesions cause an arrest in G₁. In an unknown fashion, the lesions stabilize p53, which activates the *WAF1* gene, giving rise to p21. The p21 protein binds to cyclin/Cdk, blocking the kinase activity needed for cells to progress from G₁ to S.

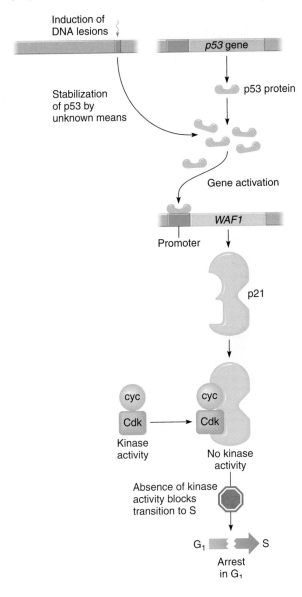

Breast Cancer Tumor Suppressor Genes. In the United States, more than 185,000 new cases of breast cancer are diagnosed each year, representing more than 31 percent of all new cancers in women, and more than 46,000 women die each year from this cancer. In developed countries, there is a 1 in 10 chance that a woman will be diagnosed with breast cancer in her lifetime. The average age of onset is 55. Approximately 5 percent of breast cancers are hereditary. As with hereditary retinoblastoma, this form of the cancer has an earlier age of onset than the sporadic form, and the cancer often is bilateral.

Among the several genes that play a role in familial breast cancer, two genes—*BRCA1* (OMIM 113705) and *BRCA2* (OMIM 600185)—have been hypothesized to be tumor suppressor genes. (Some studies have led to the alternative hypothesis that these two genes are mutator genes; see next section.) It is believed that most hereditary breast cancer in the United States results from mutations in *BRCA1* or *BRCA2*, with most of the mutations occurring in *BRCA1*.

The *breast cancer* susceptibility gene *BRCA1* is at chromosome location 17q21. Mutations of the *BRCA1* gene also lead to susceptibility to ovarian cancer. The *BRCA1* gene encompasses more than 100 kb of DNA; it is transcribed in numerous tissues, including breast and ovary, to produce a 7.8-kb mRNA that is translated to produce a 190-kDa protein with 1,863 amino acids. Various functions have been reported for the protein, including roles in mRNA transcription and cellular responses to DNA damage.

BRCA2 is at chromosome location 13q12-q13. Unlike *BRCA1*, *BRCA2* does not have an associated high risk of ovarian cancer. *BRCA2* encompasses approximately 70 kb of DNA and encodes a 3,418-amino-acid DNA-binding protein that has some similarity to the *BRCA1*-encoded protein. The *BRCA2* protein has been shown to be part of a complex that plays a role in the timely progression of cells through mitosis.

KEYNOTE

Tumor suppressor genes, like proto-oncogenes, are involved in the regulation of cell growth and division. Whereas the normal products of proto-oncogenes have a stimulatory role in those processes, the normal products of tumor suppressor genes have an inhibitory role. Therefore, when both alleles of a tumor suppressor gene are inactivated or lost, the inhibitory activity is lost, and unprogrammed cell proliferation can occur. Inactivation of tumor suppressor genes is involved in the development of a wide variety of human cancers, including breast, colon, and lung cancer.

Mutator Genes

A **mutator gene** is any gene that, when mutant, increases the spontaneous mutation frequencies of other genes. In a cell, the normal (unmutated) forms of mutator genes are involved in such important activities as DNA replication and DNA repair. Mutations of these genes can significantly impair those processes and make the cell error prone, so that it accumulates mutations. For an illustration of how a mutation in a mutator gene can result in cancer, we consider hereditary nonpolyposis colon cancer (HNPCC; OMIM 120435).

HNPCC is an autosomal dominant genetic disease in which there is an early onset of colorectal cancer. Unlike hereditary (or familial) adenomatous polyposis (FAP; see next section), no adenomas (benign tumors or polyps) are seen in HNPCC; hence its name. HNPCC accounts for perhaps 5 to 15 percent of colorectal cancers.

Four human genes named *hMSH2, hMLH1, hPMS1,* and *hPMS2* have been identified, any one of which gives a phenotype of hereditary predisposition to HNPCC when it is mutated. Tumor formation requires only one mutational event to inactivate the remaining normal allele. Thus, because of the high probability of such an event, HNPCC appears dominant in pedigrees. All four genes encode proteins involved in mismatch repair, a process for correcting mismatched base pairs left after DNA replication. (Mismatch repair is described in detail in Chapter 19, pp. 410–411.) Homologues of these genes are known in yeast, *E. coli,* and other organisms. Mutations in these genes make the DNA replication error prone, and mutation rates are significantly higher than in normal cells.

 You are a researcher at a cancer clinic investigating the origins of a rare form of bladder cancer in the iActivity *Tracking Down the Causes of Cancer* on the website.

The Multistep Nature of Cancer

The development of most cancers is a stepwise process involving an accumulation of mutations in a number of genes. It appears that perhaps six or seven independent mutations are needed over several decades of life for cancer to be induced. The multiple mutational events typically involve both the change of proto-oncogenes to oncogenes and the inactivation of tumor suppressor genes, with a resulting breakdown of the multiple cellular mechanisms that regulate growth and differentiation.

As an example, Figure 18.11 illustrates Bert Vogelstein's molecular model of multiple mutations leading to hereditary FAP, a form of colorectal cancer (OMIM 175100). Patients with FAP inherit the loss of a chromosome 5 tumor suppressor gene called *APC* (adenomatous polyposis coli). The same gene can be lost early in carcinogenesis in sporadic tumors. Once both alleles of *APC* are lost in a colon cell, increased cell growth results. If hypomethylation (decreased methylation) of the DNA occurs, a benign tumor called an *adenoma class I* (a small polyp from the colon or rectum epithelium) can develop. Then, if a mutation converts the chromosome 12 *ras* proto-oncogene into an oncogene, the cells can progress to a larger benign tumor known as *adenoma class II* (a larger polyp). Next, if both copies of the chromosome 18 tumor suppressor gene *DCC* (deleted in colon cancer) are lost, the cells progress into *adenoma class III* (large benign polyps). Deletion of both copies of the chromosome 17 tumor suppressor gene *p53* results in the progression to a carcinoma (an epithelial cancer); with other gene losses,

Figure 18.11

A multistep molecular event model for the development of hereditary adenomatous polyposis (FAP), a colorectal cancer.

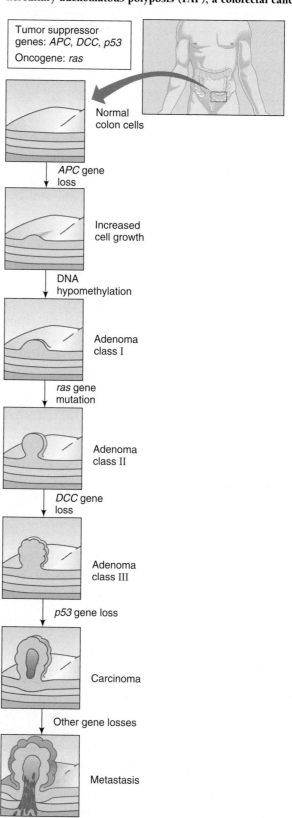

the cancer metastasizes. Note that this is only one path whereby adenomatous polyposis can occur; others are possible. However, in all paths observed, deletions of *APC* and mutations of *ras* usually occur earlier in carcinogenesis than do deletions of *DCC* and *p53*. Progressive changes in function of oncogenes and tumor suppressor genes are also thought to occur for other cancers.

K E Y N O T E

> The development of most cancers involves an accumulation of mutations in a number of genes over a significant period of life. This multistep nature of cancer typically involves mutational events that convert proto-oncogenes to oncogenes and inactivate tumor suppressor genes, thereby breaking down the multiple mechanisms that regulate growth and differentiation.

Chemicals and Radiation as Carcinogens

Several natural and artificial agents increase the frequency with which cells become cancerous. These agents, mostly chemicals and types of radiation, are known as **carcinogens.** Because of the obvious human relevance, there is a vast amount of information about carcinogenesis spanning many areas of biology; only an overview is given here.

Although we have focused much attention in this chapter on viruses as causes of cancer, chemicals are responsible for more human cancers than are viruses. Chemical carcinogenesis was discovered in the eighteenth century by Sir Percival Pott, an English surgeon who correlated the incidence of scrotal skin cancer in some of his patients with occupational exposure to coal soot when they worked as chimneysweeps as children. From the beginning of industrial development in the eighteenth century to the present day, workers in many areas have been exposed to carcinogenic agents and have developed occupationally related cancers. For example, radiologists using X rays and radium (sources of ionizing radiation) and farmers exposed to the sun's ultraviolet (UV) light (nonionizing radiation) have developed skin cancers; asbestos and insulation workers exposed to asbestos have developed bronchial and lung cancers; and workers exposed to vinyl chloride have developed liver cancer.

Chemical Carcinogens

Chemical carcinogens include both natural and synthetic chemicals. Two major classes of chemical carcinogens are recognized. *Direct-acting carcinogens* are chemicals that bind to DNA and act as mutagens. The second class, procarcinogens, must be converted metabolically to become active carcinogens; almost all of these so-called *ultimate*

carcinogens also bind to DNA and act as mutagens. In both cases, the mutations typically are point mutations (single base-pairs in the DNA). (The mutagenicity of ultimate carcinogens can be demonstrated in a number of screening tests, including the Ames test described in Chapter 19, pp. 407–408.) Thus, direct-acting and most ultimate carcinogens bring about transformation of cells and the formation of tumors by binding to and causing changes in DNA. Direct-acting carcinogens include alkylating agents. Examples of procarcinogens are polycyclic aromatic hydrocarbons (multiringed organic compounds found in the smoke from wood, coal, and cigarettes, for example), azo dyes and natural metabolites (such as aflatoxin, produced from fungal contamination of food), and nitrosamines (produced by nitrites in food). Most chemical carcinogens are procarcinogens.

The metabolic conversion of procarcinogens to ultimate carcinogens is carried out by normal cellular enzymes that function in a variety of pathways that involve hydrolysis, oxidation, and reduction, for example. If a procarcinogen interacts with the active site of one of the enzymes, then it can be modified by the enzyme to give rise to the derivative ultimate carcinogen.

Chemical carcinogens are responsible for most cancer deaths in the United States, with the top two causes of cancer—tobacco smoke and diet—being responsible for 50 to 60 percent of cancer-related deaths.

Smoking, mostly of cigarettes, is responsible for 30 percent of cancer deaths, making tobacco smoke (or rather the chemicals in the smoke) the most lethal carcinogen that exists. Tobacco smoke can cause a number of types of cancers, including lung, upper respiratory tract, esophageal, stomach, and liver cancer, and it can increase the risk of cancer in other organs, including breast, bladder, and vulva. The risk of developing cancer as a result of smoking tobacco is influenced by such factors as the amount of tobacco smoked, its tar content, and how long the person has been smoking. The younger a person is when he or she begins smoking, the greater the risk of developing cancer later in life. Moreover, secondhand smoke increases the risk of cancer, making tobacco smoke an environmental concern for everyone. One of the main types of carcinogens found in tobacco smoke is the polycyclic hydrocarbons. Once converted in the cell to their ultimate carcinogen derivatives, they react with negatively charged molecules, such as DNA, and this can result in mutations.

Radiation

We may be able to avoid or minimize environmental exposure to chemical carcinogens, but avoiding exposure to radiation in its various forms is more difficult. We are exposed to radiation from the sun, cellular telephones, radioactive radon gas, electric power lines, and some household appliances, for example. Only about 2 percent of all cancer deaths are caused by radiation, and most of the cancers involved are the highly aggressive melanoma skin cancers that can be induced by exposure to the sun's UV light. Ionizing radiation, such as that emitted by X-ray machines, decay of some radioactive materials, and radon gas, for example, has the potential to be carcinogenic, although the risk to the public generally is very low. Ionizing radiation most commonly causes leukemia and thyroid cancer.

Radiation causes mutations in DNA. The mutagenic effects of ultraviolet light and X rays are discussed in Chapter 19 (pp. 402–403). We discuss ultraviolet light as a carcinogen in more detail in the following paragraphs.

UV light is emitted by the sun, along with visible light and infrared radiation. The UV light that reaches Earth is classified into two types, based on its wavelength: ultraviolet A (UVA, spanning 320–400 nm) and ultraviolet B (UVB, spanning 290–320 nm). The intensity of UVA and UVB reaching an individual on Earth depends on a number of factors, including time of day, altitude, and materials in the atmosphere, such as dust and other particles. Generally, the ambient level of UVA is one to three orders of magnitude higher than that of UVB.

UV light causes several forms of skin cancer, the most dangerous of which are directly related to long-term exposure to UV light radiation. Both UVA and UVB play a role in carcinogenesis. Sunburn is caused mainly by UVB, which also induces skin cancer because the radiation in the wavelength range of UVB is mutagenic (see Chapter 19, pp. 402–403, for a discussion of this). UVA plays a role in skin cancer by increasing the carcinogenic effects of UVB.

The risks of UV light–induced skin cancers can be minimized by reducing one's exposure to the sun (or to UV light in any form) and by applying an effective sunscreen when out in the sun. The effectiveness of a sunscreen is indicated in terms of its sun protection factor (SPF). The SPF value of a sunscreen is the ratio of the energy required to produce a minimal skin reddening or minimal sunburn through the sunscreen compared with the energy required to produce the same reaction in the absence of the sunscreen. Theoretically, an individual who burns after 20 minutes of sun exposure can extend the period of time until a burn begins to 5 hours with an SPF 15 sunscreen if the sunscreen stays in place. However, one must be careful in interpreting what the different values mean. For example, a sunscreen with SPF 15 will block 93 percent of UVB, and a sunscreen with SPF 50 will block 98 percent of UVB. Most sunscreens provide little or no protection against UVA, which can lead to harm if using a sunscreen with a high SPF encourages long exposure to the sun.

Fortunately, many skin cancers are easy to detect and can be removed surgically.

KEYNOTE

Various types of radiation and many chemicals increase the frequency with which cells become cancerous. These agents are known as carcinogens. All carcinogens act by causing changes in the genome of the cell. A few chemical carcinogens act directly on the genome; the majority act indirectly. The latter are metabolically converted by cellular enzymes to ultimate carcinogens that bind to DNA and cause mutations. All carcinogenic forms of radiation act directly.

Summary

To understand the development of cancer (neoplasia), it is necessary to understand how normal cell division is controlled. We know that a normal eukaryotic cell moves through the cell cycle in steps that are tightly controlled by a number of molecular factors. Healthy cells grow and divide only when the balance of stimulatory and inhibitory signals received from outside favors cell proliferation and the signals are appropriately transduced to the cytoplasm and nucleus. A cancerous cell does not respond to the usual signals and reproduces without constraints.

Three classes of genes have been shown to be mutated frequently in cancer: proto-oncogenes (the mutant forms are called oncogenes), tumor suppressor genes, and mutator genes. The products of proto-oncogenes stimulate cell proliferation; the products of wild-type tumor suppressor genes inhibit cell proliferation; and the products of wild-type mutator genes are involved in the replication and repair of DNA. Mutant forms of these three types of genes all have the potential to contribute to the transformation of a cell to a tumorous state.

Some forms of cancer are caused by tumor viruses. Both DNA and RNA tumor viruses are known. DNA tumor viruses transform cells to the cancerous state through the action of one or more genes that are essential parts of the viral genome and that usually stimulate the transition from G_0 or G_1 to S. All RNA tumor viruses are retroviruses—RNA viruses that replicate via a DNA intermediate—but not all retroviruses cause cancer. When a retrovirus infects a cell, the RNA genome is released from the virus particle, and reverse transcriptase makes a cDNA copy of the genome, called the proviral DNA, that integrates into the host cell's genome. Expression of the proviral genes leads to the production of progeny viruses that exit the cell and infect other cells.

Tumor-causing RNA retroviruses contain cancer-inducing genes called oncogenes. Tumor-causing retroviruses have picked up normal cellular genes, called proto-oncogenes, while simultaneously losing some of their genetic information. Proto-oncogenes in normal cells function in various ways to regulate cell proliferation and differentiation. However, in the retrovirus, these genes have been modified or their expression regulated differently so that the oncogene protein product, now synthesized under viral control, is altered. The oncogene products, which include growth factors and protein kinases, are directly responsible for transforming cells to the cancerous state. Cellular proto-oncogenes may also mutate, resulting in a stimulatory effect on cell division. In their mutated state, these proto-oncogenes are called cellular oncogenes. Since only one allele of a proto-oncogene must be mutated to cause changes in cell growth and division, the mutations are dominant mutations.

Tumor suppressor genes, like proto-oncogenes, are involved in the regulation of cell growth and division. In this case, the normal products of tumor suppressor genes have inhibitory roles. Therefore, when both alleles of a tumor suppressor gene are inactivated or lost, the inhibitory activity is lost, and unprogrammed cell proliferation can occur. In familial (hereditary) cancers, one mutation is inherited and the other mutation occurs later in somatic cells, leading to loss of heterozygosity. In other words, the inheritance of one gene mutation predisposes a person to cancer. In sporadic (nonhereditary) cancers, both mutations occur in the somatic cells.

The development of most cancers involves an accumulation of mutations in a number of genes over a significant period of life. This multistep nature of cancer typically involves mutational events that activate oncogenes and inactivate tumor suppressor genes, thereby breaking down the multiple mechanisms that regulate growth and differentiation. Mutations of mutator genes can also contribute to the development of cancer by adversely affecting the normal maintenance of the genome's integrity through accurate DNA replication and efficient DNA repair, with the result that the cell accumulates mutations.

Various types of radiation and many chemicals, collectively known as carcinogens, increase the frequency with which cells become cancerous. Practically all carcinogens act by causing changes in the genome of the cell. A few chemical carcinogens act directly on the genome; the majority act indirectly by being converted by cellular enzymes to active derivatives called ultimate carcinogens. By understanding what carcinogens exist in the environment, we can position ourselves to minimize their effects on us and thereby perhaps decrease our risk of cancer.

Analytical Approaches for Solving Genetics Problems

Q18.1 An investigator has found a retrovirus capable of infecting human nerve cells. This is a complete virus, able to reproduce itself in the cell, and it contains no oncogenes. People who are infected suffer a debilitating encephalitis. The investigator has shown that when he infects nerve cells in culture with the complete virus, the nerve cells are killed as the virus reproduces. But if he infects cultured nerve cells with a virus in which he has created deletions in the *env* or *gag* genes, no cell death occurs. The investigator is interested in finding ways to bring about nerve cell growth or regeneration in people who have suffered nerve damage. For example, in a patient with a severed spinal cord, nerve regeneration might relieve paralysis. The investigator has cloned the human nerve growth factor gene and wants to insert it into the genome of the retrovirus from which he has deleted parts of the *env* and *gag* genes. He would then use the engineered retrovirus to infect cultured nerve cells. Adult nerve cells do not normally produce large amounts of nerve growth factor. If he is successful in inducing growth in them without causing any cell death, he would like to move on to clinical trials on injured patients. When the investigator applied for grant support to do this work, his application was denied on the grounds that there were inadequate safeguards in the plan. Why might this work be dangerous? What comparisons can you draw between the virus the investigator wants to create and, for example, avian myeloblastosis virus (AMV; see Figure 18.7a)?

A18.1 In engineering the retrovirus in the way he plans, the investigator probably would be creating a new cancer virus in which the cloned nerve growth factor gene would be the oncogene. Of course, it is an advantage that the engineered virus would not be able to reproduce itself, but we know that many "wild" cancer viruses are also defective and reproduce with the help of other viruses. If the engineered virus were to infect cells carrying other viruses (for example, wild-type versions of itself) that could supply the *env* and *gag* functions, the new virus could be reproduced and spread. Presumably, infection of normal nerve cells in vivo by the engineered retrovirus would sometimes result in abnormally high levels of nerve growth factor and thus perhaps in the production of nervous system cancers.

In avian myeloblastosis virus, the *pol* and *env* genes are partially deleted. Thus, like our investigator's virus, AMV needs a helper virus to reproduce. In AMV, the *myb* oncogene has been inserted; it encodes a nuclear protein presumably involved in control of gene expression. In our new virus, the oncogene would be the cloned nerve growth factor gene.[1]

Questions and Problems

*18.1 What is the difference between a hereditary cancer and a sporadic cancer?

---

[1] Now that the safety of retroviral vectors has been established, they are being used successfully to deliver genes to treat certain human diseases.

18.2 Distinguish between a transducing retrovirus and a nontransducing retrovirus.

*18.3 In what ways is the mechanism of cell transformation by transducing retroviruses fundamentally different from transformation by DNA tumor viruses? Even though the mechanisms are different, how are both able to cause neoplastic growth?

*18.4 Although there has been a substantial increase in our understanding of the genetic basis for cancer, the vast majority of cases of many types of cancer are not hereditary.
a. How might studying a hereditary form of a cancer provide insight into a similar, more frequent sporadic form?
b. The incidence of cancer in several members of an extended family might reasonably raise concern as to whether there is a genetic predisposition for cancer in the family. What does the term *genetic predisposition* mean? What might be the basis of a genetic predisposition to a cancer that appears as a dominant trait? What issues must be addressed before concluding that a genetic predisposition for a specific type of cancer exists in a particular family?

18.5 Cellular proto-oncogenes and viral oncogenes are related in sequence, but they are not identical. What is the fundamental difference between the two?

*18.6 Material that has been biopsied from tumors is useful for discerning both the type of tumor and the stage to which a tumor has progressed. It has been known for a long time that biopsied tissues with more differentiated cellular phenotypes are associated with less advanced tumors. Explain this finding in terms of the multistep nature of cancer.

18.7 An autopsy of a cat that died from feline sarcoma revealed neoplastic cells in the muscle and bone marrow but not in the brain, liver, or kidney. To gather evidence for the hypothesis that the virus FeSV (see Figure 18.7c) contributed to the cancer, Southern blot analysis (see Chapter 13, pp. 277–278) was performed on DNA isolated from these tissues and on a cDNA clone of the FeSV viral genome. The DNA was digested with the enzyme *Hind*III, separated by size on an agarose gel, and transferred to a membrane. The resulting Southern blot was hybridized with a ^{32}P-labeled probe made from a 1.0-kb *Hind*III fragment of the feline *fes* proto-oncogene cDNA. The autoradiogram revealed a 3.4-kb band in each lane, with an additional 1.2-kb band in the lanes with muscle and bone marrow DNA. Only a 1.2-kb band was seen in the lane loaded with *Hind*III-cut FeSV cDNA. Explain these results, including the size of the bands seen. Do these results support the hypothesis?

*18.8 The sequences of proto-oncogenes are highly conserved among a large number of animal species. Based on this fact, what hypothesis can you make about the functions of proto-oncogenes?

18.9 Explain why HIV-1, the causative agent of AIDS, is considered a nononcogenic retrovirus even though numerous types of cancers are frequently seen in patients with AIDS.

18.10 List two ways in which cancer can be induced by a retrovirus.

*18.11 Proto-oncogenes produce a diverse set of gene products.
a. What types of gene products are made by proto-oncogenes? Do these gene products share any features?
b. Which of the following mutations might result in an oncogene?
   i. a deletion of the entire coding region of a proto-oncogene
   ii. a deletion of a silencer that lies 5′ to the coding region
   iii. a deletion of an enhancer that lies 3′ to the coding region
   iv. a deletion of a 3′ splice site acceptor region
   v. the introduction of a premature stop codon
   vi. a point mutation (single base-pair change in the DNA)
   vii. a translocation that places the coding region near a constitutively transcribed gene
   viii. a translocation that places the gene near constitutive heterochromatin

*18.12 You have a culture of normal cells and a culture of cells dividing uncontrollably (isolated from a tumor). Experimentally, how might you determine whether uncontrolled growth was the result of an oncogene or a mutated pair of tumor suppressor alleles?

18.13 What are the four main ways in which a proto-oncogene can be changed into an oncogene?

*18.14 After a retrovirus that does not carry an oncogene infects a particular cell, northern blots indicate that the amount of mRNAs transcribed from a particular proto-oncogene became elevated approximately 13-fold compared with uninfected control cells. Propose a hypothesis to explain this result.

18.15 Explain how progression through the cell cycle is regulated by the phosphorylation of the retinoblastoma protein pRB. What phenotypes might you expect in cells where
a. pRB was constitutively phosphorylated?
b. pRB was never phosphorylated?

c. a severely truncated pRB protein was produced that could not be phosphorylated?

d. a normal pRB protein was produced at higher than normal levels?

e. a normal pRB protein was produced at lower than normal levels?

18.16 Mutations in the *p53* gene appear to be a major factor in the development of human cancer.

a. Explain what the normal cellular functions of the *p53* gene product are and how alterations in these functions can lead to cancer.

b. Suppose cells in a cancerous growth are shown to have a genetic alteration that results in diminished *p53* gene function. Why can we not immediately conclude that the mutation *caused* the cancer? How would the effect of the mutation be viewed in light of the current, multistep model of cancer?

*18.17 What is apoptosis? Why is the cell death associated with apoptosis desirable, and how is it regulated?

18.18 What mechanisms ensure that cells with heavily damaged DNA are unable to replicate?

18.19 How do radiation and chemical carcinogens induce cancers?

19

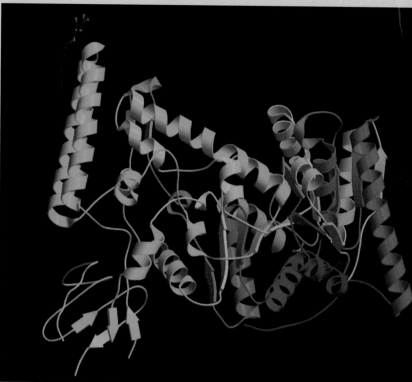

DNA Mutation and Repair

PRINCIPAL POINTS

Changes in heritable traits result from random mutation rather than from adaptation to environmental influences.

Mutation is the process by which the sequence of base pairs in a DNA molecule is altered. The alteration can be as simple as a single base pair substitution, insertion, or deletion or as complex as rearrangement, duplication, or deletion of whole sections of a chromosome. Mutations may occur spontaneously, such as by the effects of natural radiation or errors in replication, or they may be induced experimentally by the application of mutagens.

Mutations at the level of the chromosome are called chromosomal mutations (see Chapter 21). Mutations in the sequences of genes at the level of the base pair are called gene mutations.

The consequences to an organism of a gene mutation depend on a number of factors, especially the extent to which the amino acid–coding information for a protein is changed. For example, missense mutations cause the substitution of one amino acid for another, and nonsense mutations cause premature termination of polypeptide synthesis.

The effects of a gene mutation can be reversed either by reversion of the base pair sequence to its original state or by a mutation at a site distinct from that of the original mutation. The latter is called a suppressor mutation.

High-energy radiation may cause genetic damage by producing chemicals that interact with DNA or by causing unusual bonds between DNA bases. Mutations result if the genetic damage is not repaired. Ionizing radiation may also break chromosomes.

Gene mutations may also be caused by exposure to a variety of chemicals called chemical mutagens. A number of chemicals in the environment are chemical mutagens, and they can cause genetic diseases in humans and other organisms.

In prokaryotes and eukaryotes, a number of enzymes repair different kinds of DNA damage. Not all DNA damage is repaired; therefore, mutations do appear, but at low frequencies. At high dosages of mutagens, repair systems cannot correct all the damage, and cancer (in the case of eukaryotic, multicellular organisms) or cell death can result.

Geneticists have made great progress in understanding how cellular processes take place by studying mutants that have defects in those processes. A number of screening procedures have been developed to help find mutants of interest after mutagenizing cells or organisms and to detect mutations generated by specific molecular targeting methods.

i EVERY DAY WE ARE EXPOSED TO THOUSANDS OF chemicals in our air, water, and soil. Many of these chemicals are harmless or even beneficial. Others can have damaging effects, including illness, birth defects, cancer, and death. And some are even more insidious, creating changes, or mutations, in our DNA. What types of mutations can occur in our DNA? How can we identify chemicals that cause these changes? And what effects do DNA mutations have on our health? In the iActivity for this chapter, you will investigate the possible health hazards associated with contaminated groundwater.

DNA can be changed in a number of ways, such as by spontaneous changes, including errors in the replication process or by the action of particular chemicals or radiation. There are two broad types of changes to the genetic material: **chromosomal mutations**, changes involving whole chromosomes or sections of them (the topic of Chapter 21), and **point mutations**, changes in one or a few base pairs.

A point mutation will not change the phenotype of the organism unless it occurs within a gene or in the sequences regulating the gene. Thus, the point mutations that have been of particular interest to geneticists are **gene mutations**, those that affect the function of genes. A gene mutation can alter the phenotype by changing the function of a protein, as illustrated in Figure 19.1.

In this chapter, you will learn about some of the mechanisms that cause point mutations, some of the repair systems that can repair genetic damage, and some of the methods used to detect genetic mutants. As you learn about the specifics of point mutations, you should be aware that mutations are a major source of genetic variation in a species and therefore are important elements of the evolutionary process.

Adaptation Versus Mutation

Although we currently know that heritable traits result from mutations, this was not always the case. In the early part of the twentieth century, there were two opposing schools of thought. Some geneticists believed that variation among organisms resulted from random mutations that sometimes happened to be adaptive. Others believed that variations resulted from *adaptation*—that is, that the environment induced an adaptive heritable change. The adaptation theory was based on Lamarckism, which is the doctrine of the inheritance of acquired characteristics. Some observations with bacteria fueled the debate. Wild-type *E. coli*, for example, is sensitive to the virulent bacteriophage T1. If a culture of wild-type *E. coli* started from a single cell is plated in the presence of an excess of

Figure 19.1

Concept of a mutation in the protein-coding region of a gene. (Note that not all mutations lead to altered proteins, and not all mutations are in protein-coding regions.)

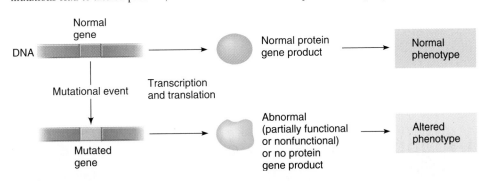

phage T1, most of the bacteria are killed. However, a very few survive and produce colonies because they are resistant to infection by T1. The resistance trait is heritable and results from a change in the bacterial cell surface that prevents the phages from absorbing to it. Supporters of the adaptation theory argued that the resistance trait arose as a result of the presence of the T1 phage in the environment. In the opposite camp, supporters of the mutation theory argued that mutations occur randomly, so that at any time in a large enough population of cells, some cells have undergone a mutation that makes them resistant to T1 (in this example), even though they have never been exposed to T1. Thus, when T1 is subsequently added, the T1-resistant bacteria are selected for.

The acquisition of resistance to T1 was used by Salvador Luria and Max Delbrück in 1943 to demonstrate that the mutation model is correct and the adaptation model incorrect. The test they designed is known as the *fluctuation test* and is based on the idea that the different models would give rise to observably different proportions of resistant bacteria within a sample of cultures.

For example, consider a dividing population of wild-type *E. coli* that starts with a single cell (Figure 19.2). Assume that phage T1 is added at generation 4, when there are 16 cells. (This number is for illustration.) If the adaptation theory is correct, a certain proportion of the generation 4 cells will be induced *at that time* to become resistant to T1 (Figure 19.2a). Most importantly, *that proportion would be the same for all identical cultures because adaptation would not commence until T1 was added.* However, if the mutation theory is correct, then the number of generation 4 cells that are resistant to T1 depends on when in the culturing process the random mutational event occurs that confers resistance to T1. If the mutational event occurs in generation 3 in our example, then 2 of the 16 cells in generation 4 will be T1-resistant (Figure 19.2b, left). However, if the mutational event occurs instead at generation 1, then 8 of the 16 generation 4 cells will be T1-resistant (Figure 19.2b, right). The key point is that if the mutation theory is correct, there should be a *fluctuation in the number of T1-resistant cells in generation 4 because the mutation to T1 resistance occurs randomly in the population and does not require the presence of T1.*

Luria and Delbrück observed a large range in the number of resistant colonies among identical cultures. This high degree of fluctuation in the number of resistant bacteria was taken as proof that resistance resulted from random mutation rather than adaptation.

Figure 19.2

Representation of a dividing population of T1 phage–sensitive wild-type *E. coli*. At generation 4, T1 phage is added. **(a)** If the adaptation theory is correct, cells mutate only when T1 phage is added, so the proportion of resistant cells in duplicate cultures is the same. **(b)** If the mutation theory is correct, cells mutate independently of when T1 phage is added, so the proportion of resistant cells in duplicate cultures is different. *Left:* If one cell mutates to resistance to T1 phage infection at generation 3, then 2 of the 16 cells at generation 4 are resistant to T1. *Right:* If one cell mutates to resistance to T1 phage infection at generation 1, then 8 of the 16 cells at generation 4 are resistant to T1.

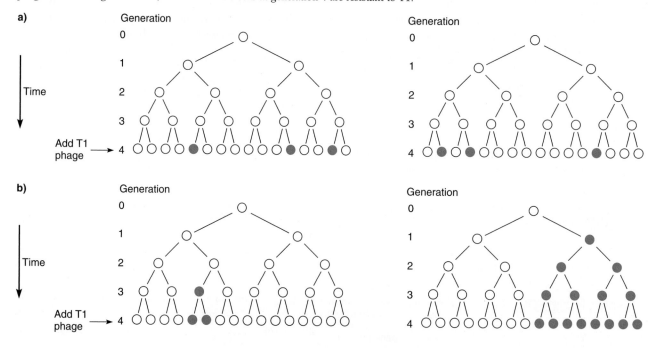

Heritable adaptive traits result from random mutation rather than by adaptation as a result of induction by environmental influences.

Mutations Defined

Mutation is the process by which the sequence of base pairs in a DNA molecule is altered. A **mutation,** then, is a DNA base pair change or chromosome change.

A cell with a mutation is a mutant cell. If a mutant cell gives rise only to somatic cells (in multicellular organisms), the mutant characteristic affects only the individual in which the mutation occurs and is not passed on to the succeeding generation. This type of mutation is called a **somatic mutation.** In contrast, mutations in the germ line of sexually reproducing organisms may be transmitted by the gametes to the next generation, producing an individual with the mutation in both its somatic and germ-line cells. Such mutations are called **germ-line mutations.**

Two different terms are used to give a quantitative measure of the occurrence of mutations. **Mutation rate** is the probability of a particular kind of mutation as a function of time, such as number per nucleotide pair per generation or number per gene per generation. **Mutation frequency** is the number of occurrences of a particular kind of mutation expressed as the proportion of cells or individuals in a population, such as number per 100,000 organisms or number per 1 million gametes.

Types of Point Mutations

Point mutations—those that change only one or a few base pairs—can be divided into two general categories: base pair substitutions and base pair insertions or deletions. A **base pair substitution mutation** involves a change in the DNA such that one base pair is replaced by another base pair. There are two general types of base pair substitution mutations. A **transition mutation** (Figure 19.3a) is a mutation from one purine-pyrimidine base pair to the other purine-pyrimidine base pair. The four types of transition mutations are AT to GC, GC to AT, TA to CG, and CG to TA. A **transversion mutation** (Figure 19.3b) is a mutation from a purine-pyrimidine base pair to a pyrimidine-purine base pair. The eight types of transversion mutations are AT to TA, TA to AT, GC to CG, CG to GC, AT to CG, CG to AT, GC to TA, and CG to AT.

Base pair substitutions in protein-coding genes can also be defined according to their effects on amino acid sequences in proteins. Depending on how a base pair substitution is translated via the genetic code, the mutations can result in no change to the protein, an insignificant change, or a noticeable change.

A **missense mutation** (Figure 19.3c) is a gene mutation in which a base pair change in the DNA causes a change in an mRNA codon so that a different amino acid is inserted into the polypeptide. A phenotypic change may or may not result, depending on the amino acid change involved. In Figure 19.3c, an AT-to-GC transition mutation changes the DNA from $\begin{smallmatrix}5'\text{-AAA-}3'\\3'\text{-TTT-}5'\end{smallmatrix}$ to $\begin{smallmatrix}5'\text{-GAA-}3'\\3'\text{-CTT-}5'\end{smallmatrix}$, altering the mRNA codon from 5'-AAA-3' (lysine) to 5'-GAA-3' (glutamic acid).

A **nonsense mutation** (Figure 19.3d) is a gene mutation in which a base pair change in the DNA results in a change of an mRNA codon from one for an amino acid to one for a stop (nonsense) codon (UAG, UAA, or UGA). For example, in Figure 19.3d, an AT-to-TA transversion mutation changes the DNA from $\begin{smallmatrix}5'\text{-AAA-}3'\\3'\text{-TTT-}5'\end{smallmatrix}$ to $\begin{smallmatrix}5'\text{-TAA-}3'\\3'\text{-ATT-}5'\end{smallmatrix}$, and this changes the mRNA codon from 5'-AAA-3' (lysine) to 5'-UAA-3', which is a nonsense codon. A nonsense mutation causes premature chain termination, so instead of complete polypeptides, polypeptide fragments (often nonfunctional) are released from the ribosomes (Figure 19.4).

animation

ⓐ Nonsense Mutations and Nonsense Suppressor Mutations

A **neutral mutation** (Figure 19.3e) is a base pair change in a gene that changes a codon in the mRNA such that the resulting amino acid substitution produces no detectable change in the *function* of the protein translated from that message. A neutral mutation is a subset of missense mutations in which the new codon codes for a different amino acid that is chemically equivalent to the original and therefore does not affect the protein's function. In Figure 19.3e, an AT-to-GC transition mutation changes the codon from 5'-AAA-3' to 5'-AGA-3', which substitutes the basic amino acid arginine for the basic amino acid lysine. Since arginine and lysine have similar properties, the protein's function may not be altered significantly.

A **silent mutation** (Figure 19.3f) is also a subset of missense mutations that occurs when a base pair change in a gene alters a codon in the mRNA such that the *same* amino acid is inserted in the protein. The protein in this case obviously has wild-type function. For example, in Figure 19.3f, a silent mutation results from an AT-to-GC transition mutation that changes the codon from 5'-AAA-3' to 5'-AAG-3', both of which specify lysine.

If one or more base pairs are added to or deleted from a protein-coding gene, the reading frame of an mRNA can change downstream of the mutation. An addition or deletion of one base pair, for example, shifts the mRNA's downstream reading frame by one base, so that incorrect amino acids are added to the polypeptide chain after the mutation site. This type of mutation is called a **frameshift mutation** (Figure 19.3g). A frameshift mutation usually

Figure 19.3

Types of base pair substitution mutations. Transcription of the segment shown produces an mRNA with the sequence 5′ … UCUCAAAAAUUUACG … 3′, which encodes … -Ser-Gln-Lys-Phe-Thr- ….

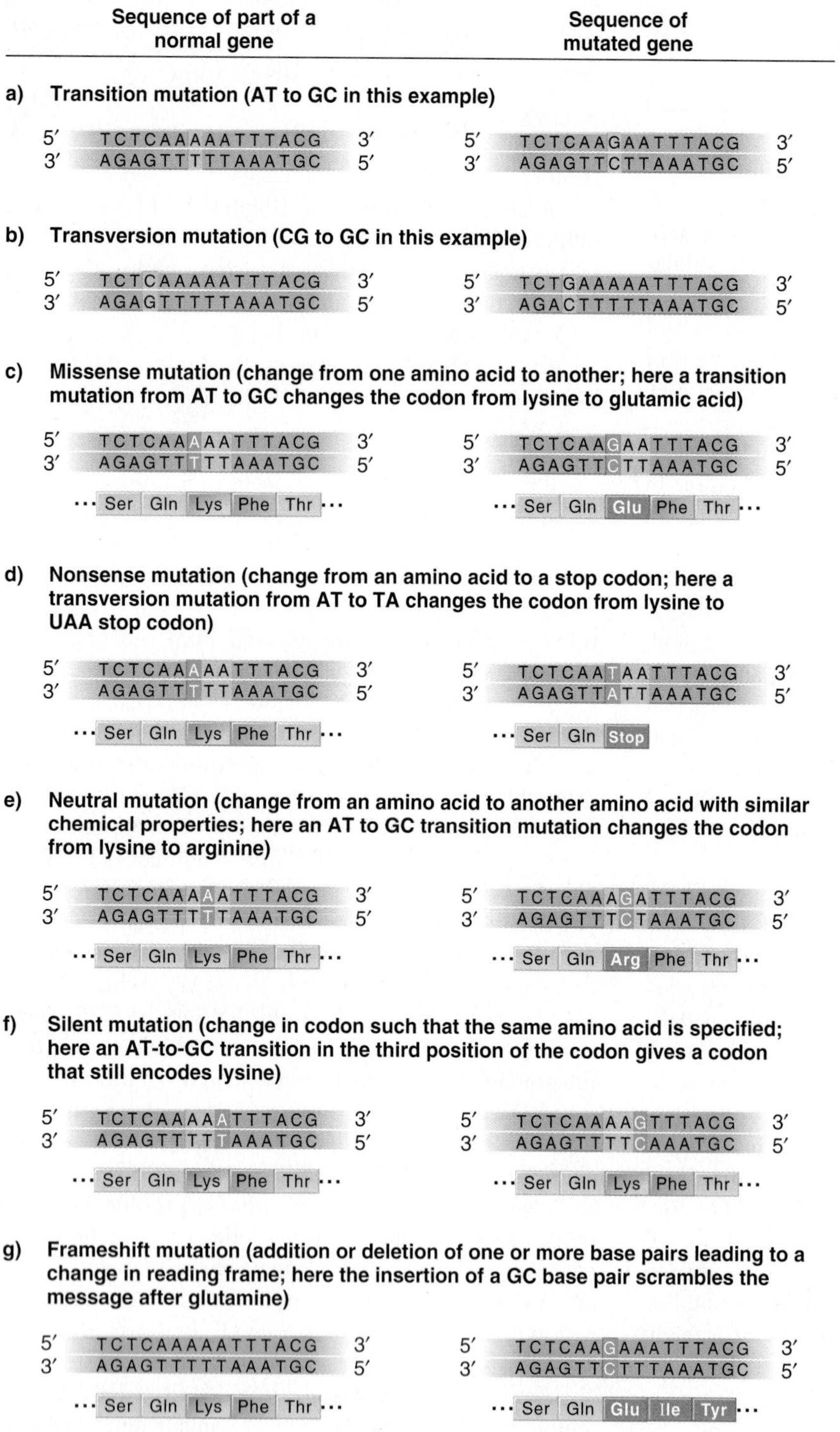

Sequence of part of a normal gene	Sequence of mutated gene

a) Transition mutation (AT to GC in this example)

5′ TCTCAAAAATTTACG 3′   5′ TCTCAAGAATTTACG 3′
3′ AGAGTTTTTAAATGC 5′   3′ AGAGTTCTTAAATGC 5′

b) Transversion mutation (CG to GC in this example)

5′ TCTCAAAAATTTACG 3′   5′ TCTGAAAAATTTACG 3′
3′ AGAGTTTTTAAATGC 5′   3′ AGACTTTTTAAATGC 5′

c) Missense mutation (change from one amino acid to another; here a transition mutation from AT to GC changes the codon from lysine to glutamic acid)

5′ TCTCAAAAATTTACG 3′   5′ TCTCAAGAATTTACG 3′
3′ AGAGTTTTTAAATGC 5′   3′ AGAGTTCTTAAATGC 5′

··· Ser Gln Lys Phe Thr ···   ··· Ser Gln Glu Phe Thr ···

d) Nonsense mutation (change from an amino acid to a stop codon; here a transversion mutation from AT to TA changes the codon from lysine to UAA stop codon)

5′ TCTCAAAAATTTACG 3′   5′ TCTCAATAATTTACG 3′
3′ AGAGTTTTTAAATGC 5′   3′ AGAGTTATTAAATGC 5′

··· Ser Gln Lys Phe Thr ···   ··· Ser Gln Stop

e) Neutral mutation (change from an amino acid to another amino acid with similar chemical properties; here an AT to GC transition mutation changes the codon from lysine to arginine)

5′ TCTCAAAAATTTACG 3′   5′ TCTCAAAGATTTACG 3′
3′ AGAGTTTTTAAATGC 5′   3′ AGAGTTTCTAAATGC 5′

··· Ser Gln Lys Phe Thr ···   ··· Ser Gln Arg Phe Thr ···

f) Silent mutation (change in codon such that the same amino acid is specified; here an AT-to-GC transition in the third position of the codon gives a codon that still encodes lysine)

5′ TCTCAAAAATTTACG 3′   5′ TCTCAAAAGTTTACG 3′
3′ AGAGTTTTTAAATGC 5′   3′ AGAGTTTTCAAATGC 5′

··· Ser Gln Lys Phe Thr ···   ··· Ser Gln Lys Phe Thr ···

g) Frameshift mutation (addition or deletion of one or more base pairs leading to a change in reading frame; here the insertion of a GC base pair scrambles the message after glutamine)

5′ TCTCAAAAATTTACG 3′   5′ TCTCAAGAAATTTACG 3′
3′ AGAGTTTTTAAATGC 5′   3′ AGAGTTCTTTAAATGC 5′

··· Ser Gln Lys Phe Thr ···   ··· Ser Gln Glu Ile Tyr ···

results in a nonfunctional protein. Often, frameshift mutations generate new stop codons, resulting in a shortened protein, or they result in read-through of the normal stop codon, resulting in longer-than-normal proteins. In Figure 19.3g, an insertion of a GC base pair scrambles the message after the codon specifying glutamine. Since each codon consists of three bases, a frameshift mutation is produced by the insertion or deletion of any number of base pairs in the DNA that is not divisible by three.

Figure 19.8

Deamination of cytosine to uracil.

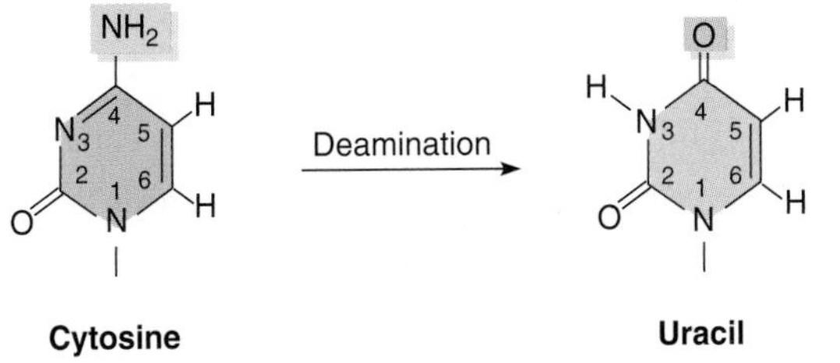

Chemical Mutagens. Chemical mutagens include both naturally occurring chemicals and synthetic substances. These mutagens can be grouped into different classes based on their mechanism of action. Here we discuss base analogues, base-modifying agents, and intercalating agents and explain how they induce mutations. Mutations induced by base analogues and intercalating agents depend on replication, whereas base-modifying agents can induce mutations at any stage of the cell cycle.

Base Analogues. **Base analogues** are bases that are very similar to the bases normally found in DNA. Base analogues exist in alternate chemical states: a normal state (the state in which the chemical is typically found) and a rare state. These alternate chemical states are called **tautomers.** In each of the two states, the base analogue pairs with a different base in DNA. Because base analogues are so similar to the normal nitrogen bases, they are occasionally incorporated into DNA in place of the normal bases. When this happens, the analogue can cause a mutation by pairing with an incorrect base during replication.

5-Bromouracil (5BU) is a base analogue mutagen. 5BU has a bromine residue instead of the methyl group of thymine. In the normal state, 5BU resembles thymine and pairs only with adenine in DNA (Figure 19.10a). In its rare state, it pairs only with guanine (Figure 19.10b). 5BU induces mutations by switching between its two chemical states once the base analogue has been incorporated into the DNA (Figure 19.10c).

If 5BU is incorporated in its normal state, it pairs with adenine. If it changes into its rare state during replication, it pairs with guanine instead. In the next round of replication, the 5BU-G base pair is resolved into a CG base pair instead of the TA base pair. By this process, a transition mutation is produced, from TA to CG. 5BU can also induce a mutation from CG to TA if it is first incorporated into DNA in its rare state and then switches to the normal state during replication (see Figure 19.10c). Thus, 5BU-induced mutations can be reverted by a second treatment of 5BU.

ical (light-induced chemical) changes in the DNA. This mutagenic activity is attenuated to some extent because UV radiation has lower-energy wavelengths than X rays and therefore has very limited penetrating power. However, at high enough doses, UV radiation can kill cells.

One of the effects of UV radiation on DNA is the formation of abnormal chemical bonds between adjacent pyrimidine molecules in the same strand or between pyrimidines on opposite strands of the double helix. This bonding is induced mostly between adjacent thymines, forming what are called *thymine dimers* (Figure 19.9), usually designated T^T. (C^C, C^T, and T^C pairs are also produced by UV radiation but in much lower amounts.) This unusual pairing produces a bulge in the DNA strand and disrupts the normal pairing of Ts with corresponding As on the opposite strand. Many thymine dimers are repaired (see pp. 408–410). All kinds of pyrimidine dimers have the potential to cause problems during DNA replication, and if enough of them remain unrepaired in the cell, cell death may result.

K E Y N O T E

Radiation may cause genetic damage by producing chemicals that affect the DNA (as in the case of X rays) or by causing the formation of unusual bonds between DNA bases, such as thymine dimers (as in the case of ultraviolet light). If radiation-induced genetic damage is not repaired, mutations or cell death may result. Radiation may also break chromosomes.

Figure 19.9

Production of thymine dimers by ultraviolet light irradiation. The two components of the dimer are covalently linked in such a way that the DNA double helix is distorted at that position.

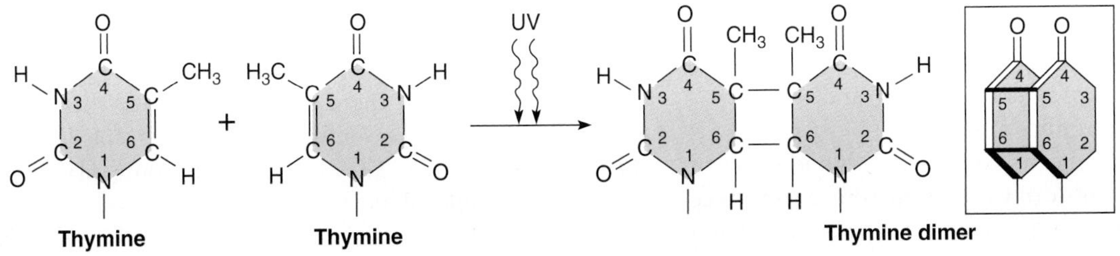

Figure 19.10

Mutagenic effects of the base analogue 5-bromouracil (5BU). (a) In its normal state, 5BU pairs with adenine. **(b)** In its rare state, 5BU (indicated by white letters on magenta) pairs with guanine. **(c)** The two possible mutation mechanisms. 5BU induces transition mutations when it incorporates into DNA in one state, then shifts to its alternate state during the next round of DNA replication.

a) Base pairing of 5-bromouracil in its normal state

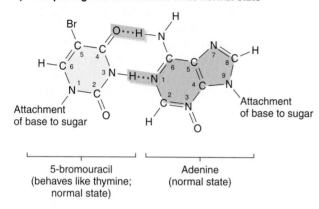

5-bromouracil (behaves like thymine; normal state)   Adenine (normal state)

b) Base pairing of 5-bromouracil in its rare state

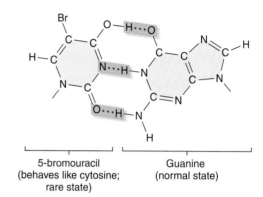

5-bromouracil (behaves like cytosine; rare state)   Guanine (normal state)

c) Mutagenic action of 5BU

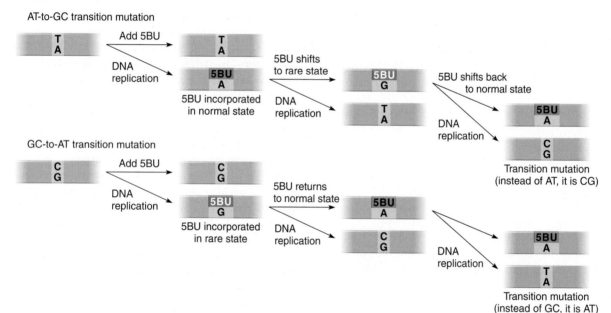

Not all base analogues are mutagens. For example, one of the approved drugs given to patients with AIDS, AZT (azidothymidine), is an analogue of thymidine, but it is not a mutagen because it does not cause base pair changes.

Base-Modifying Agents. A number of chemicals act as mutagens by modifying the chemical structure and properties of the bases. Alkylating agents, for example, introduce alkyl groups (such as $-CH_3$ and $-CH_2CH_3$) onto the bases at a number of locations. Most mutations caused by alkylating agents result from the addition of an alkyl group to the 6 oxygen of guanine to produce O^6-alkylguanine. After treatment with methylmethane sulfonate (MMS), some guanines are methylated to produce O^6-methylguanine. The methylated guanine pairs with thymine rather than cytosine, giving GC-to-AT transitions (Figure 19.11).

Intercalating Agents. Intercalating mutagens—such as proflavin, acridine, and ethidium bromide (commonly used to stain DNA in gel electrophoresis experiments)—insert themselves (*intercalate*) between adjacent bases in one or both strands of the DNA double helix (Figure 19.12). They are generally thin, platelike hydrophobic molecules.

i WHILE STUDYING INHERITANCE IN CORN PLANTS IN the 1940s, American geneticist Barbara McClintock made a startling discovery. She found that certain genetic elements could move from one location to another within a single chromosome or even between chromosomes. McClintock found that these "jumping genes," or transposable elements, can create gene mutations, affect gene expression, and produce various types of chromosome mutations.

In this chapter, you will learn about the structure and function of transposable elements in both prokaryotes and eukaryotes. Then, in the iActivity, you will have the opportunity to further explore how a transposable element in *E. coli* moves from one location to another.

Certain genetic elements in the chromosomes of prokaryotes and eukaryotes can move from one location to another in the genome. These mobile genetic elements are known as **transposable elements,** a term that reflects the *transposition* (change in position) events associated with them. Their discovery was a great surprise that altered our classic picture of genes and genomes. In this chapter, we learn about the nature of transposable elements and about the genetic changes they cause.

General Features of Transposable Elements

Transposable elements are normal and ubiquitous components of the genomes of prokaryotes and eukaryotes. Transposable elements fall into two general classes with respect to how they move. One class encodes proteins that move the DNA element directly to a new position or replicate the DNA to produce a new element that integrates elsewhere in the genome. Transposable elements of this class are found in both prokaryotes and eukaryotes. Members of the other class are related to retroviruses (see Chapter 18) in that they encode a reverse transcriptase for making DNA copies of their RNA transcripts, which subsequently integrate at new sites in the genome. Transposable elements of this class are found only in eukaryotes.

In prokaryotes, transposable elements can move to new positions on the same chromosome (because there is only one chromosome) or onto plasmids or phage chromosomes; in eukaryotes, transposable elements may move to new positions within the same chromosome or to a different chromosome. In both prokaryotes and eukaryotes, transposable elements insert into new chromosome locations with which they have no sequence homology; therefore, transposition is a different process from homologous recombination and is called *nonhomologous recombination.* Transposable elements are important because of the genetic changes they cause, and that is one reason they have been studied intensively. For example, they can produce mutations by inserting into genes, they can increase or decrease gene expression by inserting into gene regulatory sequences (such as by disrupting promoter function

or stimulating a gene's expression through the activity of promoters on the element), and they can produce various kinds of chromosomal mutations because of the mechanics of transposition. In fact, transposable elements have made important contributions to the evolution of the genomes of both prokaryotes and eukaryotes through the chromosome rearrangements they have caused.

Transposable Elements in Prokaryotes

We discuss two examples of transposable elements in prokaryotes: insertion sequence (IS) elements and transposons (Tn).

Insertion Sequences

An **insertion sequence (IS),** or **IS element,** is the simplest transposable element found in prokaryotes. An IS element contains only genes required to mobilize the element and insert the element into a chromosome at a new location. IS elements are normal constituents of bacterial chromosomes and plasmids.

animation

a Insertion Sequences in Prokaryotes

Properties of IS Elements. *E. coli* contains a number of IS elements, including IS*1*, IS*2*, and IS*10R*, each present in 0 to 30 copies per genome and each with a characteristic length and unique nucleotide sequence. IS*1* (Figure 20.1), for example, is 768 bp long and is present in 4 to 19 copies on the *E. coli* chromosome. Among prokaryotes as a whole, the IS elements range in size from 768 bp to more than 5,000 bp and are normal cell constituents; that is, they are found in most cells.

IS elements end with perfect or nearly perfect terminal inverted repeats (IRs) of 9 to 41 bp. This means that essentially the same sequence is found at each end of an IS but in opposite orientations. The inverted repeats of IS*1*, for example, consist of 23 bp of not quite identical sequences (see Figure 20.1).

Figure 20.1

The insertion sequence (IS) transposable element IS*1*. The 768-bp IS element has inverted repeat (IR) sequences at the ends. Shown below the element are the sequences for the 23-bp terminal inverted repeats.

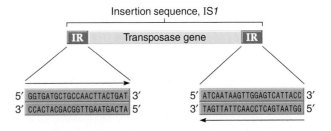

Insertion sequence, IS*1*

| IR | Transposase gene | IR |

5′ GGTGATGCTGCCAACTTACTGAT 3′
3′ CCACTACGACGGTTGAATGACTA 5′

5′ ATCAATAAGTTGGAGTCATTACC 3′
3′ TAGTTATTCAACCTCAGTAATGG 5′

IS Transposition. When an IS element transposes (which is a rare event), a copy of the IS element inserts into a new chromosome location while the original IS element remains in place. That is, transposition requires the precise replication of the original IS element using the replication enzymes of the host cell. The actual transposition also requires an enzyme encoded by the IS element called **transposase.** The transposase recognizes the IR sequences of the IS element to initiate transposition.

IS elements insert into the chromosome at sites with which they have no sequence homology. Genetic recombination between nonhomologous sequences is called *nonhomologous recombination* or *illegitimate recombination.* The sites into which IS elements insert are called *target sites.* The process of IS element insertion into a chromosome is shown in Figure 20.2. First, a staggered cut is made in the target site, and the IS element is then inserted, becoming joined to the single-stranded ends. DNA polymerase and DNA ligase fill in the gaps, producing an integrated IS element with two direct repeats of the target site sequence flanking the IS element. "Direct" in this case means that the two sequences are repeated in the same

orientation (see Figure 20.2). The direct repeats are called *target site duplications.* The sizes of target site duplications vary with the IS element but tend to be small (4 to 13 bp).

When IS elements integrate at random points along the chromosome, they often cause mutations by disrupting the coding sequence of a gene or by disrupting a gene's regulatory region. Promoters within the IS elements themselves may also have effects by altering the expression of nearby genes. Additionally, the presence of an IS element in the chromosome can cause chromosome mutations such as deletions and inversions in the adjacent DNA.

Transposons

Like an IS element, a **transposon (Tn)** contains genes for the insertion of the DNA segment into the chromosome and mobilization of the element to other locations on the chromosome. A transposon is more complex than an IS element, however, in that it contains additional genes.

There are two types of prokaryotic transposons: composite transposons and noncomposite transposons. *Com-*

Figure 20.2

Schematic of the integration of an IS element into chromosomal DNA. As a result of the integration event, the target site becomes duplicated to produce direct target repeats. Thus, the integrated IS element is characterized by its inverted repeat (IR) sequences flanked by direct target site duplications. Integration involves making staggered cuts in the host target site. After insertion of the IS, the gaps that result are filled in with DNA polymerase and DNA ligase. (Note: The base sequences given for the IRs are for illustration only and are not the actual sequences found, in length or sequence.)

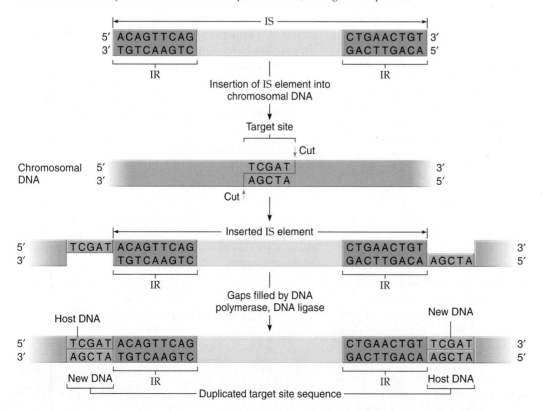

posite transposons are complex transposons with a central region containing genes (e.g., antibiotic resistance genes) flanked on both sides by IS elements (also called *IS modules*). Composite transposons may be thousands of base pairs long. The IS elements are both of the same type and are called IS*L* (for "left") and IS*R* (for "right"). The Tn*10* transposon, for example, is 9,300 bp long and consists of 6,500 bp of central, nonrepeating DNA containing the tetracycline resistance gene flanked at each end by 1,400-bp IS elements IS*10L* and IS*10R* arranged in an inverted orientation (Figure 20.3). Cells containing Tn*10* are resistant to tetracycline because of the tetracycline resistance gene. Transposition of composite transposons occurs because of the function of the IS elements they contain. One or both IS elements encodes the transposase.

Noncomposite transposons, exemplified by the 4,957-bp Tn*3*, also contain genes such as those for drug resistance, but they do not terminate with IS elements. However, they do have the repeated sequences at their ends that are required for transposition. Enzymes for transposition are encoded by genes in the central region of this type of transposon.

A number of models have been generated for transposition of transposons. Figure 20.4 shows a *cointegration* model involving the transposition of a transposon from one genome to another (e.g., from a plasmid to a bacterial chromosome or vice versa). Similar events can occur between two locations on the same chromosome. First, the donor DNA containing the transposable element fuses with the recipient DNA. Because of the way this occurs, the transposable element becomes duplicated, with one copy located at each junction between donor and recipient DNA. This fused product is called a *cointegrate*. Next, the cointegrate is resolved into two products, each with one copy of the transposable element. Since the transposable element becomes dupli-

cated, the process is called *replicative transposition*. Tn3 and related noncomposite transposons move by replicative transposition.

A second type of transposition mechanism involves the movement of a transposable element from one location to another on the same or different DNA *without* replication of the element. This mechanism is called *conservative (nonreplicative) transposition* or *simple insertion*. In other words, the element is lost from the original position when it transposes. Tn*10*, for example, transposes by conservative transposition.

As with the movement of IS elements, transposition of transposons can also cause mutations. Insertion of a transposon into the reading frame of a gene disrupts it, causing a loss of function of that gene. Insertion into a gene's controlling region can cause changes in the level of expression of the gene, depending on the promoter elements in the transposon and how they are oriented with respect to the gene. Deletion and insertion events also result from the activities of the transposons and from crossing-over between duplicated transposons in the genome.

You are a researcher in a genetics lab investigating the way a transposon called the Tn*10* transposon is transposed in the iActivity *The Genetics Shuffle* on the website.

KEYNOTE

Transposable elements are unique DNA segments that can insert themselves at one or more sites in a genome. The presence of transposable elements in a cell usually is detected by the changes they bring about in the expression and activities of the genes at or near the chromosomal sites into which they integrate. In prokaryotes, the two major types of transposable elements are insertion sequence (IS) elements and transposons (Tn).

Figure 20.3

Structure of the composite transposon Tn*10*. The general features of composite transposons are evident: a central region carrying a gene or genes, such as for drug resistance, flanked by either direct or inverted IS elements. The IS elements themselves have terminal inverted repeats.

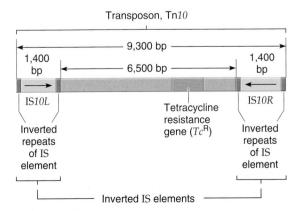

Transposable Elements in Eukaryotes

In the 1940s and 1950s, Barbara McClintock did a series of elegant genetic experiments with corn (*Zea mays*) that led her to hypothesize the existence of what she called "controlling elements," which modify or suppress gene activity in corn and are mobile in the genome. Decades later, the controlling elements she studied were shown to be transposable elements. Barbara McClintock was awarded the 1983 Nobel Prize in Physiology or Medicine for her "discovery of mobile genetic elements." A fascinating and moving biographical sketch of Barbara McClintock is given in Box 20.1.

Transposable elements have been identified in many eukaryotes, and they have been studied mostly in

Box 20.1 Barbara McClintock (1902–1992)

Barbara McClintock's remarkable life spanned the history of genetics in the twentieth century. Barbara McClintock was born in Hartford, Connecticut, to Sara Handy McClintock, an accomplished pianist as well as a poet and painter, and Thomas Henry McClintock, a physician. Both parents were quite unconventional in their attitudes toward child rearing: They were interested in what their children would and could be rather than what they should be.

During her high school years, Barbara discovered science, and she loved to learn and figure things out. After high school, Barbara attended Cornell University, where she flourished both socially and intellectually. She enjoyed her social life, but her comfort with solitude and the tremendous joy she experienced in knowing, learning, and understanding were to be the defining themes of her life. The decisions she made during her university years were consistent with her adamant individuality and self-containment. In Barbara's junior year, after a particularly exciting undergraduate course in genetics, her professor invited her to take a graduate course in genetics. After that she was treated much like a graduate student. By the time she had finished her undergraduate course work, there was no question in her mind: She had to continue her studies of genetics.

At Cornell, genetics was taught in the plant-breeding department, which at the time did not take female graduate students. To circumvent this obstacle, McClintock registered in the botany department with a major in cytology and a minor in genetics and zoology. She began to work as a paid assistant to Lowell Randolph, a cytologist. McClintock and Randolph did not get along well and soon dissolved their working relationship, but as McClintock's colleague and lifelong friend Marcus Rhoades later wrote, "Their brief association was momentous because it led to the birth of maize cytogenetics." McClintock discovered that the metaphase or late prophase chromosomes in the first microspore mitosis were far better for cytological discrimination than were root tip chromosomes. In a few weeks, she prepared detailed drawings of the maize chromosomes, which she published in *Science*.

This was McClintock's first major contribution to maize genetics and laid the groundwork for a veritable explosion of discoveries that connected the behavior of chromosomes to the genetic properties of an organism, defining the new field of cytogenetics. McClintock was awarded a Ph.D. in 1927 and appointed an instructor at Cornell, where she continued to work with maize. The Cornell maize genetics group was small, including Professor R. A. Emerson, the founder of maize genetics, McClintock, George Beadle, C. R. Burnham, Marcus Rhoades, and Lowell Randolph, together with a few graduate students. By all accounts, McClintock was the intellectual driving force of this talented group.

In 1929, a new graduate student, Harriet Creighton, joined the group and was guided by McClintock. Their work showed, for the first time, that genetic recombination is a reflection of the physical exchange of chromosome segments. A paper on the work by Creighton and McClintock, published in 1931, was perhaps McClintock's first seminal contribution to the science of genetics.

Although McClintock's fame was growing, she had no permanent position. Cornell had no female professors in fields other than home economics, so her prospects were dismal. She had already attained international recognition, but as a woman she

Barbara McClintock in 1947.

had little hope of securing a permanent academic position at a major research university. R. A. Emerson obtained a grant from the Rockefeller Foundation to support her work for two years, allowing her to continue to work independently. McClintock was discouraged and resentful of the disparity between her prospects and those of her male counterparts. Her extraordinary talents and accomplishments were widely appreciated, but she was also seen as difficult by many of her colleagues, in large part because of her quick mind and intolerance of second-rate work and thinking.

In 1936, Lewis Stadler convinced the University of Missouri to offer McClintock an assistant professorship. She accepted the position and began to follow the behavior of maize chromosomes that had been broken by X-irradiation. However, soon after her arrival at Missouri, she understood that hers was a special appointment. She found herself excluded from regular academic activities, including faculty meetings. In 1941, she took a leave of absence from Missouri and departed with no intention of returning. She wrote to her friend Marcus Rhoades, who was planning to go to Cold Spring Harbor for the summer to grow his corn. An invitation for McClintock was arranged through Milislav Demerec (member and later the director of the genetics department of the Carnegie Institution of Washington, then the dominant research laboratory at Cold Spring Harbor), who offered her a year's research appointment. Though hesitant to commit herself, McClintock accepted. When Demerec later offered her an appointment as a permanent member of the Carnegie research staff, McClintock accepted, still unsure whether she would stay. Her dislike of making commitments was a given; she insisted that she would never have become a scientist in today's world of grants because she could not have committed herself to a written research plan. It was the unexpected that fascinated her, and she was always ready to pursue an observation that didn't fit. Nevertheless, McClintock did stay at Carnegie until 1967.

At Carnegie, McClintock continued her studies on the behavior of broken chromosomes. In 1944, she was elected to the National Academy of Sciences and in 1945 to the presidency of the Genetics Society of America. In these same two years, McClin-

Figure 20.7

The *Ty* transposable element of yeast.

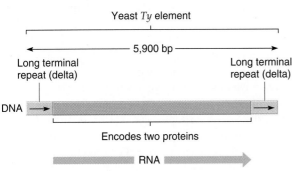

Apart from their role in hybrid dysgenesis, *P* elements are very important vectors for transferring genes into the germ line of *Drosophila* embryos, allowing genetic manipulation of the organism. Figure 20.9 illustrates an experiment in which the wild-type *rosy*[+] (*ry*[+]) gene was introduced into a strain homozygous for a mutant *rosy* allele. The wild-type *rosy* gene was introduced into the middle of a *P* element by recombinant DNA techniques and cloned in a plasmid. The recombinant plasmids were then microinjected into *rosy* embryos in the regions that would become the germ-line cells. *P* element–encoded transposase then catalyzed the movement of the *P* element, along with the wild-type *rosy* gene it contained, to the *Drosophila* genome in some of the germ-line cells. When the flies resulting from these embryos produced gametes, they contained the wild-type *rosy* gene, so descendants of these individuals had normal eye color.

the P male contains transposons called *P elements,* but the M strain has no *P* elements. Some *P* elements are autonomous elements like *Ac* in corn; that is, they encode a transposase, which can catalyze their own transposition and the transposition of shorter *P* elements that are nonautonomous elements like *Ds* in corn. *P* elements encode a repressor that blocks the transcription of the transposase gene, thereby preventing transposition of *P* elements. Thus, in the P cytotype, the *P* elements are stable in the chromosomes. However, in hybrids from M ♀ × P ♂, the cytoplasm of the egg is derived from the M cytotype, which lacks *P* repressors. Transposase is produced, activating *P* element transposition, which leads to the hybrid dysgenesis phenomenon.

KEYNOTE

> Transposable genetic elements in eukaryotes typically are transposons. Whereas most transposons move by a DNA-to-DNA mechanism, some eukaryotic transposons, such as yeast *Ty* elements, transpose via an RNA intermediate (using a transposon-encoded reverse transcriptase), thereby resembling retroviruses.

Figure 20.8

Hybrid dysgenesis, exemplified by the production of sterile flies, results from (a) a cross of M ♀ × P ♂ but not from (b) a cross of P ♀ × M ♂.

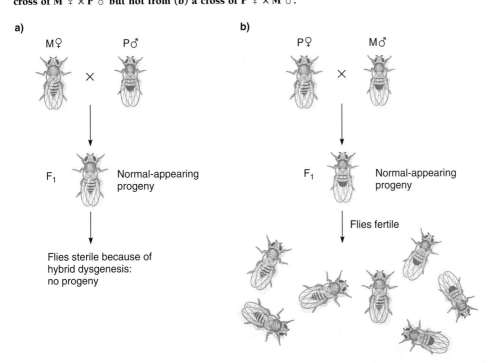

Figure 20.9

Illustration of the use of *P* elements to introduce genes into the *Drosophila* genome.

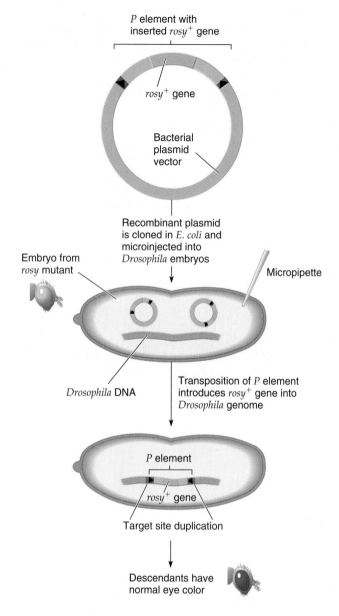

P element with inserted *rosy*⁺ gene

rosy⁺ gene

Bacterial plasmid vector

Recombinant plasmid is cloned in *E. coli* and microinjected into *Drosophila* embryos

Embryo from *rosy* mutant

Micropipette

Drosophila DNA

Transposition of *P* element introduces *rosy*⁺ gene into *Drosophila* genome

P element

rosy⁺ gene

Target site duplication

Descendants have normal eye color

Human Retrotransposons

We have learned that yeast *Ty* elements are retrotransposons, moving around the genome via RNA intermediates. There is good evidence to indicate that retrotransposons are present also in mammalian genomes.

In Chapter 8, we discussed the different repetitive classes of DNA sequences found in the genome. Of relevance here are the SINEs (short interspersed sequences) and LINEs (long interspersed sequences) found in the moderately repetitive class of sequences. SINEs are 100- to 300-bp repeated sequences interspersed between unique-sequence DNA 1,000 to 2,000 bp long. LINEs are repeated sequences more than 5,000 bp long interspersed among unique-sequence DNA up to approximately 35,000 bp long. Both SINEs and LINEs occur in DNA families, where family members are related by sequence.

In humans, a very abundant SINE family is the *Alu family*. The repeated sequence in this family is about 300 bp long and is repeated 300,000 to 500,000 times in the genome, amounting to up to 3 percent of the total genomic DNA. The name for the family comes from the fact that the sequence contains a restriction site for the enzyme *Alu*I (pronounced "al-you-one"). Over evolutionary time, members of the family have diverged, so the sequences of the individuals in the family are related but not identical.

Each Alu sequence is flanked by direct repeats of 7 to 20 bp, so they resemble transposable elements. Even though most of the moderately repetitive DNA in the genome is not transcribed, at least some members of the Alu family can be transcribed. Thus, a model is that transcriptionally active Alu sequences are actually retrotransposons that move via an RNA intermediate, much like the yeast *Ty* element.

Evidence that Alu sequences can transpose has come from a molecular study of a young male patient with neurofibromatosis (OMIM 162200), a genetic disease caused by an autosomal dominant mutation. Individuals with neurofibromatosis develop tumorlike growths (neurofibromas) over the body (see Chapter 4, p. 89). DNA analysis showed that an Alu sequence is present in the neurofibromatosis gene of the patient. Neither parent of the patient has neurofibromatosis, and neither has an Alu sequence in the neurofibromatosis gene. Since individual members of the Alu family are not identical in sequence, it was possible to track down the same Alu sequence in the patient's parents. From this analysis, it was concluded that an Alu sequence probably inserted into the neurofibromatosis gene by retrotransposition in the germ line of the father.

One mammalian LINEs family, LINEs-1 (also called *L1 elements*), is also thought to consist of retrotransposons. In humans, there are 50,000 to 100,000 copies of the L1 element, making up about 5 percent of the genome. The L1 elements vary in length; the full-length ones are autonomous transposable elements, with a gene that probably encodes a reverse transcriptase. Interestingly, in 1991, two unrelated cases of hemophilia (OMIM 306700) in children were shown to result from insertions of an L1 element into the factor VIII gene, the product of which is required for normal blood clotting. Molecular analysis showed that the insertion was not present in either set of parents, indicating that the L1 element had newly transposed. More broadly, these results show that L1 elements in humans can transpose and that they can cause disease by insertional mutagenesis.

Summary

Bacteria and eukaryotic cells contain a variety of transposable elements that have the property of moving from one site to another in the genome. We discussed two important types of transposable elements in bacteria: insertion sequence (IS) elements and transposons (Tn). The simplest type of transposable element is an IS element. An IS element typically consists of terminal inverted repeat sequences flanking a coding region, the products of which provide transposition activity. Tn elements are more complex in that they contain other genes. There are two types of prokaryotic transposons. Composite transposons consist of a central region flanked on both sides by IS elements. The central region contains genes, such as those for drug resistance. The IS elements contain the genes encoding the proteins required for transposition. Noncomposite transposons consist of a central region containing genes (such as for drug resistance), but they do not end with IS elements. Instead, short repeated sequences are found at their ends that are required for transposition. In these transposons, the transposition functions are encoded by genes in the central region.

Transposable elements in eukaryotes resemble bacterial transposons in general structure and transposition properties. Corn transposons often occur as families, each family containing an autonomous element (an element capable of transposing by itself) and one or more nonautonomous elements (elements that can transpose only if the autonomous element of the family is also present in the genome). Some eukaryotic transposons, such as yeast *Ty* elements and human (and maybe other mammalian) SINE and LINE family members, transpose via an RNA intermediate (using a transposon-encoded reverse transcriptase in the case of *Ty* and LINEs). These types of transposons resemble retroviruses in genome organization and other properties and therefore are called retrotransposons.

The presence of bacterial transposons (IS and Tn elements) and eukaryotic transposons in a cell usually is detected by the changes they bring about in the expression and activities of the genes at or near the chromosomal sites into which they integrate. Gene expression can be increased or decreased if the element inserts into a promoter or other regulatory sequence, mutant alleles of a gene can be produced if an element inserts within the coding sequence of the gene, and various chromosome rearrangements or chromosome breakage events can occur as a result of the transposition event.

Analytical Approaches for Solving Genetics Problems

Q20.1 Imagine that you are a corn geneticist and are interested in a gene you call *zma,* which is involved in formation of the tiny hairlike structures on the upper surfaces of leaves. You have a cDNA clone of this gene. In a particular strain of corn that contains many copies of *Ac* and *Ds* but no other transposable elements, you

observe a mutation of the *zma* gene. You want to figure out whether this mutation involves the insertion of a transposable element into the *zma* gene. How would you proceed? Suggest at least two approaches, and say how your expectations for an inserted transposable element would differ from your expectations for an ordinary gene mutation.

A20.1 One approach would be to make a detailed examination of leaf surfaces in mutant plants. Since there are many copies of *Ac* in the strain, if a transposable element has inserted into *zma,* it should be able to leave again, so that the mutation of *zma* would be unstable. The leaf surfaces should then show a patchy distribution of regions with and without the hairlike structures. A simple point mutation would be expected to be more stable.

A second approach would be to digest the DNA from mutant plants and the DNA from normal plants with a particular restriction endonuclease, run the digested DNAs on a gel and prepare a Southern blot, and probe the blot using the cDNA. If a transposable element has inserted into the *zma* gene in the mutant plants, then the probe should bind to different-molecular-weight fragments in mutant DNAs as compared with normal DNAs. This would not be the case if a simple point mutation had occurred.

Questions and Problems

20.1 Distinguish between prokaryotic insertion elements and transposons. How do composite transposons differ from noncomposite transposons?

20.2 What properties do bacterial and eukaryotic transposable elements have in common?

20.3 An IS element became inserted into the *lacZ* gene of *E. coli.* Later, a small deletion occurred in this gene, which removed 40 base pairs, starting to the left of the IS element. Ten *lacZ* base pairs were removed, including the left copy of the target site, and the 30 leftmost base pairs of the IS element were removed. What will be the consequence of this deletion?

*****20.4** An understanding of the molecular structures of *Ac* and *Ds* elements, together with an understanding of the basis for their transposition, has allowed for a clearer, molecular-based interpretation of Barbara McClintock's observations. Ponder the significance of this interpretation on the acceptance of her work, and use your understanding of transposition of *Ac* and *Ds* elements in corn to propose an explanation for the following.

Two different true-breeding strains of corn with colorless kernels A and B are crossed with each of two different true-breeding strains with purple kernels C and D. The F_1 from each cross is selfed, with the following results:

P:	A × C	A × D
F_1:	all purple	all purple
F_2:	3 purple : 1 colorless	3 purple : 1 colorless

P:	B × C	B × D
F_1:	all purple	all purple
F_2:	3 purple : 1 spotted*	3 purple : 1 colorless

*spotted = kernels with purple spots in a colorless background

*20.5 Consider two theoretical yeast transposons, A and B. Each contains an intron. Each transposes to a new location in the yeast genome and then is examined for the presence of the intron. In the new locations, you find that A has no intron, but B does. What can you conclude about the mechanisms of transposon movement for A and B from these facts?

*20.6 When certain strains of *Drosophila melanogaster* are crossed, hybrid dysgenesis can result in mutations, chromosomal aberrations, and sterility. A curious feature of hybrid dysgenesis is that it has been seen when females from certain laboratory strains (M cytotype) are mated with males from wild populations (P cytotype), but not when males from the same laboratory strains are mated with females from wild populations.

a. What occurs during the development of the F_1 hybrids to cause hybrid dysgenesis?

b. Why does it occur only in F_1 hybrids from M females × P males and not in F_1 hybrids from P females × M males?

c. Why does hybrid dysgenesis affect only the F_1 germ line and not F_1 somatic cells?

d. Some strains of P cytotype males are more potent at causing hybrid dysgenesis than others. Why might this be the case?

e. How might you test your explanation in (c)? Assume that you have a plasmid vector with an intact *P* element cloned into it.

20.7 In the experiment described in Figure 20.9, a wild-type ry^+ (*rosy*$^+$) gene was introduced into a strain homozygous for a mutant *ry* allele by using *P* element–mediated gene transformation.

a. If a transformed male was mated with an M strain female (see problem 20.6), would you expect the *P* element construct used in the experiment to transpose to another site in the genome of the F_1?

b. If a transformed male was mated to a P strain female (see problem 20.6), would you expect the *P* element construct used in the experiment to transpose to another site in the genome of the F_1?

c. Crosses of transformed flies to well-defined laboratory strains are more often useful for experimentation than crosses to animals from wild populations. Based on your answers to (a) and (b), what problems could arise from using transformants obtained by the approach in Figure 20.9?

20.8 After the discovery that *P* elements could be used to develop transformation vectors in *Drosophila melanogaster*, attempts were made to use them for the development of germ-line transformation in several different insect species. Savakis and his colleagues have succeeded in using a different transposable element found in *Drosophila*, the *Minos* element, to develop germ-line transformation for both *Drosophila melanogaster* and the medfly, *Ceratitus capitata*, a major agricultural pest present in Mediterranean climates.

a. What is the value of developing a transformation vector for an insect pest?

b. What basic information about the *Minos* element would need to be gathered before it could be used for germ-line transformation?

21

Chromosomal Mutations

PRINCIPAL POINTS

Chromosomal mutations are variations from the normal condition in chromosome number or chromosome structure. Chromosomal mutations can occur spontaneously, or they can be induced by chemicals or radiation.

Deletion is the loss of a DNA segment, duplication is the addition of one or more extra copies of a DNA segment, inversion is a reversal of orientation of a DNA segment in a chromosome, and translocation is the movement of a DNA segment to another chromosomal location in the genome.

Variations in the chromosome number of a cell or an organism include aneuploidy, monoploidy, and polyploidy. In aneuploidy, there are one, two, or more whole chromosomes greater or fewer than the diploid number; in monoploidy, each body cell of the organism has only one set of chromosomes; and in polyploidy, more than two sets of chromosomes are present.

A change in chromosome number or chromosome structure can have serious, even lethal consequences to the organism. In eukaryotes, abnormal phenotypes typically result from abnormal chromosome segregation during meiosis, from gene disruptions where chromosomes break, or from altered gene expression levels when the number of copies of one or more genes (gene dosage) is altered or when the rearrangement separates a gene from its regulatory sequence. Some human tumors, for example, have chromosomal mutations associated with them, either a change in the number of chromosomes or a change in chromosome structure.

i YOU LIE ON A TABLE IN A SOFTLY LIT ROOM, watching a black and white monitor. You see the image of a long needle being inserted into your uterus as you simultaneously feel the pressure of the needle against your abdomen. The doctor collects some of the amniotic fluid that surrounds your 16-week-old fetus. When the procedure is done, you get up, get dressed, and go home. Six weeks later, you go back to the clinic, where a counselor gently informs you that your unborn child has an extra chromosome 21, which causes Down syndrome.

Down syndrome is just one of a number of human disorders that are the result of variations in the normal set of chromosomes. In this chapter, you will learn about the causes and effects of different chromosomal mutations. After you have read the chapter, you can try the iActivity, in which you can use your understanding of chromosomal mutations to help a couple who are trying to conceive a child.

Chromosomal mutations—changes in normal chromosome structure or chromosome number—affect both prokaryotes and eukaryotes as well as viruses. The association of genetic defects with changes in chromosome structure or chromosome number indicates that not all genetic defects result from simple mutations of single genes. The study of normal and mutated chromosomes and their behavior is called cytogenetics. Your goal in this chapter is to learn about the various types of chromosomal mutations in eukaryotes and about some of the human disease syndromes that result from chromosomal mutations.

Types of Chromosomal Mutations

Chromosomal mutations (or **chromosomal aberrations**) are *variations from the normal (wild-type) condition in chromosome structure or chromosome number.* In Bacteria, Archaea, and Eukarya, chromosomal mutations arise spontaneously or can be induced experimentally by certain chemicals or radiation. Chromosomal mutations are detected by genetic analysis, that is, by observing changes in the linkage arrangements of genes. In eukaryotes, chromosomal mutations can also often be detected under the microscope during mitosis and meiosis. In this chapter, we limit our discussion to chromosomal changes in eukaryotes.

We often have the impression that reproduction in humans usually occurs without significant problems affecting chromosome structure or number. After all, most babies appear normal, as does the majority of the adult population. However, chromosomal mutations are more common than we once thought, and they contribute significantly to spontaneously aborted pregnancies and stillbirths. For example, major chromosomal mutations are present in about half of spontaneous abortions, and a visible chromosomal mutation is present in

about 6 out of 1,000 live births. Other studies have shown that about 11 percent of men with serious fertility problems and about 6 percent of people institutionalized with mental deficiencies have chromosomal mutations. Chromosomal mutations are significant causes of developmental disorders.

KEYNOTE

Chromosomal mutations are variations from the wild-type condition in chromosome number or chromosome structure. Chromosomal mutations can occur spontaneously, or they can be induced by chemicals or radiation.

Variations in Chromosome Structure

There are four common types of chromosomal mutations involving changes in chromosome structure: deletions and duplications (both of which involve a change in the amount of DNA on a chromosome), inversions (which involve a change in the orientation of a chromosomal segment), and translocations (which involve a change in the location of a chromosomal segment).

All four classes of chromosomal structure mutations begin with one or more breaks in the chromosome. If a break occurs within a gene, then the function of that gene may be lost. Wherever the break occurs, broken ends remain without the specialized sequences found at the ends of chromosomes (the telomeres) that prevent their degradation. The broken end of a chromosome is "sticky" and can adhere to other broken chromosome ends. This stickiness can help us understand the formation of the types of chromosomal structure mutations we will discuss.

We have learned a lot about changes in chromosome structure from the study of **polytene chromosomes,** a special type of chromosome found in certain insects such as *Drosophila.* Polytene chromosomes consist of chromatid bundles resulting from repeated cycles of chromosome duplication without nuclear or cell division. Polytene chromosomes can be a thousand times the size of corresponding chromosomes found at meiosis or in the nuclei of ordinary somatic cells and are easily detectable under the microscope. In each polytene chromosome, the homologous chromosomes are tightly paired; therefore, the observed number of polytene chromosomes per cell is reduced to half the diploid number of chromosomes. The polytene chromosomes are joined together at their centromeres by a proteinaceous structure called the chromocenter.

As a result of the intimate pairing of the multiple copies of chromatids, characteristic banding patterns are easily seen, enabling cytogeneticists to identify any segment of a polytene chromosome. In *Drosophila melanogaster,* for example, more than 5,000 bands and interbands can be counted in the four polytene chromosomes. Each band contains an average of 30,000 base pairs (30 kb) of DNA, enough to encode several average-sized proteins. DNA-cloning and -sequencing studies have shown that many bands contain up to seven genes. Genes are also found in the interbands. Polytene chromosomes are mentioned throughout this chapter because it is easy to see the different types of chromosomal mutations in *Drosophila* salivary gland polytene chromosomes.

Deletion

A **deletion** is a chromosomal mutation in which part of a chromosome is missing (Figure 21.1). A deletion is initiated when a break occurs in a chromosome. Breaks can be induced by agents such as heat, radiation (especially ionizing radiation; see Chapter 19), viruses, chemicals, or transposable elements (see Chapter 20) or by errors in recombination. Because a segment of chromosome is missing, deletion mutations cannot revert to the wild-type state.

The consequences of a deletion depend on the genes or parts of genes that have been removed. In diploid organisms, an individual heterozygous for a deletion may be normal. However, if the homologue contains recessive genes with deleterious effects, the consequences can be severe. If the deletion involves the loss of the centromere of a chromosome, the result is an acentric chromosome, which is usually lost during meiosis. This leads to the deletion of an entire chromosome from the genome, which may have very serious or lethal consequences, depending on the chromosome deleted and the organism. For example, no known living humans have one whole chromosome of a homologous pair of autosomes deleted from the genome.

In organisms in which karyotype analysis (analysis of the chromosome complement; see Chapter 3) is practical, deletions can be detected by that procedure if the losses are large enough. In that case, a mismatched pair of homologous chromosomes is seen, with one shorter than the other. In heterozygous individuals, deletions result in unpaired loops when the two homologous chromosomes pair at meiosis. Figure 21.2 diagrams such a loop in polytene chromosomes, which are paired during interphase in salivary gland cells.

A number of human disorders are caused by deletions of chromosome segments. In many cases, the abnormalities are found in heterozygous individuals; homozygotes for deletions usually die if the deletion is large. This tells us that, in humans at least, the number of copies of genes is important for normal development and function. One human disorder caused by a heterozygous deletion is *cri-du-chat syndrome* (OMIM 123450), which results from an observable deletion of part of the short arm of chromosome 5, one of the larger human chromosomes (Figure 21.3). About 1 infant in 50,000 live births has cri-du-chat ("cry of the cat") syndrome.

Another example is *Prader-Willi syndrome* (OMIM 176270), which results from heterozygosity for a deletion of part of the long arm of chromosome 15. Many individuals with the syndrome go undiagnosed, so its frequency of occurrence is not known accurately, although it is estimated to affect between 1 in 10,000 and 1 in 25,000 people, predominantly males. Infants with this syndrome are weak because their sucking reflex is poor, making feeding difficult. As a result, growth is poor. By age 5 or 6, for reasons not yet understood, children with Prader-Willi syndrome become compulsive eaters, and this produces obesity and related health problems. If untreated, afflicted individuals may feed themselves to death. Other phenotypes associated with the syndrome include poor sexual development in males, behavioral problems, and mental retardation.

Figure 21.2

Cytological effects at meiosis of heterozygosity for a deletion. Paired *Drosophila* salivary gland polytene chromosomes in a strain with a heterozygous deletion, showing the unpaired region; numbers refer to bands, which are known to include genes.

Paired polytene chromosomes

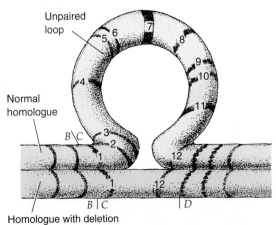

Figure 21.1

A deletion of a chromosome segment (here, *D*).

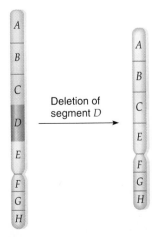

Figure 21.3

Cri-du-chat syndrome results from the deletion of part of one of the copies of human chromosome 5. **(a)** Karyotype of individual with cri-du-chat syndrome. **(b)** A child with cri-du-chat syndrome.

a)

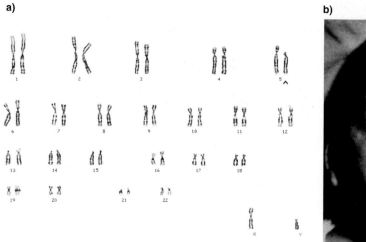

b)

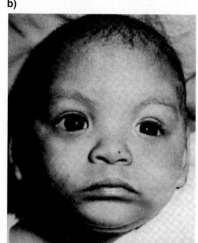

Duplication

A **duplication** is a chromosomal mutation that results in the doubling of a segment of a chromosome (Figure 21.4). The size of the duplicated segment may vary widely, and duplicated segments may occur at different locations in the genome or in a tandem configuration (i.e., adjacent to each other). Heterozygous duplications result in unpaired loops similar to those described for chromosome deletions and therefore may be detected cytologically.

Duplications of particular genetic regions can have unique phenotypic effects, as in the *Bar* mutant on the

X chromosome of *Drosophila melanogaster*. In strains homozygous for the *Bar* mutation (not to be confused with the Barr body), the number of facets of the compound eye is fewer than for the normal eye (shown in Figure 21.5a), giving the eye a bar-shaped (slitlike) rather than oval appearance (shown in Figure 21.5b). *Bar* resembles an incompletely dominant mutation because females heterozygous for *Bar* have more facets and hence a somewhat larger bar-shaped eye than do females homozygous for *Bar*. Males hemizygous for *Bar* have very small eyes, like those of homozygous *Bar* females. The *Bar* trait is the result of a duplication of a small segment (16A) of the X chromosome (see Figure 21.5b).

Duplications have played an important role in the evolution of multiple genes with related functions (a multigene family). For example, hemoglobin molecules contain two copies each of two different subunits, the α-globin polypeptide and the β-globin polypeptide. At different developmental stages from the embryo to the adult, a human individual has different hemoglobin molecules assembled from different types of α-globin and β-globin polypeptides. The genes for each of the α-globin types of polypeptides are clustered together on one chromosome, while the genes for each of the β-globin types of polypeptides are clustered together on another chromosome. (See Chapter 17 for further discussion.) The sequences of the α-globin genes are all very similar, as are the sequences of the β-globin genes. It is thought that each assembly of genes evolved from a different ancestral gene by duplication and subsequent sequence divergence.

Figure 21.4

Duplication, with a chromosome segment (here, *BC*) repeated.

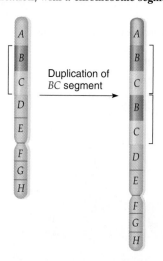

Figure 21.5

Chromosome constitutions of *Drosophila* strains, showing the relationship between duplications of region 16A of the X chromosome and the production of reduced-eye phenotypes. (a) Wild type. **(b)** Homozygous *Bar* mutant.

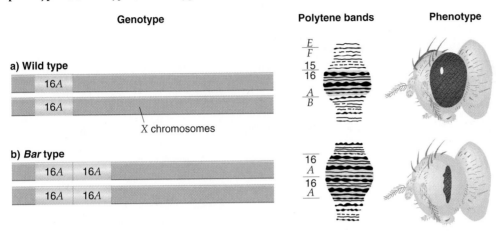

Inversion

An **inversion** is a chromosomal mutation that results when a segment of a chromosome is excised and then reintegrated in an orientation 180° from the original orientation (Figure 21.6). There are two types of inversions. A **pericentric inversion** includes the centromere (Figure 21.6a), whereas a **paracentric inversion** does not include the centromere (Figure 21.6b).

In general, genetic material is not lost when an inversion takes place, although there can be phenotypic consequences when the break points (inversion ends) occur within genes or within regions that control gene expression. Homozygous inversions can be seen because of the non-wild-type linkage relationships that result for the genes within the inverted segment and the genes that flank the inverted segment. For example, if the order of genes on the normal chromosome is *ABCDEFGH* and the *BCD* segment is inverted (shown now in bold), the gene order will now be *A**DCB**EFGH*, with *D* now more closely linked to *A* than to *E* and *B* now more closely linked to *E* than to *A* (see Figure 21.6b).

The meiotic consequences of a chromosome inversion depend on whether the inversion occurs in a homozygote or a heterozygote. If the inversion is homozygous (say, *A**DCB**EFGH/A**DCB**EFGH,* where the *BCD* segment is the inverted segment in each chromosome), then meiosis is normal and there are no problems related to gene duplications or deletions. However, crossing-over within inversion heterozygotes (say, *ABCDEFGH/A**DCB**EFGH,* where one chromosome has an inverted *BCD* segment) has serious genetic consequences. Moreover, the recombinant chromosomes are

different for crossing-over in paracentric as opposed to pericentric inversion heterozygotes. In paracentric inversion heterozygotes, the homologous chromosomes attempt to pair, so that the best possible base pairing occurs. Because of the inverted segment on one homologue, pairing of homologous chromosomes requires the formation of loops containing the inverted segments, called inversion loops (Figure 21.7a; also see Figure 21.8). Inversion heterozygotes, then, may be identified by these loops.

Paracentric inversion heterozygotes can be identified in genetic experiments because viable recombinants are

Figure 21.6

Inversions. (a) Pericentric inversion. **(b)** Paracentric inversion.

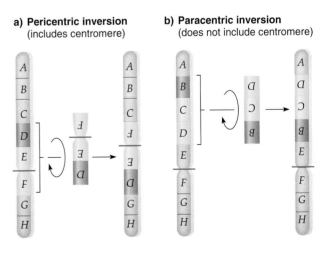

a) Pericentric inversion
(includes centromere)

b) Paracentric inversion
(does not include centromere)

Figure 21.7

Consequences of a paracentric inversion. (a) Photomicrograph of an inversion loop in polytene chromosomes of a strain of *Drosophila melanogaster* that is heterozygous for a paracentric inversion. **(b)** Meiotic products resulting from a single crossover within a heterozygous, paracentric inversion loop. Crossing-over occurs at the four-strand stage involving two nonsister homologous chromatids.

a)

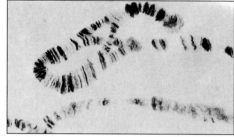

b) **Products of meitoic crossover**

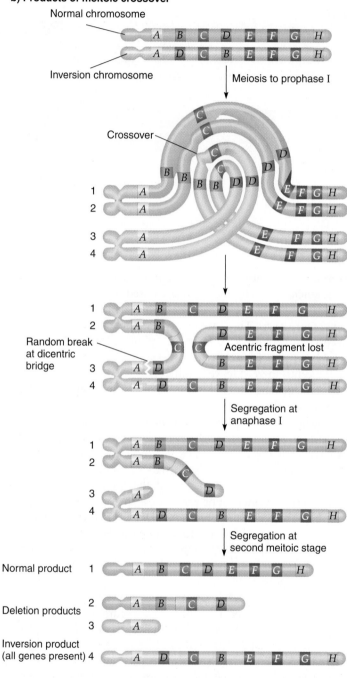

reduced significantly or suppressed. That is, the frequency of crossing-over is not lower in inversion heterozygotes than in normal cells, but gametes or zygotes derived from recombined chromatids are inviable. The inviability results from an unbalanced set of genes in the gamete: One or more genes are missing or one or more

genes are present in two copies instead of one (see Figure 21.7b). If no crossovers occur in the inversion loop of an inversion heterozygote, then all resulting gametes receive a complete set of genes (two gametes with a normal gene order, *ABCDEFGH*, and two gametes with the inverted segment, *ADCBEFGH*), and they are all viable. Figure

Figure 21.8

Meiotic products resulting from a single crossover within a heterozygous, pericentric inversion loop. Crossing-over occurs at the four-strand stage involving two non-sister homologous chromatids.

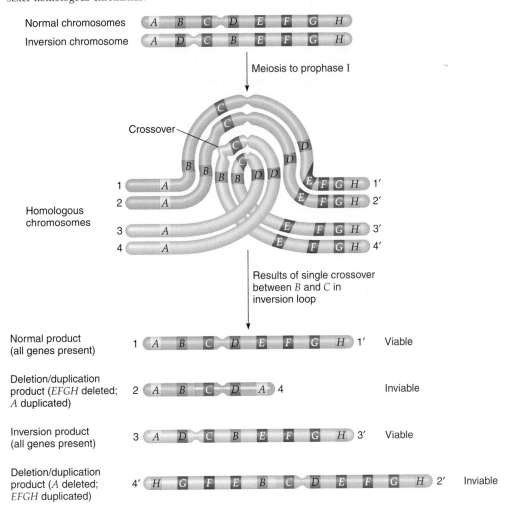

21.7b shows the effects of a single crossover in the inversion loop of an individual heterozygous for a paracentric inversion. During the first meiotic anaphase, the two centromeres migrate to opposite poles of the cell. Because of the crossover between genes *B* and *C* in the inversion loop, one recombinant chromatid becomes stretched across the cell as the two centromeres begin to migrate in anaphase, forming a **dicentric bridge,** or **dicentric chromosome**—that is, a chromosome with two centromeres. With continued migration, the dicentric bridge breaks because of tension. The other recombinant product of the crossover event is a chromosome without a centromere (an acentric fragment). This acentric fragment is unable to continue through meiosis and is usually lost (it is not found in the gametes).

In the second meiotic division, each daughter cell receives a copy of each chromosome. Two of the gametes have complete sets of genes and are viable: the gamete with the normal order of genes (*ABCDEFGH*) and the gamete

with the inverted segment of genes (*ADCBEFGH*). The other two gametes are inviable because they are unbalanced: Many genes are deleted. Thus, the only gametes that can give rise to viable progeny are those containing the chromosomes that did not involve crossing-over. However, in many cases in female animals, the dicentric chromosomes or acentric fragments arising as a result of inversion are shunted to the polar bodies, so the reduction in fertility may not be so great.

The consequences of a single crossover in the inversion loop of an individual heterozygous for a pericentric inversion are shown in Figure 21.8. The crossover event and the ensuing meiotic divisions result in two viable gametes with the nonrecombinant chromosomes *ABCDEFGH* (normal) and *ADCBEFGH* (inversion) and in two recombinant gametes that are inviable, each as a result of the deletion of some genes and the duplication of other genes.

Some crossover events within an inversion loop do not affect gamete viability. For example, a double crossover

animation

a Crossing–over in an Inversion Heterozygote

close together and involving the same two chromatids (a two-strand double crossover; see Chapter 5) produces four viable gametes. Also, recent studies with mammals show that inverted segments may remain unpaired. Since crossing-over cannot occur between unpaired segments, the generation of inviable gametes is avoided.

Translocation

A **translocation** is a chromosomal mutation in which there is a change in position of chromosome segments and the gene sequences they contain to a different location in the genome (Figure 21.9). There is no gain or loss of genetic material involved in a translocation. Two simple kinds of translocations occur. One kind involves a change in position of a chromosome segment within the same chromosome; this is called an intrachromosomal (within a chromosome) translocation (Figure 21.9a). The other kind involves the transfer of a chromosome segment from one chromosome into a nonhomologous chromosome; this is called an interchromosomal (between chromosomes) translocation (Figure 21.9b and c). If a translocation involves the transfer of a segment from one chromosome to another, it is a nonreciprocal translocation (shown in Figure 21.9a and b); if it involves the exchange of segments between the two chromosomes, it is a reciprocal translocation (shown in Figure 21.9c).

Translocations typically affect the products of meiosis. In many cases, some of the gametes produced are unbalanced in that they have duplications or deletions, and in many cases they are inviable. In other cases, such as familial Down syndrome, resulting from a duplication stemming from a translocation, gametes

are viable (see later in the chapter). We focus here on reciprocal translocations.

In strains homozygous for a reciprocal translocation, meiosis takes place normally because all chromosome pairs can pair properly and crossing-over does not produce any abnormal chromatids. In strains heterozygous for a reciprocal translocation, however, all homologous chromosome parts pair as best they can. Since one set of normal chromosomes (N) and one set of translocated chromosomes (T) are involved, the result is a crosslike configuration in meiotic prophase I (Figure 21.10). These crosslike figures consist of four associated chromosomes, each chromosome being partially homologous to two other chromosomes in the group.

Segregation at anaphase I may occur in three different ways. (We are ignoring crossing-over in this discussion.) In one way, called alternate segregation, alternate centromeres segregate to the same pole: N_1 and N_2 to one pole, T_1 and T_2 to the other pole (Figure 21.10, left). This produces two gametes, each of which is viable because it contains a complete set of genes—no more, no less. One of these gametes has two normal chromosomes, and the other has two translocated chromosomes. In the second way, called adjacent 1 segregation, adjacent nonhomologous centromeres migrate to the same pole: N_1 and T_2 to one pole, N_2 and T_1 to the other pole (Figure 21.10, middle). Both gametes produced contain gene duplications and deletions and are often inviable. Adjacent 1 segregation occurs about as frequently as alternate segregation. In the third way, called adjacent 2 segregation, different pairs of adjacent homologous centromeres migrate to the same pole: N_1 and T_1 to one pole, N_2 and T_2 to the other pole (Figure 21.10, right). Both products have gene duplications and deletions and are

animation

a Meiosis in a Translocation Heterozygote

Figure 21.9

Translocations. (**a**) Nonreciprocal intrachromosomal translocation. (**b**) Nonreciprocal interchromosomal translocation. (**c**) Reciprocal interchromosomal translocation.

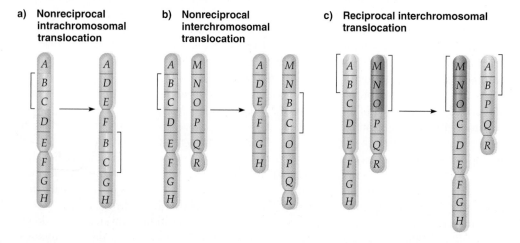

Figure 21.10

Meiosis in a translocation heterozygote in which no crossover occurs.

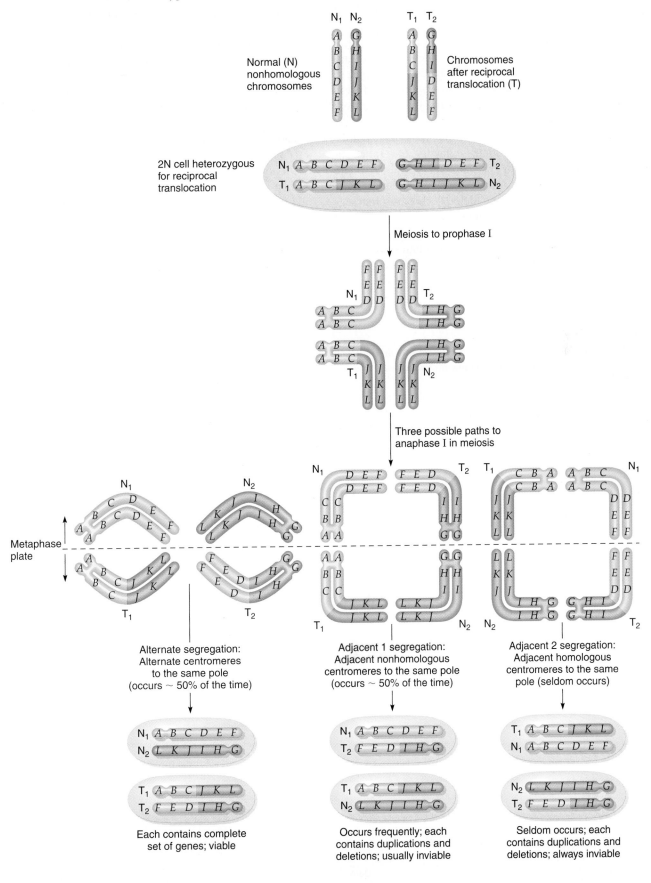

always inviable. Adjacent 2 segregation seldom occurs. In sum, of the six theoretically possible gametes, the two from alternate segregation are functional, the two from adjacent 1 segregation usually are inviable (because of gene duplications and deficiencies), and the two from adjacent 2 seldom occur and are inviable if they do. And because alternate segregation and adjacent 1 segregation occur with about equal frequency, the term *semisterility* is applied to this condition. (This term is also used for inversion heterozygotes.)

In practice, animal gametes that have large duplicated or deleted chromosome segments may function, but the zygotes formed by such gametes typically die. The gametes may function normally and viable offspring may result if the duplicated and deleted chromosome segments are small. In plants, pollen grains with duplicated or deleted chromosome segments typically do not develop completely and hence are nonfunctional.

Some human tumors are associated with consistent chromosome translocations. Examples are *chronic myel-ogenous leukemia* (*CML*; OMIM 151410; reciprocal translocation involving chromosomes 9 and 22) and Burkitt lymphoma (BL; OMIM 113970; reciprocal translocation involving chromosomes 8 and 14). CML is described here.

CML is an invariably fatal cancer involving uncontrolled replication of myeloblasts (stem cells of white blood cells). Ninety percent of patients with chronic myelogenous leukemia have a chromosomal mutation in the leukemic cells called the *Philadelphia chromosome* (*Ph*[1]), so named because the discovery was made in Philadelphia. The Philadelphia chromosome results from a reciprocal translocation involving the movement of part of the long arm of chromosome 22 (the second smallest human chromosome) to chromosome 9 and the movement of a small part from the tip of the long arm of chromosome 9 to chromosome 22 (Figure 21.11). This reciprocal translocation event apparently converts proto-oncogenes to oncogenes (see Chapter 18) that cause the transition from a differentiated cell to a tumor cell with an uncontrolled pattern of growth. Specifically, the *c-abl*

Figure 21.11

Origin of the Philadelphia chromosome in chronic myelogenous leukemia (CML) by a reciprocal translocation involving chromosomes 9 and 22. The arrows show the sites of the breakage points.

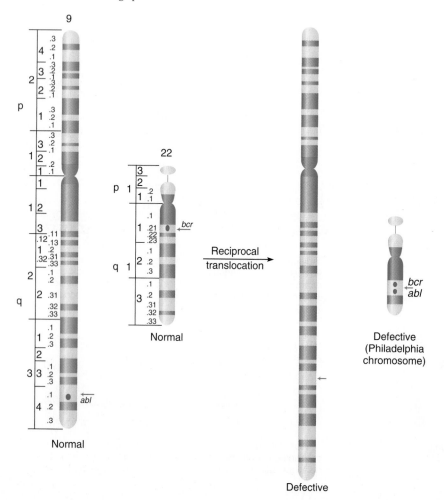

("c-able"; cellular Abelson) oncogene, normally located on chromosome 9, is translocated to chromosome 22 in patients with CML (see Figure 21.11). The translocation event positions c-*abl* within the *bcr* (break-point cluster region) gene. This hybrid oncogene arrangement somehow causes a new gene product, producing leukemia.

 As a genetic counselor, you must determine if there are any chromosomal abnormalities that could be affecting a couple's ability to have children in the iActivity *Deciphering Karyotypes* on the website.

KEYNOTE

Chromosomal mutations may involve parts of individual chromosomes rather than whole chromosomes or sets of chromosomes. The four major types of structural alterations are deletions and duplications (both of which involve a change in the amount of DNA on a chromosome), inversions (which involve no change in the amount of DNA on a chromosome but rather a change in the arrangement of a chromosomal segment), and translocations (which also involve no change in the amount of DNA but involve a change in chromosomal location of one or more DNA segments).

Fragile Sites and Fragile X Syndrome

When human cells are grown in culture, some of the chromosomes develop narrowings or unstained areas (gaps) called *fragile sites*. More than 40 fragile sites have been identified since the first one was discovered in 1965. One human condition related to a fragile site is *fragile X syndrome*, in which the X chromosome is prone to breakage at a site in the long arm of the X chromosome at position Xq27.3, as shown in Figure 21.12. After Down syndrome, fragile X syndrome is the leading genetic cause of mental retardation in the United States, with an incidence of about 1 in 1,250 males and 1 in 2,500 females (heterozygotes). As with all recessive X-linked traits, males predominantly exhibit this type of mental retardation.

The fragile X chromosome is inherited as a typical Mendelian gene. Male offspring of carrier females have a 50 percent chance of receiving a fragile X chromosome. However, only 80 percent of males with a fragile X chromosome are mentally retarded; the rest are normal. These phenotypically normal males are called *normal transmitting males* and carry a *premutation* because they can pass on the fragile X chromosome to their daughters. (A premutation can be considered a silent mutation.) The sons of those daughters frequently show symptoms of mental retardation. Female offspring of carrier (heterozygous) females also have a 50 percent chance of inheriting a frag-

Figure 21.12

Fragile site on the X chromosome. (**a**) Scanning electron micrograph and (**b**) diagram of a human X chromosome showing the location of the fragile site responsible for fragile X syndrome. (From Gerald Stine, *The New Human Genetics.* Copyright © 1989. Reproduced by permission of The McGraw-Hill Companies.)

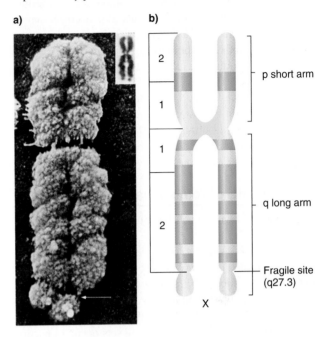

ile X chromosome. Up to 33 percent of the carrier females show mild mental retardation.

Modern molecular techniques brought to bear on this disease have resulted in an understanding of the disease at the DNA level. There is a repeated 3-bp sequence, CGG, in a gene called *FMR-1* (*f*ragile X *m*ental *r*etardation-1; OMIM 309550) located at the fragile X site. Normal individuals have an average of 29 CGG repeats (the range is from 6 to 54) in the coding region of the *FMR-1* gene. Phenotypically normal transmitting males and their daughters, as well as some carrier females, have a significantly larger number of CGG repeats, ranging from 55 to 200 copies. These individuals do not show symptoms of fragile X syndrome, and the increased number of repeats they have is the premutation previously mentioned. Males and females with fragile X syndrome have even larger numbers of the CGG repeats, ranging from 200 to 1,300 copies; these are considered the full mutations. In other words, the triplet repeat CGG in the *FMR-1* gene becomes duplicated (amplified) tandemly; below a certain threshold number of copies (about 200 or fewer), there are no clinical symptoms, and above that threshold number of copies (greater than 200), clinical symptoms are seen. Interestingly, amplification of the CGG repeats does not occur in males, only in females. Therefore, a phenotypically normal transmitting male (who has the premutation) transmits his X chromosome to his daughter. During a slipped mispairing process during DNA replication in his daughter, perhaps, the triplets may amplify, and she can transmit the amplified X to her offspring. Thus, affected males inherit the mutation from their grandfather. The function of the *FMR-1* gene, in which the triplet repeat amplification occurs, is unknown, so we do not yet understand how such amplification produces mental retardation. A protein is made by the gene but has unidentified function.

Triplet repeat amplification has also been shown to cause other human diseases, such as myotonic dystrophy (MD; OMIM 160900), spinobulbar muscular atrophy (also called Kennedy disease; OMIM 313200), and Huntington disease (HD; OMIM 143100; see Chapter 4). Each of these cases differs from fragile X syndrome in that the amplification can occur in both sexes at each generation. For each, there is a threshold number of triplet repeat copies above which symptoms of the disease are produced.

Variations in Chromosome Number

When an organism or cell has one complete set of chromosomes or an exact multiple of complete sets, that organism or cell is said to be euploid, and the condition is called **euploidy.** Thus, eukaryotic organisms that are normally diploid (such as humans and fruit flies) and eukaryotic organisms that are normally haploid (such as yeast) are euploids. Chromosome mutations that result in variations in the number of chromosome sets occur in nature, and the resulting organism or cells are also euploid. Chromosome mutations resulting in variations in the number of individual chromosomes are examples of **aneuploidy.** An aneuploid organism or cell has a chromosome number that is not an exact multiple of the haploid set of chromosomes. Both euploid and aneuploid variations affecting whole chromosomes are discussed in this section.

Changes in One or a Few Chromosomes

Generation of Aneuploidy. Changes in chromosome number can occur in both diploid and haploid organisms. Nondisjunction (see Chapter 3, pp. 59–60) of one or more chromosomes during meiosis I or meiosis II typically is responsible for generating gametes with abnormal numbers of chromosomes. Referring to Figure 3.5 and considering just one particular chromosome, it can be seen that nondisjunction at meiosis I produces four abnormal gametes: two gametes with a chromosome duplicated and two gametes with the corresponding chromosome missing. Nondisjunction at meiosis I can produce in a male a gamete with both the X and Y chromosome and in a female a gamete with both sets of homologues (and thus possible heterozygotes in a gamete). Fusion of a gamete type with two copies of a chromosome with a normal gamete produces a zygote with three copies of the particular chromosome instead of the normal two and, unless nondisjunction involved other chromosomes, two copies of all other chromosomes. Similarly, fusion of a gamete

type with no copies of a chromosome with a normal gamete produces a zygote with only one copy of the particular chromosome instead of the normal two and two copies of all other chromosomes. Nondisjunction in meiosis II (see Figure 3.5) results in two normal gametes and two abnormal gametes, that is, a single gamete with two sister chromosomes and one gamete with that same chromosome missing. Fusion of these gametes with normal gametes gives the zygote types just discussed. Unlike the case with nondisjunction at meiosis I, some normal gametes are produced by nondisjunction at meiosis II; specifically, two of the four gametes are normal. More complicated gametic chromosome compositions result when more than one chromosome is involved in nondisjunction or when nondisjunction occurs in both meiotic divisions. Furthermore, nondisjunction can occur in mitosis, giving rise to somatic cells with unusual chromosome complements. In most cases, autosomal aneuploidy is lethal in animals, so in mammals it is detected mainly in aborted fetuses. Aneuploidy is tolerated more by plants, especially in species that are considered polyploid.

Types of Aneuploidy. In aneuploidy, one or more chromosomes are lost from or added to the normal set of chromosomes. In diploid organisms, there are four main types of aneuploidy (Figure 21.13):

1. **Nullisomy** (a nullisomic cell) involves a loss of one homologous chromosome pair; that is, the cell is 2N − 2. (This can arise, for example, if nondisjunction occurs for the same chromosome in meiosis in both parents, producing gametes with no copies of that chromosome and one copy of all other chromosomes in the set.)

2. **Monosomy** (a monosomic cell) involves a loss of a single chromosome; that is, the cell is 2N − 1. (This can arise, for example, if nondisjunction in meiosis in a parent produces a gamete with no copies of a particular chromosome and one copy of all other chromosomes in the set.)

3. **Trisomy** (a trisomic cell) involves a single extra chromosome; that is, the cell has three copies of a particular chromosome and two copies of other chromosomes. A trisomic cell is 2N + 1. (This can arise, for example, if nondisjunction in meiosis in a parent produces a gamete with two copies of a particular chromosome and one copy of all other chromosomes in the set.)

4. **Tetrasomy** (a tetrasomic cell) involves an extra chromosome pair, resulting in the presence of four copies of one particular chromosome and two copies of other chromosomes. A tetrasomic cell is 2N + 2. (This can arise, for example, if nondisjunction occurs for the same chromosome in meiosis in both parents, producing gametes with two copies of that chromosome and one copy of all other chromosomes in the set.)

Figure 21.13

Normal (theoretical) set of metaphase chromosomes in a diploid (2N) organism (top) and examples of aneuploidy (bottom).

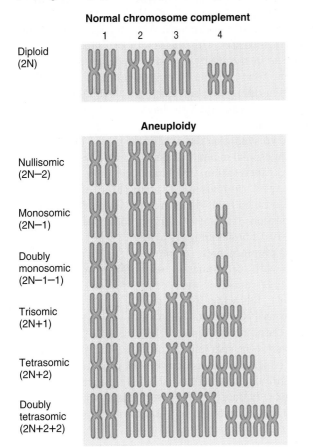

Normal chromosome complement

Diploid (2N)

Aneuploidy

Nullisomic (2N−2)

Monosomic (2N−1)

Doubly monosomic (2N−1−1)

Trisomic (2N+1)

Tetrasomic (2N+2)

Doubly tetrasomic (2N+2+2)

ratio among the progeny is 5 wild type : 1 mutant (*a*). This ratio is seen in many actual crosses of this kind.

In the following sections, we examine some examples of aneuploidy as they are found in the human population. Table 21.1 summarizes various aneuploid abnormalities for both autosomes and sex chromosomes in the human population. Examples of aneuploidy of the X and Y chromosomes are discussed in Chapter 3. Recall that in mammals, aneuploidy of the sex chromosomes is more often found in adults than aneuploidy of the autosomes because of a dosage compensation mechanism (lyonization) by which excess X chromosomes are inactivated.

In humans, autosomal monosomy is rare. Presumably, monosomic embryos do not develop significantly and are lost early in pregnancy. In contrast, autosomal trisomy accounts for about one-half of chromosomal abnormalities producing fetal deaths. In fact, only a few autosomal trisomies are seen in live births. Most of these (trisomy-8, -13, and -18) result in early death. Only in trisomy-21 (Down syndrome) does survival to adulthood occur.

Trisomy-21. Trisomy-21 (OMIM 190685) occurs when there are three copies of chromosome 21 (Figure 21.15a). Trisomy-21 occurs with a frequency of about 3,510 per 1 million conceptions and about 1,430 per

Figure 21.14

Meiotic segregation possibilities in a trisomic individual. Shown is segregation in an individual of genotype +/+/a, when two chromosomes migrate to one pole and one goes to the other pole, and assuming no crossing-over between the *a* locus and its centromere.

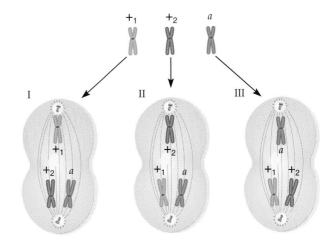

	Gametes produced after 2nd meiotic division	
	haploid	disomic
I	$+_1$	$+_2/a$
II	$+_2$	$+_1/a$
III	*a*	$+_1/+_2$

In sum: 2 +/ *a* : 2 + : 1 +/ : 1 *a*

Aneuploidy may involve the loss or the addition of more than one specific chromosome or chromosome pair. For example, a *double monosomic* has two separate chromosomes present in only one copy each; that is, it is 2N − 1 − 1. A *double tetrasomic* has two chromosomes present in four copies each; that is, it is 2N + 2 + 2. In both cases, meiotic nondisjunction involved two different chromosomes in one parent's gamete production.

Most forms of aneuploidy have serious consequences in meiosis. Monosomics, for example, produce two kinds of haploid gametes, N and N − 1. Alternatively, the odd, unpaired chromosome in the 2N − 1 cell may be lost during meiotic anaphase and not be included in either daughter nucleus, thereby producing two N − 1 gametes. There are more segregation possibilities for trisomics in meiosis. Consider a trisomic of genotype +/+/a in an organism that can tolerate trisomy, and assume no crossing-over between the *a* locus and its centromere. As shown in Figure 21.14, random segregation of the three types of chromosomes produces four genotypic classes of gametes: 2 +/a : 2 + : 1 +/+ : 1 *a*. In a cross of a +/+/a trisomic to an *a/a* individual, the predicted phenotypic

Table 21.1	Aneuploid Abnormalities in the Human Population	
Chromosomes	**Syndrome**	**Frequency at Birth**
Autosomes		
Trisomic 21	Down	14.3/10,000
Trisomic 13	Patau	2/10,000
Trisomic 18	Edwards	2.5/10,000
Sex chromosomes, females		
XO, monosomic	Turner	4/10,000 females
XXX, trisomic XXXX, tetrasomic XXXXX, pentasomic	Viable; most are fertile	14.3/10,000 females
Sex chromosomes, males		
XYY, trisomic	Normal	25/10,000 males
XXY, trisomic XXYY, tetrasomic XXXY, tetrasomic	Klinefelter	40/10,000

1 million live births. Individuals with trisomy-21 have **Down syndrome** (Figure 21.15b) and have such abnormalities as low IQ, epicanthal folds over the eyes, short and broad hands, and below-average height.

The relationship between the age of the mother and the probability of her having a trisomy-21 individual is indicated in Table 21.2. (There is no correlation with age of the father.) During the development of a female fetus before birth, the primary oocytes in the ovary undergo meiosis but stop at prophase I. In a fertile female, each month at ovulation the nucleus of a secondary oocyte (see Chapter 1) begins the second meiotic division but progresses only to metaphase, when division again stops. If a sperm penetrates the secondary oocyte, the second meiotic division is completed. The probability of nondisjunction increases with the length of time the primary oocyte is in the ovary. It is important, then, that older mothers-to-be consider testing (amniocentesis or chorionic villus sampling; see Chapter 10, pp. 218–219) to determine whether the fetus has a normal complement of chromosomes.

Down syndrome can also result from a different sort of chromosomal mutation called centric fusion or **Robertsonian translocation,** which produces three copies of the long arm of chromosome 21. This form of Down syndrome is called familial Down syndrome. A Robertsonian translocation is a type of nonreciprocal translocation in which two nonhomologous acrocentric chromosomes (chromosomes with centromeres near their ends) break at their centromeres, following which the long arms become attached to a single centromere (Figure 21.16). The short arms also join to form the reciprocal product, which typically contains nonessential genes and usually is lost within a few cell divisions. In humans, when a Robertsonian translocation joins the long arm of chromosome 21 with the long arm of chromosome 14 (or 15), the heterozygous carrier is phenotypically normal because there are two copies of all major chromosome arms and hence two copies of all essential genes.

animation
a Down Syndrome Caused by a Robertsonian Translocation

Figure 21.15
Trisomy-21 (Down syndrome). (a) Karyotype. **(b)** Individual.

a)

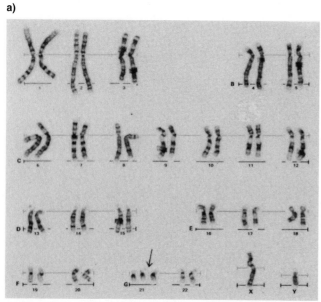

b)

Table 21.2	Relationship Between Age of Mother and Risk of Trisomy-21
Age of Mother	**Risk of Trisomy-21 in Child**
16–26	7.7/10,000
27–34	4/10,000
35–39	29/10,000
40–44	100/10,000
45–47	333/10,000
All mothers combined	14.3/10,000

There is a high risk of Down syndrome among the offspring of pairings between heterozygous carriers and normal individuals (Figure 21.17). The normal parent produces gametes with one copy each of chromosomes 14 and 21. The heterozygous carrier parent produces three reciprocal pairs of gametes, each pair produced as a result of different segregation of the three chromosomes involved. As Figure 21.17 shows, the zygotes produced by pairing these gametes with gametes of normal chromosomal constitution theoretically are as follows: One-sixth have normal chromosomes 14 and 21, one-sixth are heterozygous carriers like the parent and are phenotypically normal, one-sixth are inviable because of monosomy for chromosome 14, one-sixth are inviable because of monosomy for chromosome 21, one-sixth are inviable because of trisomy for chromosome 14, and one-sixth are trisomy-21 and therefore produce a Down syndrome individual. (These latter individuals actually have the normal diploid number of 46 chromosomes, but because of the Robertsonian translocation, they have three copies

Figure 21.16

Robertsonian translocation. Production of a Robertsonian translocation (centric fusion) by breakage of two acrocentric chromosomes at their centromeres (indicated by arrows) and fusion of the two large chromosome arms and of the two small chromosome arms.

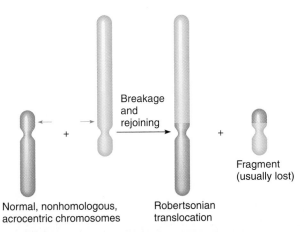

Figure 21.17

The three segregation patterns of a heterozygous Robertsonian translocation involving the human chromosomes 14 and 21. Fusion of the resulting gametes with gametes from a normal parent produces zygotes with various combinations of normal and translocated chromosomes.

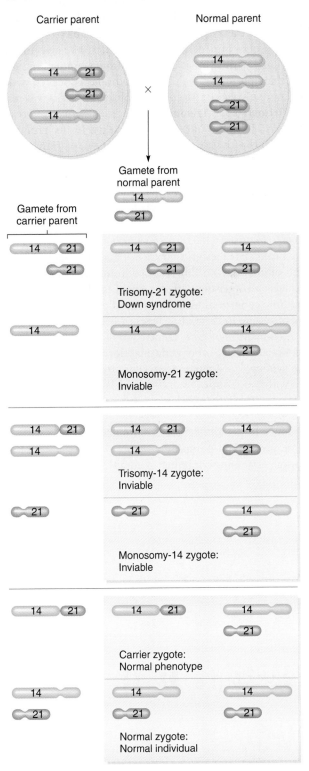

of the long arm of chromosome 21, which is sufficient to create the Down syndrome symptoms. Similarly, the trisomy-14 zygotes shown in Figure 21.17 have 46 chromosomes, but they have three copies of the long arm of chromosome 14. Apparently, the dosage of the genes involved on this larger chromosome is more critical, because trisomy-14 individuals are inviable.) In sum, one-half of the zygotes produced are inviable, and theoretically one-third of the viable zygotes give rise to an individual with familial Down syndrome, a much higher risk than for nonfamilial Down syndrome associated with the mother's age. In practice, the observed risk is lower.

Trisomy-13. **Trisomy-13** produces *Patau syndrome* (Figure 21.18). About 2 in 10,000 live births produce individuals with trisomy-13. Characteristics of individuals with trisomy-13 include cleft lip and palate, small eyes, polydactyly (extra fingers and toes), mental and developmental retardation, and cardiac anomalies, among many other abnormalities. Most die before the age of 3 months.

Trisomy-18. **Trisomy-18** produces *Edwards syndrome* (Figure 21.19). It occurs in about 2.5 in 10,000 live births. For reasons that are not known, about 80 percent of infants with Edwards syndrome are female. Individuals with trisomy-18 are small at birth and have multiple congenital malformations affecting almost every organ system in the body. Clenched fists, elongated skull, low-set malformed ears, mental and developmental retardation, and many other abnormalities are associated with the syndrome. Ninety percent of infants with trisomy-18 die within 6 months, often from cardiac problems.

Changes in Complete Sets of Chromosomes

Monoploidy and **polyploidy** involve variations from the normal state in the number of complete sets of chromosomes. Because the number of complete sets of chromosomes is involved in each case, monoploids and polyploids are both euploids. Monoploidy and polyploidy are lethal for most animal species but are less consequential in plants. Both have played significant roles in plant speciation and diversification.

Changes in complete sets of chromosomes can result when the first or second meiotic division is abortive (lack of cytokinesis) or when meiotic nondisjunction occurs for all chromosomes, for example. If such nondisjunction occurs at meiosis I, one-half of the gametes have no chromosome sets, and the other half have two chromosome sets (refer to Figure 3.5b). If such nondisjunction occurs at meiosis II, one-half of the gametes have the normal one set of chromosomes, one-quarter have two sets of chromosomes, and one-quarter have no chromosome sets (see Figure 3.5b). Fusion of a gamete with two chromosome sets with a normal gamete produces a polyploid zygote, in this case one with three sets of chromosomes, which is a *triploid* (3N). Similarly, fusion of two gametes, each with two chromosome sets, produces a *tetraploid* (4N) zygote. Polyploidy of somatic cells can also occur following mitotic nondisjunction of complete chromosome sets. Monoploid (haploid) individuals, by contrast, typically develop from unfertilized eggs.

Monoploidy. A monoploid individual has only one set of chromosomes instead of the usual two sets (Figure 21.20a). Monoploidy is sometimes called haploidy, although the term *haploidy* typically is used to describe the chromosome complement of gametes.

Monoploidy is seen only rarely in normally adult diploid organisms. Because of the presence of recessive lethal mutations (usually counteracted by dominant alleles in heterozygous individuals) in the chromosomes of many diploid eukaryotic organisms, many monoploids probably do not survive. Certain species produce mono-

Figure 21.18

Trisomy-13 (Patau syndrome). (a) Karyotype. **(b)** Individual.

a)

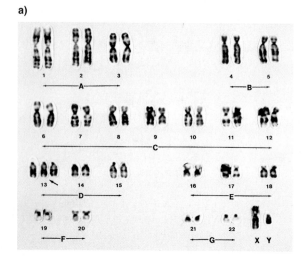

b)

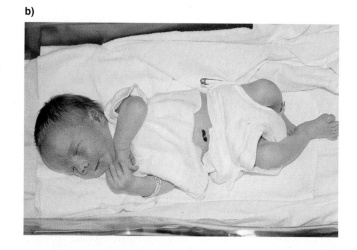

Figure 21.19

Trisomy-18 (Edwards syndrome). (a) Karyotype. (b) Individual.

a)

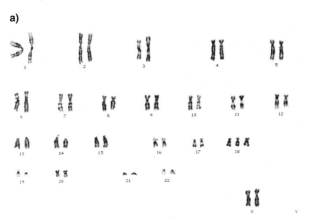

b)

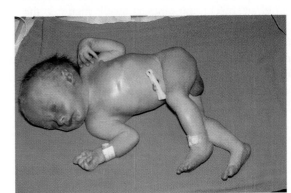

ploid organisms as a normal part of their life cycle. Some male wasps, ants, and bees, for example, are monoploid because they develop from unfertilized eggs.

Monoploids are used in plant-breeding experiments. Normal haploid cells produced by meiosis in plant anthers can be isolated and induced to grow to produce monoploid cultures for study. The single chromosome set of these monoploids then can be doubled using the chemical colchicine (which inhibits the formation of the mitotic spindle, thereby resulting in nondisjunction for all chromosomes) to produce completely homozygous diploid breeding lines.

Cells of a monoploid individual are very useful for producing mutants because there is only one dose of each of the genes. Thus, mutants can be isolated directly.

Polyploidy. Polyploidy is the chromosomal constitution of a cell or organism having three or more sets of chromosomes (Figure 21.20b). Polyploids may arise spontaneously or be induced experimentally (for instance, by using colchicine). They often occur as a result of a breakdown of the spindle apparatus in one or more meiotic divisions or in mitotic divisions. Almost all plants and animals probably have some polyploid tissues. For example, the endosperm of plants is triploid, the liver of mammals and perhaps other vertebrates is polyploid, and the giant abdominal neuron of the sea hare *Aplysia* has about 75,000 copies of the genome. Plants that are completely polyploid include wheat, which is hexaploid (6N), and the strawberry, which is octaploid (8N). Some animal species, such as the North American sucker (a freshwater fish), salmon, and some salamanders, are polyploid.

There are two general classes of polyploids: those that have an *even* number of chromosome sets and those that have an *odd* number of sets. Polyploids with an even number of chromosome sets have a better chance of being at least partially fertile because there is the potential for homologues to be segregated equally during meio-sis. Polyploids with an odd number of chromosome sets always have an unpaired chromosome for each chromosome type, so the probability of producing a balanced gamete is extremely low; such organisms usually are sterile or have increased zygote abortion.

In triploids, the nucleus of a cell has three sets of chromosomes. As a result, triploids are very unstable in meiosis because, as in trisomics, two of the three homologous

Figure 21.20

Variations in number of complete chromosome sets.
(a) Monoploidy (only one set of chromosomes instead of two).
(b) Polyploidy (three or more sets of chromosomes).

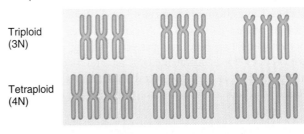

Normal chromosome complement

| | 1 | 2 | 3 |

Diploid (2N)

a) Monoploidy
(only one set of chromosomes)

Monoploid (N)

b) Polyploidy
(more than the normal number of sets of chromosomes)

Triploid (3N)

Tetraploid (4N)

chromosomes go to one pole, and the other goes to the other pole. The segregation of each chromosome from its homologues in the triploid is random, so the probability of producing balanced gametes that contain either a haploid or a diploid set of chromosomes is small; many of the gametes are unbalanced, with one copy of one chromosome, two copies of another, and so on. In general, the probability of a triploid producing a haploid gamete is $(\frac{1}{2})^n$, where n is the number of chromosomes.

In humans, the most common type of polyploidy is triploidy. Triploidy is always lethal. Triploidy is seen in 15 to 20 percent of spontaneous abortions and about 1 in 10,000 live births, but most affected infants die within 1 month. Triploid infants have many abnormalities, including a characteristically enlarged head. Tetraploidy in humans is also always lethal, usually before birth. It is seen in about 5 percent of spontaneous abortions. Very rarely, a tetraploid human is born, but such an individual does not survive long.

Polyploidy is less consequential to plants. One reason for this is that many plants undergo self-fertilization, so if a plant is produced with an even polyploid number of chromosome sets (e.g., 4N), it can still produce fertile gametes and reproduce.

Two types of polyploidy are encountered in plants. In **autopolyploidy,** all the sets of chromosomes originate in the same species. The condition probably results from a defect in meiosis that leads to diploid or triploid gametes. If a diploid gamete fuses with a normal haploid gamete, the zygote and the organism that develops from it will have three sets of chromosomes; it will be triploid. The cultivated banana is an example of a triploid autopolyploid plant. Because it has an odd number of chromosome sets, the gametes have a variable number of chromosomes and few fertile seeds are set, thereby making most bananas seedless and highly palatable. Because of the triploid state, cultivated bananas are propagated vegetatively (by cuttings). In general, the development of "seedless" fruits such as grapes and watermelons relies on odd-number polyploidy. Triploidy has also been found in grasses, garden flowers, crop plants, and forest trees.

In **allopolyploidy,** the sets of chromosomes involved come from different, though usually related, species. This situation can arise if two different species interbreed to produce an organism with one haploid set of each parent's chromosomes (one set from each species) and then both chromosome sets double. For example, fusion of haploid gametes of two diploid plants that can cross may produce an $N_1 + N_2$ hybrid plant that has a haploid set of chromosomes from plant species 1 and a haploid set from plant species 2. However, because of the differences between the two chromosome sets, pairing of chromosomes does not occur at meiosis, and no viable gametes are produced. As a result, the hybrid plants are sterile. Rarely, through a division error, the two sets of chromo-

somes double, producing tissues of $2N_1 + 2N_2$ genotype. (That is, the cells in the tissue have a diploid set of chromosomes from plant species 1 and a diploid set from plant species 2.) Each diploid set can function normally in meiosis, so that gametes produced from the $2N_1 + 2N_2$ plant are $N_1 + N_2$. Fusion of two gametes like this can produce fully fertile, allotetraploid, $2N_1 + 2N_2$ plants.

A classic example of allopolyploidy resulted from crosses made between cabbages (*Brassica oleracea*) and radishes (*Raphanus sativus*) by G. Karpechenko in 1928. Both parents have a chromosome number of 18, and the F_1 hybrids also have 18 chromosomes, 9 from each parent. These hybrids are morphologically intermediate between cabbages and radishes. The F_1 plants are mostly sterile because of the failure of chromosomes to pair at meiosis. However, a few seeds are produced through meiotic errors, and some of those seeds are fertile. The somatic cells of the plants produced from these seeds have 36 chromosomes, that is, full diploid sets of chromosomes from both the cabbage and the radish. These plants are completely fertile and belong to a breeding species named *Raphanobrassica*, a fusion of the two parental genus names. Morphologically, these plants look a lot like the F_1 hybrids.

Finally, all commercial grains, most crops, and many common commercial flowers are polyploid. In fact, polyploidy is the rule rather than the exception in agriculture and horticulture. For example, the cultivated bread wheat, *Triticum aestivum*, is an allohexaploid with 42 chromosomes. This plant species is descended from three distinct species, each with a diploid set of 14 chromosomes. Meiosis is normal because only homologous chromosomes pair, so the plant is fertile.

KEYNOTE

Variations in the chromosome number of a cell or an organism give rise to aneuploidy, monoploidy, or polyploidy. In aneuploidy, a cell or organism has one, two, or a few whole chromosomes more or less than the basic number of the species under study. In monoploidy, an organism that is usually diploid has only one set of chromosomes. And in polyploidy, an organism has three or more complete sets of chromosomes. Any or all of these abnormal conditions may have serious consequences to the organism.

Summary

In this chapter, we have considered several kinds of chromosomal mutations. Chromosomal mutations are variations from the normal (wild-type) condition in chromosome structure or chromosome number. They may occur spontaneously, or their frequency can be increased by exposure to radiation or chemical mutagens. There are four major types of chromosomal structural mutations: (1) deletion, in which a DNA segment is lost; (2) duplication, in which there are one or more extra copies of a DNA segment; (3) inversion, in which there is a reversal of orientation of a DNA segment in a chromosome; and (4) translocation, in which a DNA segment has moved to a new location in the genome.

The consequences of these kinds of structural mutations depend on the specific mutation involved. First, each kind of mutation involves one or more breaks in a chromosome. If a break occurs within a gene, then a gene mutation has been produced. In deletions, genes may be lost, and multiple mutant phenotypes may result. In some cases, deletions and duplications result in lethality or severe defects as a consequence of a deviation from the normal gene dosage. Second, chromosomal mutations in the heterozygous condition can result in production of some gametes that are inviable because of duplications or deficiencies. Commonly, this is seen after meiotic crossovers that produce inversions and translocations in heterozygotes.

Variations in chromosome number involve departure from the normal diploid (or haploid) state of the organism. For diploids, the three classes of such mutations are (1) aneuploidy, in which one to a few whole chromosomes are lost from or added to the normal chromosome set; (2) monoploidy, in which only one set of chromosomes is present in a usually diploid organism; and (3) polyploidy, in which a cell or organism has three or more sets of chromosomes. The consequences of these chromosomal mutations depend on the organism. In general, plants are more tolerant than animals of variations in the number of chromosome sets; for example, wheat is hexaploid. Although some animal species are naturally polyploid, monoploidy and polyploidy usually are lethal, probably because gene expression problems occur when abnormal numbers of gene copies are present. Even in viable individuals, viable gametes may not result because of segregation problems during meiosis.

Analytical Approaches for Solving Genetics Problems

Q21.1 Diagram the meiotic pairing behavior of the four chromatids in an inversion heterozygote *a b c d e f g*/ *a′ b′ f′ e′ d′ c′ g*. Assume that the centromere is to the left of gene *a*. Next, diagram the early anaphase configuration if a crossover occurred between genes *d* and *e*.

A21.1 This question requires a knowledge of meiosis (see Figure 1.20, p. 19) and the ability to draw and manipulate an appropriate inversion loop. Part (a) of the following figure shows the diagram for the meiotic pairing.

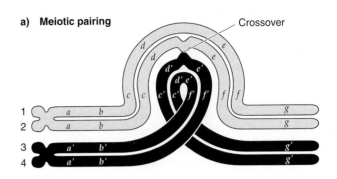

Note that the lower pair of chromatids (*a′*, *b′*, etc.) must loop over for all the genes to align; this looping is characteristic of the pairing behavior expected for an inversion heterozygote.

Once the first diagram has been constructed, answering the second part of the question is straightforward. We diagram the crossover, then trace each chromatid from the centromere end to the other end. It is convenient to distinguish maternal and paternal genes, perhaps by *a′* versus *a*, and so on, as we did in part (a) of the figure. The result of the crossover between *d* and *e* is shown in part (b).

In anaphase I of meiosis, the two centromeres, each with two chromatids attached, migrate toward the opposite poles of the cell. At anaphase, the noncrossover chromatids (top and bottom chromatids in the figure) segregate to the poles normally. As a result of the single crossover between the other two chromatids, however, unusual chromatid configurations are produced, and these configurations are found by tracing the chromatids from left to right. If we begin by tracing the second chromatid from the top, we get $\circ\ a\ b\ c\ d\ e′\ f′\ b′\ a′\ \circ$, a dicentric chromatid (where $\circ$ is a centromere); in other words, we have a single chromatid attached to two centromeres. This chromatid also has duplications and deletions for some of the genes. Thus, during anaphase, this so-called dicentric chromosome becomes stretched between the two poles of the cells as the centromeres separate, and the chromosome eventually breaks at a random location. The other product of the single crossover event is an acentric fragment (without a centromere) that can be traced starting from the right with the second chromatid from the top. This chromatid (*g f e d′ c′ g′*) contains neither a complete set of genes nor a centromere; it is an acentric fragment that will be lost as meiosis continues.

Thus, the consequence of a crossover event within the inversion in an inversion heterozygote is the production of gametes with duplicated or deleted genes. These gametes often are inviable. However, viable gametes are produced from the noncrossover chromatids: One of these chromatids (1 in part [b] of the figure) has the normal gene sequence, and the other (3 in part [b] of the figure) has the inverted gene sequence.

Q21.2 *Eyeless* is a recessive gene (*ey*) on chromosome 4 of *Drosophila melanogaster*. Flies homozygous for *ey* have tiny eyes or no eyes at all. A male fly trisomic for chromosome 4 with the genotype +/+/*ey* is crossed with a normal diploid, eyeless female of genotype *ey/ey*. What expected genotypic and phenotypic ratios would result from random assortment of the chromosomes to the gametes?

A21.2 To answer this question, we must apply our understanding of meiosis to the unusual situation of a trisomic cell. Regarding the *ey/ey* female, only one gamete class can be produced, namely, eggs of genotype *ey*. Gamete production with respect to the trisomy for chromosome 4 occurs by a random segregation pattern in which two chromosomes migrate to one pole and the other chromosome migrates to the other pole during meiosis I. (This pattern is similar to the meiotic segregation pattern shown in secondary nondisjunction of XXY cells; see Chapter 3.) Three types of segregation are possible in the formation of gametes in the trisomy, as shown in part (a) of the following figure. The union of these sperm at random with eggs of genotype *ey* occurs as shown in part (b).

a) Segregation

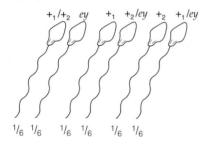

b) Union

		Eggs	Phenotype
		ey	
	+/+	+/+/*ey*	+
	ey	*ey/ey*	*ey*
Sperm	+	+/*ey*	+
	+/*ey*	+/*ey/ey*	+
	+	+/*ey*	+
	+/*ey*	+/*ey/ey*	+

c) Summary of genotypes and phenotypes

Ratios:	Genotypes	Phenotypes
	1/6 +/+/*ey*	5/6 wild type
	1/3 +/*ey/ey*	1/6 eyeless
	1/3 +/*ey*	
	1/6 *ey/ey*	

The resulting genotypic and phenotypic ratios are listed in part (c).

Questions and Problems

*21.1 A normal chromosome has the following gene sequence:

$$A B C D \quad E F G H$$

Determine the chromosomal mutation illustrated by each of the following chromosomes.

a) $A B C F E \quad D G H$

b) $A D \quad E F B C G H$

c) $A B C D \quad E F E F G H$

d) $A B C D \quad E F F E G H$

e) $A B D \quad E F G H$

*21.2 Distinguish between pericentric and paracentric inversions.

21.3 What would be the effect on protein structure if a small inversion were to occur within the amino acid–coding region of a gene?

*21.4 Inversions are known to affect crossing-over. The following homologues with the indicated gene order are given (the filled and open circles are homologous centromeres):

$$\bullet A B C D E$$
$$\circ A D C B E$$

a. Diagram the alignment of these chromosomes during meiosis.

b. Diagram the results of a single crossover between homologous genes *B* and *C* in the inversion.

c. Considering the position of the centromere, what is this sort of inversion called?

21.5 Single crossovers within the inversion loop of inversion heterozygotes give rise to chromatids with duplications and deletions. What happens if, within the inversion loop, there is a two-strand double crossover in such an inversion heterozygote when the centromere is outside the inversion loop?

*21.6 The following gene arrangements in a particular chromosome are found in *Drosophila* populations in different geographic regions. Assuming that the arrangement in part (a) is the original arrangement, in what sequence did the various inversion types probably arise?

a. $A B C D E F G H I$

b. $H E F B A G C D I$

c. $A B F E D C G H I$

d. $A B F C G H E D I$

e. $A B F E H G C D I$

*21.7 A particular plant species that had been subjected to radiation for a long time in order to produce chromosome mutations was then inbred for many generations until it was homozygous for all of these mutations. It was then crossed to the original unirradiated plant, and the meiotic process of the F₁ hybrids was examined. It was noticed that the following structures occurred, at low frequency, in anaphase I of the hybrid: a cell with a dicentric chromosome (bridge) and a fragment.

a. What kind of chromosome mutation occurred in the irradiated plant? In your answer, indicate where the centromeres are.

b. Explain in words and with a clear diagram where crossover(s) occurred and how the bridge chromosome of the cell arose.

*21.8 Mr. and Mrs. Lambert have not yet been able to produce a viable child. They have had two miscarriages and one severely defective child who died soon after birth. Studies of banded chromosomes of father, mother, and child showed that all chromosomes were normal except for pair number 6. The number 6 chromosomes of mother, father, and child are shown in the following figure:

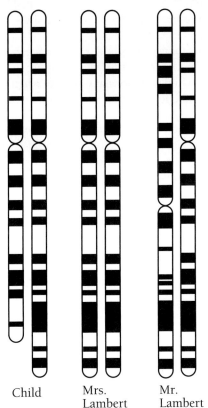

Child   Mrs. Lambert   Mr. Lambert

a. Does either parent have an abnormal chromosome? If so, what is the abnormality?

b. How did the chromosomes of the child arise? Be specific as to what events in the parents gave rise to these chromosomes.

c. Why is the child not phenotypically normal?

d. What can be predicted about future conceptions by this couple?

21.9 Mr. and Mrs. Simpson have been trying for years to have a child but have been unable to conceive. They consulted a physician, and tests revealed that Mr. Simpson had a markedly low sperm count. His chromosomes were studied, and a testicular biopsy was done. His chromosomes proved to be normal, except for pair 12. The following figure shows Mrs. Simpson's normal pair of number 12 chromosomes and Mr. Simpson's number 12 chromosomes.

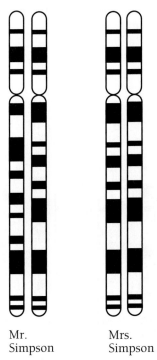

Mr. Simpson   Mrs. Simpson

a. What is the nature of the abnormality in Mr. Simpson's chromosomes of pair number 12?

b. What abnormal feature would you expect to see in the testicular biopsy (cells in various stages of meiosis can be seen)?

c. Why is Mr. Simpson's sperm count low?

d. What can be done about Mr. Simpson's low sperm count?

*21.10 Chromosome I in maize has the gene sequence *ABCDEF*, whereas chromosome II has the sequence *MNOPQR*. A reciprocal translocation resulted in *ABCPQR* and *MNODEF*. Diagram the expected pachytene (see Chapter 1, p. 20) configuration of the F₁ of a cross of homozygotes of these two arrangements.

21.11 Diagram the pairing behavior at prophase of meiosis I (see Chapter 1, pp. 18–20) of a translocation heterozygote that has normal chromosomes of gene order *abcdefg* and *tuvwxyz* and has the translocated chromosomes *abcdvwxyz* and *tuefg*. Assume that the centromere is at the left end of all chromosomes.

*21.12 Mr. and Mrs. Denton have been trying for several years to have a child. They have experienced a series of miscarriages, and last year they had a child with multiple congenital defects. The child died within days of birth. The birth of this child prompted the Dentons' physician to order a chromosome study of parents and child. The results of the study are shown in the following figure. Chromosome banding was done, and all chromosomes were normal in these individuals except some copies of number 6 and number 12. The number 6 and number 12 chromosomes of mother, father, and child are shown in the figure (the number 6 chromosomes are the larger pair).

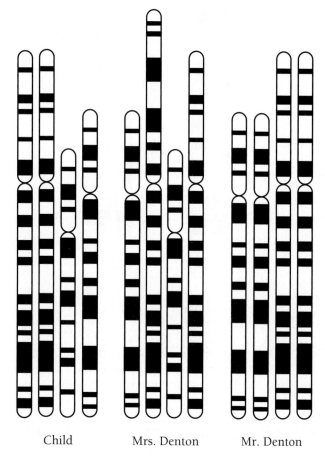

Child   Mrs. Denton   Mr. Denton

a. Does either parent have an abnormal karyotype? If so, which parent has it, and what is the nature of the abnormality?
b. How did the child's karyotype arise (what pairing and segregation events took place in the parents)?
c. Why was the child phenotypically defective?
d. What can this couple expect to occur in subsequent conceptions?
e. What medical help, if any, can be offered to them?

21.13 Irradiation of *Drosophila* sperm produces translocations between the X chromosome and autosomes, between the Y chromosome and autosomes, and between different autosomes. Translocations between the X and Y chromosomes are not produced. Explain the absence of X-Y translocations.

21.14 Define aneuploidy, monoploidy, and polyploidy.

21.15 If a normal diploid cell is 2N, what is the chromosome content of the following?
a. a nullisomic
b. a monosomic
c. a double monosomic
d. a tetrasomic
e. a double trisomic
f. a tetraploid
g. a hexaploid

*21.16 In humans, how many chromosomes would be typical of nuclei of cells that are
a. monosomic?
b. trisomic?
c. monoploid?
d. triploid?
e. tetrasomic?

*21.17 An individual with 47 chromosomes, including an additional chromosome 15, is said to be
a. triplet
b. trisomic
c. triploid
d. tricycle

*21.18 A color-blind man marries a homozygous normal woman, and after four joyful years of marriage, they have two children. Unfortunately, both children have Turner syndrome, although one has normal vision and one is color-blind. The type of color blindness involved is a sex-linked recessive trait.
a. For the color-blind child, did nondisjunction occur in the mother or in the father? Explain your answer.
b. For the child with normal vision, in which parent did nondisjunction occur? Explain your answer.

21.19 Assume that x is a new mutant gene in corn. A female x/x plant is crossed with a triplo-10 individual (trisomic for chromosome 10) carrying only dominant alleles at the x locus. Trisomic progeny are recovered and crossed back to the x/x female plant.
a. What ratio of dominant to recessive phenotypes is expected if the x locus is not on chromosome 10?
b. What ratio of dominant to recessive phenotypes is expected if the x locus is on chromosome 10?

21.20 Why are polyploids with even multiples of the chromosome set generally more fertile than polyploids with odd multiples of the chromosome set?

21.21 One plant species (N = 11) and another plant species (N = 19) produced an allotetraploid. Select the correct answer from the key regarding the following statements:
I. The chromosome number of this allotetraploid is 30.
II. The number of linkage groups of this allotetraploid is 30.
Key:
A. Statement I is true, and Statement II is true.
B. Statement I is true, but Statement II is false.
C. Statement I is false, but Statement II is true.
D. Statement I is false, and Statement II is false.

*21.22   According to Mendel's first law, genes *A* and *a* segregate from each other and appear in equal numbers among the gametes. But Mendel did not know that his plants were diploid. In fact, because plants are frequently tetraploid, he could have been unlucky enough to have started with peas that were 4N rather than 2N. Let us assume that Mendel's peas were tetraploid, that every gamete contains two alleles, and that the distribution of alleles to the gamete is random. Suppose we have a cross of *AAAA* × *aaaa*, where *A* is dominant, regardless of the number of *a* alleles present in an individual.

a.   What will be the genotype of the F_1?

b.   If the F_1 is selfed, what will be the phenotypic ratios in the F_2?

21.23   The root tip cells of an autotetraploid plant contain 48 chromosomes. How many chromosomes were contained by the gametes of the diploid from which this plant was derived?

*21.24   How many chromosomes would be found in somatic cells of an allotetraploid derived from two plants, one with N = 7 and the other with N = 10?

21.25   Plant species A has a haploid complement of four chromosomes. A related species B has five. In a geographic region where A and B are both present, C plants are found that have some characters of both species and somatic cells with 18 chromosomes. What is the chromosome constitution of the C plants likely to be? With what plants would they have to be crossed to produce fertile seed?

A Darwin's finch.

22

Population Genetics

PRINCIPAL POINTS

Population genetics is the subdiscipline of genetics that seeks to understand the causes of observed levels of genetic variation in populations and in so doing to explain the underlying genetic basis for evolutionary change. It includes an empirical aspect, which measures and quantifies the genetic variation in populations, and a theoretical or statistical side, which attempts to explain the variation in terms of mathematical models of the forces that can change gene frequencies.

The genetic structure of a population is described by the total of all alleles (the gene pool). In the case of diploid, sexually interbreeding species, the genetic structure is also characterized by the distribution of alleles into genotypes.

The Hardy-Weinberg law states that in a large, randomly mating population free from evolutionary forces, the allelic frequencies do not change, and the genotypic frequencies stabilize after one generation. In the case of two alleles, A and a, with frequencies p and q, the genotypic proportions at equilibrium are p^2, $2pq$, and q^2.

The Hardy-Weinberg principle and all other population genetics principles apply to alleles defined in a number of ways, including segregating factors that influence phenotypes (such as Mendel observed), protein variants, and any of a variety of differences at the DNA level, including single nucleotide polymorphisms (SNPs), insertions, and deletions.

The genetic structure of a species can vary both geographically and temporally.

The classical and neutral mutation models generate testable hypotheses and are used to explain how much genetic variation should exist within natural populations and what processes are responsible for the observed variation.

Mutation, genetic drift, migration, and natural selection are processes that can alter the allelic frequencies of a population.

For most loci, recurrent mutation changes allelic frequencies at such a slow rate that its effects are negligible. On the other hand, mutation is the initial source of all variation. This implies that once a novel mutation enters a population, other forces predominate in determining its changes in frequency.

Natural selection is differential reproduction of genotypes. It is measured by Darwinian fitness, which is the relative reproductive success of genotypes. Natural selection can produce a number of different effects on the gene pool of a population.

Genetic drift is random change in allelic frequencies caused by random sampling of gametes that occurs in each generation. Genetic drift produces genetic change within populations, genetic differentiation among populations, and loss of genetic variation within populations.

Nonrandom mating affects the genotypic frequencies of a population. Assortative mating can promote polymorphism, whereas inbreeding leads to an increase in homozygosity.

Migration, also called gene flow, involves movement of alleles among populations. Migration can alter the allelic frequencies of a population, and it tends to reduce genetic divergence among populations.

Principles of population genetics can be applied to the management of rare and endangered species. Genetic diversity is best maintained by establishing a population with adequate founders, expanding the population rapidly, avoiding inbreeding, and maintaining an equal sex ratio and equal family size.

i SOON AFTER MENDEL'S PRINCIPLES WERE rediscovered, geneticists began to look not only at the genetic makeup of individuals but also at the genetic makeup of populations. Population genetics allows scientists to determine whether evolution is occurring in groups of individuals as well as to determine the forces that cause populations to evolve. In this chapter, you will learn about changes in the genetic makeup of populations, how such changes are measured, and the factors that cause these changes. Then, in the iActivity, you will explore the genetics of a type of mussel that is rapidly spreading through North American waterways.

The science of genetics can be broadly divided into four major subdisciplines: transmission genetics, molecular genetics, population genetics, and quantitative genetics. Each of these four areas focuses on a different aspect of heredity. **Transmission genetics** is concerned primarily with genetic processes that occur within individuals and how genes are passed from one individual to another. Thus, the unit of study for transmission genetics is the *individual.* In **molecular genetics,** we are interested largely in the molecular nature of heredity: how genetic information is encoded within the DNA and how biochemical processes of the cell translate the genetic information into influencing the phenotype. Consequently, in molecular genetics we focus on the *cell.* **Population genetics,** the subject of this chapter, is the field of genetics that studies heredity in groups of individuals for traits determined by

one or only a few genes. **Quantitative genetics,** the subject of Chapter 23, also considers the heredity of traits in groups of individuals, but the traits of concern are determined by many genes simultaneously. The latter two fields are based on Mendelian principles applied to groups of organisms, and they are amenable to mathematical treatment. In fact, these areas provide the oldest and richest examples of the success of mathematical theory in biology. The impetus for the development of these areas came after the rediscovery of Mendel's work and its great implications for Darwinian theory. In fact, the fusion of Mendelian theory with Darwinian theory is called the neo-Darwinian synthesis and was championed by Sir Ronald Fisher, Sewall Wright, and J. B. S. Haldane (Figure 22.1). The neo-Darwinian synthesis has flourished to become the foundation of a large part of modern biology.

Population geneticists investigate the patterns of genetic variation found among individuals within groups (the **genetic structure** of populations) and how these patterns vary geographically and change over time. In this discipline, our perspective shifts away from the individual and the cell and focuses instead on a Mendelian population. A **Mendelian population** is a group of *interbreeding* individuals who share a common set of genes. The genes shared by the individuals of a Mendelian population are called the **gene pool.** To understand the genetics of the evolutionary process, we study the gene pool of a Mendelian population rather than the genotypes of its individual members. An understanding of the genetic structure of a population is also a key to our

Figure 22.1

The major architects of neo-Darwinian theory. (a) Sir Ronald Fisher. **(b)** Sewall Wright. **(c)** J. B. S. Haldane.

a)

b)

c)

understanding of the importance of genetic resources and the importance of genes for the conservation of species and biodiversity.

The advent of rapid and inexpensive methods for DNA sequencing has resulted in an explosion in the quantity of data showing genetic variation within populations at the DNA sequence level. This is genetic variation at its most fundamental level, and our abilities to discern the forces that act on this variation have increased dramatically in recent years. These data have also opened up new possibilities for the kinds of questions that population genetics can address. By studying mitochondrial DNA sequence differences, for example, we get a picture of the female lineages of a species, including relative amounts of movement, times of origin of population groups, and periods of population expansion. Similarly, DNA sequence variation in the Y chromosome reveals patterns of past movements of males of a species.

Questions frequently studied by population geneticists include the following:

1. How much genetic variation is found in natural populations, and what processes control the amount of variation observed?
2. What processes are responsible for producing genetic divergence among populations?
3. How do biological characteristics of a population, such as breeding system, fecundity, and age structure, influence the gene pool of the population?

To answer these questions, population geneticists often develop mathematical models and equations to describe what happens to the gene pool of a population under various conditions. An example is the set of equations that describes the influence of random mating on the allelic and genotypic frequencies of an infinitely large population, a model called the **Hardy-Weinberg law,** which we discuss later in this chapter. It is important to note that while the models are simple and require numerous assumptions, many of which seem unrealistic, such models are useful because they strip a process to its essence and allow us to test particular attributes of a system in isolation. With such models, we can examine what happens to the genetic structure of a population when we deliberately violate one assumption after another and then in combination. Once we understand the results of the simple models, we can incorporate more realistic conditions into the equations. In the end, we will see that many attributes of genetic variation in populations can be fitted to surprisingly simple models.

KEYNOTE

Population genetics seeks to understand the underlying causes of the observed levels of genetic variation in populations. The field includes both an empirical side, measuring variation in natural populations, and a theoretical or statistical aspect, which attempts to explain the observed variation with quantitative modeling.

Genetic Structure of Populations

Genotypic Frequencies

To study the genetic structure of a Mendelian population, population geneticists must first describe the gene pool of the population quantitatively. This is done by calculating genotypic frequencies and allelic frequencies within the population. A frequency is a proportion, and it always ranges between 0 and 1. If 43 percent of the people in a group have red hair, the frequency of red hair in the group is 0.43. To calculate the **genotypic frequencies** at a specific locus, we count the number of individuals with one particular genotype and divide this number by the total number of individuals in the population. We do this for each of the genotypes at the locus. The sum of the genotypic frequencies should be 1. Consider a locus that determines the pattern of spots in the scarlet tiger moth, *Panaxia dominula* (Figure 22.2). Three genotypes are present in most populations, and each genotype produces a different phenotype. E. B. Ford collected moths at one locality in England and found the following numbers of genotypes: 452 *BB*, 43 *Bb*, and 2 *bb*, for a total of 497 moths. The genotypic frequencies (where f = frequency of) are therefore

$$f(BB) = 452 / 497 = 0.909$$
$$f(Bb) = 43 / 497 = 0.087$$
$$f(bb) = 2 / 497 = \underline{0.004}$$
$$\text{Total} \quad 1.000$$

Figure 22.2

Panaxia dominula, **the scarlet tiger moth.** The top two moths are normal homozygotes (*BB*), those in the middle two rows are heterozygotes (*Bb*), and the bottom moth is the rare homozygote (*bb*).

Allelic Frequencies

Although genotypic frequencies at a single locus are useful for examining the effects of certain evolutionary processes on a population, population geneticists are more likely to use frequencies of alleles to describe a gene pool. The use of **allelic frequencies** offers several advantages over genotypic frequencies. For example, in sexually reproducing organisms, genotypes break down to alleles when gametes are formed, and alleles, not genotypes, are passed from one generation to the next. Consequently, only alleles have continuity over time, and the gene pool evolves through changes in the frequencies of alleles.

Allelic frequencies can be calculated in two ways: from the observed numbers of different genotypes at a particular locus or from the genotypic proportions. First, we can calculate the allelic frequencies directly from the *numbers* of genotypes. In this method, we count the number of alleles of one type at a particular locus and divide it by the total number of alleles at that locus in the population. This method is called *gene counting* and works for a wide variety of cases, including X-linked genes and mitochondrial genes. Expressing the gene-counting method as a formula, we get

$$\text{Allelic frequency} = \frac{\text{Number of copies of a given allele}}{\text{Sum of counts of all alleles in the population}}$$

For example, imagine a population of 1,000 diploid individuals with 353 *AA*, 494 *Aa*, and 153 *aa* individuals. Each *AA* individual has two *A* alleles, whereas each *Aa* heterozygote possesses only a single *A* allele. Therefore, the number of *A* alleles in the population is 2 times the number of *AA* homozygotes plus the number of *Aa* heterozygotes, or $(2 \times 353) + 494 = 1,200$. Since every diploid individual has two alleles, the total number of alleles in the population is twice the number of individuals, or $2 \times 1,000$. Using the formula just given, the allelic frequency is $1,200/2,000 = 0.60$. When two alleles are present at a locus, we can use the following formula to calculate allelic frequencies:

$$p = f(A) = \frac{(2 \times \text{count of } AA) + (\text{count of } Aa)}{2 \times \text{total number of individuals}}$$

The second method of calculating allelic frequencies goes through the step of first calculating genotypic frequencies as demonstrated previously. In this example, $f(AA) = 0.353$, $f(Aa) = 0.494$, and $f(aa) = 0.153$. From these genotypic frequencies, we calculate the allelic frequencies as follows:

$$p = f(A) = (\text{frequency of the } AA \text{ homozygote})$$
$$+ (\tfrac{1}{2} \times \text{frequency of the } Aa \text{ heterozygote})$$

$$q = f(a) = (\text{frequency of the } aa \text{ homozygote})$$
$$+ (\tfrac{1}{2} \times \text{frequency of the } Aa \text{ heterozygote})$$

The frequencies of two alleles, $f(A)$ and $f(a)$, are commonly symbolized as p and q. The allelic frequencies for a locus, like the genotypic frequencies, should always add up to 1. Therefore, once p is calculated, q can be easily obtained by subtraction: $1 - p = q$.

Allelic Frequencies with Multiple Alleles. Suppose we have three alleles—A^1, A^2, and A^3—at a locus, and we want to determine the allelic frequencies. Here, we use the same rule that we used with two alleles: We add up the number of alleles of each type and divide by the total number of alleles in the population:

$$p = f(A^1) = \frac{(2 \times \text{count of } A^1A^1) + (A^1A^2) + (A^1A^3)}{(2 \times \text{total number of individuals})}$$

$$q = f(A^2) = \frac{(2 \times \text{count of } A^2A^2) + (A^1A^2) + (A^2A^3)}{(2 \times \text{total number of individuals})}$$

$$r = f(A^3) = \frac{(2 \times \text{count of } A^3A^3) + (A^1A^3) + (A^2A^3)}{(2 \times \text{total number of individuals})}$$

To illustrate the calculation of allelic frequencies when more than two alleles are present, we will use data from a study of allelic frequencies at a locus that codes for the enzyme phosphoglucomutase (PGM). There are three alleles at this locus; each allele codes for a different molecular variant of the enzyme. In one population sample, the following numbers of genotypes were collected:

$$
\begin{array}{rcl}
A^1A^1 & = & 4 \\
A^1A^2 & = & 41 \\
A^2A^2 & = & 84 \\
A^1A^3 & = & 25 \\
A^2A^3 & = & 88 \\
A^3A^3 & = & 32 \\
\hline
\text{Total} & = & 274
\end{array}
$$

The frequencies of the alleles are calculated as follows:

$$p = f(A^1) = \frac{(2 \times 4) + (41) + (25)}{(2 \times 274)} = 0.135$$

$$q = f(A^2) = \frac{(2 \times 84) + (41) + (88)}{(2 \times 274)} = 0.542$$

$$r = f(A^3) = \frac{(2 \times 32) + (88) + (25)}{(2 \times 274)} = 0.323$$

As seen in these calculations, we add twice the number of homozygotes that possess the allele and one times the count of each of the heterozygotes that have the allele. We then divide by twice the number of individuals in the population, which represents the total number of alleles present. Notice that for each allelic frequency, we do not add all the heterozygotes in the top part of the equation because some of the heterozygotes do not have the allele; for example, in calculating the allelic frequency of A^1, we do not add the number of A^2A^3 heterozygotes in the top part of the equation, since A^2A^3 individuals do not have an

A^1 allele. We can use the same procedure for calculating allelic frequencies when four or more alleles are present.

The second method for calculating allelic frequencies (from genotypic frequencies) can also be used here. This calculation may be quicker if we have already determined the frequencies of the genotypes. The frequency of the homozygote is added to half the heterozygote frequencies because half of each heterozygote's alleles are one allele and half are another allele. If three alleles (A^1, A^2, and A^3) are present in the population, the allelic frequencies are

$$p = f(A^1) = f(A^1A^1) + \frac{f(A^1A^2)}{2} + \frac{f(A^1A^3)}{2}$$

$$q = f(A^2) = f(A^2A^2) + \frac{f(A^1A^2)}{2} + \frac{f(A^2A^3)}{2}$$

$$r = f(A^3) = f(A^3A^3) + \frac{f(A^1A^3)}{2} + \frac{f(A^2A^3)}{2}$$

Allelic Frequencies at an X-Linked Locus. Calculating allelic frequencies at an X-linked locus is slightly more complicated because males have only a single X-linked allele (in mammals and flies, for example). However, we can apply the same principles we used for autosomal loci. Remember that each homozygous female carries two X-linked alleles; heterozygous females have only one of that particular allele, and we will consider the case in which all males have only a single X-linked allele. To determine the number of alleles at an X-linked locus, we multiply the number of homozygous females by 2, then add the number of heterozygous females and the number of hemizygous males. We next divide by the total number of alleles in the population. When determining the total number of alleles, we add twice the number of females (because each female has two X-linked alleles) to the number of males (who have a single allele at X-linked loci). Using this reasoning, the frequencies of two alleles at an X-linked locus (X^A and X^a) are determined with the following equations:

$$p = f(X^A) = \frac{\begin{array}{c}(2 \times X^AX^A \text{ females}) + (X^AX^a \text{ females}) \\ + (X^AY \text{ males})\end{array}}{\begin{array}{c}(2 \times \text{number of females}) \\ + (\text{number of males})\end{array}}$$

$$q = f(X^a) = \frac{\begin{array}{c}(2 \times X^aX^a \text{ females}) + (X^AX^a \text{ females}) \\ + (X^aY \text{ males})\end{array}}{\begin{array}{c}(2 \times \text{number of females}) \\ + (\text{number of males})\end{array}}$$

If the population has the same number of males and females, then the allelic frequencies at an X-linked locus (averaged across sexes) can be determined from the genotypic frequencies as follows:

$$p = f(X^A) = \frac{2}{3}\left[f(X^AX^A) + \frac{1}{2}f(X^AX^a)\right] + \frac{1}{3}f(X^AY)$$

$$q = f(X^a) = \frac{2}{3}\left[f(X^aX^a) + \frac{1}{2}f(X^AX^a)\right] + \frac{1}{3}f(X^aY)$$

This formula assumes that the genotypic frequencies were calculated separately for each sex, so that $f(X^A Y) + f(X^a Y) = 1$. Be sure that you understand the logic behind the gene-counting method; do not just memorize the formulas. If you understand fully the basis of the calculations, you will not need to remember the exact equations and will be able to determine allelic frequencies for any situation.

K E Y N O T E

The genetic structure of a population is determined by the total of all alleles (the gene pool). In the case of diploid, sexually interbreeding individuals, the structure is also characterized by the distribution of alleles into genotypes. The genetic structure can be described in terms of allelic and genotypic frequencies. Except for rare mutations, individuals are born and die with the same set of alleles; what changes genetically over time (evolves) is the hereditary makeup of a group of individuals, reproductively connected in a Mendelian population.

The Hardy-Weinberg Law

The Hardy-Weinberg law serves as a foundation for population genetics because it offers a simple explanation for how the Mendelian principle of segregation influences allelic and genotypic frequencies in a population. The Hardy-Weinberg law is named after the two individuals who independently discovered it in the early 1900s (Box 22.1).

The Hardy-Weinberg law is divided into three parts: a set of assumptions and two major results. A simple statement of the law follows:

Part 1: In an infinitely large, randomly mating population, free from mutation, migration, and natural selection (note that there are five assumptions);

Part 2: The frequencies of the alleles do not change over time; and

Part 3: As long as mating is random, the genotypic frequencies remain in the proportions p^2 (frequency of *AA*), $2pq$ (frequency of *Aa*), and q^2 (frequency of *aa*), where *p* is the allelic frequency of *A* and *q* is the allelic frequency of *a*. The sum of the genotypic frequencies should be equal to 1 (i.e., $p^2 + 2pq + q^2 = 1$).

In short, the Hardy-Weinberg law explains what happens to the allelic and genotypic frequencies of a population as the alleles are passed from generation to generation in the absence of evolutionarily forces. In other words, if the assumptions listed in part 1 are met, alleles would be expected to combine into genotypes based on simple laws of probability, and the population is in Hardy-Weinberg equilibrium. Thus, genotypic frequencies can be predicted from allelic frequencies.

Assumptions of the Hardy-Weinberg Law

Part 1 of the Hardy-Weinberg law presents certain conditions, or assumptions, that must be present for the law to apply. First, the population must be infinitely large. If a population is limited in size, chance deviations from expected ratios can cause changes in allelic frequency, a phenomenon called **genetic drift.** The assumption of infinite size in part 1 is unrealistic: No population has an infinite number of individuals. But large populations look very similar to populations that are mathematically infinitely large. At this point, it is important to understand that populations need not be infinitely large for the Hardy-Weinberg law to provide

---

| **Box 22.1** | **Hardy, Weinberg, and the History of Their Contribution to Population Genetics** |

Godfrey H. Hardy (1877–1947), a mathematician at Cambridge University, often met R. C. Punnett, the Mendelian geneticist, at the faculty club. One day in 1908, Punnett told Hardy of a problem in genetics that he attributed to a strong critic of Mendelism, G. U. Yule (Yule later denied having raised the problem). Supposedly, Yule said that if the allele for short fingers (brachydactyly) was dominant (which it is) and its allele for normal-length fingers was recessive, then short fingers ought to become more common with each generation. In time, almost everyone in Britain should have short fingers. Punnett believed that the argument was incorrect, but he could not prove it.

Hardy was able to write a few equations showing that, given any particular frequency of alleles for short fingers and alleles for normal fingers in a population, the relative number of peo-

ple with short fingers and people with normal fingers will stay the same generation after generation if no natural selection is involved that favors one phenotype or the other in producing offspring. Hardy published a short paper describing the relationship between genotypes and phenotypes in populations, and within a few weeks a paper was published by Wilhelm Weinberg (1862–1937), a German physician of Stuttgart, that clearly stated the same relationship. The Hardy-Weinberg law signaled the beginning of modern population genetics.

To be complete, we should note that in 1903, American geneticist W. E. Castle, of Harvard University, was the first to recognize the relationship between allelic and genotypic frequencies, although it was Hardy and Weinberg who clearly described the relationship in mathematical terms. Still, the law is sometimes called the Castle-Hardy-Weinberg law.

an excellent approximation of genotypic frequencies. In fact, all of the assumptions can tolerate some deviation; the Hardy-Weinberg law still holds.

A second condition of the Hardy-Weinberg law is that mating is random. **Random mating** is mating between genotypes occurring in proportion to the frequencies of the genotypes in the population. More specifically, the probability of a mating between two genotypes is equal to the product of the two genotypic frequencies.

The Hardy-Weinberg requirement of random mating often is misinterpreted. Many students assume, incorrectly, that the population must be interbreeding randomly for all traits for the Hardy-Weinberg law to hold. If this were true, human populations would never obey the Hardy-Weinberg law because humans do not mate randomly. Humans mate preferentially for height, IQ, skin color, socioeconomic status, and other traits. However, although mating is nonrandom for some traits, most humans still mate randomly—for example, for the M-N blood types (see Chapter 4, p. 81). The principles of the Hardy-Weinberg law apply to any locus for which random mating occurs, even if mating is nonrandom for other loci.

Finally, for the Hardy-Weinberg law to work, the population must be free from mutation, migration, and natural selection (described in detail later). In other words, the gene pool must be closed to the addition or subtraction of alleles. Later we discuss these other evolutionary processes and their effect on the gene pool of a population. This condition (that no evolutionary processes act on the population) applies only to the locus in question; a population may be subject to evolutionary processes acting on some genes while still meeting the Hardy-Weinberg assumptions at other loci.

Predictions of the Hardy-Weinberg Law

If the conditions of the Hardy-Weinberg law are met, the population will be in genetic equilibrium, and two results are expected. First, the frequencies of the alleles will not change from one generation to the next, and therefore the gene pool is not evolving at this locus. Second, the genotypic frequencies will be in the proportions p^2, $2pq$, and q^2 after one generation of random mating. Also, the genotypic frequencies will remain constant in these proportions as long as all the conditions of the Hardy-Weinberg law continue to be met. When the genotypes are in these proportions, the population is said to be in *Hardy-Weinberg equilibrium*. An important use of the Hardy-Weinberg law is that it provides a mechanism for determining the genotypic frequencies from the allelic frequencies when the population is in equilibrium. This in turn provides a way to evaluate which assumptions are being violated when the expected theoretical distribution of genotypes does not match an empirically determined distribution.

To summarize, the Hardy-Weinberg law makes several predictions about the allelic frequencies and the genotypic frequencies of a population when certain conditions are satisfied. The necessary conditions are that the population is large, randomly mating, and free from mutation, migration, and natural selection. When these conditions are met, the Hardy-Weinberg law states that allelic frequencies will not change and genotypic frequencies will be determined by the allelic frequencies, occurring in the proportions p^2, $2pq$, and q^2.

Derivation of the Hardy-Weinberg Law

The Hardy-Weinberg law states that when a population is in equilibrium, the genotypic frequencies will be in the proportions p^2, $2pq$, and q^2. To understand why, consider a hypothetical population in which the frequency of allele A is p and the frequency of allele a is q. In producing gametes, each genotype passes on both alleles that it possesses with equal frequency; therefore, the frequencies of A and a in the gametes are also p and q. If one thinks of the gametes as being in a pool, the random formation of zygotes involves reaching into the pool and drawing two gametes at random. The genotypes that then form after repeatedly drawing two gametes at a time will be in frequencies predicted by the probabilities of drawing the particular allele-bearing gametes (see product rule, Chapter 2, p. 37). Table 22.1 shows the combinations of gametes when mating is random. This table illustrates the relationship between the allelic frequencies and the genotypic frequencies, which forms the basis of the Hardy-Weinberg law. We see that when gametes pair randomly, the genotypes will occur in the proportions p^2 (AA), $2pq$ (Aa), and q^2 (aa). These genotypic proportions result from the expansion of the square of the allelic frequencies $(p + q)^2 = p^2 + 2pq + q^2$, and the genotypes reach these proportions after one generation of random mating.

The Hardy-Weinberg law also states that allelic and genotypic frequencies remain constant generation after generation if the population remains large, randomly mating, and free from mutation, migration, and natural selection (i.e., evolutionary processes). This result can also be understood by considering a hypothetical, randomly mat-

Table 22.1	Possible Combinations of *A* and *a* Gametes from Gametic Pools for a Population

		Male gametes	
		A(p)	a(q)
Female gametes	A(p)	AA (p^2)	Aa (pq)
	a(q)	Aa (pq)	aa (q^2)

In sum, p^2 (AA) + $2pq$ (Aa) + q^2 (aa) = 1.00.

ing population, as illustrated in Table 22.2. In Table 22.2, all possible matings are given. By definition, random mating means that the frequency of mating between two genotypes is equal to the product of the genotypic frequencies. For example, the frequency of an $AA \times AA$ mating is equal to p^2 (the frequency of AA) $\times p^2$ (the frequency of AA) $= p^4$. The frequencies of the offspring produced from each mating are also presented in Table 22.2.

We see that the sum of the probabilities of $AA \times Aa$ (or $2p^3q$) and $Aa \times AA$ (or $2p^3q$) matings is $4p^3q$, and we know from Mendelian principles that these crosses produce ½ AA and ½ Aa offspring. Therefore, the probability of obtaining AA offspring from these matings is $4p^3q \times ½ = 2p^3q$. The frequencies of offspring produced by each type of mating are presented in the table. At the bottom of the table, the total frequency for each genotype is obtained by addition. As we can see, after random mating, the genotypic frequencies are still p^2, $2pq$, and q^2, and the allelic frequencies remain at p and q. The frequencies of the population can thus be represented in the zygotic and gametic stages as follows:

Zygotes	Gametes
$p^2 (AA) + 2pq (Aa) + q^2 (aa)$	$p (A) + q (a)$

Each generation of zygotes produces A and a gametes in proportions p and q. The gametes unite to form AA, Aa, and aa zygotes in the proportions p^2, $2pq$, and q^2, and the cycle is repeated indefinitely as long as the assumptions of the Hardy-Weinberg law hold. This short proof gives the theoretical basis for the Hardy-Weinberg law.

The Hardy-Weinberg law indicates that at equilibrium, the genotypic frequencies depend on the fre-

quencies of the alleles. This relationship between allelic frequencies and genotypic frequencies for a locus with two alleles is represented in Figure 22.3. Several aspects of this relationship should be noted: (1) The maximum frequency of the heterozygote is 0.5, and this maximum value occurs only when the frequencies of A and a are both 0.5; (2) if allelic frequencies are between 0.33 and 0.66, the heterozygote is the most numerous genotype; and (3) when the frequency of one allele is low, the homozygote for that allele is the rarest of the genotypes.

This point is also illustrated by the distribution of genetic diseases in humans, which are frequently rare and recessive. For a rare recessive trait, the frequency of the gene causing the trait is much higher than the frequency of the trait itself because most of the rare alleles are in nonaffected heterozygotes (i.e., carriers). Albinism, for example, is a rare recessive condition in humans. In *tyrosinase-negative albinism,* affected individuals have no tyrosinase activity, which is required for normal production of melanin pigment. Among North American whites, the frequency of tyrosinase-negative albinism is roughly 1 in 40,000, or 0.000025. Since albinism is a recessive condition, the genotype of affected individuals is aa. If we assume that the population meets the assumptions of the Hardy-Weinberg law, the frequency of the aa genotypes equals q^2. If $q^2 = 0.000025$, then $q = 0.005$ and $p = 1 - q = 0.995$. The heterozygote frequency is therefore $2pq = 2 \times 0.995 \times 0.005 = 0.00995$ (almost 1 percent). Thus, although the frequency of albinism is low (1 in 40,000), individuals heterozygous for albinism are much more common (almost 1 in 100). When an allele is rare, almost all copies of that allele are in heterozygotes, and recessive phenotypes often are very rare.

Table 22.2	**Algebraic Proof of Genetic Equilibrium in a Randomly Mating Population for One Gene Locus with Two Alleles**

Type of Mating ♀ ♂	Mating Frequency	Offspring Frequencies Contributed to the Next Generation by a Particular Mating		
		AA	**Aa**	**aa**
$p^2 (AA) \times p^2 (AA)$	p^4	p^4	—	—
$p^2 (AA) \times 2pq (Aa)$ ⎤[a] $2pq (Aa) \times p^2 (AA)$ ⎦	$4p^3q$	$2p^3q$	$2p^3q$	—
$p^2 (AA) \times q^2 (aa)$ ⎤ $q^2 (aa) \times p^2 (AA)$ ⎦	$2p^2q^2$	—	$2p^2q^2$	—
$2pq (Aa) \times 2pq (Aa)$	$4p^2q^2$	p^2q^2	$2p^2q^2$	p^2q^2
$2pq (Aa) \times q^2 (aa)$ ⎤ $q^2 (aa) \times 2pq (Aa)$ ⎦	$4pq^3$	—	$2pq^3$	$2pq^3$
$q^2 (aa) \times q^2 (aa)$	q^4	—	—	q^4
Totals	$(p^2 + 2pq + q^2)^2 = 1$	$p^2(p^2 + 2pq + q^2) = p^2$	$2pq(p^2 + 2pq + q^2) = 2pq$	$q^2(p^2 + 2pq + q^2) = q^2$

Genotypic frequencies $= (p + q)^2 = p^2 + 2pq + q^2 = 1$ in each generation afterward.

Allelic frequencies $= p(A) + q(a) = 1$ in each generation afterward.

[a]For example, matings between AA and Aa will occur at $p^2 \times 2pq = 2p^3q$ for $AA \times Aa$ and at $p^2 \times 2pq = 2p^3q$ for $Aa \times AA$ for a total of $4p^3q$. Two progeny types, AA and Aa, result in equal proportions from these matings. Therefore, offspring frequencies are $2p^3q$ (i.e., ½ $\times 4p^3q$) for AA and for Aa.

Figure 22.3

Relationship of the frequencies of the genotypes *AA*, *Aa*, and *aa*, to the frequencies of alleles *A* and *a* [in values of *p* (top abscissa) and *q* (bottom abscissa), respectively] in populations that meet the assumptions of the Hardy-Weinberg law. Any single population is defined by a single vertical line such as $p = 0.3$ and $q = 0.7$.

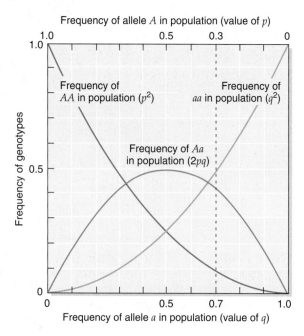

KEYNOTE

> The Hardy-Weinberg law describes what happens to allelic and genotypic frequencies of a large population when gametes fuse randomly and there is no mutation, migration, or natural selection. If these conditions are met, allelic frequencies do not change from generation to generation, and the genotypic frequencies stabilize after one generation in the proportions p^2, $2pq$, and q^2, where p and q equal the frequencies of the alleles in the population.

Extensions of the Hardy-Weinberg Law to Loci with More Than Two Alleles

When two alleles are present at a locus, the Hardy-Weinberg law tells us that at equilibrium, the frequencies of the genotypes is p^2, $2pq$, and q^2, which is the square of the allelic frequencies $(p + q)^2$. If three alleles are present (e.g., alleles *A*, *B*, and *C*) with frequencies equal to p, q, and r, the frequencies of the genotypes at equilibrium are also given by the square of the allelic frequencies:

$$(p + q + r)^2 = p^2 (AA) + 2pq (AB) + q^2 (BB) + 2pr (AC) + 2qr (BC) + r^2 (CC)$$

In the blue mussel found along the Atlantic coast of North America, three alleles are common at a locus cod-

ing for the enzyme leucine aminopeptidase (LAP). For a population of mussels inhabiting Long Island Sound, the frequencies of the three alleles were as follows:

Allele	Frequency
LAP^{98}	$p = 0.52$
LAP^{96}	$q = 0.31$
LAP^{94}	$r = 0.17$

If the population were in Hardy-Weinberg equilibrium, the expected genotypic frequencies would be as follows:

Genotype	Expected Frequency	
LAP^{98}/LAP^{98}	$p^2 = (0.52)^2$	$= 0.27$
LAP^{98}/LAP^{96}	$2pq = 2(0.52)(0.31)$	$= 0.32$
LAP^{96}/LAP^{96}	$q^2 = (0.31)^2$	$= 0.10$
LAP^{96}/LAP^{94}	$2qr = 2(0.31)(0.17)$	$= 0.11$
LAP^{94}/LAP^{98}	$2pr = 2(0.52)(0.17)$	$= 0.18$
LAP^{94}/LAP^{94}	$r^2 = (0.17)^2$	$= 0.03$

Extensions of the Hardy-Weinberg Law to Sex-Linked Alleles

If alleles are X-linked, females may be homozygous or heterozygous, but males carry only a single allele for each X-linked locus. For X-linked alleles in females, the Hardy-Weinberg frequencies are the same as those for autosomal loci: p^2 ($X^A X^A$), $2pq$ ($X^A X^B$), and q^2 ($X^B X^B$). In males, however, the frequencies of the genotypes are p ($X^A Y$) and q ($X^B Y$), the same as the frequencies of the alleles in the population. For this reason, recessive X-linked traits are more frequent among males than among females. To illustrate this concept, consider red-green color blindness, which is an X-linked recessive trait. We actually know that many different defective alleles cause red-green color blindness, but for now let's lump them together. The frequency of the color blindness allele varies among human ethnic groups; the frequency among African Americans is 0.039. At equilibrium, the expected frequency of color-blind males in this group is $q = 0.039$, but the frequency of color-blind females is only $q^2 = (0.039)^2 = 0.0015$.

When random mating occurs within a population, the equilibrium genotypic frequencies are reached in one generation. However, if the alleles are X-linked and the sexes differ in allelic frequency, the equilibrium frequencies are approached over several generations. This is because males receive their X chromosome from their mother only, whereas females receive an X chromosome from both the mother and the father. Consequently, the frequency of an X-linked allele in males is the same as the frequency of that allele in their mothers, whereas the frequency in females is the average of that in mothers and fathers. With random mating, the allelic frequencies in the two sexes oscillate back and forth each generation, and the difference in allelic frequency between the sexes is reduced by half each generation, as shown in Figure 22.4. Once the allelic frequencies of the males and

Figure 22.4

Representation of the gradual approach to equilibrium of an X-linked gene with an initial frequency of 1.0 in females and 0 in males.

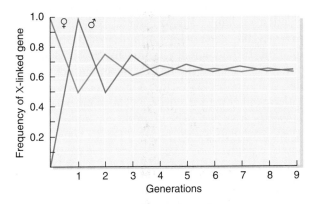

females are equal, the frequencies of the genotypes are in Hardy-Weinberg proportions after one more generation of random mating.

Testing for Hardy-Weinberg Proportions

When we take a sample from a population and calculate genotypic frequencies, it is rarely the case that the match to Hardy-Weinberg proportions is exact. To test whether the fit to Hardy-Weinberg is acceptable, we ask, "What is the chance that we would get this big a departure by chance alone?" If the observed genetic structure does not match the expected structure based on the law, we can begin to ask about which of the assumptions are being violated. To determine whether the genotypes of a population are in Hardy-Weinberg proportions, we first compute the allelic frequencies from the observed genotypic frequencies. Once we have obtained the allelic frequencies, we can calculate the expected genotypic frequencies (p^2, $2pq$, and q^2) and compare these frequencies with the actual observed frequencies of the genotypes using a chi-square test (see Chapter 2). The chi-square test gives us the probability that the difference between what we observed and what we expect under the Hardy-Weinberg law is due to chance.

Note that while there are three classes in the chi-square analysis, there is only one degree of freedom because the frequencies of alleles in a population have no theoretically expected values. Thus, p must be estimated from the observations themselves. So one degree of freedom is lost for every parameter (p in this case) that must be calculated from the data. Another degree of freedom is lost because, for the fixed number of individuals, once all but one of the classes have been determined, the last class has no degree of freedom and is set automatically. Therefore, with three genotypic classes, two degrees of freedom are lost, leaving one degree of freedom.

Using the Hardy-Weinberg Law to Estimate Allelic Frequencies

An important application of the Hardy-Weinberg law is the calculation of allelic frequencies when one or more alleles is recessive. For example, we have seen that albinism in humans results from an autosomal recessive gene. Normally, this trait is rare, but among the Hopi Indians of Arizona, albinism is remarkably common (Figure 22.5). A 1969 study, for example, showed a frequency for the trait of 26/6,000, or 0.0043, which is much higher than the frequency of albinism in most populations. Although we have calculated the frequency of the trait, we cannot directly determine the frequency of the gene for albinism because we cannot distinguish between heterozygous individuals and those homozygous for the normal allele. Recall that our computation of the allelic frequency involves counting the number of alleles:

$$p = \frac{(2 \times \text{number of homozygotes}) + (\text{number of heterozygotes})}{2N}$$

But because heterozygotes for a recessive trait such as albinism cannot be identified, this is impossible. Nevertheless, we can determine the allelic frequency from the Hardy-Weinberg law if we assume that the population is in equilibrium. At equilibrium, the frequency of the homozygous recessive genotype is q^2. For albinism among the Hopi, $q^2 = 0.0043$, and q can be obtained by taking the square root of the frequency of the affected genotype. Therefore, $q = \sqrt{0.0043} = 0.065$, and $p = 1 - q = 0.935$. Following the Hardy-Weinberg law, the frequency of heterozygotes in

Figure 22.5

Three Hopi girls, photographed about 1900. The middle child has albinism, an autosomal recessive disorder that occurs with high frequency among the Hopi Indians of Arizona.

the population is $2pq = 2 \times 0.935 \times 0.065 = 0.122$. Thus, one out of eight Hopis, on average, carries an allele for albinism.

But if the conditions of the Hardy-Weinberg law do not apply, then our estimate of allelic frequency is inaccurate. Also, once we calculate allelic frequencies with the Hardy-Weinberg assumptions, we cannot then test the population to determine whether the genotypic frequencies are in the Hardy-Weinberg expected proportions. To do so would involve circular reasoning, for we assumed Hardy-Weinberg proportions in the first place to calculate the allelic frequencies.

Genetic Variation in Space and Time

The genetic structure of populations can vary in space and time. This means that the frequencies and distribution of alleles can vary in samples of the same species in different areas or samples from the same area collected at different times. Most populations of plants and animals that have widespread geographic distributions show differences in allelic frequencies between component populations. In some cases, the spatial variation shows clear patterns or trends across space. When allelic frequencies change in a systematic way across a geographic transect, we call this an allelic frequency **cline.** Often clines are associated with changes in a physical attribute in the environment, such as temperature.

Because of the importance of geographic variation in allelic frequencies, population geneticists have devised many statistical tools for quantifying the spatial patterns of genetic variation. The simplest of these quantifies the partitioning of total genetic variance into component parts. At the simplest level, we can think of the genetic variance that exists within each local population as one component fraction of the variance and the genetic variance that results from differences between distinct local populations as another component fraction of the variance. By this kind of measure, for example, we generally find that only about 15 percent of the genetic variance in humans is found between different populations, whereas 85 percent of the total genetic variance in humans is shared across populations. Geographic patterns of genetic variation may also be of immense importance for conservation. In terms of the future evolution of a species and conservation of the genetic resources of a species, attention has to be paid to the fact that there is a spatial component to genetic variation. Conservation of genetic diversity demands some quantitative knowledge of the geographic patterns of variability.

KEYNOTE

The genetic structure of a species can vary both geographically and temporally.

Genetic Variation in Natural Populations

One of the most significant questions addressed in population genetics is how much genetic variation exists within natural populations. Genetic variation within populations is important for several reasons. First, it determines the potential for evolutionary change and adaptation. The amount of variation also provides us with important clues about the relative importance of various evolutionary processes because some processes increase variation while others decrease it. The manner in which new species arise may depend on the amount of genetic variation harbored within populations. In addition, the ability of a population to persist over time can be influenced by how much genetic variation it has to draw on should environments change. For all these reasons, population geneticists are interested in measuring genetic variation, attempting to understand the evolutionary processes that affect it, and understanding the effects of human environmental disturbance that may alter it.

Measuring Genetic Variation at the Protein Level

For many years, population geneticists were constrained in quantifying how much variation exists within natural populations. Naturalists recognized that plants and animals in nature frequently differ in phenotype, but the genetic basis of most traits is too complex to assign specific genotypes to individuals. A few traits and alleles that behaved in a Mendelian fashion, such as spot patterns in butterflies and shell color in snails, provided observable genetic variation, but these isolated cases were too few to provide any general estimate of genetic variation. Then, in 1966, population geneticists began to use protein electrophoresis for the study of polymorphism in natural populations. As in the electrophoretic separation of DNA molecules, protein electrophoresis works by separating proteins as they move through a gel matrix. Once they are separated, a specific stain is added to the gel to visualize the protein bands. In one common form of protein electrophoresis, proteins are separated on the basis of a combination of charge, which varies depending on the amino acids, and folding conformation of the protein. So if alleles of the same gene produce protein products with different charge, the proteins can be separated. For population geneticists, electrophoresis provides a technique for quickly determining the genotypes of many individuals at many loci. The amount of genetic variation within a population was commonly measured with two parameters, the **proportion of polymorphic loci** and **heterozygosity.**

A polymorphic locus is any locus that has more than one allele present within a population. The proportion of polymorphic loci (P) is calculated by dividing the number of polymorphic loci by the total number of loci examined. For example, suppose we found that of 33 loci in a population of green frogs, 18 were polymorphic. The proportion of polymorphic loci would be 18/33 = 0.55. Heterozygosity (*H*) is the proportion of an individual's loci that are heterozygous. Suppose we analyzed the genotypes of green frogs from one population at a particular locus and found that the frequency of heterozygotes was 0.09. Heterozygosity for this locus would be 0.09. We would average this heterozygosity with those for other loci and obtain an estimate of heterozygosity for the population. Note that protein electrophoresis misses much of the variation that is detected when the DNA sequence of the same gene is determined because nucleotides may vary at the DNA level but leave no trace of variation at the protein level because of the degeneracy of the genetic code. This means that estimates of heterozygosity and proportion of polymorphic loci are likely to be underestimates of the amount of total genetic variability.

The proportion of polymorphic loci and heterozygosity for many species have been surveyed with electrophoresis. Most species have large amounts of genetic variation in their proteins. This finding ruled out the **classical model** for genetic variation, which stated that most natural populations have little genetic variation. If the classical model is wrong, then what maintains so much genetic variation within populations? An alternative model, the **neutral mutation model,** acknowledges the presence of extensive variation in proteins but proposes that this variation is neutral with regard to natural selection. This does not mean that the proteins detected by electrophoresis have no function but rather that the different genotypes are physiologically equivalent. Therefore, natural selection does not act on the neutral alleles, and random processes such as mutation and genetic drift shape the patterns of genetic variation that we see in natural populations. The neutral mutation model proposes that variation at some loci affects fitness, and natural selection eliminates variation at these loci. The neutral mutation model had strong support that was seriously eroded only after full DNA sequences of alleles were available and more complicated models were found to provide better explanations of the data.

KEYNOTE

In population genetics, one often encounters competing models that seek to explain the amount of variation in natural populations. The classical model predicted that there was little genetic variation in natural populations. Protein electrophoresis revealed abundant genetic variation in natural populations, thus disproving the classical model. The neutral mutation model proposes that the genetic variation detected by electrophoresis is neutral with regard to natural selection.

iActivity You use protein electrophoresis to measure genetic variation in mussel populations in the iActivity *Measuring Genetic Variation* on the website.

Measuring Genetic Variation at the DNA Level

The development of the polymerase chain reaction (PCR; see Chapter 13, pp. 281–283) made it easy for population geneticists to obtain quantities of fragments of genes from large numbers of individuals. These fragments could then be separated directly on gels to determine size differences, cut with restriction enzymes to reveal differences, or directly sequenced. In this section, we will consider a technique for detecting genetic variation using restriction enzymes (see Chapter 13, pp. 267–268). Recall that restriction enzymes make double-stranded cuts in DNA at specific base sequences (see Table 13.1). For example, the restriction enzyme *Bam*HI recognizes the sequence $\begin{smallmatrix}5'\text{-GGATCC-}3'\\3'\text{-CCTAGG-}5'\end{smallmatrix}$.

Suppose that two individuals differ in one or more nucleotides at a particular DNA sequence and that the differences occur at a site recognized by a restriction enzyme (Figure 22.6). One individual has a DNA molecule with the restriction site, but the other individual does not. If the DNA from these two individuals is digested with the restriction enzyme and the resulting fragments are separated on a gel, the two individuals produce different patterns of fragments, as shown in Figure 22.6. The different patterns on the gel are restriction fragment length polymorphisms, or RFLPs (pronounced "riff-lips"; see Chapter 14, p. 290). They indicate that the DNA sequences of the two individuals differ. RFLPs are inherited in the same way that alleles coding for other traits are inherited, only the RFLPs do not produce any outward phenotypes; their phenotypes are the fragment patterns produced on a gel when the DNA is cut by the restriction enzyme.

RFLPs can be used to provide information about how DNA sequences differ among individuals. RFLPs involve only a small part of the DNA, specifically the few nucleotides recognized by the restriction enzyme. However, if we assume that restriction sites occur randomly in the DNA, which is not an unreasonable assumption because the sites are not expressed as traits, the presence or absence of restriction sites can be used to estimate the overall differences in sequence.

To illustrate the use of RFLPs for estimating genetic variation, suppose we isolate DNA from five wild mice and amplify a polymorphic DNA region we want to test by PCR using oligonucleotide primers complementary to each end of the region. Next, we cut the amplified DNA fragments with the restriction enzyme *Bam*HI and separate the fragments using agarose gel electrophoresis. A typical set of restriction patterns that might be obtained is shown in Figure 22.7. Thus, a mouse could be +/+ (the polymorphic restriction site is present on both chromosomes), +/– (the restriction site is present on one chromosome and absent on the other), or –/– (the restriction site is absent on both chromosomes). For the ten chromosomes present among these particular five mice, four have the restriction site and six do not. Heterozygosity at the nucleotide level can be estimated from restriction site patterns.

Figure 22.6

DNA from individual 1 and individual 2 differ in one nucleotide, found within the sequence recognized by the restriction enzyme *Bam*HI. Individual 1's DNA contains the *Bam*HI restriction sequence and is cleaved by the enzyme. Individual 2's DNA lacks the *Bam*HI restriction sequence and is not cleaved by the enzyme. When placed on an agarose gel and separated by electrophoresis, the DNAs from 1 and 2 produce different patterns on the gel, called restriction fragment length polymorphisms.

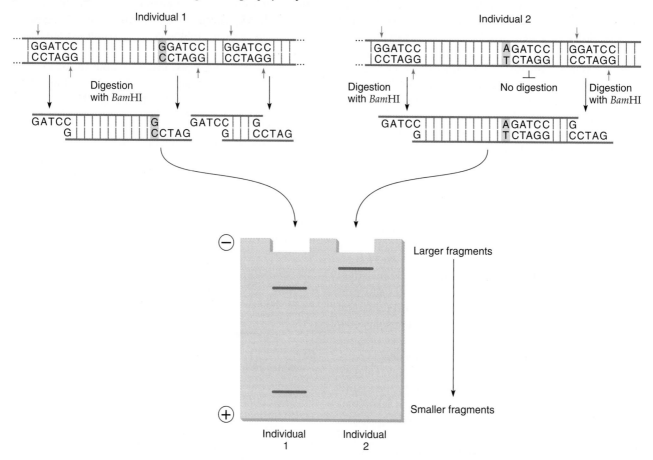

Nucleotide heterozygosity has been studied for a number of different organisms. A few examples are shown in Table 22.3. Nucleotide heterozygosity typically varies from 0.001 to 0.02 in eukaryotic organisms. Recent estimates of nucleotide heterozygosity across the entire human genome average around 0.0008. This means that an individual is heterozygous (contains different nucleotides on the two homologous chromosomes) at about 1 in every 50 to 1,000 nucleotides.

One disadvantage of using RFLPs for examining genetic variation is that this method reveals variation at only a small subset of the nucleotides that make up a gene. With RFLP analysis, we are taking a small sample of the nucleotides (those recognized by the restriction enzyme) and using this sample to estimate the overall level of variation. DNA sequencing, which was described in Chapter 13 (pp. 279–281), provides a method for detecting all nucleotide differences in a set of DNA molecules.

DNA Sequence Variation. We saw earlier that protein electrophoresis misses genetic differences that do not change the protein charge or conformation. The best method for identifying all genetic variation is to obtain the DNA sequence of the gene from each individual. In the first study to apply this approach, Martin Kreitman sequenced 11 copies (obtained from different fruit flies) of the alcohol dehydrogenase gene in *Drosophila melanogaster*. Among the 11 copies, he found different nucleotides at 43 positions within the 2,659-bp segment. Furthermore, only 3 of the 11 copies were identical at all nucleotides examined; thus, there were 8 different alleles (at the nucleotide level) among the 11 copies of this gene. This suggests that populations harbor a tremendous amount of genetic variation in their DNA sequences.

Different regions of a gene apparently are subject to different evolutionary processes, and this is reflected in their different levels of nucleotide diversity. Table 22.4 shows nucleotide diversity estimates for different parts of

| Table 22.5 | Spontaneous Mutation Frequencies at Specific Loci for Various Organisms[a] |

Organism	Trait	Mutations per 100,000 Gametes[b]
T2 Bacteriophage (virus)	To rapid lysis ($r^+ \rightarrow r$)	7
	To new host range ($h^+ \rightarrow h$)	0.001
E. coli K12 (bacterium)	To streptomycin resistance	0.00004
	To phage T1 resistance	0.003
	To leucine independence	0.00007
	To arginine independence	0.0004
	To tryptophan independence	0.006
	To arabinose dependence	0.2
Salmonella typhimurium (bacterium)	To threonine resistance	0.41
	To histidine dependence	0.2
	To tryptophan independence	0.005
Diplococcus pneumoniae (bacterium)	To penicillin resistance	0.01
Neurospora crassa	To adenine independence	0.0008–0.029
	To inositol independence	0.001–0.010
	(One inos allele, JH5202)	1.5
Drosophila melanogaster males	y^+ to yellow	12
	bw^+ to brown	3
	e^+ to ebony	2
	ey^+ to eyeless	6
Corn	Wx to waxy	0.00
	Sh to shrunken	0.12
	C to colorless	0.23
	Su to sugary	0.24
	Pr to purple	1.10
	I to i	10.60
	R^r to r^r	49.20
Mouse	a^+ to nonagouti	2.97
	b^+ to brown	0.39
	c^+ to albino	1.02
	d^+ to dilute	1.25
	ln^+ to leaden	0.80
	Reverse mutations for above genes	0.27
Chinese hamster somatic cell tissue culture	To azaguanine resistance	0.0015
	To glutamine independence	0.014
Humans	Achondroplasia	0.6–1.3
	Aniridia	0.3–0.5
	Dystrophia myotonica	0.8–1.1
	Epiloia	0.4–1
	Huntington disease	0.5
	Intestinal polyposis	1.3
	Neurofibromatosis	5–10
	Osteogenesis imperfecta	0.7–1.3
	Pelger's anomaly	1.7–2.7
	Retinoblastoma	0.5–1.2

[a]Mutations to independence for nutritional substances are from the auxotrophic condition (e.g., *leu*) to the prototrophic condition (e.g., *leu*⁺).
[b]Mutation frequency estimates of viruses, bacteria, *Neurospora*, and Chinese hamster somatic cells are based on particle or cell counts rather than gametes.
Source: From *Genetics*, 3d ed. by Monroe W. Strickberger. Copyright © 1985. Adapted by permission of Prentice Hall, Inc., Upper Saddle River, NJ.

reverse mutation. At this point, no further change in allelic frequency occurs, despite the fact that forward and reverse mutations continue to take place. With some simple algebra, population genetics theorists have shown that the equilibrium frequency for *a* is

$$\hat{q} = \frac{u}{u + v}$$

and the equilibrium value for $\hat{p}$ is

$$\hat{p} = \frac{v}{u + v}$$

Now consider how rapidly this process of pure mutation changes allele frequencies. In a population with the initial allelic frequencies $p = 0.9$ and $q = 0.1$ and mutation rates $u = 5 \times 10^{-5}$ and $v = 2 \times 10^{-5}$, we can calculate the change in allelic frequency in the first generation:

$$\Delta p = vq - up$$
$$= (2 \times 10^{-5} \times 0.1) - (5 \times 10^{-5} \times 0.9)$$

$$\Delta p = -0.000043$$

The frequency of *A* decreases by only four-thousandths of 1 percent. Because mutation rates are so low, the change in allelic frequency due to mutation pressure is exceedingly slow. To change the frequency from 0.50 to 0.49, 2,000 generations are required, and to change it from 0.1 to 0.09, 10,000 generations are necessary. If some reverse mutation occurs, the rate of change is even slower. In practice, mutation by itself changes the allelic frequencies at such a slow rate that populations are rarely in mutational equilibrium. Other processes have more profound effects on allelic frequencies, and mutation alone rarely determines the allelic frequencies of a population.

KEYNOTE

When we study what happens when we violate the assumption of the Hardy-Weinberg equilibrium of the absence of mutation, we see that mutation is the only way in which novel genetic material can come to exist within a species. The larger a population, the more potential there is for a novel mutation to arise.

Random Genetic Drift

Another major assumption of the Hardy-Weinberg law is that the population is infinitely large. Real populations are not infinite in size, but frequently they are large enough that chance factors have small effects on allelic frequencies. Some populations are small, however, and in this case, chance factors may produce large changes in allelic frequencies. Random change in allelic frequency due to chance is called **random genetic drift,** or simply *genetic drift.* Ronald Fisher and Sewall Wright (see Figure 22.1a and b) championed the importance of genetic drift.

Changes in allelic frequency resulting from random events can have important evolutionary implications in small populations. In addition, such changes can have important consequences for the conservation of a rare or endangered species. To see how chance can play a big role in altering the genetic structure of a population, imagine a small group of humans inhabiting a South Pacific island. Suppose that this population consists of only ten individuals, five of whom have green eyes and five of whom have brown eyes. For this example, we assume that eye color is determined by a single locus (actually a number of genes control eye color) and that the allele for green eyes is recessive to brown (*BB* and *Bb* code for brown eyes and *bb* codes for green). The frequency of the allele for green eyes is 0.6 in the island population. A typhoon strikes the island, killing 50 percent of the population, and it is the five with brown eyes. After the typhoon, the allelic frequency for green eyes is 1.0. Evolution has occurred in this population: The frequency of the green-eye allele has changed from 0.6 to 1.0, simply as a result of chance.

Now imagine that the population consists of 1,000 individuals, 50 percent with green eyes and 50 percent with brown eyes. A typhoon strikes the island and kills half the population. In a population of 1,000 individuals, the probability that, by chance, all 500 people who perish will have brown eyes is extremely remote. This example illustrates an important characteristic of genetic drift: Chance factors are likely to produce rapid changes in allelic frequencies only in small populations.

Random factors producing mortality in natural populations, such as the typhoon in the preceding example, is only one of several ways in which genetic drift arises. Chance deviations from expected ratios of gametes and zygotes also produce genetic drift. We have seen the importance of chance deviations from expected ratios in the genetic crosses we studied in earlier chapters. For example, when we cross a heterozygote with a homozygote (*Aa* × *aa*), we expect 50 percent of the progeny to be heterozygous and 50 percent to be homozygous. We do not expect to get exactly 50 percent every time, however, and if the number of progeny is small, the observed ratio may differ greatly from the expected. Recall that the Hardy-Weinberg law is based on random mating and expected ratios of progeny resulting from each type of mating (see Table 22.2). If the actual number of progeny differs from the expected ratio due to chance, genotypes may not be in Hardy-Weinberg proportions. Simply put, random genetic drift may result in changes in allelic frequencies.

Chance deviations from expected proportions arise from a general phenomenon called **sampling error.** Imagine that a population produces an infinitely large pool of gametes, with alleles in the proportions p and q. If random mating occurs and all the gametes unite to form viable zygotes, the proportions of the genotypes will be equal to p^2, $2pq$, and q^2, and the frequencies of the alleles in these zygotes will remain p and q. If the number of progeny is limited, however, the gametes that unite to form the progeny constitute a sample from the infinite pool of potential gametes. Just by chance, or by "error," this sample may deviate from the larger pool; the smaller the sample, the larger the potential deviation.

Bottlenecks and Founder Effects. All genetic drift arises from sampling error, but there are several ways in which sampling error occurs in natural populations. First, as already discussed, genetic drift arises when population size remains continuously small over many generations. Undoubtedly, this situation is common, particularly where populations occupy marginal habitats or when competition for resources limits population growth. In such populations, genetic drift plays an important role in the evolution of allelic frequencies. Moreover, many species are spread out over a large geographic range. This can result in a species consisting of numerous populations of small size, each undergoing drift independently. In addition, human intervention, such as the clear-cutting of forests, can result in the fragmentation of previously large continuous populations into small subdivided ones, again, each showing genetic drift.

Another way genetic drift arises is through **founder effect.** Founder effect occurs when a population is initially established by a small number of breeding individuals. Although the population may subsequently grow in size and later consist of a large number of individuals, the gene pool of the population is derived from the genes present in the original founders. Chance may play a significant role in determining which genes were present among the founders, and this has a profound effect on the gene pool of subsequent generations.

Many examples of founder effect come from the study of human populations. Consider the inhabitants of Tristan da Cunha, a small, isolated island in the South Atlantic. This island was first permanently settled by William Glass and his family in 1817. Three forms of genetic drift occurred in the evolution of the island's population. First, founder effect took place at the initial settlement. By 1855, the population of Tristan da Cunha consisted of about 100 individuals, but 26 percent of the genes of the population in 1855 were contributed by William Glass and his wife. Even in 1961, these original two settlers contributed 14 percent of all the genes in the 300 individuals of the population. The particular genes that Glass and other original founders carried heavily influenced the subsequent gene pool of the population. Second, population size remained small throughout the history of the settlement, and sampling error continually occurred.

A third form of sampling error, called **bottleneck effect,** also played an important role in the population of Tristan da Cunha. Bottleneck effect is a form of genetic drift that occurs when a population is drastically reduced in size. During such a population reduction, some genes may be lost from the gene pool as a result of chance. (Our earlier green eye/brown eye island population

change was an example of bottleneck effect.) Bottleneck effect can be viewed as a type of founder effect because the population is refounded by the few individuals that survive the reduction.

Two severe bottlenecks occurred in the history of Tristan da Cunha. The first took place around 1856 and was precipitated by the death of William Glass and the arrival of a missionary who encouraged the inhabitants to leave the island. At this time, many islanders emigrated to America and South Africa, and the population dropped from 103 individuals at the end of 1855 to 33 in 1857. A second bottleneck occurred in 1885 when 15 adult males drowned after their boat capsized, leaving only 4 adult males on the island, 1 of whom was insane and 2 of whom were old. Many of the widows and their families left the island during the next few years, and the population size dropped from 106 to 59. Both bottlenecks had a major effect on the gene pool of the population. All the genes contributed by several settlers were lost, and the relative contributions of others were altered by these events. Thus, the gene pool of Tristan da Cunha has been influenced by genetic drift in the form of founder effect, small population size, and bottleneck effect.

Effects of Genetic Drift. Genetic drift produces changes in allelic frequencies, and these changes have several effects on the genetic structure of populations. First, genetic drift causes the allelic frequencies of a population to change over time. This is illustrated in Figure 22.8. The different lines represent allelic frequencies in several populations over a number of generations. Although all populations

Figure 22.8

The effect of genetic drift on the frequency (q) of an allele in four populations. Each population begins with q equal to 0.5, and the effective population size for each is 20. The mean frequency of the allele for the four replicates is indicated by the red line. These results were obtained by a computer simulation.

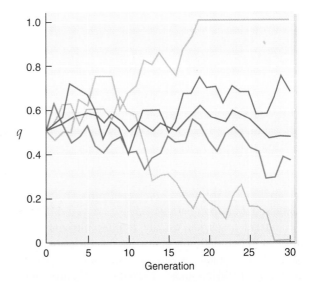

begin with an allelic frequency equal to 0.50, the frequency in each population changes over time as a result of sampling error. In each generation, the allelic frequency may increase or decrease, and over time, the frequencies wander randomly or drift (hence the name *genetic drift*). Sometimes, within 30 generations, just by chance, the allelic frequency reaches a value of 0.0 or 1.0. At this point, one allele is lost from the population and the population is said to be *fixed* for the remaining allele in a one-locus, two-allele example. In fact, in the absence of other factors, genetic drift will always lead to the loss or fixation of an allele. The time it takes for fixation to occur depends on factors such as population size and initial allelic frequency.

Once an allele has reached fixation, no further change in allelic frequency can occur unless the other allele is reintroduced through mutation or migration. If the initial allelic frequencies are equal, which allele becomes fixed is strictly random. On the other hand, if initial allelic frequencies are not equal, the rare allele is more likely to be lost. During this process of genetic drift and fixation, the number of heterozygotes in the population also decreases, and after fixation, the population heterozygosity is zero. As heterozygosity decreases and alleles become fixed, populations lose genetic variation; thus, the second effect of genetic drift is a reduction in genetic variation within populations. Since genetic drift causes random change in allelic frequency, the allelic frequencies in separate, individual populations do not change in the same direction. Therefore, populations diverge in their allelic frequencies through genetic drift.

KEYNOTE

Genetic drift, or chance changes in allelic frequency caused by sampling error, can have important evolutionary and survival implications for small populations. Genetic drift leads to loss of genetic variation within populations, genetic divergence among populations, and random fluctuation in the allelic frequencies of a population over time.

Migration

One of the assumptions of the Hardy-Weinberg law is that the population is isolated and not influenced by other populations. Many populations are not completely isolated, however, and exchange genes with other populations of the same species. Individuals migrating into a population may introduce new alleles to the gene pool and alter the frequencies of existing alleles. Thus, **migration** has the potential to disrupt Hardy-Weinberg equilibrium and may influence the evolution of allelic frequencies within populations.

The term *migration* usually implies movement of organisms. In population genetics, however, we are interested in the movement of genes, which may or may not occur when organisms move. Movement of genes takes place only when organisms or gametes migrate and contribute their genes to the gene pool of the recipient population. This process is also called **gene flow.**

Gene flow has two major effects on a population. First, it introduces new alleles to the population. Since mutation generally is a rare event, a specific mutant allele may arise in one population and not in another. Gene flow spreads unique alleles to other populations and, like mutation, is a source of genetic variation for the population. Second, when the allelic frequencies of migrants and the recipient population differ, gene flow changes the allelic frequencies within the recipient population. Through exchange of genes, different populations remain similar, and thus migration is a homogenizing force that tends to prevent populations from accumulating genetic differences among them.

To illustrate the effect of migration on allelic frequencies, we consider a simple model in which gene flow occurs in only one direction, from population x to population y. Suppose that the frequency of allele A in population x (p_x) is 0.8 and the frequency of A in population y (p_y) is 0.5. Each generation, some individuals migrate from population x to population y, and these migrants are a random sample of the genotypes in population x. After migration, population y actually consists of two groups of individuals: the migrants, with $p_x = 0.8$, and the residents, with $p_y = 0.5$. The migrants now make up a proportion of population y, which we designate m. The frequency of A in population y after migration (p'_y) is

$$p'_y = mp_x + (1 - m)p_y$$

We see that the frequency of A after migration is determined by the proportion of A alleles in the two groups that now make up population y. The first component, mp_x, represents the A alleles in the migrants; we multiply the proportion of the population that consists of migrants (m) by the allelic frequency of the migrants (p_x). The second component represents the A alleles in the residents and equals the proportion of the population consisting of residents ($1 - m$) multiplied by the allelic frequency of the residents (p_y). Adding these two components together gives us the allelic frequency of A in population y after migration.

The change in allelic frequency in population y as a result of migration (Δp) equals the original frequency of A subtracted from the frequency of A after migration:

$$\Delta p = p'_y - p_y$$

Since by the previous equation, $p'_y = mp_x + (1 - m)p_y$, the change in allelic frequency can be written as

$$\Delta p = mp_x + (1 - m)p_y - p_y$$

Multiplying $(1 - m)$ by p_y in this equation, we obtain

$$\Delta p = mp_x + p_y - mp_y - p_y$$
$$\Delta p = mp_x - mp_y$$
$$\Delta p = m(p_x - p_y)$$

This final equation indicates that the change in allelic frequency from migration depends on two factors: the proportion of the migrants in the final population and the difference in allelic frequency between the migrants and the residents. If no differences exist in the allelic frequency of migrants and residents ($p_x - p_y = 0$), then we can see that the change in allelic frequency is zero. Populations must differ in their allelic frequencies for migration to affect the makeup of the gene pool. With continued migration, p_x and p_y become increasingly similar, and as a result, the change in allelic frequency due to migration decreases. Eventually, allelic frequencies in the two populations will be equal, and no further change will occur. However, this is true only when other factors besides migration do not influence allelic frequencies.

Migration reduces divergence among populations, effectively increasing the size of the individual populations. Extensive gene flow has been shown to occur, for example, among populations of the monarch butterfly (Figure 22.9). Even a small amount of gene flow can reduce the effect of genetic drift. Calculations have shown that a single migrant moving between two populations every other generation will prevent the two populations from becoming fixed for different alleles.

KEYNOTE

Migration of individuals into a population may alter the makeup of the population gene pool if the genes carried by the migrants differ from those of the resident population. Migration, also called gene flow, tends to reduce genetic divergence among populations and increases the effective size of the population.

Natural Selection

Mutation, genetic drift, and migration alter the gene pool of a population, and they certainly influence the evolution of a species. However, mutation, migration, and genetic drift do not result in adaptation. Adaptation is the process by which traits evolve that make organisms more suited to their immediate environment; these traits increase the organism's chances of surviving and reproducing. Adaptation is responsible for the many extraordinary traits seen in nature: wings that enable a hummingbird to fly backward, leaves that enable the pitcher plant to capture and devour insects, and brains that allow humans to speak, read, and love. These biological features and countless other exquisite traits are the products of adaptation (Figure 22.10). Genetic drift, mutation, and migration all influence the pattern and process of adaptation, but adaptation arises chiefly from natural selection. **Natural selection** is the dominant force in the evolution of many traits and has shaped much of the phenotypic variation observed in nature.

animation

a Hardy-Weinberg and Natural Selection

Charles Darwin and Alfred Russell Wallace (Figure 22.11) independently developed the concept of natural selection in the mid-nineteenth century. For his innumerable contributions to our understanding of natural selection, Darwin often is regarded as the father of evolutionary theory. What is amazing about this theory is that Darwin had no clue about how genetic transmission worked. All that was necessary for his theory was that somehow offspring resemble their parents. Knowing the details of genetic transmission of traits makes the theory much deeper and richer and provides much more satisfying tests of how evolutionary change works at the genetic level.

Natural selection can be defined as differential reproduction of genotypes. It simply means that individuals with certain genes produce more offspring than others; therefore, those genes increase in frequency in the next generation, as further addressed in Chapter 23. Through natural selection, traits that contribute to survival and

Figure 22.9

Extensive gene flow occurs among populations of the monarch butterfly. (a) The butterflies overwinter in Mexico and then **(b)** migrate north during spring and summer to breeding grounds as far away as northern Canada. Extensive gene flow occurs during the migration period, with the result that monarch populations display little genetic divergence.

a)

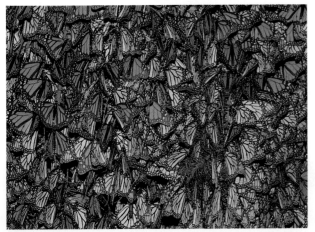

b)

Figure 22.10

Natural selection and adaptation. Natural selection produces organisms that are finely adapted to their environment. This lizard displays cryptic coloration, which allows it to blend in with its natural surroundings.

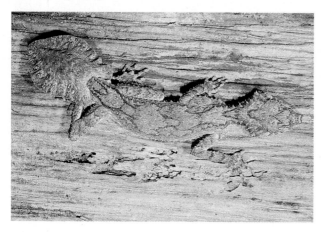

reproduction increase over time. In this way, organisms adapt to their environment.

Selection in Natural Populations. A classic example of selection in natural populations is the evolution of melanic (dark) forms of moths in association with industrial pollution, a phenomenon known as industrial melanism. One of the best-studied cases involves the peppered moth, *Biston betularia*. The common phenotype of this species, called the *typical* form, is a greyish-white color with black mottling over the body and wings.

Before 1848, all peppered moths collected in England possessed this *typical* phenotype, but in 1848, a single black moth was collected near Manchester, England. This new phenotype, called *carbonaria*, presumably arose by mutation and rapidly increased in frequency around Manchester and in other industrial regions. By 1900, the *carbonaria* phenotype had reached a frequency of more than 90 percent in several populations. High frequencies of *carbonaria* appeared to be associated with industrial regions, whereas the *typical* phenotype remained common in more rural districts.

H. B. D. Kettlewell investigated color polymorphism in the peppered moth, demonstrating that the increase in the *carbonaria* phenotype occurred as a result of strong selection against the *typical* form in polluted woods. Peppered moths are nocturnal; during the day, they rest on the trunks of lichen-covered trees. Birds often prey on the moths during the day, but because the lichens that cover the trees are naturally grey in color, the *typical* form of the peppered moth is well camouflaged against this background (Figure 22.12a). In industrial areas, however, extensive pollution beginning with the industrial revolution in the mid-nineteenth century had killed most of the lichens and covered the tree trunks with black soot. Against this black background, the *typical* phenotype was conspicuous and was readily consumed by birds. In contrast, the *carbonaria* form was well camouflaged against the blackened trees and had a higher rate of survival than the *typical* phenotype in polluted areas (Figure 22.12b). Because *carbonaria* survived better in polluted woods, more *carbonaria* genes were transmitted to the next generation; thus, the *carbonaria* phenotype increased in frequency in industrial areas. In rural areas, where pollution was absent, the *carbonaria* phenotype was conspicuous and the *typical* form was camouflaged; in these regions, the frequency of the *typical* form remained high.

Figure 22.11

(a) Charles Darwin and (b) Alfred Russel Wallace. Darwin and Wallace deserve equal credit for developing the theory of evolution through natural selection.

a)

b)

Figure 22.12

***Biston betularia*, the peppered moth, and its dark form *carbonaria* (a) on the trunk of a lichen-covered tree in the unpolluted countryside and (b) on the trunk of a tree with dark bark.** On the lichened tree, the dark form of the moth is readily seen, whereas the light form is well camouflaged. On the dark tree, the dark form of the moth is well camouflaged.

a)

b)

Fitness and Coefficient of Selection. Darwin described natural selection primarily in terms of survival. However, what is most important in the process of natural selection is the relative number of genes that are contributed to future generations. Therefore, we measure natural selection by assessing reproduction. Natural selection is measured in terms of **Darwinian fitness,** which is defined as the relative reproductive ability of a genotype.

Darwinian fitness is often symbolized as W. Since it is a measure of the relative reproductive ability, population geneticists usually assign a fitness of 1 to a genotype that produces the most offspring. The fitnesses of the other genotypes are assigned relative to this. For example, suppose that the genotype G^1G^1 on the average produces eight offspring, G^1G^2 produces an average of four offspring, and G^2G^2 produces an average of two offspring. The G^1G^1 genotype has the highest reproductive output, so its fitness is 1 ($W_{11} = 1.0$). Genotype G^1G^2 produces on the average four offspring for the eight produced by the most fit genotype, so the fitness of G^1G^2 (W_{12}) is $4/8 = 0.5$. Similarly, G^2G^2 produces two offspring for the eight produced by G^1G^1, so the fitness of G^2G^2 (W_{22}) is $2/8 = 0.25$. Table 22.6 illustrates the calculation of relative fitness values.

Darwinian fitness tells us how well a genotype is doing in terms of natural selection. A related measure is the **selection coefficient,** which is a measure of the relative intensity of selection against a genotype. The selection coefficient is symbolized by s and equals $1 - W$. In our example, the selection coefficient for G^1G^1 is $s = 0$; for G^1G^2, $s = 0.5$; and for G^2G^2, $s = 0.75$.

Effect of Selection on Allelic Frequencies. Natural selection produces a number of different effects: At times, natural

selection eliminates genetic variation, while at other times, it maintains variation; it can change allelic frequencies or prevent allelic frequencies from changing; it can produce genetic divergence between populations or maintain genetic uniformity. Which of these effects occurs depends primarily on the relative fitness of the genotypes and on the frequencies of the alleles in the population.

The change in allelic frequency that results from natural selection can be calculated by constructing a table such as Table 22.7. This table method can be used for any type of single-locus trait, whether the trait is dominant, codominant, recessive, or overdominant. To use the table method, we begin by listing the genotypes (A^1A^1, A^1A^2, and A^2A^2) and their initial frequencies. If random mating has just taken place, the genotypes are in Hardy-Weinberg proportions, and the initial frequencies are p^2, $2pq$, and q^2. We then list the fitness for each of the genotypes, W_{11}, W_{12}, and W_{22}. Now, suppose that selection occurs and only some of the genotypes survive. The contribution of each genotype to the next generation is equal to the initial frequency of the genotype multiplied by its fitness. For A^1A^1, this is $p^2 \times W_{11}$. Notice that the contributions of the three genotypes do not add up to 1. We calculate the relative contribution of each genotype by dividing each by the mean fitness of the population. The *mean fitness of the population* equals $p^2 W_{11} + 2pq\ W_{12} + q^2 W_{22} = \overline{W}$. This gives us the relative frequencies of the genotypes after selection. We then calculate the new allelic frequency (p') from the genotypes after selection, using our familiar formula $p' = $ (frequency of A^1A^1) + ($1/2 \times$ frequency of A^1A^2). Finally, the change in allelic frequency resulting from selection equals $p' - p$. A sample calculation using some actual allelic frequencies and fitness values is presented in Table 22.8 (p. 480).

Table 22.6	Computation of Fitness Values and Selection Coefficients of Three Genotypes		
	Genotypes		
	G^1G^1	G^1G^2	G^2G^2
Number of breeding adults in one generation	16	10	20
Number of offspring produced by all adults of the genotype in the next generation	128	40	40
Average number of offspring produced per breeding adult	$^{128}/_{16} = 8$	$^{40}/_{10} = 4$	$^{40}/_{20} = 2$
Fitness W (relative number of offspring produced)	$^8/_8 = 1$	$^4/_8 = 0.5$	$^2/_8 = 0.25$
Selection coefficient ($s = 1 - W$)	$1 - 1 = 0$	$1 - 0.5 = 0.5$	$1 - 0.25 = 0.75$

Natural selection results in a characteristic pattern of change in the genetic structure of a population, depending on the starting genotypic frequencies. In some cases, a form of natural selection called *directional selection* results in the elimination or at least great reduction in one of the alleles.

Selection Against a Recessive Trait. The directional effect is the one that is most often associated with natural selection. The case of the peppered moth discussed earlier falls into the category of selection against a recessive allele. We will discuss this type of case in more detail because many important traits and most new mutations are recessive and have reduced fitness. When a trait is completely recessive, both the heterozygote and the dominant homozygote have a fitness of 1, whereas the recessive homozygote has reduced fitness, as shown here:

Genotype	Fitness (W)
AA	1
Aa	1
aa	$1 - s$

If the genotypes are initially in Hardy-Weinberg proportions, the contribution of each genotype to the next generation is the frequency times the fitness:

AA	$p^2 \times 1 = p^2$
Aa	$2pq \times 1 = 2pq$
aa	$q^2 \times (1 - s) = q^2 - sq^2$

The mean fitness of the population is $p^2 + 2pq + q^2 - sq^2$. Since $p^2 + 2pq + q^2 = 1$, the mean fitness becomes $1 - sq^2$, and the normalized genotypic frequencies after selection are

AA	$\dfrac{p^2}{1 - sq^2}$
Aa	$\dfrac{2pq}{1 - sq^2}$
aa	$\dfrac{q^2 - sq^2}{1 - sq^2}$

Table 22.7	General Method of Determining Change in Allelic Frequency Caused by Natural Selection		
	Genotypes		
	A^1A^1	A^1A^2	A^2A^2
Initial genotypic frequencies	p^2	$2pq$	q^2
Fitness[a]	W_{11}	W_{12}	W_{22}
Frequency after selection	$p^2 W_{11}$	$2pq W_{12}$	$q^2 W_{22}$
Relative genotypic frequency after selection[b]	$P' = \dfrac{p^2 W_{11}}{W}$	$H' = \dfrac{2pq W_{12}}{W}$	$Q' = \dfrac{q^2 W_{22}}{W}$

Allelic frequency after selection: $p' = P' + \frac{1}{2}(H')$
$q' = 1 - p'$
Change in allelic frequency caused by selection: $\Delta p = p' = p$

[a]For simplicity, fitness in this example is considered the probability of survival. Change in allelic frequency caused by differences in the number of offspring produced by the genotypes is calculated in the same manner.

[b]$W = p^2 W_{11} + 2pq W_{12} + q^2 W_{22}$.

To obtain q', the frequency after selection, we add the frequency of the aa homozygote and half the frequency of the heterozygote:

$$q' = \frac{q^2 - sq^2}{1 - sq^2} + \frac{1}{2} \times \frac{2pq}{1 - sq^2}$$

$$= \frac{q^2 - sq^2}{1 - sq^2} + \frac{pq}{1 - sq^2}$$

$$= \frac{q^2 - sq^2 + pq}{1 - sq^2}$$

$$= \frac{q^2 + pq - sq^2}{1 - sq^2}$$

$$= \frac{q(q + p) - sq^2}{1 - sq^2}$$

Since $q + p = 1$,

$$q' = \frac{q - sq^2}{1 - sq^2}$$

Therefore, the change in the frequency of a after one generation of selection is

$$\Delta q = q' - q$$

$$= \frac{q - sq^2}{1 - sq^2} - q$$

$$= \frac{q - sq^2}{1 - sq^2} - \frac{q(1 - sq^2)}{(1 - sq^2)}$$

$$= \frac{q - sq^2 - q(1 - sq^2)}{1 - sq^2}$$

$$= \frac{q - sq^2 - q + sq^3}{1 - sq^2}$$

$$= \frac{-sq^2 + sq^3}{1 - sq^2}$$

$$= \frac{-sq^2(1 - q)}{1 - sq^2}$$

$\Delta q = -spq^2/(1 - sq^2)$ because $1 - q = p$. When $\Delta q = 0$, no further change occurs in allelic frequencies. Notice that there is a minus sign to the left of spq^2; because the values of s, p, and q are always positive or zero, Δq is negative or zero. Thus, the value of q decreases with selection.

Selection also depends on the actual frequency of the allele in the population. Figure 22.13 shows the magnitude of change in allelic frequency for each generation in three populations with different initial allelic frequencies. Population 1 begins with allelic frequency $q = 0.9$; population 2 begins with $q = 0.5$; and population 3 begins with $q = 0.1$. In this example, the homozygous recessive genotype (aa) has a fitness of 0 and the other two genotypes (AA and Aa) have a fitness of 1 (recessive lethal condition). When the frequency of q is high, as in population 1,

Figure 22.13

Effectiveness of selection against a recessive lethal genotype at different initial allelic frequencies. The three populations had initial frequencies of 0.9, 0.5, and 0.1.

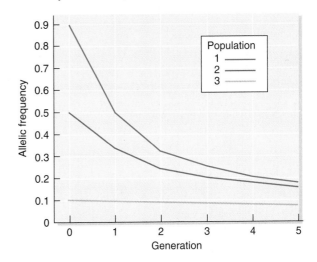

the change in allelic frequency is large; in the first generation, q drops from 0.9 to 0.47. However, when q is small, as in population 3, the change in q is much less; here q drops from 0.1 to 0.091. So as q becomes smaller, the change in q becomes less. Because of this diminishing change in frequency, it is almost impossible to eliminate a recessive trait from the population entirely. This is also understood if one realizes that the final recessive alleles in a population will almost always find themselves in the heterozygote condition. However, this result applies only to completely recessive traits; if the fitness of the heterozygote is also reduced, the change in allelic frequency will be more rapid because now selection also acts against the heterozygote in addition to the homozygote.

Heterozygote Superiority. Natural selection does not always result in a directional change in allele frequency and a decrease in genetic variation. Some forms of selection result in the maintenance of genetic variation, a balancing of selection, if you will. The simplest type of balancing selection is called **heterosis** (also called **overdominance** or **heterozygote superiority**). An equilibrium of allelic frequencies arises when the heterozygote has higher fitness than either of the homozygotes. In this case, both alleles are maintained in the population because both are favored in the heterozygote genotype. Allelic frequencies will change as a result of selection until the equilibrium point is reached and then will remain stable. The allelic frequencies at which the population reaches equilibrium depend on the relative fitness of the two homozygotes. If the selection coefficient of AA is s and the selection coefficient of aa is t, it can be shown algebraically that at equilibrium,

$$\hat{p} = f(A) = \frac{t}{s + t}$$

Table 22.8	General Method of Determining Change in Allelic Frequency Caused by Natural Selection When Initial Allelic Frequencies Are $p = 0.6$ and $q = 0.4$		

| | Genotypes | | |
	A^1A^1	A^1A^2	A^2A^2
Initial genotypic frequencies	p^2 $(0.6)^2 = 0.36$	$2pq$ $2(0.6)(0.4) = 0.48$	q^2 $(0.4)^2 = 0.16$
Fitness	$W_{11} = 0$	$W_{12} = 0.4$	$W_{22} = 1$
Frequency after selection	$p^2\, W_{11} =$ $(0.36)(0) = 0$	$2pq\, W_{12} =$ $(0.48)(0.4) = 0.19$	$q^2\, W_{22} =$ $(0.16)(1) = 0.16$
Relative genotypic frequency after selection[a]	$P' = \dfrac{p^2\, W_{11}}{\overline{W}}$ $P' = 0/0.35 = 0$	$H' = \dfrac{2pq\, W_{12}}{\overline{W}}$ $H' = 0.19/0.35$ $= 0.54$	$Q' = \dfrac{q^2\, W_{22}}{\overline{W}}$ $Q' = 0.16/0.35$ $= 0.46$

Allelic frequency after selection: $p' = P' + \frac{1}{2}(H')$
$p' = 0 + \frac{1}{2}(0.54) = 0.27$
$q' = 1 - p' = 1 - 0.27 = 0.73$
Change in allelic frequency caused by selection: $\Delta p = p' - p$
$\Delta p = 0.27 - 0.6 = -0.33$

[a]$\overline{W} = p^2\, W_1 + 2pq\, W_{12} = q^2\, W_{22}$
$= 0 + 0.19 + 0.16$
$= 0.35$

and

$$\hat{q} = f(a) = \frac{s}{s + t}$$

Notice that if selection against both homozygotes is the same (i.e., $s = t$), then the equilibrium allelic frequency is 0.5. As selection against the homozygotes becomes less symmetrical, the equilibrium allelic frequency shifts in the direction of the most fit homozygote.

The most famous example of heterozygote superiority operating in nature is provided by human sickle-cell anemia. Sickle-cell anemia results from a mutation in the gene coding for β-hemoglobin. In some populations, there are three hemoglobin genotypes: *Hb-A/Hb-A*, *Hb-A/Hb-S*, and *Hb-S/Hb-S*. Individuals with the *Hb-A/Hb-A* genotype have completely normal red blood cells; *Hb-S/Hb-S* individuals have sickle-cell anemia; and *Hb-A/Hb-S* individuals have sickle-cell trait, a mild form of sickle-cell anemia. In an environment in which malaria is common, the heterozygotes are at a selective advantage over the two homozygotes. Apparently, the abnormal hemoglobin mixture in the heterozygotes provides an unfavorable environment for the growth or maintenance of the malarial parasite in the red cell. The heterozygotes therefore have greater resistance to malaria and thus higher fitness than *Hb-A/Hb-A* individuals. The *Hb-S/Hb-S* individuals, meanwhile, are at a serious selective disadvantage because they have sickle-cell anemia. As a result, in malaria-infested areas in which the *Hb-S* gene is also found, an equilibrium state is established in which a significant number of *Hb-S* alleles are found in heterozygotes because of the selective advantage of this genotype. The distributions of malaria and the *Hb-S* allele are illus-

trated in Figure 22.14. Thus, despite the problems faced by *Hb-S* homozygotes, natural selection cannot eliminate this allele from the population because the allele has beneficial effects in the heterozygote state.

KEYNOTE

Natural selection involves differential reproduction of genotypes and is measured in terms of Darwinian fitness, the relative reproductive contribution of a genotype. The effects of selection depend on the relative fitness of the different genotypes. Directional selection results in a directional change in allele frequency, with the disfavored allele being eliminated from the population in cases where it is dominant or codominant but persisting in the population at low frequencies if it is invisible in the heterozygote. In either case, directional selection decreases the amount of genetic variation in a population. Balancing selection, exemplified here as heterozygote superiority, results in the maintenance of genetic variation in the population.

Balance Between Mutation and Selection

As we have seen, natural selection can reduce the frequency of a deleterious recessive allele. As the frequency of the allele becomes low, the change in frequency diminishes with each generation. When the allele is rare, the change in frequency is very slight. Opposing this reduction in the allele's frequency due to selection is mutation pressure, which continually produces new alleles and tends to increase the frequency. Eventually, a

Figure 22.14

The distribution of malaria caused by the parasite *Plasmodium falciparum* coincides with distribution of the *Hb-S* allele for sickle-cell anemia. The frequency of *Hb-S* is high in areas where malaria is common because *Hb-A/Hb-S* heterozygotes are resistant to malarial infection.

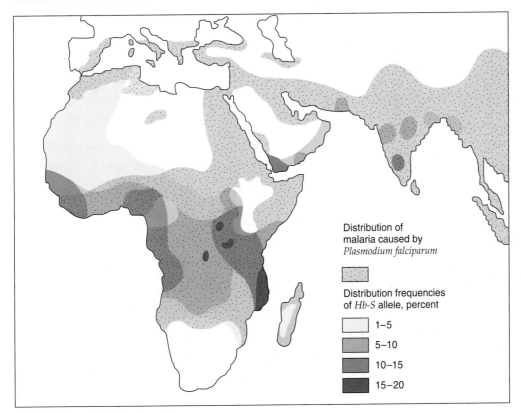

Distribution of
malaria caused by
Plasmodium falciparum

Distribution frequencies
of *Hb-S* allele, percent

1–5

5–10

10–15

15–20

balance, or equilibrium, is reached, in which the input of new alleles by recurrent mutation is exactly counterbalanced by the loss of alleles through natural selection. When equilibrium is obtained, the allele's frequency remains stable, despite the fact that selection and mutation continue, unless the equilibrium is perturbed by some other process.

Consider a population in which selection occurs against a deleterious recessive allele, *a*. As we saw earlier (pp. 478–479), the amount *a* will change in one generation (Δq) as a result of selection is

$$\Delta q = -\frac{spq^2}{1 - sq^2}$$

For a rare recessive allele, q^2 will be near 0, and the denominator in this equation, $1 - sq^2$, will be approximately 1, so that the decrease in frequency caused by selection is given by

$$\Delta q = -spq^2$$

At the same time, the frequency of the *a* allele increases as a result of mutation from *A* to *a*. Provided the frequency of *a* is low, the reverse mutation of *a* to *A* essentially can be ignored. Equilibrium between selection and

mutation occurs when the decrease in allelic frequency produced by selection is the same as the increase produced by mutation:

$$spq^2 = up$$

We can predict the frequency of *a* at equilibrium ($\hat{q}$) by simplifying this equation:

$$sq^2 = u$$
$$q^2 = u/s$$

and

$$\hat{q} = \sqrt{u/s}$$

If the recessive homozygote is lethal ($s = 1$), the equation becomes

$$\hat{q} = \sqrt{u}$$

As an example of the balance between mutation and selection, consider a recessive gene for which the mutation rate is 10^{-6} and $s = 0.1$. At equilibrium, the frequency of the gene will be $\hat{q} = \sqrt{10^{-6}/0.1} = 0.0032$. Most recessive deleterious traits remain within a population at low frequency because of equilibrium between mutation and selection.

For a dominant allele A, the frequency at equilibrium ($\hat{p}$) is

$$\hat{p} = u/s$$

If the mutation rate is 10^{-6} and $s = 0.1$, the frequency of the dominant gene at equilibrium is $10^{-6}/0.1 = 0.00001$, which is considerably less than the equilibrium frequency for a recessive allele with the same fitness and mutation rate. This is because selection cannot act on a recessive allele in the heterozygote state, whereas both the homozygote and the heterozygote for a deleterious dominant allele have reduced fitness. For this reason, detrimental dominant alleles generally are less common than recessive ones.

Assortative Mating

A fundamental assumption of the Hardy-Weinberg law is that members of the population mate randomly. But many populations do not mate randomly for some traits, and when nonrandom mating occurs, the genotypes do not exist in the proportions predicted by the Hardy-Weinberg law. One form of nonrandom mating is **positive assortative mating,** which occurs when individuals with similar phenotypes mate preferentially. Positive assortative mating is common in natural populations. For example, humans mate assortatively for height; tall men and tall women marry more frequently and short men and short women marry more frequently than would be expected on a random basis. **Negative assortative mating** occurs when phenotypically dissimilar individuals mate more often than randomly chosen individuals. If humans exhibited negative assortative mating for height, tall men and short women would marry preferentially and short men and tall women would marry preferentially. Neither positive nor negative assortative mating affects the allelic frequencies of a population, but they may influence the genotypic frequencies if the phenotypes for which assortative mating occurs are genetically determined.

Inbreeding

Another important departure from random mating is **inbreeding.** Inbreeding involves preferential mating between close relatives. Inbreeding often is measured in terms of the coefficient of inbreeding (F). The greater the value of F, the greater the reduction in heterozygosity relative to that expected from the Hardy-Weinberg expectation. Thus,

$$F = \frac{\text{expected heterozygosity} - \text{observed heterozygosity}}{\text{expected heterozygosity}}$$

In random mating, $F = 0$ because observed heterozygosity and expected heterozygosity are equal. Regular systems of inbreeding exist, however, such as self-fertilization, sib mating, and mating between first cousins. After one generation of mating in such systems, the value of F would be 0.5, 0.25, and 0.06, respectively.

The most extreme case of inbreeding is self-fertilization, which occurs in many plants and a few animals, such as some snails. The effects of self-fertilization are illustrated in Table 22.9. Assume that we begin with a population consisting entirely of Aa heterozygotes and that all individuals in this population reproduce by self-fertilization. After one generation of self-fertilization, the progeny will consist of $\frac{1}{4}$ AA, $\frac{1}{2}$ Aa, and $\frac{1}{4}$ aa. Now only half of the population consists of heterozygotes. When this generation undergoes self-fertilization, the AA homozygotes will produce only AA progeny, and the aa homozygotes will produce only aa progeny. When the heterozygotes reproduce, however, only half of their progeny will be heterozygous like the parents, and the other half will be homozygous ($\frac{1}{4}$ AA and $\frac{1}{4}$ aa). This means that in each generation of self-fertilization, the percentage of heterozygotes decreases by 50 percent. After a large number of generations, there will be no heterozygotes and the population will be divided equally between the two homozygous genotypes.

The result of continued self-fertilization is to increase homozygosity at the expense of heterozygosity. The frequencies of alleles A and a remain constant, while the fre-

Table 22.9	Relative Genotype Distributions Resulting from Self-Fertilization Over Several Generations Starting with an *Aa* Individual		
	Frequencies of Genotypes		
Generation	**AA**	**Aa**	**aa**
0	0	1	0
1	1/4	1/2	1/4
2	1/4 + 1/8 = 3/8	1/4	1/4 + 1/8 = 3/8
3	3/8 + 1/16 = 7/16	1/8	3/8 + 1/16 = 7/16
4	7/16 + 1/32 = 15/32	1/16	7/16 + 1/32 = 15/32
5	15/32 + 1/64 = 31/64	1/32	15/32 + 1/64 = 31/64
n	$[1 - (1/2)^n]/2$	$(1/2)^n$	$[1 - (1/2)^n]/2$
∞	1/2	0	1/2

quencies of the three genotypes change significantly. When less intensive inbreeding occurs, similar but less pronounced effects occur.

KEYNOTE

> Inbreeding involves preferential mating between close relatives. Continued inbreeding increases homozygosity within a population and in most species results in reduced fitness.

Summary of the Effects of Evolutionary Forces on the Genetic Structure of a Population

Let us now review the major effects of the different evolutionary processes on (1) changes in allelic frequency within a population; (2) genetic divergence between populations; and (3) increases and decreases in genetic variation within populations.

Changes in Allelic Frequency Within a Population

Mutation, migration, genetic drift, and selection all have the potential to change the allelic frequencies of a population over time. However, mutation usually occurs at such a low rate that the change resulting from mutation pressure alone frequently is negligible. Genetic drift produces substantial changes in allelic frequency when population size is small. Furthermore, mutation, migration, and selection may lead to equilibria where these processes continue to act, but the allelic frequencies no longer change. Nonrandom mating does not change allelic frequencies, but it does affect the genotypic frequencies of a population: Inbreeding leads to increases in homozygosity, and if there are deleterious recessive alleles in the population (which there almost always are), then inbreeding leads to reduced fitness.

Genetic Divergence Among Populations

Several evolutionary processes lead to genetic divergence between populations. Since genetic drift is a random process, allelic frequencies in different populations may drift in different directions; thus, genetic drift can produce genetic divergence among populations. Migration among populations has just the opposite effect, increasing effective population size and equalizing allelic frequency differences among populations. If the population size is small, different mutations may arise in different populations, so mutation may contribute to population differentiation. Natural selection can increase genetic differences between populations by favoring different alleles in different populations, or it can prevent divergence by keeping allelic frequencies uniform among populations. Nonrandom mating, by itself, will not generate genetic differences between populations, although it may contribute to the effects of other processes by increasing or decreasing effective population size.

Increases and Decreases in Genetic Variation Within Populations

Migration and mutation tend to increase genetic variation within populations by introducing new alleles to the gene pool. Genetic drift produces the opposite effect, decreasing genetic variation within small populations through loss of alleles. Since inbreeding leads to increases in homozygosity, it also diminishes genetic variation within populations; **outbreeding** (preferential mating between nonrelated individuals), on the other hand, increases genetic variation by increasing heterozygosity. Natural selection can increase or decrease genetic variation; if one particular allele is favored, other alleles decrease in frequency and can be eliminated from the population by selection. Alternatively, natural selection can increase genetic variation within populations through overdominance and other forms of balancing selection.

In natural populations, these evolutionary processes never act in isolation but combine and interact in complex ways. In most natural populations, the combined effects of these processes and their interaction determine the pattern of genetic variation observed in the gene pool over time.

The Role of Genetics in Conservation Biology

The current rate at which species are being driven to extinction is greater than it has ever been in recorded history. It is estimated that there are approximately 2 million known species and as many as 30 million that are yet to be described. As we alter the environment and reduce the amount of suitable habitat for species, their numbers frequently decline. It is important to consider the consequences on the gene pool of such species because the variability of the gene pool may affect the chances of long-term survival of the species. Many of the genetic principles and processes discussed in this chapter relate to this conservation problem. As we have seen, populations have genetic structure; conserving this structure may warrant special attention. For example, **population viability analysis** is designed to estimate how large a population must be to keep from going extinct for a particular period of time with a certain degree of certainty. If one wants to ensure that a population has the potential to evolve over long periods of time, an adequate gene pool must be maintained. Clearly, determining the genetic structure of a population and how genetic variation within populations affects the probability of extinction requires a great deal of study. The problem is particularly acute for rare and already endangered species. The effects of unintentional inbreeding of species in zoos and game management programs are diminishing as population genetics principles are being used to manage genetic structures of populations more carefully. We have seen that inbreeding, genetic drift, and selection can all decrease genetic variation, and populations may need to be maintained at certain genetic effective sizes to ensure that ample amounts of variation remain. Fortunately, we now have tools to obtain quantitative assessments of the relationships among geographic populations and of the amount of genetic variability in each. These data provide essential information for management policies. However, there is some controversy regarding the utility of genetic information in conservation biology. Some argue that the central problem is loss of habitat, and unless the rate of habitat destruction can be slowed, other efforts, such as genetic analysis, merely distract from the primary goal. It is hoped that a fusion of efforts in population ecology and genetics will provide the necessary information to better understand the best course of action for maintaining the diversity of life in our ecosystems.

K E Y N O T E

Rare and endangered species risk losing genetic variability and hence the ability to adapt to changing environmental conditions. Management practices are applied in an attempt to maintain genetic diversity by keeping adequate population size and avoiding inbreeding.

Speciation

Our discussion of population genetics so far has considered only processes of changing allelic frequencies in populations of interbreeding organisms. Population subdivision may be weak, or it may be extreme to the point that two populations never interbreed. If this occurs for a long period, eventually there will be fixation of different alleles in the subpopulations such that if they were to come together again, they would fail to mate or the hybrids would have low fitness. Recently, there has been great progress in identifying genes that play a role in the reproductive isolation of closely related species. Before we see how genetics provides exactly the tools we need to understand the basis for reproductive isolation, let's back up and consider some general principles about speciation.

Barriers to Gene Flow

If a species is a group of reproductively compatible organisms, then the process of speciation is equivalent to the erection of barriers to gene flow. Eventually, geographically isolated populations, by a combination of drift and natural selection, will diverge to the point that they can no longer reproduce with one another. The barriers to gene flow come in two major categories: those that result in poor fitness of hybrid offspring (postzygotic barriers) and those that keep the two species from mating in the first place (prezygotic barriers). In animals, it appears that mutations in many genes can result in infertility, especially male infertility. As a result, the earliest genetic change in our pair of isolated subpopulations results in hybrid offspring that are sterile. The adults from the two subpopulations still recognize one another and mate, but the offspring are a dead end. Thus, postzygotic isolation generally arises first. *Postzygotic isolation* may include *hybrid sterility, hybrid inviability,* or failure of the hybrids to perform well in future generations, or *hybrid breakdown.*

In the face of a postmating barrier, it seems that mating adults could increase their fitness if they could recognize the "wrong" species and avoid these unproductive matings. If the populations harbor genetic variation for mate recognition, then the alleles that allow the adults to discriminate successfully will increase in frequency. This simple model (called *reinforcement*) provides a mechanism whereby postzygotic isolation leads to prezygotic isolation. The genes that result in *prezygotic isolation* may keep the species apart in many different ways, including the following:

1. **Temporal isolation.** If the mating season or activity periods change such that they no longer overlap, the opportunity for mating is removed.
2. **Ecological isolation.** If the ecological niche of the two species is distinct, such that, for example, the two species' dietary preferences keep them geographically isolated, even on a small spatial scale, again the opportunity for mating is removed.

Consider now the cases in which the two species freely overlap and have plenty of opportunity for mating. There still are genetic means to prevent the formation of zygotes:

1. **Behavioral incompatibility.** If the two species recognize each other as distinct and avoid each other as mates, both benefit.
2. **Mechanical isolation.** The two species may not be able to discriminate one from another, but if their genitalia do not fit together, zygotes cannot be formed.
3. **Gametic isolation.** Even if they mate and gametes come in contact with one another, there still remains a highly complex process of gametic fusion that can fail. In the case of plants, pollen from the wrong species often lands on the stigma surface. There is a chemical communication between the pollen (or pollen tube) and the stigma and style that has to be correct, or the pollen fails to germinate or the pollen tube fails to grow.

Once prezygotic isolation is partially achieved, there is a snowball effect in which the rate of divergence accelerates. Individuals who engage in interspecific matings suffer an increasing disadvantage until at last the barrier to gene flow is complete. There is good empirical support for every step of this process, and the study of the genetic basis for species isolation remains an active field.

Genetic Basis for Speciation

Because of the argument that younger species tend to rely more on postzygotic isolation, these species are sought after by evolutionary geneticists seeking to understand the nature of the genes that partially isolate the species. It is often the case that male hybrids are sterile but female hybrids are fertile. This is especially true when the males are heterogametic (making two different kinds of gametes, X- and Y-bearing sperm) and the females are homogametic (producing only X-bearing eggs). This rule is so pervasive that J. B. S. Haldane noted it, and we call the phenomenon **Haldane's rule.**

The observation of Haldane's rule immediately begs the question, What is the genetic basis for hybrid male sterility? In crosses between many species of *Drosophila*, such as *D. simulans* and *D. mauritiana*, the F_1 females are viable and fertile, whereas the F_1 males are sterile. One can backcross the females to *D. simulans*, and a few of the backcross males (who are now roughly ¾ *D. simulans*) are fertile. In experiments such as this, *Drosophila* population geneticists have learned that many genes are involved in the fertility of male hybrids.

Another example of prezygotic isolation occurs in species of abalone, a subtidal marine gastropod that sheds its gametes into the seawater. Sperm and eggs of more than one species may co-occur (although some species exhibit temporal and ecological isolation), so the only way to avoid interspecific matings is for the eggs to allow penetration only by conspecific sperm. This is accomplished by means of molecules in the sperm and the egg. The sperm protein lysin is able to disaggregate the egg glycoprotein VERL in a species-specific manner. Because the sperm and egg components must track one another and because they must be able to respond quickly if they come in contact with new species, these molecules are undergoing rapid adaptive evolution. You will learn more about the inferences of adaptive change in proteins in Chapter 24.

Summary

Population genetics is the study of the genetic structure of populations and species and how the structure changes, or evolves, over time. The gene pool of a population is the total of all genes within the population, and it is described in terms of allelic and genotypic frequencies. The Hardy-Weinberg law describes what happens to allelic and genotypic frequencies of a large, randomly mating population free from evolutionary processes; when these conditions are met, allelic frequencies do not change, and genotypic frequencies stabilize after one generation in the proportions p^2, $2pq$, q^2, where p and q equal the allelic frequencies of the population.

The classical and neutral mutation models have generated testable hypotheses that help explain how much genetic variation should exist within natural populations and what processes are responsible for the variation observed. Protein electrophoresis showed that most populations of plants and animals contain large amounts of genetic variation, proving that the classical model was wrong. The current view is that genetic variation is maintained in populations by a combination of forces and that some particular genes may be heavily influenced by natural selection, resulting in fixation of an advantageous allele or recurrent loss of deleterious mutations. Variation in other genes appears to fit the neutral model, suggesting that variation is maintained by a balance between genetic drift and mutation.

Mutation, genetic drift, migration, and natural selection are processes that can alter the allelic frequencies of a population. Although mutation is the source of all variation in a population, it usually changes allelic frequencies at a very slow rate, so slow that almost any of the other forces swamp the effects of recurrent mutation on allelic frequencies. Genetic drift, chance change in allelic frequencies due to small effective population size, leads to a loss of genetic variation within a population, genetic divergence among populations, and random change of allelic frequency within a population. Migration tends to reduce genetic divergence among populations and increases effective population size. Natural selection is the differential reproduction of genotypes. The relative reproductive contribution of genotypes is measured in terms of Darwinian fitness. The effects of natural selection depend on the fitness of the genotypes, the degree of dominance, and the frequencies of the alleles in the population. Nonrandom mating affects the effective population size and genotypic frequencies of a population; the allelic frequencies are unaffected. One type of nonrandom mating, inbreeding, leads to an increase in homozygosity.

New techniques of molecular genetics, including analysis of restriction fragment length polymorphisms and RNA and DNA sequences, have supported prior insights obtained from analyses of proteins with respect to evolutionary processes. Different parts of a gene are found to have different levels of polymorphism; the parts of the gene that have the least effect on fitness appear to evolve at the highest rates. This suggests that natural selection generally removes deleterious mutations. Exceptions occur in genes where it is advantageous to have high levels of variability, such as genes in the immune system that increase the variety of pathogens that can be recognized.

Analytical Approaches for Solving Genetics Problems

Q22.1 In a population of 2,000 gaboon vipers, a genetic difference with respect to venom exists at a single locus. The alleles are incompletely dominant. The population shows 100 individuals homozygous for the *t* allele (genotype *tt*, nonpoisonous), 800 heterozygous (genotype *Tt*, mildly poisonous), and 1,100 homozygous for the *T* allele (genotype *TT*, lethally poisonous).
a. What is the frequency of the *t* allele in the population?
b. Are the genotypes in Hardy-Weinberg equilibrium?

A22.1 This question addresses the basics of calculating allelic frequencies and relating them to the genotype frequencies expected of a population in Hardy-Weinberg equilibrium.
a. The *t* frequency can be calculated from the information given because the trait is an incompletely dominant one. There are 2,000 individuals in the population under study, meaning a total of 4,000 alleles at the *T/t* locus. The number of *t* alleles is given by

$$(2 \times tt \text{ homozygotes}) + (1 \times Tt \text{ heterozygotes})$$
$$= (2 \times 100) + (1 \times 800) = 1,000$$

This calculation is straightforward because both alleles in the nonpoisonous snakes are *t*, whereas only one of the two alleles in the mildly poisonous snakes is *t*. Since the total number of alleles under study is 4,000, the frequency of *t* alleles is $1,000/4,000 = 0.25$. This system is a two-allele system, so the frequency of *T* must be 0.75.
b. For the genotypes to be in Hardy-Weinberg equilibrium, the distribution must be $p^2 (TT) + 2pq (Tt) + q^2 (tt)$ genotypes, where *p* is the frequency of the *T* allele and *q* is the frequency of the *t* allele. In part (a), we established that the frequency of *T* is 0.75 and the frequency of *t* is 0.25. Therefore, $p = 0.75$ and $q = 0.25$. Using these values, we can determine the expected genotypic frequencies if this population is in Hardy-Weinberg equilibrium:

$$(0.75)^2 \ TT + 2(0.75)(0.25) \ Tt + (0.25)^2 \ tt$$

This expression gives $0.5625 \ TT + 0.3750 \ Tt + 0.0625 \ tt$. Thus, with 2,000 individuals in the population, we would expect 1,125 *TT*, 750 *Tt*, and 125 *tt*. These values are close to the values given in the question, suggesting that the population is indeed in genetic equilibrium.

To check this result, we should perform a chi-square analysis (see Chapter 2), using the given numbers (not frequencies) of the three genotypes as the observed numbers and the calculated numbers as the expected numbers. The chi-square analysis is as follows, where $d = (\text{observed} - \text{expected})$:

Genotype	Observed	Expected	d	d^2	d^2/e
TT	1,100	1,125	−25	625	0.556
Tt	800	750	+50	2,500	3.334
tt	100	125	−25	625	5.000
Totals	2,000	2,000	0		8.890

Thus, the chi-square value (i.e., the sum of all the d^2/e values) is 8.89. For the reasons discussed in the text for a similar example, there is only one degree of freedom. Looking up the chi-square value in the chi-square table (Table 2.5), we find a *P* value of approximately 0.0025. So about 25 times out of 10,000 we would expect chance deviations of the magnitude observed. In other words, our hypothesis that the population is in Hardy-Weinberg equilibrium is not substantiated. In this case, our guess that it was in equilibrium was inaccurate. Nonetheless, the population is not greatly removed from an equilibrium state.

Q22.2 Approximately one normal allele in 30,000 mutates to the X-linked recessive allele for hemophilia in each human generation. Assume for the purposes of this problem that one *h* allele in 300,000 mutates back to the normal alternative in each generation. (Note that in reality it is difficult to measure the reverse mutation of a human recessive allele that is essentially lethal, such as the allele for hemophilia.) The mutation frequencies are indicated in the following diagram, where $u = 10v$:

$$h^+ \xrightarrow{u} h$$
$$h^+ \xleftarrow{v} h$$

What allelic frequencies would prevail at equilibrium under mutation pressures alone in these circumstances?

A22.2 This question seeks to test your understanding of the effects of mutation on allelic frequencies. In the chapter, we discussed the consequences of mutation pressure. The conclusion was that if *A* mutates to *a* at *n* times the frequency with which *a* mutates back to *A*, then at equilibrium, the value of *q* will be $\hat{q} = u/(u + v)$ or $\hat{q} = nv/(n + 1)v$. Applying this general derivation to this particular problem, we simply use the values given. We are told that the forward mutation rate is 10 times the reverse mutation rate, or $u = 10v$. At equilibrium, the value of *q* will be $\hat{q} = u/(u + v)$. Since $u = 10v$, this equation becomes $\hat{q} = 10v/11v$, so $q = 10/11$, or 0.909. Therefore, at equilibrium brought about by mutation pressures, the frequency of *h* (the hemophilia allele) is 0.909, and the frequency of h^+ (the normal allele) is $\hat{p}$, that is, $(1 - \hat{q}) = (1 - 0.909) = 0.091$.

Questions and Problems

*22.1 In the European land snail *Cepaea nemoralis*, multiple alleles at a single locus determine shell color. The allele for brown (C^B) is dominant to the allele for pink (C^P) and to the allele for yellow (C^Y). The dominance hierarchy among these alleles is $C^B > C^P > C^Y$. In one population sample of *Cepaea*, the following color phenotypes were recorded:

Brown	236
Pink	231
Yellow	33
Total	500

Assuming that this population is in Hardy-Weinberg equilibrium (large, randomly mating, and free from evolutionary processes), calculate the frequencies of the C^B, C^Y, and C^P alleles.

22.2 Three alleles are found at a locus coding for malate dehydrogenase (MDH) in the spotted chorus frog. Chorus frogs were collected from a breeding pond, and each frog's genotype at the MDH locus was determined with electrophoresis. The following numbers of genotypes were found:

M^1M^1	8
M^1M^2	35
M^2M^2	20
M^1M^3	53
M^2M^3	76
M^3M^3	62
Total	254

a. Calculate the frequencies of the M^1, M^2, and M^3 alleles in this population.
b. Using a chi-square test, determine whether the MDH genotypes in this population are in Hardy-Weinberg proportions.

22.3 In a large interbreeding population, 81 percent of the individuals are homozygous for a recessive character. In the absence of mutation or selection, what percentage of the next generation would be homozygous recessives? Homozygous dominants? Heterozygotes?

*22.4 Let *A* and *a* represent dominant and recessive alleles whose respective frequencies are *p* and *q* in a given interbreeding population at equilibrium (with $p + q = 1$).
a. If 16 percent of the individuals in the population have recessive phenotypes, what percentage of the total number of recessive genes exist in the heterozygous condition?
b. If 1.0 percent of the individuals are homozygous recessive, what percentage of the recessive genes occur in heterozygotes?

*22.5 A population has eight times as many heterozygotes as homozygous recessives. What is the frequency of the recessive gene?

22.6 In a large population of range cattle, the following ratios are observed: 49 percent red (*RR*), 42 percent roan (*Rr*), and 9 percent white (*rr*).
a. What percentage of the gametes that give rise to the next generation of cattle in this population will contain allele *R*?
b. In another cattle population, only 1 percent of the animals are white and 99 percent are either red or roan. What is the percentage of *r* alleles in this case?

22.7 In a gene pool, the alleles *A* and *a* have initial frequencies of *p* and *q*, respectively. Show that the allelic frequencies and zygotic frequencies do not change from generation to generation as long as there is no selection, mutation, or migration, the population is large, and the individuals mate at random.

*22.8 The *S-s* antigen system in humans is controlled by two codominant alleles, *S* and *s*. In a group of 3,146 individuals, the following genotypic frequencies were found: 188 *SS*, 717 *Ss*, and 2,241 *ss*.
a. Calculate the frequency of the *S* and *s* alleles.
b. Determine whether the genotypic frequencies are in Hardy-Weinberg equilibrium by using the chi-square test.

22.9 Refer to problem 22.8. A third allele is sometimes found at the *S* locus. This allele S^u is recessive to both the *S* and the *s* alleles and can be detected only in the homozygous state. If the frequencies of the alleles *S*, *s*, and S^u are *p*, *q*, and *r*, respectively, what would be the expected frequencies of the phenotypes *S–*, *Ss*, *s–*, and S^uS^u?

22.10 In a large interbreeding human population, 60 percent of individuals belong to blood group O (genotype *i/i*). Assuming negligible mutation and no selective advantage of one blood type over another, what percentage of the grandchildren of the present population will be type O?

*22.11 A selectively neutral, recessive character appears in 40 percent of the males and in 16 percent of the females in a large, randomly interbreeding population. What is the gene's frequency? What proportion of females are heterozygous for it? What proportion of males are heterozygous for it?

22.12 Suppose you found two distinguishable types of individuals in wild populations of some organism in the following frequencies:

	Type 1	Type 2
Females	99%	1%
Males	90%	10%

The difference is known to be inherited. Are these data compatible with the trait being X-linked?

*22.13 Red-green color blindness is caused by a sex-linked recessive gene. About 64 women out of 10,000 are color-blind. What proportion of men would be expected to show the trait if mating is random?

22.14 About 8 percent of the men in a population are red-green color-blind (because of a sex-linked recessive gene). Answer the following questions with respect to color blindness, assuming random mating in the population.
a. What percentage of women would be expected to be color-blind?
b. What percentage of women would be expected to be heterozygous?
c. What percentage of men would be expected to have normal vision two generations later?

22.15 List some of the basic differences in the classical and neutral mutation models of genetic variation.

*22.16 Two alleles of a locus, *A* and *a*, can be interconverted by mutation:

$$A \xrightarrow{\;\;u\;\;}[\;\;v\;\;] a$$

where *u* is a mutation rate of 6.0×10^{-7} and *v* is a mutation rate of 6.0×10^{-8}. What will be the frequencies of *A* and *a* at mutational equilibrium, assuming no selective difference, no migration, and no random fluctuation caused by genetic drift?

22.17 The land snail *Cepaea nemoralis* is native to Europe but has been accidentally introduced into North America at several localities. These introductions occurred when a few snails were inadvertently transported on plants, building supplies, soil, or other cargo. The snails subsequently multiplied and established large, viable populations in North America.

Assume that today the average size of *Cepaea* populations found in North America is equal to the average size of *Cepaea* populations in Europe. What predictions can you make about the amounts of genetic variation present in European and North American populations of *Cepaea*? Explain your reasoning.

*22.18 A population of 80 adult squirrels resides on campus, and the frequency of the *Est*[1] allele among these squirrels is 0.70. Another population of squirrels is found in a nearby woods, and there the frequency of the *Est*[1] allele is 0.5. During a severe winter, 20 of the squirrels from the woods population migrate to campus in search of food and join the campus population. What will be the allelic frequency of *Est*[1] in the campus population after migration?

22.19 Upon sampling three populations and determining genotypes, you find the following three genotypic distributions. What does each of these distributions imply with regard to selective advantages of population structure?

Population	AA	Aa	aa
1	0.04	0.32	0.64
2	0.12	0.87	0.01
3	0.45	0.10	0.45

22.20 The frequency of two adaptively neutral alleles in a large population is 70 percent *A* and 30 percent *a*. The population is wiped out by an epidemic, leaving only four individuals, who produce many offspring. What is the probability that the population several years later will be 100 percent *AA*? (Assume no mutations.)

*22.21 A completely recessive gene, through changed environmental circumstances, becomes lethal in a certain population. It was previously neutral, and its frequency was 0.5.
a. What was the genotype distribution when the recessive genotype was not selected against?
b. What will be the allelic frequency after one generation in the altered environment?
c. What will be the allelic frequency after two generations?

22.22 Human individuals homozygous for a certain recessive autosomal gene die before reaching reproductive age. Despite this removal of all affected individuals, there is no indication that homozygotes occur less frequently in succeeding generations. To what might you attribute the continued appearance of recessives?

*22.23 A completely recessive gene (Q^1) has a frequency of 0.7 in a large population, and the Q^1Q^1 homozygote has a relative fitness of 0.6.
a. What will be the frequency of Q^1 after one generation of selection?
b. If there is no dominance at this locus (the fitness of the heterozygote is intermediate to the fitnesses of the homozygotes), what will the allelic frequency be after one generation of selection?
c. If Q^1 is dominant, what will the allelic frequency be after one generation of selection?

22.24 As discussed earlier in this chapter, the gene for sickle-cell anemia exhibits heterozygote advantage. An individual who is an *Hb-A/Hb-S* heterozygote has increased resistance to malaria and therefore has greater fitness than the *Hb-A/Hb-A* homozygote, who is susceptible to malaria, and the *Hb-S/Hb-S* homozygote, who has sickle-cell anemia. Suppose that the fitness values of the genotypes in Africa are as presented here:

$$Hb\text{-}A/Hb\text{-}A = 0.88$$
$$Hb\text{-}A/Hb\text{-}S = 1.00$$
$$Hb\text{-}S/Hb\text{-}S = 0.14$$

Give the expected equilibrium frequencies of the sickle-cell gene (*Hb-S*).

*22.25 Achondroplasia, a type of dwarfism in humans, is caused by an autosomal dominant gene. The mutation rate for achondroplasia is about 5×10^{-5}, and the fitness of achondroplastic dwarfs has been estimated to be about 0.2, compared with unaffected individuals. What is the equilibrium frequency of the achondroplasia gene based on this mutation rate and fitness value?

22.26 The frequencies of the L^M and L^N blood group alleles are the same in each of the populations I, II, and III, but the genotypes' frequencies are not the same, as shown in the following table. Which of the populations is most likely to show each of the following characteristics: random mating, inbreeding, genetic drift? Explain your answers.

	$L^M L^M$	$L^M L^N$	$L^N L^N$
I	0.50	0.40	0.10
II	0.49	0.42	0.09
III	0.45	0.50	0.05

*22.27 DNA was collected from 100 people randomly sampled from a given human population and was digested with the restriction enzyme *Bam*HI; the fragments were separated by electrophoresis and then transferred to a membrane filter using the Southern blot technique. The blots were probed with a particular cloned sequence. Three different patterns of hybridization were seen on the blots. Some DNA samples (56 of them) showed a single band of 6.3 kb, others (6) showed a single band at 4.1 kb, and others (38) showed both the 6.3- and 4.1-kb bands.

a. Interpret these results in terms of *Bam*HI sites.
b. What are the frequencies of the restriction site alleles?
c. Does this population appear to be in Hardy-Weinberg equilibrium for the relevant restriction sites?

*22.28 Fifty tiger salamanders from one pond in west Texas were examined for genetic variation by using the technique of protein electrophoresis. The genotype of each salamander was determined for five loci (AmPep, ADH, PGM, MDH, and LDH-1). No variation was found at AmPep, ADH, and LDH-1; in other words, all individuals were homozygous for the same allele at these loci. The following numbers of genotypes were observed at the MDH and PGM loci.

MDH Genotypes	Number of Individuals	PGM Genotypes	Number of Individuals
AA	11	DD	35
AB	35	DE	10
BB	4	EE	5

Calculate the proportion of polymorphic loci and the heterozygosity for this population.

22.29 What factors cause genetic drift?

*22.30 What are the primary effects of the following evolutionary processes on the allelic and genotypic frequencies of a population?
a. mutation
b. migration
c. genetic drift
d. inbreeding

22.31 Explain how overdominance leads to an increased frequency of sickle-cell anemia in areas where malaria is widespread.

Quantitative Genetics

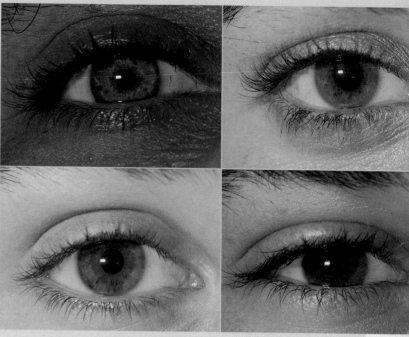

Various human eye colors.

23

PRINCIPAL POINTS

Discontinuous traits exhibit only a few distinct phenotypes. Continuous traits display a range of phenotypes.

Continuous traits have many phenotypes because they are encoded by many genotypes (are polygenic) and because environmental factors cause each genotype to produce a range of phenotypes.

Continuous traits are studied by using samples of populations and statistical concepts such as the mean, variance, and correlation of characters.

The polygene or multiple-gene hypothesis of inheritance proposes that quantitative traits are determined by a number of genes, whose effects add together to determine the phenotype.

Quantitative trait loci (QTLs) that determine continuous traits can be identified through marker-based mapping. QTLs provide estimates of the number and relative importance of genes influencing quantitative genetic variation.

Variation among individuals can be partitioned into genetic and environmental components. However, genotypes may behave differently in different environments, so caution must be exercised when designing and interpreting experiments that measure genetic and environmental contributions to phenotypic variation.

The broad-sense heritability of a trait is the proportion of the phenotypic variance that results from genetic differences among individuals. The narrow-sense heritability is the proportion of the phenotypic variance that results from additive genetic variance. Both measures depend on a particular population in a particular environment.

The amount that a trait changes in one generation as a result of selection for the trait is called the response to selection. The magnitude of the response to selection depends on the selection differential and the narrow-sense heritability.

Genetic correlations arise when two traits are influenced by the same genes. When a trait is selected, genetically correlated traits will also exhibit a response to selection.

i JUST LIKE SNOWFLAKES, NO TWO FINGERPRINTS ARE alike, even those of identical twins. Yet, research shows that the patterns of ridges on our fingers and palms are inherited. What factors would produce such a variety of phenotypes? Is the trait encoded by more than one locus? Are there environmental factors involved? Is there a relationship between a person's fingerprints and another trait, such as hair color or blood type? In this chapter, you will learn the answer to questions such as these. Then, in the iActivity, you can apply what you've learned as you investigate whether a relationship exists between fingerprint patterns and high blood pressure.

The mutations used in early genetic studies and later in molecular studies were characterized by the presence of a few distinct phenotypes. The effects of variant alleles and a single gene locus were observable at the level of the organism, so the phenotype could be used as a quick assay for the genotype. The seed coats of pea plants, for example, were either grey or white, the seed pods were green or yellow, and the plants were tall or short. In each trait, the different phenotypes were distinct, and each phenotype was easily separated from all other phenotypes. Traits such as these, with only a few distinct phenotypes, are called **discontinuous traits.**

For discontinuous traits, a simple relationship usually exists between the genotype and the phenotype. In most cases, each genotype produces only a single phenotype, and frequently, each phenotype results from a single genotype. However, we saw in Chapter 4 that the relationship between the genotype and the phenotype is not always so simple. Variable **penetrance** and **expressivity,** as well as **pleiotropy** and **epistasis,** can be quite common. In addition, single genotypes can give rise to a range of phenotypes as the genotype interacts with variable environments during development to give rise to a **norm of reaction.** As a result of these and other factors, there are not many traits with phenotypes that fall into a few distinct categories. Many traits (probably most), such as birth weight and adult height in humans, protein content in corn, and number of eggs laid by *Drosophila,* exhibit a wide range of possible phenotypes. Traits such as these, with a continuous distribution of phenotypes, are called **continuous traits.** The distribution of a continuous trait—birth weight in humans—is illustrated in Figure 23.1. Since the phenotypes of continuous traits must be described in quantitative terms, such traits are also known as **quantitative traits,** and the study of the inheritance of quantitative traits is the field of **quantitative genetics.**

Numerous traits have continuous distributions; quantitative genetics therefore plays an important role in our understanding of evolution, conservation, and other areas of applied biology. In agriculture, for example, crop yield, rate of weight gain, milk production, and fat content are all continuous traits that are studied with the techniques of quantitative genetics. Geneticists also use these methods to study continuous traits found in humans, such as blood pressure, antibody titer, fingerprint pattern, and birth weight (see Figure 23.1).

The Nature of Continuous Traits

Why Some Traits Have Continuous Phenotypes

Continuous traits by definition have a continuous range of phenotypes. To understand the inheritance of continuous traits, we must first determine why some traits have a range of phenotypes.

Multiple phenotypes of a trait arise in several ways. Frequently, a range of phenotypes occurs because numerous genotypes exist among the individuals of a group; this happens when the trait is influenced by a large number of loci. For example, when a single locus with two alleles determines a trait, three genotypes are present: *AA, Aa,* and *aa.* With two loci, each with two alleles, the number of genotypes is $3^2 = 9$ (*AA BB, Aa BB, AA Bb, Aa Bb, AA bb, aa BB, aa Bb, Aa bb,* and *aa bb*). In general, the number of genotypes is 3^n, where *n* equals the number of loci with two alleles; if more than two alleles are present at a locus, the number of genotypes is even greater. As the number of loci influencing a trait increases, the number of genotypes quickly becomes large. Traits encoded by many loci are called **polygenic traits.** If each genotype in a polygenic trait encodes a separate phenotype, many phenotypes will be present. And because many phenotypes are present and the differences between phenotypes are slight, the trait appears to be continuous.

Figure 23.1

Distribution of birth weight of babies (males and females) born to teenagers in Portland, Oregon, in 1992.

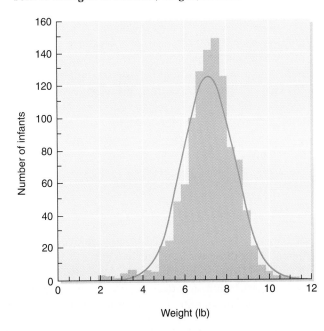

More frequently, several genotypes of a polygenic trait produce the same phenotype. For example, with dominance, only two phenotypes are produced by the three genotypes at a locus (*AA* and *Aa* produce one phenotype and *aa* produces a different phenotype). If dominance occurs among the alleles at each of three loci, the number of genotypes is $3^3 = 27$, but the number of phenotypes is only $2^3 = 8$. When epistatic interactions occur among the alleles at different loci, as discussed in Chapter 4, several genotypes may code for the same phenotype. In many polygenic traits, multiple genotypes code for the same phenotype, and the relationship between genotype and phenotype is obscured.

Another reason that a trait may have a range of phenotypes is that environmental factors also affect the trait. When environmental factors exert an influence on the phenotype, each genotype is capable of producing a range of phenotypes (the *norm of reaction*). Which phenotype is expressed depends both on the genotype and on the specific environment in which the genotype is found. For most continuous traits, both multiple genotypes and environmental factors influence the phenotype; such a trait is a **multifactorial trait.**

When multiple genes and environmental factors influence a trait, one does not find the simple relationship between genotype and phenotype that exists in discontinuous traits. To understand the action and role of each individual gene controlling a continuous trait, quantitative genetics uses special statistical and analytical procedures.

K E Y N O T E

> Discontinuous traits exhibit only a few distinct phenotypes and can be described in qualitative terms. Continuous traits, on the other hand, display a spectrum of phenotypes and must be described in quantitative terms. The relationship between the genotype and the phenotype is complex. One genotype may give rise to a range of phenotypes, and many genotypes can give rise to the same phenotype. Numerous phenotypes are present in a continuous trait because the trait can be encoded by many loci, producing many genotypes, and because environmental factors can cause each genotype to produce a range of phenotypes. Understanding how variation among individuals in a particular trait is determined during development is the major underlying theme of quantitative genetics.

Questions Studied in Quantitative Genetics

Not only are the methods used in quantitative genetics different from those we have previously studied, but the fundamental nature of the questions asked also is different. In transmission genetics, we frequently determine the probability of inheriting a discontinuous trait. How-

ever, continuous traits are determined by numerous genes, and no simple relationship exists between the genotype and the phenotype. Furthermore, individuals differ in the quantity of a trait, not its presence or absence, so it makes no sense to ask about the probability of inheriting a continuous trait as we did for simple discontinuous traits. The following questions frequently are studied by quantitative geneticists:

1. To what degree does the observed variation in phenotype result from differences in genotype and to what degree does this variation reflect the influence of different environments? In our study of discontinuous traits, this question assumed little importance because the differences in phenotype were assumed to reflect differences in genes.

2. How many genes determine the phenotype of a trait? When only a few loci are involved and the trait is discontinuous, the number of loci involved often can be determined by examining the phenotypic ratios in genetic crosses. With complex, continuous traits, however, determining the number of loci involved is more difficult.

3. Are the contributions of the determining genes equal? Or do a few genes have major effects on the trait, while other genes only modify the phenotype slightly?

4. To what degree do alleles at the different loci interact with one another? Are the effects of alleles additive?

5. When selection occurs for a particular phenotype, how rapidly does the trait change? Do other traits change at the same time?

6. What is the best method for selecting and mating individuals to produce desired phenotypes in the progeny?

Statistical Tools

One of the fundamental questions addressed in the study of quantitative traits is how much of the variation that exists among individuals in populations is genetically determined and how much is environmentally induced. Thus, at the heart of the field of quantitative genetics is the enduring question of **nature versus nurture,** or genes versus environment. In quantitative genetics, we phrase the problem in terms of variation: How much of the variation in some aspect of the phenotype (V_P) results from genetic variation (V_G), and how much from environmental variation (V_E)? This relationship can be expressed as

$$V_P = V_G + V_E$$

To work this equation, we have to learn how to measure variation in phenotype and how to partition the variation into genetic and environmental components. For this we need to understand some statistical methodology, much of which was developed specifically to deal with these genetic issues.

Samples and Populations

Suppose we want to describe some aspect of a trait in a large group of individuals. For example, we might be interested in the average birth weight of infants born in New York City during 1987. One way to do this is to collect information on the weight of each of the thousands of babies born in New York City in 1987. An alternative method would be to collect information on a subset of the group—say, birth weights on 100 infants born in New York City during 1987—and then use the average obtained on this subset as an estimate of the average for the entire city. Biologists and other scientists commonly use this sampling procedure in data collection, and statistics are necessary for analyzing such data. The group of ultimate interest (in our example, all infants born in New York City during 1987) is called the **population,** and the subset used to give us information about the population (our set of 100 babies) is called a **sample.** For a sample to give us confidence in information about the population, it must be large enough that chance differences between the sample and the population are not misleading. If our sample consisted of only a single baby and that infant was unusually large, then our estimate of the average birth weight of all babies would not be very accurate. The sample must also be a random subset of the population. If all the babies in our sample came from a hospital for premature infants, then we would grossly underestimate the true average birth weight of the population. Although this might seem obvious, a great many errors in judgment are made because data are not collected randomly. Figure 23.1 shows a distribution of birth weight in humans.

KEYNOTE

To describe and study a large group of individuals, scientists frequently examine a subset of the group. This subset is called a sample, and the sample provides information about the larger group, which is called the population. The sample must be of reasonable size, and it must be a random subset of the larger group to provide accurate information about the population.

Distributions

When we studied discontinuous traits, we were able to describe the phenotypes found among a group of individuals by stating the proportion of individuals falling into each phenotypic class. As we discussed earlier, continuous traits exhibit a range of phenotypes, and describing the phenotypes found within a group of individuals is more complicated. One means of summarizing the phenotypes of a continuous trait is with a **frequency distribution,** which is a description of the population in terms of the proportion of individuals that fall within a certain range of phenotypes.

To make a frequency distribution, we construct classes that consist of individuals falling within a specified range of the phenotype, and then we count the number of individuals in each class. Table 23.1 presents a frequency distribution constructed from the data in Wilhelm Johannsen's study of the inheritance of seed weight in the dwarf bean, *Phaseolus vulgaris.* As shown in the table, 5,494 beans from the F_2 progeny of a cross were weighed and classified into nine groups, or classes, each of which covered a 100-mg range of weight. A frequency distribution such as this can be displayed graphically by plotting the phenotypes in a frequency histogram, as shown in Figure 23.2 for Johannsen's beans. In the histogram, the phenotypic classes are indicated along the horizontal axis, and the number present in each class is plotted on the vertical axis. A curve that traces the outline of the histogram shows a shape that is characteristic of the frequency distribution.

Many continuous phenotypes exhibit a symmetrical, bell-shaped distribution similar to the one shown in

Figure 23.2

Frequency histogram for bean weight in *Phaseolus vulgaris* plotted from data in Table 23.1. A normal curve has been fitted to the data and is superimposed on the frequency histogram.

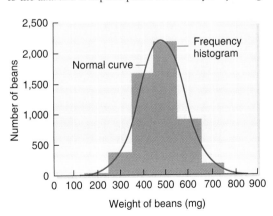

Table 23.1	Weight of 5,494 F₂ Beans (Seeds of *Phaseolus vulgaris*) Observed by Johannsen in 1903

Weight (mg)	50–150	150–250	250–350	350–450	450–550	550–650	650–750	750–850	850–950
(Midpoint of range)	(100)	(200)	(300)	(400)	(500)	(600)	(700)	(800)	(900)
Number of beans	5	38	370	1,676	2,255	928	187	33	2

Figure 23.2. This type of distribution is called a **normal distribution.** The normal distribution is a theoretical distribution that has specific properties and is produced when a large number of independent factors influence the measurement. Since many continuous traits are multifactorial (influenced by multiple genes and multiple environmental factors), observing a nearly normal distribution for these traits is not surprising.

The Mean

A frequency distribution of a phenotypic trait can be summarized in the form of two convenient statistics, the mean and the variance. The **mean,** also known as the average, gives us information about where the center of the distribution of the phenotypes in a sample is located along a continuous range of possibilities. The mean of a sample ($\bar{x}$) is calculated by simply adding up all the individual measurements (Σx_i, where Σ is "sum" and x_i is the individual measurements $x_1, x_2, x_3, \ldots, x_n$) and dividing by the number of measurements we added (n).

$$\text{Mean} = \bar{x} = \frac{\Sigma x_i}{n}$$

Table 23.2 presents a sample calculation of the mean body lengths (and other statistics, discussed next) of ten spotted salamanders from Penobscot County, Maine.

The mean frequently is used in quantitative genetics to characterize the phenotypes of a group of individuals. For example, Edward M. East examined the inheritance of

Figure 23.3

Graphs showing three distributions with the same mean but different variances.

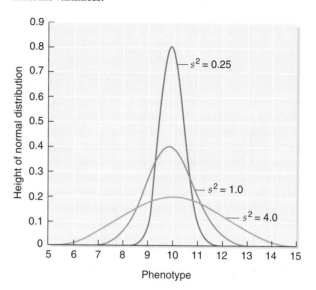

flower length in several strains of the tobacco plant. He crossed a short-flowered strain of tobacco with a long-flowered strain. Within each strain, however, flower length varied, so East reported that the mean phenotype of the short strain was 40.4 mm and the mean phenotype of the long strain was 93.1 mm. The F_1 progeny, which consisted of 173 plants, had a mean flower length of 63.5 mm. In this situation, the mean provides a convenient way for quickly characterizing and comparing the phenotypes of parents and offspring.

The Variance and the Standard Deviation

A second statistic that provides key information about a distribution is the **variance.** The variance is a measure of how much the individual measurements spread out around the mean—how variable the measurements are. Two distributions may have the same mean but very different variances, as shown in Figure 23.3. The variance, symbolized as s^2, is defined as the average squared deviation from the mean.

$$\text{Variance} = s^2 = \frac{\Sigma(x_i - \bar{x})^2}{n - 1}$$

The variance can be calculated by first subtracting the mean from each individual measurement. This difference is then squared and all the squared values are added up. The sum of these squared values is then divided by the number of original measurements minus 1. (For mathematical reasons, which we will not discuss here, the variance is obtained by dividing by $n - 1$ instead of n.)

The **standard deviation** often is preferred to the variance because the standard deviation shares the same

Table 23.2	Sample Calculations of the Mean, Variance, and Standard Deviation for Body Length of Ten Spotted Salamanders from Penobscot County, Maine

Body Length (x_i) (mm)	($x_i - \bar{x}$)	($x_i - \bar{x}$)^2
65	(65 − 57.1) = 7.9	7.9² = 62.41
54	(54 − 57.1) = −3.1	−3.1² = 9.61
56	(56 − 57.1) = −1.1	−1.1² = 1.2
60	(60 − 57.1) = 2.9	2.9² = 8.41
56	(56 − 57.1) = −1.1	1.1² = 1.21
55	(55 − 57.1) = −2.1	2.1² = 4.41
53	(53 − 57.1) = −4.1	−4.1² = 16.81
55	(55 − 57.1) = −2.1	−2.1² = 4.41
58	(58 − 57.1) = 0.9	0.9² = 0.81
59	(59 − 57.1) = 1.9	1.9² = 3.61
$\Sigma x_i = 571$		$\Sigma(x_i - \bar{x})^2 = 112.9$

$$\text{Mean} = \bar{x} = \frac{\Sigma x_i}{n} = \frac{571}{10} = 57.1$$

$$\text{Variance} = s_x^2 = \frac{\Sigma(x_i - \bar{x})^2}{n - 1} = \frac{112.9}{9} = 12.54$$

$$\text{Standard deviation} = s_x = \sqrt{12.54} = 3.54$$

units as the original measurements (whereas the variance is in the units squared). The standard deviation is simply the square root of the variance:

$$\text{Standard deviation} = s = \sqrt{s^2}$$

Sample calculations for the variance and the standard deviation are presented in Table 23.2. A broad curve implies high variability in the quantity measured and a correspondingly large standard deviation. A narrow curve, in contrast, indicates little variability in the quantity measured and a correspondingly small standard deviation.

Once we know the mean and the standard deviation, a theoretical normal distribution is completely specified. It always has the shape indicated in Figure 23.4, where 66 percent of the individual observations have values within one standard deviation above or below ($\pm\,1s$) the mean of the distribution, about 95 percent of the values fall within two standard deviations ($\pm\,2s$) of the mean, and more than 99 percent fall within three standard deviations ($\pm\,3s$). We can infer many things about our data and experiments using these objectively determined percentages.

The variance and the standard deviation provide valuable information about the phenotypes of a group of individuals. In our discussion of the mean, we saw how East used the mean to describe flower length of parents and offspring in crosses of the tobacco plant. When East crossed a strain of tobacco with short flowers to a strain with long flowers, the F_1 offspring had a mean flower length of 63.5 mm, which was intermediate to the phe-

notypes of the parents. When he intercrossed the F_1, the mean flower length of the F_2 offspring was 68.8 mm, approximately the same as the mean phenotype of the F_1. However, the F_2 progeny differed from the F_1 in an important attribute that is not apparent if we examine only the means of the phenotype: The F_2 were more variable in phenotype than the F_1. The variance in the flower length of the F_2 was 42.4 mm^2, whereas the variance in the F_1 was only 8.6 mm^2. This finding indicated that more genotypes were present among the F_2 progeny than among the F_1. Thus, the mean and the variance are both necessary for fully describing the distribution of phenotypes among a group of individuals.

Correlation

A difficulty encountered when thinking about the phenotype is that it is somewhat artificial to pick out traits and study them in isolation. Organisms are composites of a multitude of traits. Some of these traits like height and weight may actually be two members of a more general trait called size. Genes and environmental factors that affect the development of size may affect both traits. Genes that affect height may have pleiotropic effects on weight. In other words, two or more traits are often associated or *correlated*. This means that if one variable changes, the other is also likely to change. For example, arm length and leg length are correlated in most animals including humans—individuals with long arms have correspondingly long legs and vice versa. The correlation coefficient is a statistic that measures the strength of the association between two variables in the same individual or experimental unit. Suppose we have two variables, x and y (x might equal arm length and y might equal leg length), and we wish to calculate the correlation between them. We begin by obtaining the **covariance** of x and y which is a measure of how much variation is shared by both traits. The covariance is computed by taking the same deviations from the mean used in calculating the variance, but for each trait x and y. Instead of squaring these values as for the variance, the product (i.e., the crossproduct) of the two is taken for each pair of x and y values and the products are added together. The sum is then divided by $n - 1$ to give the covariance of x and y, where n equals the number of xy pairs:

$$\text{cov}_{xy} = \frac{\sum (x_i - \bar{x})(y_i - \bar{y})}{n - 1}$$

An algebraically equivalent equation, which is easier to compute, is

$$\text{cov}_{xy} = \frac{\sum x_i y_i - \frac{1}{n}(\sum x_i \sum y_i)}{n - 1}$$

where $\sum x_i y_i$ is the sum of each value of x multiplied by each corresponding value of y, $\sum x_i$ is the sum of all x values, and $\sum y_i$ is the sum of all y values.

Figure 23.4

Normal distribution curve showing the proportions of the data in the distribution that are included within certain multiples of the standard deviation.

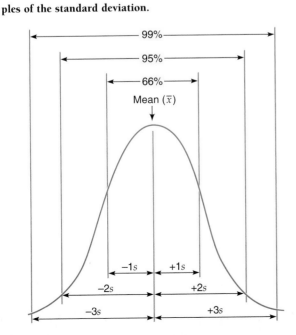

The correlation coefficient r can then be obtained by dividing the covariance by the product of the standard deviations of x and y.

$$\text{Correlation coefficient} = r = \frac{\text{cov}_{xy}}{s_x s_y}$$

where s_x equals the standard deviation of x, and s_y equals the standard deviation of y. Table 23.3 gives a sample calculation of the correlation coefficient between two variables.

The correlation coefficient is a standardized measure of covariance that can range from −1 to +1. The sign of the correlation coefficient, whether it is positive or negative, indicates the direction of the correlation. If the correlation coefficient is positive, then an increase in one variable tends to be associated with an increase in the other variable. If seed size and seed number are positively correlated in sunflowers, for example, plants with larger seeds also tend to produce more seeds. Positive correlations are illustrated in Figure 23.5b, c, d, and f. A negative correlation coefficient indicates that an increase in one variable is associated with a decrease in the other. If seed size and seed number are negatively correlated,

plants with large seeds tend to produce fewer seeds on the average than plants with smaller seeds. Figure 23.5e represents a negative correlation. The absolute value of the correlation coefficient (its magnitude if the sign is ignored) provides information about the strength of the association. When the correlation coefficient is close to −1 or to +1, the correlation is strong, meaning that a change in one variable is almost always associated with a corresponding change in the other variable. For example, the x and y variables in Figure 23.5f are strongly associated and have a correlation coefficient of 0.9. On the other hand, a correlation coefficient near 0 indicates a weak relationship between the variables, as is illustrated in Figure 23.5b. The condition of no correlation is shown in Figure 23.5a.

Several important points about correlation coefficients warrant emphasis. First, a correlation between variables means only that the variables are associated; correlation *does not imply that a cause-effect relationship exists*. The classic example of a noncausal correlation between two variables is the positive correlation that exists between number of ministers and liquor consumption in cities with population size over 10,000. One should not conclude

Figure 23.5

Scatter diagrams showing the correlation of x and y variables. Diagrams b), c), d), and f) show positive correlations, whereas diagram e) shows a negative correlation. The absolute value of the correlation coefficient (r) indicates the strength of the association. For example, diagram f) illustrates a relatively strong correlation and diagram b) illustrates a relatively weak correlation. In diagram a), the x and y variables are not correlated.

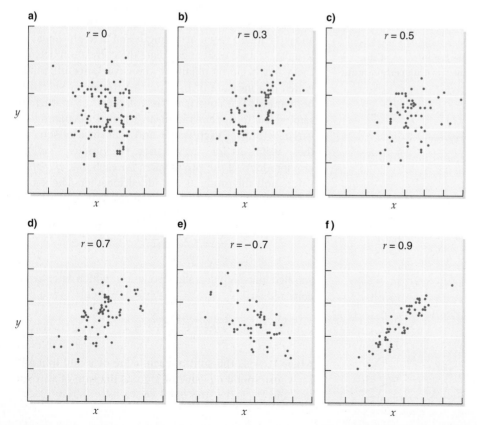

| Table 23.3 | Sample Calculation of the Correlation Coefficient for Body Length and Head Width of Tiger Salamanders |

Body Length (mm)			Head Width (mm)			
x_i	$x_i - \bar{x}$	$(x_i - \bar{x})^2$	y_i	$y_i - \bar{y}$	$(y_i - \bar{y})^2$	$x_i y_i$
72.00	−7.92	62.67	17.00	−0.75	0.56	1224
62.00	−17.92	321.01	14.00	−3.75	14.06	868
86.00	6.08	37.01	20.00	2.25	5.06	1720
76.00	−3.92	15.34	14.00	−3.75	14.06	1064
64.00	−15.92	253.34	15.00	2.75	7.56	960
82.00	2.08	4.34	20.00	2.25	5.06	1640
71.00	−8.92	79.51	15.00	−2.75	7.56	1065
96.00	16.08	258.67	21.00	3.25	10.56	2016
87.00	7.08	50.17	19.00	1.25	1.56	1653
103.00	23.08	532.84	23.00	5.25	27.56	2369
86.00	6.08	37.01	18.00	0.25	0.06	1548
74.00	−5.92	35.01	17.00	−0.75	0.56	1258
$\Sigma x_i =$ 959.00		$\Sigma(x_i - x)^2 =$ 1,686.92	$\Sigma y_i =$ 213.00		$\Sigma(y_i - y)^2 =$ 94.25	$\Sigma x_i y_i =$ 17,385

$\bar{x} = \Sigma x_i/n = 959/12 = 79.92$

$\bar{y} = \Sigma y_i/n = 213/12 = 17.75$

Variance of $x = s_x^2 = \Sigma(x_i - x)^2/n - 1 = 1,686.92/11 = 153.35$

Standard deviation of $\bar{x} = s_x = \sqrt{s_x^2} = \sqrt{153.35} = 12.38$

Variance of $y = s_y^2 = \Sigma(y_i - x)^2/n - 1 = 94.25/11 = 8.57$

Standard deviation of $y = s_y = \sqrt{s^2} = \sqrt{8.57} = 2.93$

Covariance $= \text{cov}_{xy} = (\Sigma x_i y_i - 1/n(\Sigma x_i \Sigma y_i))/n - 1$

$$\text{cov}_{xy} = \frac{(17,385 - 1/12(959 \times 213))}{12 - 1}$$

$\text{cov}_{xy} = 32.97$

Correlation coefficient $= r = \text{cov}_{xy}/(s_x s_y) = 32.97/(12.38 \times 2.93)$

$r = 0.91$

from this correlation that ministers are consuming all the alcohol. Alcohol consumption and number of ministers are associated because both are positively correlated with a third factor, population size; larger cities contain more ministers and also have higher alcohol consumption. Assuming that two factors are causally related because they are correlated may lead to erroneous conclusions.

Another important point is that *correlation is not the same thing as identity*. Correlation only means that a change in one variable is associated with a corresponding change in the other variable. Two variables can be highly correlated, and yet have very different values. For example, the height of college-age males is correlated with the height of their fathers; tall fathers tend to produce tall sons and short fathers tend to produce short sons. This correlation results from the fact that genes influence human height. However, most college-age males today are taller than their fathers, probably because better diet and health care have increased the average height of all individuals in recent years. Thus, fathers and sons exhibit a correlation in height, but they are not the same height.

K E Y N O T E

The correlation coefficient is a measure of how strongly two variables are associated. A positive correlation coefficient indicates that the two variables change in the same direction; an increase in one variable is usually associated with a corresponding increase in the other variable. When the correlation coefficient is negative, the variables are inversely related; an increase in one variable is most often associated with a decrease in the other. The absolute value of the correlation coefficient provides information about the strength of the association. Correlation does not imply that a cause-effect relationship exists between the two variables.

Polygenic Inheritance

In the examples of polygenic inheritance described in the following sections, geneticists did not know at first how these traits were inherited, although it was apparent that their pattern of inheritance differed from that of discontinuous traits.

Inheritance of Ear Length in Corn

In a study of ear length in corn, *Zea mays,* reported in 1913, Rollins Emerson and E. East used two pure-breeding strains of corn, each of which displayed little variation in ear length. The two varieties were black Mexican sweet corn (which had short ears of mean length 6.63 cm) and Tom Thumb popcorn (which had long ears of mean length 16.80 cm).

Emerson and East crossed the two strains and then interbred the F_1 plants. Figure 23.6 presents the results in photographs and histograms. Note that the F_1s have a mean ear length of 12.12 cm, which is approximately intermediate between the mean ear lengths of the two parental lines. The parental plants are pure-breeding, so each is homozygous for whatever genes control the lengths of their ears. Since the two parental plants differ in ear length, though, each must be genetically different. When two pure-breeding strains are crossed, the F_1 plants are heterozygous for all genes, and all plants should have the same genotype. Therefore the range of ear length phenotypes seen in the F_1 plants must result from factors other than genetic differences; these other factors probably are environmental.

In the F_2, the mean ear length of 12.89 cm is about the same as the mean for the F_1 population, but the F_2 population has a much larger variation around the mean than the F_1 population has. This variation is easy to see in Figure 23.6b; it can also be shown by calculating the standard deviation s. The standard deviation of the long-eared parent is 1.887, and that of the short-eared parent is 0.816. In the F_1, s = 1.519, and in the F_2, s = 2.252. These numbers confirm that the F_2 has greater variability, something we could conclude by looking at the data.

If the environment was responsible for variation in the parental and the F_1 generations, it would likely have a similar effect on the F_2. However, we have no reason to suppose that the environment would have a greater influence on the F_2 than on the other two generations, so there must be another explanation for the greater variation in ear length in the F_2 generation. A more reasonable hypothesis is that the increased variability of the F_2 results from the presence of greater genetic variation in the F_2.

Setting aside the environmental influence for the moment, the data reveal four observations that apply generally to quantitative inheritance studies similar to this one:

1. The mean value of the quantitative trait in the F_1 is approximately intermediate between the means of the two true-breeding parental lines.

Figure 23.6

Inheritance of ear length in corn. (a) Representative corn ears from the parental, F_1, and F_2 generations from an experiment in which two pure-breeding corn strains that differed in ear length were crossed and then the F_1s interbred. (b) Histograms of the distributions of ear length (in centimeters) from the experiment represented in part (a); the vertical axes represent the percentages of the different populations found at each ear length. (From E. W. Sinnott and L. C. Dunn, *Principles of Genetics* [New York: McGraw-Hill, 1925].)

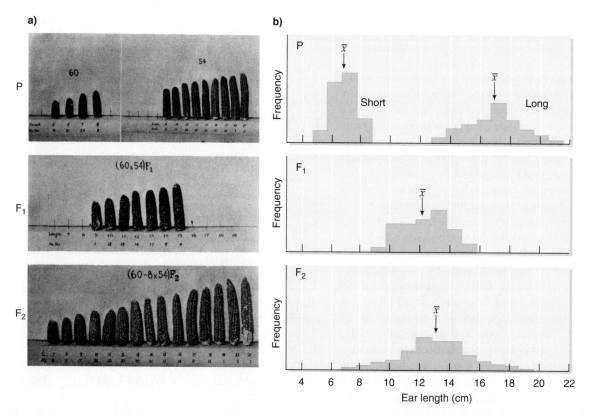

2. The mean value for the trait in the F_2 is approximately equal to the mean for the F_1 population.

3. The F_2 shows more variability around the mean than the F_1 does.

4. The extreme values for the quantitative trait in the F_2 extend further into the distribution of the two parental values than do the extreme values of the F_1.

The data cannot be explained in terms of the standard single gene locus Mendelian genetic principles that govern the inheritance of discontinuous traits. That is, if a single gene were responsible for the two phenotypes of the original parents (AA = homozygous, long; aa = homozygous, short), then the F_1 data could be explained if we assume incomplete dominance. However, crossing the F_1 heterozygote (Aa) should produce a 1:2:1 ratio of AA, Aa, and aa, or long, intermediate, and short phenotypes. The data do not fall into such discrete classes.

K E Y N O T E

> For a quantitative trait, the F_1 progeny of a cross between two phenotypically distinct, pure-breeding parents has a phenotype intermediate between the parental phenotypes. The F_2 shows more variability than the F_1, with a mean phenotype close to that of the F_1. The extreme phenotypes of the F_2 extend well beyond the range of the F_1 and into the ranges of the two parental values.

Polygene Hypothesis for Quantitative Inheritance

The simplest explanation for the data obtained from Emerson and East's experiments on corn ear length and from other experiments with quantitative traits is that quantitative traits are controlled by not one but many genes. This explanation, called the **polygene** or **multiple-gene hypothesis for quantitative inheritance,** is one of the landmarks of genetic thought.

The polygene hypothesis can be traced back to 1909 and the classic work of Hermann Nilsson-Ehle, who studied the color of wheat kernels. Nilsson-Ehle crossed true-breeding lines of plants with red kernels and plants with white kernels. The F_1 had grains that were all the same shade of an intermediate color between red and white. At this point, he could not rule out incomplete dominance as the basis for the F_1 results. However, when he intercrossed the F_1s, a number of the F_2 progeny showed a ratio of approximately 15 red kernels (all shades) to 1 white kernel, clearly a deviation from a 3 : 1 ratio expected for a monohybrid cross. He recognized four discrete shades of red, in addition to white, among the progeny. He counted the relative number of each class and found a 1:4:6:4:1 phenotypic ratio of wheat with dark red, medium red, intermediate red, light red, and white kernels, respectively. Note that $\frac{1}{16}$ of the F_2 had a kernel phenotype

animation

a Polygene Hypothesis for Wheat Kernel Color

as extreme as the original red parent, and $\frac{1}{16}$ had a kernel phenotype as extreme as the original white parent.

How can the data be interpreted in genetic terms? In Chapter 4 (Analytical Approaches for Solving Genetics Problems, problem 4.3c, pp. 94–95), the explanation for the 15:1 ratio was that two allelic pairs are involved in determining the phenotypes segregating in the cross. Because several of the F_2 populations from the wheat crosses exhibited a 15 red : 1 white ratio, we can apply that explanation to the kernel trait.

Let us hypothesize that there are two pairs of independently segregating alleles that control the production of red pigment: Alleles R (red) and C (crimson) result in red pigment, and alleles r and c result in the lack of pigment. Nilsson-Ehle's parental cross and the F_1 genotypes can then be shown as follows:

$$
\begin{array}{ccc}
P & RR\,CC & \times \quad rr\,cc \\
 & (\text{dark red}) & (\text{white}) \\
 & \downarrow & \\
F_1 & & Rr\,Cc \\
 & & (\text{intermediate red})
\end{array}
$$

When the F_1 is interbred, the distribution of genotypes in the F_2 is that typical of dihybrid inheritance, that is, $\frac{1}{16}\,RR\,CC + \frac{2}{16}\,Rr\,CC + \frac{1}{16}\,rr\,CC + \frac{2}{16}\,RR\,Cc + \frac{4}{16}\,Rr\,Cc + \frac{2}{16}\,rr\,Cc + \frac{1}{16}\,RR\,cc + \frac{2}{16}\,Rr\,cc + \frac{1}{16}\,rr\,cc$. If R and C are dominant to r and c, the 9:3:3:1 phenotypic ratio characteristic of dihybrid inheritance will result. For the wheat kernel color phenotype, then, dominance is not the simple answer because the observed phenotypic ratio approximates 1:4:6:4:1. Note that these numbers in the phenotypic ratio are the same as the coefficients in the **binomial expansion** of $(a+b)^4$. The following calculation demonstrates how we can arrive at the coefficients and their associated terms of this expansion. In essence, we multiply $(a + b)$ by $(a + b)$, then multiply the product by $(a + b)$, and so on:

$$
\begin{array}{l}
\quad a \;+\; b \\
\times\; a \;+\; b \\
\hline
=\; a^2 \;+\; ab \\
\qquad\;+\; ab \;+\; b^2 \\
\hline
=\; a^2 \;+\; 2ab \;+\; b^2 \qquad \text{—this is } (a+b)^2 \\
\times\; a \;+\; b \\
\hline
=\; a^3 \;+\; 2a^2b \;+\; ab^2 \\
\qquad\;+\; a^2b \;+\; 2ab^2 \;+\; b^3 \\
\hline
=\; a^3 \;+\; 3a^2b \;+\; 3ab^2 \;+\; b^3 \qquad \text{—this is } (a+b)^3 \\
\times\; a \;+\; b \\
\hline
=\; a^4 \;+\; 3a^3b \;+\; 3a^2b^2 \;+\; ab^3 \\
\qquad\;+\; a^3b \;+\; 3a^2b^2 \;+\; 3ab^3 \;+\; b^4 \\
\hline
=\; a^4 \;+\; 4a^3b \;+\; 6a^2b^2 \;+\; 4ab^3 \;+\; b^4 \quad \text{—this is } (a+b)^4
\end{array}
$$

A simple explanation for the wheat kernel color phenotypic distribution, then, is that each dose of a gene controlling pigment production allows the synthesis of a certain amount of pigment. Therefore, the intensity of red coloration is a function of the number of *R* or *C* alleles in the genotype; *RR CC* (term a^4 in the expansion) would be dark red, and *rr cc* (b^4 in the expansion) would be white. Table 23.4 summarizes this situation with regard to the five phenotypic classes observed by Nilsson-Ehle. In other words, the genes represented by capital letters code for products that add to the phenotypic characteristic; for example, each allele, *R* or *C*, causes more red pigment to be added to the wheat kernel color phenotype. Alleles that contribute to the phenotype are called **contributing alleles.** The alleles that do not have any effect on the phenotypes of the quantitative trait, such as the *r* and *c* alleles in the wheat kernel color trait, are called **noncontributing alleles.** Thus, the inheritance of red kernel color in wheat is an example of a multiple-gene or polygene series of as many as four contributing alleles.

We must be cautious in interpreting the genetic basis of this particular quantitative trait, though. Some F_2 populations show only three phenotypic classes, with a 3:1 ratio of red to white, whereas other F_2 populations show a 63:1 ratio of red to white, with discrete classes of color between the dark red and the white. These results indicate that the genetic basis for the quantitative trait can vary with the strain of wheat involved. The 3:1 case could be explained by a single gene system with two contributing alleles, and the 63:1 case could indicate a polygene series with six contributing alleles. The number of discrete classes in the latter case would be seven, with the proportion of each class following the coefficients in the binomial expansion of $(a + b)^6$, that is, 1:6:15:20:15:6:1.

The multiple-gene hypothesis that fits the wheat kernel color example so well has been applied to other examples of quantitative inheritance, including ear length in corn. In its basic form, the multiple-gene hypothesis proposes that a number of the attributes of quantitative inheritance can be explained on the basis of the action and segregation of a number of allelic pairs that each have a small but additive effect on the phenotype. These allelic pairs with small effects are called *polygenes*. For the most part, the multiple-gene hypothesis is satisfactory as a working hypothesis for interpreting many quantitative traits. The whole picture of quantitative traits is very complicated, though, and there are still many gaps in our understanding of quantitative inheritance. Fortunately, with the application of molecular techniques, we are slowly getting a better understanding of the nature of the genes involved.

KEYNOTE

Quantitative traits are based on polygenes, genes in a multiple-gene series. The multiple-gene hypothesis assumes that contributing and noncontributing alleles in the series operate so that as the number of contributing alleles increases, there is an additive effect on the phenotype.

Quantitative Trait Loci

The genes underlying quantitative traits—**quantitative trait loci (QTLs)**—cannot be determined through pedigree analysis because environmental effects and the action of other segregating genes tend to obscure the effect of any single gene. However, when important segments of the genome can be identified and correlated with phenotypic variation, we can begin to zero in on important QTLs. The basic approach to QTL identification follows the techniques of marker-based mapping outlined in Chapter 5.

QTL identification is essentially an exercise in correlating segments of the genome with phenotypic differences between individuals. Typically, inbred lines that have been selected for differing phenotypes (i.e., are homozygous for different alleles) are crossed to generate a recombinant inbred strain. The F_1 is expected to be heterozygous at most loci. By further crossing the F_1 to parental lines or to itself, the amount of phenotypic variation is increased as segregation is increased. The F_2 is then analyzed to determine whether any of the marker genotypes are correlated with phenotypic variation. If a marker locus is unlinked to a QTL, the average phenotype is the same for all of its genotypes. If the marker locus is linked to a QTL, then genotypes will differ in the value of the quantitative trait examined.

In this way, we can begin to determine not only how many genes underlie quantitative traits but also how big their effects are and exactly where they reside in the genome. Some of the most exciting applications of this work have led to identification of QTLs that are responsible for differences among species. Some of the most important species differences between plants are the suites of traits that attract pollinators, including color, shape, odor, and nectar rewards. Monkeyflowers have diverged along this axis, with hummingbird-pollinated species such as *Mimulus cardinalis* exhibiting red coloration, deep nectar tubes with lots of nectar, and reflexed petals, whereas species such as *M. lewisii* have little nectar reward, broad petals, and pink flowers characteristic of bee-pollinated flowers (Figure 23.7). Crosses between these species have revealed that pollinator attraction and efficiency are controlled by a number

Table 23.4	Genetic Explanation for the Number and Proportions of F_2 Phenotypes for the Quantitative Trait Red Kernel Color in Wheat		
Genotype	**Number of Contributing Alleles for Red**	**Phenotype**	**Fraction of F_2**
RR CC	4	Dark red	$\frac{1}{16}$
RR Cc or *Rr CC*	3	Medium red	$\frac{4}{16}$
RR cc or *rr CC* or *Rr Cc*	2	Intermediate red	$\frac{6}{16}$
rr Cc or *Rr cc*	1	Light red	$\frac{4}{16}$
rr cc	0	White	$\frac{1}{16}$

Figure 23.7

Mimulus lewisii **(A, C) and** *M. cardinalis* **(B, D) flowers.** Flowers are shown from the front (A, B) as an approaching pollinator views them. In side views (C, D), the relative positions of the stigma and anthers are shown.

of genes, some of which have large effects. Most of the floral traits scored appear to be controlled in part by at least one gene that influences 25 percent or more of the phenotypic variation in the F_2 population. Figure 23.8 (p. 502) illustrates the distribution of QTLs for various floral traits identified through marker analysis. Notice that many of the QTLs for different traits are physically close, suggesting that floral characters are genetically correlated with one another.

K E Y N O T E

Marker-based mapping approaches can be used to correlate segments of the genome with phenotypic variation in quantitative traits when more traditional pedigree analyses are not possible. Quantitative trait loci (QTLs) provide estimates of the number of genes and size of effects influencing variation in continuous traits.

Heritability

Heritability is the proportion of a population's phenotypic variation that is attributable to genetic factors. The term frequently is misused. For example, in humans, when individuals in a family resemble each other in some aspect of the phenotype, be it stature or intelligence, a genetic basis often is assumed to be responsible for the similarity. But the resemblance among family members could just as easily result from their shared environment rather than their shared genes. Carefully planned quantitative genetic experiments are the only way to distinguish between these alternatives in any organism being studied.

The concept of heritability is used to examine the relative contributions of genes and environment to variation in a specific trait. Polygenic traits such as weight of cattle,

Figure 23.8

QTL maps for 12 floral traits in monkeyflowers. Boxes show the position of markers correlated to phenotypic traits. Taller boxes indicate QTLs that explain ≥ 25 percent of the variance in a trait. A – H are linkage groups.

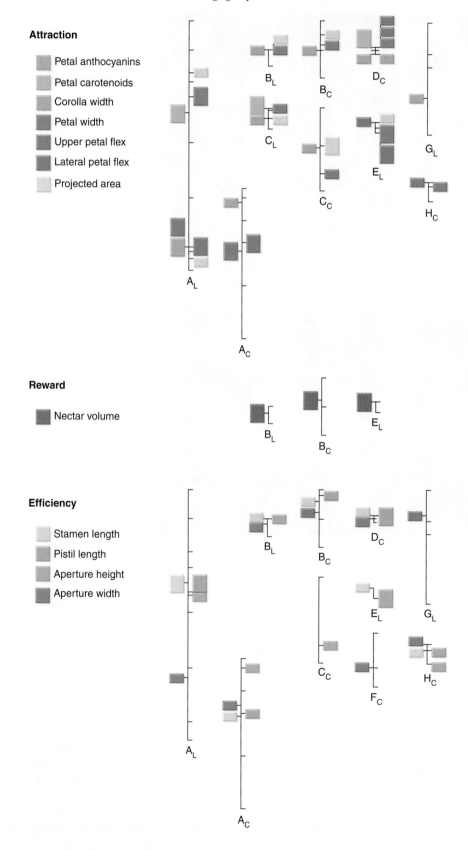

Attraction

Petal anthocyanins
Petal carotenoids
Corolla width
Petal width
Upper petal flex
Lateral petal flex
Projected area

Reward

Nectar volume

Efficiency

Stamen length
Pistil length
Aperture height
Aperture width

number of eggs laid by chickens, and amount of fleece produced by sheep are important for breeding programs and agricultural management. Similarly, the extent to which genetic and environmental factors contribute to human variation in traits such as blood pressure and birth weight is important for health care.

This section deals with two types of heritability: *broad-sense heritability* and *narrow-sense heritability*. To assess heritability, we must first measure the variation in the trait and partition that variance into components attributable to different causes.

Components of the Phenotypic Variance

We can consider variance among individuals as a stick that can be broken into various pieces. The length of each piece corresponds to the amount of variation contributed by that causal factor. The **phenotypic variance,** represented by V_P, is a measure of all of the variability of a trait (i.e., the whole stick). It is calculated by computing the variance of the trait for a group of individuals, as outlined in this chapter's section on statistics. Some of the phenotypic variation may arise because of genetic differences between individuals (different genotypes within the group); this is **genetic variance,** symbolized by V_G. Additional variation often results from the contribution of different environments to differences in phenotypes between individuals; this is **environmental variance,** symbolized by V_E, and by definition it includes any nongenetic source of variation. Temperature, nutrition, and parental care are examples of environmental factors that may cause differences during development between individuals. Thus, we have two pieces of our stick that correspond to the basic nature-nurture issue discussed earlier:

$$V_P = V_G + V_E$$

One hundred percent of the variation among individuals is accounted for by genetic and environmental influences; however, the partitioning of phenotypic variance is more complicated than this. The sum of the genetically caused variance and environmentally caused variance may not add up to the total phenotypic variance. This is because the genetically caused variance and environmentally caused variance may covary, and another term ($COV_{G,E}$) is needed. For example, let's say milk production in cows is influenced by genes, but it is also influenced by the amount of feed a farmer provides. The farmer knows his cows and provides the offspring of good milking cows more feed and poor milking cows less feed. In this way, the variance in milk production is increased beyond that which would be expected on the basis of genes and environment operating independently.

In addition, we may want to know whether we can expect offspring to resemble their parents. Just knowing that there is a genetic component to the variation does not answer this question. A reason for this is that there is another source of phenotypic variance called genotype-by-environment interaction, or $G \times E$. Variance caused by $G \times E$ exists when the relative effects of the genotypes differ among environments. For example, in a cold environment, genotype *AA* of a plant may result in a plant that is 40 cm tall and genotype *Aa* may result in a plant that is 35 cm tall. However, when the genotypes are moved to a warm climate, the plant produced by genotype *Aa* may now be 60 cm, and the *AA* plant may be 50 cm tall. In this example, both genotypes produce taller plants in the warm environment, so there is an environmental effect on variance. There is also a genetic effect, but it is not the kind that results in consistent resemblance among relatives. The genetic effect depends on the environment. The relative performance of the genotypes switches in the two environments. Therefore, both environmental differences (temperature) and genetic differences (genotypes) contribute to the phenotypic variance. However, the effects of genotype and environment cannot simply be added together. An additional component that accounts for how genotype and environment interact must be considered, and this is $V_{G \times E}$.

The phenotypic variance, composed of differences arising from genetic variation, environmental variation, genetic-environmental covariation, and genetic-environmental interaction, can be represented by the following equation:

$$V_P = V_G + V_E + 2COV_{G,E} + V_{G \times E}$$

The relative contributions of these four factors to the phenotypic variance depend on the genetic composition of the population, the specific environment, and the manner in which the genes interact with the environment.

KEYNOTE

Variation among individuals can be partitioned into genetic and environmental components. The fact that genotypes might not be distributed randomly across environments and that genotypes may behave differently in different environments means that the results of an experiment determining the relative importance of genetic and environmental factors may depend in nonobvious ways on the environment in which the experiment is performed.

The pieces of the stick of variation can be further broken to reveal more precise components of causal influence. Genetic variance, V_G, can be subdivided into components arising from different types of gene action and interactions between genes. Some of the genetic variance occurs as a result of the average effects of the different alleles on the phenotype. For example, an allele g may, on the average, contribute 2 cm in height to a plant, and the allele G may contribute 4 cm. In this case, the gg homozygote would contribute $2 + 2 = 4$ cm in height, the Gg heterozygote would contribute $2 + 4 = 6$ cm in height, and the GG homozygote would contribute $4 + 4 = 8$ cm in height. To determine the genetic contribution to height, we would then add the effects of alleles at this locus to the effects of alleles at other loci that might influence the phenotype. Genes such as these are said to have additive effects, and variation resulting from this sort of gene action is called **additive genetic variance,** symbolized V_A. The genes that determine kernel color in wheat (see pp. 499–500) are strictly additive in this way.

Some genes may exhibit dominance; thus, **dominance variance** (symbolized V_D) is another source of genetic variance. If dominance is present, the individual effects of the alleles are not strictly additive; we must also consider how alleles at a locus interact. In the presence of dominance by the G allele, the heterozygote Gg would contribute 8 cm in height to the phenotype, the same amount as the GG homozygote. Thus, a population consisting of both genotypes would have genetic variation, but there would be no corresponding phenotypic variation. As the degree of dominance diminishes, the genotypic differences more clearly become phenotypic differences as the dominance variance turns into additive genetic variance.

The presence of epistasis adds another source of genetic variation, called epistatic or **interaction variance** (symbolized V_I). So we can partition the genetic variance as follows:

$$V_G = V_A + V_D + V_I$$

and the total phenotypic variance up to this point can then be summarized as

$$V_P = V_A + V_D + V_I + V_E + 2\text{COV}_{G,E} + V_{G \times E}$$

It is very difficult to design experiments that can analyze all these components simultaneously, and assumptions about some of them usually must be made. For example, it is often assumed that there is no genotype-by-environment covariation ($2\text{COV}_{G,E}$) or G × E variance, but the well-trained geneticist would always remember that the results of such an experiment must be presented with appropriate caution.

Broad-Sense and Narrow-Sense Heritability

Geneticists are interested in how much of the phenotypic variance, V_P, can be attributed to genetic variance, V_G. This quantity is called the **broad-sense heritability** and can be thought of as how much of the stick of variation is made up of genetic variance. Broad-sense heritability is expressed as a proportion:

$$\text{Broad-sense heritability} = h_B^2 = \frac{V_G}{V_P}$$

(The B in the heritability term, h^2, signifies broad-sense.) Heritability of a trait can range from 0 to 1. A broad-sense heritability of 0 indicates that none of the variation in phenotype among individuals results from genetic differences. A broad-sense heritability of 0.5 means that 50 percent of the phenotypic variation arises from genetic differences among individuals, and a broad-sense heritability of 1 suggests that all the phenotypic variance is genetically based. Broad-sense heritability includes genetic variation from all types of genes and gene actions. It ignores the fact that the genetic variance may be of the additive, dominance, or interactive sort. It also assumes that genotype-by-environment interaction ($V_{G \times E}$) is not important. Thus, the usefulness of broad-sense heritability is questionable.

Since the additive genetic variance allows one to make accurate predictions about the resemblance between offspring and parents, quantitative geneticists frequently determine the **narrow-sense heritability,** which is the proportion of the phenotypic variance that results from additive genetic variance:

$$\text{Narrow-sense heritability} = h_N^2 = \frac{V_A}{V_P}$$

(The N in the heritability term, h^2, signifies narrow-sense.)

Understanding Heritability

Heritability estimates have a number of significant limitations. Unfortunately, these limitations often are ignored. As a result, heritability is one of the most misunderstood and widely abused concepts in genetics. The following are some of the important qualifications and limitations of heritability.

1. **Broad-sense heritability does not indicate the extent to which a trait is genetic.** What broad-sense heritability does measure is the *proportion of the phenotypic variance* among individuals in a population that results from genetic differences. The proportion of the phenotypic variance among individuals in a population that results from genetic differences may seem like the same thing as the extent to which a trait is genetic, but these two statements actually mean different things. Genes often influence the development of a trait, and thus the trait may be said to be genetic. For example, our ability to learn about the game of football depends on the proper development of our nervous system, which is clearly controlled by genes. Thus, we could say that knowledge of the game of football is a genetically influenced trait. However, the differences we see among individuals in their knowledge of football usually is not caused by genes but by different environments and

personal interests. Since broad-sense heritability measures the proportion of the phenotypic variation among individuals that results from genetic differences, broad-sense heritability for knowledge of football would be zero, despite the fact that genes do influence our ability to have knowledge of football.

2. **Heritability does not indicate what proportion of an individual's phenotype is genetic.** Since it is based on the variance, which can be calculated only for a group of individuals, heritability is characteristic of a population. An individual does not have heritability; a population does.

3. **Heritability is not fixed for a trait.** Thus, there is no universal heritability for a trait such as human height. Rather, the heritability value for a trait depends on the genetic makeup and the specific environment of the population. We could calculate the heritability of stature (height) among Hopi Indians living in Arizona, for example, but heritability calculated for other populations and other environments might be very different.

4. **High heritability for a trait does not imply that population differences in the same trait are genetically determined.** Therefore, heritability cannot be used to draw conclusions about the nature of differences between populations. Let's say we had two groups of humans and determined that variation in book-reading ability within each group had a high heritability. One group was raised in a book-rich environment, and most individuals could read well. The other group was raised in a book-poor environment, and individuals read poorly. What conclusions would you draw about the genetic differences between the two populations? Can book-reading ability be enhanced by social intervention programs, or is it hopeless because book-reading ability is "genetic"?

5. **Traits shared by members of the same family do not necessarily have high heritability.** When members of the same family share a trait, the trait is said to be **familial.** Familial traits may arise because family members share genes or because they are exposed to the same environmental factors. Thus, familiality is not the same as heritability.

Heritability values for a number of traits in different species are given in Table 23.5. These heritability values are based on various populations and have been determined using a variety of methods, including the parent-offspring method. Estimates of heritability are rarely precise, and most measured heritability values have large standard errors. Therefore, heritability values calculated for human traits must be viewed with special caution, given the difficulties of separating genetic and environmental influences.

Table 23.5	Heritability Values for Some Traits in Humans, Domesticated Animals, and Natural Populations[a]	
Organism	**Trait**	**Heritability**
Humans	Stature	0.65
	Serum immunoglobulin (IgG) level	0.45
Cattle	Milk yield	0.35
	Butterfat content	0.40
	Body weight	0.65
Pigs	Back-fat thickness	0.70
	Litter size	0.05
Poultry	Egg weight	0.50
	Egg production (to 72 weeks)	0.10
	Body weight (at 32 weeks)	0.55
Mice	Body weight	0.35
Drosophila	Abdominal bristle number	0.50
Jewelweed	Germination time	0.29
Milkweed bugs	Wing length (females)	0.87
	Fecundity (females)	0.50
Spring peepers (frogs)	Size at metamorphosis	0.69
Wood frogs	Development rate (mountain population)	0.31
	Size at metamorphosis (mountain population)	0.62

[a]The estimates given in this table apply to particular populations in particular environments; heritability values for other populations may differ.

K E Y N O T E

Broad-sense heritability of a trait represents the pro-portion of the phenotypic variance that results from genetic differences between individuals. Narrow-sense heritability is more limited: It measures the proportion of the phenotypic variance that results from additive genetic variance.

Response to Selection

Two fields of study in which quantitative genetics has played a particularly important role are plant and animal breeding and evolutionary biology. Both fields are concerned with genetic change within groups of organisms. In the case of plant and animal breeding, genetic change can lead to improvement in yield, hardiness, size, and other agriculturally important qualities; in the case of evolutionary biology, genetic change occurs in natural populations as a result of the processes discussed in Chapter 22. **Evolution** can be defined as genetic change that takes place over time within a group of organisms. Therefore, both evolutionary biologists and plant and animal breeders are interested in the process of evolution, and both use the methods of quantitative genetics to predict the rate and magnitude of genetic change.

The essential element of natural selection is that individuals with certain genotypes leave more offspring than others. In this way, groups of individuals (populations) change, or evolve, over time and become better adapted to their particular environment. Humans bring about evolution in domestic plants and animals through the similar process of **artificial selection.** In artificial selection, humans, not nature, select the individuals that are to survive and reproduce. If the selected traits have a genetic basis, then the genetic structure of the selected population will change over time and evolve, just as traits in natural populations evolve as a result of natural selection. Artificial selection can be a powerful tool in bringing about rapid evolutionary change, as evidenced by the extensive variation observed in domesticated plants and animals. For example, all breeds of domestic dogs are derived from one species that was domesticated some 10,000 years ago. The large number of breeds that exist today, encompassing a tremendous variety of sizes, shapes, colors, and even behaviors, have been produced by artificial selection and selective breeding.

Both the process of natural selection, as described by Charles Darwin, and artificial selection, practiced by plant and animal breeders, depend on the presence of genetic variation. Only if genetic variation is present within a population of individuals can that population change genetically and evolve. Furthermore, the amount and type of genetic variation present are crucial in determining how fast evolution will occur. Both evolutionary biologists and breeders are interested in determining how much genetic variation for a particular trait exists within a population. As we have seen, quantitative genetics often is used to answer this question.

Estimating the Response to Selection

When natural or artificial selection is imposed on a phenotype, the phenotype changes from one generation to the next if genetic variation underlying the trait is present in the population. The amount that the phenotype changes in one generation is called the **selection response,** R. To illustrate the concept of selection response, suppose a geneticist wants to produce a strain of *Drosophila melanogaster* with large body size. To increase body size in fruit flies, the geneticist would first examine flies from a genetically diverse population and would measure the body size of these *unselected flies*. Suppose that our geneticist found the mean body weight of the unselected flies to be 1.3 mg. After determining the mean body weight of this population, the geneticist would select flies that were endowed with large bodies (assume that the mean body weight of these selected flies was 3.0 mg). He would then place the large, selected flies in a separate culture vial and allow them to interbreed. After the F_1 offspring of these selected parents emerged, the geneticist would measure the body weights of the F_1 flies and compare them with the body weights of the original, unselected population.

What our geneticist has done in this procedure is to apply selection for large body size to the population of fruit flies. If genetic variation underlies the variation in body size of the original population, the offspring of the selected flies will resemble their parents, and the mean body size of the F_1 generation will be greater than the mean body size of the original population. If the F_1 flies have a mean body weight of 2.0 mg, which is considerably larger than the mean body weight of 1.3 mg observed in the original, unselected population, a response to selection has occurred.

The amount of change that occurs in one generation, or the selection response, depends on two things: the narrow-sense heritability and the selection differential, S. The **selection differential** is defined as the difference between the mean phenotype of the selected parents and the mean phenotype of the population before selection. In our example of body size in fruit flies, the original population had a mean weight of 1.3 mg, and the mean weight of the selected parents was 3.0 mg, so the selection differential was 3.0 mg − 1.3 mg = 1.7 mg. The selection response is related to the selection differential and the heritability by the following formula, known as the breeder's equation:

$$R = h^2S$$

When the geneticist applied artificial selection to body size in fruit flies, the difference in the mean body weight of the F_1 flies and the original population was 2.0 mg − 1.3 mg = 0.7 mg, which is the response to selection. We now have values for two of the three parameters in the preceding equation: the selection response (0.7 mg) and the selection differential (1.7 mg). By rearranging the formula for the selection response, we can solve for the narrow-sense heritability:

$$\text{Narrow-sense heritability} = h_N^2 = \frac{\text{selection response}}{\text{selection differential}}$$

$$= \frac{0.7 \text{ mg}}{1.7 \text{ mg}} = 0.41$$

Measuring the response to selection provides another means for determining the narrow-sense heritability, and heritabilities for many traits are determined in this way.

A trait will continue to respond to selection, generation after generation, as long as genetic variation for the trait remains within the population. The results from an actual selection experiment on phototaxis in *Drosophila pseudoobscura* are presented in Figure 23.9. Phototaxis is a behavioral response to light. In this study, flies were scored for the number of times they moved toward light in a total of 15 light-dark choices. Two different response-to-selection experiments were carried out. In one, attraction to light was selected, and in the other, avoidance of light was selected. As can be seen in Figure 23.8, the fruit flies responded to selection for positive and negative phototactic behavior for a number of generations. Eventually, however, the response to selection tapered off, and finally, no further directional change in phototactic behavior occurred. One possible reason for this lack of response in later generations is that no more genetic variation for phototactic behavior existed within the population. In other words, all flies at this point were homozygous for all the alleles affecting the behavior. If this were the case, phototactic behavior could not undergo further evolution in this population unless input of additional genetic variation occurred. More often, some variation still exists for the trait, even after the selection response levels off, but the population fails to respond to selection because the genes for the selected trait have detrimental effects on other traits.

KEYNOTE

The amount that a trait changes in one generation as a result of selection for the trait is called the selection response or the response to selection. The magnitude of selection response depends on both the intensity of selection, called the selection differential, and the narrow-sense heritability.

Genetic Correlations

The phenotypes of two or more traits may be associated or correlated; this means that the traits do not vary independently. For example, fair skin, blond hair, and blue eyes are often found together in the same individual. The association is not perfect—we sometimes see individuals with dark hair, fair skin, and blue eyes—but the traits are found together with enough regularity for us to say that they are correlated. The **phenotypic correlation** between two traits can be computed by measuring two phenotypes on a number of individuals and then calculating a correlation coefficient for the two traits (see the section on statistics for a discussion of correlation coefficients). One reason for a phenotypic correlation among traits is that the traits are influenced by a common set of genes. Indeed, this is the most likely reason for the association among hair color, eye color, and skin color in humans. Rarely do genes affect only a single trait. More commonly, each gene influences a number of traits, and this is particularly true for the polygenes that influence continuous traits. When genes affect multiple phenotypes, we say they are *pleiotropic*, a concept we introduced in Chapter 10. If a gene is pleiotropic, it simultaneously affects two or more traits, and thus the phenotypes of those two traits will be correlated. For example, the genes that affect growth rates in humans also influence both weight and height, so these two phenotypes tend to be

Figure 23.9

Selection for phototaxis in *Drosophila pseudoobscura*. The upper graph is the line selected for avoidance of light. The lower graph is the line selected for attraction to light. The phototactic score is the number of times the fly moved toward the light out of a total of 15 light-dark choices.

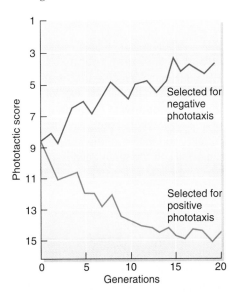

correlated. When pleiotropy is present (i.e., some of the genes influencing two traits are the same), we say that a **genetic correlation** exists for the traits.

Pleiotropy is not the only cause of phenotypic correlations among traits. Environmental factors may also influence several traits simultaneously and may cause nonrandom associations between phenotypes. For example, adding fertilizer to the soil often causes plants both to grow taller and to produce more flowers. If we measured plant height and counted the number of flowers on a group of responsive plants, some of which received fertilizer and some of which did not, we would find that the two traits are correlated; those plants receiving fertilizer would be tall and would have many flowers, while those without fertilizer would be short and have few flowers. However, this phenotypic correlation does not result from any genetic correlation (pleiotropy), but from the common effect of the environmental factor, the fertilizer, on both traits.

Genetic correlations may be positive or negative. A positive correlation means that genes causing an increase in the magnitude of one trait bring about a simultaneous increase in the magnitude of the other. In chickens, body weight and egg weight have a positive genetic correlation. If breeders select for heavier chickens, thereby favoring the genes for large body size, the size of the chickens will increase and the mean weight of the eggs produced by these chickens will also increase. This increase in egg weight occurs because the genes that produce heavier chickens have a similar effect on egg weight.

Other traits exhibit negative genetic correlations. In this case, genes that cause an increase in one trait tend to produce a corresponding decrease in another trait. Egg weight and egg number in chickens have a negative genetic correlation, for instance. When breeders select for chickens that produce larger eggs, the average egg size increases, but the number of eggs laid by each chicken decreases.

Genetic correlations are important for predicting the changes in phenotype that result from selection. If two traits are genetically correlated, they evolve together. Any genetic change in one trait occurring as a result of selection will simultaneously alter the other trait, since both are influenced by the same genes. Such correlation often presents problems for plant and animal breeders. As an example, milk yield and butterfat content have a negative genetic correlation in cattle. The same genes that cause an increase in milk production bring about a decrease in butterfat content of the milk. Thus, when breeders select for increased milk yield, the amount of milk produced by the cows may go up, but the butterfat content will decrease at the same time. Negative correlations among traits often place practical constraints on the ability of breeders to select for desirable traits. Knowing about the presence of genetic correlations before undertaking an expensive breeding program is essential in order to avoid the production of associated undesirable traits in the selected stock.

An organism's ability to adapt to a particular environment is also strongly influenced by genetic correlations among traits, and thus genetic correlations are of considerable interest to evolutionary biologists. As an illustration, consider two traits in tadpoles: developmental rate and size at metamorphosis. Most tadpoles are found in small ponds and pools, where fish (potential predators) are absent and food is abundant. A major liability in using this type of aquatic habitat is that ponds often dry up, frequently before the tadpoles have developed sufficiently to metamorphose into frogs and leave the water. One might expect, then, that natural selection would favor a maximum rate of development in tadpoles, so that the tadpoles could quickly metamorphose into frogs. However, many species of tadpoles fail to develop at maximum rates, contrary to what we might expect to evolve under natural selection. One reason for the slow rate of development is that a negative genetic correlation may exist between developmental rate and body size at metamorphosis. Genes that accelerate development also tend to cause metamorphosis at a smaller size, at least in some populations. Thus, selection for fast metamorphosis will also produce smaller frogs, and size is extremely important in determining the survival of young frogs. Small frogs tend to lose water more rapidly in the terrestrial environment, are more likely to be eaten by predators, and have more difficulty finding sufficient food. The negative genetic correlation between developmental rate and body size at metamorphosis places constraints on the frogs' ability to evolve rapid development and on their ability to evolve large body size at metamorphosis. Knowing about such genetic correlations is important for understanding how animals adapt or fail to adapt to a particular environment. Table 23.6 presents some genetic correlations that have been detected in studies of quantitative genetics.

iActivity
You are a researcher trying to determine whether fingerprint patterns can be correlated to high blood pressure in the iActivity *Your Fate in Your Hands* on the website.

KEYNOTE

Genetic correlations arise when two traits are influenced by the same genes. When a trait is selected, any genetically correlated traits will also exhibit a selection response. Thus, the evolution of a population, the outcome of an artificial breeding program, or the ability of a population to respond evolutionarily and avoid extinction, depends on the simultaneous integration and correlation of many aspects of the phenotype.

Summary

Quantitative genetics is the field of genetics that studies the inheritance of continuous, or quantitative, traits— those with a range of phenotypes. Continuous traits usually result from the influence of multiple genes and

Table 23.6	Genetic Correlations Between Traits in Humans, Domesticated Animals, and Natural Populations[a]	
Organism	**Traits**	**Genetic Correlation**
Humans	IgG, IgM	0.07
Cattle	Butterfat content, milk yield	−0.38
Pigs	Weight gain, back-fat thickness	0.13
	Weight gain, efficiency	0.69
Chickens	Egg weight, egg production	−0.31
	Body weight, egg weight	0.42
	Body weight, egg production	−0.17
Mice	Body weight, tail length	0.29
Jewelweed	Seed weight, germination time	−0.81
Milkweed bugs	Wing length, fecundity	−0.57
Wood frogs	Developmental rate, size at metamorphosis	−0.86
Drosophila	Early life fecundity, resistance to starvation	−0.91

[a]The estimates given in this table apply to particular populations in particular environments; genetic correlations for other individuals may differ.

environmental factors. Statistics such as the mean, variance, standard deviation, and correlation of characters can be used to describe continuous traits.

Polygenic traits are caused by genes at multiple loci, each of which follows the principles of Mendelian inheritance. The multiple-gene hypothesis assumes that the effects of the alleles at a locus are small and additive, but at present this is at best a working model and much more remains to be known. Specific quantitative trait loci (QTLs) that determine variation in continuous traits can be identified through marker-based mapping.

The broad-sense heritability is the proportion of the phenotypic variance in a population that results from genetic differences. The narrow-sense heritability indicates the proportion of the phenotypic variance that results from additive genetic variance.

The response to selection is the amount a trait changes in one generation as a result of selection; it depends on the narrow-sense heritability and the selection differential. Evolutionary responses that result from the effects of natural selection depend on narrow-sense heritability.

The field of quantitative genetics strives to clarify the relationship between a complex genetic architecture and complex phenotypic architecture. Are most continuous traits controlled by many genes with small effects, or are they controlled by few genes with major effects? How important are epistatic interactions and pleiotropic effects? These are not easy questions to answer. But in the answers will lie extremely important findings that will affect the way in which we view the relationship between the genotype and phenotype, which in turn affects our understanding of developmental biology and evolution.

Analytical Approaches for Solving Genetics Problems

Q23.1   Assume that genes *A*, *B*, *C*, and *D* are members of a multiple-gene series that control a quantitative trait. Each of these genes has a duplicate, cumulative effect in that each contributes 3 cm of height to the organism when it is present. Each gene assorts independently. In addition, gene *L* is always present in the homozygous state, and the *LL* genotype contributes a constant 40 cm of height. The alleles *a*, *b*, *c*, and *d* do not contribute anything to the height of the organism. If we ignore height variation caused by environmental factors, an organism with genotype *AA BB CC DD LL* would be 64 cm high, and one with genotype *aa bb cc dd LL* would be 40 cm. A cross is made of *AA bb CC DD LL* × *aa BB cc DD LL* and is carried into the F_2 by selfing of the F_1.

a. How does the height of the F_1 individuals compare with the height of each of the parents?

b. Compare the mean of the F_1 with the mean of the F_2, and comment on your findings.

c. What proportion of the F_2 population would show the same height as the *AA bb CC DD LL* parent?

d. What proportion of the F_2 population would show the same height as the *aa BB cc DD LL* parent?

e. What proportion of the F_2 population would breed true for the height shown by the *aa BB cc DD LL* parent?

f. What proportion of the F_2 population would breed true for the height characteristic of F_1 individuals?

A23.1   This question explores our understanding of the basic genetics involved in a multiple-gene series that in this case controls a quantitative trait. The approach we will take is essentially the same as the approach used with a series of independently assorting genes that control distinctly different traits. That is, we make predictions on the basis of genotypes and relate the results to phenotypes, or we make predictions on the basis of phenotypes and relate the results to genotypes.

a. Each allele represented by a capital letter contributes 3 cm of height to the base height of 40 cm, which is controlled by the ever-present *LL* homozygosity. Therefore, the *AA bb CC DD LL* parent, which has six capital-letter alleles from the *A–D*, multiple-gene series, is 40 + (6 × 3) = 58 cm high. Similarly, the *aa BB cc DD LL* parent has four capital-letter alleles and therefore is 40 + 12 = 52 cm high. The F_1 from a cross between these two individuals would be heterozygous for the *A*, *B*, and *C* loci and homozygous for *D* and *L*, that is, *Aa Bb Cc DD LL*. This progeny has five capital-letter alleles apart from *LL* and therefore is 40 + 15 = 55 cm high.

b. The F_2 is derived from a self of the *Aa Bb Cc DD LL* F_1. All the F_2 individuals will be *DD LL*, making them at least 40 + 6 = 46 cm high. Now we must deal with the heterozygosity at the other three loci. What we need to calculate is the relative proportions of individuals with all the various possible numbers of capital-letter alleles. This calculation is equivalent to determining the relative distribution of three independently assorting traits, each showing incomplete dominance. In other words, we must calculate directly the relative frequencies of all possible genotypes for the three loci and collect those with no, one, two, three, four, five, and six capital-letter alleles. Thus, the probability of getting an individual with two capital-letter alleles for each locus is $\frac{1}{4}$, the probability of getting an individual with one capital-letter allele for each locus is $\frac{1}{2}$, and the probability of getting an individual with no capital-letter alleles for each locus is $\frac{1}{4}$. So the probability of getting an F_2 individual with six capital-letter alleles for the *A*, *B*, and *C* loci is $(\frac{1}{4})^3 = \frac{1}{64}$, and the same probability is obtained for an individual with no capital-letter alleles. This analysis gives us a clue about how we should consider all the possible combinations of genotypes that have the other numbers of capital-letter alleles. That is, the simplest approach is to compute the coefficients in the binomial expansion of $(a + b)^6$, as explained on page 499. The expansion gives a 1:6:15:20:15:6:1 distribution of zero, one, two, three, four, five, and six capital-letter alleles, respectively. Since each capital-letter allele in the *A*, *B*, and *C* set contributes 3 cm of height over the 46-cm height given by the *DD LL* genotype common to all, the F_2 individuals would fall into the following distribution:

Number of Capital-Letter Alleles	Height Added to Basic Height of 46 cm for Common *DD LL* Genotype (cm)	Height of Individuals (cm)	Frequency
6	18	64	1
5	15	61	6
4	12	58	15
3	9	55	20
2	6	52	15
1	3	49	6
0	0	46	1

The distribution is clearly symmetrical, giving an average of 55 cm, the same height shown in F_1 individuals.

c. The *AA bb CC DD LL* parent was 58 cm, so we can read the proportion of F_2 individuals that show this same height directly from the table in part (b). The answer is $\frac{15}{64}$.

d. The *aa BB cc DD LL* parent was 52 cm, and from the table in part (b), the proportion of F_2 individuals that show this same height is $\frac{15}{64}$.

e. We are asked to determine the proportion of the F_2 population that would breed true for the height shown by the *aa BB cc DD LL* parent, which was 52 cm. To breed true, the organism must be homozygous. We have also established that *DD LL* is a constant genotype for the F_2 individuals, giving a basic height of 46 cm. Therefore, for a height of 52 cm, two additional, active, capital-letter alleles must be present apart from those at the *D* and *L* loci. With the requirement for homozygosity, there are only three genotypes that give a 52-cm height; they are *AA bb cc DD LL*, *aa BB cc DD LL*, and *aa bb CC DD LL*. The probability of each combination occurring in the F_2 is $\frac{1}{64}$, so the answer to the problem is $\frac{1}{64} + \frac{1}{64} + \frac{1}{64} = \frac{3}{64}$. (Note that the individual probability for each genotype can be calculated. That is, probability of *AA* = $\frac{1}{4}$, probability of *bb* = $\frac{1}{4}$, probability of *cc* = $\frac{1}{4}$, and probability of *DD LL* = 1, giving an overall probability for *AA bb cc DD LL* of $\frac{1}{64}$.)

f. We are asked to determine the proportion of the F_2 population that would breed true for the height characteristic of F_1 individuals. Again, the basic height given by *DD LL* is 46 cm. The F_1 height is 55 cm, so three capital-letter alleles must be present in addition to *DD LL* to give that height because 3 × 3 cm = 9 cm, and 9 cm + 46 cm = 55 cm. However, because an individual must be homozygous to be true-breeding, the answer to this question is none, because 3 is an odd number, meaning that at least one locus must be heterozygous to get the 55-cm height.

Q23.2 Five field mice collected in Texas had weights of 15.5 g, 10.3 g, 11.7 g, 17.9 g, and 14.1 g. Five mice collected in Michigan had weights of 20.2 g, 21.2 g, 20.4 g, 22.0 g, and 19.7 g. Calculate the mean weight and the variance in weight for mice from Texas and mice from Michigan.

A23.2 To answer this question, we use the formulas given in the chapter section on statistical tools. The formula for the mean is

$$\bar{x} = \frac{\Sigma x_i}{n}$$

The symbol Σ means to add, and the x_i represents all the individual values. We begin by summing up all the weights of the mice from Texas:

$$\Sigma x_i = 15.5 + 10.3 + 11.7 + 17.9 + 14.1 = 69.5$$

Next, we divide this summation by n, which represents the number of values added together. In this case, we added together five weights, so $n = 5$. The mean for the Texas mice is therefore

$$\frac{\Sigma x_i}{n} = \frac{69.5}{5} = 13.9$$

To calculate the variance in weight among the Texas mice, we use the formula

$$s^2 = \frac{\Sigma (x_i - \bar{x})^2}{n - 1}$$

We must take each individual weight and subtract it from the mean weight of the group. Each value obtained from this subtraction is then squared, and all squared values are added up, as shown below:

15.5 − 13.9 = 1.6	$(1.6)^2 = 2.56$
10.3 − 13.9 = −3.6	$(-3.6)^2 = 12.96$
11.7 − 13.9 = −2.2	$(-2.2)^2 = 4.84$
17.9 − 13.9 = 4.0	$(4.0)^2 = 16.0$
14.1 − 13.9 = 0.2	$(0.2)^2 = 0.04$
	36.4

The sum of all the squared values is 36.4. All that remains for us to do is to divide this sum by $n - 1$, which is $5 - 1 = 4$:

$$s^2 = \frac{\Sigma (x_i - \bar{x})^2}{n - 1} = \frac{36.4}{4} = 9.1$$

The mean and the variance for the Texas mice are 13.9 and 9.1.

We now repeat these steps for the mice from Michigan.

$$\Sigma x_i = 20.2 + 21.2 + 20.4 + 22.0 + 19.7 = 103.5$$

$$\frac{\Sigma x_i}{n} = \frac{103.5}{4} = 20.7$$

$$s^2 = \frac{\Sigma (x_i - \bar{x})^2}{n - 1}$$

20.2 − 20.7 = −0.5	$(-0.5)^2 = 0.25$
21.2 − 20.7 = 0.5	$(0.5)^2 = 0.25$
20.4 − 20.7 = −0.3	$(-0.3)^2 = 0.09$
22.0 − 20.7 = 1.3	$(1.3)^2 = 1.69$
19.7 − 20.7 = −1.0	$(-1.0)^2 = 1.0$
	3.28

$$s^2 = \frac{\Sigma (x_i - \bar{x})^2}{n - 1} = \frac{3.28}{4} = 0.82$$

The mean and the variance for the Michigan mice are 20.7 and 0.82.

We conclude that the Michigan mice are much heavier than the Texas mice, and the Michigan mice also exhibit less variance in weight.

Questions and Problems

*23.1 The following measurements of head width and wing length were made on a series of steamer ducks:

Specimen	Head Width (cm)	Wing Length (cm)
1	2.75	30.3
2	3.20	36.2
3	2.86	31.4
4	3.24	35.7
5	3.16	33.4
6	3.32	34.8
7	2.52	27.2
8	4.16	52.7

a. Calculate the mean and the standard deviation of head width and of wing length for these eight birds.
b. Calculate the correlation coefficient for the relationship between head width and wing length in this series of ducks.
c. What conclusions can you make about the association between head width and wing length in steamer-ducks?

23.2 Answer the following questions.
a. In a family of six children, what is the probability that three will be girls and three will be boys?
b. In a family of five children, what is the probability that one will be a boy and four will be girls?
c. What is the probability that in a family of six children, all will be boys?

*23.3 In flipping a coin, there is a 50 percent chance of obtaining heads and a 50 percent chance of obtaining tails on each flip. If you flip a coin ten times, what is the probability of obtaining exactly five heads and five tails?

*23.4 The F_1 generation from a cross of two pure-breeding parents that differ in a size character usually is no more variable than the parents. Explain.

23.5 If two pure-breeding strains, differing in a size trait, are crossed, is it possible for F_2 individuals to have phenotypes that are more extreme than either grandparent (i.e., be larger than the largest or smaller than the smallest in the parental generation)? Explain.

23.6 Two pairs of genes with two alleles each, *A/a* and *B/b*, determine plant height additively in a population. The homozygote *AA BB* is 50 cm tall, and the homozygote *aa bb* is 30 cm tall.
a. What is the F_1 height in a cross between the two homozygous stocks?
b. What genotypes in the F_2 will show a height of 40 cm after an $F_1 \times F_1$ cross?
c. What will be the F_2 frequency of the 40-cm plants?

*23.7 Three independently segregating genes (*A, B, C*), each with two alleles, determine height in a plant. Each capital-letter allele adds 2 cm to a base height of 2 cm.
a. What are the heights expected in the F_1 progeny of a cross between homozygous strains *AA BB CC* (14 cm) and *aa bb cc* (2 cm)?
b. What is the distribution of heights (frequency and phenotype) expected in an $F_1 \times F_1$ cross?
c. What proportion of F_2 plants will have heights equal to the heights of the original two parental strains?
d. What proportion of the F_2 will breed true for the height shown by the F_1?

23.8 Repeat problem 23.7, but assume that each capital-letter allele doubles the existing height; for example, *Aa bb cc* = 4 cm, *AA bb cc* = 8 cm, *AA Bb cc* = 16 cm, and so on.

23.9 Assume that three equally and additively contributing pairs of alleles control flower length in nasturtiums. A completely homozygous plant with 10-mm flowers is crossed with a completely homozygous plant with 30-mm flowers. F_1 plants all have flowers about 20 mm long. F_2 plants show a range of lengths from 10 to 30 mm, with about $1/64$ of the F_2 having 10-mm flowers and $1/64$ having 30-mm flowers. What distribution of flower length would you expect to see in the offspring of a cross between an F_1 plant and the 30-mm parent?

*23.10 In a particular experiment, the mean internode length in spikes (the floral structures) of the barley variety *asplund* was found to be 2.12 mm. In the variety *abed binder,* the mean internode length was found to be 3.17 mm. The mean of the F_1 of a cross between the two varieties was approximately 2.7 mm. The F_2 gave a continuous range of variation from one parental extreme to the other. Analysis of the F_3 generation showed that in the F_2, 8 out of the total 125 individuals were of the *asplund* type, giving a mean of 2.19 mm. Eight other individuals were similar to the parent *abed binder,* giving a mean internode length of 3.24 mm. Is the internode length in spikes of barley a discontinuous or a quantitative trait? Why?

23.11 Assume that the difference between a type of oat yielding about 4 g per plant and a type yielding 10 g is the result of three equal and cumulative multiple-gene pairs, *AA BB CC*. If you cross the type yielding 4 g with the type yielding 10 g, what will be the phenotypes of the F_1 and the F_2? What will be their distribution?

*23.12 Assume that in squashes the difference in fruit weight between a 3-lb type and a 6-lb type results from three allelic pairs, *A/a*, *B/b*, and *C/c*. Each capital-letter allele contributes a half pound to the weight of the squash. From a cross of a 3-lb plant (*aa bb cc*) with a 6-lb plant (*AA BB CC*), what will be the phenotypes (weights) of the F_1 and the F_2? What will be their distribution?

23.13 Refer to the assumptions stated in problem 23.12 to determine the range in fruit weight of the offspring in the following squash crosses.
a. *AA Bb CC* × *aa Bb Cc*
b. *AA bb Cc* × *Aa BB cc*
c. *aa BB cc* × *AA BB cc*

*23.14 Assume that the difference between a corn plant 10 dm (decimeters) high and one 26 dm high results from four pairs of equal and cumulative multiple alleles, with the 26-dm plants being *AA BB CC DD* and the 10-dm plants being *aa bb cc dd*.
a. What will be the size and genotype of an F_1 from a cross between these two true-breeding types?
b. Determine the limits of height variation in the offspring from the following crosses:
i. *Aa bb cc dd* × *Aa bb Cc dd*
ii. *aa BB cc dd* × *Aa Bb Cc dd*
iii. *AA BB Cc DD* × *aa BB cc Dd*
iv. *Aa Bb Cc Dd* × *Aa bb Cc Dd*

23.15 Refer to the assumptions given in problem 23.14. For this problem, two 14-dm corn plants, when crossed, give nothing but 14-dm offspring (case A). Two other 14-dm plants give one 18-dm, four 16-dm, six 14-dm, four 12-dm, and one 10-dm offspring (case B). Two other 14-dm plants, when crossed, give one 16-dm, two 14-dm, and one 12-dm offspring (case C). What genotypes for each of these 14-dm parents (cases A, B, and C) would explain these results? Would it be possible to get a plant taller than 48 dm by selection in any of these families?

*23.16 Pigmentation in the imaginary river bottom dweller *Mucus yuccas* is a quantitative character controlled by a set of five independently segregating polygenes with two alleles each: *A/a, B/b, C/c, D/d,* and *E/e.* Pigment is deposited at three different levels, depending on the threshold of gene products produced by the capital-letter alleles. Greyish-brown pigmentation is seen if at least four capital alleles are present; light tan pigmentation is seen if two or three capital alleles are present; and whitish-blue pigmentation is seen if these thresholds are not met. If an *AA BB CC DD EE* animal is crossed with a true-breeding *aa bb cc dd ee* animal and the progeny are selfed, what kinds of phenotypes are expected in the F_1 and F_2?

23.17 Since monozygotic twins share all their genetic material, and dizygotic twins share, on average, half of their genetic material, twin studies sometimes can be useful for evaluating the genetic contribution to a trait. Consider the following two instances.

An intelligence quotient (IQ) assesses intellectual performance on a standardized test that involves reasoning ability, memory, and knowledge of an individual's language and culture. IQ scores are transformed so that the population mean score is 100 and 95 percent of the individuals have scores in the range between 70 and 130. Observations in the United States and England found that monozygotic twins had an average difference of 6 IQ points, dizygotic twins had an average difference of 11 points, and random pairs of individuals had an average difference of 21 points.

In a large sample of pairs of twins in the United States where one twin was a smoker, 83 percent of monozygotic twins both smoked, whereas 62 percent of dizygotic twins both smoked.

From these data, can you infer the genetic determination of IQ or smoking?

23.18 A quantitative geneticist determines the following variance components for leaf width in a population of wildflowers growing along a roadside in Kentucky:

> Additive genetic variance (V_A) = 4.2
> Dominance genetic variance (V_D) = 1.6
> Interaction genetic variance (V_I) = 0.3
> Environmental variance (V_E) = 2.7
> Genetic-environmental variance ($V_{G\times E}$) = 0.0

a. Calculate the broad-sense heritability and the narrow-sense heritability for leaf width in this population of wildflowers.

b. What do the heritabilities obtained in part (a) indicate about the genetic nature of leaf width variation in this plant?

*23.19 Assume that all genetic variance affecting seed weight in beans is genetically determined and is additive. From a population in which the mean seed weight was 0.88 g, a farmer selected two seeds, each weighing 1.02 g. He planted these seeds and crossed the resulting plants with each other, then collected and weighed their seeds. The mean weight of their seeds was 0.96 g. What is the narrow-sense heritability of seed weight?

*23.20 Members of the inbred rat strain SHR are salt-sensitive: They respond to a high-salt environment by developing hypertension. Members of a different inbred rat strain, TIS, are not salt-sensitive. Imagine that you placed a population consisting only of SHR rats in an environment that was variable in regard to distribution of salt, so that some rats would be exposed to more salt than others. What would be the heritability of blood pressure in this population?

*23.21 In Kansas, a farmer is growing a variety of wheat called TK138. He calculates the narrow-sense heritability for yield (the amount of wheat produced per acre) and finds that the heritability of yield for TK138 is 0.95. The next year he visits a farm in Poland and observes that a Russian variety of wheat growing there, UG334, has only about 40 percent as much yield as TK138 grown on his farm in Kansas. Since he found the heritability of yield in his wheat to be very high, he concludes that the American variety of wheat (TK138) is genetically superior to the Russian variety (UG334), and he tells the Polish farmers that they can increase their yield by using TK138. What is wrong with his conclusion?

*23.22 A scientist wants to determine the narrow-sense heritability of tail length in mice. He measures tail length among the mice of a population and finds a mean tail length of 9.7 cm. He then selects the ten mice in the population with the longest tails; mean tail length in these selected mice is 14.3 cm. He interbreeds the mice with the long tails and examines tail length in their progeny. The mean tail length in the F_1 progeny of the selected mice is 13 cm. Calculate the selection differential, the response to selection, and the narrow-sense heritability for tail length in these mice.

23.23 Suppose that the narrow-sense heritability of wool length in a breed of sheep is 0.92, and the narrow-sense heritability of body size is 0.87. The genetic correlation between wool length and body size is –0.84. If a breeder selects for sheep with longer wool, what will be the most likely effects on wool length and on body size?

*23.24 The heights of nine college-age males and the heights of their fathers are presented on the next page.

Height of Son (inches)	Height of Father (inches)
70	70
72	76
71	72
64	70
66	70
70	68
74	78
70	74
73	69

a. Calculate the mean and the variance of height for the sons and do the same for the fathers.

b. Calculate the correlation coefficient for the relationship between the height of father and height of son.

c. Determine the narrow-sense heritability of height in this group by regression of the son's height on the height of father.

*23.25 The narrow-sense heritability of egg weight in a particular flock of chickens is 0.60. A farmer selects for increased egg weight in this flock. The difference between the mean egg weight of the unselected chickens and the mean egg weight of the selected chickens is 10 g. How much should egg weight increase in the offspring of the selected chickens?

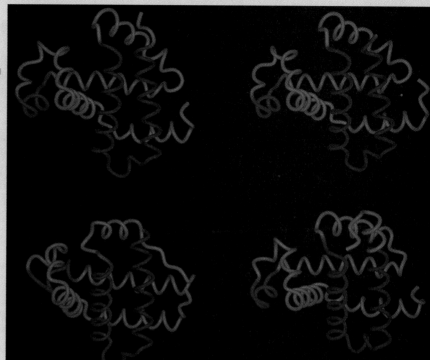

Polypeptides encoded by four different globin genes.

24

Molecular Evolution

PRINCIPAL POINTS

Rates of molecular evolution can be measured by comparing DNA sequences. Rates of change vary both within and between genes. Regions with the lowest rates of change usually are those that are most functionally constrained and most subject to natural selection.

Mutations are changes in nucleotide sequences, whereas substitutions are mutations that have passed through the filter of selection. Synonymous substitution rates indicate the actual mutation rate operating within a genome. Even small amounts of selection, such as those associated with different codons, can have dramatic effects over the course of evolution.

Mutations are rare events, and most changes in amino acid sequence tend to be removed through natural selection.

Relative rate tests suggest that although the molecular clocks of some genes run at a steady rate over long periods of time, it is unreasonable to assume that all lineages in a gene tree (a tree depicting the relationship of a single gene within and between species) accumulate substitutions at the same rate.

Gene trees do not always correspond to species trees (trees depicting the relationships of species based on morphological or paleontological analyses or molecular data from several genes) because some genetic polymorphisms within populations predate speciation events.

In eukaryotic organisms, genes frequently occur in multiple copies with identical or very similar sequences. A group of such genes is called a multigene family. Duplications of genes, in whole or in part, are the principal raw material from which proteins with new functions are made.

515

Increases in the number of taxa being considered dramatically increases the possible number of phylogenetic trees that can describe the relationship between those taxa. The distance matrix approach (statistical analysis that groups taxa on the basis of their overall similarity) and the parsimony approach (premised on the concept that the tree that invokes the fewest number of mutations is most likely to be correct because mutations are rare events) are two different ways to choose which of the many possible trees are most likely to represent the true evolutionary relationship.

The functional domains of many proteins correspond to regions encoded in single exons at the level of their genes. Many genes appear to have been derived by "mixing and matching" such functional domains of already useful proteins through exon shuffling.

i IN A HOT, HARSH REGION OF THE DESERT, YOU SIFT carefully through the sediment looking for fossils. Suddenly, you glimpse a fragment of bone. Eventually, you unearth a leg bone and part of a jaw, which lab tests show to be from an ancient hominid. Is this the distant predecessor of modern humans or merely an unrelated species that became extinct hundreds of thousands of years ago? How could you find out? In this chapter, you will learn how population geneticists can apply molecular genetics techniques to answer questions about how species evolve. Then, in the iActivity, you will have the opportunity to use some of the same tools and techniques to determine whether we are the direct descendants of Neanderthals.

While individuals are the entities affected by natural selection, it is populations and genes that change over evolutionary time. **Molecular evolution** is evolution at the molecular level of DNA and protein sequences. The study of molecular evolution uses the theoretical foundation of population genetics to address two essentially different sets of questions: how DNA and protein molecules evolve and how genes and organisms are evolutionarily related.

Aside from the differences in the questions asked, population genetics and molecular evolution differ primarily in the time frame of their perspective. Population genetics (see Chapter 22) focuses on the changes in gene frequencies that occur from generation to generation, whereas molecular evolution typically considers the much longer time frames associated with speciation. Very small departures from the conditions needed to maintain

Hardy-Weinberg equilibrium have small effects on gene frequencies in the short term but can take on great significance on evolutionary time scales. Moreover, random effects, such as those associated with small amounts of sampling error, along with exceptionally small differences in fitness can become the predominant process of genomic change when applied cumulatively over hundreds or thousands of generations.

The field of molecular evolution is multidisciplinary. It routinely invokes data and insights from genetics, ecology, evolutionary biology, statistics, and even computer science. However, before the widespread development of the tools of molecular biology in the 1970s and 1980s, researchers interested in the study of how biologically important molecules change over time had little data available to study. The ability to clone, sequence, and hybridize DNA removed the species barrier in population genetics studies. It also opened a window on a world that had been only dimly perceived where genes evolve by the accumulation of **mutations** (see Chapter 19), **duplication** (see Chapter 21), and **transposition** (see Chapter 20). For the first time, studies of evolution had an abundance of parameters that could be measured and theories that could be tested. Molecular analyses made it clear that genomes are historical records that can be unraveled to identify the dynamics behind evolutionary processes and to reconstruct the chronology of change. The same approaches that allow these documents of evolutionary history to be put in order and deciphered also facilitate classification of the living world into true **phylogenetic relationships**—the hierarchical genealogical relationships between separated populations or species across the vast distances of evolutionary time. Previously unimagined relationships between organisms became apparent, and even the kingdoms of life at the root of all systematics had to be rearranged.

In this chapter, we present the principles of molecular evolution and show how molecules with new functions arise and how the phylogenetic relationships of molecules and organisms are determined.

Patterns and Modes of Substitutions

Nucleotide Substitutions in DNA Sequences

Substitutions in Protein and DNA Sequences. An important question in the study of evolution is how the patterns and rates of substitution differ between different parts of the same gene. These studies began in earnest in the 1970s and 1980s, when the best molecular data available came from the amino acid sequences in proteins from a variety of organisms. It quickly became apparent that some amino acid differences were more likely to be observed between two **homologous proteins**—proteins that share a common ancestor—than were others. Specifically, amino acids were most likely to be replaced with amino acids that had similar chemical characteristics (see the groupings in

among the different mitochondrial protein-coding genes but in all cases is much higher than the average nonsynonymous rate of nuclear genes. It is not entirely clear why animal mtDNA undergoes such rapid evolutionary change, but the reason is likely to be related to a higher error rate during mtDNA replication and repair. (Unlike nuclear DNA polymerases, mitochondrial DNA polymerases have no proofreading ability.) Higher concentrations of mutagens, such as oxygen free radicals (i.e., O_2^-) resulting from the metabolic processes carried out in mitochondria, may also play a role in higher rates of substitution. It may also be that the selection pressure that normally eliminates many mutations in nuclear genes is relaxed in the mitochondria because most cells contain several dozen mitochondria. Regardless, changes in the proteins, tRNAs, and rRNAs encoded by the mitochondrial genome appear to be less detrimental to individual fitness than similar changes in the proteins, tRNAs, and rRNAs encoded by nuclear genes.

Mammalian mtDNA also differs from nuclear DNA in that the overwhelming majority of mtDNA is inherited clonally from the mother. Mitochondria are located in the cytoplasm, and only the mother's egg cell contributes cytoplasm to a zygote. As a consequence, mtDNA does not undergo meiosis, and all offspring should be identical to the maternal genotype for mtDNA sequences (the offspring are clones for mtDNA genes). This pattern of inheritance allows matriarchal lineages (descendants from one female) to be traced and provides a means for examining family structure in some populations. This, in conjunction with the rapid and regular rate of accumulation of nucleotide sequence differences, has allowed mtDNA to become a valuable tool for comparing closely related lineages. An example of geographic variation in mtDNA sequences in pocket gophers living in the southeastern United States is shown in Figure 24.2.

Molecular Clocks

As described earlier, the differences in the nucleotide and amino acid replacement rates between nuclear genes can be striking but are likely to result primarily from differences in the selective constraint on each individual protein. However, rates of molecular evolution for loci with similar functional constraints can be quite uniform over long periods of evolutionary time. In fact, when Emile Zuckerkandl and Linus Pauling performed the very first comparative studies of protein sequences in the 1960s,

Figure 24.2

Lineage relationships among mtDNA types in pocket gophers. The lowercase letters are different mtDNA types grouped according to similarity and superimposed on a geographic map of the collection sites. The tick marks across the connecting lines are the numbers of inferred mutational steps.

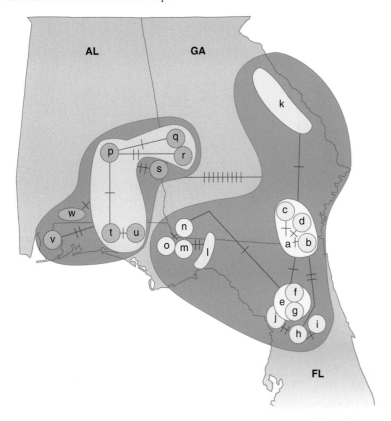

the evidence that substitution rates were constant within homologous proteins over many tens of millions of years prompted the researchers to liken the accumulation of amino acid changes to the steady ticking of a **molecular clock.** The molecular clock may run at different rates in different proteins, but the number of differences between two homologous proteins appeared to be very well correlated with the amount of time since speciation caused them to diverge independently, as shown in Figure 24.3. This observation immediately stimulated intense interest in using biological molecules in evolutionary studies. A steady rate of change between two sequences should facilitate not only the determination of phylogenetic relationships between species but also the times of their divergence in much the same way that radioactive decay was used to date geologic times.

Despite its great promise, however, Zuckerkandl and Pauling's molecular clock hypothesis has been controversial. Classic evolutionists argued that the erratic tempo of morphological evolution was inconsistent with a steady rate of molecular change. Disagreements regarding divergence times have also placed in question the uniformity of evolutionary rates at the heart of the idea.

Relative Rate Test. Most divergence dates used in molecular evolution studies come from interpretations of the notoriously incomplete fossil record and are of questionable accuracy. To avoid any questions regarding speciation dates, V. M. Sarich and A.C. Wilson devised a simple way to estimate the overall rate of substitution in different lineages that does not depend on specific knowledge of divergence times. For example, to determine the **relative rate** of substitution in the lineages for species 1 and 2 in Figure 24.4, we need to have a less related species 3 as an **outgroup.** Outgroups usually can be readily agreed upon; for instance, in this example, if species 1 and 2 are humans and gorillas, respectively, then species 3 could be another primate, such as a baboon. In the evolutionary relationship portrayed in Figure 24.4, the point in time when species 1 and 2 diverged is marked with the letter A.

Figure 24.3

The molecular clock runs at different rates in different proteins. One reason is that the neutral substitution rate differs among proteins. Fibrinogen appears to be unconstrained and has a high neutral substitution rate, whereas cytochrome *c* has a lower neutral substitution rate and may be more constrained. Data are from a wide variety of organisms.

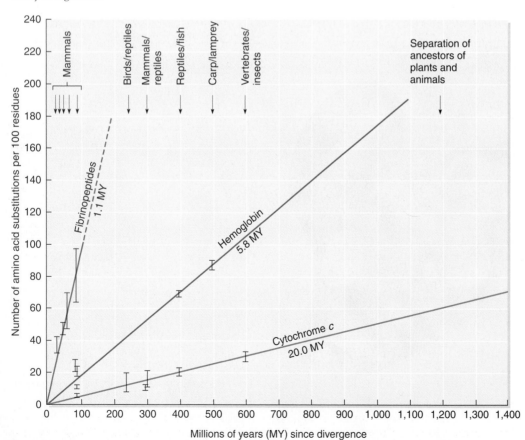

Figure 24.4

Phylogenetic tree used in a relative rate test. Species 3 is an outgroup known to have been evolving independently before the divergence of species 1 and 2. The letter A denotes the common ancestor of species 1 and 2.

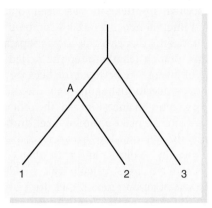

The number of substitutions between any two species is assumed to be the sum of the number of substitutions along the branches of the tree connecting them such that

$$d_{13} = d_{A1} + d_{A3}$$
$$d_{23} = d_{A2} + d_{A3}$$
$$d_{12} = d_{A1} + d_{A2}$$

where d_{13}, d_{23}, and d_{12} are easily obtained measures of the differences between species 1 and 3, species 2 and 3, and species 1 and 2, respectively. Simple algebraic manipulation of those statements allows the amount of divergence that has taken place in species 1 and species 2 since they last shared a common ancestor to be calculated using these equations:

$$d_{A1} = \frac{(d_{12} + d_{13} - d_{23})}{2}$$

$$d_{A2} = \frac{(d_{12} + d_{23} - d_{13})}{2}$$

By definition, the time since species 1 and 2 began diverging independently is the same, so the molecular clock hypothesis predicts that values for d_{A1} and d_{A2} should also be the same.

Exponentially increasing amounts of DNA sequence data from a very wide variety of species are available for testing the molecular clock's premise that the rate of evolution for any given gene is constant over time in all evolutionary lineages. Substitution rates in rats and mice have been found to be largely the same. In contrast, molecular evolution in humans and apes appears to have been only half as rapid as that which has occurred in Old World monkeys since their divergence. Indeed, relative rate tests performed on homologous genes in rats and humans suggest that rodents have accumulated substitutions at twice the rate of primates since they last shared a common ancestor during the time of the mammalian

radiation 80 to 100 million years ago. The rate of the molecular clock clearly varies among taxonomic groups, and such departures from constancy of the clock rate pose a problem in using molecular divergence to date the times of existence of recent common ancestors. Before such inferences can be made, it is necessary to demonstrate that the species being examined have a uniform clock, such as the one observed within rodents.

Causes of Variation in Rates. Several possible explanations have been put forward to account for the differences in evolutionary rates revealed by the relative rate tests. For instance, generation times in monkeys are shorter than in humans, and the generation time of rodents is much shorter still. The number of germ-line DNA replications, occurring once per generation, should be more closely correlated with substitution rates than simple divergence times. Differences may also result in part from a variety of other differences between two lineages since the time of their divergence, such as average repair efficiency, average exposure to mutagens, and the opportunity to adapt to new ecological niches and environments.

KEYNOTE

Relative rate tests suggest that substitutions do not always accumulate at the same rate in different evolutionary lineages. Primates—humans in particular—appear to be evolving more slowly than other mammals since the time of the mammalian radiation 80 to 100 million years ago. Faster rates of change may result from shorter average generation times or from a variety of other factors.

Molecular Phylogeny

Because evolution is defined as genetic change in the face of selective dynamics, genetic relationships are of primary importance in the deciphering of evolutionary relationships. The greatest promise of the molecular clock hypothesis is the implication that molecular data can be used to decipher the phylogenetic relationships among all living things. Quite simply, organisms with high degrees of molecular similarity are expected to be more closely related than those that are dissimilar. Before the tools of molecular biology were available to provide molecular data for such analyses, evolutionary biologists relied entirely on comparison of phenotypes to infer genetic similarities and differences. They assumed that if the phenotypes were similar, the genes that coded for the phenotypes were also similar; if the phenotypes were different, the genes were different. Thus, phenotypes were used for evolutionary studies. Originally, the phenotypes examined consisted largely of gross anatomical features. Later, behavioral, ultrastructural, and biochemical characteristics were also studied. Comparisons of such traits were used successfully to construct evolutionary trees for many groups of plants and animals and are still the basis of many evolutionary studies today.

However, relying on the study of such traits has limitations. Sometimes, similar phenotypes can evolve in organisms that are distantly related—a process called *convergent evolution*. For example, if a naive biologist tried to construct an evolutionary tree on the basis of whether wings were present or absent in an organism, he might place birds, bats, and flying insects in the same evolutionary group because all have wings. In this particular case, it is fairly obvious that these three groups of organisms are not closely related; they differ in many features other than the possession of wings, and the wings themselves are very different in their design. But this extreme example shows that phenotypes can be misleading about evolutionary relationships, and phenotypic similarities do not necessarily reflect genetic similarities.

Another problem with relying on phenotypes to determine evolutionary relationships is that many organisms do not have easily studied phenotypic features suitable for comparison. For example, the study of relationships among bacteria has always been problematic because bacteria have few obvious traits that correlate with the degree of their genetic relatedness. A third problem arises when we try to compare distantly related organisms. What phenotypic features should be compared, for example, in an analysis of bacteria and mammals, where there are so few characteristics in common?

Earlier in this chapter, we saw that molecular approaches can generate useful information about DNA sequences and how they evolve. Even though the relative rate of molecular evolution may vary from one lineage to another, and molecularly inferred divergence times must be treated with caution, molecular approaches to generating phylogenies usually can be relied on to group organisms correctly. Many have argued that molecular phylogenies are more reliable even when alternative data are available because the effects of natural selection generally are less pronounced at the DNA sequence level. When differences between molecular and morphological phylogenies are found, they usually create valuable opportunities to examine the effect of natural selection acting at the level of phenotypic differences, ranging from the molecular to the gross anatomical.

Phylogenetic Trees

Because of the long history of evolutionary studies even before molecular data were available, the general approaches used to elucidate relationships between species are fairly well established. A central principle in all phylogenetic reconstructions is the idea of a **phylogenetic tree** that graphically describes the relationship among different species. All living things on Earth, both in the present and in the past, share a single, common ancestor that lived roughly 4 billion years ago. Every phylogenetic tree portrays at least some portion of that ancestry with **branches** that connect two (occasionally more) adjacent **nodes.** Terminal nodes indicate the taxa for which molecular information has been obtained for analysis, and internal nodes represent common ancestors before the branching that gave rise to two separate groups of organisms. Branch lengths often are scaled to reflect the amount of divergence between the taxa they connect. Where it is possible to distinguish one internal node as representing a common ancestor to all the other nodes on a tree, it is possible to make a **rooted tree.** Unrooted trees specify only the relationship between nodes and say nothing about the evolutionary path that was taken. Roots for unrooted trees usually can be determined through the use of an outgroup. As in the example of an outgroup used for the relative rate test described earlier, outgroups are taxa that have unambiguously separated earlier from the other taxa being studied. In the case of humans and gorillas, when baboons are used as an outgroup, the root of the tree can be placed somewhere along the branch connecting baboons to the common ancestor of humans and gorillas. When only three taxa are being considered, there are three possible rooted trees but only one unrooted tree (Figure 24.5).

Number of Possible Trees. The number of possible rooted and unrooted trees quickly becomes staggering as more taxa are considered (Table 24.3). The actual number of possible rooted (N_R) and unrooted (N_U) trees for any number of taxa (n) can be determined with the following equations:

$$N_R = \frac{(2n - 3)!}{2^{n-2}(n - 2)!}$$

$$N_U = \frac{(2n - 5)!}{2^{n-3}(n - 3)!}$$

The value for n can be extremely large (conceivably every species or even every individual organism that has ever lived).

Gene Versus Species Trees. A phylogenetic tree based on the divergence observed within a single homologous gene is more appropriately called a **gene tree** than a **species tree.** Such trees may represent the evolutionary history of a gene but not necessarily that of the species in which it is found. Species trees usually are best obtained from analyses that use data from multiple genes. While this may sound counterintuitive, divergence within genes typically occurs before the splitting of populations that occurs when new species are created. For the locus being considered in Figure 24.6, some individuals in species 1 may actually be more similar to individuals in species 2 than they are to other members of their own population. The differences between gene and species trees tend to be

Figure 24.5

Rooted and unrooted phylogenetic trees. The relationship between three taxa (labeled 1, 2, and 3) can be described by only one unrooted tree but three different rooted trees. The letter A denotes a common ancestor of all three species in the rooted tree.

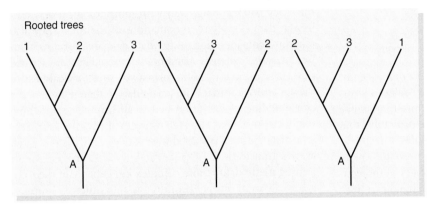

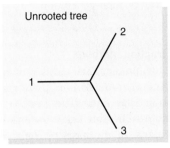

particularly important when considering loci where diversity within populations is advantageous, as in the major histocompatibility complex (MHC) described earlier. If MHC alleles alone were used to determine species trees, many humans would be grouped with gorillas rather than other humans because the polymorphism they carry is older than the split in the two lineages.

KEYNOTE

> Sequence polymorphisms often predate speciation events. As a result, it is possible for phylogenetic trees made from a single gene to not always reflect the relationships between species. Species trees are best constructed by considering multiple genes.

Despite the staggering number of rooted and unrooted trees that can be generated even when using a small number of taxa (see Table 24.3), only one of the possible trees

represents the true phylogenetic relationship between the taxa being considered. Since the true tree usually is known only when artificial data are used in computer simulations, most phylogenetic trees generated with molecular data are called **inferred trees**. Distinguishing which of all the possible trees is most likely to be the true tree can be a daunting task and is typically left to high-speed computers. The computer algorithms used in these searches usually use

Figure 24.6

Shared polymorphism. Transspecies, or shared, polymorphism may occur if the ancestor was polymorphic for two or more alleles and if alleles persist to the present in both species.

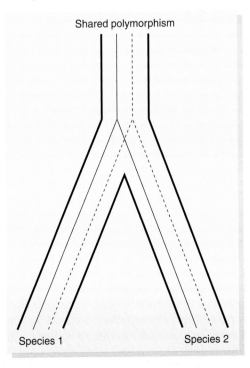

Table 24.3	Numbers of Rooted and Unrooted Trees That Describe the Possible Relationships Between Different Numbers of Taxa	
Number of Taxa	**Number of Rooted Trees**	**Number of Unrooted Trees**
2	1	1
3	3	2
4	15	3
5	105	15
10	34,459,425	2,027,025
20	8.20×10^{21}	2.22×10^{20}
30	4.95×10^{38}	8.69×10^{36}

one of two different kinds of approaches: distance matrix approaches and parsimony-based approaches. A basic understanding of the logic behind these approaches should lead to an understanding of exactly what information phylogenetic trees convey and what sort of molecular data are most useful for their generation.

Reconstruction Methods

At least two fundamentally different approaches are commonly used to determine phylogenetic relationships using molecular data. The first to be used, distance matrix approaches, are based on statistical principles that group things on the basis of their overall similarity to each other. This statistical approach is used for many kinds of data analysis in addition to the study of molecular evolution. In contrast, parsimony approaches group organisms in ways that minimize the number of substitutions that must have occurred since they last shared a common ancestor and are most useful in molecular evolution studies.

Distance Matrix Approaches to Phylogenetic Tree Reconstruction.

The oldest distance matrix method is also the simplest of all methods for tree reconstruction. Originally proposed in the early 1960s to help with the evolutionary analysis of morphological characters, the **unweighted pair group method with arithmetic mean (UPGMA)** is largely statistically based and requires data that can be condensed to a measure of genetic distance between all the pairs of taxa being considered. To illustrate the construction of a phylogenetic tree using the UPGMA method, consider a group of four taxa called A, B, C, and D. Assume that the pairwise distances between each of the taxa are as given in the following matrix:

Taxa	A	B	C
B	d_{AB}	–	–
C	d_{AC}	d_{BC}	–
D	d_{AD}	d_{BD}	d_{CD}

In this matrix, d_{AB} represents the distance (perhaps as calculated by the Jukes-Cantor model) between taxa A and B, d_{AC} is the distance between taxa A and C, and so on. UPGMA begins by clustering the two taxa with the smallest distance separating them into a single, composite taxon. In this case, assume that the smallest value in the distance matrix corresponds to d_{AB}, in which case taxa A and B are the first to be grouped together (AB). After the first clustering, a new distance matrix is computed, with the distance between the new taxon (AB) and taxa C and D being calculated as $d_{(AB)C} = \frac{1}{2}(d_{AC} + d_{BC})$ and $d_{(AB)D} = \frac{1}{2}(d_{AD} + d_{BD})$. The taxa separated by the smallest distance in the new matrix are then clustered together to make another new composite taxon. The process is repeated until all taxa have been grouped together. If scaled branch lengths are

to be used on the tree to represent the evolutionary distance between taxa, branch points are positioned at a distance halfway between the taxa being grouped (i.e., at $d_{AB}/2$ for the first clustering).

A strength of distance matrix approaches in general is that they work equally well with morphological and molecular data as well as combinations of the two. They also take into consideration all the data available for a particular analysis, whereas parsimony approaches (described next) discard many "noninformative" sites. A weakness of the UPGMA approach in particular is that it assumes a constant rate of evolution across all lineages, something that the relative rate tests tell us is not always the case. Several distance matrix–based alternatives to UPGMA, such as the transformed distance method and the neighbor-joining method, are more complex but capable of incorporating different rates of evolution within different lineages.

Parsimony-Based Approaches to Phylogenetic Tree Reconstruction.

While the distance-based methods of tree reconstruction are grounded in statistics, parsimony-based approaches rely more heavily on the biological principle that mutations are rare events. Parsimony approaches assume that the tree that invokes the fewest number of mutations is likely to be the best, and that tree is deemed a tree of **maximum parsimony.**

As mentioned earlier, the parsimony-based approach does not use all sites when considering molecular data. Instead, it focuses only on positions within a multiple alignment that favors one tree over an alternative in terms of the number of substitutions they invoke. Not all positions within a multiple alignment favor one tree over an alternative from the perspective of parsimony. Consider the following alignment of four nucleotide sequences:

	Site					
Sequence	**1**	**2**	**3**	**4**	**5***	**6***
I	G	C	G	A	T	G
II	G	T	G	T	T	G
III	G	T	T	G	C	A
IV	G	T	C	C	C	A

In such an alignment, only the fifth and sixth sites (marked with asterisks) qualify as **informative sites** from a parsimony perspective. As shown in Figure 24.7, only three possible unrooted trees can be drawn that describe the relationship between four taxa. The unrooted tree that groups sequences I and II separate from sequences III and IV would require only one mutation to have occurred in the branch that connects both groupings. Either of the two alternative trees that group the taxa differently would require two mutations and therefore do not represent the most parsimonious arrangement of the sequences. In contrast, all three of the possible unrooted trees for site 1 are indistinguishable from the perspective of parsimony because no mutations must be invoked for

Figure 24.7

Three different unrooted trees describing all possible relationships between four taxa. Using the sequences (uppercase letters) and sites shown in the text, all three trees for each of the six sites are shown. Dark lines are drawn on branches along which substitutions must have occurred, and inferred ancestral states are shown in lowercase letters. The sequence for site 1 requires no substitutions regardless of which tree is used, site 2 requires one for all trees, site 3 requires two for all trees, and site 4 requires three for all trees. Only sites 5 and 6 (within the dashed line box) have one tree with a fewer number of substitutions than the alternative trees; that makes them informative sites.

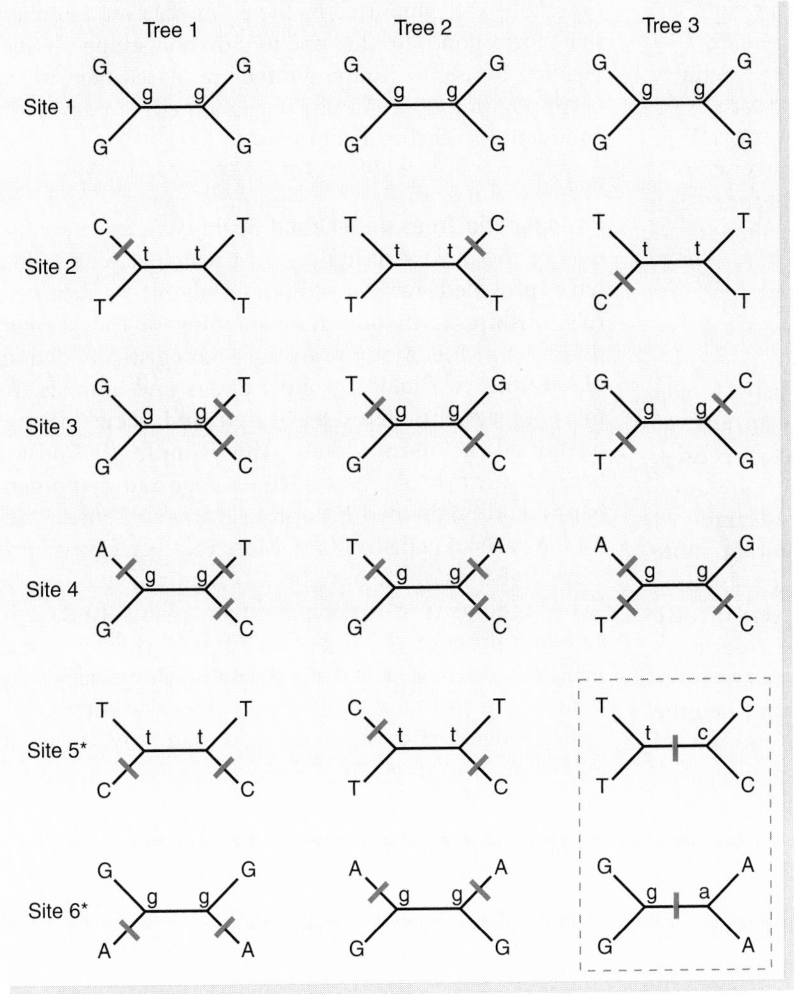

any of them. Similarly, site 2 is uninformative because one mutation occurs in all three of the possible trees. Likewise, site 3 is uninformative because all three trees require two mutations, and site 4 is uninformative because all three trees require three mutations. In general, for a site to be informative regardless of how many sequences are aligned, it has to have at least two different nucleotides, and each of these nucleotides has to be present at least twice.

Maximum parsimony trees are determined by first identifying all informative sites within an alignment and then determining which of all possible unrooted trees invokes the fewest number of mutations for each of those sites. The tree or trees that invoke the fewest num-

ber of mutations when all sites within an alignment are considered is the most parsimonious tree. A very useful by-product of the parsimony approach is the generation of inferred ancestral sequences at each node of a tree (see Figure 24.7). These inferred ancestral sequences go a long way toward making a nonissue of the infamous "missing links" of the fossil record and, when analyzed carefully, can give remarkably clear insights into the nature of long-dead organisms and even the environment in which they lived. Of course, the parsimony approach described here assumes that all nucleotides are just as likely to mutate into any of the three alternative nucleotides. More complicated parsimony algorithms take the difference in transition and transversion frequencies into account, although none is particularly reliable when rates of substitutions between branches of a tree differ dramatically.

K E Y N O T E

> The number of possible trees that describe the relationship between even a small number of taxa can be very large. Distance matrix methods rely on statistical relationships between taxa to group them. Parsimony approaches are more biologically based and assume that the tree that invokes the fewest number of mutations is most likely to be the best. No method can guarantee that it will yield the true phylogenetic tree, but when multiple substitutions are not likely to have occurred and evolutionary rates within all lineages are fairly equal, distance matrix and parsimony methods both work well.

Bootstrapping and Tree Reliability. Obviously, longer sequence alignments require a longer time to analyze than shorter ones when the parsimony approach is used. However, because of the relationship between the number of taxa and the corresponding number of unrooted trees illustrated in Table 24.3, the addition of more sequences has a much more dramatic effect on the time required to find a preferred tree. Once data sets involve 30 or more species, the number of possible trees is so large that it is simply not possible to examine all possible trees and assess the fit of the data to each, even when the fastest computers are used. Alternative trees are not all independent of each other, however, and many parsimony algorithms use shortcuts to avoid having to perform an exhaustive search. However, neither the distance matrix nor the maximum parsimony methods are certain to yield the correct tree. Numerous variations on each approach have been suggested, and intensive simulation studies have been performed to compare the statistical reliability of almost all tree construction methods. The results of these simulations are easy to summarize: Data sets that allow one method to infer the correct phylogenetic relationship generally work well with all the currently popular methods. However, if many changes have occurred in the simulated data sets or rates of change vary among branches, then none of the methods works very reliably. As a general rule, if a data set yields similar trees when analyzed by the fundamentally different distance matrix and parsimony methods, that tree can be considered fairly reliable.

It is also possible for portions of inferred trees to be determined with varying degrees of confidence. **Bootstrap tests** allow a rough quantification of those confidence levels. The basic approach of the bootstrap test is very straightforward:

- a subset of the original data is drawn (columns of the sequence alignment are randomly selected from the original set);
- each column could be sampled more than one time (sampling with replacement);
- a tree is then inferred from this new data set.

This process is repeated hundreds or thousands of times, and portions of the inferred tree that have the same groupings in many of the repetitions are those that are especially well supported by the entire data set. Numbers that correspond to the fraction of bootstrapped trees yielding the same grouping often are placed next to the corresponding nodes in phylogenetic trees to convey the confidence in each part of the tree.

Phylogenetic Trees on a Grand Scale

One of the most striking cases in which sequence data have provided new information about evolutionary relationships is in our understanding of the primary divisions of life. Many years ago, biologists divided all of life into two major groups: plants and animals. As more organisms were discovered and their features examined in more detail, this simple dichotomy became unworkable. It was later recognized that organisms could be divided into prokaryotes and eukaryotes on the basis of cell structure. More recently, several primary divisions of life have been recognized, such as the five kingdoms (prokaryotes, protista, plants, fungi, and animals) proposed by Robert Whittaker. However, negative evidence, such as the absence of internal membranes (the primary distinction of prokaryotes), has been recognized as a notoriously bad way to group organisms taxonomically.

The Tree of Life. In the mid-1980s, RNA and DNA sequences were used to uncover the primary lines of evolutionary history among all organisms. In one study, Carl Woese, Norm Pace, and colleagues constructed an evolutionary tree of life based on the nucleotide sequences of the 16S rRNA, which all organisms (as well as mitochondria and chloroplasts) possess. As illustrated in Figure 24.8, their evolutionary tree revealed three major evolutionary groups: the Bacteria (the traditional prokaryotes as well as mitochondria and chloroplasts), the Eukarya, and the Archaea (including thermophilic bacteria and many other little-known organisms). Bacteria and Archaea, although both prokaryotic in that they have no internal membranes, were found to be as different genetically as Bacteria and Eukarya. The deep evolutionary differences that separate the Bacteria and the Archaea were not obvious on the basis of phenotype, and the fossil record was silent on the issue. The differences became clear only after their nucleotide sequences were compared. Sequences of other genes, including those for 5S rRNAs, large rRNAs, and the genes coding for some

Figure 24.8

An evolutionary tree of life revealed by comparison of 16S rRNA sequences.
(Reprinted with permission from N. Pace, "A Molecular View of Microbial Diversity in the Biosphere," in *Science* 276 (1997):735. Copyright © 1997 American Association for the Advancement of Science.)

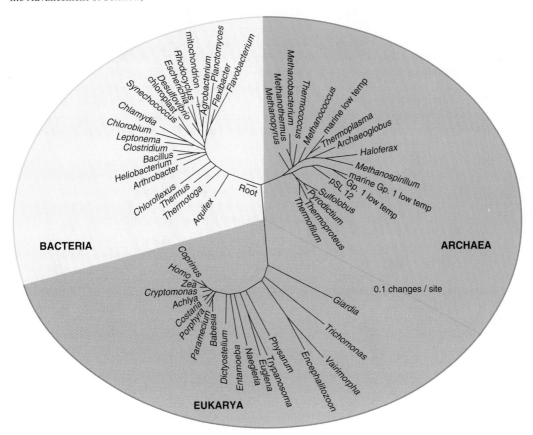

fundamentally important proteins, support the idea that three major evolutionary groups exist among living organisms. DNA sequence data also suggest that the genes of eukaryotic organelles such as mitochondria and chloroplasts actually have separate, independent origins from their nuclear counterparts (Box 24.1).

Human Origins. Another field in which DNA sequences are being used to study evolutionary relationships is human evolution. In contrast to the extensive variation observed in size, body shape, facial features, and skin color, genetic differences among human populations are small. For example, analysis of mtDNA sequences shows that the mean difference in sequence between two human populations is about 0.33 percent. Other primates exhibit much larger differences. For example, the two subspecies of orangutan differ by as much as 5 percent. This indicates that all human groups are closely related. Nevertheless, some genetic differences occur among different human groups. Surprisingly, the greatest differences are not found among populations located on different conti-

nents but between human populations residing in Africa. All other human populations show fewer differences than we find among the African populations. Many experts interpret these findings to mean that humans experienced their origin and early evolutionary divergence in Africa. It is hypothesized that after a number of genetically differentiated populations had evolved in Africa, a small group of humans may have migrated out of Africa and given rise to all other human populations. This hypothesis has been called the out-of-Africa theory. Sequence data from both mitochondrial DNA and the nuclear Y chromosome (the male sex chromosome) are consistent with this hypothesis. Further interpretation of this data suggests that all people alive today have mitochondria that came from a "mitochondrial Eve" and that all men have Y chromosomes derived from a "Y chromosome Adam" roughly 200,000 years ago. While the out-of-Africa theory is not universally accepted, DNA sequence data are playing an increasingly important role in the study of human evolution and in the study of the evolution of many lineages.

Box 24.1	The Endosymbiont Theory

The tree of life in Figure 24.8 suggests that the differences between Bacteria, Eukarya, and the Archaea result from independent evolution that has been taking place far longer than the time since plants and animals diverged. Analyses such as these have also shed light on the long-standing question of how the compartmental organization of eukaryotic cells could have evolved from the simpler condition still found in bacteria and archaea. The most important clue to providing a satisfying answer to that question came with the realization that mitochondria, chloroplasts, and the 16S ribosomal DNA of the nucleus were evolving independently even before the first eukaryotes appeared. In fact, the closest living relative of mitochondria today actually appears to be the bacteria *Rickettsia prowazekii*, the causative agent of epidemic typhus. A logical inference was that mitochondria and chloroplasts were free-living organisms that at some point in the past became engulfed by a prokaryote-like organism. The endosymbiosis (*endo* meaning "internal," *symbiosis* meaning "cooperative relationship") that resulted became the eukaryotes we see today. In other words, a merger of at least two or three evolutionary lineages gave rise to significantly different forms of life.

The endosymbiont theory was originally suggested by Russian biologist, C. Mereschkovsky in the early 1900s, based on microscopic examinations of eukaryotes. More recent molecular analyses, especially those of Lynn Margulis in 1981, have led to general acceptance of this model for the origin of these eukaryotic organelles. Numerous additional similarities between prokaryotes, mitochondria, and chloroplasts corroborate the 16S rRNA-based phylogenies. For instance, all organisms in the Bacteria branch of the tree of life (see Figure 24.8) have circular chromosomes, similar genomic arrangements and replication processes, similar sizes, and similar drug sensitivities, all features that distinguish them from what is associated with the nucleus of eukaryotic cells. Mitochondria and chloroplasts share these properties.

In time, the endosymbionts in eukaryotic cells have become very specialized, with the nucleus being the predominant site at which heritable information is stored, mitochondria being the primary site for oxidative phosphorylation, and chloroplasts being the site at which photosynthesis occurs. Many of the genes essential for organelle function have moved to the nucleus, and the relationship between organelles and their host cells has become an obligatory and elaborate one in which no compartment can live independently.

Sources: Andersson, S. G. E., Zomorodipour, A., Andersson, J. O., Sicheritz-Ponten, T., Alsmark, U. C., Podowski, R. M., Naslund, A. K., Eriksson, A. S., Winkler, H. H., and Kurland, C. G. 1998. "The genome sequence of *Rickettsia prowazekii* and the origin of mitochondria," *Nature* 396:133–140.

Margulis, L., 1981. *Symbiosis in Cell Evolution: Life and Its Environment in the Early Earth.* San Francisco: W. H. Freeman.

You have joined a team of molecular geneticists and anthropologists who have developed a technique for extracting and analyzing ancient DNA from Neanderthal fossils in the iActivity *Were Neanderthals Our Ancestors?* on the website.

Acquisition and Origins of New Functions

A long-standing question of deep interest to those who study molecular evolution is the issue of how genes with new functions arise. As early as 1932, J. B. S. Haldane suggested that new genes arise from the process of mutating redundant copies of already existing genes. Although other means, such as transposition (see Chapter 21), have since been described, Haldane's argument still does a good job of describing the origin of most new genes.

Multigene Families

In eukaryotic organisms, we often find tandemly arrayed, multiple copies of genes, all having identical or very similar sequences. These **multigene families** are sets of related genes that have evolved from some ancestral gene through gene duplication. The globin gene family that encodes the proteins used to make up the oxygen-carrying hemoglobin molecule in our blood has become a classic example of such a multigene family. The organization and expression pattern of this multigene family in humans was discussed in Chapter 17 (pp. 355–356). Briefly, the globin multigene family is comprised of seven α-like genes found on chromosome 16 and six β-like genes found on chromosome 11.

Globin genes are also found in other animals, and globin-like genes are even found in plants, suggesting that this is a very ancient gene family. Almost all functional globin genes in animal species have the same general structure, consisting of three exons separated by two introns. However, the numbers of globin genes and their order varies among species, as is shown for the β-like genes in Figure 24.9. Since all globin genes have similarities in structure and sequence, it appears that an ancestral globin gene (perhaps most like the present-day myoglobin gene) duplicated and diverged to produce an ancestral α-like gene and an ancestral β-like gene. These two genes then underwent repeated duplications, giving rise to the various α-like and β-like genes found in vertebrates today.

Repeated gene duplication, such as that giving rise to the globin gene family, appears to be a frequent evolutionary occurrence. Indeed, the number of copies of globin genes varies even within some human populations. For example, most humans have two α-globin genes on chromosome 16, as shown in Figure 17.19. However,

Figure 24.9

Organization of the globin gene families in several mammalian species.

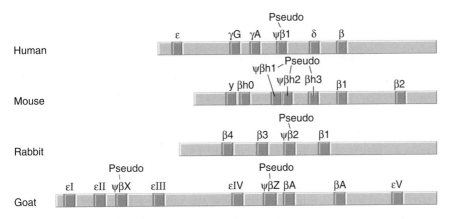

some individuals have a single α-globin gene on chromosome 16; and other individuals have three or even four copies of the α-globin gene on one of their chromosomes. These observations suggest that duplication and deletion of genes in multigene families are part of a constant process that continues to operate today. Gene duplications and deletions often arise as a result of misalignment of sequences during crossing-over, a process called *unequal crossing-over.* Duplications can also arise through transposition.

Gene Duplication and Gene Conversion

Following gene duplication, one of the copies of a gene may undergo changes in sequence as though free from functional constraint—as long as the other copy continues to function. As you might expect from the previous discussions in this chapter, most changes to the copy normally would be selectively disadvantageous or even render it a nonfunctional pseudogene. On rare occasions, however, the changes may lead to subtle alterations of function or pattern of expression that are advantageous to the organism, and the change sweeps through a population. This "tinkering" approach to evolution becomes even more of a "win/no-lose" scenario when misalignments between pseudogene copies and the functional copy occur during subsequent recombination events and the inactivating changes are corrected by **gene conversion,** a nonreciprocal recombination process resulting in a replacement of one sequence by another and therefore producing loss of one of the variant sequences. In this way, gene conversion events can give an organism multiple chances to create a gene with a new function from the duplicate of an already functional gene. Like gene duplication, gene conversion also continues to operate to this day, although it is usually most apparent when helpful substitutions to a gene copy are "corrected." For example, the two genes on the X chromosome that allow most humans to distinguish between

red and green light are 98 percent identical at the nucleotide level, and most spontaneous occurrences of deficiencies in green vision occur as a result of gene conversions between the two.

Arabidopsis Genome Results

The extent to which organisms use gene duplication to generate proteins with new functions and thereby evolve is becoming increasingly clear as more and more genome-sequencing projects are coming to their conclusions. For example, with only about 125 million nucleotides in its genome, *Arabidopsis thaliana* (thale-cress) was the first plant genome to be completely sequenced (see Chapter 15, p. 311). Its short generation time and small size make it a favorite organism of plant geneticists, but it was a particularly appealing choice for genome sequencers because studies had indicated that its genome had undergone much less duplication than that seen in other more commercially important plants. But when the sequencing was completed at the end of year 2000, a little over half of the 25,500 *Arabidopsis* genes were found to be duplicates. Phylogenetic analyses such as the distance matrix and parsimony methods described earlier revealed only about 11,600 distinct families of one or more genes. Even in this unusually nonredundant genome, the process of evolving through gene duplication followed by tinkering holds sway.

K E Y N O T E

Gene duplication events appear to have occurred frequently in the evolutionary history of all organisms. Copies of genes provide the raw material for evolution in that they are free to accumulate substitutions that sometimes give rise to proteins with new, advantageous functions.

Domain (Exon) Shuffling. It should also be pointed out that an increase in the number of copies of a DNA segment can also occur for segments of a genome that are smaller than complete genes. Numerous examples of genes that contain internal duplication of one or more protein domains have been found, such as the human serum albumin gene, which is made up almost entirely of three perfect copies of a 195-amino-acid domain. Elongation of a gene through internal duplication of functional domains does not lead to new proteins with significantly different functions very quickly, however. Most complex proteins are assemblages of several different protein domains that perform varied functions, such as acting as a substrate-binding site or a membrane-spanning region. Perhaps not coincidentally, the beginnings and ends of exons often correspond to the beginnings and ends of domains within complex proteins.

In 1978, Walter Gilbert proposed that the first genes had a limited number of protein domains within their repertoire and that most, if not all, of the gene families seen in living things today came through **domain shuffling:** the duplication and rearrangement of those domains (usually encoded by individual exons) in different combinations. Domain (or exon) shuffling is a controversial idea that presupposes that introns were a feature of the most primitive life on Earth, even though they are now found mostly in Eukarya and not in the simpler Bacteria and Archaea. Still, numerous striking examples of complex genes that are made of bits and pieces of other genes are known, and it is clear that at least some genes with novel functions have been created in this way.

K E Y N O T E

Internal duplications within genes are not uncommon, and many exons correspond to discrete functional domains within proteins. Some genes with novel functions seem to have been created through a process of domain (or exon) shuffling, in which regions between and within genes are recombined in new ways.

Summary

The mathematical theory developed by population geneticists is applied to long time frames in the study of molecular evolution to decipher the sometimes cryptic historical record preserved in the DNA sequences of organisms. It provides insights into which portions of genes are functionally important, the evolutionary relationship between widely varying groups of organisms, and the mechanisms by which genes with novel functions arise.

Rates of evolution vary widely within and between genes. In most cases, the primary cause of these differences is variation in the level to which changes affect the function of genes or their proteins. Portions of a genome that have the least impact on fitness appear to evolve the fastest. Alignments between two or more sequences allow estimates to be made of the number of substitutions that have accumulated since they diverged from a single sequence in a common ancestor. Many genes accumulate substitutions at a constant rate for long periods of evolutionary time, although relative rate tests show that some lineages (like that of humans) tend to evolve more slowly than others.

Sequence alignments also can be used as a starting point in phylogenetic reconstructions of very diverse groups of organisms. Caution must be exercised when interpreting relationships based on data from only one gene, but two fundamentally different approaches often yield trees that describe similar relationships when substitution rates are low and rates of evolution are fairly constant for all lineages. The distance matrix methods are based on statistical principles that group taxa together in a fashion that depends on their overall similarity to each other. Parsimony-based methods rely on the biological principle that mutations are rare events and the assumption that a tree that invokes the fewest number of mutations is most likely to represent the actual evolutionary relationship of sequences. The large number of possible trees makes phylogenetic reconstruction computationally intensive, but statistical methods are available to determine the robustness of any tree. Molecular phylogenies have numerous advantages over trees generated with morphological data sets and have provided new insights into the very deepest branches to the tree of life.

Gene duplication, in whole or in part, seems to have played a fundamental role in the evolution of proteins with novel functions. Gene duplications often arise as a result of misalignment of sequences during crossing-over and typically result in tandemly arrayed families of genes on chromosomes. Duplicates of functional genes are free of selective constraint and can accumulate substitutions without affecting the fitness of an organism. When a duplicated gene undergoes a mutation that confers a useful new function to an organism, it is again placed under selective constraint as the organism begins to derive an increased fitness from it. Multigene families are common in complex organisms such as eukaryotes, and more than half of all genes within any given genome are likely to have evolved in such a way.

Analytical Approaches for Solving Genetics Problems

Q24.1 Consider the following five-way multiple alignment of hypothetical homologous sequences. Generate a distance matrix that describes the pairwise relationship of all the sequences presented. Use the UPGMA method to generate a tree that describes the relationship between these sequences.

```
            10         20         30
A: GCCAACGTCC ATACCACGTT GTTTAGCACC
B: GCCAACGTCC ATACCACGTT GTCAAACACC
C: GGCAACGTCC ATACCACGTT GTTATACACC
D: GCTAACGTCC ATATCACGCT GTCATGTACC
E: GCTGGTGTCC ATATCACGTT ATCATGTACC

            40         50
A: GGTTCTCGTC CGATCACCGA
B: GGTTCTCGTC CGATCACCGA
C: GGTTCTCGTC AGGTCACCGA
D: GGTCCTCGTC AGATCCCCAA
E: GGTACTCGTC CGATCACCGA
```

A24.1 A distance matrix is made by determining the number of differences observed in all possible pairwise comparisons of the sequences. The number of differences between sequence A and B (d_{AB}), for instance, is 3. The complete distance matrix is shown here:

Taxa	A	B	C	D
B	3	–	–	–
C	6	5	–	–
D	11	10	11	–
E	11	10	13	9

The smallest distance separating any of the two sequences in the multiple alignment corresponds to d_{AB}, so taxon A and taxon B are grouped together. A new distance matrix is then made in which the composite group (AB) takes their place. Distances between the remaining taxa and the new group are determined by taking the average distance between its two members (A and B) and all other remaining taxa [i.e., $d_{(AB)C} = \frac{1}{2}(d_{AC} + d_{BC})$, so $d_{(AB)C} = \frac{1}{2}(6 + 5) = 5.5$], and the resulting matrix looks like this:

Taxa	AB	C	D
C	5.5	–	–
D	10.5	11	–
E	10.5	13	9

The smallest distance separating any two taxa in this new matrix is the distance between (AB) and C, so a new combined taxon, (AB)C, is created. Another distance matrix using this new grouping then looks like this:

Taxa	(AB)C	D
D	10.75	–
E	11.75	9

In this last matrix, the smallest distance is between taxa D and E ($d_{DE} = 9$), so they are grouped together as (DE). One way to symbolically represent the final clustering of taxa is ((AB)C)(DE). Alternatively, a tree such as the following can be used:

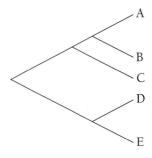

Q24.2 Using the same five sequences from problem 24.1, which positions within the alignment correspond to informative sites for parsimony analyses?

A24.2 The following positions are informative sites for parsimony analyses: 3, 14, 23, 25, 26, 27, and 41. They are the only ones that have at least two different nucleotides, with each of those nucleotides being present at least twice.

Questions and Problems

*24.1 The following sequence is that of the first 45 codons from the human gene for preproinsulin:

```
ATG GCC CTG TGG ATG CGC CTC CTG CCC CTG CTG GCG
CTG CTG GCC CTC TGG GGA CCT GAC CCA GCC GCA GCC
TTT GTG AAC CAA CAC CTG TGC GGC TCA CAC CTG GTG
GAA GCT CTC TAC CTA GTG TGC GGG GAA
```

Using the genetic code (Figure 12.7), determine what fraction of mutations at the first, second, and third positions of these 45 codons will be synonymous. At which position is natural selection likely to have the greatest effect and are nucleotides most likely to be conserved?

*24.2 The following sequences represent an optimum alignment of the first 50 nucleotides from the human and sheep preproinsulin genes:

```
                10         20         30
Human: ATGGCCCTGT GGATGCGCCT CCTGCCCCTG
Sheep: ATGGCCCTGT GGACACGCCT GGTGCCCCTG

                40         50
Human: CTGGCGCTGC TGGCCCTCTG
Sheep: CTGGCCCTGC TGGCACTCTG
```

Estimate the number of substitutions that have occurred in this region since humans and sheep last shared a common ancestor, using the Jukes-Cantor model.

*24.3 Using the alignment in problem 24.2 and assuming that humans and sheep last shared a common ancestor 80 million years ago, estimate the rate at which the sequence of the first 50 nucleotides in their preproinsulin genes has been accumulating substitutions.

*24.4 Would the mutation rate be greater or less than the observed substitution rate for a sequence of a gene such as the one shown in problem 24.2? Why?

*24.5 If the rate of nucleotide evolution along a lineage is 1.0 percent per million years, what is the rate of substitution per nucleotide per year? What would be the observed rate of divergence between two species evolving at that rate since they last shared a common ancestor?

*24.6 The average synonymous substitution rate in mammalian mitochondrial genes is approximately ten times the average value for synonymous substitutions in nuclear genes. Why would it be better to use comparisons of mitochondrial sequences to study human migration patterns and nuclear genes when studying the phylogenetic relationships of mammalian species that diverged 80 million years ago?

*24.7 Why might mitochondrial DNA sequences accumulate substitutions at a faster rate than nuclear genes in the same organism?

*24.8 Why might substitution rates differ from one species to another, and how would such differences depart from Zuckerkandl and Pauling's assumptions for molecular clocks?

*24.9 Suppose we examine the rates of nucleotide substitution in two nucleotide sequences isolated from humans. In the first sequence (sequence A), we find a nucleotide substitution rate of 4.88×10^{-9} substitutions per site per year. The substitution rate is the same for synonymous and nonsynonymous substitutions. In the second sequence (sequence B), we find a synonymous substitution rate of 4.66×10^{-9} substitutions per site per year and a nonsynonymous substitution rate of 0.70×10^{-9} substitutions per site per year. Referring to Table 24.1, what might you conclude about the possible roles of sequence A and sequence B?

*24.10 What evolutionary process might explain a coding region in which the rate of amino acid replacement is greater than the rate of synonymous substitution?

*24.11 Natural selection does not always act just at the level of amino acid sequences in proteins. Ribosomal RNAs, for instance, are functionally dependent on extensive and specific intramolecular secondary structures that form when complementary nucleotide sequences within a single rRNA interact. Would the regions involved in such pairing accumulate mutations at the same rate as unpaired regions? Why?

*24.12 The following three-way alignment is of the nucleotide sequences from the beginning of the human, rabbit, and duck α-globin genes:

```
                10          20          30
Human:  ATGGTGCTGT CTCCTGCCGA CAAGACCAAC
Rabbit: ATGGTGCTGT CTCCCGCTGA CAAGACCAAC
Duck:   ATGGTGCTGT CTGCGGCTGA CAAGACCAAC

                40          50
Human:  GTCAAGGCCG CCTGGGACAA
Rabbit: ATCAAGACTG CCTGGGAAAA
Duck:   GTCAAGGGTG TCTTCTCCAA
```

Given that humans and rabbits are known to be more closely related to each other than they are to ducks (an outgroup), use the relative rate test to determine whether there has been a change in the rate of substitution in this region of the human and rabbit genomes since they last shared a common ancestor.

*24.13 What are some of the advantages of using DNA sequences to infer evolutionary relationships?

*24.14 As suggested by the popular movie *Jurassic Park*, organisms trapped in amber have proved to be a good source of DNA from tens and even hundreds of millions of years ago. However, when using such sequences in phylogenetic analyses, it is usually not possible to distinguish between samples that come from evolutionary dead ends and those that are the ancestors of organisms still alive today. Why would the former be no more useful than simply including the DNA sequence of another living species in an analysis?

*24.15 In the phylogenetic analysis of a group of closely related organisms, the conclusions drawn from one locus were found to be at odds with those from several others. What might account for the discordant locus?

*24.16 What is the chance of randomly picking the one rooted phylogenetic tree that describes the true relationship between a group of six organisms? Are the odds better or worse for randomly picking from among all the possible unrooted trees for those organisms?

*24.17 Increasing the amount of sequence information available for analysis usually has little effect on the length of time computer programs use to generate phylogenetic trees with the parsimony approach. Why doesn't the amount of sequence information affect the total number of possible rooted and unrooted trees?

*24.18 When bootstrapping is used to assess the robustness of branching patterns in a tree of maximum parsimony, why is it more important to use sequences that have as many informative sites as possible than to simply use longer sequences?

*24.19 What are the advantages of gene duplication (in whole or in part) in generating genes with new functions? Suggest an alternative way in which genes with new functions could arise.

Glossary

10-nm nucleofilament *See* **nucleofilament.**

acrocentric chromosome A chromosome with a centromere near the end such that it has one long arm plus a stalk and a satellite.

activator In eukaryotic transcription, a transcription factor that controls which promoters are transcribed actively.

additive genetic variance Genetic variance that arises from the additive effects of genes on the phenotype; represented by V_A.

adenine (A) A purine base found in RNA and DNA. In double-stranded DNA, adenine pairs with the pyrimidine thymine.

allele One of two or more alternative forms of a single gene locus. Different alleles of a gene each have a unique nucleotide sequence, but their activities are all concerned with the same biochemical and developmental process, even though their individual phenotypes may differ.

allele-specific oligonucleotide (ASO) hybridization A procedure, using PCR primers, to distinguish alleles that differ by one base pair.

allelic frequencies The frequencies of alleles at a locus occurring among individuals in a population.

allelomorph (allele) A term coined by William Bateson; literally means "alternative form"; later shortened by others to *allele*.

allopolyploidy Polyploidy involving two or more genetically distinct sets of chromosomes.

alternation of generations The two distinct reproductive phases of green plants in which stages alternate between haploid cells and diploid cells (gametophyte cells and sporophyte cells).

Ames test A test developed by Bruce Ames in the early 1970s that investigates new or old environmental chemicals for carcinogenic effects. It uses the bacterium *Salmonella typhimurium* as a test organism for mutagenicity of compounds.

amino acid One of the building blocks of polypeptides. There are 20 different amino acids.

aminoacyl-tRNA A tRNA molecule covalently bound to an amino acid. This complex brings the amino acid to the ribosome so that it can be used in polypeptide synthesis.

aminoacyl-tRNA synthetase An enzyme that catalyzes the addition of a specific amino acid to a tRNA molecule. Since there are 20 amino acids, there are also 20 synthetases.

amniocentesis A procedure in which a sample of amniotic sac fluid is withdrawn from the amniotic sac of a developing fetus and cells are cultured and examined for chromosomal abnormalities.

anaphase The stage in mitosis or meiosis during which the sister chromatids (mitosis) or homologous chromosomes (meiosis) separate and migrate toward the opposite poles of the cell.

anaphase II The stage of meiosis II during which the centromeres (and therefore the chromatids) are pulled to the opposite poles of the spindle. The separated chromatids are now called chromosomes in their own right.

aneuploidy The abnormal condition in which one or more whole chromosomes of a normal set of chromosomes are missing or are present in more than the usual number of copies. Aneuploidy also is the abnormal condition in which one or more parts of a chromosome (or chromosomes) are duplicated or deleted.

antibody A protein molecule that recognizes and binds to a foreign substance introduced into the organism.

anticodon A three-nucleotide sequence that pairs with a codon in mRNA by complementary base pairing.

antigen Any large molecule that stimulates the production of specific antibodies or binds specifically to an antibody.

antiparallel Showing opposite polarity; in the case of double-stranded DNA, this means that the chemical polarity of one chain is opposite to the chemical polarity of the other chain.

applied research Research done with an eye toward making products that can be commercialized or at least made available to humankind for practical benefit.

artificial selection Human determination as to which individuals will survive and reproduce. If the selected traits have a genetic basis, they will change and evolve.

attenuation A regulatory mechanism in certain bacterial biosynthetic operons that controls gene expression by causing RNA polymerase to terminate transcription.

autonomously replicating sequences (ARSs) Specific sequences (e.g., in baker's yeast, *Saccharomyces cerevisiae*) that, when included as part of an extrachromosomal, circular DNA molecule, confer on that molecule the ability to replicate autonomously.

autopolyploidy Polyploidy involving more than two chromosome sets of the same species.

autosome A chromosome other than a sex chromosome.

auxotroph (auxotrophic mutant) A mutant that requires a nutritional supplement to grow. An auxotrophic mutation affects an organism's ability to make a particular molecule essential for growth.

auxotrophic mutation (nutritional mutation, biochemical mutation) A mutation that affects an organism's ability to make a particular molecule essential for growth.

back mutation *See* **reverse mutation.**

bacteria Spherical, rod-shaped or spiral-shaped, single-cellular or multicellular, filamentous prokaryotic organisms.

bacterial artificial chromosome (BAC) Vector for cloning large DNA fragments (up to about 200 kb) in *E. coli*. BACs contain the origin of replication of the *F* factor, a multiple cloning site, and a selectable marker.

bacteriophages Viruses that attack bacteria.

Barr body A highly condensed mass of chromatin found in the nuclei of normal females but not in the nuclei of normal male cells. It represents a cytologically condensed and inactivated X chromosome.

basal transcription factor Protein required for the initiation of transcription by a eukaryotic RNA polymerase.

base A purine or pyrimidine that is part of a nucleotide building block of DNA or RNA.

base analogue A chemical whose molecular structure is extremely similar to a base normally found in DNA.

base pair substitution mutation A change in a gene such that one base pair is replaced by another base pair (e.g., AT replaced by GC pair).

basic research Research done to further knowledge for knowledge's sake.

behavioral incompatibility Concerning barriers to gene flow, the phenomenon in which two species recognize and avoid each other as mates.

bidirectional replication The DNA synthesis that takes place in both directions away from the origin of the replication point.

binomial expansion The expansion of $(a + b)^n$, where n is any whole number.

biochemical mutation *See* **auxotrophic mutation.**

bioinformatics A field that fuses biology with mathematics and computer science to find genes within a genomic sequence, align sequences in databases to determine the degree of matching, predict the structure and function of gene products, describe the interactions between genes and gene products at a global level within the cell and between organisms, and postulate phylogenetic relationships for sequences.

biparental inheritance Plant zygotes that show traits indicating chloroplast chromosomes from both parents are present and active.

bivalent A pair of homologous, synapsed chromosomes during the first meiotic division.

bootstrap test A test that allows for a rough quantification of those confidence levels attached to the branching patterns of a phylogenetic tree chosen by the parsimony approach.

bottleneck effect A form of genetic drift that occurs when a population is drastically reduced in size. Some genes may be lost from the gene pool as a result of chance.

branch A graphic representation in a phylogenetic tree of the divergence of an extant or ancestral taxon from an ancestral organism.

branch-point sequence The consensus sequence in mammalian cells, YNCURA (where Y is a pyrimidine, R is a purine, and N is any base), to which the free 5′ end of the intron loops and binds to the A nucleotide in the sequence during intron splicing.

broad-sense heritability A quantity representing the proportion of the phenotypic variance that consists of genetic variance.

C value The amount of DNA found in the haploid set of chromosomes.

CAAT box One of the eukaryotic promoter elements found in approximately 75 base pairs upstream of the initiation site, but it can function at a number of other locations and in either orientation with respect to the start point. The consensus sequence is 5′-GGCCAATCT-3′.

cancer A disease characterized by the uncontrolled and abnormal division of eukaryotic cells and by the spread of the disease (metastasis) to disparate sites in the organism.

5′ capping The addition of a methylated guanine nucleotide (a "cap") to the 5′ end of a pre-mRNA molecule. The cap is retained on the mature mRNA molecule.

carcinogen Any of several natural and artificial agents, mostly chemicals and types of radiation, that increase the frequency with which cells become cancerous.

catabolite repression (glucose effect) The inactivation of an inducible bacterial operon in the presence of glucose even though the operon's inducer is present.

cDNA DNA copies made from an RNA template catalyzed by the enzyme reverse transcriptase.

cDNA library The collection of molecular clones that contains cDNA copies of the entire mRNA population of a cell. *See also* **cDNA.**

cell cycle The cyclical process of cellular growth and reproduction in unicellular and multicellular eukaryotes. The cycle includes nuclear division, or mitosis, and cell division, or cytokinesis.

cell division A process whereby one cell divides to produce two cells.

cell-free, protein-synthesizing system A system, isolated from cells, that contains ribosomes, mRNA, tRNAs with amino acids attached, and all the necessary protein factors for the in vitro synthesis of polypeptides.

cellular oncogene (c-*onc*) A gene, present in a functional state in a cancerous cell, that is responsible for the cancerous state.

CEN sequence DNA sequence of the centromere.

centimorgan (cM) The map unit, named in honor of T. H. Morgan.

centromere A specialized region of a chromosome, seen as a constriction under the microscope. This region is important in the activities of the chromosomes during cellular division.

chain-terminating codon One of three codons for which no normal tRNA molecule exists with an appropriate anticodon. A nonsense codon in an mRNA specifies the termination of polypeptide synthesis.

character An observable phenotypic feature of the developing or fully developed organism that is the result of gene action.

charged tRNA The product of an amino acid added to a tRNA.

charging The process of attaching the correct amino acid to the tRNA.

chiasma A cross-shaped structure formed during crossing-over and visible during the diplonema stage of meiosis.

chi-square (χ^2) test A statistical procedure that determines what constitutes a significant difference between observed results and results expected on the basis of a particular hypothesis; a goodness-of-fit test.

chloroplast The cellular organelle found only in green plants that is the site of photosynthesis in the cells containing it.

chorionic villus sampling A procedure in which a sample of chorionic villus tissue of a developing fetus is examined for chromosomal abnormalities.

chromatid One of the two visibly distinct longitudinal subunits of all replicated chromosomes that becomes visible between early prophase and metaphase of mitosis.

chromatin The piece of DNA-protein complex that is studied and analyzed. Each chromatin fragment reflects the general features of chromosomes but not the specifics of any individual chromosome.

chromosomal aberration *See* **chromosomal mutation.**

chromosomal mutation The variation from the wild-type condition in chromosome number or structure.

chromosome The genetic material of the cell, complexed with protein and organized into a number of linear structures. It literally means "colored body" because the threadlike structures are visible under the light microscope only after they are stained with dyes.

chromosome theory of inheritance The theory that the chromosomes are the carriers of the genes. The first clear formulation of the theory was made by both Sutton and Boveri, who independently recognized that the transmission of chromosomes from one generation to the next closely paralleled the pattern of transmission of genes from one generation to the next.

chromosome walking A process to identify adjacent clones in a genomic library. In chromosome walking, a piece of DNA is used to probe a genomic library to find an overlapping clone, then a piece of that clone is used as a probe to screen the library again for an overlapping clone, and so on.

cis-dominance The phenomenon of a gene or DNA sequence controlling only genes that are on the same contiguous piece of DNA.

cis-trans (complementation) test A test that determines whether two different mutations are within the same cistron (gene).

classical model A hypothesis of genetic variation proposing that natural populations contain little genetic variation as a result of strong selection for one allele.

cline A systematic change in allelic frequencies across a geographic transect.

clonal selection A process whereby cells that already have antibodies on their surface specific to an antigen are stimulated to proliferate and secrete that antibody.

clone contig map A set of ordered, overlapping clones comprising all the DNA of an entire chromosome, or part of a chromosome, without any gaps.

cloning *See* **molecular cloning.**

cloning vector (cloning vehicle) A double-stranded DNA molecule that is able to replicate autonomously in a host cell and with which a DNA fragment (or fragments) can be bonded to form a recombinant DNA molecule for cloning.

coding sequence The part of an mRNA molecule that specifies the amino acid sequence of a polypeptide during translation.

codominance The situation in which the heterozygote exhibits the phenotypes of both homozygotes.

codon A group of three adjacent nucleotides in an mRNA molecule that specifies either one amino acid in a polypeptide chain or the termination of polypeptide synthesis.

codon usage bias A disproportionate use of one or a few synonymous codons within a codon family for a particular gene or across a genome.

coefficient of coincidence A number that expresses the extent of interference throughout a genetic map; the ratio of observed double-crossover frequency to expected double-crossover frequency. Interference is equal to 1 minus the coefficient of coincidence.

combinatorial gene regulation Transcriptional control (i.e., whether a gene is active or inactive) achieved by combining a few regulatory proteins (negative and positive) binding to particular DNA sequences.

comparative genome analysis Comparison of the nucleotide sequence from two or more organisms, usually to determine which regions are evolutionarily conserved and likely to be functionally important.

comparative genomics The comparison of entire genomes of different species, with the goal of enhancing our understanding of the functions of each genome.

complement DNA library *See* **cDNA library.**

complementary base pairing The hydrogen bonding between a particular purine and a particular pyrimidine in double-stranded nucleic acid molecules (DNA-DNA, DNA-RNA, or RNA-RNA). The major specific pairings are guanine with cytosine and adenine with thymine or uracil.

complementary base pairs The specific A-T and G-C base pairs in double-stranded DNA. The bases are held together by hydrogen bonds between the purine and pyrimidine bases in each pair.

complementary DNA *See* **cDNA.**

complementation test *See* **cis-trans test.**

complete dominance The case in which one allele is dominant to the other, so that at the phenotypic level, the heterozygote is essentially indistinguishable from the homozygous dominant.

complete medium For a microorganism, the complete medium supplies all the building blocks and vitamins needed to synthesize macromolecules required for growth and reproduction.

complete recessiveness The situation in which an allele is phenotypically expressed only when it is homozygous.

concerted evolution (molecular drive) A poorly understood evolutionary process that produces uniformity of sequence in multiple copies of a gene.

conditional mutant A mutant organism that is normal under one set of conditions but becomes seriously impaired or dies under other conditions.

conditional mutation A mutation that has a normal phenotype under one condition and a mutant phenotype under another condition.

conjugation A process having a unidirectional transfer of genetic information through direct cellular contact between a donor ("male") and a recipient ("female") bacterial cell.

consensus sequence The sequence indicating which nucleotide is found most frequently at each position.

conservative model A DNA replication scheme in which the two parental strands of DNA remain together and serve as a template for the synthesis of a new daughter double helix.

constitutive gene A gene whose products are essential to the normal functioning of the cell, no matter what the life-supporting environmental conditions are. These genes are always active in growing cells.

constitutive heterochromatin Condensed chromatin that is always genetically inactive and is found at homologous sites on chromosome pairs.

continuous trait *See* **quantitative (continuous) trait.**

contributing allele An allele that contributes to the phenotype.

controlling site A specific sequence of nucleotide pairs adjacent to the gene where the transcription of a gene occurs in response to a particular molecular event.

coordinate induction The simultaneous transcription and translation of two or more genes brought about by the presence of an inducer.

core enzyme The portion of the *E. coli* RNA polymerase that is the active enzyme and can be written as $\alpha_2\beta\beta'\sigma$.

cotransduction The simultaneous transduction of two or more bacterial genes, a good indication that the bacterial genes are closely linked.

cotranslational transport The movement of a protein into the endoplasmic reticulum simultaneously with its synthesis.

coupling An arrangement in which the two wild-type alleles are on one homologous chromosome and the two recessive mutant alleles are on the other.

covariance A statistic used to calculate the correlation coefficient between two variables. The covariance is calculated by taking the sum of $(x - \bar{x})(y - \bar{y})$ over all pairs of values for the variables x and y, where $\bar{x}$ is the mean of the x values and $\bar{y}$ is the mean of all y values.

crisscross inheritance A type of gene transmission passed from a male parent to a female child to a male grandchild.

cross *See* **cross-fertilization.**

cross-fertilization (cross) The fusion of male gametes from one individual and female gametes from another.

crossing-over A term introduced by Morgan and E. Cattell in 1912 to describe the process of reciprocal chromosomal interchange by which recombinants arise.

cytokinesis The division of the cytoplasm during cell division. The two new nuclei compartmentalize into separate daughter cells, and the mitotic cell division process is completed.

cytosine (C) A pyrimidine base found in RNA and DNA. In double-stranded DNA, cytosine pairs with the purine guanine.

dark repair system *See* **excision repair system.**

Darwinian fitness The relative reproductive ability of a genotype.

daughter chromosome A former sister chromatid that detached when the centromeres separated at the beginning of mitotic anaphase or meiotic anaphase II.

degeneracy Multiple coding; more than one codon per amino acid.

degradation control Regulation of the RNA breakdown rate in the cytoplasm.

deletion A chromosomal mutation resulting in the loss of a segment of the genetic material and the genetic information contained therein from a chromosome.

deoxyribonuclease (DNase) An enzyme that catalyzes the degradation of DNA to nucleotides.

deoxyribonucleic acid (DNA) A polymeric molecule consisting of deoxyribonucleotide building blocks that in a double-stranded, double-helical form is the genetic material of most organisms.

deoxyribonucleotide The basic building block of DNA, consisting of a sugar (deoxyribose), a base, and a phosphate.

deoxyribose The pentose (five-carbon) sugar found in DNA.

development The process of regulated growth that results from the interaction of the genome with cytoplasm and the environment. It involves a programmed sequence of phenotypic events that are typically irreversible.

diakinesis The stage of prophase I that follows diplonema and during which the four chromatids of each tetrad are most condensed and the chiasmata often terminalize.

dicentric bridge *See* **dicentric chromosome.**

dicentric chromosome A chromosome with two centromeres. For example, as a result of the crossover between genes B and C in the inversion loop, one recombinant chromatid becomes stretched across the cell as the two centromeres begin to migrate, forming a dicentric bridge.

dideoxy (Sanger) sequencing A method of rapid sequencing of DNA molecules developed by Fred Sanger. This technique incorporates the use of dideoxy nucleotides in a DNA polymerase-catalyzed DNA synthesis reaction.

dideoxynucleotide A modified nucleotide that has a 3′-H on the deoxyribose sugar rather than a 3′-OH. If a dideoxy nucleoside triphosphate (ddNTP) is used in a DNA synthesis reaction, the ddNTP can be incorporated into the growing chain. However, no further DNA synthesis can occur because no phosphodiester bond can be formed with an incoming DNA precursor.

differentiation An aspect of development that involves the formation of different types of cells, tissues, and organs through the processes of specific regulation of gene expression.

dihybrid cross A cross between two dihybrids of the same type. Individuals that are heterozygous for two pairs of alleles at two different loci are called dihybrid.

dioecious Referring to plant species that have male and female sex organs on different individuals.

diploid (2N) A eukaryotic cell with two sets of chromosomes.

diplonema The stage of prophase I in which the chromosomes begin to repel one another and tend to move apart.

discontinuous trait A heritable trait in which the mutant phenotype is sharply distinct from the alternative, wild-type phenotype.

disjunction The process in anaphase during which sister chromatid pairs undergo separation.

dispersed repeated DNA Repeated sequences in the genome that are distributed at irregular intervals.

dispersive model A DNA replication scheme in which the parental double helix is cleaved into double-stranded DNA segments that act as templates for the synthesis of new double-stranded DNA segments. Somehow, the segments reassemble into complete DNA double helices, with parental and progeny DNA segments interspersed.

DNA *See* **deoxyribonucleic acid.**

DNA chip *See* **DNA microarray.**

DNA fingerprinting *See* **DNA typing.**

DNA helicase An enzyme that catalyzes the unwinding of the DNA double helix during replication in *E. coli;* product of the *dnaB* gene.

DNA ligase (polynucleotide ligase) An enzyme that catalyzes the formation of a covalent bond between free single-stranded ends of DNA molecules during DNA replication and DNA repair.

DNA marker A genetic marker detected using molecular tools that focus on the DNA itself rather than on the gene product or associated phenotype.

DNA microarray An ordered grid of DNA molecules of known sequence—probes—fixed at known positions on a solid substrate, either a silicon chip, glass, or, less commonly, a nylon membrane. Labeled free DNA molecules—targets—are added to the fixed probes to analyze identities or quantities of target molecules. DNA microarrays allow for the simultaneous analysis of thousands of DNA target molecules.

DNA molecular testing A type of genetic testing that focuses on the molecular nature of mutations associated with disease.

DNA polymerase An enzyme that catalyzes the synthesis of DNA.

DNA polymerase I An *E. coli* enzyme that catalyzes DNA synthesis, originally called the Kornberg enzyme.

DNA primase *See* **primase.**

DNA profiling *See* **DNA typing.**

DNA typing The use of restriction fragment length polymorphisms for DNA analysis to identify an individual.

docking protein An integral membrane protein of the endoplasmic reticulum (ER) to which the nascent polypeptide-signal recognition particle (SRP)-ribosome complex binds to facilitate the binding of the polypeptide's signal sequence and associated ribosome to the ER.

domain shuffling A controversial idea that genes with new functions are made by the duplication and rearrangement of protein domains (usually encoded by individual exons) in different combinations.

dominance variance Genetic variance that arises from the dominance effects of genes; represented by V_D.

dominant Referring to an allele or phenotype that is expressed in either the homozygous or the heterozygous state.

dominant lethal allele An allele that exhibits a lethal phenotype when present in the heterozygous condition.

dosage compensation A mechanism in mammals that compensates for X chromosomes in excess of the normal complement. *See also* **Barr body.**

double crossover Two crossovers occurring in a particular region of a chromosome in a meiosis.

Down syndrome *See* **trisomy-21.**

duplication A chromosomal mutation that results in the doubling of a segment of a chromosome.

effective population size The effective number of adults contributing gametes to the next generation.

effector (effector molecule) A small molecule involved in controlling expression of a regulated gene.

endosymbiont hypothesis The hypothesis that mitochondria and chloroplasts originated as free-living prokaryotes that invaded primitive eukaryotic cells and established a mutually beneficial (symbiotic) relationship.

enhancer (enhancer element) In eukaryotes, a type of DNA sequence element having a strong, positive effect on transcription by RNA polymerase II.

environmental sex determination The process by which the environment plays a major role in determining the sex of an organism.

environmental variance Any nongenetic source of phenotypic variation among individuals; represented by V_E.

episome An autonomously replicating plasmid (a circular, double-stranded DNA molecule) that is capable of integrating into the host cell's chromosome.

epistasis A form of gene interaction in which one gene interferes with the phenotypic expression of another nonallelic gene so that the phenotype is governed by the former gene and not by the latter gene when both genes are present in the genotype.

essential gene A gene that when mutated can result in a lethal phenotype.

euchromatin Chromatin that is condensed during division but becomes uncoiled during interphase.

eukaryote Literally meaning "true nucleus," an organism that has cells in which the genetic material is located in a membrane-bounded nucleus. Eukaryotes can be unicellular or multicellular.

euploidy The condition in which an organism or cell has one complete set of chromosomes or an exact multiple of complete sets.

evolution Genetic change that takes place over time within a group of organisms.

excision repair (dark repair) system An enzyme-catalyzed, light-independent process of repair of ultraviolet light–induced pyrimidine dimers in DNA that involves removal of the dimers and synthesis of a new piece of DNA complementary to the undamaged strand.

exon The part of an mRNA molecule that specifies the amino acid sequence of a polypeptide during translation. *See also* **coding sequence.**

exon shuffling *See* **domain shuffling.**

expressivity The degree to which a particular genotype is expressed in the phenotype.

F₁ generation (first filial generation) The offspring that result from the first experimental crossing of two parental strains in animals or plants.

F₂ generation The second filial generation, produced by selfing the F₁.

facultative heterochromatin Chromatin that may become condensed throughout the cell cycle and may contain genes that are inactivated when the chromatin becomes condensed.

familial Referring to a trait shared by members of a family.

F-duction Transfer of the few host genes carried on an F′ plasmid in conjugation between an F′ and an F⁻ cell. If the genes are different between the two cell types, the recipient becomes partially diploid for the genes on the F′.

fine-structure mapping High-resolution mapping of allelic sites within a gene.

first filial generation *See* **F₁ generation.**

fitness *See* **Darwinian fitness.**

formylmethionine (fMet) A specially modified amino acid involving the addition of a formyl group to the amino group of methionine. It is the first amino acid incorporated into a polypeptide chain in prokaryotes and in eukaryotic cellular organelles.

forward mutation A mutational change from a wild-type allele to a mutant allele.

founder effect A form of genetic drift. A phenomenon that occurs when a population is initially established by a small number of breeding individuals that has come by migration from a large population.

F-pili (sex pili) Hairlike cell surface components produced by cells containing the F factor, which allow the physical union of F⁺ and F⁻ cells or Hfr and F⁻ cells to take place.

frameshift mutation A mutational addition or deletion of a base pair in a gene that disrupts the normal reading frame of an mRNA, which is read in groups of three bases.

frequency distribution A means of summarizing the phenotypes of a continuous trait whereby the population is described in terms of the proportion of individuals that have each phenotype.

functional genomics The comprehensive analysis of the functions of genes and nongene sequences in entire genomes.

G banding A particular staining technique that generates specific banding patterns for each chromosome, enabling each chromosome in the karyotype to be distinguished clearly.

gamete A mature reproductive cell that is specialized for sexual fusion. Each gamete is haploid and fuses with a cell of similar origin but of opposite sex to produce a diploid zygote.

gametogenesis The formation of male and female gametes by meiosis.

gametophyte The haploid sexual generation in the life cycle of plants that produces the gametes.

GC box A eukaryotic promoter element with the consensus sequence 5′-GGGCGG-3′ that can be found in either orientation upstream of the transcription initiation site. The GC boxes appear to help the RNA polymerase near the transcription start point.

gene (Mendelian factor) The determinant of a characteristic of an organism. Genetic information is coded in the DNA, which is responsible for species and individual variation. A gene's nucleotide sequence specifies a polypeptide or RNA and is subject to mutational alteration.

gene conversion A nonreciprocal recombination process resulting in a sequence becoming identical to another.

gene expression The process by which a gene produces its product and the product carries out its function.

gene flow The movement of genes that takes place when organisms migrate and then reproduce, contributing their genes to the gene pool of the recipient population.

gene locus *See* **locus.**

gene marker *See* **genetic marker.**

gene mutation A heritable alteration of the genetic material, usually from one allelic form to another.

gene pool The total genetic information encoded in all the genes in a breeding population existing at a given time.

gene regulatory elements Base pair sequences associated with a gene, which are involved in the regulation of gene expression.

gene segregation *See* **principle of segregation (first law).**

gene tree A phylogenetic tree based on the divergence observed within a single homologous gene. Gene trees are not always a good representation of the relationship between species because polymorphisms in any given gene may have arisen before speciation events.

GeneChip® array *See* **DNA microarray.**

generalized transduction A type of transduction in which any gene may be transferred between bacteria.

genetic code The base pair information that specifies the amino acid sequence of a polypeptide.

genetic correlation An association between the genes that determine two traits.

genetic counseling The procedures whereby the risks of prospective parents having a child who expresses a genetic disease are evaluated and explained to them. The genetic counselor typically makes predictions about the probabilities of particular traits (deleterious or not) occurring among children of a couple.

genetic drift Random change in allelic frequency.

genetic engineering The alteration of the genetic constitution of cells or individuals by directed and selective modification, insertion, or deletion of an individual gene or genes. In some cases, novel gene combinations are made by joining DNA fragments from different organisms.

genetic map (linkage map) A representation of the genetic distance separating nonallelic gene loci in a linkage structure.

genetic mapping The use of genetic crosses to locate genes on chromosomes relative to one another.

genetic marker (gene marker) Any genetically controlled phenotypic difference used in genetic analysis, particularly in the detection of genetic recombination events.

genetic recombination A process by which parents with different genetic characters give rise to progeny such that the genes in which the parents differed are associated in new combinations. For example, from *A B* and *a b* the recombinants *A b* and *a B* are produced.

genetic structure The patterns of genetic variation found among individuals within groups.

genetic testing Analysis to determine whether an individual who has symptoms or is at a high risk of developing a genetic disease because of a family history of a heritable disease actually has a particular gene mutation.

genetic variance Genetic sources of phenotypic variation among individuals of a population; includes dominance genetic variance, additive genetic variance, and epistatic genetic variance; represented by V_G.

genetics The science of heredity, involving the structure and function of genes and the way genes are passed from one generation to the next.

genome The total amount of genetic material in a chromosome set; in eukaryotes, this is the amount of genetic material in the haploid set of chromosomes of the organism.

genomic imprinting Phenomenon in which the expression of certain genes is determined by whether the gene is inherited from the female or male parent.

genomic library A collection of molecular clones that contains at least one copy of every DNA sequence in the genome.

genomics The development and application of new mapping, sequencing, and computational procedures to analyze the entire genome of organisms.

genotype The complete genetic makeup of an organism.

genotypic frequencies The frequencies or percentages of different genotypes found within a population.

genotypic sex determination The process by which the sex chromosomes play a decisive role in the inheritance and determination of sex.

germ-line mutation A mutation that occurs in the germ line of a sexually reproducing organism and that may be transmitted by the gametes to the next generation, giving rise to an individual with the mutant state in both its somatic and germ-line cells.

glucose effect *See* **catabolite repression.**

Goldberg-Hogness box (TATA box; TATA element) A sequence found approximately at position −25 from the transcription initiation site. The Goldberg-Hogness box has the seven-nucleotide consensus sequence TATAAAA and has a general function in the initiation of transcription.

group I intron self-splicing *See* **self-splicing.**

guanine (G) A purine base found in RNA and DNA. In double-stranded DNA, guanine pairs with the pyrimidine cytosine.

Haldane's rule The rule of male hybrids being sterile and female hybrids being fertile.

haploid (N) A cell or individual with one copy of each nuclear chromosome.

haplosufficient In a heterozygote for a gene showing normal dominance, half the amount of protein produced by the homozygote may be sufficient for normal cell function. In this case, the gene is said to be haplosufficient.

Hardy-Weinberg law (Hardy-Weinberg equilibrium, Hardy-Weinberg law of genetic equilibrium) An extension of Mendel's laws of inheritance that describes the expected relationship between allelic frequencies in natural populations and the frequencies of individuals of various genotypes in the same populations.

harlequin chromosomes 5-bromodeoxyuridine (5-BUdR), a thymidine analogue, is incorporated into DNA during replication. When both DNA strands contain 5-BUdR, the chromatid stains less intensely than when only one DNA strand contains the analogue. When cells are grown in the presence of 5-BUdR for two replication cycles, the two sister chromatids stained differentially are called harlequin chromosomes.

helix-destabilizing proteins *See* **single-strand DNA-binding (SSB) proteins.**

hemizygous The condition of X-linked genes in males. Males that have an X chromosome with an allele for a particular gene but do not have another allele of that gene in the gene complement are hemizygous.

hereditary trait A characteristic under control of the genes that is transmitted from one generation to another.

heritability The proportion of phenotypic variation in a population attributable to genetic factors.

hermaphroditic In animals (e.g., nematodes), referring to species in which each individual has both testes and ovaries; in plants, referring to species that have both stamens and pistils on the same flower.

heterochromatin Chromatin that remains condensed throughout the cell cycle and is genetically inactive.

heteroduplex DNA A region of double-stranded DNA with different sequence information on the two strands.

heterogametic sex The sex that has sex chromosomes of different types (e.g., XY) and therefore produces two kinds of gametes with respect to the sex chromosomes.

heterosis The phenomenon in which the heterozygous genotypes with respect to one or more characters are superior in comparison with the corresponding homozygous genotypes in terms of growth, survival, phenotypic expression, and fertility.

heterozygosity The proportion of individuals heterozygous at a locus; the state of being heterozygous; *see also* **heterozygous.**

heterozygote superiority *See* **heterosis.**

heterozygous In a diploid organism, having different alleles of one or more genes and therefore producing gametes of different genotypes.

***Hfr* (high-frequency recombination)** A male cell in *E. coli* with the *F* factor integrated into the bacterial chromosome. When the *F* factor promotes conjugation with a female (F^-) cell, bacterial genes are transferred to the female cell with high frequency.

highly repetitive DNA A DNA sequence that is repeated between 10^5 and 10^7 times in the genome.

histone One of a class of basic proteins that are complexed with DNA in chromosomes and play a major role in determining the structure of eukaryotic nuclear chromosomes.

holandric Referring to traits controlled by genes on the Y chromosome.

homeobox A 180-bp consensus sequence found in the protein-coding sequences of genes that regulate development.

homeodomain The 60-amino-acid part of proteins that corresponds to the homeobox sequence of genes. All homeodomain-containing proteins appear to be located in the nucleus.

homeotic mutations Mutations that alter the identity of particular segments, transforming them into copies of other segments.

homogametic sex The gender in the species, most often the female, that produces only the X sex chromosome.

homologous Referring to chromosomes or chromosome regions that are identical with respect to the genetic loci they contain.

homologous chromosomes The members of a chromosome pair that are identical in the arrangement of genes they contain and in their visible structure.

homologous proteins Proteins that show a common ancestor.

homologue Each individual member of a pair of homologous chromosomes.

homozygous In a diploid organism, having the same alleles of one or more genes and therefore producing gametes of identical genotypes.

homozygous dominant A diploid organism that has the same dominant allele for a given gene locus on both members of a homologous pair of chromosomes.

homozygous recessive A diploid organism that has the same recessive allele for a given gene locus on both members of a homologous pair of chromosomes.

Human Genome Project (HGP) A project to obtain the sequence of the complete 3 billion (3×10^9) nucleotide pairs of the human genome and to map all human genes.

hybrid dysgenesis The appearance of a series of defects, including mutations, chromosomal aberrations, and sterility, when certain strains of *Drosophila melanogaster* are crossed.

hypersensitive site (hypersensitive region) A region of DNA around transcriptionally active genes that is highly sensitive to digestion by DNase I.

hypothetico-deductive method of investigation Research method involving making observations, forming hypotheses to explain the observations, making experimental predictions based on the hypotheses, and finally testing the predictions. The last step produces new observations, so a cycle is set up leading to a refinement of the hypotheses and perhaps eventually to the establishment of a law or accepted principle.

imaginal disk In the *Drosophila* blastoderm, undifferentiated cells that will develop into an adult tissue or organ.

immunoglobulins Specialized proteins (antibodies) secreted by B cells that circulate in the blood and lymph and are responsible for humoral immune responses.

inbreeding Preferential mating between close relatives.

incomplete (partial) dominance The condition resulting when one allele is not completely dominant to another allele, so that the heterozygote has a phenotype between that shown in individuals homozygous for either individual allele involved. An example of partial dominance is the frizzle chicken.

indel A gap in a sequence alignment where it is not possible to determine whether an insertion occurred in one sequence or a deletion occurred in another.

induced mutation A mutation that results from treatment with mutagens.

inducer A chemical or environmental agent that brings about the transcription of a bacterial operon.

induction The synthesis of a gene product (or products) in response to the action of an inducer, that is, a chemical or environmental agent.

inferred tree One of the many possible phylogenetic trees that might describe the relationship between taxa.

infinitely many alleles model A specific form of the neutral theory in which it is assumed that each subsequent neutral mutation generates a new allele of a gene, never previously seen in the population.

informative sites Positions within a multiple alignment that favor one possible phylogenetic tree over an alternative from the perspective of parsimony. In general, such sites have at least two different nucleotides or amino acids, and each must be present at least twice within the multiple alignment.

initiation factors Proteins involved in the initiation of translation.

insertion sequence (IS element) The simplest transposable genetic element found in prokaryotes. It is a mobile segment of DNA that contains genes required for insertion of the DNA segment into a chromosome and for mobilization of the element to different locations.

interaction variance Genetic variance that arises from epistatic interactions among genes; represented by V_I.

interference The phenomenon in which the presence of one crossover interferes with the formation of another crossover nearby.

intergenic suppressor A mutation whose effect is to suppress the phenotypic consequences of another mutation in a gene distinct from the gene in which the suppressor mutation is located.

internal control region (ICR) Promoter sequence, recognized by RNA polymerase III, that is located within the gene sequence (e.g., in tRNA genes and 5S rRNA genes of eukaryotes).

interspersed repeated DNA *See* **dispersed repeated DNA.**

intervening sequence (IVS) *See* **intron.**

intragenic suppressor A mutation whose effect is to suppress the phenotypic consequences of another mutation within the same gene in which the suppressor mutation is located.

intron A nucleotide sequence in eukaryotes that must be excised from a structural gene transcript to convert the transcript into a mature messenger RNA molecule containing only coding sequences that can be translated into the amino acid sequence of a polypeptide.

inversion A chromosomal mutation that results when a segment of a chromosome is excised and then reintegrated in an orientation 180° from the original orientation.

IS element *See* **insertion sequence (IS element).**

karyotype A complete set of all the metaphase chromatid pairs in a cell (literally, "nucleus type").

kinetochores Specialized structures on each face of the centromere of each eukaryotic chromosome that become attached to special microtubules called kinetochore microtubules.

Klinefelter syndrome A human clinical syndrome that results from disomy for the X chromosome in a male, which results in a 47,XXY male. Many of the affected males are mentally deficient, have underdeveloped testes, and are taller than average.

lagging strand In DNA replication, the DNA strand that is synthesized discontinuously in the 5′→3′ direction away from the replication fork.

leader sequence One of three main parts of the mRNA molecule. The leader sequence is located at the 5′ end of the mRNA molecule and contains the coded information that the ribosome and special proteins read to tell them where to begin the synthesis of the polypeptide.

leading strand In DNA replication, the DNA strand synthesized continuously in the 5′→3′ direction toward the replication fork.

leptonema The stage during meiosis in prophase I at which the chromosomes have begun to coil and are visible.

lethal allele An allele that results in the death of an organism.

light repair *See* **photoreactivation.**

LINEs (long interspersed repeated sequences) The dispersed families of repeated sequences that are about 5,000 base pairs or more in length.

linkage The condition in which genes are located on the same chromosome.

linkage map *See* **genetic map.**

linked genes Genes that are located on the same chromosome.

linker *See* **restriction site linker.**

locus (*plural*, loci) The position of a gene on a genetic map; the specific place on a chromosome where a gene is located.

lod score method The lod (logarithm of odds) score method is a statistical analysis, usually performed by computer programs, based on data from pedigrees. It is used to test for linkage between two loci in humans.

looped domain A loop of supercoiled DNA that serves to compact the chromosome.

Lyon hypothesis *See* **lyonization.**

lyonization A mechanism in mammals that allows them to compensate for X chromosomes in excess of the normal complement. The excess X chromosomes are cytologically condensed and inactivated, and they do not play a role in much of the development of the individual. The name derives from the discoverer of the phenomenon, Mary Lyon.

lysogenic A term describing a bacterium that contains a temperate phage in the prophage state. The bacterium is said to be lysogenic for that phage. On induction, phage reproduction is initiated, progeny phages are produced, and the bacterial cell lyses.

lysogenic pathway A path, besides the lytic cycle, that a phage can follow. The chromosome does not replicate; instead, it inserts itself physically into a specific region of the host cell's chromosome in a way that is essentially the same as *F* factor integration.

lysogeny The phenomenon of the insertion of a temperate phage chromosome into a bacterial chromosome, where it replicates when the bacterial chromosome replicates. In the lysogenic state, the phage genome is repressed and is said to be in the prophage state.

lytic cycle A type of phage life cycle in which the phage takes over the bacterium and directs its growth and reproductive activities to express the phage's genes and to produce progeny phages.

macromolecule A large molecule (such as DNA, RNA, and proteins) that has a molecular weight of at least a few thousand daltons.

map unit (mu) A unit of measurement used for the distance between two gene pairs on a genetic map. A crossover frequency of 1 percent between two genes equals 1 map unit. *See also* **centimorgan.**

mapping function A mathematical formula used to correct the observed recombination values for the incidence of multiple crossovers.

maternal effect The phenotype in an individual that is established by the maternal nuclear genome as the result of mRNA or proteins that are deposited in the oocyte before fertilization. These inclusions direct early development in the embryo.

maternal inheritance A phenomenon in which the mother's phenotype is expressed exclusively.

mating types A genic system in which two sexes are morphologically indistinguishable but carry different alleles and will mate.

maximum parsimony Describing the phylogenetic tree (or trees) that invokes the fewest number of mutations and therefore is most likely to represent the true evolutionary relationship between species or their genes.

mean The average of a set of numbers, calculated by adding all the values represented and dividing by the number of values.

megasporogenesis The formation in flowering plants of megaspores and the production of the embryo sac (the female gametophyte).

meiosis Two successive nuclear divisions of a diploid nucleus that result in the formation of haploid gametes or meiospores having one-half the genetic material of the original cell.

meiosis I The first meiotic division that results in the reduction of the number of chromosomes. This division consists of four stages: prophase I, metaphase I, anaphase I, and telophase I.

meiosis II The second meiotic division, resulting in the separation of the chromatids.

Mendelian factor *See* **gene.**

Mendelian population An interbreeding group of individuals sharing a common gene pool; the basic unit of study in population genetics.

Mendel's first law *See* **principle of segregation.**

Mendel's second law *See* **principle of independent assortment.**

messenger RNA (mRNA) The RNA molecule that contains the coded information for the amino acid sequence of a protein.

metacentric chromosome A chromosome that has the centromere approximately in the center of the chromosome.

metaphase A stage in mitosis or meiosis in which chromosomes become aligned along the equatorial plane of the spindle.

metaphase II The stage of meiosis II during which the centromeres line up on the equator of the second-division spindles (in each of two daughter cells formed from meiosis I).

metaphase plate The plane where the chromosomes become aligned during metaphase.

metastasis The spreading of malignant tumor cells throughout the body so that tumors develop at new sites.

microsatellite *See* **STR.**

migration Movement of organisms from one location to another.

minimal medium For a microorganism, the medium that contains the simplest set of ingredients that the microorganism can use to synthesize all the molecules required for growth and reproduction.

minisatellite *See* **VNTR.**

missense mutation A gene mutation in which a base pair change in the DNA causes a change in an mRNA codon, with the result that a different amino acid is inserted into the polypeptide in place of the one specified by the wild-type codon.

mitochondrion Organelle found in the cytoplasm of all aerobic animal and plant cells; the principal source of energy in the cell.

mitosis The process of nuclear division in haploid or diploid cells, producing daughter nuclei that contain identical chromosome complements and are genetically identical to one another and to the parent nucleus from which they arose.

moderately repetitive DNA A DNA sequence that is reiterated from a few to as many as 10^5 times in the genome.

molecular clock A controversial hypothesis that for any given gene, mutations accumulate at an essentially constant rate in all evolutionary lineages as long as the gene retains its original function.

molecular cloning The generation of many copies of a DNA molecule (e.g., a recombinant DNA molecule) by replication in a suitable host.

molecular drive *See* **concerted evolution.**

molecular evolution A population genetics–based field of study of how macromolecules evolve and how genes and organisms are evolutionarily related.

molecular genetics A subdivision of the science of genetics involving how genetic information is encoded within the DNA and how biochemical processes of the cell translate the genetic information into the phenotype.

monoecious Referring to plants in which male and female gametes are produced in the same individual.

monohybrid cross A cross between two true-breeding individuals that differ with respect to the alleles of one locus.

monoploidy The condition of having only one set of chromosomes instead of the normal two sets.

monosomy An aberrant, aneuploid state in a normally diploid cell or organism in which one chromosome is missing, leaving one chromosome with no homologue.

mRNA splicing A process whereby an intervening sequence between two coding sequences in an RNA molecule is excised and the coding sequences ligated (spliced) together.

mRNA transport control Regulating the number of transcripts that exit the nucleus to the cytoplasm.

multifactorial trait A trait influenced by multiple genes and environmental factors.

multigene family A set of related genes that have evolved from some ancestral gene through gene duplication.

multiple alleles Many alternative forms of a single gene.

multiple cloning site *See* **polylinker.**

multiple crossovers More than one crossover occurring in a particular region of a chromosome in a meiosis.

multiple-gene hypothesis for quantitative inheritance *See* **polygene hypothesis for quantitative inheritance.**

mutagen Any physical or chemical agent that induces mutations.

mutagenesis The creation of mutations.

mutant allele Any alternative to the wild-type allele of a gene. Mutant alleles may be dominant or recessive to wild-type alleles.

mutation Any detectable and heritable change in the genetic material not caused by genetic recombination.

mutation frequency The number of occurrences of a particular kind of mutation expressed as the proportion of cells or individuals in a population.

mutation rate The probability of a particular kind of mutation as a function of time.

mutator gene A gene that when mutant increases the spontaneous mutation frequencies of other genes.

narrow-sense heritability The proportion of the phenotypic variance that results from additive genetic variance.

natural selection Differential reproduction of genotypes.

nature versus nurture The relative importance of genetic and environmental factors in determining phenotypes, especially behaviors.

negative assortative mating Mating that occurs between dissimilar individuals more often than it does between randomly chosen individuals.

neutral mutation A base pair change in the gene that changes a codon in the mRNA such that there is no change in the function of the protein translated from that message.

neutral mutation model A model that replaced the classical model by acknowledging the presence of extensive genetic variation in proteins but proposing that this variation is neutral with regard to natural selection.

neutral theory of molecular evolution The process of changes in DNA and protein sequences in a population driven solely by mutation and random genetic drift.

nitrogenous base A nitrogen-containing base that, along with a pentose sugar and a phosphate, is one of the three parts of a nucleotide, the building block of RNA and DNA.

node A graphic representation of an ancestral organism in a phylogenetic tree.

noncontributing allele An allele that does not have any effect on the phenotype of a quantitative trait.

nondisjunction A failure of homologous chromosomes or sister chromatids to separate at anaphase.

nonhistone A type of acidic or neutral protein found in chromatin.

nonhomologous chromosomes Chromosomes that contain dissimilar genes and do not pair during meiosis.

non-Mendelian inheritance (cytoplasmic inheritance) The inheritance of characters determined by genes not located on the nuclear chromosomes but on mitochondrial or chloroplast chromosomes. Such genes show inheritance patterns distinctly different from those of nuclear genes.

nonparental ditype (NPD) One of three types of tetrads possible when two genes are segregating in a cross. The NPD tetrad contains four nuclei, all of which have recombinant (nonparental) genotypes, that is, two of each possible type.

nonsense codon *See* **chain-terminating codon.**

nonsense mutation A gene mutation in which a base pair change in the DNA causes a change in an mRNA codon from an amino acid–coding codon to a chain-terminating (nonsense) codon. As a result, polypeptide chain synthesis is terminated prematurely and is therefore nonfunctional or, at best, partially functional.

nonsynonymous Referring to mutations at nucleotide positions that change the resulting amino acid sequence of a protein.

nontranscribed spacer (NTS) sequence Sequence that is not transcribed and that is found between transcription units in rDNA. Important sequences that control transcription of the rDNA are within the NTS.

norm of reaction The extent to which the phenotype produced by a genotype varies with the environment.

normal distribution A probability distribution in statistics, graphically displayed as a bell-shaped curve.

northern blot analysis A technique similar to Southern blotting except that RNA rather than DNA is separated and transferred to a filter for hybridization with a probe.

nuclear matrix A filamentous structural framework of protein inside the nuclear envelope.

nuclease An enzyme that catalyzes the degradation of a nucleic acid by breaking phosphodiester bonds. Nucleases specific for DNA are called deoxyribonucleases (DNases), and nucleases specific for RNA are called ribonucleases (RNases).

nucleic acid A class of compounds that includes DNA and RNA. The building block of a nucleic acid is the nucleotide, which consists of sugar, nitrogenous base, and phosphate.

nucleofilament A fiber seen in chromatin. It is approximately 10 nm in diameter and consists of DNA wrapped around nucleosome cores.

nucleoid Central region in a bacterial cell in which the chromosome is compacted.

nucleolus An organelle within the eukaryotic nucleus; the site of transcription of the ribosomal RNA genes and assembly of the ribosomal subunits.

nucleoside Combination of a sugar and a nitrogenous base. When a phosphate is added, the nucleoside becomes a nucleotide.

nucleoside phosphate *See* **nucleotide.**

nucleosome The basic structural unit of eukaryotic nuclear chromosomes, consisting of two molecules each of the four core histones (H2A, H2B, H3, and H4, the histone octamer), a single molecule of the linker histone H1.

nucleotide A monomeric molecule of RNA or DNA that consists of three distinct parts: a pentose (ribose in RNA, deoxyribose in DNA), a nitrogenous base, and a phosphate group.

nucleotide excision repair (NER) system *See* **excision repair (dark repair) system.**

nucleus A discrete structure within the cell that is bounded by a nuclear membrane. It contains most of the genetic material of the cell.

null hypothesis A hypothesis that states that there is no real difference between the observed data and the predicted data.

nullisomy The aberrant, aneuploid state in a normally diploid cell or organism in which there is a loss of one pair of homologous chromosomes.

nutritional mutant *See* **auxotrophic mutant.**

nutritional mutation *See* **auxotrophic mutation.**

Okazaki fragments The short, single-stranded DNA fragments in discontinuous DNA replication that are synthesized during DNA replication and subsequently covalently joined to make a continuous strand.

oligomer A short DNA molecule.

oligonucleotide array *See* **DNA microarray.**

oligonucleotide hybridization analysis A way to type a single nucleotide polymorphism (SNP) by hybridizing a short oligonucleotide with a sequence corresponding to the common allele at the SNP locus.

oncogene A gene whose action promotes unregulated cell proliferation. Oncogenes are altered forms of proto-oncogenes.

oncogenesis Tumor (cancer) initiation in an organism.

one gene–one enzyme hypothesis The hypothesis, based on Beadle and Tatum's studies in biochemical genetics, that each gene controls the synthesis of one enzyme.

one gene–one polypeptide hypothesis Updated version of the one gene–one enzyme hypothesis, which states that each gene controls the synthesis of a polypeptide chain.

oogenesis The development in the gonad of the female germ cell (egg cell) of animals.

open reading frame In a segment of DNA, a potential protein-coding sequence identified by an initiator codon in frame with a chain-terminating codon.

operator The controlling site adjacent to a promoter that is responsible for controlling the transcription of genes that are contiguous to the promoter.

operon A cluster of genes whose expressions are regulated together by operator-regulator protein interactions plus the operator region itself and the promoter.

optimal alignment In the analysis of changes at the nucleotide or amino acid level between two or more gene sequences, an approximation of the true alignment of sequences, where gaps are inserted to maximize the similarity among the sequences being aligned.

ordered tetrads A structure resulting from meiosis in which the four meiotic products are in an order reflecting exactly the orientation of the four chromatids at the metaphase plate in meiosis I.

origin A specific site on the chromosome at which the double helix denatures into single strands and continues to unwind as the replication fork migrates.

origin of replication A specific DNA sequence that is necessary for the initiation of DNA replication in prokaryotes.

outbreeding Preferential mating between nonrelated individuals.

outgroup A species that is least related to the others in a group of species because it has diverged from the group before all the others diverged from each other.

overdominance *See* **heterosis.**

ovum A mature egg cell. In the second meiotic division, the secondary oocyte produces two haploid cells; the large cell rapidly matures into the ovum.

P generation The parental generation, the immediate parents of an F_1.

pachynema The stage in meiosis (mid-prophase I) during which the homologous pairs of chromosomes exchange chromosome regions.

paracentric inversion An inversion in which the inverted segment occurs on one chromosome arm and does not include the centromere.

parental ditype (PD) One of three types of tetrads possible when two genes are segregating in a cross. The PD tetrad contains four nuclei, all of which are parental genotypes, with two of one parent and two of the other parent.

parental genotypes (parental classes, parentals) Genotypes among progeny of crosses that have combinations of genetic markers like one or other of the parents in the parental generation.

parental imprinting *See* **genomic imprinting.**

partial dominance *See* **incomplete (partial) dominance.**

partial reversion A reversion of a mutation such that a non-wild-type amino acid is inserted into the polypeptide. Complete or partial function may be restored, depending on the change.

particulate factor The term Mendel used to describe a factor that carries hereditary information and is transmitted from par-

ents to progeny through the gametes. We now know these factors by the name *genes*.

pedigree analysis A family tree investigation that involves the careful compilation of phenotypic records of the family over several generations.

penetrance The frequency with which a dominant or homozygous recessive gene manifests itself in the phenotype of an individual.

pentose sugar A five-carbon sugar that, along with a nitrogenous base and a phosphate group, is one of the three parts of a nucleotide, the building block of RNA and DNA.

peptide bond A covalent bond in a polypeptide chain that joins the α-carboxyl group of one amino acid to the α-amino group of the adjacent amino acid.

peptidyl transferase The enzyme that catalyzes the formation of the peptide bond in protein synthesis.

pericentric inversion An inversion in which the inverted segment includes parts of both chromosome arms and therefore includes the centromere.

phage lysate The progeny phages released after lysis of phage-infected bacteria.

phage vector A phage that carries pieces of bacterial DNA between bacterial strains in the process of transduction.

phages Shortened form of **bacteriophages.**

phenocopy An abnormal phenotypic modification resulting from special environmental conditions. It mimics a similar phenotype caused by gene mutation.

phenotype The observable properties of an organism, produced by the genotype and its interaction with the environment.

phenotypic correlation An association between two traits.

phenotypic variance A measure of a trait's variability; represented by V_P.

phosphate group A component that, along with a pentose sugar and a nitrogenous base, is one of the three parts of a nucleotide, the building block of RNA and DNA. Because phosphate groups are acidic in nature, DNA and RNA are called nucleic acids.

phosphodiester bond A covalent bond in RNA and DNA between a sugar and a phosphate. Phosphodiester bonds form the repeating sugar-phosphate array of the backbone of DNA and RNA.

photoreactivation (light repair) One way by which pyrimidine dimers can be repaired. The dimers are reverted directly to the original form by exposure to visible light in the wavelength range of 320 to 370 nm.

phylogenetic relationship The hierarchical relationship between species or populations across vast distances of evolutionary time.

phylogenetic tree A graphic representation of the phylogeny of a group of species or genes.

physical map Map of physically identifiable regions or markers on genomic DNA, constructed without genetic recombination analysis.

pistil The female reproductive organ in a flowering plant that typically consists of the stigma, the style, and the ovary.

plaque A round, clear area in a lawn of bacteria on solid medium that results from the lysis of cells by repeated cycles of phage lytic growth.

plasma membrane Lipid bilayer that surrounds the cytoplasm of both animal and plant cells.

plasmid An extrachromosomal genetic element consisting of double-stranded DNA that replicates autonomously from the host chromosome.

pleiotropic Referring to genes or mutations that result in multiple phenotypic effects **(pleiotropy).**

pleiotropy Multiple phenotypic effects resulting from a single mutant gene.

point mutants Organisms whose phenotypes result from an alteration of a single nucleotide pair.

point mutation A mutation caused by a substitution of one base pair for another.

polarity A term referring to a bacterial operon that codes for a polygenic mRNA. It is the phenomenon whereby certain nonsense mutations not only result in the loss of activity of the enzyme encoded by the gene in which they are located but also reduce significantly or abolish the synthesis of enzymes coded by structural genes on the operator distal side of the mutation. The mutations are called *polar mutations.*

poly(A) polymerase (PAP) The enzyme that catalyzes the production of the 3′ poly(A) tail.

poly(A) site The 3′ end of mRNA to which 50 to 250 adenine nucleotides are added as part of mRNA posttranscriptional modification.

poly(A) tail A sequence of 50 to 250 adenine nucleotides that is added as a posttranscriptional modification at the 3′ ends of most eukaryotic mRNAs.

polycistronic mRNA *See* **polygenic mRNA.**

polygene (multiple-gene) hypothesis for quantitative inheritance The hypothesis that quantitative traits are controlled by many genes.

polygenic (polycistronic) mRNA A single mRNA transcript in prokaryotic operons of two or more adjacent structural genes that specifies the amino acid sequences of the corresponding polypeptides.

polygenic trait A trait encoded by many loci.

polylinker (multiple cloning site) A region of clustered unique restriction sites in a cloning vector.

polymerase chain reaction (PCR) A method used to replicate defined DNA sequences selectively and repeatedly from a DNA mixture.

polynucleotide A linear sequence of nucleotides in DNA or RNA.

polypeptide A polymeric, covalently bonded linear arrangement of amino acids joined by peptide bonds.

polyploidy The condition of a cell or organism that has more than its normal number of sets of chromosomes.

polyribosome (polysome) The complex between an mRNA molecule and all the ribosomes that are translating it simultaneously.

polytene chromosome A special type of chromosome consisting of a bundle of numerous chromatids that have arisen by repeated cycles of replication of single chromatids without nuclear division. This type of chromosome is characteristic of various tissues of Diptera.

population A group of interbreeding individuals that share a set of genes.

population genetics A branch of genetics that describes in mathematical terms the consequences of Mendelian inheritance on the population level.

population viability analysis Analysis of the survival probabilities of different genotypes in the population.

positional cloning The isolation of a gene associated with a genetic disease on the basis of its approximate chromosomal position.

positive assortative mating Mating that occurs more frequently between individuals who are phenotypically similar than it does between randomly chosen individuals.

precursor mRNA (primary transcripts; pre-mRNA) The initial transcript of a gene that is modified or processed to produce the mature, functional mRNA molecule. In eukaryotes, for example, the transcript is modified at both the 5′ and the 3′ ends, and in a number of cases RNA sequences that do not code for amino acids are present and must be excised.

precursor rRNA (pre-rRNA) A primary transcript of adjacent rRNA genes (16S, 23S, and 5S rRNA genes in prokaryotes; 18S, 5.8S, and 28S rRNA genes in eukaryotes) plus flanking and spacer DNA that must be processed to release the mature rRNA molecules.

precursor tRNA (pre-tRNA) A primary transcript of a tRNA gene whose bases must be extensively modified and that must be processed to remove extra RNA sequences to produce the mature tRNA molecule. In some cases, the primary transcript may contain the sequences of two or more tRNA molecules.

Pribnow box A part of the promoter sequence in prokaryotic genomes that is located at about ten base pairs upstream from the transcription starting point. The consensus sequence for the Pribnow box is TATAAT. The Pribnow box often is called the TATA box.

primary oocyte In female animals, a cell produced from secondary oogonia in the ovary.

primary transcripts *See* **precursor mRNA, rRNA, and tRNA.**

primase Enzyme that, as part of the primosome, catalyzes the formation of the RNA primer to initiate DNA synthesis.

primer *See* **RNA primer.**

primosome A complex of *E. coli* primase, helicase, and perhaps other polypeptides that together become functional in catalyzing the initiation of DNA synthesis.

principle of independent assortment (second law) The law that the factors (genes) for different traits assort independently of one another. In other words, genes on different chromosomes behave independently in the production of gametes.

principle of segregation (first law) The law that two members of a gene pair (alleles) segregate (separate) from each other during the formation of gametes. As a result, one-half the gametes carry one allele and the other half carry the other allele.

probability The ratio of the number of times a particular event is expected to occur to the number of trials during which the event could happen.

proband In human genetics, an affected person with whom the study of a character in a family begins. *See also* **propositus; proposita.**

probe array Generic term for a **DNA microarray, DNA chip,** or **GeneChip® array.**

product rule The rule that the probability of two independent events occurring simultaneously is the product of each of their probabilities.

prokaryote A cellular organism whose genetic material is not located within a membrane-bound nucleus (*see also* **eukaryote**).

promoter (promoter sequence) A specific regulatory nucleotide sequence in the DNA to which RNA polymerase binds for the initiation of transcription.

promoter elements (modules) Consensus sequences found in the promoter region of the transcription initiation site. The elements are the TATA box (or Goldberg-Hogness box), CAT element, and the GC element.

proofreading In DNA synthesis, the process of recognizing a base pair error during the polymerization events and correcting it. Proofreading is a property of the DNA polymerase in prokaryotic cells.

prophage A temperate bacteriophage integrated into the chromosome of a lysogenic bacterium. It replicates with the replication of the host cell's chromosome.

prophase The first stage in mitosis or meiosis, during which the chromosomes (already replicated) condense and become visible under the microscope.

prophase I The first stage of meiosis. There are several stages of prophase I, including leptonema, zygonema, pachynema, diplonema, and diakinesis.

prophase II The stage of meiosis II during which there is chromosome contraction.

proportion of polymorphic loci A ratio calculated by determining the number of polymorphic loci and dividing by the total number of loci examined.

proposita In human genetics, an affected female person with whom the study of a character in a family begins. *See also* **proband.**

propositus In human genetics, an affected male person with whom the study of a character in a family begins. *See also* **proband.**

protein A high-molecular-weight, nitrogen-containing organic compound of complex shape and composition.

protein degradation control Regulation of the protein degradation rate.

proteome The complete set of proteins in a cell.

proteomics The cataloging and analysis of proteins to determine when a protein is expressed, how much is made, and with what other proteins the protein may interact.

proto-oncogene A gene that in normal cells functions to control the normal proliferation of cells and that when mutated or changed in any other way becomes an oncogene.

prototroph A strain that is a wild type for all nutritional requirement genes and thus requires no supplements in its minimal growth medium.

prototrophic strain *See* **prototroph.**

pseudodominance The unexpected expression of a recessive trait, caused by the absence of a dominant allele.

Punnett square A matrix that describes all the possible gametic fusions that will give rise to the zygotes that will produce the next generation.

pure-breeding *See* **true-breeding (pure-breeding) strain.**

purine A type of nitrogenous base. In DNA and RNA, the purines are adenine and guanine.

pyrimidine A type of nitrogenous base. Cytosine is a pyrimidine in DNA and RNA, thymine is a pyrimidine in DNA, and uracil is a pyrimidine in RNA.

quantitative genetics Study of the inheritance of quantitative traits.

quantitative (continuous) trait A trait that shows a continuous variation in phenotype over a range.

quantitative trait loci (QTLs) The genes underlying quantitative traits.

radiation hybrid (RH) A rodent cell line that carries a small fragment of the genome of another organism, such as a human.

random genetic drift *See* **genetic drift.**

random mating Mating between genotypes occurring in proportion to the frequencies of the genotypes in the population.

rDNA repeat units The tandem arrays of rRNA genes, 18S-5.8S-28S, repeated many times along the chromosome.

recessive Referring to an allele or phenotype that is expressed only in the homozygous state.

recessive lethal allele An allele that causes lethality when it is homozygous.

reciprocal cross Repeating a particular genetic cross but with the sexes of the two parents switched. In the garden pea example, reciprocal crosses for smooth and wrinkled seeds are smooth female X wrinkled male and wrinkled female X smooth male.

recombinant chromosome A chromosome that emerges from meiosis with a combination of genes different from a parental combination of genes.

recombinant DNA molecule A new type of DNA sequence that has been constructed or engineered in the test tube from two or more distinct DNA sequences.

recombinant DNA technology A collection of experimental procedures that allow molecular biologists to splice a DNA fragment from one organism into DNA from another organism and to clone the new recombinant DNA molecule. It includes the development and application of particular molecular techniques, such as biotechnology or genetic engineering. This technology is important, for example, in the production of antibiotics, hormones, and other medical agents used in the diagnosis and treatment of certain genetic diseases.

recombinants The individuals or cells that have nonparental combinations of genes as a result of the processes of genetic recombination.

recombination *See* **genetic recombination.**

regulated gene A gene whose activity is controlled in response to the needs of a cell or organism.

regulatory proteins Proteins active in the activation or repression of transcription of the gene.

relative rate In molecular evolution studies, a simple way to estimate the overall rate of substitution in different lineages that does not depend on specific knowledge of divergence times.

release factor *See* **termination factor.**

replica plating The procedure for transferring the pattern of colonies from a master plate to a new plate. In this procedure, a velveteen pad on a cylinder is pressed lightly onto the surface of the master plate, thereby picking up a few cells from each colony to inoculate onto the new plate.

replication bubble Opposing replication forks caused by the local denaturing of DNA during replication.

replication fork A Y-shaped structure formed when a double-stranded DNA molecule unwinds to expose the two single-stranded template strands for DNA replication.

replicon (replication unit) The stretch of DNA in eukaryotes from the origin of replication to the two termini of replication on each side of the origin.

repressible operon Operon for which gene activity is repressed when a chemical is added. The tryptophan operon is an example of a repressible operon.

repressor *See* **repressor gene.**

repressor gene A regulatory gene whose product is a protein that controls the transcriptional activity of a particular operon.

repressor molecule The protein product of a repressor gene.

repulsion An arrangement in which each homologous chromosome carries the wild-type allele of one gene and the mutant allele of the other one.

restriction endonuclease (restriction enzyme) An enzyme important for analyzing DNA and constructing recombinant DNA molecules because of its ability to cleave double-stranded DNA molecules at specific nucleotide pair sequences.

restriction enzyme *See* **restriction endonuclease.**

restriction fragment length polymorphisms (RFLPs) The different restriction maps that result from different patterns of distribution of restriction sites. They are detected by the presence of restriction fragments of different lengths on gels.

restriction map A genetic map of DNA showing the relative positions of restriction enzyme cleavage sites.

restriction site linker (linker) A short, double-stranded oligodeoxyribonucleotide, about 8 to 12 nucleotide pairs long, that is synthesized by chemical means and contains the cleavage site for a specific restriction enzyme within its sequence.

retrovirus A single-stranded RNA virus that replicates via a double-stranded DNA intermediate. The DNA integrates into the host's chromosome, where it can be transcribed.

reverse mutation (reversion) A mutational change from a mutant genotype back to a wild-type (or partially wild-type) genotype.

reverse transcriptase An enzyme (an RNA-dependent DNA polymerase) that makes a complementary DNA copy of an mRNA strand.

reversion *See* **reverse mutation.**

ribonuclease (RNase) An enzyme that catalyzes the degradation of RNA to nucleotides.

ribonucleic acid (RNA) A usually single-stranded polymeric molecule consisting of ribonucleotide building blocks. RNA is chemically very similar to DNA. The three major types of RNA in cells are ribosomal RNA (rRNA), transfer RNA (tRNA), and messenger RNA (mRNA), each of which performs an essential role in protein synthesis (translation). In some viruses, RNA is the genetic material.

ribonucleotide The basic building block of RNA, consisting of a sugar (ribose), a base, and a phosphate.

ribose The pentose sugar of the nucleotide building block of RNA.

ribosomal DNA (rDNA) The regions of the DNA that contain the genes for the rRNAs in prokaryotes and eukaryotes.

ribosomal protein One of the proteins that along with rRNA molecules make up the ribosomes of prokaryotes and eukaryotes.

ribosomal RNA (rRNA) The RNA molecules of discrete sizes that along with ribosomal proteins make up ribosomes of prokaryotes and eukaryotes.

ribosome A complex cellular particle composed of ribosomal protein and rRNA molecules that is the site of amino acid polymerization during protein synthesis.

ribozyme A self-cleaving RNA with enzyme activity that functions catalytically.

RNA *See* **ribonucleic acid.**

RNA editing Posttranscriptional insertion and deletion of nucleotides in an mRNA molecule.

RNA enzyme An RNA molecule that has catalytic function.

RNA polymerase An enzyme that catalyzes the synthesis of RNA molecules from a DNA template in a process called *transcription.*

RNA polymerase I An enzyme in eukaryotes located in the nucleolus. It catalyzes the transcription of the 18S, 5.8S, and 28S rRNA genes.

RNA polymerase II An enzyme in eukaryotes found only in the nucleoplasm of the nucleus. It catalyzes the transcription of mRNA-coding genes.

RNA polymerase III An enzyme in eukaryotes found only in the nucleoplasm. It catalyzes the transcription of the tRNA and 5S rRNA genes.

RNA primer A preexisting polynucleotide chain in DNA replication to which new nucleotides can be added.

RNA processing control The second level of control of gene expression in eukaryotes. This level involves regulating the production of mature RNA molecules from precursor RNA molecules.

RNA splicing *See* **mRNA splicing.**

RNA synthesis *See* **transcription.**

Robertsonian translocation A type of nonreciprocal translocation in which the long arms of two nonhomologous acrocentric chromosomes become attached to a single centromere.

rooted tree A phylogenetic tree in which one internal node is represented as a common ancestor to all the other nodes on the tree.

rRNA transcription units *See* **ribosomal DNA.**

sample The subset used to give information about a population. It must be of reasonable size and it must be a random subset of the larger group to provide accurate information about the population.

sampling error The phenomenon in which chance deviations from expected proportions arise in small samples.

secondary oocyte A large cell produced by the primary oocyte. In the ovaries of female animals, the diploid primary oocyte goes through meiosis I and unequal cytokinesis to produce two cells; the large cell is called the secondary oocyte.

segmentation gene In *Drosophila*, genes that determine the segments of the embryo and adult.

selection The favoring of particular combinations of genes in a given environment.

selection coefficient A measure of the relative intensity of selection against a genotype.

selection differential In natural and artificial selection, the difference between the mean phenotype of the selected parents and the mean phenotype of the population before selection.

selection response The amount by which a phenotype changes in one generation when selection is applied to a group of individuals.

self-fertilization (selfing) The union of male and female gametes from the same individual.

selfing *See* **self-fertilization.**

self-splicing The excision of introns from some precursor RNA molecules that occurs by a protein-independent reaction in some organisms.

semiconservative model A DNA replication scheme in which each daughter molecule retains one of the parental strands.

semidiscontinuous Concerning DNA replication, when one new strand is synthesized continuously and the other discontinuously.

sense codon An mRNA molecule that specifies an amino acid in the corresponding polypeptide (as opposed to a nonsense codon).

sequence tagged site (STS) A short segment of DNA that defines a unique position in the human genome; an STS usually is detected by the **polymerase chain reaction (PCR).**

sex chromosome A chromosome in eukaryotic organisms that is represented differently in the two sexes. In many organisms, one sex possesses a pair of visibly different chromosomes. One is an X chromosome, and the other is a Y chromosome. Commonly, the XX sex is female and the XY sex is male.

sex-influenced trait A trait that appears in both sexes, but either the frequency of occurrence in the two sexes is different or there is a different relationship between genotype and phenotype.

sex-limited trait A genetically controlled character that is phenotypically exhibited in only one of the two sexes.

sex-linked *See* **X-linked.**

sexual reproduction Reproduction involving the fusion of haploid gametes produced by meiosis.

Shine-Dalgarno sequence In prokaryotic mRNAs, a sequence upstream of the AUG initiation codon that is complementary to a short sequence at the 3′ end of the 16S rRNA found in the 30S ribosomal subunit. Formation of complementary base pairs between the mRNA and 16S rRNA allows the ribosome to locate the start codon for correct initiation of translation.

short tandem repeat *See* **STR.**

shuttle vector A cloning vector that can replicate in two or more host organisms. Shuttle vectors are used for experiments in which recombinant DNA is to be introduced into organisms other than *E. coli.*

signal hypothesis The hypothesis that the secretion of proteins from a cell occurs through the binding of a hydrophobic amino terminal extension to the membrane and the subsequent removal and degradation of the extension in the cisternal space of the endoplasmic reticulum.

signal peptidase An enzyme in the cisternal space of the endoplasmic reticulum that catalyzes removal of the signal sequence from the polypeptide.

signal recognition particle (SRP) In eukaryotes, a complex of a small RNA molecule with six proteins that can temporarily halt protein synthesis by recognizing the signal sequence of a nascent polypeptide destined to be translocated through the endoplasmic reticulum, binding to it and thereby blocking further translation of the mRNA.

signal sequence The hydrophobic amino terminal extension found on proteins that are secreted from a cell. The amino terminus (extension) is removed and degraded in the cisternal space of the endoplasmic reticulum.

silencer *See* **silencer element.**

silencer element In eukaryotes, a transcriptional regulatory element that decreases RNA transcription rather than stimulating it as enhancer elements do.

silent mutation A mutational change resulting in a protein with a wild-type function because of an unchanged amino acid sequence.

simple telomeric sequence A simple, tandemly repeated DNA sequence at or very close to the extreme end of a chromosomal DNA molecule.

SINEs (short interspersed repeated sequences) One class of interspersed and highly repeated sequences that consists of dispersed families with unit lengths of fewer than 500 base pairs and repeated for as many as hundreds of thousands of copies in the genome.

single nucleotide polymorphism (SNP) A DNA marker resulting from a base pair difference at one particular site in a genome. One base pair at a position showing a difference is the "common" one, present in most individuals, and another base pair at that position is a less common variant.

single-strand DNA-binding (SSB) proteins (helix-destabilizing proteins) Proteins that help the DNA unwinding process by stabilizing the single-stranded DNA.

sister chromatid A chromatid derived from replication of one chromosome during interphase of the cell cycle.

small nuclear ribonucleoprotein particles (snRNPs) The complex formed by small nuclear RNA and protein in which the processing of pre-mRNA molecules occurs.

small nuclear RNA (snRNA) Found only in eukaryotes, one of four major classes of RNA molecules produced by transcription. snRNAs are used in the processing of pre-mRNA molecules.

somatic mutation A mutation affecting only somatic cells (in multicellular organisms) and therefore affecting only the individual in which the mutation occurs (i.e., it is not passed on to the succeeding generation).

Southern blot technique A technique invented by E. M. Southern in which DNA fragments are transferred from a gel to a nitrocellulose filter, facilitating the analysis of genes and gene transcripts.

spacer sequence Any of the transcribed sequences found between and flanking coding RNA sequences. Spacer sequences are removed during processing of pre-rRNA and pre-tRNA to produce mature molecules.

specialized transduction A type of transduction in which only specific genes are transferred.

species tree A phylogenetic tree depicting the evolutionary history of a group of species, usually best obtained from analyses of multiple genes.

sperm cells (spermatozoa) The male gametes; the spermatozoa produced by the testes in male animals.

spermatogenesis Development of the male animal germ cell within the male gonad.

spliceosomes The splicing complexes formed by the association of several snRNPs bound to the pre-mRNA.

spontaneous mutation A mutation that occurs without the use of chemical or physical mutagenic agents.

sporophyte The diploid, asexual generation in the life cycle of plants that produces haploid spores by meiosis.

stamen The male reproductive organ in a flowering plant that usually consists of a stalk, called a filament, bearing a pollen-producing anther.

standard deviation The square root of the variance. It measures the extent to which each measurement in the data set differs from the mean value and is used as a measure of the extent of variability in a population.

steroid hormone response element (HRE) The DNA sequence to which steroid hormones bind to activate a gene.

stop codon *See* **chain-terminating codon.**

STR (short tandem repeat) A region in the genome where there are identical DNA segments two, three, or four base pairs long tandemly arranged head to tail. Also called a **microsatellite.**

structural gene A gene that codes for an mRNA molecule and hence for a polypeptide chain.

structural genomics The genetic mapping, physical mapping, and sequencing of entire genomes.

submetacentric chromosome A chromosome that has the centromere nearer one end than the other. Such chromosomes appear J-shaped at anaphase.

substitution A mutational change of one base pair for another.

sum rule The rule that the probability of either of two mutually exclusive events occurring is the sum of their individual probabilities.

supercoiled Describing DNA when the double helix has been twisted in space about its own axis.

suppressor gene A gene that causes suppression of a mutation in another gene.

suppressor mutation A mutation at a second site that totally or partially restores a function lost because of a primary mutation at another site.

synapsis The intimate association of homologous chromosomes brought about by the formation of a zipperlike structure along the length of the chromatids called the synaptonemal complex.

synaptonemal complex A complex structure spanning the region between meiotically paired (synapsed) chromosomes that is concerned with crossing-over rather than with chromosome pairing.

synonymous Referring to mutations at nucleotide positions that do not change the resulting amino acid sequence of a protein.

tandemly repeated DNA Repeated sequences in the genome that are clustered together, so that the sequence repeats many times in a row.

TATA box A part of the promoter sequence in prokaryotic genomes that is located at about ten base pairs upstream from the transcription starting point. The consensus sequence for the TATA box is TATAAT. The TATA box was originally named the Pribnow box.

TATA element *See* **Goldberg-Hogness box.**

tautomers Alternate chemical forms in which DNA (or RNA) bases are able to exist.

telocentric chromosome A chromosome that has the centromere more or less at one end.

telomere-associated sequence A repeated, complex DNA sequence extending from the molecular gene of chromosomal DNA, suspected to mediate a telomere-specific interaction.

telophase A stage in mitosis or meiosis during which the migration of the daughter chromosomes to the two poles is completed.

telophase II The last stage of meiosis II, during which a nuclear membrane forms around each set of chromosomes and cytokinesis takes place.

temperate phages Phages that have a choice between lytic and lysogenic pathways.

temperature-sensitive mutants Temperature-sensitive mutants function normally until the temperature is raised past some threshold level, at which time some temperature-sensitive defect is manifested.

template strand The unwound single strand of DNA on which new strands are made (following complementary base-pairing rules).

termination factor (release factor; RF) One of the specific proteins in polypeptide synthesis (translation) that read the chain termination codons and then initiate a series of specific events to terminate polypeptide synthesis.

terminator *See* **transcription terminator sequence.**

testcross A cross of an individual of unknown genotype, usually expressing the dominant phenotype, with a homozygous recessive individual to determine the unknown genotype.

testis-determining factor Gene product in placental mammals that sets the switch toward male sexual differentiation.

tetrad analysis Genetic analysis of all the products of a single meiotic event. Tetrad analysis is possible in organisms in which the four products of a single nucleus that has undergone meiosis are grouped together in a single structure.

tetrasomy The aberrant, aneuploid state in a normally diploid cell or organism in which an extra chromosome pair results in the presence of four copies of one chromosome type and two copies of every other chromosome type.

tetratype (T) One of the three types of tetrads possible when two genes are segregating in a cross. The T tetrad contains two parental and two recombinant nuclei, one of each parental type and one of each recombinant type.

three-point testcross A test involving three genes within a short section of the chromosome. It is used to map genes for their order in the chromosome and for the distance between them.

thymine (T) A pyrimidine base found in DNA but not in RNA. In double-stranded DNA, thymine pairs with adenine.

topoisomerase An enzyme that catalyzes the supercoiling of DNA.

totipotency The capacity of a nucleus to direct events through all the stages in development and therefore produce a normal adult.

totipotent Referring to a nucleus that directs events through all the stages in development to produce a normal adult.

trailer sequence The sequence of the mRNA molecule beginning at the end of the amino acid–coding sequence and ending at the 3′ end of the mRNA. The trailer sequence is not translated and varies in length from molecule to molecule.

trans-dominant The phenomenon of a gene or DNA sequence controlling genes that are on a different piece (strand) of DNA.

transconjugant In bacteria, a recipient inheriting donor DNA in the process of conjugation.

transcription The transfer of information from a double-stranded DNA molecule to a single-stranded RNA molecule; also called RNA synthesis.

transcription terminator sequence (terminator) A transcription regulatory sequence located at the distal end of a gene that signals the termination of transcription.

transcriptional control The first level of control of gene expression in eukaryotes. This level involves regulating whether a gene is to be transcribed and the rate at which transcripts are produced.

transcriptome The set of mRNA transcripts in a cell.

transducing phage The phage that is the vehicle by which genetic material is shuttled between bacteria.

transducing retrovirus A retrovirus that has picked up an oncogene from the cellular genome.

transductant In bacteria, a recipient inheriting donor DNA in the process of transduction.

transduction A process by which bacteriophages mediate the transfer of bacterial genetic information from one bacterium (the donor) to another (the recipient); a process whereby pieces of bacterial DNA are carried between bacterial strains by a phage.

transfer RNA (tRNA) One of the four classes of RNA molecules produced by transcription and involved in protein synthesis; molecules that bring amino acids to the ribosome, where they are matched to the transcribed message on the mRNA.

transformant The genetic recombinant generated by the transformation process.

transformation (a) A process in which genetic information is transferred by means of extracellular pieces of DNA in bacteria. (b) The failure of cells to remain constrained in their growth properties, giving rise to tumors.

transforming principle In Frederick Griffith's experiment on transformation, the unknown agent responsible for the change in genotype.

transgene A gene introduced into the genome of an organism by genetic manipulation to alter its genotype.

transgenic cell A cell or organism that has had its genotype altered by having a gene introduced into its genome by artificial means.

transgenic organism An organism that has had its genotype altered by the introduction of a gene into its genome by genetic manipulation.

transition mutation A specific type of base pair substitution mutation that involves a change in the DNA from one purine-pyrimidine base pair to the other purine-pyrimidine base pair at a particular site (e.g., A T to GC).

transitions *See* **transition mutation.**

translation (protein synthesis) The conversion in the cell of the mRNA base sequence information into an amino acid sequence of a polypeptide.

translational control The regulation of protein synthesis by ribosome synthesis among mRNAs.

translocation (a) A chromosomal mutation involving a change in position of a chromosome segment (or segments) and the gene sequences it contains. This process is also called transposition. (b) In polypeptide synthesis, the movement of the ribosome, one codon at a time, along the mRNA toward the 3′ end.

transmission genetics (classical genetics) A subdivision of the science of genetics dealing primarily with how genes are passed from one generation to another and how genes recombine.

transposable element A genetic element of chromosomes of both prokaryotes and eukaryotes that has the capacity to mobilize itself and move from one location to another in the genome.

transposase An enzyme encoded by the IS element of a transposon that catalyzes transposition activity of a transposable element.

transposition The movement of genetic material from one genomic location to another, usually through the action of a transposon or some other kind of transposable element.

transposon (Tn) A mobile DNA segment that contains genes for the insertion of the DNA segment into a chromosome and for mobilization of the element to other locations on the chromosomes.

transversion mutation A specific type of base pair substitution mutation that involves a change in the DNA from a purine-pyrimidine base pair to a pyrimidine-purine base pair at the same site (e.g., A T to T A or GC to T A).

transversions *See* **transversion mutation.**

trihybrid cross A cross between individuals of the same type that are heterozygous for three pairs of alleles at three different loci.

trisomy An aberrant, aneuploid state in a normally diploid cell or organism in which there are three copies of a particular chromosome instead of two copies.

trisomy-13 A human clinical condition characterized by various abnormalities. It is caused by the presence of an extra copy of chromosome 13; also called Patau syndrome.

trisomy-18 A human clinical condition characterized by various abnormalities. It is caused by the presence of an extra copy of chromosome 18; also called Edwards syndrome.

trisomy-21 A human clinical condition characterized by various abnormalities. It is caused by the presence of an extra copy of chromosome 21. Individuals with trisomy-21 have Down syndrome.

true reversion A point mutation from mutant back to wild type in which the change codes for the original amino acid of the wild type.

true-breeding (pure-breeding) strain A strain allowed to self-fertilize for many generations to ensure that the traits to be studied are inherited and unchanging.

tumor suppressor gene A gene in a normal cell that encodes a product that suppresses uncontrolled cell proliferation.

tumor virus A virus that induces cells to dedifferentiate and divide to produce a tumor.

Turner syndrome A human clinical syndrome that results from monosomy for the X chromosome in the female, which gives a 45,X female. These females fail to develop secondary sexual characteristics, tend to be short, have weblike necks, have poorly developed breasts, are usually infertile, and exhibit mental deficiencies.

uniparental inheritance A phenomenon, usually exhibited by extranuclear genes, in which all progeny have the phenotype of only one parent.

unique-sequence (single-copy) DNA A DNA sequence that has one to a few copies per genome.

unweighted-pair group method with arithmetic mean (UPGMA) A statistically based approach that groups taxa together on the basis of their overall pairwise similarities to each other; often called cluster analysis.

uracil (U) A pyrimidine base found in RNA but not in DNA.

variable number of tandem repeats *See* **VNTR.**

variance A statistical measure of how values vary from the mean.

variance of gene frequency The variance in the frequency of an allele among a group of populations.

viral oncogene A viral gene that transforms a cell it infects to a cancerous state. *See also* **cellular oncogene; oncogenesis.**

virulent phage A phage like T4, which always follows the lytic cycle when it infects bacteria.

visible mutation A mutation that affects the morphology or physical appearance of an organism.

VNTR (variable number of tandem repeats) A region in the genome where there are identical segments of DNA five to a few tens of base pairs long tandemly arranged head to tail. Also called a **minisatellite.**

wild type A strain, organism, or gene of the type that is designated as the standard for the organism with respect to genotype and phenotype.

wild-type allele The allele designated as the standard ("normal") for a strain of organism.

wobble hypothesis A theory proposed by Francis Crick that the base at the 5′ end of the anticodon (3′ end of the codon) is not as constrained as the other two bases. This feature allows less exact base pairing, so that the 5′ end of the anticodon can potentially pair with one of three different bases at the 3′ end of the codon.

X chromosome A sex chromosome present in two copies in the homogametic sex and in one copy in the heterogametic sex.

X chromosome–autosome balance system A genotypic sex determination system. The main factor in sex determination is the ratio between the numbers of X chromosomes and the number of sets of autosomes. Sex is determined at the time of fertilization, and sex differences are assumed to result from the action during development of two sets of genes located in the X chromosomes and in the autosomes.

X chromosome nondisjunction An event occurring when the two X chromosomes fail to separate in meiosis, producing eggs with two X chromosomes or with no X chromosomes instead of the usual one X chromosome.

X-linked Referring to genes located on the X chromosome.

X-linked dominant trait A trait caused by a dominant mutant gene carried on the X chromosome.

X-linked recessive trait A trait caused by a recessive mutant gene carried on the X chromosome.

Y chromosome A sex chromosome that when present is found in one copy in the heterogametic sex, along with an X chromosome, and is not present in the homogametic sex. Not all organisms with sex chromosomes have a Y chromosome.

Y chromosome mechanism of sex determination The system in which the Y chromosome determines the sex of an individual. Individuals with a Y chromosome are genetically male, and individuals without a Y chromosome are genetically female.

yeast artificial chromosome (YAC) A cloning vector in which DNA fragments several hundred kilobase pairs long can be cloned in yeast. A YAC is a linear vector with a yeast telomere at each end, a centromere, a sequence for autonomous replication in yeast, a selectable marker for yeast, and a polylinker.

Y-linked (holandric ["wholly male"]) trait A trait caused by a mutant gene carried on the Y chromosome but with no counterpart on the X.

zygonema The stage during meiosis in prophase I at which homologous chromosomes begin to pair in a highly specific way (like a zipper).

zygote The cell produced by the fusion of male and female gametes.

Cook, P. R. 1999. The organization of replication and transcription. *Science* 284:1790–1795.

Cramer, P., Bushnell, D. A., Fu, J., Gnatt, A. L., Maier-Davis, B., Thompson, N. E., Burgess, R. R., Edwards, A. M., David, P. R., and Kornberg, R. D. 2000. Architecture of RNA polymerase II and implications for the transcription mechanism. *Science* 288:640–649.

Crick, F. H. C. 1979. Split genes and RNA splicing. *Science* 204:264–271.

Eick, D., Wedel, A., and Heumann, H. 1994. From initiation to elongation: Comparison of transcription by prokaryotic and eukaryotic RNA polymerases. *Trends Genet.* 10:292–296.

Geiduschek, E. P., and Tocchini-Valentini, G. P. 1988. Transcription by RNA polymerase III. *Annu. Rev. Biochem.* 57:873–914.

Gelles, J., and Landick, R. 1998. RNA polymerase as a molecular motor. *Cell* 93:13–16.

Gesteland, R. F., and Atkins, J. F., eds. 1993. *The RNA World.* Cold Spring Harbor, NY: Cold Spring Harbor Laboratory Press.

Grabowski, P. J., Seiler, S. R., and Sharp, P. A. 1985. A multicomponent complex is involved in the splicing of messenger RNA precursors. *Cell* 42:355–367.

Green, M. R. 1986. Pre-mRNA splicing. *Annu. Rev. Genet.* 20:671–708.

———. 1991. Biochemical mechanisms of constitutive and regulated pre-mRNA splicing. *Annu. Rev. Cell Biol.* 7:559–599.

Guarente, L., and Birmingham-McDonogh, O. 1992. Conservation and evolution of transcriptional mechanisms in eukaryotes. *Trends Genet.* 8:27–32.

Guthrie, C. 1992. Messenger RNA splicing in yeast: Clues to why the spliceosome is a ribonucleoprotein. *Science* 253:157–163.

Hahn, S. 1998. The role of TAFs in RNA polymerase II transcription. *Cell* 95:579–582.

Halle, J.-P., and Meisterernst, M. 1996. Gene expression: Increasing evidence for a transcriptosome. *Trends Genet.* 12:161–163.

Hochschild, A., and Dove, S. L. 1998. Protein-protein contacts that activate and repress prokaryotic transcription. *Cell* 92:597–600.

Horowitz, D. S., and Krainer, A. R. 1994. Mechanisms for selecting 5′ splice sites in mammalian pre-mRNA splicing. *Trends Genet.* 10:100–105.

Jeffreys, A. J., and Flavell, R. A. 1977. The rabbit beta-globin gene contains a large insert in the coding sequence. *Cell* 12:1097–1108.

Katagiri, F., and Chua, N.-H. 1992. Plant transcription factors: Present knowledge and future challenges. *Trends Genet.* 8:22–27.

Korzheva, N., Mustaev, A., Kozlov, M., Malhotra, A., Nikiforov, V., Goldfarb, A., and Darst, S. A. 2000. A structural model of transcription elongation. *Science* 289:619–625.

Lang, W. H., Morrow, B. E., Ju, Q., Warner, J. R., and Reeder, R. H. 1994. A model for transcription termination by RNA polymerase I. *Cell* 79:527–534.

Maquat, L. E., and Carmichael, G. G. 2001. Quality control of mRNA function. *Cell* 104:173–176.

Marmur, J., Greenspan, C. M., Palecek, E., Kahan, F. M., Levine, J., and Mandel, M. 1963. Specificity of the complementary RNA formed by *Bacillus subtilis* infected with bacteriophage SP8. *Cold Spring Harbor Symp. Quant. Biol.* 28:191–199.

Nilsen, T. W. 1994. RNA-RNA interactions in the spliceosome: Unraveling the ties that bind. *Cell* 78:1–4.

Nomura, M. 1973. Assembly of bacterial ribosomes. *Science* 179:864–873.

Nomura, M., Morgan, E. A., and Jaskunas, S. R. 1977. Genetics of bacterial ribosomes. *Annu. Rev. Genet.* 11:297–347.

O'Hare, K. 1995. mRNA 3′ ends in focus. *Trends Genet.* 11:253–257.

Orphanides, G., and Reinberg, D. 2000. RNA polymerase II elongation through chromatin. *Nature* 407:471–475.

Pabo, C. O., and Sauer, R. T. 1992. Transcription factors: Structural families and principles of DNA recognition. *Annu. Rev. Biochem.* 61:1053–1093.

Padgett, R. A., Grabowski, P. J., Konarska, M. M., and Sharp, P. A. 1985. Splicing messenger RNA precursors: Branch sites and lariat RNAs. *Trends Biochem. Sci.* 10:154–157.

Proudfoot, N. 2000. Connecting transcription to messenger RNA processing. *Trends Biochem. Sci.* 25:290–293.

Roeder, R. G. 1991. The complexities of eukaryotic transcription initiation: Regulation of preinitiation complex assembly. *Trends Biochem. Sci.* 16:402–408.

Rogers, J. H. 1989. How were introns inserted into nuclear genes? *Trends Genet.* 5:213–216.

Schmidt, F. J. 1985. RNA splicing in prokaryotes: Bacteriophage T4 leads the way. *Cell* 41:339–340.

Sharp, P. A. 1994. Split genes and RNA splicing. Nobel lecture. *Cell* 77:805–815.

Sollner-Webb, B. 1988. Surprises in polymerase III transcription. *Cell* 52:153–154.

Srivastava, A. K., and Schlessinger, D. 1990. Mechanism and regulation of bacterial ribosomal RNA processing. *Annu. Rev. Microbiol.* 44:105–129.

Stragier, P. 1991. Dances with sigmas. *EMBO J.* 10:3559–3566.

Struhl, K. 1996. Chromatin structure and RNA polymerase II connection: Implications for transcription. *Cell* 84:179–182.

Symons, R. H. 1992. Small catalytic RNAs. *Annu. Rev. Biochem.* 61:641–671.

Thompson, C. C., and McKnight, S. L. 1992. Anatomy of an enhancer. *Trends Genet.* 8:232–236.

Tilghman, S. M., Curis, P. J., Tiemeier, D. C., Leder, P., and Weissman, C. 1978. The intervening sequence of a mouse β-globin gene is transcribed within the 15S β-globin mRNA precursor. *Proc. Natl. Acad. Sci. USA* 75:1309–1313.

Tilghman, S. M., Tiemeier, D. C., Seidman, J. G., Peterlin, B. M., Sullivan, M., Maizel, J. V., and Leder, P. 1978. Intervening sequence of DNA identified in the structural portion of a mouse beta-globin gene. *Proc. Natl. Acad. Sci. USA* 78:725–729.

Wahle, E., and Keller, W. 1992. The biochemistry of 3′-end cleavage and polyadenylation of messenger RNA precursors. *Annu. Rev. Biochem.* 61:419–440.

Weinstock, R., Sweet, R., Weiss, M., Cedar, H., and Axel, R. 1978. Intragenic DNA spacers interrupt the ovalbumin gene. *Proc. Natl. Acad. Sci. USA* 75:1299–1303.

Weis, L., and Reinberg, D. 1992. Transcription by RNA polymerase II: Initiator-directed formation of transcription-competent complexes. *FASEB J.* 6:3300–3309.

White, R. J., and Jackson, S. P. 1992. The TATA-binding protein: A central role in transcription by RNA polymerases I, II, and III. *Trends Genet.* 8:284–288.

Wolffe, 1994. Transcription: In tune with the histones. *Cell* 77:13–16.

Zaug, A. J., and Cech, T. R. 1986. The intervening sequence RNA of *Tetrahymena* is an enzyme. *Science* 231:470–475.

Chapter 12: Gene Expression: Translation

Ban, N., Nissen, P., Hansen, J., Moore, P. B., and Steitz, T. A. 2000. The complete atomic structure of the large ribosomal subunit at 2.4Å resolution. *Science* 289:905–920.

Blobel, G., and Dobberstein, B. 1975. Transfer of proteins across membranes. I. Presence of proteolytically processed and unprocessed nascent immunoglobulin light chains on membrane-bound ribosomes of murine myeloma. *J. Cell Biol.* 67:835–851.

Brenner, S., Jacob, F., and Meselson, M. 1961. An unstable intermediate carrying information from genes to ribosomes for protein synthesis. *Nature* 190:576–581.

Carter, A. P., Clemons, W. M., Brodersen, D. E., Morgan-Warren, R. J., Wimberly, B. T., and Ramakrishnan, V. 2000. Functional insights from the structure of the 30S ribosomal subunit and its interactions with antibiotics. *Nature* 407:340–348.

Crick, F. H. C. 1966. Codon-anticodon pairing: The wobble hypothesis. *J. Mol. Biol.* 19:548–555.

Crick, F. H. C., Barnett, L., Brenner, S., and Watts-Tobin, R. J. 1961. General nature of the genetic code for proteins. *Nature* 192:1227–1232.

Garen, A. 1968. Sense and nonsense in the genetic code. *Science* 160:149–159.

Horowitz, S., and Gorovsky, M. A. 1985. An unusual genetic code in nuclear genes of *Tetrahymena*. *Proc. Natl. Acad. Sci. USA* 82:2452–2455.

Jackson, R. J., and Standart, N. 1990. Do the poly(A) tail and 3′ untranslated region control mRNA translation? *Cell* 62:15–24.

Khorana, H. G. 1966–67. Polynucleotide synthesis and the genetic code. *Harvey Lect.* 62:79–105.

Kozak, M. 1983. Comparison of initiation of protein synthesis in procaryotes, eucaryotes, and organelles. *Microbiol. Rev.* 47:145.

———. 1989. Context effects and inefficient initiation at non-AUG codons in eukaryotic cell-free translation systems. *Mol. Cell. Biol.* 9:5073–5080.

McCarthy, J. E. G., and Brimacombe, R. 1994. Prokaryotic translation: The interactive pathway leading to initiation. *Trends Genet.* 10:402–407.

Meyer, D. I. 1982. The signal hypothesis: A working model. *Trends Biochem. Sci.* 7:320–321.

Morgan, A. R., Wells, R. D., and Khorana, H. G. 1966. Studies on polynucleotides. LIX. Further codon assignments from amino acid incorporation directed by ribopolynucleotides containing repeating trinucleotide sequences. *Proc. Natl. Acad. Sci. USA* 56:1899–1906.

Nierhaus, K. H. 1990. The allosteric three-site model for the ribosomal elongation cycle: Features and future. *Biochemistry* 29:4997–5008.

Nirenberg, M., and Leder, P. 1964. RNA code words and protein synthesis. *Science* 145:1399–1407.

Nirenberg, M., and Matthaei, J. H. 1961. The dependence of cell-free protein synthesis in *E. coli* upon naturally occurring or synthetic polyribonucleotides. *Proc. Natl. Acad. Sci. USA* 47:1588–1602.

Nissen, P., Hansen, J., Ban, N., Moore, P. B., and Steitz, T. A. 2000. The structural basis of ribosome activity in peptide bond formation. *Science* 289:920–930.

Noller, H. F., Hoffarth, V., and Zimniak, L. 1992. Unusual resistance of peptidyl transferase to protein extraction procedures. *Science* 256:1416–1419.

Pfeffer, S. R., and Rothman, J. E. 1987. Biosynthetic protein transport and sorting by the endoplasmic reticulum and Golgi. *Annu. Rev. Biochem.* 56:829–852.

Rogers, S., Wells, R., and Rechsteiner, M. 1986. Amino acid sequences common to rapidly degraded proteins: The PEST hypothesis. *Science* 234:364–368.

Ryan, K. R., and Jensen, R. E. 1995. Protein translocation across mitochondrial membranes: What a long, strange trip it is. *Cell* 83:517–519.

Schekman, R. 1985. Protein localization and membrane traffic in yeast. *Annu. Rev. Cell Biol.* 1:115–143.

Schnell, D. J. 1995. Shedding light on the chloroplast protein import machinery. *Cell* 83:521–524.

Shine, J., and Dalgarno, L. 1974. The 3′-terminal sequence of *Escherichia coli* 16S ribosomal RNA: Complementarity to nonsense triplet and ribosome binding sites. *Proc. Natl. Acad. Sci. USA* 71:1342–1346.

Silver, P. A. 1991. How proteins enter the nucleus. *Cell* 64:489–497.

Watson, J. D. 1963. The involvement of RNA in the synthesis of proteins. *Science* 140:17–26.

Wimberly, B. T., Brodersen, D. E., Clemons, W. M., Morgan-Warren, R. J., Carter, A. P., Vonrhein, C., Hartsch, T., and Ramakrishnan, V. 2000. Structure of the 30S ribosomal subunit. *Nature* 407:327–339.

Zheng, N., and Gierasch, L. M. 1996. Signal sequences: The same yet different. *Cell* 86:849–852.

Chapter 13: Recombinant DNA Cloning Technology

Arber, W. 1965. Host-controlled modification of bacteriophage. *Annu. Rev. Microbiol.* 19:365–378.

Arber, W., and Dussoix, D. 1962. Host specificity of DNA produced by *Escherichia coli*. I. Host controlled modification of bacteriophage lambda. *J. Mol. Biol.* 5:18–36.

Boyer, H. W. 1971. DNA restriction and modification mechanisms in bacteria. *Annu. Rev. Microbiol.* 25:153–176.

Danna, K., and Nathans, D. 1971. Specific cleavage of simian virus 40 DNA by restriction endonuclease of *Haemophilus influenzae*. *Proc. Natl. Acad. Sci. USA* 68:2913–2917.

Erlich, H. A., and Arnheim, N. 1992. Genetic analysis using the polymerase chain reaction. *Annu. Rev. Genet.* 26:479–506.

Feinberg, A. P., and Vogelstein, B. 1983. A technique for radiolabeling DNA restriction endonuclease fragments to high specific activity. *Anal. Biochem.* 132:6–13.

———. 1984. Addendum: A technique for radiolabeling DNA restriction endonuclease fragments to high specific activity. *Anal. Biochem.* 137:266–267.

Luria, S. E. 1953. Host-induced modification of viruses. *Cold Spring Harbor Symp. Quant. Biol.* 18:237–244.

Maxam, A. M., and Gilbert, W. 1977. A new method for sequencing DNA. *Proc. Natl. Acad. Sci. USA* 74:560–564.

Mullis, K. B. 1990. The unusual origin of the polymerase chain reaction. *Sci. Am.* 262 (April):56–65.

Sanger, F., and Coulson, A. R. 1975. A rapid method for determining sequences in DNA by primed synthesis with DNA polymerase. *J. Mol. Biol.* 94:441–448.

Southern, E. M. 1975. Detection of specific sequences among DNA fragments separated by gel electrophoresis. *J. Mol. Biol.* 98:503–517.

Watson, J. D., Gilman, M., Witkowski, J., and Zoller, M. 1992. *Recombinant DNA*, 2nd ed. New York: Scientific American Books, Freeman.

White, T. J., Arnheim, N., and Erlich, H. A. 1989. The polymerase chain reaction. *Trends Genet.* 5:185–188.

Chapter 14: Applications of Recombinant DNA Technology

Anderson, W. F. 1992. Human gene therapy. *Science* 256:808–813.

Cavazzana-Calvo, M., Havein-Bey, S., de Saint Basile, G., Gross, F., Yvon, E., Nusbaum, P., Selz, F., Hu, C., Certain, S., Casanova, J.-L., Bousso, P., Le Deist, F., and Fischer, A. 2000. Gene therapy of human severe combined immunodeficiency (SCID)-X1 disease. *Science* 288:669–672.

Chien, C.-T., Bartel, P. L., Sternglanz, R., and Fields, S. 1991. The two-hybrid system: A method to identify and clone genes for proteins that interact with a protein of interest. *Proc. Natl. Acad. Sci. USA* 88:9578–9582.

Collins, F. 1992. Cystic fibrosis: Molecular biology and therapeutic implications. *Science* 256:774–779.

Culver, K. V., and Blaese, R. M. 1994. Gene therapy for cancer. *Trends Genet.* 10:174–178.

Eisenstein, B. I. 1990. The polymerase chain reaction: A new method of using molecular genetics for medical diagnosis. *N. Engl. J. Med.* 322:178–183.

Fields, S., and Sternglanz, R. 1994. The two-hybrid system: An assay for protein-protein interactions. *Trends Genet.* 10:286–292.

Geisbrecht, B. V., Collins, C. S., Reuber, B. E., and Gould, S. J. 1998. Disruption of a PEX1-PEX6 interaction is the most common cause of the neurological disorders Zellweger syndrome, neonatal adrenoleukodystrophy, and infantile Refsum disease. *Proc. Natl. Acad. Sci. USA* 95:8630–8635.

Gilliam, T. C., Tanzi, R. E., Haines, J. L., Bonner, T. I., Faryniarz, A. G., Hobbs, W. J., MacDonald, M. E., Cheng, S. V., Folstein, S. E., Conneally, P. M., Wexler, N. S., and Gusella, J. F. 1987. Localization of the Huntington's disease gene to a small segment of chromosome 4 flanked by *D4S10* and the telomere. *Cell* 50:565–571.

Green, E. D., and Olson, M. V. 1990. Chromosomal region of the cystic fibrosis gene in yeast artificial chromosomes: A model for human genome mapping. *Science* 250:94–98.

Harris, J. D., and Lemoine, N. R. 1996. Strategies for targeted gene therapy. *Trends Genet.* 12:400–405.

Huntington's Disease Collaborative Research Group. 1993. A novel gene containing a trinucleotide repeat that is expanded and unstable on Huntington's disease chromosomes. *Cell* 72:971–983.

Johnston, M., Flick, J. S., and Pexton, T. 1994. Multiple mechanisms provide rapid and stringent glucose repression of *GAL* gene expression in *Saccharomyces cerevisiae*. *Mol. Cell. Biol.* 14:3834–3841.

Kay, M. A., and Woo, S. L. C. 1994. Gene therapy for metabolic disorders. *Trends Genet.* 10:253–257.

Kerem, B.-S., Rommens, J. M., Buchanan, J. A., Markiewicz, D., Cox, T. K., Chakravarti, A., Buchwald, M., and Tsui, L.-C. 1989. Identification of the cystic fibrosis gene: Genetic analysis. *Science* 245:1073–1080.

Klee, H., Horsch, R., and Rogers, S. 1987. Agrobacterium-mediated plant transformation and its further applications to plant biology. *Annu. Rev. Plant Physiol.* 38:467–486.

Knowlton, R. G., Cohen-Haguenauer, O., Van Cong, N., Frézal, J., Brown, V. A., Barker, D., Braman, J. C., Schumm, J. W., Tsui, L.-C., Buchwald, M., and Donis-Keller, H. 1985. A polymorphic DNA marker linked to cystic fibrosis is located on chromosome 7. *Nature* 318:380–385.

Koenig, M., Hoffman, E. P., Bertelson, C. J., Monaco, A. P., Feener, C., and Kunkel, L. M. 1987. Complete cloning of the Duchenne muscular dystrophy (DMD) cDNA and preliminary genomic organization of the DMD gene in normal and affected individuals. *Cell* 50:509–517.

Krings, M., Stone, A., Schmitz, R. W., Krainitzki, H., Stoneking, M., and Pääbo, S. 1997. Neanderthal DNA sequences and the origin of modern humans. *Cell* 90:19–30.

Morgan, R. A., and Anderson, W. F. 1993. Human gene therapy. *Annu. Rev. Biochem.* 62:191–217.

Mulligan, R. C. 1993. The basic science of gene therapy. *Science* 260:926–932.

Pääbo, S. 1993. Ancient DNA. *Sci. Am.* 269 (November):86–92.

Riordan, J. R., Rommens, J. M., Kerem, B., Alon, N., Rozmahel, R., Grzelczak, Z., Zielenski, J., Lok, S., Plavsic, N., Chou, J. L., Drumm, M. L., Ianuzzi, M. C., Collins, F. S., and Tsui, L.-C. 1989. Identification of the cystic fibrosis gene: Cloning and characterization of complementary DNA. *Science* 245:1066–1073.

Rommens, J. M., Ianuzzi, M. C., Kerem, B., Drumm, M. L., Melmer, G., Dean, M., Rozmahel, R., Cole, J. L., Kennedy, D., Hidaka, N., Zsiga, M., Buchwald, M., Riordan, J. R., Tsui, L.-C., and Collins, F. S. 1989. Identification of the cystic fibrosis gene: Chromosome walking and jumping. *Science* 245:1059–1065.

Rozsa, F. W., Shimizu, S., Lichter, P. R., Johnson, A. T., Othman, M. I., Scott, K., Downs, C. A., Nguyen, T. D., Polansky, J., and Richards, J. E. 1998. *GLC1A* mutations point to regions of potential functional importance on the TIGR/MYOC protein. *Mol. Vis.* 4:20.

Ryner, L. C., Goodwin, S. F., Castrillon, D. H., Anand, A., Villella, A., Baker, B. S., Hall, J. C., Taylor, B. J., and Wasserman, S. A. 1996. Control of male sexual behavior and sexual orientation in *Drosophila* by the *fruitless* gene. *Cell* 87:1079–1089.

Stafford, H. A. 2000. Crown gall disease and *Agrobacterium tumefaciens*: A study of the history, present knowledge, missing information, and impact on molecular genetics. *Botanical Rev.* 66:99–118.

Wicking, C., and Williamson, B. 1991. From linked marker to gene. *Trends Genet.* 7:288–293.

Wolfenbarger, L. L., and Phifer, P. R. 2000. The ecological risks and benefits of genetically engineered plants. *Science* 290:2088–2093.

Chapter 15: Genome Analysis

Adams, M. D., et al. 2000. The genome sequence of *Drosophila melanogaster*. *Science* 287:2185–2215.

Allzadeh, A. A., et al. 2000. Distinct types of diffuse large B-cell lymphoma identified by gene expression profiling. *Nature* 403:503–511.

Arabidopsis Genome Initiative. 2000. Analysis of the genome sequence of the flowering plant *Arabidopsis thaliana*. *Nature* 408:796–815.

Bevan, M., and Murphy, G. 1999. The small, the large and the wild. The value of comparison in plant genomics. *Trends Genet.* 15:211–214.

Blattner, F. R., et al. 1997. The complete genome sequence of *Escherichia coli* K-12. *Science* 277:1453–1463.

Bult, C. J., et al. 1996. Complete genome sequence of the methanogenic archaeon, *Methanococcus jannaschii*. *Science* 273:1058–1073.

Chipping Forecast, the. 1999. *Nat. Genet.* 21(suppl):1–60.

Cho, R. J., Campbell, M. J., Winzeler, E. A., Steinmetz, L., Conway, A., Wodicka, L., Wolfsberg, T. G., Gabriellan, A. E., Landsman, D., Lockhart, D. J., and Davis, R. W. 1998. A genome-wide transcriptional analysis of the mitotic cell cycle. *Mol. Cell* 2:65–73.

Chu, S., DeRisi, J., Eisen, M., Mulholland, J., Botstein, D., Brown, P. O., and Herskowitz, I. 1998. The transcriptional program of sporulation in budding yeast. *Science* 282:699–705.

Davies, K. 2001. After the genome: DNA and human disease. *Cell* 104:465–467.

DeRisi, J. L., Iyer, V. R., and Brown, P. O. 1997. Exploring the metabolic and genetic control of gene expression on a genomic scale. *Science* 278:680–686.

Dib, C., Fauré, S., Fizames, C., Samson, D., Drouot, N., Vignal, A., Millasseau, P., Marc, S., Hazan, J., Seboun, E., Lathrop, M., Gyapay, G., Morissette, J., and Weissenbach, J. 1996. A comprehensive genetic map of the human genome based on 5,264 microsatellites. *Nature* 380:152–154.

Dujon, B. 1996. The yeast genome project: What did we learn? *Trends Genet.* 12:263–270.

Fleischmann, R. D., et al. Whole-genome random sequencing and assembly of *Haemophilus influenzae* Rd. *Science* 269:496–512.

Fraser, C. M., et al. 1995. The minimal gene complement of *Mycoplasma genitalium*. *Science* 270:397–403.

Gavin, A.-C., et al. 2002. Functional organization of the yeast proteome by systematic analysis of protein complexes. *Nature* 415:141–147.

Goff, S. A., et al. 2002. A draft sequence of the rice genome (*Oryza sativa* L. ssp. *japonica*). *Science* 296:92–104.

Goffeau, A., et al. 1996. Life with 6000 genes. *Science* 274:546–567.

Kornberg, T. B., and Krasnow, M. A. 2000. The *Drosophila* genome sequence: Implications for biology and medicine. *Science* 287:2218–2220.

Nature 15 February 2001: An issue with a special section on "The Human Genome," an analysis of the draft sequence of the human genome.

Ried, T., Baldini, A., Rand, T. C., and Ward, D. C. 1992. Simultaneous visualization of seven different DNA probes by *in situ* hybridization using combinatorial fluorescence and digital imaging microscopy. *Proc. Natl. Acad. Sci. USA* 89:1388–1392.

Rubin, G. M., and Lewis, E. B. 2000. A brief history of *Drosophila*'s contributions to genome research. *Science* 287:2216–2218.

Science 16 February 2001: An issue focused on "The Human Genome," an analysis of the draft sequence of the human genome.

Smith, V., Botstein, D., and Brown, P. O. 1995. Genetic footprinting: A genomic strategy for determining a gene's function given its sequence. *Proc. Natl. Acad. Sci. USA* 92:6479–6483.

Tang, C. M., Hood, D. W., and Moxon, E. R. 1997. *Haemophilus* influence: The impact of whole genome sequencing on microbiology. *Trends Genet.* 13:399–404.

The *C. elegans* Sequencing Consortium. 1998. Genome sequence of the nematode *C. elegans*: A platform for investigating biology. *Science* 282:2012–2018.

White, R., and Lalouel, J.-M. 1988. Chromosome mapping with DNA markers. *Sci. Am.* 258 (February):40–48.

Young, R. A. 2000. Biomedical discovery with DNA arrays. *Cell* 102:9–15.

Yu, J., et al. 2002. A draft sequence of the rice genome (*Oryza sativa* L. ssp. *indica*). *Science* 296:79–92.

Chapter 16: Regulation of Gene Expression in Bacteria and Bacteriophages

Bell, C. E., Frescura, P., Hochschild, A., and Lewis, M. 2000. Crystal structure of the λ repressor C-terminal domain provides a model for cooperative operator binding. *Cell* 101:801–811.

Bertrand, K., Korn, L., Lee, F., Platt, T., Squires, C. L., Squires, C., and Yanofsky, C. 1975. New features of the structure and regulation of the tryptophan operon of *Escherichia coli*. *Science* 189:22–26.

Bertrand, K., and Yanofsky, C. 1976. Regulation of transcription termination in the leader region of the tryptophan operon of *Escherichia coli* involves tryptophan as its metabolic product. *J. Mol. Biol.* 103:339–349.

Fisher, R. F., Das, A., Kolter, R., Winkler, M. E., and Yanofsky, C. 1985. Analysis of the requirements for transcription pausing in the tryptophan operon. *J. Mol. Biol.* 182:397–409.

Gilbert, W., and Muller-Hill, B. 1966. Isolation of the *lac* repressor. *Proc. Natl. Acad. Sci. USA* 56:1891–1898.

Jacob, F., and Monod, J. 1961. Genetic regulatory mechanisms in the synthesis of proteins. *J. Mol. Biol.* 3:318–356.

Lee, F., and Yanofsky, C. 1977. Transcription termination at the *trp* operon attenuators of *Escherichia coli* and *Salmonella typhimurium*: RNA secondary structure and regulation of termination. *Proc. Natl. Acad. Sci. USA* 74:4365–4369.

Lewis, M., Chang, G., Horton, N. C., Kercher, M. A., Pace, H. C., Schumacher, M. A., Brennan, R. G., and Lu, P. 1996. Crystal structure of the lactose operon repressor and its complexes with DNA and inducer. *Science* 271:1247–1254.

Matthews, K. S. 1996. The whole lactose repressor. *Science* 271:1245–1246.

Niu, W., Kim, Y., Tau, G., Heyduk, T., and Ebright, R. H. 1996. Transcription activation at class II CAP-dependent promoters: Two interactions between CAP and RNA polymerase. *Cell* 87:1123–1134.

Ptashne, M. 1967. Isolation of the λ phage repressor. *Proc. Natl. Acad. Sci. USA* 57:306–313.

———. 1984. Repressors. *Trends Biochem. Sci.* 9:142–145.

———. 1992. *A Genetic Switch*, 2nd ed. Oxford: Cell Press and Blackwell Scientific Publications.

Ptashne, M., and Gilbert, W. 1970. Genetic repressors. *Sci. Am.* 222 (June):36–44.

Yanofsky, C. 1981. Attenuation in the control of expression of bacterial operons. *Nature* 289:751–758.

———. 1987. Operon-specific control by transcription attenuation. *Trends Genet.* 3:356–360.

Yanofsky, C., and Kolter, R. 1982. Attenuation in amino acid biosynthetic operons. *Annu. Rev. Genet.* 16:113–134.

Chapter 17: Regulation of Gene Expression in Eukaryotes

Bachvarova, R. F. 1992. A maternal tail of poly(A): The long and the short of it. *Cell* 69:895–897.

Beachy, P. A. 1990. A molecular view of the *Ultrabithorax* homeotic gene of *Drosophila*. *Trends Genet.* 6:46–51.

Beato, M., Herrlich, P., and Schützm, G. 1995. Steroid hormone receptors: Many actors in search of a plot. *Cell* 83:851–857.

Beelman, C. A., and Parker, R. 1995. Degradation of mRNA in eukaryotes. *Cell* 81:179–182.

Carpousis, A. J., Vanzo, N. F., and Raynal, L. C. 1999. mRNA degradation. A tale of poly(A) and multiprotein machines. *Trends Genet.* 15:24–28.

Cedar, H. 1988. DNA methylation and gene activity. *Cell* 53:3–4.

Chen, C.-Y. A., and Shyu, A.-B. 1995. AU-rich elements: Characterization and importance of mRNA degradation. *Trends Biochem. Sci.* 20:465–470.

Davis, M. M., Calame, K., Early, P. W., Livant, D. L., Joho, R., Weissman, I. L., and Hood, L. 1980. An immunoglobulin heavy chain gene is formed by at least two recombinational events. *Nature* 283:733–739.

Davis, M. M., Kim, S. K., and Hood, L. 1980. Immunoglobulin class switching: Developmentally regulated DNA rearrangements during differentiation. *Cell* 22:1–2.

Efstratiadis, A., Posakony, J. W., Maniatis, T., Lawn, R. M., O'Connell, C., Spritz, R. A., DeRiel, J. K., Forget, B. G., Weissman, S. M., Slighton, J. L., Blechtl, A. E., Smithies, O., Baralle, F. E., Shoulders, C. C., and Proudfoot, N. J. 1980. The structure and evolution of the human β-globin gene family. *Cell* 21:653–668.

Gasser, S. M. 2001. Positions of potential: Nuclear organization and gene expression. *Cell* 104:639–642.

Gellert, M. 1992. V(D)J recombination gets a break. *Trends Genet.* 8:408–412.

Green, M. R. 1989. Pre-mRNA processing and mRNA nuclear export. *Curr. Opin. Cell Biol.* 1:519–525.

Gross, D. S., and Garrard, W. T. 1987. Poising chromatin for transcription. *Trends Biochem. Sci.* 12:293–297.

———. 1988. Nuclease hypersensitive sites in chromatin. *Annu. Rev. Biochem.* 57:159–197.

Grunstein, M. 1992. Histones as regulators of genes. *Sci. Am.* 267 (October):68–74B.

Gurdon, J. B. 1968. Transplanted nuclei and cell differentiation. *Sci. Am.* 219 (December):24–35.

Gurdon, J. B., Laskey, R. A., and Reeves, R. 1975. The developmental capacity of nuclei transplanted from keratinized skin cells of adult frogs. *J. Embryol. Exp. Morph.* 34:93–112.

Hanna-Rose, W., and Hansen, U. 1996. Active repression mechanisms of eukaryotic transcription repressors. *Trends Genet.* 12:229–234.

Hochstrasser, M. 1996. Protein degradation or regulation: Ub the judge. *Cell* 84:813–815.

Holstege, F. C. P., Jennings, E. G., Wyrick, J. J., Lee, T. I., Hengartner, C. J., Green, M. R., Golub, T. R., Lander, E. S., and Young, R. A. 1998. Dissecting the regulatory circuitry of a eukaryotic genome. *Cell* 95:717–728.

Johnston, M., Flick, J. S., and Pexton, T. 1994. Multiple mechanisms provide rapid and stringent repression of *GAL* gene expression in *Saccharomyces cerevisiae*. *Mol. Cell. Biol.* 14:3834–3841.

Jones, P. A. 1999. The DNA methylation paradox. *Trends Genet.* 15:34–37.

Karlsson, S., and Nienhuis, A. W. 1985. Development regulation of human globin genes. *Annu. Rev. Biochem.* 54:1071–1078.

Kornberg, R. D. 1999. Eukaryotic transcriptional control. *Trends Genet.* 15:M46–M49.

Landschulz, W. H., Johnson, P. F., and McKnight, S. L. 1988. The leucine zipper: A hypothetical structure common to a new class of DNA binding protein. *Science* 240:1759–1763.

Lucas, P. C., and Granner, D. K. 1992. Hormone response domains in gene transcription. *Annu. Rev. Biochem.* 61:1131–1173.

O'Malley, B. W., and Schrader, W. T. 1976. The receptors of steroid hormones. *Sci. Am.* 234 (February):32–43.

Pabo, C. O., and Sauer, R. T. 1992. Transcription factors: Structural families and principles of DNA recognition. *Annu. Rev. Biochem.* 61:1053–1093.

Pankratz, M. J., and Jäckle, H. 1990. Making stripes in the *Drosophila* embryo. *Trends Genet.* 6:287–292.

Paranjape, S. M., Kamakaka, R. T., and Kadonaga, J. T. 1994. Role of chromatin structure in the regulation of transcription by RNA polymerase II. *Annu. Rev. Biochem.* 63:265–297.

Parthun, M. R., and Jaehning, J. A. 1992. A transcriptionally active form of GAL4 is phosphorylated and associated with GAL80. *Mol. Cell. Biol.* 12:4981–4987.

Ptashne, M. 1989. How gene activators work. *Sci. Am.* 243 (January):41–47.

Rhodes, D., and Klug, A. 1993. Zinc fingers. *Sci. Am.* 259 (February):56–65.

Rivera-Pomar, R., and Jäckle, H. 1996. From gradients to stripes in *Drosophila* embryogenesis: Filling in the gaps. *Trends Genet.* 12:478–483.

Ross, J. 1996. Control of messenger RNA stability in higher eukaryotes. *Trends Genet.* 12:171–175.

Scott, M. P., Tamkun, J. W., and Hartzell III, G. W. 1989. The structure and function of the homeodomain. *Biochim. Biophys. Acta* 989:25–48.

Struhl, K. 1999. Fundamentally different logic of gene regulation in eukaryotes and prokaryotes. *Cell* 98:104.

Studitsky, V. M., Clark, D. J., and Felsenfeld, G. 1994. A histone octamer can step around a transcribing polymerase without leaving the template. *Cell* 76:371–382.

Tsai, M.-J., and O'Malley, B. W. 1994. Molecular mechanisms of action of steroid/thyroid receptor superfamily members. *Annu. Rev. Biochem.* 63:451–486.

Varshavsky, A. 1996. The N-end rule: Functions, mysteries, uses. *Proc. Natl. Acad. Sci. USA* 93:12142–12149.

Verdine, G. L. 1994. The flip side of DNA methylation. *Cell* 76:197–200.

Wang, T. Y., Kostraba, N. C., and Newman, R. S. 1976. Selective transcription of DNA mediated by nonhistone proteins. *Prog. Nucleic Acid Res. Mol. Biol.* 19:447–462.

Wilmut, I., Schnieke, A. E., McWhir, J., Kind, A. J., and Campbell, K. H. S. 1997. Viable offspring derived from fetal and adult mammalian cells. *Nature* 385:810–813.

Wolffe, A. P. 1994. Transcription: In tune with the histones. *Cell* 77:13–16.

Wolffe, A. P., and Pruss, D. 1996. Targeting chromatin disruption: Transcription regulators that acetylate histones. *Cell* 84:817–819.

Chapter 18: Genetics of Cancer

Bishop, J. M. 1987. The molecular genetics of cancer. *Science* 235:305–311.

Brown, M. A., and Solomon, E. 1997. Studies on inherited cancers: Outcomes and challenges of 25 years. *Trends Genet.* 13:202–206.

Cavenee, W. K., and White, R. L. 1995. The genetic basis of cancer. *Sci. Am.* 272 (March):72–79.

Cleaver, J. E. 1994. It was a very good year for DNA repair. *Cell* 76:1–4.

Fishel, R., Lescoe, M. K., Rao, M. R. S., Copeland, N. G., Jenkins, N. A., Garber, J., Kane, M., and Kolodner, R. 1994. The human mutator gene homolog *MSH2* and its association with hereditary nonpolyposis colon cancer. *Cell* 75:1027–1038.

Hartwell, L. H., and Kastan, M. B. 1994. Cell cycle control and cancer. *Science* 266:1821–1828.

Jiricny, J. 1994. Colon cancer and DNA repair: Have mismatches met their match? *Trends Genet.* 10:164–168.

Kamb, A. 1995. Cell-cycle regulators and cancer. *Trends Genet.* 11:136–140.

Kingston, R. E., Baldwin, A. S., and Sharp, P. A. 1985. Transcription control by oncogenes. *Cell* 41:3–5.

Leach, F. S., et al. 1993. Mutations of a *mutS* homolog in hereditary nonpolyposis colorectal cancer. *Cell* 75:1215–1225.

Levine, A. J. 1997. p53, the cellular gatekeeper for growth and division. *Cell* 88:323–331.

Mancini, M. A., Shan, B., Nickerson, J. A., Penman, S., and Lee, W.-H. 1994. The retinoblastoma gene product is a cell cycle–dependent, nuclear matrix–associated protein. *Proc. Natl. Acad. Sci. USA* 91:418–422.

Marmorstein, L. Y., Kinev, A. V., Chan, G. K. T., Bochar, D. A., Beniya, H., Epstein, J. A., Yen, T. J., and Shiekhatter, R. 2001. A human *BRCA2* complex containing a structural DNA binding component influences cell cycle progression. *Cell* 104:247–257.

Rabbitts, T. H. 1994. Chromosomal translocations in human cancer. *Nature* 372:143–149.

Ratner, L., Josephs, S. F., and Wong-Staal, F. 1985. Oncogenes: Their role in neoplastic transformation. *Annu. Rev. Microbiol.* 39:419–449.

Rebbeck, T. R., Couch, F. J., Kant, J., Calzone, K., DeShano, M., Peng, Y., Chen, K., Garber, J. E., and Weber, B. L. 1996. Genetic heterogeneity in hereditary breast cancer: Role of *BRCA1* and *BRCA2*. *Am. J. Hum. Genet.* 59:547–553.

Vousden, K. H. 2000. p53: Death star. *Cell* 103:691–694.

Weber, B. L. 2002. Cancer genomics. *Cancer Cell* 1:37–47.

Weinberg, R. A., 1995. The retinoblastoma protein and cell cycle protein. *Cell* 81:323–330.

Weinstein, R. A. 1997. The cat and mouse games that genes, viruses, and cells play. *Cell* 88:573–575.

Welcsh, P. L., Owens, K. N., and King, M.-C. 2000. Insights into the functions of *BRCA1* and *BRCA2*. *Trends Genet.* 16:69–74.

Wooster, R., et al. 1995. Identification of the breast cancer susceptibility gene *BRCA2*. *Nature* 378:789–792.

Wooster, R., and Stratton, M. R. 1995. Breast cancer susceptibility: A complex disease unravels. *Trends Genet.* 11:3–5.

Chapter 19: DNA Mutation and Repair

Ames, B. N., Durston, W. E., Yamasaki, E. and Lee, F. 1973. Carcinogens are mutagens: A simple test system combining liver homogenates for activation and bacteria for detection. *Proc. Natl. Acad. Sci. USA* 70:2281–2285.

Boyce, R. P., and Howard-Flanders, P. 1964. Release of ultraviolet light–induced thymine dimers from DNA in *E. coli* K12. *Proc. Natl. Acad. Sci. USA* 51:293–300.

Cleaver, J. E. 1994. It was a very good year for DNA repair. *Cell* 76:1–4.

Devoret, R. 1979. Bacterial tests for potential carcinogens. *Sci. Am.* 241(August):40–49.

Fishel, R., Lescoe, M. K., Rao, M. R. S., Copeland, N. G., Jenkins, N. A., Garber, J., Kane, M., and Kolodner, R. 1993. The human mutator gene homolog *MSH2* and its association with hereditary nonpolyposis colon cancer. *Cell* 75:1027–1038.

Lederberg, J., and Lederberg, E. M. 1952. Replica plating and indirect selection of bacterial mutants. *J. Bacteriol.* 63:399–406.

Luria, S. E., and Delbrück, M. 1943. Mutations of bacteria from virus sensitivity to virus resistance. *Genetics* 28:491–511.

Modrich, P. 1987. DNA mismatch correction. *Annu. Rev. Biochem.* 56:435–466.

Morgan, A. R. 1993. Base mismatches and mutagenesis: How important is tautomerism? *Trends Biochem. Sci.* 18:160–163.

Setlow, R. B., and Carrier, W. L. 1964. The disappearance of thymine dimers from DNA: An error-correcting mechanism. *Proc. Natl. Acad. Sci. USA* 51:226–231.

Tessman, I., Liu, S.-K., and Kennedy, A. 1992. Mechanism of SOS mutagenesis of UV-irradiated DNA: Mostly error-free processing of deaminated cytosine. *Proc. Natl. Acad. Sci. USA* 89:1159–1163.

Chapter 20: Transposable Elements

Bhattacharyya, M. K., Smith, A. M., Ellis, T. H. N., Hedley, C., and Martin, C. 1990. The wrinkled-seed character of pea described by Mendel is caused by a transposon-like insertion in a gene encoding starch-branching enzyme. *Cell* 60:115–122.

Boeke, J. D., and Devine, S. E. 1998. Yeast retrotransposons: Finding a nice quiet neighborhood. *Cell* 93:1087–1089.

Boeke, J. D., Garfinkel, D. J., Styles, C. A., and Fink, G. R. 1985. Ty elements transpose through an RNA intermediate. *Cell* 40:491–500.

Bucheton, A. 1990. I transposable elements and I-R hybrid dysgenesis in *Drosophila*. *Trends Genet.* 6:16–21.

Cohen, S. N., and Shapiro, J. A. 1980. Transposable genetic elements. *Sci. Am.* 242(February):40–49.

Davies, D., Goryshin, I. Y., Reznikoff, W. S., and Rayment, I. 2000. Three-dimensional structure of the Tn5 synaptic complex transposition intermediate. *Science* 289:77–84.

Engles, W. R. 1983. The *P* family of transposable elements in *Drosophila*. *Annu. Rev. Genet.* 17:315–344.

Federoff, N. V. 1989. About maize transposable elements and development. *Cell* 56:181–191.

Foster, T. J., Davis, M. A., Roberts, D. E., Takashita, K., and Kleckner, N. 1981. Genetic organization of transposon Tn10. *Cell* 23:201–213.

Iida, S., Meyer, J., and Arber, W. 1983. Prokaryotic IS elements. In *Mobile Genetic Elements*, J. A. Shapiro (ed.) pp. 159–221. New York: Academic Press.

Kingsman, A. J., and Kingsman, S. M. 1988. Ty: A retroelement moving forward. *Cell* 53:333–335.

Kleckner, N. 1981. Transposable elements in prokaryotes. *Annu. Rev. Genet.* 15:341–404.

McClintock, B. 1939. The behavior in successive nuclear divisions of a chromosome broken at meiosis. *Proc. Natl. Acad. Sci. USA* 25:405–416.

———. 1950. The origin and behavior of mutable loci in maize. *Proc. Natl. Acad. Sci. USA* 36:344–355.

———. 1951. Chromosome organization and genic expression. *Cold Spring Harbor Symp. Quant. Biol.* 16:13–47.

———. 1953. Induction of instability at selected loci in maize. *Genetics* 38:579–599.

———. 1956. Controlling elements and the gene. *Cold Spring Harbor Symp. Quant. Biol.* 21:197–216.

———. 1961. Some parallels between gene control systems in maize and in bacteria. *Am. Naturalist* 95:265–277.

———. 1965. The control of gene action in maize. *Brookhaven Symp. Biol.* 18:162–184.

———. 1984. The significance of responses of the genome to challenge. Nobel lecture. *Science* 226:792–801.

Chapter 21: Chromosomal Mutations

Barr, M. L., and Bertram, E. G. 1949. A morphological distinction between neurones of the male and female, and the behavior of the nucleolar satellite during accelerated nucleoprotein synthesis. *Nature* 163:676–677.

Borst, P., and Greaves, D. R. 1987. Programmed gene rearrangements altering gene expression. *Science* 235:658–667.

Caskey, C. T., Pizzuti, A., Fu, Y.-H., Fenwick, R. G., and Nelson, D. L. 1992. Triplet repeat mutations in human disease. *Science* 256:784–789.

Dalla-Favera, R., Martinotti, S., Gallo, R., Erickson, J., and Croce, C. 1983. Translocation and rearrangements of the c-myc oncogene locus in human undifferentiated B-cell lymphomas. *Science* 219:963–997.

DeKlein, A., van Kessel, A. G., Grosveld, G., Bartram, C. R., Hagemeijer, A., Bootsma, D., Spurr, N. K., Heisterkamp, N., Groffen, J., and Stephenson, J. R. 1982. A cellular oncogene is translocated to the Philadelphia chromosome in chronic myelocytic leukemia. *Nature* 300:765–767.

Huntington's Disease Collaborative Research Group. 1993. A novel gene containing a trinucleotide repeat that is expanded and unstable in Huntington's disease chromosome. *Cell* 72:971–983.

Kremer, E., Pritchard, M., Lynch, M., Yu, S., Holman, K., Baker, E., Warren, S. T., Schlessinger, D., Sutherland, G. R., and Richards, R. I. 1991. Mapping of DNA instability at the fragile X to a trinucleotide repeat sequence p(CGG)n. *Science* 252:1711–1714.

Lyon, M. F. 1961. Gene action in the X-chromosomes of the mouse (*Mus musculus* L). *Nature* 190:372–373.

Penrose, L. S., and Smith, G. F. 1966. *Down's Anomaly.* Boston: Little, Brown.

Richards, R. I., and Sutherland, G. R. 1992. Dynamic mutations: A new class of mutations causing human disease. *Cell* 70:709–712.

———. 1992. Fragile X syndrome: The molecular picture comes into focus. *Trends Genet.* 8:249–255.

Shaw, M. W. 1962. Familial mongolism. *Cytogenetics* 1:141–179.

Sutherland, G. R., Baker, E., and Richards, R. I. 1998. Fragile sites still breaking. *Trends Genet.* 14:501–506.

Tarleton, J. C., and Saul, R. A. 1993. Molecular genetic advances in fragile X syndrome. *J. Pediatr.* 122:169–185.

Verkerk, A. J. M. H., Piertti, M., Sutcliff, J. S., Fu, Y.-H., Kuhl, D. P. A., Pizzuti, A., Reiner, O., Richards, S., Victoria, M. F., Zhang, F., Eussen, B. E., van Ommen, G.-J. B., Blonden, L. A. J., Riggins, G. J., Chastain, J. L., Kunst, C. B., Galjaard, H., Caskey, C. T., Nelson, D. L., Oostra, B. A., and Warrent, S. T. 1991. Identification of a gene (*FMR-1*) containing a CGG repeat coincident with a breakpoint cluster region exhibiting length variation in fragile X syndrome. *Cell* 65:905–914.

Chapter 22: Population Genetics

Avise, J. C. 1986. Mitochondrial DNA and the evolutionary genetics of higher animals. *Phil. Trans. Roy. Soc. Lond., Ser. B* 321:325–342.

Buri, P. 1956. Gene frequency in small populations of mutant *Drosophila. Evolution* 10:367–402.

Crow, J. F. 1986. *Basic Concepts in Population, Quantitative, and Evolutionary Genetics.* New York: Freeman.

Darwin, C. 1860. *On the Origin of Species by Means of Natural Selection, or the Preservation of Favoured Races in the Struggle for Life.* New York: Appleton.

Dobzhansky, T. 1951. *Genetics and the Origin of Species,* 3rd ed. New York: Columbia University Press.

Fisher, R. A. 1930. *The Genetical Theory of Natural Selection.* Oxford: Clarendon Press.

Ford, E. B. 1971. *Ecological Genetics,* 3rd ed. London: Chapman & Hall.

Gillespie, J. H. 1991. *The Causes of Molecular Evolution.* Oxford: Oxford University Press.

Glass, B., Sacks, M. S., Jahn, E. F., and Hess, C. 1952. Genetic drift in a religious isolate: An analysis of the causes of variation in blood group and other gene frequencies in a small population. *Am. Nat.* 86:145–159.

Hardy, G. H. 1908. Mendelian proportions in a mixed population. *Science* 28:49–50.

Hartl, D. L., and Clark, A. G. 1995. *Principles of Population Genetics,* 3rd ed. Sunderland, MA: Sinauer.

Hedrick, P. H. 2000. *Genetics of Populations.* Boston: Science Books International.

Hillis, D. M., and Moritz, C. 1990. *Molecular Systematics.* Sunderland, MA: Sinauer.

Kettlewell, H. B. D. 1961. The phenomenon of industrial melanism in the Lepidoptera. *Annu. Rev. Entomol.* 6:245–262.

Kreitman, M. 1983. Nucleotide polymorphism at the alcohol dehydrogenase locus of *Drosophila melanogaster. Nature* 304:412–417.

Lewontin, R. C. 1974. *The Genetic Basis of Evolutionary Change.* New York: Columbia University Press.

———. 1985. Population genetics. *Annu. Rev. Genet.* 19:81–102.

Lewontin, R. C., Moore, J. A., Provine, W. B., and Wallace, B. 1981. *Dobzhansky's Genetics of Natural Populations I-XLIII.* New York: Columbia University Press.

Li, W.-H. 1997. *Molecular Evolution.* Sunderland, MA: Sinauer.

Maniatis, T., Fritsch, E. F., Lauer, L., and Lawn, R. M. 1980. The molecular genetics of human hemoglobin. *Annu. Rev. Genet.* 14:145–178.

Nei, M. 1987. *Molecular Evolutionary Genetics.* New York: Columbia University Press.

Powell, J. R. 1997. *Progress and Prospects in Evolutionary Biology: The Drosophila model.* New York: Oxford University Press.

Selander, R. K., and Kaufman, D. W. 1975. Self-fertilization and genetic population structure in a colonizing land snail. *Proc. Natl. Acad. Sci. USA* 70:1186–1190.

Soulé, M. E., ed. 1986. *Conservation Biology: The Science of Scarcity and Diversity.* Sunderland, MA: Sinauer.

Weir, B. S. 1996. *Genetic Data Analysis II.* Sunderland, MA: Sinauer.

Chapter 23: Quantitative Genetics

Blumer, M. G. 1980. *The Mathematical Theory of Quantitative Genetics.* Oxford: Clarendon Press.

Darwin, C. 1860. *On the Origin of Species by Means of Natural Selection, or the Preservation of Favoured Races in the Struggle for Life.* New York: Appleton.

Dobzhansky, T., and Pavlovsky, O. 1969. Artificial and natural selection for two behavioral traits in *Drosophila pseudoobscura. Proc. Natl. Acad. Sci. USA* 62:75–80.

East, E. M. 1910. A Mendelian interpretation of variation that is apparently continuous. *Am. Nat.* 44:65–82.

———. 1916. Studies on size inheritance in *Nicotiana. Genetics* 1:164–176.

East, E. M., and Jones, D. F. 1919. *Inbreeding and Outbreeding.* Philadelphia: Lippincott.

Falconer, D. S. 1989. *Introduction to Quantitative Genetics.* New York: Wiley.

Hill, W. G., ed. 1984. *Quantitative Genetics,* Parts I and II. New York: Van Nostrand Reinhold.

Lander, E. S., and Botstein, D. 1989. Mapping Mendelian factors underlying quantitative traits using RFLP linkage maps. *Genetics* 121:185–199.

Mather, K. 1943. Polygenic inheritance and natural selection. *Biol. Rev.* 18:32–64.

Nilsson-Ehle, H. 1909. Kreuzungsuntersuchungen an Hafer und Weizen. *Lunds Univ. Aarskr. N. F. Atd., Ser. 2*, 5 (2):1–122.

Paterson, A. H., Lander, E. S., Hewitt, J. D., Person, S., Lincoln, S. E., and Tanksley, S. D. 1988. Resolution of quantitative traits into Mendelian factors by using a complete RFLP linkage map. *Nature* 335:721–726.

Selander, R. K., and Kaufman, D. W. 1975. Self-fertilization and genetic population structure in a colonizing land snail. *Proc. Natl. Acad. Sci. USA* 70:1186–1190.

Thoday, J. M. 1961. Location of polygenes. *Nature* 191:368–370.

Weir, B. S., Eisen, E. J., Goodman, M. M., and Namkoong, G., eds. 1988. *Proceedings of the Second International Conference on Quantitative Genetics.* Sunderland, MA: Sinauer.

Chapter 24: Molecular Evolution

Haldane, J. B. S. 1932. *The Causes of Evolution.* London: Longmans and Green.

Jukes, T. H., and Cantor, C. R. 1969. Evolution of protein molecules. In *Mammalian Protein Metabolism,* H. N. Munro (ed.) pp. 21–123. New York: Academic Press.

Klein, J., and Figueroa, F. 1986. Evolution of the major histocompatibility complex. *CRC Crit. Rev. Immunol.* 6:295–386.

Perutz, M. F. 1983. Species adaptation in a protein molecule. *Mol. Biol. Evol.* 1:1–28.

Sarich, V. M., and Wilson, A. C. 1967. Immunological time scale for hominid evolution. *Science* 158:1200–1203.

Solutions to Selected Questions and Problems

Chapter 1 Genetics: An Introduction

1.1 c

1.3 c

1.6 a. Yes, if a sexual mating system exists in that species. In that case, two haploid cells can fuse to produce a diploid cell, which can then go through meiosis to produce haploid progeny. The fungi *Neurospora crassa* and *Saccharomyces cerevisiae* exemplify this positioning of meiosis in the life cycle.

b. No, because a diploid cell cannot be formed and meiosis occurs only starting with a diploid cell.

1.8 c

1.10 a. metaphase

b. anaphase

1.14 a. The chance that a gamete would have a particular maternal chromosome is $\frac{1}{2}$. Therefore, the chance of obtaining a gamete with all three maternal chromosomes is $(\frac{1}{2})^3 = \frac{1}{8}$.

b. The set of gametes with some maternal and paternal chromosomes is composed of all gametes *except* those that have only maternal or only paternal chromosomes. That is, p(gamete with both maternal and paternal chromosomes) $= 1 - p$ (gamete with only maternal or only paternal chromosomes). From part (a), the chance of a gamete having chromosomes from only one parent is $\frac{1}{8}$. Using the sum rule, p (gamete with both maternal and paternal chromosomes) $= 1 - (\frac{1}{8} + \frac{1}{8}) = \frac{3}{4}$.

1.15 One of the long chromosomes and the short chromosome might be members of a heteromorphic pair—that is, X and Y chromosomes, respectively.

1.17 False. Owing to the randomness of independent assortment and to crossing-over, both of which characterize meiosis, the probability of any two sperm cells being genetically identical is extremely remote.

1.19 a. $17 + 26 = 43$ chromosomes

b. Similar chromosomes pair in meiosis. The pairing pattern seen in the hybrid indicates that some of the chromosomes in arctic and red foxes share evolutionary similarity but others do not. Unpaired chromosomes will not segregate in an orderly manner, giving rise to unbalanced meiotic products with either extra or missing chromosomes. This can lead to sterility for two reasons. First, meiotic products that are missing chromosomes may not have genes necessary to form gametes. Second, even if gametes are able to form, a zygote generated from them will not have the chromosome set from the hybrid, the red, or the arctic fox. The zygote will have missing or extra genes, causing it to be inviable.

1.20 The probability of a given homologue going to one particular pole is $\frac{1}{2}$. The probability of all five paternal chromosomes going to the same pole is $(\frac{1}{2})^5 = \frac{1}{32}$. The same answer applies for all maternal chromosomes going to one pole.

Chapter 2: Mendelian Genetics

2.1 a. red

b. 3 red, 1 yellow

c. all red

d. $\frac{1}{2}$ red, $\frac{1}{2}$ yellow

2.3 The F_2 genotypic ratio (if C is colored, c is colorless) is $\frac{1}{4}$ $CC : \frac{1}{2}$ $Cc : \frac{1}{4}$ cc. If we consider just the colored plants, there is a 1:2 ratio of CC homozygotes to Cc heterozygotes. Therefore, if a colored plant is picked at random, the probability that it is CC is $\frac{1}{3}$ (i.e., $\frac{1}{3}$ of the colored plants are homozygous), and the probability that it is Cc is $\frac{2}{3}$. Only if a Cc plant is selfed will more than one phenotypic class be found among its progeny; therefore, the answer is $\frac{2}{3}$.

2.4 a. Parents are Rr (rough) and rr (smooth); F_1 are Rr (rough) and rr (smooth).

b. $Rr \times Rr \rightarrow \frac{3}{4}$ rough, $\frac{1}{4}$ smooth

2.6 Progeny ratio approximates 3:1, so the parent is heterozygous. Of the dominant progeny, there is a 1:2 ratio of homozygous to heterozygous, so $\frac{1}{3}$ will breed true.

2.7 Black is dominant to brown. If B is the allele for black and b for brown, then female X is Bb and female Y is BB.

2.10

	Parents		Progeny		Female Parent
	female ×	male	grey	white	Genotype
a.	grey ×	white	81	82	Gg
b.	grey ×	grey	118	39	Gg
c.	grey ×	white	74	0	GG
d.	grey ×	grey	90	0	GG or Gg

2.11 a. To obtain a white babbit in a cross between a pair of F_1 babbits, both babbits must be Bb in genotype, and a bb offspring must be produced by these parents. Among the black F_1 progeny, there is a $\frac{1}{3}$ chance of picking a BB individual and a $\frac{2}{3}$ chance of picking a Bb individual. Therefore,

P(white offspring) $= P$(both F_1 babbits are Bb and a bb offspring is produced)

$\qquad = P$(both F_1 babbits are Bb) $\times P$(bb offspring)

$\qquad = (\frac{2}{3} \times \frac{2}{3}) \times (\frac{1}{4})$

$\qquad = \frac{1}{9}$

b. If he crosses an F_1 male (Bb or BB) to the parental female (Bb), two types of crosses are possible. The crosses and probabilities are (1) Bb (F_1 male) $\times$ Bb (parental female), $P = \frac{2}{3} \times 1 = \frac{2}{3}$; and (2) BB (F_1 male) $\times$ Bb (parental female), $P = \frac{1}{3} \times 1 = \frac{1}{3}$. Only the first cross can produce white progeny, $\frac{1}{4}$ of the time. Therefore, the chance that this strategy will yield white progeny is

$P = \frac{2}{3}$ (chance of $Bb \times Bb$ cross) $\times \frac{1}{4}$ (chance of bb offspring) $= \frac{1}{6}$

c. The best strategy is as follows. Remate the initial two black babbits (both are *Bb*) to obtain a white male offspring (*P* = ¼ (white *bb*) × ½ (male) = ⅛). Retain this male and breed it back to its mother. This cross would be *Bb* × *bb* and give ½ white (*bb*) and ½ black (*Bb*) offspring. The progeny of this cross could be used to develop a "breeding colony" consisting of black (*Bb*) females and white (*bb*) males. These would consistently produce half white and half black offspring.

2.13 Try fitting the data to a model in which catnip sensitivity or insensitivity is controlled by a pair of alleles at one gene. Hypothesize that since sensitivity is seen in all the progeny of the initial mating between catnip-sensitive Cleopatra and catnip-insensitive Antony, sensitivity is dominant. Let *S* represent the sensitive allele and *s* the insensitive allele. Then the initial cross would have been *S*– × *ss*, and the progeny are *Ss*. If two of the *Ss* kittens mate, one would expect 3 *Ss* (sensitive):1 *ss* (insensitive) kittens. In the mating with Augustus, the cross would be *Ss* × *ss* and should give a 1 *Ss* (sensitive):1 *ss* (insensitive) progeny ratio. The observed progeny ratios are not far off from these expectations.

An alternative hypothesis is that sensitivity (*s*) is recessive and insensitivity (*S*) is dominant. For Antony and Cleopatra to have sensitive (*ss*) offspring, they would need to be *Ss* and *ss*, respectively. When two of their *ss* progeny mate, only sensitive, *ss* offspring should be produced. Since this is not observed, this hypothesis does not explain the data.

2.15 a. *WW Dd* × *ww dd*

   b. *Ww dd* × *Ww dd*

   c. *ww DD* × *WW dd*

   d. *Ww Dd* × *Ww dd*

   e. *Ww Dd* × *Ww dd*

2.16 The cross is *Aa Bb Cc* × *Aa Bb Cc*.

   a. Considering the *A* gene alone, the probability of an offspring showing the *A* trait from *Aa* × *Aa* is ¾. Similarly, the probability of showing the *B* trait from *Bb* × *Bb* is ¾ and the *C* trait from *Cc* × *Cc* is ¾. Therefore, the probability of a given progeny being phenotypically *A B C* (using the product rule) is ¾ × ¾ × ¾ = ²⁷⁄₆₄.

   b. Considering the *A* gene, the probability of an *AA* offspring from *Aa* × *Aa* is ¼. The same probability is the case for a *BB* offspring and for a *CC* offspring. Therefore, the probability of an *AA BB CC* offspring (from the product rule) is ¼ × ¼ × ¼ = ¹⁄₆₄.

2.20 a. The F₁ has the genotype *Aa Bb Cc^h* and is all agouti, black. The F₂ is ²⁷⁄₆₄ agouti, black; ⁹⁄₆₄ agouti, black, Himalayan; ⁹⁄₆₄ agouti, brown; ⁹⁄₆₄ black; ³⁄₆₄ agouti, brown, Himalayan; ³⁄₆₄ black, Himalayan; ³⁄₆₄ brown; ¹⁄₆₄ brown, Himalayan.

   b. F₂ animals that are black and agouti (and either Himalayan or have full-body pigmentation) have the genotypes *A– B– C–* and *A– B– c^h c^h*. Among the *A–* animals, ⅔ are *Aa*. Among the *B–* animals, ⅓ are *BB*. Among the *C–* and *c^h c^h* animals, ½ are *Cc^h*. Thus, among all of the black, agouti F₂, ⅔ × ⅓ × ½ = ⅑ are *Aa BB Cc^h*. If one limits the animals under consideration to those that are black, agouti, and not Himalayan (i.e., *A– B– C–*), then ⅔ of the *C–* animals are *Cc^h*, and the proportion of *Aa BB Cc^h* animals is ⅔ × ⅓ × ⅔ = ⁴⁄₂₇.

   c. From the cross *Aa Bb Cc^h* × *Aa Bb Cc^h*, ¼ of the progeny will be *bb* and show brown pigment. This will be the case regardless of whether the animals are pigmented over their

entire body or are Himalayan. Thus, ¼ of the Himalayan mice will show brown pigment.

   d. From the cross *Aa Bb Cc^h* × *Aa Bb Cc^h*, ¾ of the progeny will be *B–* and show black pigment. This will be the case regardless of whether the animals are agouti or non-agouti. Thus, ¾ of the agouti mice will show black pigment.

2.24 Mating type C is determined only by the genotype *aa bb*. Thus, C must be genotype *aa bb*. Crosses of the other strains to C, then, are testcrosses and can tell us the genotypes of the strains. Therefore, A is *Aa Bb*, B is *aa Bb*, and D is *Aa bb*.

2.25 a. The cross is *Ww Rr* × *W r*. Workers are females, and in the progeny of the cross they are ¼ *WW Rr* (black, use wax) : ¼ *Ww Rr* (black, use wax) : ¼ *WW rr* (black, use resin) : ¼ *Ww rr* (black, use resin). In sum, all workers are black-eyed, and ½ use wax and ½ use resin.

   b. ¼ each *W R, W r, w R, w r.*

   c. Here the question is whether the female is heterozygous, not whether her progeny will be wingless. Madonna is *Cc*, so there is a ¼ chance that her granddaughter will be heterozygous and that wingless males will be found in the hive (i.e., ½ that her daughter is heterozygous × ½ that the daughter of the daughter is heterozygous).

   d. The chance that the F₄ generation great-great-granddaughter will be heterozygous is ½ × ½ × ½ × ½ = ¹⁄₁₆.

2.26 a. Mother must be homozygous *Aa* in order to have children who have the trait.

   b. Father is homozygous *aa* because he expresses the trait.

   c. The cross is *Aa* × *aa*, so children will be *aa* if they have the trait (II.2 and II.5) and *Aa* if they do not have the trait (II.1, II.3, and II.4).

   d. From the cross *Aa* × *aa*, the prediction is that ½ of the progeny will be *Aa* (normal) and ½ will be *aa* (expressing the trait). There are five children, two of whom have the trait and three of whom are normal. Thus, the ratio fits as well as it could for five children.

Chapter 3: Chromosomal Basis of Inheritance, Sex Linkage, and Sex Determination

3.3 Fathers always give their X chromosome to their daughters, so the woman must be heterozygous for the color blindness trait and is *c⁺c*. Her husband received his X chromosome from his mother and has normal color vision, so he is *c⁺Y*. The cross is therefore *c⁺c* × *c⁺Y*. All daughters will receive the paternal X bearing the *c⁺* allele and have normal vision. Sons will receive the maternal X, so half will be *cY* and be color-blind, and half will be *c⁺Y* and have normal color vision.

3.4 Let *c* and *c⁺* be the color-blind and normal vision alleles, respectively, and let *a* and *a⁺* be the albino and normal pigmentation alleles, respectively. Then the cross can be represented as *c⁺c⁺ aa* × *cY a⁺a⁺*. As all the offspring will be *a⁺a*, all will have normal pigmentation. The offspring will either be *c⁺c* or *c⁺Y*, and have normal color vision. The daughters will, however, be carriers for the color-blindness trait.

3.5 The parentals are *ww vg⁺vg⁺* and *w⁺Y vgvg*.

   a. The F₁ males are all *wY vg⁺vg*, white eyes, long wings. The F₁ females are all *ww⁺ vg⁺vg* red eyes, long wings.

   b. The F₂ females are ⅜ red, long; ⅜ white, long; ⅛ red, vestigial; ⅛ white, vestigial. The same ratios of the respective phenotypes apply for males.

   c. The cross of F₁ male with the parental female is *wY vg⁺vg* × *ww vg⁺vg⁺*. All progeny have white eyes and long wings. The cross of an F_1 female with the parental male is $ww^+ \, vg^+vg \times w^+Y \, vgvg$. Female progeny: All have red eyes, half have long wings, and half have vestigial wings. Male progeny: $\frac{1}{4}$ red, long; $\frac{1}{4}$ red, vestigial; $\frac{1}{4}$ white, long; $\frac{1}{4}$ white, vestigial.

3.8 The simplest hypothesis is that brown-colored teeth are determined by a sex-linked dominant mutant allele. Man A was BY and his wife was bb. All sons will be bY normals and cannot pass on the trait. All the daughters receive the X chromosome with the B mutant allele.

3.10 a. To have produced a child with cystic fibrosis, both parents must have been heterozygous for the autosomal recessive mutant gene. Therefore, the probability of their next child having cystic fibrosis is $\frac{1}{4}$ (i.e., $\frac{1}{4}$ of the progeny from $Aa \times Aa$ will be aa).

b. Nonaffected children will be either AA or Aa. Therefore, $\frac{2}{3}$ of the nonaffected children will be Aa heterozygotes.

3.12 a. Since one is concerned with only a single trait, one can consider just part of the cross: $AA \times aY$. The progeny will either be AY or Aa, and all are $A-$. Therefore, the chance of obtaining an $A-$ individual in the F_1 is 1.

b. As shown in (a), there is no chance ($P = 0$) of obtaining an aY individual in the F_1.

c. The F_1 progeny will be $A- \, Bb \, Cc \, Dd$. Half will be female, so $P = \frac{1}{2}$.

d. Two: $Aa \, Bb \, Cc \, Dd$ (females) and $AY \, Bb \, Cc \, Dd$ males

e. For the X chromosome trait, the F_1 cross is $AY \times Aa$. Half of the female offspring ($\frac{1}{4}$ of the total) will be heterozygous Aa individuals. For each of the autosomal traits, $\frac{1}{2}$ of the offspring will be heterozygous (e.g., $Bb \times Bb$ gives $\frac{1}{2} \, Bb$ individuals). Therefore, the chance that an F_2 individual will be heterozygous for all four traits is $\frac{1}{4} \times \frac{1}{2} \times \frac{1}{2} \times \frac{1}{2} = \frac{1}{32}$.

f. Before determining the probabilities, consider that at any autosomal gene, there is a $\frac{1}{4}$ chance of obtaining either type of homozygote (e.g., BB, bb) and a $\frac{1}{2}$ chance of obtaining a heterozygote. At the A gene, the cross is $AY \times Aa$, so there is a $\frac{1}{4}$ chance of obtaining an AY male, a $\frac{1}{4}$ chance of obtaining an aY male, a $\frac{1}{4}$ chance of obtaining an Aa female, and a $\frac{1}{4}$ chance of obtaining an AA female (there will be a $\frac{1}{2}$ chance of obtaining an $A-$ female). Then the chance of obtaining (1) an $A- \, bb \, CC \, dd$ (female) is $P = (\frac{1}{2} \times \frac{1}{4} \times \frac{1}{4} \times \frac{1}{4}) = \frac{1}{128}$, (2) an $aY \, BB \, Cc \, Dd$ (male) is $P = (\frac{1}{4} \times \frac{1}{4} \times \frac{1}{2} \times \frac{1}{2}) = \frac{1}{64}$, (3) an $AY \, bb \, CC \, dd$ (male) is $P = (\frac{1}{4} \times \frac{1}{4} \times \frac{1}{4} \times \frac{1}{4}) = \frac{1}{256}$, and (4) an $aa \, bb \, Cc \, Dd$ (female) is $P = (0 \times \frac{1}{4} \times \frac{1}{2} \times \frac{1}{2}) = 0$.

3.15 This problem raises the issue that the precise mode of inheritance of a trait often cannot be determined when a pedigree is small and the frequency of the trait in a population is unknown. For example, pedigree A could easily fit an autosomal dominant trait (AA and Aa = affected): The affected father would be heterozygous for the trait (Aa), the mother would be unaffected (aa), and half of their offspring would be affected. However, if the trait were autosomal recessive (i.e., aa = affected individuals) and the mother were heterozygous (Aa) and the father homozygous (aa), half of the offspring would still be affected. One could also fit the pedigree to an X-linked recessive trait: The mother would be heterozygous ($X^A X^a$), the father would be hemizygous ($X^a Y$), and half of the progeny would be affected (either $X^A X^a$ or $X^a Y$). An X-linked dominant trait would not fit the pedigree because it would require all the daughters of the affected father to be affected (because they all receive their father's X), and not allow a son to be affected (because he does not receive his father's X). Pedigrees B and C can be solved by similar analytical reasoning.

	Pedigree A	Pedigree B	Pedigree C
Autosomal recessive	Yes	Yes	Yes
Autosomal dominant	Yes	Yes	No
X-linked recessive	Yes	Yes	No
X-linked dominant	No	No	No

3.17 a. Y-linked inheritance can be excluded because females are affected. X-linked recessive inheritance can also be excluded because an affected mother (I.2) has a normal son (II.5). Autosomal recessive inheritance can also be excluded because in such a case two affected parents (such as II.1 and II.2) could not have unaffected offspring, which they do.

b. The two remaining mechanisms of inheritance are X-linked dominant and autosomal dominant. Genotypes can be written to satisfy both mechanisms of inheritance. Of these two, X-linked dominant inheritance may be more likely because II.6 and II.7 have only affected daughters, indicating criss-cross inheritance. If the trait were autosomal dominant, one would expect half of the daughters and half of the sons to be affected.

3.20 Answer (a) is untrue because if the affected father is heterozygous, only half of his offspring should be affected. Answer (b) is untrue because if the mother is heterozygous, half of her offspring, regardless of sex type, should be affected. Answer (c) is untrue because if the parents are each heterozygous, $\frac{1}{4}$ of their offspring should be homozygous recessive and normal. Answer (d) is the most likely. However, if the mutation is new in either the child or his or her parents, his or her grandparent could have been unaffected.

3.21 a. False. Two affected individuals will always have affected children ($aa \times aa$ can give only aa offspring).

b. False. An autosomal trait is inherited independent of sex type.

c. Need not be true. The trait could be masked by normal dominant alleles through many generations before two heterozygotes marry and produce affected, homozygous offspring.

d. Could be true. If the trait is rare, an unaffected individual marrying into the pedigree is likely to be homozygous for a normal allele. The trait is recessive, and the children receive the dominant, normal allele from the unaffected parent, so the children will be normal. This answer would not be true if the unaffected individual was heterozygous. In this case, half of the children would be affected.

3.23 Since hemophilia is an X-linked trait, the most likely explanation is that random inactivation of X chromosomes (lyonization; see p. 63) produces individuals with different proportions of cells with the normal allele. Thus, if some women had only 40% of their cells with an active h^+ allele and 60% with the h (hemophilia) allele, but other women had 60% of their cells with an active h^+ allele and 40% with the h allele, there would be a significant difference in the amount of clotting factor these two individuals would make.

Chapter 4: Extensions of Mendelian Genetic Analysis

4.2 Six possible genotypes: w/w, $w/w1$, $w/w2$, $w1/w1$, $w1/w2$, and $w2/w2$.

4.6 The woman's genotype is $I^A I^B$, and the man's genotype is $I^A i$.

a. $\frac{1}{2} \times \frac{1}{2} = \frac{1}{4}$

b. Zero. A blood group O baby is not possible.

c. $1/2$ (probability of male) × $1/4$ (probability of AB) × $1/2$ (probability of male) × $1/4$ (probability of B) = $1/64$ probability that all four conditions will be fulfilled

4.8 Blood type O, because genotype is *i/i*

4.10 Half will be C^R/C^W and therefore will resemble the parents.

4.12 **a.** The cross is $FF\ G^NG^N \times ff\ G^OG^O$. The F_1 is $Ff\ G^OG^N$, which is fuzzy with round leaf glands.

b. As the alleles at the G gene show incomplete dominance, there will be a modified 9 : 3 : 3 : 1 ration in the F_2. The progeny will be $3/16$ fuzzy, oval-glanded ($F-G^OG^O$), $6/16$ fuzzy, round-glanded ($F-G^OG^N$), $3/16$ fuzzy, no-glanded ($F-G^NG^N$), $1/16$ smooth, oval-glanded ($ff\ G^OG^O$), $2/16$ smooth, round-glanded ($ff\ G^OG^N$), and $1/16$ smooth, no-glanded ($ff\ G^NG^N$).

c. The cross can be written as $Ff\ G^NG^O \times ff\ G^OG^O$. The progeny will be $1/4$ fuzzy, oval-glanded ($Ff\ G^OG^O$), $1/4$ fuzzy, round-glanded ($Ff\ G^NG^O$), $1/4$ smooth, oval-glanded ($ff\ G^OG^O$), and $1/4$ smooth, round-glanded ($ff\ G^NG^O$).

4.18 To show no segregation among the progeny, the chosen plant must be homozygous. The genotypes comprising the $9/16$ colored plants are 1 *A/A B/B* : 2 *A/a B/B* : 2 *A/A B/b* : 4 *A/a B/b*. Only one of these genotypes is homozygous; the answer is $1/9$.

4.19 There are two independently assorting genes involved. Because of epistasis, only two phenotypes are seen in the F_2; that is, *A/– B/–* genotypes are runner, and the other three genotypes are bunch. Thus, the original cross was *A/A b/b* × *a/a B/B*, giving an F_1 of *A/a B/b*.

4.20 **a.** The cross is *A/a B/b* × *A/a B/b*, which gives $9/16$ *A/– B/–* : $3/16$ *A/– b/b* : $3/16$ *a/a B/–* : $1/16$ *a/a b/b*. The *A/– B/–* rabbits are not deaf because they produce both substances needed for hearing. The other three genotypic classes result in deafness because one or the other or both of the enzymes needed for hearing are not produced. Therefore, the phenotypic ratio is 9 hearing rabbits : 7 deaf rabbits.

b. Epistasis. In this case, it is duplicate recessive epistasis.

c. The cross is *a/a B/b* × *Aa/Bb*, which gives $3/8$ *A/a B/–* : $1/8$ *A/a b/b* : $3/8$ *a/a B/–* : $1/8$ *a/a b/b*. Only the *A/a B–* rabbits can hear; the rest are deaf. Thus, the phenotypic ratio is 3 hearing rabbits : 5 deaf rabbits.

4.23 **a.** If A^Y governs yellow and a^+ governs nonyellow (agouti), then A^Y/A^Y are lethal, A^Y/a^+ are yellow, and a^+/a^+ are agouti. Let c^+ determine colored coat and *c* determine albino. The parental genotypes, then, are $A^Y/a^+\ c^+/c$ (yellow) and $A^Y/a^+\ c/c$ (white).

b. The proportion is 2 yellow : 1 agouti : 1 albino. None of the yellows breed true because they are all heterozygous, with homozygous A^Y/A^Y individuals being lethal.

4.25 **a.** *Y/Y R/R* (crimson) × *y/y r/r* (white) gives *Y/y R/r* F_1 plants, which have magenta-rose flowers. Selfing the F_1 gives an F_2 as follows: $1/16$ crimson (*Y/Y R/R*), $2/16$ orange-red (*Y/Y R/r*), $1/16$ yellow (*Y/Y r/r*), $2/16$ magenta (*Y/y R/R*), $4/16$ magenta-rose (*Y/y R/r*), $2/16$ pale yellow (*Y/y r/r*), and $4/16$ white (*y/y R/R*, *y/y R/r*, and *y/y r/r*). Progeny of the F_1 backcrossed to the crimson parent are $1/4$ crimson (*Y/Y R/R*), $1/4$ magenta-rose (*Y/y R/r*), $1/4$ magenta (*Y/y R/R*), and $1/4$ orange-red (*Y/Y R/r*).

b. The cross can be denoted as *Y/Y R/r* × *Y/y r/r*. The progeny will be $1/4$ orange-red (*Y/Y R/r*), $1/4$ magenta-rose (*Y/y R/r*), $1/4$ yellow (*Y/Y r/r*), and $1/4$ pale yellow (*Y/y r/r*).

c. The cross can be denoted as *Y/Y r/r* × *y/y R/r*, and the progeny will be $1/2$ *Y/y R/r* (magenta-rose) and $1/2$ *Y/y r/r* (pale yellow).

4.27 **a.** The simplest approach is to calculate the proportion of progeny that will be black and then subtract that answer from 1. The black progeny have the genotype *A/– B/– C/–*, and the proportion of these progeny is $(3/4)^3$. Therefore, the proportion of colorless progeny is $1 - (3/4)^3 = 1 - 27/64 = 37/64$.

b. With this pathway, an individual is black only if it has the first two steps of the pathway (those provided by *A* and *B*) and lacks the inhibitor provided by *C*, i.e., if it is *A/– B/– c/c*. The chance of obtaining this genotype from a cross of *A/a B/b C/c* × *A/a B/b C/c* is $3/4 \times 3/4 \times 1/ = 9/64$. The proportion of the F_2 that is colorless is $1 - 9/64 = 55/64$. There will be a 55 colorless : 9 black ratio. Thus, the ration of black to colorless in the F_2 can be used to distinguish between hypotheses concerning these two pathways.

4.29 In males, *H/H* and *H/h* are horned, and *h/h* is hornless; in females, *H/H* is hornless. The cross is an *H/H W/W* male × *h/h w/w* female. The F_1 is *H/h W/w*, which gives horned white males and hornless white females. Interbreeding the F_1 gives the following F_2:

	Male	Female
$3/16$ *H/H W/–*	horned, white	horned, white
$6/16$ *H/h W/–*	horned, white	hornless, white
$3/16$ *h/h W/–*	hornless, white	hornless, white
$1/16$ *H/H w/w*	horned, black	horned, black
$2/16$ *H/h w/w*	horned, black	hornless, black
$1/16$ *h/h v/w*	hornless, black	hornless, black

In sum, the ratios are $9/16$ horned white : $3/16$ hornless white : $3/16$ horned black : $1/16$ hornless black males and $3/16$ horned white : $9/16$ hornless white : $1/16$ horned black : $3/16$ hornless black females.

Chapter 5: Gene Mapping in Eukaryotes

5.2 From the chi-square test, $\chi^2 = 16.10$; *P* is less than 0.01 at 3 degrees of freedom. This test reveals that the two genes do not fit a 1:1:1:1 ratio. It does not say why. Linkage might seem reasonable until it is realized that the minority classes are not reciprocal classes (both carry the *aa* phenotype). If the segregation at each locus is considered, however, the *B/–* : *b/b* ratio is about 1:1 (203:197), whereas the *A/–* : *a/a* ratio is not (240:160). The departure, then, specifically results from a deficiency of *a/a* individuals. This departure should be confirmed in other crosses that test the segregation at locus *A*. In corn, further evidence might show up as a class of ungerminated seeds or seedlings that die early.

5.4

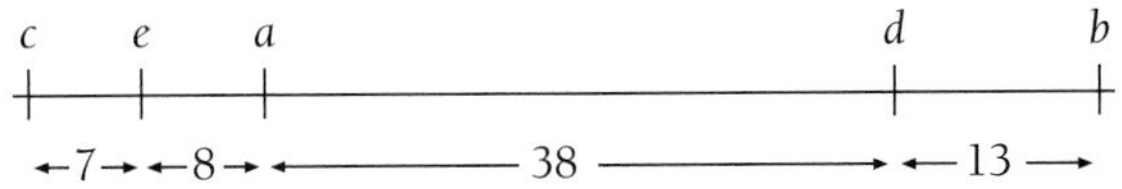

Note that the interval lengths are not strictly additive (e.g., *a–d* = 38 and *a–e* = 8, but *d–e* = 45) because of the effects of multiple crossovers, as described in the chapter.

5.6 45% $a\ b^+$, 45% $a^+\ b$, 5% $a\ b$, and 5% $a^+\ b^+$

5.7 **a.** 0.035 + 0.465 = 0.50

b. All the daughters have $a^+\ b^+$ phenotype.

5.8 Each chromosome pair segregates independently. We can compute the relative proportions of gametes produced for each homologous pair of chromosomes separately from the known map distances (*P*, parental; *R*, recombinant):

P	R	P	R	P	R
AB 0.4	Ab 0.1	CD 0.45	Cd 0.05	EF 0.35	Ef 0.15
ab 0.4	aB 0.1	cd 0.45	cD 0.05	ef 0.35	eF 0.15

To answer the question, simply multiply the probabilities of getting the particular gamete from the F_1 multiple heterozygote.

a. $A B C D E F = 0.40 \times 0.45 \times 0.35 = 0.063$, or 6.3%

b. $A B C d e f = 0.40 \times 0.05 \times 0.35 = 0.007$, or 0.7%

c. $A b c D E f = 0.10 \times 0.05 \times 0.15 = 0.00075$ or 0.075%

d. $a B C d e f = 0.10 \times 0.05 \times 0.35 = 0.00175$ or 0.175%

e. $a b c D e F = 0.40 \times 0.05 \times 0.15 = 0.003$ or 0.3%

5.9 a. 47.5% each of $D\,P\,h$ and $d\,p\,h$, 2.5% of $D\,p\,h$, and 2.5% $d\,P\,h$

b. 23.75% each of $d\,P\,H$, $d\,P\,h$, $D\,p\,h$, and $D\,p\,h$; and 1.25% each of $D\,P\,H$, $D\,P\,h$, $d\,p\,H$, and $d\,p\,h$

5.13 a. By doing a three-point mapping analysis as described in the chapter, we find that the order of genes in the chromosomes is *dp-b-hk*, with 35.5 mu between *dp* and *b* and 5.4 mu between *b* and *hk*.

b. (1) The frequency of observed double crossovers is 1.4%, and the frequency of expected double crossovers is 1.917%. The coefficient of coincidence, therefore, is 1.4/1.917 = 0.73. (2) The interference value is given by 1 − coefficient of coincidence, or 0.27.

5.16 The two X chromosomes in the female have the genotypes $a +$ and $+ b$. The lethal must be on the $+ b$ chromosomes of the female parent because there are far fewer parental $+ b$ chromosomes in the progeny males than $a +$ chromosomes. Therefore, we can diagram the cross as

$$\frac{a + +}{+ b\ l}\, \female \times \xrightarrow{+ + +}\, \male$$

All female progeny of the cross will be wild-type. At this point, however, we do not know the order of the three genes on the chromosome.

We would expect 1,000 males normally, but we get only 499. The remainder can be considered the lethal (*l*) progeny. This fact allows us to predict the entire set of eight genotypes expected among the zygotes of the cross. This is done by knowing that $a +$ is really $+ +$, where the last $+$ sign is the wild-type allele of the lethal. Thus, each genotype seen is accompanied in this cross by one carrying the *l* allele. Writing them out and recognizing the equality of the reciprocal classes, we have

405 $a + +$	and	405 $+ b\ l$
44 $+ b +$	and	44 $a + +$
48 $+ + +$	and	48 $a\ b\ l$
2 $a\ b +$	and	2 $+ + l$

This is 810 parentals ($a + +$ and $+ b\ l$); 88 single crossovers, one region ($+ b +$ and $a + +$); 96 single crossovers, one region ($+ + +$ and $a\ b\ l$); and 4 double crossovers ($a\ b +$ and $+ + l$). Traditional methods lead us to the following map:

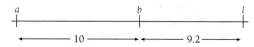

5.22 a. From applying the tetrad analysis formula,

$$\text{map distance} = \frac{\frac{1}{2}T + NPD}{\text{Total}} \times 100\%$$

the *a–b* distance is 19.6 map units, the *b–c* distance is 11 map units, and the *a–c* distance is 14 map units. The gene order is *a–c–b*.

Chapter 6: Gene Mapping in Bacteria and Bacteriophages

6.1 The whole chromosome would have to be transferred in order for the recipient to become a donor in an *Hfr* × *F⁻* cross; that is, the F factor in the *Hfr* strain is transferred to the *F⁻* cell last. This transfer takes approximately 100 minutes, and usually the conjugal unions break apart before then.

6.2

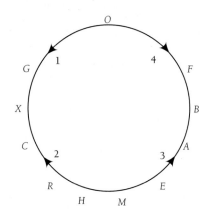

6.4 Strain A is *thy⁻ leu⁺*, and strain B is *thy⁺ leu⁻*. Transformed B should be *thy⁺ leu⁻*, and its presence should be detected on a medium containing neither threonine nor leucine.

6.5 0.07 mu. The plaques produced on *K12(λ)* are *r⁺* phages, and in the undiluted lysate there are 470×5 per milliliter (because 0.2 mL was plated), or 2,350/mL. The *r⁺* phages were generated by recombination between the two *rII* mutations. The other product of the recombination event is the double mutant, and it does not grow on *K12(λ)*. Therefore, the true number of recombinants in the population is equivalent to twice the number of *r⁺* phages because for every wild-type phage produced, there ought to be a doubly mutant recombination produced. Therefore, there are 4,700 recombinants/mL. The total number of phages in the lysate is $627 \times$ (dilution factor) × (1 mL divided by the sample size plated) per milliliter, or $672 \times 1,000 \times 10 = 6,720,000$/mL. The map distance between the mutations is $(4,700/6,720,000) \times 100\% = 0.07\% = 0.07$ mu.

6.7 Two answers are compatible with the data:

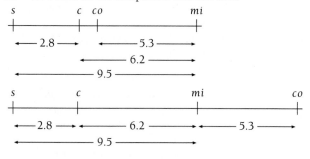

6.9 0.5% recombination (the numbers of plaques counted are so small that this value is a rather rough approximation).

6.12 a. 3 genes

b. A, D, and F are in one; B and G are in the second; and C and E are in the third.

Chapter 7: Non-Mendelian Genetics

7.1 The features of extranuclear inheritance are differences in reciprocal cross results (not related to sex), nonmappability to known nuclear chromosomes, Mendelian segregation not followed, and the indifference to nuclear substitution.

7.4 **a.** The *tudor* mutation is a maternal effect mutation. Homozygous *tudor* mothers give rise to sterile progeny, regardless of their mate.

b. The "grandchildless" phenotype results from the absence of some maternally packaged component in the egg needed for the development of the F$_1$'s germ line.

7.5 **a.** If normal cytoplasm is [N] and the male-sterile cytoplasm is [Ms], then the F$_1$ genotype is [Ms] *Rf/rf* and is male-fertile.

b. The cross is [Ms] *Rf/rf* female × [N] *rf/rf* male. Half of the progeny would be [Ms] *Rf/rf* and half would be [Ms] *rf/rf*. Thus, half of the progeny will be male-fertile and half will be male-sterile.

7.7 ½ *petite*, ½ wild-type (*grande*).

7.10 The first possibility is that the results are the consequence of a sex-linked lethal gene. The females would be homozygous for a dominant gene *L* that is lethal in males but not in females. In this case, mating the F$_1$ females of an *L/L* × +/Y cross to +/Y males should give a sex ratio of 2 females : 1 male in the progeny flies.

The second possibility is that the trait is cytoplasmically transmitted via the egg and is lethal to males. In this case, the same F$_1$ females should continue to have only female progeny when mated with +/Y males.

7.14 The parental snails were *D/d* female and *d/–* male. The F$_1$ snail is *d/d*. (Given the F$_1$ genotype, the male can be either homozygous *d* or heterozygous, but the determination cannot be made from the data given.)

Chapter 8 DNA: The Genetic Material

8.2 **a.** lived

b. died

c. lived

d. died (DNA from the *IIIS* bacteria transformed the *IIR* bacteria to a virulent form)

8.4 **a, b,** and **c.** C, N and H are all present both in proteins and in DNA, so label would be located both within and on the surface of the host cell.

8.8 **a.** 3′-T C A A T G G A C T A G C A T-5′

b. 3′-A A G A G T T C T T A A G G T-5′

8.9 The chemical properties of nucleotides allow these compounds to perform several essential functions. One key function is storage of energy in high-energy bonds: bonds that, when broken, release more energy than covalent bonds. The two nucleotide chains of DNA are held together by hydrogen bonds between the bases and by other forces. The hydrogen bonds always connect adenine to thymine and cytosine to guanine. The linking of specific bases is called complementary base pairing and is determined by the sizes and shapes of the bases. The bonds between adenine and thymine are double bonds and between guanine and cytosine are triple bonds. The double bonds would be the most difficult to break. The triple bonds would be the easiest because the electron cloud is perpendicular, so it can be attacked from more than one plane. Double bonds would have fewer planes and would be more secure.

8.11 b, c, and d

8.14 Since the DNA molecule is double-stranded, (A) = (T) and (G) = (C). If there are 80 T residues, there must be 80 A

residues. If there are 110 G residues, there must be 110 C residues. The molecule has (110 + 110 + 80 + 80) = 380 nucleotides, or 190 base pairs.

8.15 First, notice that (A) ≠ (T) and (G) ≠ (C). Thus, the DNA is not double-stranded. The bacterial virus appears to have a single-stranded DNA genome.

8.19 **a.** 200,000

b. 10,000

c. 3.4×10^4 nm

8.22 Answer is C.

8.23 **a.** Only eukaryotic chromosomes have centromeres, the sections of the chromosome found near the point of attachment of mitotic or meiotic spindle fibers. In some organisms, such as *S. cerevisiae,* they are associated with specific *CEN* sequences. In other organisms, they have a more complex repetitive structure.

b. Hexose sugars are not found in either eukaryotic or bacterial chromosomes, as *pentose* sugars are found in DNA.

c. Amino acids are found in proteins involved in chromosome compaction, such as the proteins that hold the ends of looped domains in prokaryotic chromosomes and the histone and nonhistone proteins in eukaryotic chromatin.

d. Both eukaryotic and bacterial chromosomes share supercoiling.

e. Telomeres are found only at the end of eukaryotic chromosomes and are required for replication and chromosome stability. They are associated with specific types of sequences: simple telomeric sequences and telomere-associated sequences.

f. Nonhistone proteins are found only in eukaryotic chromosomes and have structural (higher-order packaging) and possibly other functions.

g. DNA is found in both prokaryotic and eukaryotic chromosomes. (However, some viral chromosomes have RNA as their genetic material.)

h. Nucleosomes are the fundamental unit of packaging of DNA in eukaryotic chromosomes and are not found in prokaryotic chromosomes.

i. Circular chromosomes are found only in prokaryotes, not in eukaryotes.

j. Looping is found in both eukaryotic and prokaryotic chromosomes. In eukaryotic chromosomes, the 30-nm nucleofilament is packed into looped domains by nonhistone chromosomal proteins. In bacterial chromosomes such as that of *E. coli,* there are about 100 looped domains, each containing about 40 kb of supercoiled DNA.

8.25 **a.** The belt forms a right-handed helix. Although you wrapped the belt around the can axis in a counterclockwise direction from your orientation (looking down at the can), the belt was winding up and around the side of the can in a clockwise direction from its orientation. While the belt is wrapped around the can, curve the fingers of your right hand over the belt and use your index finger to trace the direction of the belt's spiral. Your right index finger will trace the spiral upward, the same direction your thumb points when you wrap your hand around the can. Therefore, the belt has formed a right-handed helix.

b. Three

c. Three. The number of helical turns is unchanged, although the twist in the belt is.

d. The belt appears more twisted because the pitch of the helix is altered, and the edges of the belt (positioned much like the complementary base pairs of a double helix) are twisted more tightly.

e. When it is twisted around the can, the length of the belt decreases by about 70–80%, depending on the initial length of the belt and the diameter of the can.

f. Yes. As the DNA of linear chromosomes is wrapped around histones to form the 10-nm nucleofilament, it becomes supercoiled. In much the same manner as you must add twists to the belt in order for it to lie flat on the surface of the can, supercoils must be introduced into the DNA for it to wrap around the histones.

g. Topoisomerases increase or reduce the level of negative supercoiling in DNA. For linear DNA to be packaged, negative supercoils must be added to it.

8.28 In unique-sequence DNA. See text p. 178.

Chapter 9: DNA Replication

9.2 Key: ^{15}N-^{15}N DNA = HH; ^{15}N-^{14}N DNA = HL; ^{14}N-^{14}N DNA = LL.

a. Generation 1: all HL; 2: ½ HL, ½ LL; 3: ¼ HL, ¾ LL; 4: ⅛ HL, ⅞ LL; 6: 1/32 HL, 31/32 LL; 8: 1/128 HL, 127/128 LL

b. Generation 1: ½ HH, ½ LL; 2: ¼ HH, ¾ LL; 3: ⅛ HH, ⅞ LL; 4: 1/16 HH, 15/16 LL; 6: 1/64 HH, 63/64 LL; 8: 1/256 HH, 255/256 LL

9.4 a. That DNA replication is semiconservative does not a priori require DNA replication to be semidiscontinuous. For example, if both of the two old strands were completely unwound and replication were initiated from the 3′ end of each, it could proceed continuously in a 5′→3′ direction along each strand. Alternatively, if DNA polymerase were able to synthesize DNA in both the 3′→5′ and 5′→3′ directions, DNA replication could proceed continuously on both DNA strands.

b. That DNA replication is semidiscontinuous does ensure that it is semiconservative. In the semidiscontinuous model, each old, separated strand serves as a template for a new strand. This is the essence of the semiconservative model.

c. That DNA polymerase synthesizes just one new strand from each "old" single-stranded template and can synthesize in only one direction (5′→3′) ensures that replication is semiconservative.

9.5 See text, pp. 188-189. DNA polymerase; intact, high-molecular-weight DNA; dATP, dGTP, dTTP, and dCTP; and magnesium ions are needed.

9.8 None is an analogue for adenine, B and D are analogues of thymine, C is an analogue of cytosine, and A is an analogue of guanine.

9.11 Since a replication fork moves at a rate of 10^4 bp per minute and each replicon has two replication forks moving in opposite directions, in one replicon, replication would be occurring at a rate of 2×10^4 bp/minute. Since all the organism's DNA is replicated in 3 minutes, the number of replicons in the diploid genome is

$$\frac{4.5 \times 10^8 \text{ bp}}{3 \text{ minutes}} \times \frac{1 \text{ replicon}}{2 \times 10^4 \text{ bp / minute}} = 7,500$$

9.13 DNA ligase catalyzes the formation of a phosphodiester bond between the 3′-OH and the 5′-monophosphate groups on either side of a single-strand DNA gap, sealing the gap. Temperature-sensitive ligase mutants would be unable to seal such gaps at the restrictive (high) temperature, leading to fragmented lagging strands and presumably cell death. If a biochemical analysis were performed on DNA synthesized after E. coli were shifted to a restrictive temperature, there would be an accumulation of DNA fragments the size of Okazaki fragments. This would provide additional evidence that DNA replication must be discontinuous on one strand.

9.14 Assume that the amount of a gene's product is directly proportional to the number of copies of the gene present in the E. coli cell. Assay the enzymatic activity of genes at various positions in the E. coli chromosome during the replication period. Then, some genes (those immediately adjacent to the origin) will double their activity very shortly after replication begins. Relate the map position of genes having doubled activity to the amount of time that has transpired since replication was initiated. If replication is bidirectional, there should be a doubling of the gene products both clockwise and counterclockwise from the origin.

9.16 Clearly, DNA replication in the Jovian bug does not occur as it does in E. coli. Assuming that the double-stranded DNA is antiparallel as it is in E. coli, the Jovian DNA polymerases must be able to synthesize DNA in the 5′→3′ direction (on the leading strand) as well as in the 3′→5′ direction (on the lagging strand). This is unlike any DNA polymerase on Earth.

9.17 See text, p. 189 and p. 196.

9.19 Assuming these cells have the typical 4-hour G$_2$ period, you would add ^{3}H thymidine to the medium, wait 4.5 hours, and prepare a slide of metaphase chromosomes. Autoradiography would then be done on this chromosome preparation. Regions of chromosomes displaying silver grains are then the late-replicating regions. Cells that were at earlier stages in the S period when they began to take up ^{3}H will not have had sufficient time to reach metaphase.

9.20 a. Harlequin chromosomes are prepared by allowing tissue culture cells to undergo two rounds of DNA replication in the presence of BUdR. After BUdR (a base analogue that replaces T) becomes incorporated in the DNA, metaphase chromosomes stained with Giemsa stain and a fluorescent dye have one darkly stained and one lightly stained chromatid. Since BUdR-DNA stains less intensely than T-DNA, the presence of two differently stained chromatids indicates that one chromatid has two BUdR-labeled strands and the other has one BUdR- and one T-labeled strand.

This supports the semiconservative model of DNA replication. After one round of replication, each DNA molecule has one BUdR- and one T-labeled strand. After two rounds of replication, each of these molecules produces one DNA molecule with both strands labeled with BUdR (the light chromatid) and one DNA molecule with one BUdR- and one T-labeled strand (the dark chromatid).

b. Chromosomes with chromatids that appear to contain segments of both T-labeled DNA and BUdR-labeled DNA might have had a sister chromatid exchange (mitotic crossing-over).

9.21 Telomerase synthesizes the simple-sequence telomeric repeats at the ends of chromosomes. The enzyme is made up of both protein and RNA, and the RNA component has a base sequence that is complementary to the telomere repeat unit. The RNA component is used as a template for the telomere repeat, so that if the RNA component were altered, the telomere repeat would be as well. Therefore, the mutant in this question is likely to have an altered RNA component.

Chapter 10: Gene Control of Proteins

10.2 The double homozygote should have PKU but not AKU. The PKU block should prevent most homogentisic acid from being formed, so that it could not accumulate to high levels.

10.4 a. The simplest approach is to calculate the proportion of the F_2 that are colored and to subtract that answer from 1. In this case, the noncolorless are the brown or black progeny. The proportion of progeny that make at least the brown pigment is given by the probability of having the following genotype: $a^+/- \, b^+/-$ $c^+/- \, (d^+/d^+, \, d^+/d, \, \text{or} \, d/d)$. The answer is $\frac{3}{4} \times \frac{3}{4} \times \frac{3}{4} = \frac{27}{64}$. Therefore, the proportion of colorless is $1 - \frac{27}{64} = \frac{37}{64}$.

b. The brown progeny have the following genotype: $a^+/- \, b^+/- \, c^+/- \, d/d$. The probability of getting individuals with this genotype is $\frac{3}{4} \times \frac{3}{4} \times \frac{3}{4} \times \frac{1}{4} = \frac{27}{256}$.

10.6 a. Since any of the normal alleles a^+, b^+, or c^+ is sufficient to catalyze the reaction leading to color, in order for color to fail to develop, all three normal alleles must be missing. That is, the colorless F_2 must be $a/a \, b/b \, c/c$. The chance of obtaining such an individual is $\frac{1}{4} \times \frac{1}{4} \times \frac{1}{4} \times = \frac{1}{64}$.

b. Now, colorless F_2 are obtained if *either* one or both steps of the pathway are blocked. That is, colorless F_2 are obtained in either of the following genotypes: $d/d \, -/- \, -/- \, -/-$ (the first or both steps blocked) or $d^+/- \, a/a \, b/b \, c/c$ (second step blocked). The chance of obtaining such individuals is

$$(\tfrac{1}{4} \times 1 \times 1 \times 1) + (\tfrac{3}{4} \times \tfrac{1}{4} \times \tfrac{1}{4} \times \tfrac{1}{4}) = \frac{(64+3)}{256} = \frac{67}{256}.$$

10.8 a. Half have white eyes (the sons), and half have fire-red eyes (the daughters).

b. All have fire-red eyes.

c. All have brown eyes.

d. All are w^+; $\frac{1}{4}$ are $bw^+/- \, st^+/-$, red; $\frac{1}{4}$ are $bw^+/- \, st^+/st$, scarlet; $\frac{1}{4}$ are $bw/bw \, st^+/-$, brown; $\frac{1}{4}$ are $bw/bw \, st/st$, the color of 3-hydroxykynurenine plus the color of the precursor to biopterin, or colorless.

10.9 Wild-type T4 will produce progeny phages at all three temperatures. Let us suppose that model (1) is correct. If cells infected with the double mutant are first incubated at 17°C, progeny phages will be produced, and the cells will lyse. The explanation is as follows: The first step, A to B, is controlled by a gene whose product is heat-sensitive. At 17°C, the enzyme works, and A is converted to B, but B cannot then be converted to mature phages because that step is cold-sensitive. When the temperature is raised to 42°C, the A-to-B step is blocked, but the accumulated B can be converted to mature phages because the enzyme involved with that step is cold-sensitive and functional at the high temperature. If model (2) is the correct pathway, then the progeny phages should be produced in a 42°-to-17°C temperature shift but not vice versa. In general, two gene product functions can be ordered by this method whenever one temperature shift allows phage production and the reciprocal shift does not, according to the following rules: (1) If a low-to-high temperature shift results in phages but a high-to-low temperature shift does not, the *hs* step precedes the *cs* step (model 1); (2) if a high-to-low temperature shift results in phages but a low-to-high temperature shift does not, the *cs* step precedes the *hs* step (model 2).

10.11 One approach to this problem is to try to sequentially fit the data to each pathway, as if each were correct. Check where each mutant could be blocked (remember, each mutant carries only one mutation), whether the mutant would be able to grow if supplemented with the single nutrient that is listed, and whether the mutant would not be able to grow if supplemented with the "no growth" intermediate. It will not be possible to fit the data for mutant 4 to pathway b or data for mutants 3 and 4 to pathway c. The data for all mutants can be fit only to pathway d. Therefore, d must be the correct pathway.

A second approach to this problem is to realize that in any linear segment of a biochemical pathway (a segment without a branch), a block early in the segment can be circumvented by any metabolites that normally appear later in the same segment. Consequently, if two (or more) intermediates can support growth of a mutant, they normally are made after the blocked step in the same linear segment of a pathway. From the data given, compounds D and E both circumvent the single block in mutant 4. This means that compounds D and E lie after the block in mutant 4 on a linear segment of the metabolic pathway. The only pathway where D and E lie in an unbranched linear segment is pathway d. Mutant 4 could be blocked between A and E in this pathway. Mutant 4 cannot be fit to a single block in any of the other pathways that are shown, so the correct pathway is d.

10.12 Use the nutrition and accumulation data to infer where each mutant is blocked. 1 accumulates ornithine, and can grow only if citrulline or arginine is added. Hence, 1 is blocked in the conversion of ornithine to citrulline. 2 and 3 are blocked in the conversion of citrulline to arginine. Mutant 4 is more complex. It can grow only if supplemented with arginine, but accumulates ornithine. 4 must be a double mutant, and is blocked in the conversion of ornithine to citrulline as well as from citrulline to arginine. This gives the following pathway:

$$\text{precursor} \xrightarrow{} \text{ornithine} \xrightarrow{1,\,4} \text{citrulline} \xrightarrow{2,\,3,\,4} \text{arginine}$$

Enzyme: $\quad\quad\quad A \quad\quad\quad\quad\quad B \quad\quad\quad\quad\quad\quad C$

Use the complementation-test data to determine whether two mutants affect the same function. Two mutants able to grow on minimal media complement each other: one provides a function missing in the other. Failure to grow indicates no complementation: both are missing the same function. 2 and 3 complement 1, so 2 and 3 affect a different step than 1. 2 and 3 fail to complement each other, so affect the same step. 4 fails to complement 1, 2, or 3, and affects both steps. This is consistent with the above diagram: 1 and 4 affect enzyme B, and 2, 3, and 4 affect enzyme C.

Start to analyze the recombination data by assigning symbols to the genes and writing out the genotypes and crosses. Let b^+ encode enzyme B and c^+ encode enzyme C. Then $1 = b^1 \, c^+$, $2 = b^+ \, c^2$, $3 = b^+ \, c^3$, and $4 = b^4 \, c^4$. The prototrophic ascospores produced in the crosses 1×2, 1×3, and 2×3 are $b^+ \, c^+$ and are half of the recombinant progeny (e.g., in 1×2, parental types are $b^1 \, c^+$ and $b^+ \, c^2$, recombinant types are $b^+ \, c^+$ and $b^1 \, c^2$). The results can be tabulated as:

Cross	Genotypes	$b^+ \, c^+$ Spores
1×2	$b^1 \, c^+ \times b^+ \, c^2$	25.0%
1×3	$b^1 \, c^+ \times b^+ \, c^3$	25.0%
1×4	$b^1 \, c^+ \times b^4 \, c^4$	$<1 \times 10^{-6}$
2×3	$b^+ \, c^2 \times b^+ \, c^3$	0.002%
2×4	$b^+ \, c^2 \times b^4 \, c^4$	0.001%
3×4	$b^+ \, c^3 \times b^4 \, c^4$	$<1 \times 10^{-6}$

Use the recombination frequencies to infer whether the b and c genes are linked, and whether two alleles at one gene are identical or can be separated by recombination. 1×2 and 1×3 give 25 percent prototrophs (show 50 percent recombinants), so b and c are unlinked. 2×3 produces very few prototrophs, as they can only arise from rare, intragenic recombination. The

0.002% prototrophs (= half the recombinants) indicate that c^2 and c^3 are 0.004 mu apart. 4 fails to recombine with 1 or 3, indicating that $b^4 = b^1$ and $c^4 = c^3$ (they occupy the same site). The 0.001% prototrophs from 2×4 arise when a c^+ allele is obtained after intragenic recombination between c^2 and c^4, and the b^+ allele (in mutant 2) assorts with it into the ascopore. When c^+ assorts with b^4, (half of the time) a prototrophic ascospore is not obtained. This is why only 0.001% prototrophs are recovered from 2×4, instead of the 0.002% prototrophs recovered from 2×3. This results in the following map:

$$b^1 = b^4 \qquad c^3 = c^4$$

a. Three distinct mutational sites exist based on the recombination analysis: one in the b gene, and two in the c gene.

b. Two polypeptide chains are affected based on the complementation analysis: one encoded by the b gene, and one encoded by the c gene.

c. The genotypes are as given in the table above.

d. There are 0.004 mu between the two mutant sites in the c gene identified by c^2 and c^3. The b and c genes are unlinked.

e. The b and c genes assort independently of one another. Thus, strains 1, 2, or 3, when mated with the wild type, will give 50 percent prototrophs and 50 percent auxotrophs. As strain 4 is a double mutant, it will only give 25 percent prototrophs. Four equally frequent genotypes would be expected from the cross $b^4 c^4 \times + +$. Only one of the four will be $+ +$.

10.14 Baby Joan, whose blood type is O, must belong to the Smith family. Mrs. Jones, being of blood type AB, could not have an O baby. Therefore, Baby Jane must be hers.

10.15a. Normal parents have affected offspring; the disease appears to be recessive. However, because patients with 50% of GSS activity have a mild form of the disease, an individual who was heterozygous (mutant/+) for a complete loss-of-function GSS mutation might show mild symptoms. Therefore, in a population the disease may show variable penetrance. The expressivity of the disease in an individual depends on the nature of the person's GSS mutation. The alleles discussed here appear to be recessive, but a complete loss-of-function allele may show partial dominance.

b. Patient 1, with 9% of normal GSS activity, has a more severe form of the disease, and patient 2, with 50% of normal GSS activity, has a less severe form of the disease. Thus, increased disease severity is associated with less GSS enzyme activity.

c. The two amino acid substitutions can disrupt different regions of the enzyme's structure (consider the effect of different amino acid substitutions on hemoglobin function, discussed in the text). As amino acids vary in their polarity and charge, different amino acid substitutions within the same structural region could have different chemical effects on protein structure. This also could lead to different levels of enzymatic function. (For a discussion of the chemical differences between amino acids, see Chapter 12.)

d. By analogy with the disease PKU, discussed in the text, 5-oxoproline is produced only when a precursor to glutathione accumulates in large amounts because of a block in a biosynthetic pathway. When GSS levels are 9% of normal, this occurs. When GSS levels are 50% of normal, sufficient GSS enzyme is present to partially complete the pathway and prevent high levels of 5-oxoproline.

e. The mutations are allelic because both the severe and the mild forms of the disease are associated with alterations in the same polypeptide that is a component of the GSS enzyme. (Note that although the data in this problem suggest that the GSS enzyme is composed of a single polypeptide, they do not exclude the possibility that GSS has multiple polypeptide subunits.)

f. If GSS is normally found in fetal fibroblasts, one could, in principle, measure GSS activity in fibroblasts obtained via amniocentesis. The GSS enzyme level in cells from at-risk fetuses could be compared with that in normal control samples to predict disease caused by inadequate GSS levels. Some variation in GSS level might be seen, depending on the alleles present. Since more than one mutation is present in the population, it is important to devise a functional test that assesses GSS activity rather than a test that identifies a single mutant allele.

10.18 If the enzyme that catalyzes the d→e reaction is missing, the mutant strain should accumulate d and be able to grow on minimal medium to which e is added. In addition, it should not be able to grow on minimal medium or on minimal medium to which X, c, or d is added but should grow if Y is added. Therefore, plate the strain on these media and test which allow growth of the mutant strain and which intermediate is accumulated if the strain is plated on minimal medium.

10.22 Autosomes are chromosomes that are found in two copies in both males and females. That is, an autosome is any chromosome except the X and Y chromosomes. Since individuals have two of each type of autosome, they have two copies of each gene on an autosome. The forms of the gene, or the alleles at the gene, can be the same or different. They can have either two normal alleles (homozygous for the normal allele), one normal allele and one mutant allele (heterozygous for the normal and mutant alleles), or two mutant alleles (homozygous for the mutant allele). A recessive mutation is one that exhibits the phenotype only when it is homozygous. Therefore, an autosomal recessive mutation is a mutation on any chromosome except the X or Y and causes a phenotype only when homozygous. Heterozygotes exhibit a normal phenotype.

Of the diseases discussed in this chapter, many are autosomal recessive. For example phenylketonuria, albinism, Tay-Sachs disease, and cystic fibrosis are autosomal recessive diseases. Heterozygotes for the disease allele are normal, but homozygotes with the disease allele are affected. For phenylketonuria and albinism, homozygotes are affected because they lack a required enzymatic function. In these cases, heterozygotes have a normal phenotype because their single normal allele provide sufficient enzyme function.

Parents contribute one of their two autosomes to their gametes, so that each offspring of a couple receives an autosome from each parent. If in a particular conception each of two heterozygous parents contributes a chromosome with the normal allele, the offspring will be homozygous for the normal allele and be normal. If in a particular conception one of the two heterozygous parents contributes a chromosome with the normal allele and the other parent contributes a chromosome with the mutant allele, the offspring will be heterozygous but be normal. If in a particular conception each parent contributes a chromosome with the mutant allele, the offspring will be homozygous for the mutant allele and develop the disease. Therefore, heterozygous parents can have both normal and affected children. Since each conception is independent, two heterozygous parents can have all normal, all affected, or any mix of normal and affected children.

Chapter 11: Gene Expression: Transcription

11.3 See text, pp. 228–230 and 230–232.

11.6

RNA Polymerase	Cellular Location	Product & Other Characteristics
I	nucleolus	28S, 18S, 5.8S rRNA: These are structural and functional components of the ribosome, which functions during translation.
II	nucleoplasm	1) mRNAs: These provide protein coding information to ribosome. 2) Some mRNAs (small nuclear RNAs): These function in a variety of nuclear processes, including RNA splicing and processing.
III	nucleoplasm	tRNAs (transfer RNAs): These carry amino acids to the ribosome during translation. Some small nuclear RNAs: These function in a variety of nuclear processes. 5S rRNA: This is a functional and structural component of the ribosome's 60S (large) subunit.

11.9 By convention, DNA sequence is given $5' \rightarrow 3'$. This is the same polarity as the antitemplate strand, the strand that is complementary to the $3' \rightarrow 5'$ template strand used for transcription. By using this convention, analyzing a DNA sequence is made easier. If a region of the sequence is transcribed, the RNA will have an identical sequence, except that U will replace T.

Approach this problem by surveying the sequence for the GC, CAAT, and TATA box consensus sequences described on p. 232 of the text. The following figure summarizes their location and the approximate site of transcription initiation. An mRNA that could be produced (a conceptual mRNA) is given in lowercase letters. This mRNA has two signals (underlined) that could be used by the ribosome to recognize the start and termination of translation. Initiation of a polypeptide chain is signaled by a 5'-AUG-3' codon (encoding a methionine). Chain termination is signaled by a UAA stop codon. Features of translation are described in Chapter 12.

```
        −80                    −65                    −50
     GC element             CAAT box              GC element
5' AGAGGGCGGT   CCGTATCGGC CAATCTGCTC   CAAGGGCGGA

                              −30
                            TATA box
TTCACACGTT  GTTATATAAA TGACTGGGCG TACCCCAGGG

+1 (approx., conceptual)
transcription initiation potential translation start
TTCGAGTATT  CTATCGTATG GTGCACCTGA     CT(...)
mRNA 5' uauu   cuaucguaug gugcaccuga     cu(...)

        potential translation stop
GCTCACAAGT ACCACTAAGC   (...)
gcucacaagu accacuaagc   (...)
```

11.14 In eukaryotes, 18S, 28S, and 5.8S rRNA genes are transcribed from the rDNA into a single pre-rRNA molecule. The pre-

mRNA is processed by removing the internal and external spacer sequences, leading to the production of the mature rRNAs. See text Figure 11.16. The eukaryotic 5S rRNA is transcribed separately to produce mature rRNA molecules that need no further processing. Eukaryotic 5S rRNAs are imported into the nucleolus where they are assembled with the other mature rRNAs and the ribosomal protein subunits to produce functional ribosomal subunits. Some *Tetrahymena* pre-rRNAs have self-splicing introns (group I) in the 28S rRNA. It is important to remember that the removal of the intron in the 28S rRNA via self-splicing is a separate process from the cleavage of spacer sequences from the mature rRNAs. See text Figure 11.17.

Protein-coding eukaryotic transcripts synthesized by RNA polymerase II are extensively processed. A 5'-5' bonded m^{7}Gppp cap is added to the 5' end of the nascent transcript when the chain is 20–30 nucleotides long. A poly(A) tail of variable length can be added to the 3' end of the transcript, and spliceosomes remove intronic sequences. The site of poly(A) addition, as well as splice-site selection, can be regulated, so that more than one alternatively processed mature mRNA is sometimes produced from a single precursor mRNA transcript.

11.16a. This would prevent splicing out of any sequences corresponding to introns and would result in abnormal sequences of most proteins. This should be lethal.

b. If splicing is not affected, this should have no phenotypic consequence.

c. This would prevent the splicing out of intron 2 material. Since the 5' cut could still be made, it is likely that this mutation would lead to the absence of functional mRNA and the absence of β-globin. (This should produce a phenotype similar to that seen when the β-globin gene is deleted. In that case, there is anemia compensated by increased production of fetal hemoglobin. The condition is called β-thalassemia.)

11.18 The first two bases of an intron typically are 5'-GU-3', which are essential for base pairing with the U1 snRNA during spliceosome assembly. A GC-to-TA mutation at the initial base pair of the first intron impairs base pairing with the U1 snRNA, so that the 5' splice site of the first intron is not identified. This causes the retention of the first intron in the *tub* mRNA and a longer mRNA transcript in *tub/tub* mutants. When the mutant *tub* mRNA is translated, the retention of the first intron could result in either the introduction of amino acids not present in the *tub*+ protein or, if the intron contained a chain termination (stop) codon, premature translation termination and the production of a truncated protein. In either case, a nonfunctional gene product is produced.

The *tub* mutation is recessive because the single *tub*+ allele in a *tub/tub*+ heterozygote produces mRNAs that are processed normally, and when these are translated, enough normal (*tub*+) product is produced to obtain a wild-type phenotype. Only the *tub* allele produces abnormal transcripts. When both copies of the gene are mutated in *tub/tub* homozygotes, no functional product is made, and a mutant, obese phenotype results.

11.19a. This could occur. Since the 5S genes are clustered, 10 copies could be removed in one deletion.

b. This could not occur. The 18S genes are part of a cluster of 18S, 5.8S, and 28S genes. The 18S genes cannot be deleted without also deleting the others.

c. This could occur. The genes are clustered.

d. This could not occur. The 18S, 5.8S, and 28S genes are clustered at a locus separate from the 5S genes.

11.21a. 3
  b. 1, 2, 3, 4
  c. 3
  d. 1, 2
  e. 1
  f. 4
  g. 1

Chapter 12: Gene Expression: Translation

12.1 b

12.6 Consider Figure 12.7, p. 254, in answering this question.

Amino Acid	tRNAs Needed	Rationale
Ile	1	3 codons can use 1 tRNA (wobble)
Phe	1	2 codons can use 1 tRNA (wobble)
Tyr	1	" " " " "
His	1	" " " " "
Gln	1	" " " " "
Asn	1	" " " " "
Lys	1	" " " " "
Asp	1	" " " " "
Glu	1	" " " " "
Cys	1	" " " " "
Trp	1	1 codon
Met	2	Single codon, but need one tRNA for initiation and one tRNA for elongation
Val	2	4 codons: 2 can use 1 tRNA (wobble)
Pro	2	" " " " "
Thr	2	" " " " "
Ala	2	" " " " "
Gly	2	" " " " "
Leu	3	6 codons: 2 can use 1 tRNA (wobble)
Arg	3	" " " " "
Ser	3	" " " " "
Total	**32**	**61 codons**

12.7 Since a dipeptide is formed, translation initiation is not affected, nor is the first step of elongation: the binding of a charged tRNA in the A site and the formation of a peptide bond. However, since *only* a dipeptide is formed, it appears that translocation is inhibited.

12.11a. 4 A : 6 C

$$AAA = (^4/_{10})(^4/_{10})(^4/_{10}) = 0.064, \text{ or } 6.4\% \text{ Lys}$$
$$AAC = (^4/_{10})(^4/_{10})(^6/_{10}) = 0.096, \text{ or } 9.6\% \text{ Asn}$$
$$ACA = (^4/_{10})(^6/_{10})(^4/_{10}) = 0.096, \text{ or } 9.6\% \text{ Thr}$$
$$CAA = (^6/_{10})(^4/_{10})(^4/_{10}) = 0.096, \text{ or } 9.6\% \text{ Gln}$$
$$CCC = (^6/_{10})(^6/_{10})(^6/_{10}) = 0.216, \text{ or } 21.6\% \text{ Pro}$$
$$CCA = (^6/_{10})(^6/_{10})(^4/_{10}) = 0.144, \text{ or } 14.4\% \text{ Pro}$$
$$CAC = (^6/_{10})(^4/_{10})(^6/_{10}) = 0.144, \text{ or } 14.4\% \text{ His}$$
$$ACC = (^4/_{10})(^6/_{10})(^6/_{10}) = 0.144 \text{ or } 14.4\% \text{ Thr}$$

In sum, 6.4% Lys, 9.6% Asn, 9.6% Gln, 36.0% Pro, 24.0% Thr, and 14.4% His.

  b. 1 A : 3 U : 1 C; the same logic is followed here, using $^1/_5$ as the fraction for A

$$AAA = 0.008, \text{ or } 0.8\% \text{ Lys}$$
$$AAU = 0.024, \text{ or } 2.4\% \text{ Asn}$$
$$AUA = 0.024, \text{ or } 2.4\% \text{ Ile}$$
$$UAA = 0.024, \text{ or } 2.4\% \text{ chain terminating}$$
$$AUU = 0.072, \text{ or } 7.2\% \text{ Ile}$$
$$UAU = 0.072, \text{ or } 7.2\% \text{ Tyr}$$
$$UUA = 0.072, \text{ or } 7.2\% \text{ Leu}$$
$$UUU = 0.216, \text{ or } 21.6\% \text{ Phe}$$
$$AAC = 0.008, \text{ or } 0.8\% \text{ Asn}$$
$$ACA = 0.008, \text{ or } 0.8\% \text{ Thr}$$
$$CAA = 0.008, \text{ or } 0.8\% \text{ Gln}$$
$$ACC = 0.008, \text{ or } 0.8\% \text{ Thr}$$
$$CAC = 0.008, \text{ or } 0.8\% \text{ His}$$
$$CCA = 0.008, \text{ or } 0.8\% \text{ Pro}$$
$$CCC = 0.008, \text{ or } 0.8\% \text{ Pro}$$
$$UUC = 0.072, \text{ or } 7.2\% \text{ Phe}$$
$$UCU = 0.072, \text{ or } 7.2\% \text{ Ser}$$
$$CUU = 0.072, \text{ or } 7.2\% \text{ Leu}$$
$$UCC = 0.024, \text{ or } 2.4\% \text{ Ser}$$
$$CUC = 0.024, \text{ or } 2.4\% \text{ Leu}$$
$$CCU = 0.024, \text{ or } 2.4\% \text{ Pro}$$
$$UCA = 0.024, \text{ or } 2.4\% \text{ Ser}$$
$$UAC = 0.024, \text{ or } 2.4\% \text{ Tyr}$$
$$CUA = 0.024, \text{ or } 2.4\% \text{ Leu}$$
$$CAU = 0.024, \text{ or } 2.4\% \text{ His}$$
$$AUC = 0.024, \text{ or } 2.4\% \text{ Ile}$$
$$ACU = 0.024, \text{ or } 2.4\% \text{ Thr}$$

In sum, 0.8% Lys, 3.2% Asn, 12.0% Ile, 2.4% chain terminating, 9.6% Tyr, 19.2% Leu, 28.8% Phe, 4.0% Thr, 0.8% Gln, 3.2% His, 4.0% Pro, and 12.0% Ser. The likelihood is that the chain would not be long because of the chance of the chain-terminating codon.

12.12

Word Size	Number of Combinations
a. 5	$2^5 = 32$
b. 3	$3^3 = 27$
c. 2	$5^2 = 25$

(The minimum word sizes must uniquely designate 20 amino acids.)

12.14 **a.** A A A A T A A A A A T A etc.
  b. T T T T A T T T T T A T etc.
  c. A A A for Phe and A U A for Tyr

12.16 No chain-terminating codons can be produced from only As and Gs. But the stop codon UAA can be made from As and Us. Therefore, the population A proteins will be longer than those from population B. Most of the population B proteins will be free in solution rather than attached to ribosomes.

12.18 The anticodon 5'-GAU-3' recognizes the codon 5'-AUC-3', which encodes Ile. The mutant tRNA anticodon 5'-CAU-3' would recognize the codon 5'-AUG-3', which normally encodes Met. The mutant tRNA therefore would compete with tRNA.Met for the recognition of the 5'-AUG-3' codon and, if successful, insert Ile into a protein where Met should be. Since a special tRNA.Met is used for initiation, only AUG codons other than the initiation AUG will be affected. Therefore, this protein will have four different N-terminal sequences, depending on which tRNA occupies the A site in the ribosome when the codon AUG is present there:

15.3 A SNP is a single nucleotide polymorphism. Since a single base pair change can alter the site recognized by a restriction endonuclease, a SNP can also be a RFLP. Since STRs and VNTRs are based on variable numbers of tandemly repeated sequences (1–4 and 5–10 bp, respectively), they usually are not SNPs.

15.5 Sequencing of archaeon genomes has shown that their genes are not uniformly similar to those of Bacteria or Eukarya. While most of the archaeon genes involved in energy production, cell division, and metabolism are similar to their counterparts in Bacteria, the genes involved in DNA replication, transcription, and translation are similar to their counterparts in Eukarya.

15.9 a. Since prokaryotic ORFs should reside in transcribed regions, they should follow a bacterial promoter, containing −35 (TTGACA) and −10 (TATAAT) consensus sequences. Within the transcribed region, but before the ORF, there should be a Shine-Dalgarno sequence (UAAGGAGG) used for ribosome binding. Nearby should be an AUG (or GUG, in some systems) initiation codon. This should be followed by a set of in-frame, sense codons. The ORF should terminate with a stop (UAG, UAA, UGA) codon.

b. Eukaryotic introns introduce nontranslated sequences into the coding region of a gene. These sequences will be spliced out of the primary mRNA transcript before it is translated.

c. Eukaryotic introns typically contain a GU at their 5′ ends, an AG at their 3′ ends, and a YNCURAY branch-point sequence 18 to 38 nucleotides upstream of their 3′ ends. To identify eukaryotic ORFs in DNA sequences, scan sequences following a eukaryotic promoter for the presence of possible introns by searching for sets of these three consensus sequences. Then try to translate sequences obtained if potential introns are removed, testing whether a long ORF with good codon usage (for the species in question) can be generated. Since alternative mRNA splicing exists at many genes, more than one possible ORF may be found in a given DNA sequence.

15.11 One approach is to use model organisms (e.g., transgenic mice) that have been developed as models to study a specific human disease. Expose them and a control population to specific environmental conditions, and then simultaneously assess disease progression and alterations in patterns of gene expression (using microarrays). This would provide a means to establish a link between environmental factors and patterns of gene expression that are associated with disease onset or progression.

15.13

S, F, C	Aligning DNA sequences within databases to determine the degree of matching
F, C	Annotation of sequences within a sequenced genome
F	Characterizing the transcriptome and proteome present in a cell at a specific developmental stage or in a particular disease state
C	Comparing the overall arrangements of genes and nongene sequences in different organisms to understand how genomes evolve
F	Describing the function of all genes in a genome
F, C	Determining the functions of human genes by studying their homologues in nonhuman organisms
F	Developing a comprehensive two-dimensional polyacrylamide gel electrophoresis map of all proteins in a cell
S	Developing a physical map of a genome
S, F, C	Developing DNA microarrays (DNA chips)
F, C	Identifying homologues to human disease genes in organisms suitable for experimentation
S	Identifying a large collection of simple tandem repeat or microsatellite sequences to use as DNA markers within one organism
F, C	Making gene knockouts and observing the phenotypic changes associated with them

Chapter 16: Regulation of Gene Expression in Bacteria and Bacteriophages

16.3 A constitutive phenotype can be the result of a *lacI⁻* or *lacOᶜ* mutation.

16.5 *lacI⁺ lacOᶜ lacP⁺ lacZ⁺ lacY⁻/lacI⁺ lacO⁺ lacP⁺ lacZ⁻ lacY⁺*. (It cannot be ruled out that one of the repressor genes is *lacI⁺*.)

16.7 Answer is in Table 16.A, p. 580.

16.9 The CAP, in a complex with cAMP, is required to facilitate RNA polymerase binding to the *lac* promoter. The RNA polymerase binding occurs only in the absence of glucose and only if the operator is not occupied by repressor (i.e., lactose is also absent). A mutation in the *CAP* gene, then, would render the *lac* operon incapable of expression because RNA polymerase would not be able to recognize the promoter.

16.11 For a wild-type *trp* operon, the absence of tryptophan results in antitermination; that is, the structural genes are transcribed and the tryptophan biosynthetic enzymes are made. This occurs because a lack of tryptophan results in the absence of, or at least a very low level of, Trp-tRNA.Trp. In turn, this causes the ribosome translating the leader sequence to stall at the Trp codons (see Figure 16.14, p. 335). When the ribosome is stalled at the Trp codons, the RNA being synthesized just ahead of the ribosome by RNA polymerase assumes a particular secondary structure. This favors continued transcription of the structural genes by the polymerase. If the two Trp codons were mutated to stop codons, then the mutant operon would function constitutively in the same way as the wild-type operon in the absence of tryptophan. The ribosome would stall in the same place, and antitermination would result in transcription of the structural genes.

For a wild-type *trp* operon, the presence of tryptophan turns off transcription of the structural genes. This occurs because the presence of tryptophan leads to the accumulation of Trp-tRNA.Trp, which allows the ribosome to read the two Trp codons and stall at the normal stop codon for the leader sequence. When stalled in that position, the antitermination signal cannot form in the RNA being synthesized; instead, a termination signal is formed, resulting in the termination of transcription. In a mutant *trp* operon with two stop codons instead of the Trp codons, the stop codons cause the ribosome to stall, even though tryptophan and Trp-tRNA.Trp are present. This results in an antitermination signal and transcription of the structural genes.

In sum, in both the presence and the absence of tryptophan, the mutant *trp* operon will not show attenuation. The structural genes will be transcribed in both cases, and the tryptophan biosynthetic enzymes will be synthesized.

16.15 The *cI* gene product is a repressor protein that acts to keep the lytic functions of the phage repressed when λ is in the lysogenic state. A *cI* mutant strain would lack the repressor and be unable to repress lysis, so that the phage would always follow a lytic pathway.

Table 16.A

	Genotype	Inducer Absent:		Inducer Present:	
		β-galactosidase	Permease	β-galactosidase	Permease
a.	$I^+ P^+ O^+ Z^+ Y^+$	−	−	+	+
b.	$I^+ P^+ O^+ Z^- Y^+$	−	−	−	+
c.	$I^+ P^+ O^+ Z^+ Y^-$	−	−	+	−
d.	$I^- P^+ O^+ Z^+ Y^+$	+	+	+	+
e.	$I^s P^+ O^+ Z^+ Y^+$	−	−	−	−
f.	$I^+ P^+ O^c Z^+ Y^+$	+	+	+	+
g.	$I^s P^+ O^c Z^+ Y^+$	+	+	+	+
h.	$I^+ P^+ O^c Z^+ Y^-$	+	−	+	−
i.	$I^- P^+ O^+ Z^+ Y^+$ / $I^+ P^+ O^+ Z^- Y^-$	−	−	+	+
j.	$I^- P^+ O^+ Z^+ Y^-$ / $I^+ P^+ O^+ Z^- Y^+$	−	−	+	+
k.	$I^s P^+ O^+ Z^+ Y^-$ / $I^+ P^+ O^+ Z^- Y^+$	−	−	−	−
l.	$I^+ P^+ O^c Z^- Y^+$ / $I^+ P^+ O^+ Z^+ Y^-$	−	+	+	+
m.	$I^- P^+ O^c Z^+ Y^-$ / $I^+ P^+ O^+ Z^- Y^+$	+	−	+	+
n.	$I^s P^+ O^+ Z^+ Y^+$ / $I^+ P^+ O^c Z^+ Y^+$	+	+	+	+
o.	$I^+ P^- O^c Z^+ Y^-$ / $I^+ P^+ O^+ Z^- Y^+$	−	−	+	+
p.	$I^+ P^- O^+ Z^+ Y^-$ / $I^+ P^+ O^c Z^- Y^-$	−	+	−	+
q.	$I^- P^- O^+ Z^+ Y^+$ / $I^+ P^+ O^+ Z^- Y^-$	−	−	−	−
r.	$I^- P^+ O^+ Z^+ Y^-$ / $I^+ P^- O^+ Z^- Y^+$	−	−	+	−

Chapter 17: Regulation of Gene Expression in Eukaryotes

17.3 a. In fragile X syndrome, the expanded CGG repeat results in hypermethylation and transcriptional silencing. In Huntington's disease, the expanded CAG repeat results in the inclusion of a polyglutamine stretch within the huntingtin protein, which causes it to have a novel, abnormal function.

b. A heterozygote with a CGG repeat expansion near one copy of the *FMR-1* gene will still have one normal copy of the *FMR-1* gene. The normal gene can produce a normal product, even if the other is silenced. (The actual situation is made somewhat more complex by the process of X inactivation in females, but in general one would expect that a mutation that caused transcriptional silencing of one allele would not affect a normal allele on a homologue.) In contrast, a novel, abnormal protein is produced by the CAG expansion in the disease allele in Huntington's disease. Since the disease phenotype is caused by the presence of the abnormal protein, the disease trait is dominant.

c. Transcriptional silencing may require significant amounts of hypermethylation and so require more CGG repeats for an effect to be seen. In contrast, protein function may be altered by a relatively smaller sized polyglutamine expansion.

17.5 The data indicate that ovalbumin synthesis depends on the presence of the hormone estrogen. These data do not address the mechanism by which estrogen achieves its effects. Theoretically, it could act (1) to increase transcription of the ovalbumin gene by binding to an intracellular receptor that, as an activated complex, stimulates transcription at the ovalbumin gene; (2) to stabilize the ovalbumin precursor mRNA; (3) to increase the processing of the precursor ovalbumin mRNA; (4) to increase the transport of the processed ovalbumin mRNA out of the nucleus; (5) to stabilize the mature ovalbumin mRNA once it has been transported into the cytoplasm; (6) to stimulate translation of the ovalbumin mRNA in the cytoplasm; or (7) to stabilize (or process) the newly synthesized ovalbumin protein. Experiments in which the levels of ovalbumin mRNA were measured have shown that the production of ovalbumin mRNA is regulated primarily at the level of transcription.

17.7 a. In bacterial operons, a common regulatory region controls the production of single mRNA from which multiple protein products are translated. These products function in a related biochemical pathway. Here, two proteins that are involved in the synthesis and packaging of acetylcholine are both produced from a common primary mRNA transcript.

b. Unlike the proteins translated from an mRNA synthesized from a bacterial operon, the protein products produced at the *VAChT/ChAT* locus are not translated sequentially from the same mRNA. Here, the primary mRNA appears to be alternatively processed to produce two distinct mature mRNAs. These mRNAs are translated starting at different points, producing different proteins.

c. At least two mechanisms are involved in the production of the different ChAT and VAChT proteins: alternative mRNA processing and alternative translation initiation. After the first exon, an alternative 3′ splice site is used in the two different mRNAs. In addition, different AUG start codons are used.

17.10 a. Based on the work of Wilmut and his colleagues, the nose cells would first be dissociated and grown in tissue culture. The cells would be induced into a quiescent state (the G_0 phase of the cell cycle) by reducing the concentration of growth serum in the media. Then, they would be fused with enucleated oocytes from a donor female and allowed to grow and divide by mitosis to produce embryos. The embryos would be implanted into a surrogate female. After the establishment of pregnancy, its progression would need to be maintained.

b. Although the nuclear genome generally would be identical to that in the original nose cell, cytoplasmic organelles presumably would derive from those in the enucleated oocyte. Therefore, the mitochondrial DNA would not derive from the original leader. In addition, because telomeres in an older individual are shorter, one might expect the telomeres in the cloned leader to be those of an older individual.

c. In mature B cells, DNA rearrangements at the heavy- and light-chain immunoglobulin genes have occurred. One would expect the cloned leader to be immunocompromised because he would be unable to make the wide spectrum of antibodies present in a normal individual.

d. The production of Dolly was significant because it demonstrated the apparent complete totipotency of a nucleus from a mature mammary epithelium cell. The other six lambs developed from the transplanted nuclei of embryonic or fetal cells, which one might expect to be less determined and have a higher degree of totipotency due to their younger developmental age.

e. There is no way to predict the psychological profile of the cloned leader based on his genetic identity. Even identical twins, who are genetically more identical than such a clone, do not always share behavioral traits.

17.12 Preexisting mRNAs (stored in the oocyte) are recruited into polysomes as development begins following fertilization.

17.16

Mutant	Class
a	segmentation gene (segment polarity)
b	maternal effect gene (anterior-posterior gradient)
c	segmentation gene (gap)
d	homeotic gene (eye to wing transformation)
e	segmentation gene (gap)

17.17 Preexisting mRNA that was made by the mother and packaged into the oocyte before fertilization is translated up to the gastrula stage. After gastrulation, new mRNA synthesis is necessary for the production of proteins needed for subsequent embryonic development.

17.18 The tissue taken from the blastula/gastrula has not yet been committed to its final differentiated state in terms of its genetic programming; that is, it has not yet been *determined*. Thus, when the tissue is transplanted into the host, it adopts the fate of nearby tissues and differentiates in the same way as they do. Presumably, cues from the tissue surrounding the transplant determine its fate. In contrast, tissues in the neurula stage are stably determined. By the time the neurula developmental stage has been reached, a developmental program has been set. In other words, the fate of neurula tissue transplants is *determined*. Upon transplantation, they will differentiate according to their own set genetic program. Tissue transplanted from a neurula to an older embryo cannot be influenced by the surrounding tissues. It will develop into the tissue type for which it has been determined—in this case, an eye.

Chapter 18: Genetics of Cancer

18.1 Hereditary cancer is associated with the inheritance of a germ-line mutation; sporadic cancer is not. Consequently, hereditary cancer runs in families. For some cancers, both hereditary and sporadic forms exist, with the hereditary form being much less frequent. For example, retinoblastoma occurs when both normal alleles of the tumor suppressor gene *RB* are inactivated. In hereditary retinoblastoma, a mutated, inactive allele is transmitted via the germ line. Retinoblastoma occurs in cells of an *RB/+* heterozygote when an additional somatic mutation occurs. In the sporadic form of the disease, retinoblastoma occurs when both alleles are inactivated somatically.

18.3 Tumor growth induced by transducing retroviruses results either from the activity of a single viral oncogene or from the activation of a proto-oncogene caused by the nearby integration of the proviral DNA. The oncogene can cause abnormal cellular proliferation via the variety of mechanisms discussed in the text. The expression of a proto-oncogene, normally tightly regulated during cell growth and development, can be altered if it comes under the control of the promoter and enhancer sequences in the retroviral LTR.

DNA tumor viruses do not carry oncogenes. They transform cells through the action of one or more genes within their genomes. For example, in a rare event, the DNA virus can be integrated into the host genome, and the DNA replication of the host cell may be stimulated by a viral protein that activates viral DNA replication. This would cause the cell to move from the G_0 to the S phase of the cycle.

For both transducing retroviruses and DNA tumor viruses, an abnormally expressed protein leads to the activation of the cell from G_0 to S and abnormal cell growth.

18.4 a. Studies of hereditary forms of cancer have led to insights into the fundamental cellular processes affected by cancer. For example, substantial insights into the important role of DNA repair and the relationship between the control of the cell cycle and DNA repair have come from analyses of the genes responsible for hereditary forms of human colorectal cancer. For breast cancer, studying the normal functions of the *BRCA1* and *BRCA2* genes promises to provide substantial insights into breast and ovarian cancer.

b. Genetic predisposition for cancer is the presence of an inherited mutation that, with additional somatic mutations during the individual's life span, can lead to cancer. For diseases such as retinoblastoma, a genetic predisposition has been associated with the inheritance of a recessive allele of the *RB* tumor suppressor gene. Retinoblastoma occurs in *RB/+* individuals when the normal allele is mutated in somatic cells and the pRB protein no longer functions. Because somatic mutation is likely, the disease appears to be dominant in pedigrees.

Although there is a substantial understanding of the genetic basis for cancer and the genetic abnormalities present in somatic cancerous cells, there are also substantial environmental risk factors for specific cancers. Environmental risk factors must be investigated thoroughly when a pedigree is evaluated for a genetic predisposition for cancer.

18.6 Tumors result from multiple mutational events that typically involve both the activation of oncogenes and the inactivation of tumor suppressor genes. The analysis of hereditary adenomatous polyposis, an inherited form of colorectal cancer, has shown that the more differentiated cells found in benign, early-stage tumors are associated with fewer mutational events, whereas the less differentiated cells found in malignant and metastatic tumors are associated with more mutational events. Although the path by which mutations accumulate varies between tumors, additional mutations that activate oncogenes and inactivate tumor suppressor genes generally result in dedifferentiation and increased cellular proliferation.

18.8 The high degree of conservation of proto-oncogenes suggests that they function in normal, essential, conserved cellular processes. Given the relationship between oncogenes and proto-oncogenes, it also suggests that cancer occurs when these processes are not correctly regulated.

18.11 a. Proto-oncogenes encode a diverse set of gene products that include growth factors, receptor and nonreceptor protein kinases, receptors lacking protein kinase activity, membrane-associated GTP-binding proteins, cytoplasmic regulators involved in intracellular signaling, and nuclear transcription factors. These gene products all function in intercellular and intracellular circuits that regulate cell division and differentiation.

b. In general, mutations that activate a proto-oncogene convert it into an oncogene. Since (i), (iii), and (viii) cause a decrease in gene expression, they are unlikely to result in an oncogene. Since (ii) and (vii) could activate gene expression, they could result in an oncogene. Mutations (iv), (v), and (vi) cannot be predicted with certainty. The deletion of a 3′ splice site acceptor would alter the mature mRNA and possibly the protein produced and may or may not affect the protein's function and regulation. Similarly, it is difficult to predict the effect of a nonspecific point mutation or a premature stop codon. The text presents examples in which these types of mutations have caused the activation of a proto-oncogene and resulted in an oncogene.

18.12 Experimentally fuse cells from the two cell lines and then test the resultant hybrids for their ability to form tumors. If the uncontrolled growth of the tumor cell line was caused by a mutated pair of tumor suppressor alleles, the normal alleles present in the normal cell line would "rescue" the tumor cell line defect. The hybrid line would grow normally and be unable to form a tumor. If the uncontrolled growth of the tumor cell line was caused by an oncogene, the oncogene would also be present in the hybrid cell line. The hybrid line would grow uncontrollably and form a tumor.

18.14 One hypothesis is that the proviral DNA has integrated near the proto-oncogene, and the expression of the proto-oncogene has come under the control of promoter and enhancer sequences in the retroviral LTR. This could be assessed by performing a whole-genome Southern blot analysis to determine whether the organization of the genomic DNA sequences near the proto-oncogene have been altered.

18.17 Apoptosis is programmed, or suicidal, cell death. Cells targeted for apoptosis are those that have large amounts of DNA damage and so are at a greater risk for neoplastic transformation. (During the development of some tissues in multicellular organisms, cell death via apoptosis is a normal process.) Apoptosis is regulated by p53, among other proteins. In cells with large amounts of DNA damage, p53 is stabilized. This leads to the activation of *WAF1*, whose product, p21, causes cells to arrest in G_1 by binding to cyclin/Cdk complexes and blocking the kinase activity needed for progression from G_1 to S. If the DNA damage cannot be repaired, apoptosis will be induced by p53.

Chapter 19: DNA Mutation and Repair

19.1 b: Mutations are heritable changes in genetic information.

19.3 e: None of these (i.e., all are classes of mutation).

19.4 c: Ultraviolet light usually causes the induction of thymine dimers and their persistence or imperfect repair.

19.6 a. The normal anticodon was 5′-CAG-3′, and the mutant one is 5′-CAC-3′. The mutation event was a CG to GC transversion.

b. Presumably Leu

c. Val

d. Leu

19.7 Acridine is an intercalating agent and so can be expected to induce frameshift mutations. 5BU is incorporated into DNA in place of T but is likely to be read as C by DNA polymerase because of a shift from its normal to its rare chemical state. Thus, 5BU-induced mutations would be expected to be point mutations, usually TA-to-CG transitions. If these expectations are realized, *lacZ-1* probably would contain a single amino acid difference from the normal β-galactosidase, although it could be a truncated normal protein caused by a nonsense point mutation. *lacZ-2* should have a completely altered amino acid sequence after some point and might also be truncated.

19.8 a. Six

b. Three. The mutation results in replacement of the UGG codon in the mRNA by UAG, which is a chain termination of the codon.

19.10 a. UAG :CAG Gln, AAG Lys, GAA Glu, UUG Leu, UCG Ser, UGG Trp, UAU Tyr, UAC Tyr, UAA chain terminating

b. UAA: CAA Gln, AAA Lys, GAA Glu, UUA Leu, UCA Ser, UGA chain terminating, UAU Tyr, UAC Tyr, UAG chain terminating

c. UGA: CGA Arg, AGA Arg, GGA Gly, UUA Leu, UCA Ser, UAA chain terminating, UGU Cys, UGC Cys, UGG Trp

19.12

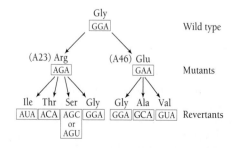

e. Prenatal monitoring of fetal chromosomes could be done, followed by therapeutic abortion of chromosomally unbalanced fetuses.

21.16 a. 45

  b. 47

  c. 23

  d. 69

  e. 48

21.17 b. An individual with three instead of two chromosomes is said to be trisomic.

21.18 a. The cross can be written as $X^+X^+ \times X^cY$. Turner syndrome children are XO. If the child is color blind, it received its father's X^c via a chromosomally normal X-bearing sperm. Therefore, the egg that was fertilized must have lacked an X, and nondisjunction occurred in the mother.

  b. If the child has normal vision, it must have received its mother's normal X^+. To be XO, this must be the only X the embryo received, so that the egg must have been fertilized by a nullo-X, nullo-Y sperm. Nondisjunction occurred in the father.

21.22 a. *AA aa*

  b. If we label the four alleles *A1, A2, a1,* and *a2,* there are six possible gamete types: *A1 A2, A1 a1, A1 a2, A2 a1, A2 a2, a1 a2,* or ⅙ *AA,* ⁴⁄₆ *Aa,* and ⅙ *aa.* The possible gamete pairings are as follows:

	⅙ *AA*	⁴⁄₆ *Aa*	⅙ *aa*
⅙ *AA*	¹⁄₃₆ *AAAA*	⁴⁄₃₆ *AAAa*	¹⁄₃₆ *AAaa*
⁴⁄₆ *Aa*	⁴⁄₃₆ *AAaa*	¹⁶⁄₃₆ *AAaa*	⁴⁄₃₆ *Aaaa*
⅙ *aa*	¹⁄₃₆ *AAaa*	⁴⁄₃₆ *Aaaa*	¹⁄₃₆ *aaaa*

Phenotypically this gives 35/36 *A* : 1/36 *a.*

21.24 The initial allopolyploid will have 17 chromosomes. After doubling, the somatic cells will have 34 chromosomes.

Chapter 22: Population Genetics

22.1 Equate the frequency of each color with the frequency expected in Hardy-Weinberg equilibrium, letting $p = f(C^B)$, $q = f(C^P)$, and $r = f(C^Y)$.

Brown: $f(C^BC^B) + f(C^BC^P) + f(C^BC^Y) = p^2 + 2pq + 2pr = 236/500 = 0.472$

Pink: $f(C^PC^P) + f(C^PC^Y) = q^2 + 2qr = 231/500 = 0.462$

Yellow: $f(C^YC^Y) = r^2 = 33/500 = 0.066$

Now solve for *p, q,* and *r,* knowing that $p + q + r = 1$.

$$r^2 = 0.066, \text{ so } r = \sqrt{0.066} = 0.26$$

There are two approaches to solve for *q.* First, because $q^2 + 2qr = 0.462$, we can substitute in $r = 0.26$, getting $q^2 + 2q(0.26) = 0.462$. We recognize this as a quadratic equation, set it equal to 0, and solve for *q;* that is, we solve the equation $q^2 + 0.52q - 0.462 = 0$.

Solving the quadratic equation for *q,* we get

$$q = \frac{-0.52 \pm \sqrt{(0.52)^2 - 4(1)(-0.462)}}{2(1)} = 0.467$$

A second approach to solve for *q* is to realize that
$$q^2 + 2qr = 0.462$$
$$r^2 = 0.066$$

Adding left and right sides of the equations together, we get

$$q^2 + 2qr + r^2 = 0.066 + 0.462$$
$$(q + r)^2 = 0.528$$
$$q + r = 0.726$$
$$q = 0.726 - r = 0.726 - 0.26 = 0.467$$

Since $p + q + r = 1, p = 1 - (q + r) = 1 - (0.26 + 0.467) = 0.273.$

22.4 a. $\sqrt{0.16} = 0.40 = 40\%$ = frequency of recessive alleles; $1 - 0.4 = 0.6 = 60\%$ = frequency of dominant alleles; $2pq = (2)(0.4)(0.6) = 0.48$ = probability of heterozygous diploids. Then $(0.48)/[(2 \times 0.16) + 0.48] = 0.48/0.80 = 60\%$ of recessive alleles are heterozygotes.

  b. If $q^2 = 1\% = 0.01$, then $q = 0.1$, $p = 0.9$, and $2pq = 0.18$ heterozygous diploids. Therefore, $(0.18)/[0.18 + 2(0.01)] = 0.18/0.20 = 0.90 = 90\%$ of recessive alleles in heterozygotes.

22.5 $2pq/q^2 = 8$, so $2p = 8q$; then $2(1 - q) = 8q$, and $2 = 10q$, or $q = 0.2$.

22.8 a. Let *p* equal the frequency of *S* and *q* equal the frequency of *s.* Then

$$q = \frac{2(188) SS + 717 Ss}{2(3,146)} = \frac{1,093}{6,292} = 0.1737$$

$$p = \frac{717 Ss + 2(2,241) ss}{2(3,146)} = \frac{5,199}{6,292} = 0.8263$$

b.

Class	Observed	Expected	*d*	d^2/e
SS	188	95	+93	91.0
Ss	717	903	−186	38.3
ss	2,241	2,148	+93	4.0
	3,146	3,146	0	133.3

There is only one degree of freedom because the three genotypic classes are completely specified by two allele frequencies: *p* and *q* (df = number of phenotypes − number of alleles = 3 − 2 = 1). The χ^2 value of 133.3 for one degree of freedom gives $P < 0.0001$. Therefore, the distribution of genotypes differs significantly from that expected if the population were in Hardy-Weinberg equilibrium.

22.11 There are several possible explanations for the difference in the frequency of the trait in males and females. Two are sex linkage and autosomal linkage with sex-influenced expression. Sex linkage can be readily examined. If the population is in Hardy-Weinberg equilibrium (which this one is), the frequency of the recessive allele causing the trait is *q,* and the gene is X-linked, then the frequency in XY males would be *q,* while the frequency in XX females would be q^2. Since the frequency in males is 0.4 and the frequency in females is $(0.4)^2 = 0.16$, the data fit a model of sex linkage with $q = 0.4$. The frequency of heterozygous XX (female) individuals is $2pq = 2(0.6)(0.4) = 0.48$. Since the trait appears to be sex linked, no heterozygous males exist.

22.13 Let *q* equal the frequency of the recessive allele, and *p* equal the frequency of the dominant allele. One expects homozygotes showing the trait to appear at a frequency of q^2

in a population at equilibrium. If $q^2 = 64/10{,}000 = 0.0064$, $q = 0.08$. Thus, one would expect 8 percent of XY male individuals to show the trait.

22.16 Let q equal the frequency of a, and p equal the frequency of A, with $q + p = 1$. As discussed on text pp. 470-472, when the population is at equilibrium, the frequency of p and q is given by:

$$q = \frac{u}{u + v} = \frac{6 \times 10^{-7}}{(6 \times 10^{-7}) + (6 \times 10^{-8})}$$

$$= \frac{6 \times 10^{-7}}{(6 \times 10^{-7}) + (0.6 \times 10^{-7})} = \frac{6}{6.6} = 0.91$$

$$p = 1 - q = 1 - 0.91 = 0.09$$

Thus, the frequencies are 0.0081 *AA*, 0.1638 *Aa*, and 0.8281 *aa*.

22.18

$$p'_x = mp_y + (1 - m)p_x$$
$$p'_x = [20/(20 + 80)](0.50)$$
$$+ \{1 - [20/(20 + 80)]\}\,(0.70) = 0.66$$

22.21 a. When selectively neutral, the genes distribute themselves according to the Hardy-Weinberg law, so 0.25 are *AA*, 0.5 are *Aa*, and 0.25 are *aa*.

   b. $q = 0.33$

   c. $q = 0.66$

22.23 a. $q = 0.63$

   b. $q = 0.64$

   c. $q = 0.66$

22.25 $q = u/s = (5 \times 10^{-5})/0.8 = 0.0000625$

22.27 a. One explanation is that a restriction fragment length polymorphism exists in the population. On some chromosomes, a RFLP having a size of 4.1 kb is found, while on others, the 6.3 kb fragment is found. Individuals having just one size band are homozygotes, while individuals with two different-sized bands are heterozygous. The difference in sizes of the fragments could result from a missing site in the 6.3 kb individuals, or the insertion of a 2.2 kb piece of DNA between two sites that are normally 4.1 kb apart.

   b. Homozygotes have two identical alleles, while heterozygotes have two different alleles. There are $(56 \times 2) + 38 = 150$, 6.3 kb alleles, and $(6 \times 2) + 38 = 50$, 4.1 kb alleles. Let q equal the frequency of 6.3 kb alleles, and p equal the frequency of 4.1 kb alleles. Then $q = 0.75$ and $p = 0.25$.

   c. For a population in Hardy-Weinberg equilibrium, one would expect $q^2 = 0.5625$ 6.3 kb homozygotes (compared to the 0.56 seen), $2pq = 0.375$ heterozygotes (compared to the 0.38 seen) and $p^2 = 0.0625$ 4.1 kb homozygotes (compared to the 0.06 seen). The population appears to be in Hardy-Weinberg equilibrium.

22.28 Since five loci were examined, and only two have more than one allele, $(2/5)(100\%) = 40\%$ of the loci are polymorphic. Heterozygosity is calculated by averaging the frequency of heterozygotes for each locus. The frequency of heterozygotes for the AmPep, ADH, and LDH-1 loci is zero. At MDH, 35 out of 50 individuals were heterozygous (0.70). At PGM, 10 out of 50 individuals were heterozygous (0.20). Thus, the average heterozygosity is

$$\frac{0 + 0 + 0 + 0.7 + 0.2}{5} = 0.18.$$

22.30 a. Mutation leads to change in allelic frequencies within a population if no other forces are acting and introduces genetic variation. If population size is small, mutation may lead to genetic differentiation between populations.

   b. Migration increases the population size and has the potential to disrupt a Hardy-Weinberg equilibrium. It can increase genetic variation and may influence the evolution of allelic frequencies within populations. Over many generations, migration reduces divergence between populations and equalizes allelic frequencies between populations.

   c. Genetic drift produces changes in allelic frequencies within a population. It can reduce genetic variation and increase the homozygosity within a population. Over time, it leads to genetic change. When several populations are compared, genetic drift can lead to increased genetic differences between populations.

   d. Inbreeding increases the homozygosity within a population and decreases its genetic variation.

Chapter 23: Quantitative Genetics

23.1 a. The mean is obtained by summing the individual values and dividing by the total number of values. The mean head width is $25.21/8 = 3.15$ cm, and the mean wing length is $281.7/8 = 35.21$ cm.

The standard deviation equals the square root of the variance (the square root of s^2). The variance is computed by squaring the sum of the difference between each measurement and the mean value, and dividing this sum by the number of measurements minus one. One has:

$$s_{headwidth} = \sqrt{s^2} = \sqrt{\frac{\Sigma(x_i - \bar{x})^2}{n - 1}} = \sqrt{\frac{1.70}{7}} = \sqrt{0.24} = 0.49 \text{ cm}$$

$$s_{winglength} = \sqrt{s^2} = \sqrt{\frac{\Sigma(x_i - \bar{x})^2}{n - 1}} = \sqrt{\frac{413.35}{7}} = \sqrt{59.05} = 7.68 \text{ cm}$$

   b. The correlation coefficient, r, is calculated from the covariance, *cov*, of two quantities. Let head width be represented by x, and wing length be represented by y. r is defined as:

$$r = \frac{cov_{xy}}{s_x s_y} = \frac{\dfrac{\Sigma x_i y_i - \dfrac{1}{n}(\Sigma x_i \Sigma y_i)}{n - 1}}{s_x s_y}$$

The first factor $(\Sigma x_i y_i)$ is obtained by taking the sum of the products of the individual measurements of head width and the corresponding measurements for wing length. The next factor $(\Sigma x_i \Sigma y_i)/n$ is the product of the sums of the two sets of measurements divided by the number of pairs of measurements. The difference between these values is then divided by $(n - 1)$, and then by the products fof the standard deviations of each measurement. One has:

$$r = \frac{\dfrac{913 - (1/8)(25.21 \times 281.7)}{7}}{0.49 \times 7.68} = \frac{3.61}{3.76} = 0.96$$

c. Head width and wing length show a strong positive correlation, nearly 1.0. This means that ducks with larger heads will almost always have longer wings, and ducks with smaller heads will almost always have smaller wings.

23.3 $252/1{,}024 = 0.246$

23.4 If each pure-breeding parent is homozygous for the genes (however many there are) controlling the size character, each parent is homogeneous in type. A cross of two pure-breeding strains will generate an F_1 heterozygous for the loci controlling the size trait. Since the F_1 consists of all heterozygotes, it is genetically as homogeneous as each of the parents. Therefore, it shows no greater variability than the parents.

23.7 a. 8 cm

b. 1 (2 cm) : 6 (4 cm) : 15 (6 cm) : 20 (8 cm) : 15 (10 cm) : 6 (12 cm) : 1 (14 cm)

c. Proportion of F_2 of *AA BB CC* is $(\frac{1}{4})^3$; proportion of F_2 of *aa bb cc* is $(\frac{1}{4})^3$; proportion of F_2 with heights equal to one or the other is $(\frac{1}{4})^3 + (\frac{1}{4})^3 = \frac{2}{64}$.

d. The F_1 height of 8 cm can be produced only by maintaining heterozygosity at no less than one gene pair (e.g., *AA Bb cc*). Thus, although 20 of 64 F_2 plants are expected to be 8 cm tall, none of these will breed true for this height.

23.10 It is a quantitative trait. Variation appears to be continuous over a range rather than falling into three discrete categories. Also, the pattern in which the F_1 mean falls between the parental means and in which F_2 individuals show a continuous range of variation from one parental extreme to the other is typical of quantitative inheritance.

23.12 The cross is *aa bb cc* (3 lb) × *AA BB CC* (6 lb), which gives an F_1 that is *Aa Bb Cc* and weighs 4.5 lb. The distribution of phenotypes in the F_2 is 1 (3 lb) : 6 (3.5 lb) : 15 (4 lb) : 20 (4.5 lb) : 15 (5 lb) : 6 (5.5 lb) : 1 (6 lb).

23.14 a. The progeny are *Aa Bb Cc Dd*, which are 18 dm high.

b. (i) The minimum number of capital-letter alleles is one and the maximum number is four, giving a height range of 12 to 18 dm. (ii) The minimum number is one and the maximum number is four, giving a height range of 12 to 18 dm. (iii) The minimum number is four and the maximum number is six, giving a height range of 18 to 22 dm. (iv) The minimum number is zero and the maximum number is seven, giving a height range of 10 to 24 dm.

23.16 The F_1 is *A/a B/b C/c D/d E/e*, and so it is greyish-brown. The F_2 phenotypes are determined by the number of capital alleles contributed from each F_1 parent. Thus, determine the chance of obtaining 0, 1, 2, or 3 capital alleles from each F_1 parent: Since the parent is heterozygous at all five loci, the chance of obtaining any specified set of five alleles in one gamete is $(1/2)^5$. The chance of obtaining a particular number of capital alleles from an F_1 is the number of ways in which that number of alleles can be obtained multiplied by $(1/2)^5$. Consideration of all the possibilities reveals that there is one way to obtain 0 capital alleles, 5 ways to obtain 1 capital allele, 10 ways of obtaining two capital alleles, and 10 ways of obtaining three capital alleles. With this information, tabulate the ways in which the progeny can be formed:

F_1 Gamete #1		F_1 Gamete #2		Capital Alleles	F_2 Progeny	
Capital Alleles	Gamete Fraction	Capital Alleles	Gamete Fraction		F_2 Fraction	Phenotype
0	$(1/2)^5$	0	$(1/2)^5$	0	$(1/2)^{10}$	whitish blue
0	$(1/2)^5$	1	$5(1/2)^5$	1	$5(1/2)^{10}$	whitish blue
1	$5(1/2)^5$	0	$(1/2)^5$	1	$5(1/2)^{10}$	whitish blue
0	$(1/2)^5$	2	$10(1/2)^5$	2	$10(1/2)^{10}$	light tan
1	$5(1/2)^5$	1	$5(1/2)^5$	2	$25(1/2)^{10}$	light tan
2	$10(1/2)^5$	0	$(1/2)^5$	2	$10(1/2)^{10}$	light tan
0	$(1/2)^5$	3	$10(1/2)^5$	3	$10(1/2)^{10}$	light tan
1	$5(1/2)^5$	2	$10(1/2)^5$	3	$50(1/2)^{10}$	light tan
2	$10(1/2)^5$	1	$5(1/2)^5$	3	$50(1/2)^{10}$	light tan
3	$10(1/2)^5$	0	$(1/2)^5$	3	$10(1/2)^{10}$	light tan

In the F_2, $11(1/2)^{10} = 11/1{,}024$ will be bluish white and $165(1/2)^{10} = 165/1{,}024$ will be light tan. The remaining $[1 - 176(1/2)^{10} =] 848/1{,}024$ F_2 progeny will be greyish brown.

A faster method to solve this problem is to use the coefficients of the binomial expansion to determine the proportion of progeny with different numbers of capital and lowercase alleles. Let n = total number of alleles, s = number of capital alleles, t = number of lowercase alleles, a = chance of obtaining a capital allele, b = chance of obtaining a lowercase allele, and $x! = (x)(x-1)(x-2)...(1)$, with $0! = 1$. Then the chance P of obtaining progeny with a specified number of each type of allele is given by:

$$P(s,t) = \frac{n!}{s!t!}a^s b^t$$

$$P(0,10) = \frac{10!}{0!10!}\left(\frac{1}{2}\right)^0\left(\frac{1}{2}\right)^{10} = \frac{1}{1{,}024}$$

$$P(1,9) = \frac{10!}{1!9!}\left(\frac{1}{2}\right)^1\left(\frac{1}{2}\right)^9 = \frac{10}{1{,}024}$$

$\left.\right\}\ \dfrac{11}{1{,}024}$ whitish blue

$$P(2,8) = \frac{10!}{2!8!}\left(\frac{1}{2}\right)^2\left(\frac{1}{2}\right)^8 = \frac{45}{1{,}024}$$

$$P(3,7) = \frac{10!}{3!7!}\left(\frac{1}{2}\right)^3\left(\frac{1}{2}\right)^7 = \frac{120}{1{,}024}$$

$\left.\right\}\ \dfrac{165}{1{,}024}$ light tan

$$1 - \frac{11 + 165}{1{,}024} = \frac{848}{1{,}024}\ \text{greyish brown}$$

23.19 The selection differential is 0.14 g; the selection response is 0.08 g; $h^2 = 0.08/0.14 = 0.57$.

23.20 Zero because there would be no genetic variability in the population to affect blood pressure. Blood pressure would vary only because of salt exposure.

23.21 Heritability is specific to a particular population and a specific environment and cannot be used to draw conclusions about the basis of population differences. Because the environments of the farms in Kansas and Poland differ and because the two wheat varieties differ in their genetic makeup, the heritability of yield calculated in Kansas cannot be applied to the wheat grown in Poland. Furthermore, the yield of TK138 probably would be different in Poland and might even be less that the yield of the Russian variety when grown in Poland.

23.22 Selection differential = 14.3 − 9.7 = 4.6 cm

Selection response = 13 − 9.7 = 3.3 cm

Narrow-sense heritability = selection response/selection differential = 3.3/4.6 = 0.72

23.24 a. Fathers: mean = 71.9, variance = 11.6 in.

Sons: mean = 70, variance = 10.25 in.

b. Correlation = 0.49 in.

c. Slope = b = 0.46 in.

Narrow-sense heritability = 2(b) = 2(0.46) = 0.92

23.25 Selection response = narrow-sense heritability × selection differential

Selection response = 0.60 × 10 g = 6 g

Chapter 24: Molecular Evolution

24.1 Each of the three codon positions can change to three different nucleotides, so a total of 135 substitutions must be considered for each position. At the first position only 9 of the 135 possible changes have no effect on the amino acid sequence of the polypeptide (0.07 are synonymous). Every change at the second codon position results in an amino acid substitution (0.00 are synonymous). A total of 98 changes at the third codon position have no effect on the protein's amino acid sequence (0.73 are synonymous). Natural selection is much more likely to act on mutations that change amino acid sequences, such as the second and first codon positions, and those are the ones where the least change is likely to be seen as sequences diverge.

24.2 $K = -\frac{3}{4}\ln[1 - 4/3(p)]$

$= -\frac{3}{4}\ln[1 - 4/3(0.12)]$

$= -\frac{3}{4}\ln(1 - 0.16)$

$= -\frac{3}{4}(-0.17)$

$= 0.13$

24.3 $r = K/2T$

$= 0.13/[2(80,000,000)]$

$= 8.1 \times 10^{-10}$

24.4 The mutation rate would be greater than or equal to the substitution rate for any locus. Mutations are any nucleotide changes that occur during DNA replication or repair, whereas substitutions are mutations that have passed through the filter of selection. Many mutations are eliminated through the process of natural selection.

24.5 1×10^{-8} substitutions/nucleotide/year; 2×10^{-8} substitutions/nucleotide/year

24.6 The high substitution rate in mammalian mitochondrial genes is useful when determining the relationships between evolutionarily closely related groups of organisms, such as members of a single species. When longer divergence times are involved, such as those associated with the mammalian radiation, more slowly evolving nuclear loci are more convenient to study because multiple substitutions are less likely to have occurred.

24.7 The increased rate of substitution in animal mtDNA may result from a higher mutation rate due to error prone replication and DNA repair, since mitochondrial DNA polymerases lack proofreading ability; from increased concentrations of mutagens, such as free radicals resulting from mitochondrial metabolic processes; and/or from relaxed selection pressure, since most cells have at least several dozen mitochondria, each of which has multiple copies of the mitochondrial genome.

24.8 Generation times of the organisms may be significantly different now or have been different at some point since they diverged. Average repair efficiency, average exposure to mutagens, and the need to adapt to new ecological niches and environments may also differ between the two species.

24.9 Sequence A is a pseudogene, and sequence B is a still functioning gene.

24.10 Diversifying selection

24.11 Regions involved in base pairing would not accumulate substitutions as quickly as those that are not. If the secondary structure of an RNA molecule is under selective constraint, then the only changes that would be found would be ones for which a compensatory change also occurred in its complementary sequence.

24.12 First identify how many differences exist between pairs of the sequences. In the annotation below, H represents human; R represents rabbit; D represents duck; an underlined base indicates a base that is different between the mammalian and duck sequence; and an "x" below the aligned sequences identifies the position of a base that is different between the indicated pair of species.

```
H:       ATGGTGCTCT CTCCTGCCGA CAAGACCAAC GTC AAGCCG CCTGGGACAA
R:       ATGGTGCTGT CTCCCGCTGA CAAGACCAAC ATCAAGACTG CCTGGGAAAA
D:       ATGGTGCTGT CTGCGGCTGA CAAGACCAAC GTCAAGGGTG TCTTCTCCAA
H vs. D:            x  x  x                    xx   x  xxxx
R vs. D:            x x                 x      xx   x  xxxxx
H vs. R:              x  x               x       x x      x
```

Based on this analysis, there are 10 substitutions between humans and ducks, 11 substitutions between rabbits and ducks, and 6 substitutions between humans and rabbits. With ducks as an outgroup, we can draw a species tree as follows:

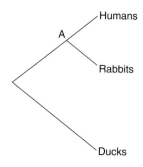

If the molecular clock hypothesis is correct, then $d_{A\text{-Humans}} = d_{A\text{-Rabbits}}$ (d in these equations represents "distance": $d_{A\text{-Humans}}$ is the distance between the point of common ancestry (A) to humans). The path by which the distances are calculated should not matter if the molecular clocks throughout the tree are uniform. Check this by inserting the numbers of substitutions determined from the analysis of the sequences into equations that trace the paths in the species tree. These equations are presented in the text, and reflect the relationships between distances in the species tree.

$d_{A\text{-Humans}} = (d_{\text{Humans-Rabbits}} + d_{\text{Humans-Ducks}} - d_{\text{Rabbits-Ducks}})/2$

$= (6 + 10 - 11)/2$

$= 2.5$

$d_{A\text{-Rabbits}} = (d_{\text{Humans-Rabbits}} + d_{\text{Rabbits-Ducks}} - d_{\text{Humans-Ducks}})/2$

$= (6 + 11 - 10)/2$

$= 3.5$

This analysis indicates that although $d_{A\text{-Humans}}$ and $d_{A\text{-Rabbits}}$ are similar, they are not identical. This suggests that there is a difference in the rate of substitution in this region of the rabbit and human genomes since they last shared a common ancestor.

24.13 Using DNA sequence information to infer evolutionary relationships has several advantages. First, DNA sequences serve as historical records that can be unraveled to identify the dynamics of evolutionary history. Second, since they do not rely on phenotypes for comparison of genetic relatedness, they circumvent problems associated with phylogenetic classification in organisms based solely on phenotypes. These include: problems in the identification of usable phenotypes in organisms which either lack phenotypes that correlate with their degree of genetic relatedness or for which there is no fossil record evidence (e.g., the eubacteria and the archaeabacteria); problems in the identification of usable phenotypes in groups of organisms where few characters are shared (e.g., the bacteria and mammals); and problems associated with convergent evolution, where distantly related organisms can share similar phenotypes. Third, they allow for an abundance of parameters that can be measured so that theories can be tested, as they provide highly accurate and reliable information, allow direct comparison of the genetic differences among organisms, are easily quantified, and can be used for all organisms.

24.14 The appeal of analyzing ancient DNA samples is that they might allow ancestral sequences to be determined (and not just inferred). However, it is almost impossible to prove that an ancient organism is from the same lineage as an extant species. Increasing the number of taxa in any analysis increases the robustness of any phylogenetic inferences, but extant taxa are almost invariably easier to obtain.

24.15 Divergence within genes typically occurs before the splitting of populations that occurs when new species are created. Preexisting polymorphisms such as those seen in the major histocompatibility locus could account for such a discordance.

24.16 These chances are related to the total number of rooted and unrooted trees that can be generated using six taxa.

$$N_{rooted} = (2n - 3)!/[2^{n-2}(n - 2)!]$$
$$= 9!/[16 \times (4)!]$$
$$= 362,880/384$$
$$= 945$$
$$N_{unrooted} = (2n - 5)!/[2^{n-3}(n - 3)!]$$
$$= 7!/[8 \times (3)!]$$
$$= 5,040/48$$
$$= 105$$

Since there are only 105 possible unrooted trees and 945 possible rooted trees, choosing a correct unrooted tree is more likely.

24.17 The equations used to determine how many rooted and unrooted trees can describe the relationship between taxa have no parameter that considers the amount of data associated with each taxon but do consider the number of taxa.

24.18 Informative sites are the only sites that are considered in parsimony analyses.

24.19 Most new genes arise from the process of mutating redundant copies of already existing genes. Copies of genes are free to accumulate substitutions, whereas the original version remains under selective constraint. One alternative might be to have tracts of noncoding sequences undergo random changes until they code for something that provides a selective advantage. The chances of a sequence randomly accumulating mutations that give it an open reading frame and appropriate promoter elements all at the same time are extremely small.

Credits

Text and Illustration Credits

Figure 1.6: From Robert H. Tamarin, *Principles of Genetics*. Copyright © 1996 McGraw-Hill. Used by permission of McGraw-Hill Companies, Inc.

Figure 1.8: Peregrine Publishing

Figure 1.9: Peregrine Publishing

Figure 1.21: From *Biological Science*, 4th Edition by William T. Keeton and James Gould with Carol Grant Gould. Copyright © 1986, 1980, 1979, 1978, 1972, 1967 by W.W. Norton & Company, Inc. Used by permission of W.W. Norton & Company, Inc.

Figure 5.1: Figure from *Genetics*, Second Edition by Ursula W. Goodenough, copyright © 1978 by Holt, Rinehart and Winston. Reproduced by permission of the publisher. This material may not be reproduced in any form or by any means without the prior written permission of the publisher.

Figure 5.12: From *General Genetics* by Srb. Owen, Edgar, © 1965 by W. H. Freeman and Company. Used with permission.

Figure 7.1b: Reprinted with permission from *The New Britton and Brown Illustrated Flora of the Northeastern United States and Adjacent Canada* (volume 2) by Henry Allen Gleason et al. Copyright 1952 by The New York Botanical Society.

Figure 7.2, flower drawings: Reprinted with permission from *The New Britton and Brown Illustrated Flora of the Northeastern United States and Adjacent Canada* (volume 2) by Henry Allen Gleason et al. Copyright 1952 by The New York Botanical Society.

Figure 8.15: From Pluta et al., *Science*, Vol. 270, 1995, pp. 1591–94. Copyright © 1995 American Association for the Advancement of Science. Reprinted by permission.

Figure 8.16: Reprinted from *Genes IV* by Benjamin Lewin. Copyright © 1990 with permission from Excerpta Medica, Inc.

Figure 8.18: From *Molecular Biology of the Cell,* Second Edition by Bruce Alberts. Copyright © 1989. Reproduced by permission of Routledge, Inc., part of The Taylor & Francis Group.

Figure 8.22: From *Molecular Biology of the Cell*, Second Edition by Bruce Alberts. Copyright © 1989. Reproduced by permission of Routledge, Inc., part of The Taylor & Francis Group.

Figure 8.23: Reprinted from *Cell*, Vol. 97, Greider, p. 419–422, Copyright © 1999 with permission from Excerpta Medica, Inc.

Figure 9.5: From *Molecular Biology of the Cell*, Second Edition by Bruce Alberts. Copyright © 1989. Reproduced by permission of Routledge, Inc., part of The Taylor & Francis Group.

Figure 9.8: Adapted from *Biology: Concepts and Connections*, Second Edition by Campbell et al., figure 8.10, p. 136. © 1997 by the Benjamin/Cummings Publishing Company. Reprinted by permission of Pearson Education, Inc.

Figure 11.9: Reprinted from *Cell*, Vol. 87, N. Proudfoot "Ending the Message Is Not So Simple" pp. 779–781, copyright 1996 with permission by Excerpta Medica, Inc.

Figure 11.12: From Maizels and Winer "RNA Editing" *Nature*, Vol. 334, 1988, p. 469. Copyright © 1988 Macmillan Magazines Limited. Reprinted by permission.

Figure 12.4: © Irving Geis. Rights owned by Howard Hughes Medical Institute. Not to be used without permission.

Figure 12.9: Peregrine Publishing

Figure 14.1: From Johnston et al., *Molecular Cell Biology*, Vol. 14, pp. 3834–3841, 1994. Copyright © 1994 American Society for Microbiology. Used with permission.

Figure 14.4a: Roza et al., *Molecular Vision*, Vol. 4, No. 20, 1998, figure 3a. Used with permission.

Figure 14.7: From *Recombinant DNA* by J. D. Watson, M. Gillman, J. Witkowski, and M. Zoller. © 1983, 1992 by J. D. Watson, M. Gillman, J. Witkowski, and M. Zoller. Used with the permission of W.H. Freeman and Company.

Figure 14.8: From *Recombinant DNA* by J. D. Watson, M. Gillman, J. Witkowski, and M. Zoller. © 1983, 1992 by J. D. Watson, M. Gillman, J. Witkowski, and M. Zoller. Used with the permission of W.H. Freeman and Company.

Figure 15.7a: Figure from http://biohpc.biotec.or/th/speakers/Viraphong/DNA_Microarray_Technology/sid007.htm. Used with permission.

Figure 15.1: Figure from "DNA Mapping" *Time*, July 3, 2000, p. 71. © 2000 Time Inc. Reprinted by permission.

Figure 15.2: Reprinted with permission from Fleischmann, et al., *Science*, July 28, 1995, Vol. 269, p. 507. Copyright © 1995 American Association for the Advancement of Science.

Figure 15.5: Reprinted from *Trends in Genetics*, Vol. 12, No. 7B. Dujon, Fig. 3, p. 267, copyright 1966 with permission from Elsevier Science.

Figure 15.6: Copyright © Stanford University. Used with permission.

Figure 15.8a: Reprinted with permission from Chu et al., *Science*, Vol. 282, No. 699, figure 1. Copyright © 1998 American Association for the Advancement of Science.

Figure 15.8b: from http://orca.ucsc.edu/mathdocs/microarray/madn/microarray_exp.jpg

Figure 15.8c: Copyright © Patrick Brown. Used with permission.

Figure 17.1: *Genetics*, Fifth Edition by Peter J. Russell p. 538, fig. 17.1. Copyright © 1998. Reprinted by permission of Pearson Education, Inc.

Figure 17.11: Adapted from Painter, *Journal of Heredity,* Vol. 25, 1934, pp. 465–467. Reprinted by permission of Oxford University Press.

Figure 17.16: Reprinted from *Genes IV* by Benjamin Lewis. Copyright © 1990 with permission from Excerpta Medica, Inc.

Figure 17.20: Illustration by Bunji Tagawa in "Transdetermination in Cells," *Scientific American,* November 1968, p. 116. Reprinted by permission.

Table 18.3: Reprinted with permission from J. Marx, *Science,* Vol. 261, 1993, pp. 1385–1387. Copyright © 1993 American Association for the Advancement of Science.

Figure 19.15: From Brooker, *Genetics: Analysis and Principles,* p. 474. Copyright © 1998. Reprinted by permission of Pearson Education, Inc.

Figure 19.19: Copyright © Stanford University. Used with permission.

Figure 20.7: "Ty-transposable element of yeast" adapted from Watson by permission of Gerald B. Fink. Reprinted by permission of Pearson Education, Inc.

Figure 21.11: Reprinted from *Cancer and Cytogenetics,* Vol. 11, O. Prakash and J. J. Yunis, "High Resolution Chromosomes of the +(922) Leukemias" pp. 361–368, copyright © 1984 with permission from Elsevier Science, Inc.

Figure 21.12: From Gerald Stine, *The New Human Genetics.* Copyright © 1989. Used with permission from McGraw Hill Companies, Inc.

Table 22.4: from *Genetics,* 3rd ed., by Monroe W. Strickberger. Copyright © 1985. Adapted by permission of Pearson Education, Inc., Upper Saddle River, NJ.

Figure 22.2: From *Ecological Genetics* by E. B. Ford. Copyright © 1975. Reprinted by permission of The Natural History Museum Picture Library.

Figure 22.3: from P. Buri in *Evolution* 10 (1956): 367. Reprinted by permission of the Society for the Study of Evolution.

Figure 22.8: Courtesy of Andrew Clark.

Figure 23.7: From *An Introduction to Genetics* by A. H. Sturtevant and G. W. Beadle, 1962. Reprinted by permission of Dover Publishers, Inc.

Figure 23.8: "Quantitative Trait Loci Affecting Differences in Floral Morphology between Two Species of Monkeyflower (Mimulus)" H. D. Bradshaw, Jr. et al., *Genetics,* Vol 149, May 1998, p. 378. © Genetics Society of America.

Table 24.1: From W. Li, C. Luo, and C. Wu, "Evolution of DNA Sequences" in *Molecular Evolutionary Genetics* Vol. 2, 1985, pp. 150–174 by R. J. MacIntyre, ed. Reprinted by permission of Kluwer Academic/Plenum Publishers.

Figure 24.2: From J. C. Avise, *Natural History and Evolution.* Copyright © 1994 Chapman and Hall. Used with permission.

Figure 24.3: From R. E. Dickerson, *Journal of Molecular Evolution,* Vol. 1, 1971, pp. 26–45. Copyright © 1971 Springer-Verlag. Used with permission.

Figure 24.6: From Hartl and Clark *Principles of Population Genetics,* Third Edition, p. 373. Copyright Sinauer Associations. Reprinted by permission from the publisher.

PHOTOGRAPH CREDITS

Chapter 1 Opener: ©Howard Hughes Medical Institute/Peter Arnold, Inc. 1.2, left: Jane Grushow/Grant Heilman Photography, Inc. 1.2, right: H. Reinhard/Okapia/Photo Researchers, Inc. 1.4, both: James A. Lake, University of California at Los Angeles/*Scientific American* (Aug. 1981): 84–97 1.5a: Pharmacia Corporation 1.5b: Ken Sherman/Bruce Coleman Inc. 1.7a: J. Forsdyke/Gene Cox/Science Photo Library/Photo Researchers, Inc. 1.7b: Grant Heilman/Grant Heilman Photography, Inc. 1.7c: Courtesy of John Sulston, Medical Research Council/Laboratory of Molecular Biology 1.7d: Dr. Jeremy Burgess/Science Photo Library/Photo Researchers, Inc. 1.7e, f: Pearson Education U.S. ELT/Scott Foresman 1.7g: Peter J. Russell 1.7h: David M. Phillips/Visuals Unlimited 1.7i: K. Aufderheidel/Visuals Unlimited 1.7j: Cabisco/Visuals Unlimited 1.7k: John Colwell/Grant Heilman Photography, Inc. 1.7l: Larry Lefever/Grant Heilman Photography, Inc. 1.7m: Hans Reinhard/Bruce Coleman Inc. 1.10: Custom Medical Stock Photo, Inc. 1.13: Published in Sumner et al., *Nature New Biology* 232 (1971): 31–32, Fig. 2. Courtesy of Dr. A. T. Sumner, MRC Human Genetics Unit, Western General Hospital, Edinburgh, Scotland. 1.16a-e: Michael Abbey/SS/Photo Researchers, Inc. 1.18a: G. F. Bahr/Armed Forces Institute of Pathology 1.18b: K. G. Murti/Visuals Unlimited 1.18c: J. R. Paulson and U.K. Laemmli, *Cell* 12 (1977): 817–28. © 1977 M.I.T. Photo courtesy of Dr. U. K. Laemmli, with permission from Elsevier Science.

Chapter 2 Opener: © Nigel Cattlin/Holt Studios Int./Earth Scenes 2.2: The Granger Collection 2.15a: G. Whiteley/Photo Researchers, Inc. 2.15b: Biophoto Associates/Photo Researchers, Inc. 2.18, left: Gershoff/Retna Ltd. USA 2.18, right: Davila/Retna Ltd. USA 2.19: From *The Journal of Heredity* 23 (Sept.

1932): 345. Reproduced by permission of Oxford University Press.

Chapter 3 Opener: © Gopal Murti/Phototake 3.8: Digamber S. Borgaonkar, Ph.D. 3.9: Digamber S. Borgaonkar, Ph.D. 3.10a: George Wilder/Visuals Unlimited 3.10b: M. Abbey/Photo Researchers, Inc. 3.11: Courtesy of Peter J. Russell 3.12a: Courtesy of the Library of Congress 3.13: National Library of Medicine

Chapter 4 Opener: © Jean Paul Ferrero/Ardea London, Ltd./Ardea Photographics 4.4a: Barbara J. Wright/Animals Animals/Earth Scenes 4.4b: Larry Lefever/Grant Heilman Photography, Inc. 4.4c: Christine Prescott-Allen/Animals Animals/Earth Scenes 4.4d: Larry Lefever/Grant Heilman Photography, Inc. 4.11a-c: Dr. James H. Tonsgard 4.12: Paul S. Conklin

Chapter 5 Opener: © Jean-Claude Revy/Phototake

Chapter 6 Opener: ©Dr. Dennis Kunkel/Phototake 6.1: Michael Gabridge/Custom Medical Stock Photo, Inc. 6.10a, b: Courtesy of Dr. Harold W. Fisher, University of Rhode Island 6.14: Bruce Iverson 6.16: Courtesy of Gunther S. Stent, University of California, Berkeley

Chapter 7 Opener: Courtesy of Dr. Roderrick Capaldi & Daciana Margineantu 7.1a: Color-Pic, Inc. 7.4: M. I. Walker/Science Source/ Photo Researchers, Inc. 7.7: Kim Taylor/Bruce Coleman Inc.

Chapter 8 Opener: © Ken Eward/Science Source/Photo Researchers, Inc. 8.1: Manfred Kage/Peter Arnold, Inc. 8.8, left and center: Agence France Presse/CORBIS 8.8, right: Photo Researchers, Inc. 8.9a, left: Cold Spring Harbor Laboratory Archives/Peter Arnold, Inc. 8.9a, right: UPI/CORBIS 8.12a–c: Richard Pastor/U.S. Food and Drug Administration 8.13a: Courtesy of Dr. Harold W. Fisher, University of Rhode Island 8.13b: From W. J. Thomas and R. W. Horne, "The Structure of Bacteriophage phiX174," *Virology* 15 (1961): 2–7. ©Academic Press. 8.14: Dr. Gopal Murti/Science Photo Library/Photo Researchers, Inc. 8.15a, b: Dr. Jack Griffith/University of North Carolina/School of Medicine 8.19: Dr. Jack Griffith/University of North Carolina/School of Medicine 8.20a: Barbara Hamkalo

Chapter 9 Opener: © Clive Freeman/The Royal Institution/Science Photo Library/Photo Researchers, Inc. 9.3: Dr. Sheldon Wolff Figure 9.13: David Dressler, Oxford University, UK, from *Proc. Natl. Acad. Sci. USA* 75(1978): 65

Chapter 10   Opener: © Ken Eward/Science Source/Photo Researchers, Inc. 10.5a: Bill Longcore/Science Source/ Photo Researchers, Inc. 10.5b: Jackie Lewis/Royal Free Hospital/Science Photo Library/Photo Researchers, Inc.

Chapter 11   Opener: Courtesy of K. Kamada & S. K. Burley. From J. L. Kim, D. B. Nikolov, and S. K. Burley, (1993) Co-crystal structure of TBP recognizing the minor groove of a TATA element, *Nature* 365, 520–527 11.13, both: Courtesy of Dr. James A. Lake/*Scientific American* Aug. 1981: 84–97 11.19b: Tripos, Inc.

Chapter 12   Opener: Photo courtesy of V. Ramakrishnan and colleagues/MRC Laboratories, Cambridge, UK

Chapter 13   Opener: © Jean-Claude Revy/Phototake

Chapter 14   Opener: © S. Miller/Custom Medical Stock Photo 14.1: Courtesy of Dr. Mark Johnston (From Johnston et al *Mol. Cell. Biol.* 14, 3834–3841, 1994) 14.4a: Courtesy of Frank Rozsa and Julia Roberts

Chapter 15   Opener: DOE Human Genome Program 15.7b: AECOM cDNA Microarray Facility, Aldo Massimi, Facility Director 15.3: Sinclair Stammers/Science Source/Photo Researchers, Inc. 15.8b: Tyson A. Clark, Graduate Student/UC Santa Cruz

Chapter 16   Opener: ©Ken Eward/Science Source/Photo Researchers, Inc.

Chapter 17   Opener: Courtesy of Stephen Paddock, James Langeland, Peter DeVries, and Sean B. Carroll of the Howard Hughes Medical Institute at the University of Wisconsin (*BioTechniques,* Jan. 1993) 17.11b: Richard Feldmann/Phototake NYC 17.18b, c: Edward B. Lewis, California Institute of Technology 17.19a-c: From Anthony Griffiths et al., *An Introduction to Genetic Analysis,* 7th ed. (New York: W.H. Freeman), 2000, p. 485

Chapter 18   Opener: Courtesy of Y. Cho et al., *Science* 265: 346–55 18.1: SIU/Visuals Unlimited 18.3: Custom Medical Stock Photo, Inc.

Chapter 19   Opener: Protein Data Bank/RCSB 19.17: Ken Greer/Visuals Unlimited

Chapter 20   Opener: Courtesy of Dr. Nina Federoff 20.5: Virginia Walbot, Stanford University UNF Box 20.1: AP/Wide World Photos

Chapter 21   Opener: ©Department of Energy/Photo Researchers, Inc. 21.3a: Dr. Laird Jackson, Thomas Jefferson University Hospital, Division of Medical Genetics 21.3b: C. Weinkove and R. McDonald, *South African Medical Journal* 43 (1969): 318; from *Syndromes of the Head and Neck,* 3rd ed., by Robert Gorlin, M. Michael Cohen, and L. Stefan Levin, Oxford University Press 21.7a: Courtesy of Grant L. Pyrah, Department of Biology 21.12a: Courtesy of Christine J. Harrison, from *The American Journal of Medical Genetics,* vol. 20 (1983): 280–85. Reprinted by permission of Wiley-Liss, Inc., a division of John Wiley & Sons, Inc. 21.15a:

National Library of Medicine 21.15b: M. Coleman/Visuals Unlimited 21.18a, b: Dr. Laird Jackson, Thomas Jefferson University Hospital, Division of Medical Genetics 21.19a, b: Dr. Laird Jackson, Thomas Jefferson University Hospital, Division of Medical Genetics

Chapter 22   Opener: © Dieter and Mary Plage/Bruce Coleman, Inc.

22.1a: CORBIS 22.1b: Hildegard Adler, photographer; courtesy of James F. Crow 22.1c: UPI/CORBIS 22.5: The Field Museum, Neg #CSA 118, Chicago 22.9a: Kevin Byron/Bruce Coleman Inc. 22.9b: W. Perry Conway/Tom Stack & Associates 22.10: Cliff B. Frith/Bruce Coleman Inc. 22.11a: AKG/Photo Researchers, Inc. 22.11b: The Granger Collection 22.12a, b: Breck P. Kent

Chapter 23   Opener: ©Camera M.D. Studios 23.6: From E. W. Sinnott and L. C. Dunn, *Principles of Genetics* (New York, McGraw-Hill), 1925 23.7: Courtesy of D. W. Schemske & H. D. Bradshaw, Jr.

Chapter 24   Opener: Photo courtesy of Mike Raymer

TRADEMARK ACKNOWLEDGMENTS

Flavr Savr is a trademark of Calgene, Inc. NutraSweet is a registered trademark of NutraSweet Company, Inc. Roundup is a registered trademark of Monsanto Company (De Corp). Roundup Ready soybeans is a trademark of Monsanto Company (De Corp).

Index

Page numbers in *italics* indicate material in figures and tables.